Lecture Notes in Computer Science　　10772

Commenced Publication in 1973
Founding and Former Series Editors:
Gerhard Goos, Juris Hartmanis, and Jan van Leeuwen

More information about this series at http://www.springer.com/series/7409

Gabriella Pasi · Benjamin Piwowarski
Leif Azzopardi · Allan Hanbury (Eds.)

Advances in Information Retrieval

40th European Conference on IR Research, ECIR 2018
Grenoble, France, March 26–29, 2018
Proceedings

 Springer

Editors
Gabriella Pasi
Department of Informatics, Systems,
   and Communication
University of Milano-Bicocca
Milan
Italy

Benjamin Piwowarski
LIP6 – UPMC/CNRS
University Pierre et Marie Curie
Paris
France

Leif Azzopardi
University of Glasgow
Glasgow
UK

Allan Hanbury
Technical University of Vienna
Vienna
Austria

ISSN 0302-9743   ISSN 1611-3349 (electronic)
Lecture Notes in Computer Science
ISBN 978-3-319-76940-0   ISBN 978-3-319-76941-7 (eBook)
https://doi.org/10.1007/978-3-319-76941-7

Library of Congress Control Number: 2018934361

LNCS Sublibrary: SL3 – Information Systems and Applications, incl. Internet/Web, and HCI

Printed on acid-free paper

This Springer imprint is published by the registered company Springer International Publishing AG
part of Springer Nature
The registered company address is: Gewerbestrasse 11, 6330 Cham, Switzerland

Preface

These proceedings contain the papers presented at ECIR 2018, the 40th European Conference on Information Retrieval. The conference is organized by the Laboratoire d'Informatique de Grenoble (LIG), in cooperation with the Information Retrieval Specialist Group of the British Computer Society (BCS-IRSG). It was supported by the Special Interest Group on Information Retrieval (ACM SIGIR), and the French Association of Information Retrieval (ARIA).

ECIR 2018 received 171 full paper and 112 short paper submissions. All submitted papers were reviewed by at least three members of the international Program Committee. In all, 39 full papers (23%) were accepted for oral presentation and 39 short papers (35%) were accepted for poster presentation. The accepted papers come from universities, research institutes, and/or commercial organizations. The traditional strong focus on students was maintained with a student mentoring program and a student grant program.

Accepted papers cover the state of the art in information retrieval including topics such as: topic modeling, deep learning, evaluation, user behavior, document representation, recommendation systems, retrieval methods, learning and classification, and micro-blogs.

The success of such a conference is made possible only by the huge effort of several people and organizations. We wish to thank our various committees – for full papers, short papers, demonstrations, tutorials, workshops and awards – for their hard work ensuring the quality of the ECIR 2018 program. We thank the researchers who submitted their work. We also wish to thank the student mentoring chairs, the student grant chairs, the sponsor chairs, the publicity chairs, the proceedings chairs, the industry day chairs, and the website chair.

We thank our sponsoring institutions, Naver Labs, Google, Université Grenoble Alpes, Grenoble INP, Grenoble Alpes Data Institute, Grenoble Alpes Métropole, Labex Persyval, ARIA, SIGIR, and Springer for their support of ECIR 2018. We are also grateful to our three keynote speakers.

Our final thanks are for the members of the local Organizing Committee for their hard work over many months and to the local and student volunteers who contributed to a smooth running of ECIR 2018.

March 2018

Gabriella Pasi
Benjamin Piwowarski
Leif Azzopardi
Allan Hanbury

Organization

General Chairs

Éric Gaussier   University of Grenoble Alpes, France
Lorraine Goeuriot   University of Grenoble Alpes, France
Georges Quénot   CNRS, France

Program Chairs

Gabriella Pasi   University of Milano-Bicocca, Italy
Benjamin Piwowarski   CNRS, France

Short Papers and Poster Chairs

Leif Azzopardi   University of Glasgow, UK
Allan Hanbury   Technical University of Vienna, Austria

Workshop Chairs

Elaine Toms   University of Sheffield, UK
Jian-Yun Nie   University of Montréal, Canada

Tutorial Chairs

Claudia Hauff   DEFT, The Netherlands
Mohand Boughanem   University of Toulouse, France

Demo Chairs

Jean-Pierre Chevallet   University of Grenoble Alpes, France
Jaap Kamps   University of Amsterdam, The Netherlands

Industry Day Chairs

Gabriella Kazai   Lumi, UK
Miguel Martinez   Signal Media, UK

Proceedings Chairs

Philippe Mulhem   CNRS, France
Michel Beigbeder   École Nationale Supérieure des Mines de Saint-Étienne, France

Publicity Chairs

Laure Soulier   University of Paris 6, France
Ingo Frommholz   University of Bedfordshire, UK

Sponsor Chairs

Massih-Reza Amini   University of Grenoble Alpes, France
Patrice Bellot   Université Aix-Marseille, France
Pavel Braslavski   Ural Federal University, Russia

Student Mentoring Chairs

Jacques Savoy   University of Neuchâtel, Switzerland
Mihai Lupu   Vienna University of Technology, Austria

Student Grant Chairs

Fabio Crestani   University of Lugano (USI), Switzerland
Josiane Mothe   University of Toulouse, France

Website Infrastructure

Christophe Brouard   University of Grenoble Alpes, France

Program Committee

Ahlers Dirk   Norwegian University of Science and Technology,
                            Norway
Albakour M-Dyaa   Signal Media, UK
Alonso Omar   Microsoft, USA
Amati Giambattista   Fondazione Ugo Bordoni, Italy
Amini Massih-Reza   LIG, France
Andersson Linda   Vienna University of Technology, Austria
Arampatzis Avi   Democritus University of Thrace, Greece
Arapakis Ioannis   Telefonica Research, Spain
Arguello Jaime   The University of North Carolina at Chapel Hill, USA
Leif Azzopardi   University of Strathclyde, UK
Balog Krisztian   University of Stavanger, Norway
Barreiro Alvaro   University of A Coruña, Spain
Basili Roberto   University of Rome Tor Vergata, Italy
Bellogin Alejandro   Universidad Autonoma de Madrid, Spain
Bellot Patrice   Aix-Marseille Université - CNRS (LSIS), France
Berrut Catherine   LIG, Université Grenoble Alpes, France
Bordogna Gloria   National Research Council of Italy - CNR, Italy
Boughanem Mohand   IRIT, Université Paul Sabatier Toulouse, France

Bozzon Alessandro	Delft University of Technology, The Netherlands
Braslavski Pavel	Ural Federal University, Russia
Buitelaar Paul	Insight Centre for Data Analytics, National University of Ireland Galway, Ireland
Cabanac Guillaume	IRIT, Université Paul Sabatier Toulouse 3, France
Cacheda Fidel	Universidad de A Coruña, Spain
Calabretto Sylvie	LIRIS, France
Calado Pável	Universidade de Lisboa, Portugal
Cambazoglu B. Barla	NTENT Inc., Spain
Can Fazli	Bilkent University, Turkey
Carpineto Claudio	Fondazione Ugo Bordoni, Italy
Castells Pablo	Universidad Autónoma de Madrid, Spain
Chevalier Max	IRIT, France
Chen Long	University of Glasgow, UK
Corney David	Signal Media, UK
Crestani Fabio	University of Lugano (USI), Switzerland
Croft Bruce	University of Massachusetts Amherst, USA
Cummins Ronan	University of Cambridge, UK
De Cock Martine	University of Washington, USA
de La Fuente Pablo	Universidad de Valladolid, Spain
de Rijke Maarten	University of Amsterdam, The Netherlands
de Vries Arjen	Radboud University, The Netherlands
Di Buccio Emanuele	University of Padua, Italy
Dietz Laura	University of New Hampshire, UK
Di Nunzio Giorgio Maria	University of Padua, Italy
Dixon Julie	University of Aberdeen, UK
Duan Huizhong	WalmartLabs, USA
Eickhoff Carsten	ETH Zurich, Switzerland
Fang Hui	University of Delaware, USA
Fernández-Luna Juan M.	University of Granada, Spain
Ferro Nicola	University of Padua, Italy
Freire Ana	Universitat Pompeu Fabra, Spain
Frommholz Ingo	University of Bedfordshire, UK
Fuhr Norbert	University of Duisburg-Essen, Germany
Gallinari Patrick	LIP6 - University of Paris 6, France
Ganesan Kavita	UIUC, USA
Giannopoulos Giorgos	Imis Institute, Athena R.C., Greece
Goeuriot Lorraine	Laboratoire d'informatique de Grenoble, France
Gonzalo Julio	UNED, Spain
Granitzer Michael	University of Passau, Germany
Grau Brigitte	LIMSI, CNRS, ENSIIE, Université Paris-Saclay, France
Gravier Guillaume	CNRS, IRISA, France
Gurrin Cathal	DCU, Ireland
Hagen Matthias	Bauhaus-Universität Weimar, Germany
Hanbury Allan	Vienna University of Technology, Austria

Martins Bruno	IST - Instituto Superior Técnico, Portugal
Mass Yosi	IBM Haifa Research Lab, Israel
Mccreadie Richard	University of Glasgow, UK
Meij Edgar	Bloomberg L.P., UK
Melucci Massimo	University of Padua, Italy
Mendoza Marcelo	Universidad Técnica Federico Santa María, Chile
Micarelli Alessandro	Roma Tre University, Italy
Mizzaro Stefano	University of Udine, Italy
Moshfeghi Yashar	University of Strathclyde, UK
Mothe Josiane	Institut de Recherche en Informatique de Toulouse, France
Moussa Samir	Signal Media, UK
Mulhem Philippe	LIG-CNRS, France
Müller Henning	HES-SO, Switzerland
Nardini Franco Maria	ISTI-CNR, Italy
Nejdl Wolfgang	L3S and University of Hannover, Germany
Nguyen Dong	University of Twente, The Netherlands
Nie Jian-Yun	University of Montreal, Canada
Nuernberger Andreas	Otto von Guericke University of Magdeburg, Germany
Oakes Michael	University of Wolverhampton, UK
Orlando Salvatore	Università Ca' Foscari Venezia, Italy
Ounis Iadh	University of Glasgow, UK
P Deepak	Queen's University Belfast, UK
Pasi Gabriella	Università degli Studi di Milano Bicocca, Italy
Pavlu Virgil	Northeastern University, UK
Pecina Pavel	Charles University in Prague, Czech Republic
Perego Raffaele	ISTI-CNR, Italy
Petras Vivien	HU Berlin, Germany
Pinel-Sauvagnat Karen	IRIT, France
Piroi Florina	Vienna University of Technology, ISIS, IFS, Austria
Piwowarski Benjamin	CNRS/Pierre et Marie Curie University, France
Plachouras Vassilis	Thomson Reuters, UK
Quénot Georges	Laboratoire d'Informatique de Grenoble, CNRS, France
Rosso Paolo	Universitat Politècnica de València, Spain
Roussinov Dmitri	University of Strathclyde, UK
Rueger Stefan	Knowledge Media Institute, UK
Russell-Rose Tony	UXLabs, UK
Said Alan	University of Skövde, Sweden
Santos Rodrygo	Universidade Federal de Minas Gerais, Brazil
Sebastiani Fabrizio	Qatar Computing Research Institute, Qatar
Sébillot Pascale	IRISA, France
Sedes Florence	IRIT P. Sabatier University, France
Semeraro Giovanni	University of Bari, Italy
Serdyukov Pavel	Yandex, Russia
Shakery Azadeh	University of Tehran, Iran
Sormunsen Eero	University of Tampere, Finland

Soulier Laure	Sorbonne Universités UPMC-LIP6, France
Tamine Lynda	IRIT, France
Tombros Anastasios	Queen Mary University of London, UK
Tonellotto Nicola	ISTI-CNR, Italy
Tsai Ming-Feng	National Chengchi University, Taiwan
Tsikrika Theodora	Information Technologies Institute, CERTH, Greece
Tzitzikas Yannis	University of Crete and FORTH-ICS, Greece
Vargas Saúl	Mendeley Ltd., UK
Velupillai Sumithra	TCS, KTH Royal Institute of Technology, Sweden
Verberne Suzan	LIACS, Leiden University, The Netherlands
Viviani Marco	Università degli Studi di Milano-Bicocca - DISCo, Italy
Vrochidis Stefanos	Information Technologies Institute, Greece
Womser-Hacker Christa	University of Hildesheim, Germany
Yang Tao	Ask.com and UCSB, USA
Yu Hai-Tao	University of Tsukuba, Japan
Zuccon Guido	Queensland University of Technology, Australia

Additional Reviewers

Avgerinakis Konstantinos	Information Technologies Institute, Greece
Banerjee Somnath	Jadavpur University, India
Basile Pierpaolo	Aldo Moro University, Italy
Biancalana Claudio	Roma Tre University, Italy
Bibaev Vitaly	Saint Petersburg Academic University, Russia
Blinov Vladislav	Ural Federal University, Russia
Campillos Llanos Leonardo	LIMSI, France
Carvalho André	Lisbon University, Portugal
Chakravarthi Bharathi Raja	Insight, Ireland
Corney David	University of London, UK
Daudert Tobias	Insight, Ireland
Delasalles Edouard	LIP6, France
Dias Charles-Emmanuel	LIP6, France
Dos Santos Ludovic	LIP6, France
Effrosynidis Dimitrios	Democritus University of Thrace, Greece
Elbeshausen Stefanie	Hildesheim University, Germany
Farnadi Golnoosh	University of California Santa Cruz, USA
Gaede Maria	IBI, Germany
Gasparetti Fabio	Roma Tre University, Italy
Gialampoukidis Ilias	ITI, Greece
Gossen Tatiana	Magdeburg University, Germany
Hansen Casper	University of Copenhagen, Denmark
Hoang Thi Bich Ngoc	IRIT, France
Ivanschitz Bernd	Research Studios, Austria

Ji Shiyu	University of California, Santa Barbara, USA
Jurgovsky Johannes	Passau University, Germany
Kamateri Eleni	ITI, Greece
Kiesel Johannes	Weimar University, Germany
Kotzyba Michael	Magdeburg University, Germany
Laclau Charlotte	University of Grenoble Alpes, France
Lamrayah Mehdi	LIP6, France
Landín Alfonso	University of A Coruña, Spain
Li Wei	DCU, Ireland
Lionakis Panagiotis	University of Crete, Crete
Low Thomas	Magdeburg University, Germany
Marra Giuseppe	DIIES, Italy
Mavropoulos Thanassis	ITI, Greece
McCreadie Richard	Glasgow University, UK
Melo Tiago	INESC-ID, Portugal
Mountantonakis Michalis	University of Crete, Crete
Muntean Cristina	ISTI, Italy
Musto Cataldo	Bari University, Italy
Narducci Fedelucio	Bari University, Italy
Niebler Thomas	University of Würzburg, Germany
Pajot Arthur	LIP6, France
Palotti Joao	Vienna University of Technology, Austria
Papantoniou Katerina	University of Crete, Crete
Parapar Javier	University of A Coruña, Spain
Perez Estruch Carlos	PRLHT, Spain
Portier Pierre-Edouard	Insa de Lyon, France
Recalde Lorena	Pompeu Fabra University, Spain
Roitero Kevin	Udine University, Italy
Rossiello Gaetano	Bari University, Italy
Rubtsova Yuliya	Novosibirsk State University, Russia
Schinas Manos	ITI, Greece
Schlör Daniel	University of Würzburg, Germany
Schlötterer Jörg	Passau University, Germany
Schwerdt Johannes	Magdeburg University, Germany
Sgontzos Konstantinos	University of Crete, Crete
Shao Jinji	University of California, Santa Barbara, USA
Spitz Andreas	Heidelberg University, Germany
Sushmita Shanu	University of Washington, USA
Syed Shahbaz	Weimar University, Germany
Symeonidis Symeon	Democritus University of Thrace, Greece
Thiel Marcus	Magdeburg University, Germany
Trani Roberto	ISTI, Italy
Tsikrika Theodora	ITI, Greece
Valcarce Daniel	University of A Coruña, Spain
Vezzani Federic	Padua University, Italy

Voelske Michael	Weimar University, Germany
Wang Bingyu	Northeastern University, USA
Watt Stuart	Turalt, Canada
Yang Liu	University of Massachusetts, USA
Yang Peilin	University of Delaware, USA
Yu Haitao	University of Tsukuba, Japan
Zablocki Éloi	LIP6, France
Zamani Hamed	University of Massachusetts, USA
Zampoglou Markos	ITI, Greece
Zayed Omnia	Insight, Ireland
Zehe Albin	University of Würzburg, Germany
Zhao Haozhen	Legal Technology Solutions, USA
Zlabinger Markus	Vienna University of Technology, Austria
Zoller Daniel	University of Würzburg, Germany
Zuccon Guido	Queensland University of Technology, Australia

Sponsors

Silver Sponsors

Grenoble Data Institute

Naver Labs

Persyval Lab

ACM SIGIR

SIGIR

Special Interest Group
on Information Retrieval

Bronze Sponsors

ARIA

Google

Grenoble Alpes Métropole

GRENOBLE • ALPES
METROPOLE

Grenoble INP

Springer

Université Grenoble Alpes

Contents

Evaluations and User Behavior

Representation

Recommendation

Retrieval

Learning/Classification

Micro-blogs

Short Papers

Demonstrations

Tutorials

Topic Modelling

Entity-Centric Topic Extraction and Exploration: A Network-Based Approach

Andreas Spitz[✉] and Michael Gertz

Heidelberg University, Heidelberg, Germany
{spitz,gertz}@informatik.uni-heidelberg.de

Abstract. Topic modeling is an important tool in the analysis of corpora and the classification and clustering of documents. Various extensions of the underlying graphical models have been proposed to address hierarchical or dynamical topics. However, despite their popularity, topic models face problems in the exploration and correlation of the (often unknown number of) topics extracted from a document collection, and rely on compute-intensive graphical models. In this paper, we present a novel framework for exploring evolving corpora of news articles in terms of topics covered over time. Our approach is based on implicit networks representing the cooccurrences of entities and terms in the documents as weighted edges. Edges with high weight between entities are indicative of topics, allowing the context of a topic to be explored incrementally by growing network sub-structures. Since the exploration of topics corresponds to local operations in the network, it is efficient and interactive. Adding new news articles to the collection simply updates the network, thus avoiding expensive recomputations of term and topic distributions.

Keywords: Networks · Topic models · Evolving networks

1 Introduction

Given a collection of time-stamped documents spanning a period of time, what topics are covered in the documents and how do they evolve? This question is common in corpus analysis, ranging from the exploration of news media to the analysis of corpora of historic documents. In most cases, answers to the above questions are provided by employing probabilistic topic models, which are based on Latent Dirichlet Allocation, LDA [4]. In these models, documents are assumed to consist of one or more topics, which are distributions of words. Once identified, the topics for a corpus can also be used for document classification and clustering. Due to this range of applications, a multitude of tools for computing topics and extensions to LDA have been proposed (for an overview, see [2]).

Despite the versatility of topic models, topics are often simply represented as lists of ranked terms, some of which are even difficult to associate with the topic. More recently, there have thus been approaches that enable the exploration and analysis of the computed topics and ranked terms (e.g., [7,11]). To utilize

© Springer International Publishing AG, part of Springer Nature 2018
G. Pasi et al. (Eds.): ECIR 2018, LNCS 10772, pp. 3–15, 2018.
https://doi.org/10.1007/978-3-319-76941-7_1

additional annotation data during topic extraction, some approaches also include named entities in the topic model [9,12], an aspect that is important for the analysis of news articles, which typically revolve around such entities. However, these approaches rely on computationally expensive graphical models, and often produce a fixed number of topics that cannot be altered after the extraction.

In this paper, we present an alternative and efficient framework for the exploration and analysis of topics, including their evolution over time. Our approach is based on *implicit networks* that are constructed from the cooccurrences of terms and named entities in the documents, which are encoded as weighted edges in the network. Based on the central role that entities play in the evolution of news stories, we conjecture that frequently cooccurring pairs of entities are indicative of a topic. Starting from such seed edges between two words in the network representation, one can then construct the *context* of a potential topic by following other highly weighted edges to adjacent terms and entities. With this approach, the exploratory character of topic discovery and the overlap between identified topics becomes obvious. Important seed edges can easily be determined or expanded, meaning that the model is not constrained to a fixed number of topics. For news articles, publication dates provide an effective means of focussing on entity and term cooccurrences in a given time frame. Thus, the evolution of topics in terms of edge weights and word contexts can be effectively explored in such an implicit network, which also supports the addition of new documents, resulting in new nodes, edges, and updated edge weights. Most importantly, it is not necessary to recompute topics and word distributions for an evolving corpus when new documents are added. As a result, a network-based framework can support a variety of entity-centric topic analysis and exploration tasks.

2   Related Work

The framework that we propose in this paper is related to two broader areas of research, namely topic modeling and network-based document representation.

Topic Modeling. Since the introduction of topic models based on Latent Dirichlet Allocation [4], numerous frameworks for topic models have been proposed. The review by Blei summarizes the diverse approaches and directions [2]. However, since topic models primarily extract topics from a document collection in the form of ranked lists of terms, it has been questioned to what extent such a representation is semantically meaningful or interpretable, beyond providing an approximate initial parameter of topics to discover. This issue has been addressed by Chang et al. [5] by proposing novel quantitative methods for evaluating the semantic meaning of topics. Our work is most closely related to entity-centric topic modeling, e.g., [9,12], which focusses on (named) entities and their inclusion in the topics. These approaches rely on the original concept of LDA and thus share the inherent problems of divining an appropriate number of topics to discover and the necessity of interpreting lists of ranked terms.

More recent works in the area of topic modeling also propose frameworks that enable the interactive construction and exploration of topics from a collection of documents, e.g., [7,11], thus providing the user with added flexibility in terms of corpus analysis. Another interesting direction that implicitly addresses the aspect of collocation is the combination of word embeddings and topic models [15]. Common to most extensions of probabilistic models for entity-centric topic modeling and exploration is the reliance on a prior (computationally expensive) construction of lists of ranked terms. In contrast, we use an underlying network document model that adds versatility to the subsequent steps.

Network-based Document Models and Analyses. Network-based representations of documents and corpora have become a popular tool for analyzing the context of entities and terms. The most prominent and well-known examples are collocation networks (for an in-depth discussion, see [6,8]), even though the derived networks are only used for representation and not for network analysis tasks. Some recent works employ network-based representations of documents for the discovery and description of events [14,16,18]. These works focus on named entities in the documents, for which an implicit network is constructed from entity and term cooccurrences, but do not consider the extraction, exploration, or temporal evolution of topics. Other recent approaches in support of more general information retrieval tasks such as ranking have been proposed by Blanco and Lioma [1] and Rousseau and Vazirgiannis [13].

A combination of term collocation patterns and topic models was recently proposed by Zuo et al. [20]. Compared to the network-based approach that we present here, their approach is tailored towards short documents and relies on LDA, thus incurring the same problems as the topic models outlined above.

3 Network Model

In the following, we describe how an implicit network representation is constructed for a collection $D = \{d_1, \ldots, d_n\}$ of news articles. We assume each document d_i to have a timestamp $t(d_i)$, indicating the article's publication time. The network for the collection D, denoted as a graph $G_D(V, E)$, consists of a set of nodes V representing words in the documents and a set of labeled edges E that describe the *cooccurrence* of pairs of words in the documents.

Network Nodes. With the focus on entity-centric topics, we distinguish different types of entities such as persons, organizations, and locations. We assume that words in the sentences have been tagged with respect to a known set of named entities NE. Stop words are removed, the remaining untagged words are denoted as the set of terms T. We then let the set of nodes be $V = NE \cup T$. Each node can be assigned occurrence statistics of the corresponding entity or term, such as document or sentence positions. To model *cooccurrence* data, on the other hand, we utilize edge attributes in the network model.

Network Edges. Edges $E \subset V \times V$ describe the cooccurrence of entities and terms in documents, requiring that at least one node of an edge $e = (v_1, v_2)$

corresponds to an entity, i.e., $v_1 \in NE$ or $v_2 \in NE$. The cooccurrence of two respective words can be limited by a maximum sentence distance $sdist$. For $sdist = 0$, only cooccurrences in the same sentence are considered, for $sdist = 1$ the sentences directly before or after a given sentence, and so on.

Our model is based on three cooccurrence statistics, namely (1) the number of word cooccurrences, (2) the publication dates of the articles in which they cooccur, and (3) the textual distances at which the words cooccur. This scheme allows us to easily integrate new documents, eventually providing a basis for exploring the evolution of topics, as outlined in Sect. 4. To include the above features, each edge has an associated list of $\langle d, t, \delta \rangle$ tuples that encode the document d, timestamp t and the smallest mention distance δ (counted in sentences). For the first cooccurrence of two words v_1 and v_2 in a document d, we add the corresponding edge to the network, as well as the pair $\langle d, t(d), \delta(v_1, v_2) \rangle$ to that edge's list. If the same words cooccur again in the same document, we simply update the distance if necessary. If a cooccurrence of the two words is found in a new document, a new tuple is added to the list. In a sense, these lists represent a time series of word cooccurrences that support subsequent explorations, and enable efficient updates of the network representation. Similarly, we store lists of tuples $\langle d, t \rangle$ as node attributes for individual word mentions.

The resulting network serves as a model for a timestamped document collection, which is represented as collocations of words. In the following, we discuss how substructures in the network can be associated with topics in the documents.

4   Network-Based Topic Exploration

How are topics reflected in an implicit network? In the following, we argue that the core of topics is formed by edges between frequently cooccurring nodes, and that topics can be grown around such edges in a well-defined manner. We propose two growth approaches that specifically allow for an interactive exploration of topics. Finally, we discuss evolving topics based on the temporal edge labels.

4.1   Edge Weighting

Given the structure of news events, it appears to be a reasonable conjecture that a high cooccurrence frequency of two entities is indicative of a topic in news. For example, interactions between politicians, parties, countries, companies, or other actors and locations all involve more than one entity, which supports a faceted exploration of topics. Extracting and linking such named entities is an established task, so the question that remains is how such important edges can be identified and filtered from spurious connections in the network. For a naive approach, assume an edge $e = (v_i, v_j)$ with tuple list $L(e) = \langle (d_1, t_1, \delta_1), \dots, (d_k, t_k, \delta_k) \rangle$. Let $w(e) = |L(e)|$ be the weight of that edge, that is, we use the length of the list to reflect the overall cooccurrence frequency. Clearly, an edge with a higher weight is more likely to be at the center of an important topic than an edge with a lower weight. Based on this intuition, we introduce a weight for the edges of the graph G_D that supports such a filtering and includes both the overall and the temporal frequency of joint mentions, as well as the cooccurrence distances.

For an edge $e = (v_1, v_2)$, let $D(e) = \{d \mid (d, \cdot, \cdot) \in L(e)\}$ denote the set of documents in which v_1 and v_2 cooccur. Similarly, let $T(e) = \{t \mid (\cdot, t, \cdot) \in L(e)\}$ denote the set of timestamps at which both words that correspond to v_1 and v_2 occur jointly in a document. Let $D(v)$ and $T(v)$ be defined analogously for single nodes v. Finally, let $\Delta(e) = \langle \delta_1, \ldots, \delta_{|L(e)|} \rangle$ be the sequence of minimum distances at which the words that correspond to the two nodes of edge e occur in documents. We then obtain a combined weight for edge $e = (v_1, v_2)$ as

$$\omega(e) = 3 \left[\frac{|D(v_1) \cup D(v_2)|}{|D(e)|} + \frac{\max\{T(e)\} - \min\{T(e)\}}{|T(e)|} + \frac{|L(e)|}{\sum_{\delta \in \Delta(e)} \exp(-\delta)} \right]^{-1}$$

Intuitively, the measure represents the harmonic mean of three individual components, namely the number of joint versus individual mentions, the temporal coverage density, and an exponentially decaying weight by mention distance [18]. The resulting weight is normalized such that $\omega \in [0, 1]$.

Using the weights computed this way allows a pruning of low frequency edges and the detection of important entity connections, from which we grow and explore topics in the following. Since the components of the edge weights can be computed during network construction (or network updates with new documents), no additional post-processing costs occur during topic exploration.

4.2   Topic Construction and Edge Growth

Assuming an ordering of edges in G_D by weight, the top-ranked edges correlate to topic *seeds* as described above. Thus, one can select the top-ranked k edges for some value of k and treat them as seeds around which the topics are grown. Note that some seed edges may share nodes, an aspect that we discuss in Sect. 4.3. To grow topic substructures around the selected edges, we introduce two types of growth patterns, *triangular growth* and *external node growth*.

Triangular Growth. Given an edge $e = (v_1, v_2)$ between entities v_1 and v_2 along with a network substructure that only contains e, v_1, and v_2, this initial substructure can be grown by adding neighbours of both entities. Formally, let $N(v_1)$ and $N(v_2)$ denote the neighbours of nodes v_1 and v_2 respectively, then $N(v_1) \cap N(v_2)$ is the set of all nodes in G_D that share v_1 and v_2 as neighbours. To rank nodes in this potentially very large set, we utilize a scoring function on the edge weights. Specifically, let $s : V \rightarrow \mathbb{R}$ such that $s(x) = \min\{\omega(x, v_1), \omega(x, v_2)\}$. Obviously, nodes with a higher score cooccur more often and more consistently with both entities of the seed edge. Ranking nodes in the shared neighbourhood according to s thus allows us to select the most related terms to the topic that is represented by the seed edge. It is then a simple matter of adding any number of such nodes (along with the edges connecting them to v_1 and v_2) to the network substructure to incrementally grow the topic. Since all new nodes can be ranked according to s, we obtain a relevance score for nodes in relation to the given edge e. In addition to the two seed words v_1 and v_2, a topic can thus be viewed as a list of ranked words that are added to the initial two words based on their cooccurrence patterns, much like a classic topic model. However, this growth strategy also results in a descriptive network substructure as illustrated in Fig. 1, where l such triangles are added to the seed edge e.

Based on the process described above, the incremental addition of words to the seed edge clearly supports different aspects of topic and cooccurrence exploration. First, instead of adding terms as described above, it is equally viable to select entities or even specific types of entities in the shared neighbourhood. For example, if v_1 and v_2 are both persons, one could restrict the growth process to add only other persons. Second, one should keep in mind that, depending on the realization of the network, the implicit network can be used as an inverted index [16], thus allowing the user to inspect articles and sentences in which two words cooccur during the incremental construction and exploration of the network substructures.

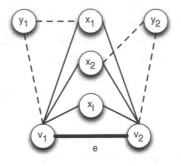

Fig. 1. Edge growth approach for seed edge e with words v_1 and v_2. Nodes $x_1, \ldots, x_l$ denote words added during triangular growth. Nodes y_1 and y_2 are (optional) additions in a subsequent external growth phase.

External Node Growth. While the construction of edge triangles and word rankings is a key component in extracting and exploring a topic in the classical sense, the node set determined in this way can also be explored based on further expansion techniques. As a topic substructure grows, the seed edge and its incident nodes are not the only available attachment points for further edges and nodes. Instead, one could also add further nodes that are connected only to one of the initial words v_1 or v_2, but also to some of the other nodes that were added in subsequent triangles. The external node growth process is also illustrated in Fig. 1, where nodes y_1 and y_2 connect to the substructure by dotted lines. While this attachment step has no analogy in classic topic extraction, it introduces additional degrees of freedom in an interactive exploration of topics.

4.3   Topic Fusion and Overlap

In the network-based extraction of topics, the substructure that is grown from the top-ranked seed edge is self-contained. However, this is not necessarily the case for substructures grown from subsequent edges in the list of top-ranked entity edges. Assume an edge $e' \neq e$ with $\omega(e') < \omega(e)$ from the list of seed edges. While new nodes are added to the substructure around e', it is possible that an edge is added that is incident to a node of the previously extracted substructure grown from e. In practice, this overlap between two topics may occur for terms or entities. In the most extreme case, even seed edges may overlap in one of their entity nodes, leading to the fusion of two topics. In classic topic models, the same word may belong to different topics with a high probability, which is analogous to partially overlapping topics in our model, where the same node can be part of different topics. In fact, we argue that topics should overlap for entities in news that participate in multiple topics, and in Sect. 5 we show how network visualization highlights such overlapping substructures during topic exploration.

4.4 Evolving Topics

An important aspect of topic modeling is the evolution of topics over time (e.g., [3,10]). In such a setting, visualizations turn out to be especially helpful in highlighting changes in a topic's relevance over time or how it compares to other topics. Our framework directly supports the exploration of temporal topic characteristics due to the timestamp information contained in the edge labels.

Key to the exploration of evolving topics based on an implicit network is the *network projection*. In general, a network projection filters nodes and edges that do not satisfy certain user-specified conditions. For example, a given network could be projected to only cooccurrences of entities, i.e., all term nodes and incident edges are removed. For studying the evolution of topics, such conditions concern the timestamps associated with the cooccurrence information of edges. In principle, given a collection of news articles spanning a time interval from t_{min} to t_{max}, the interval can be partitioned to construct corresponding networks. For example, the evolution of news topics over multiple weeks or months can be considered by focusing on multiple networks constructed for these intervals.

For a seed edge in such an interval, one can then directly compare respective sub-networks from different time intervals side-by-side, thus highlighting what nodes have gained or lost relevance for that topic in a given interval with respect to neighbouring intervals. There are numerous visualization metaphors one can consider in this setting, all of them relying on the visualization of sub-networks. Key to all these approaches is the representation of topics in the (entity-centric) context of a network structure that explicitly represents the implicit word relations in the documents. In the following, we give an example of such a process for exploring the evolution of topics over time in a document collection.

5 Evaluation

We give a description of the news article data used in our exploration of topics, before presenting the results for topic networks and a comparison to LDA topics.

5.1 News Article Data

To demonstrate the advantages of a network-based approach and investigate the evolution of topics, we focus on the extraction of topics from news articles that provide both a temporal component and a large scale. Since we are unaware of existing data sets that are sufficiently large and annotated for named entities (or even topics), we collect the articles from the RSS feeds of a variety of international outlets. For reproducibility, we make the list of article URLs available[1].

Data Collection. We collect all articles from 14 English-speaking news outlets located in the U.S. (CNN, LA Times, NY Times, USA Today, CBS News, The Washington Post, IBTimes), Great Britain (BBC, The Independent, Reuters, SkyNews, The Telegraph, The Guardian), and Australia (Sidney Morning Herald). We collect all their feeds that are related to political news from June 1, 2016, to November 30, 2016. After removing articles that have less than 200 or over 20,000 characters or more than 100 identified entities per article (e.g., lists of real estate sales), we obtain 127,485 articles with 5.4M sentences.

Data Preparation. We use manually created extraction rules for each news outlet to strip the HTML code and cleanly extract the text, before performing sentence splitting, tokenization, entity recognition, entity linking, entity classification, and stemming. For the recognition and classification of named entities, we use the Ambiverse API [2], which disambiguates entity mentions to Wikidata IDs. To classify named entities into persons, locations, and organizations, we map Wikidata entities to YAGO3 entities and classify them according to the YAGO hierarchy since a classification in Wikidata is problematic [17]. We use the class `wordnet_person_100007846` for persons, `wordnet_social_group_107950920` for organizations and `yagoGeoEntity` for locations. For sentence splitting and part-of-speech tagging, we use the Stanford POS tagger [19].

Network Construction. To construct the network, we use a modified version of the LOAD implicit network extraction code [18] that we adapted to utilize disambiguated entities and add document timestamps and outlet identifiers to edges. Terms are stemmed with the Snowball Porter stemmer[3]. We set the window size for the extraction of entity cooccurrences to $sdist = 5$. The resulting network has 27.7 k locations, 72.0 k actors, 19.6 k organizations, and 329 k terms, which are connected by 10.6 M labelled edges. To generate edge weights for the resulting network, we use the weighting scheme ω described in Sect. 4.1.

5.2   Entity-Centric Extraction of Topics

As a first step of our exploration, we consider the extraction of traditional topics as lists of words with importance weights. Based on the underlying assumption that topics are focussed on entities, we first obtain a ranking by weight of all edges in the network that connect two entities. Thus, we utilize a global ranking of edges to identify relevant seeds for topics, which stands in contrast to the local entity-centric approaches that have been used on implicit networks so far [16,18]. The top-ranked edges are then considered to form the seeds of

---

[1] The URLs of articles in our data, the extracted implicit network, and our program code are available at https://dbs.ifi.uni-heidelberg.de/resources/nwtopics/.

[2] https://www.ambiverse.com/.

[3] http://snowballstem.org/.

Table 1. Traditional topics as ranked lists of terms, extracted for the four top-ranked edges in the network generated from the subset of NewYork Times articles. For each edge, the two incident entities and their Wikidata identifiers are given.

Beirut - Lebanon		Russia - Moscow		Russia - Putin		Trump - Obama	
Q3820 - Q822		Q159 - Q649		Q159 - Q7747		Q22686 - Q76	
Term	Score	Term	Score	Term	Score	Term	Score
syrian	0.14	russian	0.28	russian	0.29	presid	0.40
rebel-held	0.12	soviet	0.06	presid	0.18	american	0.21
rebel	0.06	nato	0.06	annex	0.09	republican	0.19
cease-fir	0.05	diplomat	0.06	nato	0.08	democrat	0.19
bombard	0.05	syrian	0.06	hack	0.08	campaign	0.18
bomb	0.04	rebel	0.05	west	0.08	administr	0.17

topics. Subsequently, each such edge is grown to a topic description by adding neighbouring terms that are connected to both entities as described in Sect. 4.2. The resulting ranking can be utilized as a topic. Since only the local neighbourhood of the two entities is considered once the global ranking is obtained, this process is extremely efficient, can be parallelized by edge and computed at query time to obtain, expand, or reduce an arbitrary number of topics interactively.

As an example, we show the topics that are induced by the four highest ranked entity edges in the subset of NewYork Times articles in Table 1 (topics are ranked from left to right by the value of ω of the seed edge). We find that the topics are overall descriptive and can be interpreted within the context of news in 2016. With regard to location mentions, the example shows well a prevalent bias in news articles, which often include the location of the correspondent or news agency at the start of the article (i.e., articles about the war in Syria are often not reported from Syria itself but neighbouring Lebanon). However, even when seed edges overlap in a common entity, the resulting topics are still descriptive and nuanced, as the example of Russia shows, which is associated with aspects of Russian politics and the involvement of Putin. If topics that are focused on such synonyms (e.g., the mention of a capital instead of a country) are not of interest, filtering edges by entity type is easily possible.

Much like traditional topic models, the topics and topic qualities that we find vary strongly by news outlet, as we show in Sect. 5.4. Since the topics of each edge are independent, it is easier to discard unwanted topics than it would be for traditional topic models with interdependent topics. Overall, we find that we can replicate list-based topics with an edge-centric approach. However, a representation of topics as lists of terms is needlessly abstract and the network supports the extraction of visually descriptive topics, as we show in the following.

5.3   Topic Network Extraction and Exploration

To fully utilize the network representation, we extract topic substructures. Instead of lists of terms from nodes that surround seed edges, we extract the nodes themselves to continually grow a descriptive network structure. We proceed in the same way as described for traditional topics above by extracting a ranked list of entity-centric edges and selecting the top-ranked edges. For each edge, we then include a number of terms that are adjacent to both entities in the network. The number of adjacent nodes can be selected arbitrarily, but a value of around three term nodes per edge tends to result in a visually interpretable network. By projecting the network according to the publication date of the corresponding article, we can introduce a temporal dimension and investigate the evolution of topic networks over time or even dynamically for selected intervals.

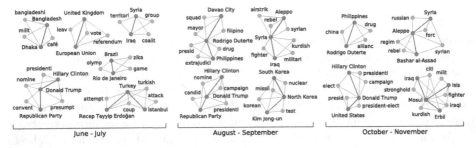

Fig. 2. Topic substructures of entities (purple) and terms (green) for the article subset of CNN. Shown are the 8 highest ranked edges and the 3 most relevant connected terms. The data is divided into three segments of two months to highlight topic evolution. (Color figure online)

In Fig. 2, we show temporal snapshots for the CNN article subnetwork. The results for other news outlets are similar (albeit with a regional or political bias). While we still find the same descriptive terms in the graph representation as we do in the case of ranked term lists, we also observe the additional structure of the underlying network. Unlike term lists, which represent isolated topics, the overlaps of edges show topic relations directly. In fact, we observe entity-entity subgraph structures that emerge from the top-ranked edges and lend further support to their topics. For example, the topics of Trump, Clinton, and the U.S. are clearly related. On the temporal axis, we find that the topics correlate well with political events. For example, the Brexit topic disappears after the referendum in June 2016 (of course, it is more pronounced in British outlets), while several war-related topics shift focus to follow ongoing campaign locations, and the US election topic is expectedly stable. Overall, the network representation adds a structure to the visualization that is easily recognizable and explorable.

5.4   Comparison to LDA Topics

It is well known that topics are subjective and a strict evaluation is difficult, especially for an exploratory approach. To relate network topics to traditional topic models, we compare their list-of-term representation to LDA topics [4]. We extract topics for each news outlet from the network as described in Sect. 5.2. For LDA topics, we group the news articles by outlet, prepare the plain text with Tidytext[4], and generate topics with the R implementation of LDA[5]. As a metric for the comparison, we compute the coverage to capture how well each of the topics that are produced by one approach are reflected in topics produced by the other approach. Formally, for two sets of topics T and U of size $k = |T| = |U|$, we let

$$coverage(T, U) := \frac{1}{k} \sum_{t \in T} \max_{u \in U} \{jaccard(u, t)\}.$$

Note that the coverage as defined above is not symmetric. Since entities are a major component of network topics, we add the tokenized labels of seed nodes to the term lists before selecting the top-ranked terms of a network topic.

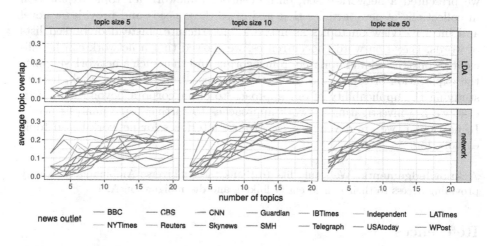

Fig. 3. Comparison of coverage between network topics and LDA topics. Topic size denotes the number of words per topic. Results are shown for the coverage of LDA topics in network topics (top row) and network topics in LDA topics (bottom row).

Figure 3 shows the comparison results for all 14 news outlets. The overall coverage increases with the number of topics and with the number of terms per topic. However, the coverage of LDA topics in network topics is much worse than the other way around. Combined with the increasing coverage for larger numbers of topics, this indicates the more narrowly focussed nature of network topics. The

---

[4] https://cran.r-project.org/web/packages/tidytext/.
[5] https://cran.r-project.org/web/packages/topicmodels/.

coverage of LDA topics in network topics increases with the number of topics since this narrows the scope of LDA topics. Overall, we find that network topics seem to be well reflected in LDA topics once the number of extracted topics for LDA is large enough to distinguish between the multitude of news topics. However, for a number of topics beyond 20, the runtime of LDA becomes an serious issue on the larger news outlets since the number of topics is a multiplicative factor in the runtime complexity of LDA, while it is an addend in the complexity of network topics. Furthermore, the number of network topics can be dynamically adjusted during the exploration phase and supports settings where the repeated extraction of traditional topics is too compute intensive.

6   Conclusions and Ongoing Work

The detection of topics in a (dynamic) collection of documents is a central task in corpus analysis, in particular for document classification and clustering. While there exist many approaches for topic modeling, they often fall short in terms of intuitive, interactive, and efficient topic exploration methods. In this paper, we presented a network-based, entity-centric framework for topic exploration in collections of news articles that addresses these needs and provides a novel and intuitive view on topics as network substructures instead of ranked lists of words. Based on word cooccurrences, we showed that a network can be efficiently constructed and updated from a corpus and that such a network supports the exploration of topics and their relationships. In our ongoing work, besides studying the applicability of our approach to other corpora beyond news articles, we are investigating network approaches to further topic analysis needs. In particular, we are developing easy-to-use interfaces for network-based topic exploration.

Acknowledgements. We would like to thank the Ambiverse Ambinauts for kindly providing access to their named entity linking and disambiguation API.

References

1. Blanco, R., Lioma, C.: Graph-based term weighting for information retrieval. Inf. Retr. **15**(1), 54–92 (2012)
2. Blei, D.M.: Probabilistic topic models. Commun. ACM **55**(4), 77–84 (2012)
3. Blei, D.M., Lafferty, J.D.: Dynamic topic models. In: ICML (2006)
4. Blei, D.M., Ng, A.Y., Jordan, M.I.: Latent Dirichlet allocation. J. Mach. Learn. Res. **3**, 993–1022 (2003)
5. Chang, J., Gerrish, S., Wang, C., Boyd-Graber, J.L., Blei, D.M.: Reading tea leaves: how humans interpret topic models. In: NIPS (2009)
6. Evert, S.: The statistics of word cooccurrences: word pairs and collocations. Ph.D. thesis, University of Stuttgart, Germany (2005)
7. Gretarsson, B., O'Donovan, J., Bostandjiev, S., Höllerer, T., Asuncion, A., Newman, D., Smyth, P.: TopicNets: visual analysis of large text corpora with topic modeling. ACM Trans. Intell. Syst. Technol. **3**(2), 23:1–23:26 (2012)

8. Gries, S.T.: 50-something years of work on collocations. Int. J. Corpus Linguist. **18**(1), 137–166 (2013)
9. Han, X., Sun, L.: An entity-topic model for entity linking. In: EMNLP (2012)
10. Hong, L., Yin, D., Guo, J., Davison, B.D.: Tracking trends: incorporating term volume into temporal topic models. In: KDD (2011)
11. Hu, Y., Boyd-Graber, J., Satinoff, B., Smith, A.: Interactive topic modeling. Mach. Learn. **95**(3), 423–469 (2014)
12. Newman, D., Chemudugunta, C., Smyth, P.: Statistical entity-topic models. In: KDD (2006)
13. Rousseau, F., Vazirgiannis, M.: Graph-of-word and TW-IDF: new approach to ad hoc IR. In: CIKM (2013)
14. Sarma, A.D., Jain, A., Yu, C.: Dynamic relationship and event discovery. In: WSDM (2011)
15. Shi, B., Lam, W., Jameel, S., Schockaert, S., Lai, K.P.: Jointly learning word embeddings and latent topics. In: SIGIR (2017)
16. Spitz, A., Almasian, S., Gertz, M.: EVELIN: exploration of event and entity links in implicit networks. In: WWW Companion (2017)
17. Spitz, A., Dixit, V., Richter, L., Gertz, M., Geiss, J.: State of the union: a data consumer's perspective on Wikidata and its properties for the classification and resolution of entities. In: Wikipedia Workshop at ICWSM (2016)
18. Spitz, A., Gertz, M.: Terms over LOAD: leveraging named entities for cross-document extraction and summarization of events. In: SIGIR (2016)
19. Toutanova, K., Klein, D., Manning, C.D., Singer, Y.: Feature-rich part-of-speech tagging with a cyclic dependency network. In: HLT-NAACL (2003)
20. Zuo, Y., Zhao, J., Xu, K.: Word network topic model: a simple but general solution for short and imbalanced texts. Knowl. Inf. Syst. **48**(2), 379–398 (2016)

Predicting Topics in Scholarly Papers

Seyed Ali Bahrainian[✉], Ida Mele, and Fabio Crestani

Faculty of Informatics, Università della Svizzera italiana (USI), Lugano, Switzerland
{bahres,ida.mele,fabio.crestani}@usi.ch

Abstract. In the last few decades, topic models have been extensively used to discover the latent topical structure of large text corpora; however, very little has been done to model the continuation of such topics in the near future. In this paper we present a novel approach for tracking topical changes over time and predicting the topics which would continue in the near future. For our experiments, we used a publicly available corpus of conference papers, since scholarly papers lead the technological advancements and represent an important source of information that can be used to make decisions regarding the funding strategies in the scientific community. The experimental results show that our model outperforms two major baselines for dynamic topic modeling in terms of predictive power.

Keywords: Topic prediction · Topic modeling
Temporal evolution of topics

1 Introduction

Scientific papers are a vehicle for advancing science and technological development. Discovering topics from scientific papers and analyzing their evolution over time is beneficial for making important decisions by governments, research organizations, funding agencies, and even researchers. As an example, research-funding organizations can adjust their granting policies based on insights produced by predictive models in order to favor topics that are trending and get increasing attention rather than those that are losing momentum and interest.

In this paper, we propose a novel evolutionary method capable of predicting the topics from the past time slices that would continue in the future time slices. Our model updates itself in an evolutionary process based on reinforcement learning which learns from the data and corrects itself over time. We use a publicly available dataset of papers from the Neural Information Processing (NIPS) conference which were published over 29 years. We compare the predictive performance of our method against two dynamic-topic-modeling baselines in terms of near-future prediction of continuing topics. The first baseline is the *Dynamic Topic Model* (DTM) by Blei et al. [4] which tracks the evolution of topics over time. DTM assumes that all topics are present in all the time slices of a sequential corpus of text. The second baseline is the *discrete Dynamic Topic*

© Springer International Publishing AG, part of Springer Nature 2018
G. Pasi et al. (Eds.): ECIR 2018, LNCS 10772, pp. 16–28, 2018.
https://doi.org/10.1007/978-3-319-76941-7_2

Model (dDTM) [1] which modifies DTM by relaxing the assumption that a topic should be present in all the time slices. Thus, dDTM tracks the evolution of intermittent topics over time, hence the word "discrete" in the name. It is noteworthy to mention that whether the topics extracted from the NIPS papers tend to be continuous or discrete, we use both DTM and dDTM as baselines to rigorously test our proposed model in the different possible scenarios.

The contributions of this paper are as follows:

– We present a novel approach for predicting the continuing topics over time.
– We conduct an analysis of the dataset of NIPS papers to show different features of our novel method, and compare it with DTM and dDTM baselines.

The remainder of this paper is organized as follows: Sect. 2 describes the related work, and Sect. 3 briefly explains the background on dynamic topic modeling. We present our topic tracking evolutionary model in Sect. 4. In Sect. 5, we show our experimental setup and results. Finally, Sect. 6 concludes the paper.

2   Related Work

In this section, we shortly review the related work regarding the use of topic models on scholarly articles.

An early work based on applying topic models to scientific articles was the one by Griffiths et al. [8]. They proposed an approach for reconstructing the official Proceedings of the National Academy of Sciences (PNAS). In a more recent work, Talley et al. [12] used topic models to create a high-level representation of scientific articles. They applied topic models to abstracts of National Institutes of Health (NIH) grant proposals to discover the research directions of various research groups and institutes. Their analysis showed unexpected overlaps in research priorities across institutes.

In 2006, Blei and Lafferty [4] introduced Dynamic Topic Model (DTM) for analyzing the evolution of topics in chronologically-ordered datasets. The model, based on Latent Dirichlet Allocation (LDA) [5], can capture the evolution of a topic over time and show various trends, for example, the changing probability of a term in a topic over time, or the popularity of that term at different time intervals. Based on DTM, Continuous-time Dynamic Topic model (cDTM) [13] and discrete Dynamic Topic Model (dDTM) [1] were introduced. cDTM can track any change in a topic and was shown to be effective for short time intervals and the changes in topics are often very small. On the other hand, dDTM provides more flexibility since it relaxes the assumption that topics need to be continuously present over all the time slices. Indeed dDTM can capture sudden variation in the topic change and also the evolution of intermittent topics over time.

All these temporal models can be used to discover latent topics discussed in sequential document collections and capture the evolution of topics by chaining the same topics over time. Although they were evaluated by computing log-likelihood on a future time slice given the data of past time slices, they were not used to predict the topics from the past that would continue in the future in terms of standard IR metrics. In this paper, we use the DTM and dDTM models as state-of-the-art baselines for temporal topic modeling and compare the performance of our topic-prediction model against them.

3   Background

3.1   Dynamic Topic Model

Topic models are defined as hierarchical Bayesian networks of discrete data where documents are distributions over topics and topics are represented as sets of words drawn from a fixed vocabulary that together represent a high-level concept [13]. These probabilistic methods can be used to have a low-dimensional representation of document corpora. Do et al. [7] state that "from an application viewpoint, topic model is a tool for extracting emergent hidden patterns from a collection of data."

LDA [5] is a well known topic model used to discover the latent topics in a document collection. Since the model is not influenced by the temporal ordering of the documents, it has the drawback of mixing together topics related to different temporal periods. To overcome this limitation, Blei and Lafferty proposed Dynamic Topic Model (DTM) [4] which divides a sequential corpus of documents into time slices. Then, it applies LDA to each of them in order to model the latent topics present in the time slices. The hyperparameters of LDA are chained together over consecutive time slices using a linear Kalman filter [9] which allows a linear evolution of the topics. To elaborate further, the parameters of each topic, $\beta_{t,k}$, are chained together in a state space model that evolves with a Gaussian noise. Subsequently, DTM draws each topic β such that:

$$\beta_{t,k}|\beta_{t-1,k} \sim \mathcal{N}(\beta_{t-1,k}, \sigma^2 I) \tag{1}$$

where $\mathcal{N}$ is a logistic normal distribution. The σ parameter, in Eq. 1, allows for variation in a topic over two subsequent time slices. By assigning small values to σ, the model ensures that one topic would not evolve to a different topic over two subsequent time slices. A similar evolution process holds for the α parameter, as α impacts the per-document topic proportions, θ, that is drawn from a Dirichlet distribution. The graphical model of DTM is illustrated in Fig. 1.

In spite of being a powerful model for statistical interpretation of a sequential corpus, DTM comes with two limitations:

(1) The assumption that topics change slowly over time which holds for some document collections (e.g., the articles from the journal *Science*) where topics evolve at a low pace, but does not hold for others (e.g., online discussions, news).

Moreover, the topic evolutions may have skips in the timeline and it is reasonable to assume that in textual streams, different topics may emerge,

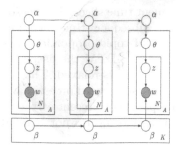

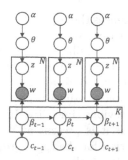

Fig. 1. Graphical model of DTM.

Fig. 2. Graphical model of dDTM. $C = c_0, \ldots, c_n$ is the resulting topic chain.

disappear, and appear again after some time. Hence the topics observed in one time slice might be completely different from the previous one and the assumption that each topic at time slice t has to be connected to a topic at time slice $t-1$ may be a limitation.

(2) As shown in Eq. 1, a topic evolves linearly with a Gaussian noise which allows only small variations in the topics over two consecutive time slices. The effect of such modeling is that if a new topic (i.e., it did not exist in the first time slice of the dataset) emerges in the data the model will not immediately respond to it. Only if the topic persists over a number of time slices, DTM would gradually take it into account in the evolution process and be able to model it. This causes DTM to be unable to immediately detect emerging topics.

In the next section we present another topic model called discrete Dynamic Topic Model (dDTM) which aims at addressing the above limitations. As demonstrated in previous works [1,10], it can model the above-mentioned real-world applications for users' posts and news articles more effectively. Differently from DTM, it uses a non-linear evolution process for the topics and does not rely on the approximations of the evolution of the topic proportions over time.

3.2 Discrete Dynamic Topic Model

The dDTM [1] estimates the topical chains based on the multinomial distributions over words (i.e., topics) in a non-linear fashion as opposed to DTM where topics evolve linearly over all time slices.

The graphical model of dDTM is illustrated in Fig. 2. As we can see, similarly to DTM, the model is based on LDA. Given a sequential dataset (i.e., stream of documents) divided into different time slices, dDTM applies LDA for computing topics in each time slice. As in DTM, the model chains together the multinomial distributions over observed words (i.e., topics), β, to estimate the latent topical chains over time. Differently from DTM, it does not estimate the α parameter

which influences the topic proportions over subsequent time slices. The reason is that in the case of streaming data, the data over two subsequent time slices might not always be similar enough, hence an estimation of topic proportions based on the first time slice might not be correct. Indeed, in dDTM the assumption that a topic would always be present in all the time slices is relaxed. Thus, it is not reasonable to adjust topic proportions of one time slice based on estimations from the previous time slice. In settings where a topic is not present at the time slice t_0, but it is present at t_1, the topic proportions of t_0 are not a reliable approximation for the topic proportions of t_1.

For chaining the topics over different time slices, dDTM uses a Hidden Markov Model (HMM) [11]. Similarly to the Kalman filter, it implements an Expectation Maximization (EM) algorithm, named Baum-Welch [6]. However, DTM using a Kalman filter makes the assumption that the evolution of a topic in the state space model is linear, and this is suitable for continuous-time settings. Whereas, the HMM used by dDTM enables to relax the assumption that every topic should be present in all time slices allowing the discovery of latent topic chains in discrete-state settings.

4   Our Model

In this section, we present a novel approach for predicting topics that will continue in the future. In our research on how to effectively predict the continuing topics over time, we devise various strategies which consider the *recency* and the *establishment* effects. The former is captured by increasing the weight of the most recent topics, whereas the latter assigns higher weights to the more established topics. Based on these two effects, we developed a dynamic method that combines recency and establishment measurements.

4.1   The K2RE Method

We call our methodology *Kalman combination of Recency and Establishment (K2RE)*. It combines the two effects of recency and establishment using a Kalman filter. In the following we first explain the two effects and then further elaborate on other details of our model.

Recency. The recency effect ranks the topics by assigning higher weights to most recent topics. Then, as an energy function it computes the correlation of each topic (whose continuation is to be predicted) with the vector of newly computed weights. We formally define the recency effect as follows: given the topics of the last n consecutive time slices of a sequential dataset, we would like to predict which topics continue in the $(n + 1)_{th}$ time slice. Let V be the vocabulary of all words occurring in the first n time slices. We construct a word vector containing probability scores corresponding to each word in V. The assigned probability scores are higher for the words appearing in the most recent topics. Thus, first we compute LDA topics (this is explained in Sect. 5.1) from the first n time slices

and then we compute the average probability of each word present in all topics according to the recency effect using the following equation:

$$P_{ref,Rec} = \sum_{n=1}^{N} \sum_{t=1}^{T} \sum_{w_i \in t} \frac{P(w_i) * 2^n}{(n * t)} \tag{2}$$

where n is the sequence number of the time slice, t is the number of topics derived from each time slice, and w_i is a word from that topic. The 2^n is the rate with which higher weights are assigned to recent topics. The resulting word vector is an average representation of the probability of all the words present in all the n time slices.

Therefore, this effect assigns higher weight to a word which has occurred in the most recent time slice of a sequential corpus. We refer to the word vector where the probability of each word is computed with Eq. 2 as the recency-reference vector.

Establishment. As for the establishment effect, given a vocabulary V made of all the words occurring in the first n time slices, we create a word vector containing probability scores corresponding to each word in V. In this case, the assigned probability scores are higher for the words which have persisted over time. For this purpose, we compute LDA topics (this is explained in Sect. 5.1) from the first n time slices and compute the average probability of each word present in all the topics according to the establishment effect using the following equation:

$$P_{ref,Est} = \sum_{n=1}^{N} \sum_{t=1}^{T} \sum_{w_i \in t} \frac{P(w_i) * 2^{-n}}{(n * t)} \tag{3}$$

where n is the time-slice sequence number, t is the number of topics derived from each time slice, and w_i is a word from that topic. The 2^{-n} is the rate with which higher weights are assigned to established topics and, as we can see, it is the opposite of Eq. 2. It weighs the words opposite to that of the recency effect. Therefore, the word vector constructed by averaging all topics in all n previous time slices based on the establishment effect will be a representation of the average occurrence of each word in V, where the most established (i.e., persisting in occurrence) words have higher weights. We refer to the word vector where the probability of each word is computed using Eq. 3 as the establishment-reference vector.

Combining Recency and Establishment. Now that we defined the recency and establishment effects, we explain how to combine them. Our model dynamically estimates the weights of each of the effects for each time slice and corrects itself over time by learning from the data. We refer to this method as a *Kalman combination of Recency and Establishment (K2RE)*. It utilizes the Kalman filter [9] to estimate the recency and establishment weights over time. By measuring the dataset changing behavior in terms of establishment and recency, K2RE adapts itself to the dynamics of the data over time.

In the following, we first present a general overview of K2RE and then elaborate on its details. Figure 3 illustrates the components of the K2RE method for integrating the scores from the recency and the establishment effects.

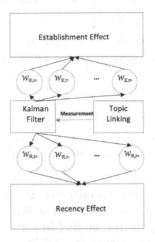

Fig. 3. Components of the K2RE method.

The linear interpolation for a time slice t is defined as:

$$K2RE_t = w_{E,t} * Score_{establishment} + w_{R,t} * Score_{recency} \qquad (4)$$

where $Score_{establishment}$ and $Score_{recency}$ are computed by the establishment and the recency effects, respectively. Furthermore, $w_{E,t}$ and $w_{R,t}$ are establishment weights and recency weights computed by the Kalman filter at time t, such that $w_{E,t} + w_{R,t} = 1$. This means that at each time slice t, each of the two effects will be given a weight that reflects its contribution to the future topics. The weights can be either equally assigned or they reflect the effect that dominates. The Kalman filter is always initialized by assigning equal probability of 0.5 to both $w_{E,t}$ and $w_{R,t}$ and then it dynamically updates the weights based on the learning from the data, according to the following system of equations:

$$f(n) = \begin{cases} X^G = A^G f_{t-1} + \varepsilon_t^G & t = 2,\ldots,T \\ z_t = H^G f_t + w_t^G & t = 1,\ldots,T \end{cases}$$

where f_t is the system state at time t, A^G denotes the transition of the dynamic system from $t-1$ to t, H^G describes how to map state f_t to an observation, z_t (i.e., measurement), and both ε_t^G and w_t^G are mutually independent Gaussian noise variables with co-variances R_t and Q_t, respectively. The dynamic system evolves over time and updates itself proportional to the Kalman gain.

Now we describe the topic-linking module which is based on the dDTM [1]. As described in Sect. 3.2, dDTM tracks the evolution of intermittent topics that

may occur discretely over time, such that a topic does not need to be necessarily present over all the time slices. In the topic-linking module shown in Fig. 3, we use a component of dDTM which links together similar topics over time. Such linking can be discrete or continuous under our model (i.e., a topic may be present over all time slices or it may skip some time slices). In particular, we use a Gaussian random walk in a Markovian state space model. The Markov assumption enforces probabilities of a hidden state at time t to be computed merely depending on the previous time slice and not based on all the previous states. We utilized this assumption to compute topic chains that capture the evolution of a topic discretely over time, so that two topics will be linked over two different time slices if they are similar according to the following criterion:

$$\beta_{t,k}|\beta_{t-m,1..k} \sim \mathcal{N}(\beta_{t-1}, \sigma^2 I) \tag{5}$$

where $\beta_{t,k}$ is topic k at time slice t, and $m \in \{1, 2, 3, \ldots, n\}$ with n being the number of previous time slices and σ is the maximum variance allowed from the mean of a topic in the previous time slice. By assigning a small value to σ, our model links two topics that are highly similar. Furthermore, we use the Baum-Welch [3] algorithm to learn the forward and backward probabilities of the transitions among the topics. The model takes as input the topics from the first n time slices and computes their continuation in the $(n+1)_{th}$ time slice.

After linking similar topics over every two consecutive time slices, the topic-linking module computes the recency rate of the topics for time slice n which is the number of topics in the time slice $n-1$ that have been present in the time slice n divided by the total number of topics in the same time slice. This measurement is given as the observation matrix to the Kalman filter for each time slice. Subsequently, the Kalman filter computes the evolution of recency and establishment weights using the Kalman filter system of equations and, using Eq. 4, a K2RE reference vector is generated.

For the purpose of computing correlation between each topic and the K2RE reference vector, we use the Pearson correlation metric that is $P_{X,Y} = \frac{COV(X,Y)}{\sigma_X \sigma_Y}$, where $P_{X,Y}$ is the Pearson correlation of two populations X and Y, COV is the covariance, and σ is the standard deviation. In our use case, the Pearson correlation indicates the level of correlation of each topic with the K2RE reference vector. The intuition behind using correlation as an energy function is that, topics are by definition a set of words that depend on one another and a change in one word may cause changes in the probabilities of other words. According to this intuition we chose the Pearson correlation as an energy function for comparing word vectors. Finally, we require a threshold for distinguishing continued and discontinued topics. To determine an effective threshold we use *10-fold cross validation*. Hence, we split the dataset in 10 folds, iteratively leaving out a chunk of the data, and compute the threshold which minimizes the Mean Squared Error (MSE) of prediction on the remaining folds. Then, using the computed threshold, we evaluate the left-out fold.

5   Experimental Setup

In this section we present our experimental setup, including a description of the dataset followed by the evaluation of our approach.

5.1   Dataset Description

Our dataset consists of all the papers of the Neural Information Processing Systems (NIPS) conference published between years 1987 and 2015 [1]. Therefore, our dataset is spread over 29 years. The total number of papers is 5993.

Topic extraction: The dataset is sorted chronologically and divided into time slices of fixed size (one year). We treat every paper as a document and applied LDA to extract the latent topics from each time slice. Since the number of topics (K) discussed in two different time slices might vary, it is important to estimate the number of topics per time slice. For this purpose, similar to the method proposed in [8], we went through a model selection process. This consists in keeping the LDA Dirichlet parameters (commonly known as α and η) fixed and assigning several values to K. We computed an LDA model for each assignment and subsequently we picked the model that satisfies:

$$\underset{K}{argmin}\ logP(W|K)$$

where W indicates all the words in the corpus. We repeated this process for each time slice to find the optimal number of topics for all the time slices.

Labeling: In our prediction tasks we assume that given the topics of the first n time slices (the first 28 years of the dataset) we would like to predict those that will persist in the $(n+1)_{th}$ time slice (the $29th$ year of the dataset). We carried out the labeling process semi-automatically. At first, using a k-nearest-neighbors implementation, for each of the topics from the first 28 time slices we identify the top 5 neighboring topics in the $29th$ time slice to simplify the annotation task. Then, given a topic from the first 28 time slices and its top 5 neighbors in the last time slice, we asked three human assessors (who were domain experts) to determine whether the topic was a continuation or not. This was done for all the topics of the last 28 time slices to have a ground truth for the prediction task. By aggregating the votes of the 3 human assessors each topic was labeled.

The assessors were given instructions on how to label the topics as 'continued' and 'not continued'. These instructions include putting more emphasis on the top 20 words in each topic to take a decision. That is due to the fact that the users of our system would look at the top words of each topic to understand it. As a result, our dataset consists of 839 topics out of which 305 topics are labeled as continued and 534 topics are labeled as discontinued.

---

[1] The dataset was downloaded from https://www.kaggle.com/.

5.2   Evaluation

In this section we present the evaluation of our method against the state-of-the-art dynamic topic modeling approaches.

First experiment. We evaluate our method using standard IR evaluation metrics, namely, precision, recall, F_1 measure and Mean Average Precision (MAP).

For performing the continuing-topic prediction task using DTM and dDTM we compute the log likelihood of each topic (whose continuation has to be predicted) under the corresponding model. Then we normalize the log likelihood scores to compute a score between 0 and 1. Furthermore, similar to the K2RE case we use 10-fold cross validation to compute the performance of all the models.

The dDTM model estimates the number of topic chains automatically. However, both dDTM and DTM require that we manually set the number of topics per each time slice. As we explained in Sect. 5.1, we estimated the number of topics in each time slice using model selection when building our dataset. Since the average number of estimated topics per time slice was very close to 30 we initialize both DTM and dDTM by setting their number of topics to 30.

Figure 4 shows an interpolated precision-recall diagram. The figure shows that our novel model outperforms the baselines in terms of the precision-recall curves and MAP.

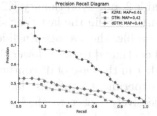

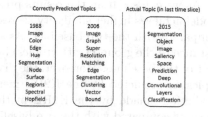

Fig. 4. Precision and recall of our model against DTM and dDTM.

Fig. 5. An example of a continued topic against the ground truth

Furthermore, Table 1 shows a comparison between all the methods with respect to precision, recall, F_1 measure, and MAP. Our results show that the K2RE method outperforms the two other state-of-the-art dynamic topic models used as baselines.

Second experiment. As a second experiment, we use a more objective evaluation without relying on human annotated data. For this purpose, we compute the average Euclidean distance between each topic that was predicted by our model as "continued" and its closest neighboring topic in the last time slice. By doing so, we can compare our model against the other models in terms of how correctly they could distinguish between those topics that continued and the ones that they did not. In this case the lower the distance measure the better each model has performed. Furthermore, by repeating the same procedure for

Table 1. Comparison of approaches: precision, recall, F_1 measure, and MAP.

	Precision(%)	Recall(%)	F_1(%)	MAP(%)
K2RE	61.56	84.78	71.33	61.53
DTM	50.03	85.66	63.17	42.84
dDTM	49.95	89.51	64.12	44.38

the topics that did not continue we can again compute an average Euclidean distance for how similar they are to the actual topics of the last time slice. In this case the higher the distance the better a model has performed.

We present the results of this experiment in Table 2.

Table 2. Comparison of approaches based on distance with ground-truth topics

	K2RE	DTM	dDTM
Ave. Euclidean dist. (Continued)	0.0305	0.0543	0.0493
Ave. Euclidean dist. (Discontinued)	0.1026	0.0744	0.0850

As we can see from the results presented in Table 2, our evolutionary K2RE model achieves higher similarity scores to the actual topics in the last time slice compared with the two baselines. As shown in the table, the average Euclidean distance to the ground truth (topics) in the case of continued topics is lower than that of the baselines. Moreover, the distance is higher in the case of discontinued topics as predicted by the K2RE method. This experiment confirms that the predictions made by the K2RE method are objectively closer to the ground truth as compared with the two baselines.

Qualitative example. Finally, we show a qualitative example of two topics which were predicted by K2RE to continue in the last time slice against the actual topic appearing in that time slice. We chose a topic about "image processing" (shown in Fig. 5) and we observed that *image processing* is an important topic for the NIPS conference over the years. Image segmentation based on color changes to image super resolution and then ten years later into deep convolutional neural networks, object detection, and image segmentation. As a further example on how our model correctly predicted a discontinued topic, we present the top-10 words of a randomly chosen topic from the year 1999: "spatial, temporal, localization, space, vector, belief, sequence, probability, robot, state". As we can see, this topic is mostly related to the navigation and localization of a robot. Our analysis of the data shows that the word "localization" did not occur in the 2015 time slice, hence such topic disappeared. A funding agency or a research organization with access to such insights (e.g., which topics continue and what are the main themes of a continuing topic over time) can make informed decisions and planning. Indeed, by looking at the image processing topic in 2014 we might be able to come up with more specific and detailed directions of research about 2015 (e.g., use of convolutional neural networks) but knowing the general trend and how past research is evolving into the future is also of strong importance.

6   Conclusion

In this paper, we introduced an evolutionary model capable of predicting topics that continue in the next time slice in a sequential corpus of documents. Our results showed that our method outperforms the state-of-the-art dynamic topic models in the prediction task. Our evolutionary K2RE model can learn the changes in the data over time and adapt itself to the changes. We used a corpus of scholarly papers to show the effectiveness of our model.

As a future work we plan to extend our model to other domains. As an example, our model could be adapted to the meetings scenario to prepare users for a future meeting [2] or be used in recommender systems for tracking user intent and context over time in order to anticipate users' information needs.

References

1. Bahrainian, S.A., Mele, I., Crestani, F.: Modeling discrete dynamic topics. In: Proceedings of the Symposium on Applied Computing, SAC 2017, pp. 858–865 (2017)
2. Bahreinian, S.A., Crestani, F.: Towards the next generation of personal assistants: systems that know when you forget. In: Proceedings of the ACM SIGIR International Conference on Theory of Information Retrieval, ICTIR 2017, pp. 169–176 (2017)
3. Baum, L.E., Sell, G.R.: Growth transformations for functions on manifolds. Pacific J. Math. **27**(2), 211–227 (1968)
4. Blei, D.M., Lafferty, J.D.: Dynamic topic models. In: Proceedings of the 23rd International Conference on Machine Learning, ICML 2006, pp. 113–120 (2006)
5. Blei, D.M., Ng, A.Y., Jordan, M.I.: Latent dirichlet allocation. J. Mach. Learn. Res. **3**, 993–1022 (2003)
6. Dempster, A.P., Laird, N.M., Rubin, D.B.: Maximum likelihood from incomplete data via the EM algorithm. J. Roy. Stat. Soc. B **39**(1), 1–38 (1977)
7. Do, T.-M.-T., Gatica-Perez, D.: By their apps you shall understand them: mining large-scale patterns of mobile phone usage. In: Proceedings of the 9th International Conference on Mobile and Ubiquitous Multimedia, MUM 2010, pp. 27:1–27:10 (2010)
8. Griffiths, T.L., Steyvers, M.: Finding scientific topics. In: Proceedings of the National Academy of Sciences (2004)
9. Kalman, R.E.: A new approach to linear filtering and prediction problems. Trans. ASME-J. Basic Eng. **82**(Series D), 35–45 (1960)
10. Mele, I., Bahrainian, S.A., Crestani, F.: Linking news across multiple streams for timeliness analysis. In: Proceedings of the 2017 ACM on Conference on Information and Knowledge Management, CIKM 2017, pp. 767–776 (2017)
11. Rabiner, L.R.: A tutorial on hidden markov models and selected applications in speech recognition. In: Readings in Speech Recognition (1990)

12. Talley, E.M., Newman, D., Mimno, D., Herr II, B.W., Wallach, H.M., Burns, G.A., Leenders, A.M., McCallum, A.: Database of NIH grants using machine-learned categories and graphical clustering. Nat. Methods 8(6), 443–444 (2011)
13. Wang, C., Blei, D., Heckerman, D.: Continuous time dynamic topic models. In: Proceedings of Uncertainty in Artificial Intelligence (2008)

Topic Lifecycle on Social Networks: Analyzing the Effects of Semantic Continuity and Social Communities

Kuntal Dey[1], Saroj Kaushik[2], Kritika Garg[3], and Ritvik Shrivastava[4]

[1] IBM Research, New Delhi, India
kuntadey@in.ibm.com
[2] IIT, Delhi, New Delhi, India
saroj@cse.iitd.ac.in
[3] Ch. Brahm Prakash GEC, New Delhi, India
kgarg.kritika@gmail.com
[4] NSIT, New Delhi, India
ritviks.it@nsit.net.in

Abstract. Topic lifecycle analysis on Twitter, a branch of study that investigates Twitter topics from their birth through lifecycle to death, has gained immense mainstream research popularity. In the literature, topics are often treated as one of (a) hashtags (independent from other hashtags), (b) a burst of keywords in a short time span or (c) a latent concept space captured by advanced text analysis methodologies, such as Latent Dirichlet Allocation (LDA). The first two approaches are not capable of recognizing topics where different users use different hashtags to express the same concept (semantically related), while the third approach misses out the user's explicit intent expressed via hashtags. In our work, we use a word embedding based approach to cluster different hashtags together, and the temporal concurrency of the hashtag usages, thus forming topics (a semantically and temporally related group of hashtags). We present a novel analysis of topic lifecycles with respect to communities. We characterize the participation of social communities in the topic clusters, and analyze the lifecycle of topic clusters with respect to such participation. We derive first-of-its-kind novel insights with respect to the complex evolution of topics over communities and time: temporal morphing of topics over hashtags within communities, how the hashtags die in some communities but morph into some other hashtags in some other communities (that, it is a community-level phenomenon), and how specific communities adopt to specific hashtags. Our work is fundamental in the space of topic lifecycle modeling and understanding in communities: it redefines our understanding of topic lifecycles and shows that the social boundaries of topic lifecycles are deeply ingrained with community behavior.

© Springer International Publishing AG, part of Springer Nature 2018
G. Pasi et al. (Eds.): ECIR 2018, LNCS 10772, pp. 29–42, 2018.
https://doi.org/10.1007/978-3-319-76941-7_3

1   Introduction

Twitter has been a key social network platform for diffusion of information via user interactions. Several research works have been carried out, that analyze the user-generated content, to identify the characteristics of information diffusion. One core research area has focused on the topics present in user-generated content, either via hashtag analysis or sophisticated text-analytics driven derivations. And based upon that, research has further focused on identifying the topics of user interest, and understanding the lifecycle of these topics - how these topics emerge, how they spread over the social network successfully (or not) and proliferate across several users, and eventually how they subside over time.

Some works in the literature have attempted to investigate lifecycles of topics. In a pioneering work, Ardon *et al.* [2] investigated the shape and rate of adoption of topics among social users, where they treated hashtags as topics. They observed that, topics (hashtags) have a five-phase lifecycle, peaking in the middle phases. They presented a detailed study of the social graphs associated with the topics, such as the degree distributions, the presence (and essence) of giant components, and geographical distributions.

Other works have also attempted to understand the lifecycle of topics; however, they have focused on the linguistic aspects more than the social aspects, and have treated the topic lifetime problem (how long a topic lasts, without focusing on *socially with whom*) in the form of a hashtag disambiguation problem. In an early work, Yang and Leskovec [21] detected similar distributions of usage of given Twitter hashtags in form of temporal usage shapes, using K-Spectral Centroid (KSC) clustering. However, this work did not investigate (a) the temporal overlap of different hashtags - whether or not a given pair of hashtags occurs at similar times, (b) the semantic concept space addressed by the corresponding tweets - if two different hashtags originate from tweets with the same meaning then it goes uncaptured, and (c) the social angle was completely missing too. In a recent work, Stilo and Velardi [20] proposed SAX, a temporal sense clustering algorithm based on the hypothesis that semantically related hashtags have similar and synchronous usage patterns. Thus, SAX overcomes a key shortcoming of KSC by considering the temporal overlap of different hashtags. However, it still does not account for the social angle; and in addition, does not attempt to consider the semantic space overlap across hashtags, which in turn leads to clustering of topically unrelated tweets also. Further, none of these approaches attempt to understand the morphing of topics and whether intricate social community interaction dynamics are associated with any such morphing.

On the contrary, we believe that, social communities (that are formed purely based upon familiarity structures), and the intricacy of interactions of users, are the core determinants of topic lifecycle - how topics are born, how they spread, and how they die and morph. However, we note that, in order to understand topics in the true sense, one needs to first acknowledge that, (a) in reality topics spread over and beyond a single hashtag: *#federer* and *#rogerfederer* are the same topics really, and (b) considering the latent semantic concept space of tweets is insufficient to account for the user's intent unless the hashtag is also considered: "I love him" is not the same as "I love him #Obama" - the former is probably a simple expression of personal love while the later is clearly a political expression. Hence, we propose a novel technique to bring related hashtags together by clustering as a combination of the semantic space (*#NFL* is National Football League for sports but National Fertilizers Limited for agriculture), hashtags and the time of expression (#USOpen is "obviously" the golf tag during the golf time but the tennis tag during the tennis time). We hypothesize that, hashtags, and topics derived using the hashtags, bear the following characteristics.

- **Hypothesis 1 - Conceptually related hashtags overlap semantically and temporally:** Different users use different hashtags at the same time for the same topic, that are semantically related and temporally overlapping. That is, one user would use *#wimbledon* while another would use *#bigW*, but their content would semantically (conceptually) overlap, and the usage would be temporally around similar (overlapping) times too.
- **Hypothesis 2 - Hashtags associate with communities at a given time:** Hashtag usages are community-level characteristics rather than individual-level. Individuals mostly tend to use the same hashtag that their community would use, for a given topic, at a given time. That is, if two users u_1 and u_2 belong to the same community, then they both are likely to use *#federer* instead of one using *#federer* and the other using *#rogerfederer*.
- **Hypothesis 3 - Hashtags are independently used across communities:** Inter-community independence of hashtag usage is an inherent property of social networks. That is, for the same topic, at the same time, while one community would use one hashtag, another community would use another hashtag. That is, community C_1 as a whole would tend to use the term *#federer* while community C_2 as a whole would tend to use the term *#rogerfederer*.
- **Hypothesis 4 - Hashtags evolve independently (atomically) within communities:** Evolution and lifecycle of hashtags (and topics) are community specific. The global (overall) lifecycle of a given topic can be derived as an aggregation of the lifecycle of topics within individual communities, along an overall span of time. For example, in a given span of 7 days, community C_1 would use the hashtag *#federer* for the first 2 days and then use the hashtag *#rogerfederer* for the next 5 days, while, community C_2 would use *#federer* for the first 4 days and then *#rogerfederer* for the next 3 days. The overall graph structure will suggest a majority usage of *#federer* for the first 2 days (since both C_1 and C_2 use this hashtag in the first 2 days), a mixed usage for the next 2 days (since C_1 uses one and C_2 uses another hashtag during this period) and a majority usage of *#rogerfederer* in the final 3 days. However, within the graph, the evolutions have a clear boundary - they are distinct, without much mixing, when investigated atomically from the standpoint of communities.

We demonstrate the effectiveness of our approach using around 20–30% of eighteen days of Twitter data. We observe that, all the four observations we have made, are novel in the literature. Our work is the first of its kind in the space of Twitter topic lifecycle analysis, and presents insights that are fundamental for understanding the underlying dynamics of topics and their lifecycles.

2   Related Work

The topic identification literature on Twitter has used three different approaches. First, hashtags have been treated as topics, such as by [8]. Second, a burst of keywords in a short span of time are identified, and each bursting keyword is treated as a topic. Works, such as [6,7,13], use this. Third, the latent semantic concepts of given tweets - often identified with sophisticated text-to-topic assignment techniques such as Latent Dirichlet Allocation (LDA) [4] - are treated as topics, and the tweets that address these spaces are said to belong to these topics. Works, such as [12], follow this. The first two approaches miss out on the latent semantic concept space addressed by the content, since they simply examine the keywords and hashtags instead of the overall content space. Thus, these approaches would not be able to identify that tweets containing hashtags *#mj*, *#michaeljackson*, *#jackson* and *#m_jackson* potentially address the same topic. The third captures the semantic space of the concept inside the text well, but miss the explicit user intent expressed via hashtags. Other works, such as [10,15,17], also use hashtag and LDA based methods for identifying topics, and analyzing their spatio-temporal evolution.

The Twitter topic lifecycle analysis literature has seen a strong work by Ardon *et al.* [2]. They observe five phases in event lifecycles: pre-growth phase, growth phase, peak phase, decay phase and post-phase. They perform the topic lifecycle analysis using individual hashtags as topics, and they further use a tool to identify places, entities *etc.* and assign these as tags (in turn, these tags become topics). Amongst other works, the K-Spectral Centroid (KSC) clustering approach by Yang and Leskovec [21] detects occurrence pattern similarity of hashtags, but does not consider any of, the time of occurrences, the semantic concept covered by the tweets having these hashtags, and the social network (friendship of users) aspects. Stilo and Velardi [20] propose SAX, that overcomes the temporal overlap aspect, but does not address the other two (semantic and social).

In general, the space of information diffusion has been extremely well-studied on Twitter. Several works, such as Bakshy *et al.* [3], Kawk *et al.* [11] and Myers *et al.* [14], have investigated this problem. Social affinity of discussions on Twitter has been observed by Narang *et al.* [17], and the geo-spatial characteristics of such discussions have been studied by Nagar *et al.* [16]. Many other works also galore. An extensive survey of the literature, towards information diffusion and topic lifecycle analysis, has been conducted by Dey *et al.* [9].

However, no work in the prior literature examines the lifecycle of a collection of hashtags with topics in the context of communities. Further, none of the works attempt to investigate along the lines of correlating social communities with information topic lifecycles. Our work, thus, is the first of its kind.

3   Our Approach

The input to our system is a collection of tweets that consisting of at least one hashtag. The aim is to (a) create topics by creating clusters of semantically related hashtags with temporal overlap, (b) create communities, and (c) analyze the hashtag and topic lifecycles with respect to the communities, in terms of how topics morph over evolutions of hashtags within and across communities, as described in Sect. 1.

The overview of our approach is as follows. We create a timeline for the hashtags, tracking the usage frequency (count) of each hashtag within each timeslot. We identify a word embedding for each hashtag using the content associated with it (since hashtags by themselves are non-dictionary words), using pre-trained embedding. Using the similarity of embeddings as the distance measure for each pair of hashtags, we perform k-means clustering of the hashtags. These clusters are further split such that, each hashtag present in a given (splitted) cluster temporally overlap in terms of occurrence. Each cluster of hashtags (after splitting) is treated as a topic. We identify modularity-based communities [18] that are present in the underlying social network. The hashtag usage of each user of a given community is aggregated to derive the hashtag usage made by the community, thereby creating a hashtag usage timeline of each community as a whole. In addition, we overlap the topic cluster memberships of these hashtags, to create a topic participation timeline for each community as a whole. These timelines are used to obtain hashtag-level and topic-level insights, in a community-agnostic manner as well as in the context of communities.

The details of our approach are provided below.

3.1   Identifying "Word Embedding" of Hashtags

We identify semantically related hashtags, using a word embedding technique followed by k-means clustering.

Step 1: Document creation
We create a document for each hashtag appearing in the dataset. Let the set of hashtags present in the document be $H = \{h_1, h_2, h_3, ...\}$. To this, we collect all the tweets t_{h_i} where a given hashtag h_i appears, and then append the tweets. Thus for each hashtag h_i, we create a document D_{h_i} as $D_{h_i} = \bigcup \{t_{h_i}\}$.

Step 2: Computing the "word embedding" of hashtags
In the next step, a word embedding model is created for each document (corresponding to a hashtag). We eliminate all the hashtags occurring in document D_{h_i}, as well as, eliminate all the mentions. We take the pre-trained Twitter-specific version of GloVe word embedding [19] as an external resource, which has been learned on 2 billion tweets containing 27 billion tokens with a 1.2 million vocabulary size. Let W_{h_i} be the set of words appearing in D_{h_i}. For each word $w_{h_i} \in W_{h_i}$, that is, each word that appears within the document of the hashtag, we look up the GloVe embedding of the word, and if found, we retain

the word along with its embedding. Finally, we compute an embedding v_{h_i} for each given hashtag h_i as a whole, using the embedding of the words that appear in the tweets containing the hashtag. We compute this as the average of all the word embeddings that appear in its document.

$$v_{h_i} = \frac{\sum\limits_{w_{h_i} \in D_{h_i}} (v_{w,h_i})}{|D_{h_i}|} \tag{1}$$

In Eq. 1, $|D_{h_i}|$ represents the total length of the document D_{h_i} as a count (total number) of the words appearing in the document, retaining words as many times as they appear. The repeating behavior of words is retained, as this implicitly provides proportionate weight the embedding bears in the context of that hashtag: a word more used along with a given hashtag will get counted more frequently. Further, in Eq. 1, v_{w,h_i} denotes the embedding of an individual word present in the pre-trained embedding. The computation is repeated for all hashtags $h_i \in H$, creating a complete embedding map, for all the words w_{h_i} under the context of all the hashtags h_i that they appear in.

3.2 Topic Cluster Creation Using Related Hashtags

Semantically Related Hashtag Cluster Creation
We use the embeddings obtained in the earlier step, to obtain semantically related clusters. In order to do this, we define a distance function for a given pair of embeddings: the value of cosine similarity of two given embeddings is treated as the distance between the pair of embeddings. Cosine similarity of two vectors v_1 and v_2 (in this case, two embedding vectors) of dimension d is given as:

$$similarity = cos(\theta) = \frac{\sum\limits_{i=1}^{d} v_{1_i}.v_{2_i}}{\sqrt{\sum\limits_{i=1}^{d} v_{1_i}^2} . \sqrt{\sum\limits_{i=1}^{d} v_{2_i}^2}} \tag{2}$$

We now perform k-means clustering, in order to create clusters T_s of conceptually (semantically) related hashtags.

Temporally Relating Hashtags for Cluster Creation
Hashtags that would be contained in the same cluster, would be semantically as well as temporally related. Hence, in the next step, we examine each semantically related cluster $t_s \in T_s$ in terms of temporal overlap. Allen [1] created an exhaustive list of temporal relationships that can exist between a pair of time periods. This includes *overlap*: partof event A and event B co-occur, *meets*: event A starts as soon as event B stops, and *disjoint*: event A and event B share no common time point. In out setting, an event is an instance of a tweet using a given hashtag.

We create a time series of the individual hashtags, as well as the semantic clusters of hashtags obtained earlier. For each timeslot, we compute whether or not a given hashtag is used. We temporally relate a pair of hashtags h_i and h_j if they either satisfy the *overlaps* relationship, or if there exists one or more hashtags h_k, such that, h_i is temporally related to h_k, and h_k *overlaps* h_j, or, they are disjont by less than a threshold number of days (2 days for our experiments). Two hashtags h_i and h_j are temporally unrelated if $\nexists h_k$ such that h_i is temporally related to h_k, and, h_k *overlaps* h_j. The *temporally related* relationship is recursive in nature, and can be expressed as

$$h_i \odot h_j \implies \left((\exists h_k) h_i \odot h_k \right) \cap (h_k \circledcirc h_j) \tag{3}$$

where $\odot$ denotes the *temporally related* relationship and $\circledcirc$ denotes the *overlaps* relationship. A given semantic cluster T_s will be split into two (or more) clusters T_{s,t_1} and T_{s,t_2}, if there are two (or more) sets temporally related hashtags.

Topic Cluster Finalization
We finalize our topics, defined as hashtag clusters, such that each hashtag cluster consists of hashtags that are both semantically and temporally related. As an example, at the end of the process, hashtags such as {#tennis, #federer, #rogerfederer, #roger} *etc.* are expected to be together in one cluster together if they occur closely in time, while hashtags such as {#politics, #trump, #donlandtrump, #donald} *etc.* are expected to be together another cluster together.

3.3 Creating Community-Level Hashtag and Topic Timelines

Using the Twitter followership network of the users that posted the tweets, we discover modularity-based communities [18]. We subsequently perform aggregation of the users hashtag usage behavior, in order to find the total usage of each hashtag by community members, and find timelines. Two timelines are found.

Hashtag-level Usage Timeline of Communities
For each given timeslot, all the usages of a given hashtag for all the community members are summed up, to find the total number of usages of the hashtag by the community (that is, its members). This gives the usage characteristics of each hashtag for each community, over each timeslot. Further, we also note the topic cluster that each hashtag belongs to, which in turn gives, for each community, for each timeslot, a triplet

$$< community, timeslot, < topic \ and \ hashtag \ usage \ characteristics >>$$

wherein, each element within $< topic \ and \ hashtag \ usage \ characteristics >$ consists of the following triplet

$$< hashtag, cluster \ of \ the \ hashtag, usage \ count \ of \ the \ hashtag >$$

Topic (Cluster)-level Usage Timeline of Communities
For each given timeslot, for each community, we sum up the usage count of all the hashtags belonging to the same topic cluster. This is useful for identifying

the participation of each given community in the topic as a whole, within the given timeslot. This is captured in form of a triplet

$$< community, timeslot, < topic\ usage\ count\ over\ all\ hashtags >>$$

wherein, each element within $< topic\ usage\ count\ over\ all\ hashtags >$ consists of the following pair

$$< cluster\ of\ the\ hashtag, usage\ count\ of\ all\ the\ hashtags\ in\ the\ cluster >$$

3.4   Topic Lifecycle Analysis: Individual Topics and Communities

We investigate two main aspects of topic lifecycles, both for our community-agnostic analysis as well as the analysis in the context of communities.

Dominant Hashtag Detection and Topic Morphing

A dominant hashtag is the one which has been most frequently used within a given timeslot, among all the hashtags. In effect, it is the most representative hashtag of a topic at a given timeslot. If a topic t_k comprises of hashtags $H = \{_k h_{1,k}\ h_2, ...,_k h_m\}$ for a given timeslot and if a function g_c counts the number of times each hashtag $_k h_i$ was used, then, the dominant hashtag for the given timeslot is defined as

$$_k h_x = \forall(i)(max(g_c(_k h_i))) \tag{4}$$

While the traditional analysis of the dominant hashtag would tend to follow a lifecycle observed by Ardon *et al.* [2], the lifecycle of the topic would be different, as over time, one dominant hashtag would take over another. The change of the dominant hashtag of a given cluster over time, captures the morphing of the corresponding topic from being captured mostly by one hashtag to another. The analysis is conducted at the level of communities also, in order to find the dominant hashtag usage made by each community at each timeslot and its evolution over time. Note that, a topic morphs, when its dominant hashtag changes from one to the other.

Topic Intensity Detection

The intensity of a topic is derived as the summation of the number of times each hashtag is used. We compute it both for the topics overall, as well as for each community. It denotes the total presence of the topic (as a summation of the presence of its constituent hashtags) within the time slot, and in the other case, for each community. If a topic t_k comprises of hashtags $H = \{_k h_{1,k}\ h_2, ...,_k h_m\}$ for a given timeslot and if a function g_c counts the number of times each hashtag $_k h_i$ was used, then, the dominant hashtag for the given timeslot is defined as

$$_k h_x = \sum_{i=1}^{m}(g_c(_k h_i)) \tag{5}$$

Note that, a topic dies, when its intensity becomes zero. Further, if a topic intensity becomes zero within a community C_1 but is non-zero in another community C_2, it indicates that C_1 is no longer discussing the topic (the topic has died within community C_1) but C_2 is still discussing it (the topic is alive within C_2).

4   Experiments

Dataset Description

Our experiments use the tweet dataset[1] by Yang and Leskovec [21]. It comprises of around 20%–30% of entire Twitter data of that period. We use the data from 11th to 30th June 2009. The corresponding social network connections data was obtained[2] (Kwak *et al.* [11]). We pre-process the data, to retain all the hashtags that occurred between 40–1,000 times within this period. This ensures that hashtags occurring frequently enough are retained, while the hashtags that associate with an excessively high number of tweets (mostly outliers) get ignored. We retain the users that posted these tweets, and use the social connections among these users to form their social network subgraph. The dataset is presented on Table 1.

Table 1. Description of available data. All the tweets are from June 2009.

Total num. of tweets	Num. hashtags retained	Num. tweets retained	Num. users retained	Avg. num. tweets per user
18,572,084	4,244	471,470	158,118	2.98

Experimental Setup

We conduct our experiments on the given data, following the steps delineated in Sect. 3. We create 1-day timeslots for our experiments. We use the BGLL algorithm [5] for discovering communities. We use the KMEANS package of Python for doing k-means clustering. We repeat our experiments at different granularities of k for finding clusters. Since we have 4,244 hashtags, we range the value of k as $k = \{200, 400, 600, 800, 1000, 1200, 1400, 1600, 1800, 2000\}$, thus exploring at different clustering granularities. We create the timeline for individual hashtags, for topics (clusters), and for the participation of communities in different topics over different hashtags across the different timeslots.

Inspecting the Topic Clusters

We examine the topic clusters derived by our process, to inspect the effectiveness of the embedding-and-clustering approach, given the relative novelty of this approach for clustering hashtags on Twitter data. We present a few randomly chosen samples of topic clusters on Table 2. Given space constraints, we have picked some of the k-values at random (k being the number of clusters in the corresponding k-means clustering), and have shown one randomly chosen topic cluster from each randomly chosen k-value. It is visibly clear that the clusters are of consistently of good quality.

---

[1] https://snap.stanford.edu/data/twitter7.html.
[2] http://an.kaist.ac.kr/traces/WWW2010.html.

Table 2. Examples of random clusters with random k-values (k of k-means clustering)

k-value	Cluster content
2,000	#Nats, #Rangers, #WhiteSox
1,800	#musician, #musiclover, #singer
1,400	#Jackson, #jackson, #Rip, #1984, #jacko, #kingofpop
1,000	#marijuana, #drugwar, #drugs, #smoking
600	#Fashion, #tshirts, #shoes, #makeup, #clothing, #sneakers, #handbags
200	#cancer, #Health, #diet, #medical, #organic, #weightloss, #firstaid, #ynw, #healthy, #nutrition, #medicine, #stemcells, #Cancer, #drugs, #alcoholism, #hiv, #FDA

Topic Lifecycles - Overall and in Context of Communities: Our Findings

Our experiments provide strong support for all the four hypothesis we propose in our work. We create the following kinds of plots to support our hypothesis.

1. *User participation plots:* These plots show the participation of given users to given hashtags (by virtue of the user using the hashtags).
2. *Hashtag lifecycle plots:* These plots show the overall lifespan of individual hashtags.
3. *Topic lifecycle plots:* These plots show the overall lifespan of the topics (clusters), aggregated across hashtags.
4. *Hashtag lifecycle plots per-community:* These plots show the lifespan of given individual hashtags, for a given community, indicating the participation of the community as a whole to these hashtags.
5. *Topic lifecycle plots per-community:* These plots show the overall lifespan of the topics (clusters), aggregated across hashtags, for a given community, indicating the participation of the community as a whole to a given topic.

For qualitative analysis, we randomly choose two topic clusters from our dataset. Cluster C1 comprises of the hashtags *#marijuana, #drugwar, #drugs, #smoking* and cluster C2 comprises of the hashtags *#freeiran, #iranrevolution, #revolution*. We randomly choose two users making sure that they are not connected with each other, and plot their hashtag usage characteristics towards cluster C1 over time in Fig. 1. We observe that, they use different hashtags for the semantic concept captured by the cluster (one uses *#freeiran* while the other uses *#iranrevolution*). On manual inspection, we see this behavior frequently repeating in the overall dataset, though we restrict to only one visual example here due to space constraints. The observation supports our **first hypothesis** - *conceptually related hashtags overlap semantically and temporally.*

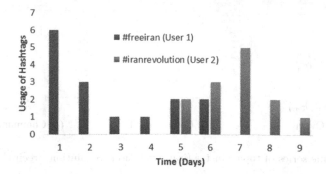

Fig. 1. Temporal overlaps of pairs of semantically related hashtags used by two random users

We capture the timeseries of the individual hashtags in Figs. 2(a) and 3(a), and the timeseries of these hashtags with respect to two randomly chosen communities, respectively in Figs. 2(b) and (c) for cluster C1, and Figs. 3(b) and (c) for cluster C2. It is visibly obvious from Figs. 2(b) and (c) that, while the overall topic sees a good mix of all the hashtags (see Fig. 2(a)), however, at given times, a given hashtag is clearly the dominant one in each community at a given time. Since the hashtag usage at the level of a given community is simply the collective (aggregate) behavior of the members of the community, it entails that hashtag

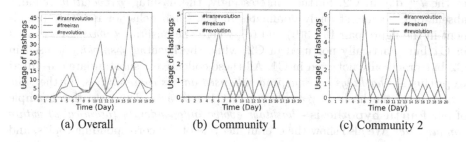

Fig. 2. Time series of hashtags (iranrevolution, revolution, freeiran cluster)

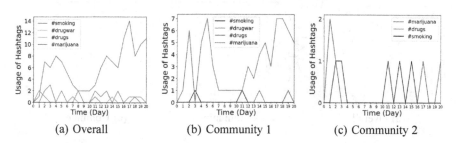

Fig. 3. Time series of hashtags (drugs, smoking, drugwars, marijuana cluster)

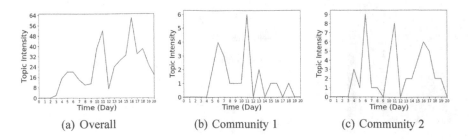

<table>
<tr><td>(a) Overall</td><td>(b) Community 1</td><td>(c) Community 2</td></tr>
</table>

Fig. 4. Time series of topic cluster (iranrevolution, revolution, freeiran cluster)

usage behavior is a community-level phenomenon. This characteristic is reflected clearly in the other cluster as well. These examples (and many others that we consistently observe, but do not report due to space constraints) corroborates our **second hypothesis** - *hashtags associate with communities at a given time*, rather than independently among users.

Inspecting the community level hashtag usage timelines carefully, and comparing the hashtag usage behavior across the community pairs, the third and fourth hypothesis become clear. For instance, comparing the hashtag usage behaviors shown in the figure pair Fig. 2(b) and (c), it can be seen that although the hashtag *#iranrevolution* follows similar dominance timelines across the two communities, the other hashtags have a different characteristics. The hashtag *#freeiran* is used from the 5th to the 8th day in C1 but mostly from the 6th to the 7th day in C2. Further, interestingly, the hashtag *#revolution* remains absent in C1 while strongly dominates in C2. Such behavior is highly prominent in the figure pair Fig. 3(b) and (c), where the hashtag *#marijuana* is used in C1 but practically not used in C2, while the hashtag *#smoking* is used in C2 but practically not used in C1. All these collectively substantiate our **third hypothesis** - *hashtags are independently used across communities*. Further, the evolution of the hashtag *#revolution* in C1 acts as a demonstrative example of our **fourth hypothesis** - *hashtags evolve independently (atomically) within communities*. We also show the overall lifecycle of the corresponding topics, and

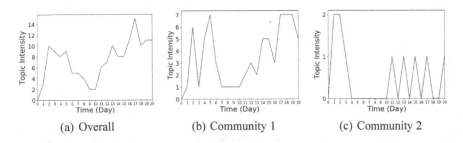

<table>
<tr><td>(a) Overall</td><td>(b) Community 1</td><td>(c) Community 2</td></tr>
</table>

Fig. 5. Time series of topic cluster (drugs, smoking, drugwars, marijuana cluster)

their evolution, at an overall level in Figs. 4(a) and 5(a), and at a per-community level in Figs. 2(b) and (c) for topic cluster C1 and Figs. 3(b) and (c) for topic cluster C2.

Note that, while we restrict our report to a small number of examples due to space constraints, we observe these characteristics to hold over a substantial volume of the data that we could manually inspect.

5 Conclusion

In this paper, we provided a novel analysis of topic lifecycles, in the context of social communities identified on Twitter. We used semantically and temporally related clusters of hashtags as topics. We used word embedding to enable hashtag clustering, thus ensuring the presence of higher order latent semantic space. We provided novel insights on peculiarities of evolution of topics, manifested via usage of hashtags over time and the underlying social communities: hashtags (and topics) that remain within communities, topics that see the use of different hashtags in different communities at similar (overlapping) points of time, and topics that morph over hashtags within some communities while keep the hashtag used unchanged on other communities. Empirically, we formed a baseline of hashtag lifecycles, and derived overall topic lifecycles by analyzing the aggregate characteristics of all hashtags in a given topic cluster. Our experiments substantiated our set of hypotheses. Our work would play a transformational role in the current understanding of information diffusion models, as well as, in understanding the social boundaries of topic lifecycles over time.

References

1. Allen, J.F.: Maintaining knowledge about temporal intervals. Commun. ACM **26**(11), 832–843 (1983)
2. Ardon, S., Bagchi, A., Mahanti, A., Ruhela, A., Seth, A., Tripathy, R.M., Triukose, S.: Spatio-temporal and events based analysis of topic popularity in twitter. In: CIKM, pp. 219–228. ACM (2013)
3. Bakshy, E., Rosenn, I., Marlow, C., Adamic, L.: The role of social networks in information diffusion. In: WWW, pp. 519–528. ACM (2012)
4. Blei, D.M., Ng, A.Y., Jordan, M.I.: Latent dirichlet allocation. J. Mach. Learn. Res. **3**, 993–1022 (2003)
5. Blondel, V.D., Guillaume, J.L., Lambiotte, R., Lefebvre, E.: Fast unfolding of communities in large networks. J. Stat. Mech: Theory Exp. **2008**(10), P10008 (2008)
6. Cataldi, M., Di Caro, L., Schifanella, C.: Emerging topic detection on twitter based on temporal and social terms evaluation. In: Tenth International Workshop on Multimedia Data Mining, p. 4. ACM (2010)
7. Cataldi, M., Schifanella, C., Candan, K.S., Sapino, M.L., Di Caro, L.: CoSeNa: a context-based search and navigation system. In: Conference on Management of Emergent Digital EcoSystems, p. 33. ACM (2009)

8. Cunha, E., Magno, G., Comarela, G., Almeida, V., Gonçalves, M.A., Benevenuto, F.: Analyzing the dynamic evolution of hashtags on twitter: a language-based approach. In: Languages in Social Media (ACL) (2011)

9. Dey, K., Kaushik, S., Subramaniam, L.V.: Literature survey on interplay of topics, information diffusion and connections on social networks. arXiv preprint arXiv:1706.00921 (2017)

10. Ifrim, G., Shi, B., Brigadir, I.: Event detection in twitter using aggressive filtering and hierarchical tweet clustering. In: SNOW-DC@ WWW, pp. 33–40 (2014)

11. Kwak, H., Lee, C., Park, H., Moon, S.: What is twitter, a social network or a news media? In: WWW, pp. 591–600. ACM (2010)

12. Lau, J.H., Collier, N., Baldwin, T.: On-line trend analysis with topic models:\# twitter trends detection topic model online. In: COLING, pp. 1519–1534 (2012)

13. Mathioudakis, M., Koudas, N.: Twittermonitor: trend detection over the twitter stream. In: SIGMOD, pp. 1155–1158. ACM (2010)

14. Myers, S.A., Zhu, C., Leskovec, J.: Information diffusion and external influence in networks. In: SIGKDD, pp. 33–41. ACM (2012)

15. Naaman, M., Becker, H., Gravano, L.: Hip and trendy: characterizing emerging trends on twitter. J. Am. Soc. Inform. Sci. Technol. 62(5), 902–918 (2011)

16. Nagar, S., Narang, K., Mehta, S., Subramaniam, L.V., Dey, K.: Topical discussions on unstructured microblogs: analysis from a geographical perspective. In: Lin, X., Manolopoulos, Y., Srivastava, D., Huang, G. (eds.) WISE 2013. LNCS, vol. 8181, pp. 160–173. Springer, Heidelberg (2013). https://doi.org/10.1007/978-3-642-41154-0_12

17. Narang, K., Nagar, S., Mehta, S., Subramaniam, L.V., Dey, K.: Discovery and analysis of evolving topical social discussions on unstructured microblogs. In: Serdyukov, P., Braslavski, P., Kuznetsov, S.O., Kamps, J., Rüger, S., Agichtein, E., Segalovich, I., Yilmaz, E. (eds.) ECIR 2013. LNCS, vol. 7814, pp. 545–556. Springer, Heidelberg (2013). https://doi.org/10.1007/978-3-642-36973-5_46

18. Newman, M.E.: Modularity and community structure in networks. Proc. Natl. Acad. Sci. 103(23), 8577–8582 (2006)

19. Pennington, J., Socher, R., Manning, C.D.: GloVe: global vectors for word representation. EMNLP 14, 1532–1543 (2014)

20. Stilo, G., Velardi, P.: Hashtag sense clustering based on temporal similarity. Comput. Linguist. 43, 181–200 (2017)

21. Yang, J., Leskovec, J.: Patterns of temporal variation in online media. In: WSDM, pp. 177–186. ACM (2011)

Health Applications

Medical Forum Question Classification
Using Deep Learning

Raksha Jalan⑩, Manish Gupta(✉), and Vasudeva Varma

International Institute of Information Technology-Hyderabad, Hyderabad, India
jalan.raksha@research.iiit.ac.in, {manish.gupta,vv}@iiit.ac.in

Abstract. With the rapid increase in the number as well as quality of online medical forums, patients are increasingly using the Internet for health information and support. Online health forums play an important role in addressing consumers health information needs. However, given the large number of queries, and limited number of experts, a significant fraction of the questions remains unanswered. Automatic question classifiers can overcome this issue by directing questions to specific experts according to their topic preferences to get quick and better responses.

In this paper, we aim to classify health forum questions where classes of questions mainly focus on capturing user intentions. We strongly believe that a good estimate of user intentions will help direct their questions to the best responders. We propose a novel approach of combining medical domain based features with deep learning models for question classification task. To further improve performance of the data-hungry deep learning models, we resort to weak supervision strategies. We propose a new variant of the existing self-training method called "Self-Training with Lookups" for weak supervision. Our results demonstrate that combining features generated from biomedical entities along with other language representation features for deep learning networks can lead to substantial improvement in modeling user generated health content. Weak supervision further enhances the accuracy. The proposed model outperforms the state-of-the-art method on a benchmark dataset of 11000 questions with a margin of 3.13%.

Keywords: Medical question classification · Deep learning models
Weak supervision

1 Introduction

With the increasing penetration of Internet to even the remote parts of the world, web based information access has become an integral part of meeting healthcare information needs. Unfortunately, search engines can still not answer a large number of healthcare queries effectively. Hence, recently, there has been a boom in the number of healthcare question-answering websites both specific

---

M. Gupta—A Principal Applied Scientist at Microsoft.

© Springer International Publishing AG, part of Springer Nature 2018
G. Pasi et al. (Eds.): ECIR 2018, LNCS 10772, pp. 45–58, 2018.
https://doi.org/10.1007/978-3-319-76941-7_4

to certain diseases, as well as general ones. To get better and quick responses to questions put up by an exponentially growing set of online information seekers, it is important to categorize questions and direct them to appropriate experts based on question types. As a first step in this direction we propose a novel approach for classifying questions posted on health forums into seven different categories each of which captures a unique user intent.

1. **Demographic** (DEMO): Questions targeted towards specific demographic sub-groups characterized by age, gender, profession, ethnicity, etc.
2. **Disease** (DISE): Questions related to a specific disease.
3. **Treatment** (TRMT): Questions related to a specific treatment or procedure.
4. **Goal-oriented** (GOAL): Questions related to achieving a health goal, such as weight management, exercise regimen, etc.
5. **Pregnancy** (PREG): Questions related to pregnancy, difficulties with conception, mother and unborn child's health during pregnancy.
6. **Family support** (FMLY): Questions related to issues of a caregiver (rather than a patient), such as support of an ill child or spouse.
7. **Socializing** (SOCL): Questions related to socializing, including hobbies and recreational activities, rather than a specific health-related issue.

Most of the research reported in health domain has been performed on small datasets. Data annotation in general is expensive and time consuming. Specially in health domain, data annotation costs are significant since most of the annotations need to be done by medical experts. In such cases, weakly supervised techniques can lead to significant improvement in performance of various models. To the best of our knowledge, we are the first to introduce weak supervision for the question classification task in medical domain. In this paper we also propose a novel method for weak supervision: "self-training with lookups" which outperforms the typical self-training method [2].

Our proposed solution uses domain-specific knowledge by generating medical features along with word embeddings and word n-grams to train deep learning models in using the weakly supervised data. For building deep networks, we experiment with multiple sequence learning architectures including Hierarchical Bidirectional LSTMs [15] and fully connected neural networks. In order to capture the strength of multiple models, we train them separately and use weighted average ensemble method to generate final class predictions.

Overall, we make the following contributions in this paper. (1) We propose a novel approach to generate features in the medical domain for the question classification task. (2) We present a new variant of the self-training method called "self-training with lookups". (3) We demonstrate the use of weakly supervised data to train deep learning models which is very important in medical domain where we generally have limited annotated data. (4) Our best performing model provides an accuracy of 71.13%, which beats the state-of-the-art method with a margin of 3.13%. Data and code is available at https://tinyurl.com/medCat18.

The paper has been organized as follows. Related work is discussed in Sect. 2. We present details of our proposed methods in Sect. 3 followed by results and analysis in Sect. 4. Finally, we conclude with a brief summary in Sect. 5.

2   Related Work

Although there is a large amount of user generated content about healthcare on different social media sites, few studies have applied deep learning or artificial intelligence techniques for knowledge discovery on a large scale of data in this particular emerging area. In this section, we first discuss about recent advances in language modeling using deep learning, and then discuss about literature on question classification in healthcare.

2.1   Language Modeling Using Deep Learning

Representation learning [7,10] and deep learning models [6,9,15,19] have been shown to be very effective for a large number of NLP and IR tasks [16]. Word embeddings [10] let you treat individual words as related units of meaning, rather than entirely distinct IDs. Benefiting from its recurrent structure, Recurrent Neural Networks (RNNs) [9] have been found to be very suitable to process variable length texts. But standard RNNs suffer from vanishing and exploding gradient problems. Long Short-Term Memory Networks (LSTMs) [6] deal with these problems by introducing memory cells and gates which allow for a better control over the gradient flow and enable better preservation of "long-range dependencies". Bidirectional LSTMs (BLSTMs) [17] utilize both the previous and future context by processing the sequence in both directions. Finally, hierarchical networks [15,19] using two level BLSTMs have also been proposed. We use BLSTMs with attention to implement our baselines. Further, considering the hierarchical structure of questions, we build Hierarchical Attention BLSTMs for modeling questions. Our final model is based on ensemble techniques where we create an ensemble of multiple deep models generated using word embeddings and TF-IDF vectors.

2.2   Question Classification in the Health Domain

Liu et al. [8] classified medical domain questions according to whether they were asked by health-care professionals or consumers using statistical and category features to train SVMs. Roberts et al. [14] classified medical questions as Patient specific, general knowledge or research using lexical, syntactic and semantic features. Guo et al. [5] classified Chinese health-related questions into six categories (Condition Management, Healthy Lifestyle, Diagnosis, Health Provider Choice, Treatment, and Epidemiology) using lexical, grammatical, CMeSH concepts, keywords and statistical features. Mrabet et al. [12] identify topics from healthcare questions based on the contained entities. The question classification task discussed in this paper was first proposed at ICHI 2016[1]. The winner team [18] used TF-IDF features to train Logistic Regression, SVMs and Random Forests. They also trained a Convolutional Neural Network (CNN) for which they

---

[1] http://www.ieee-ichi.org/healthcare_data_analytics_callenge.html.

used pre-trained Google-news vectors. Final predictions were obtained by averaging the predictions generated by all four classifiers. They did not incorporate any domain based medical features into their models. Also, the small size of the dataset without usage of any weakly supervised or semi-supervised techniques leads to low accuracy. In this paper, we experiment with multiple feature sets, and multiple deep learning architectures along with weak supervision.

3   Proposed Methods

In this section we discuss our proposed methods for the question classification task. We start with a discussion of various pre-processing steps. Next, we mention various standard sequence learning architectures which we adapt for our task. Further, we describe our medical features based approach, followed by our approach to perform weak supervision and ensemble learning.

3.1   Pre-processing

Data obtained from medical forums is usually noisy as it is user generated. Hence, we perform basic pre-processing and cleaning of the data before building classification models. We remove all hyper-links, special characters, punctuations and stopwords from the title and the body of questions. Then we perform case-folding and lemmatization. We concatenated the title and body of the question together for all questions before using it for our models. So the question dataset can be represented as $Q = \{q_1, q_2, q_3, ..., q_N\}$, where $q_i = t_i + b_i$, such that q_i corresponds to the ith question and t_i and b_i are the title and the body of the question respectively. When training instances are limited, word embeddings generated using only training instances cannot capture the semantic meaning of words effectively. Hence, in our experiments we use pre-trained embeddings and fine-tune them during training to improve the performance of classification. We experiment with multiple pre-trained word embeddings to obtain the best representation for the questions. Among embeddings such as Wikipedia-Pubmed-and-PMC-w2v vectors [11], Google-News Vectors[2] and Global Vectors for Word Representation (GloVe) [13], we found GloVe to outperform others for our task. Hence, we report results using GloVe embeddings.

3.2   Standard Sequence Learning Architectures

Next, let us discuss multiple variants of the sequence learning framework which can be leveraged for our task.

---

[2] https://code.google.com/archive/p/word2vec/.

3.2.1   Bidirectional LSTM (BLSTM)

Bidirectional LSTMs train two LSTM models together on the input sequence. The first is trained on the input sequence itself, and the other on a reversed copy of the input sequence. This provides both forward as well as backward context to the network and results in faster and more holistic learning on the problem. Each question is represented using a matrix with each row corresponding to a GloVe vector for a word in the question. This matrix is passed to BLSTM model, which generates the encoded representation of question which is then passed as input to softmax layer for classification. We use Bidirectional LSTM with 128 dimensions in each direction.

3.2.2   Bidirectional LSTMs with Attention Networks (BLSTM-A)

"Attention Mechanism" has been proposed recently, for effective modeling of long-term dependencies. Attention mechanisms allow for a more direct dependence between the state of the model at different points in time. Questions are represented using a matrix same as that for BLSTM network which is then passed to a BLSTM to generate encoded representation of questions. Attention layer is built on top of the Bidirectional-LSTM layer (128) followed by softmax layer at the top for classification.

3.2.3   Hierarchical Networks

So far, we represented a question as a sequence of words and encoded it using a matrix. Better representations for questions of larger length can be obtained by incorporating knowledge of their structure in the model's architecture. When questions contain multiple sentences, each question can be considered as a sequence of sentences and each sentence as sequence of words. In this style, questions are represented using a 3D tensor where the additional third dimension represents sentences in the question. Each of these sentences are then represented using the typical 2D matrix generated from sequence of words from the sentence and their corresponding GloVe vectors.

Hierarchical Bidirectional-LSTM (H-BLSTM): For this model, each question is encoded in a hierarchical fashion. Sentences within question are first encoded using sentence encoders (using Bidirectional-LSTMs) and then these encoded sentences are passed to a document (question in our case) encoder (again using another Bidirectional-LSTM), which encodes the question, and generates a final encoding for the question. This encoded representation is then passed to a softmax layer for classification. We used same number of LSTM dimensions (128) for both sentence encoder and document encoder.

Hierarchical Attention Networks (HAN): Not all sentences in a question are equally informative for representing a question and determining the informative sentences involves modeling the interactions of the words, not just their presence in isolation. We implemented a hierarchical network with two levels of attention mechanisms applied at the word and sentence-level, enabling it to attend differentially to more and less important content when generating the question encoding. Finally encoded representation of question is passed to a softmax layer for classification.

3.2.4 Term Frequency-Inverse Document Frequency Based Deep Network (TFIDF-DN)

Term frequency-Inverse Document Frequency (TFIDF) [3] is one of the popular feature representation methods for text which can efficiently capture statistical properties of words. We represented questions with their TF-IDF vectors and trained neural network consisting a fully connected layer followed by a dropout layer with a softmax layer at the top.

3.3   Deep Model Based on Medical Features

In order to incorporate domain knowledge and important medical signals, we use MetaMap [1] which is primarily designed to extract medical entities from biomedical documents. It gives us the advantage of capturing critical medical features based on multi-word combinations which are difficult to capture using single word based methods proposed in the previous sub-section. For example: "Heart Attack", "Chronic Fatigue Syndrome", "Chest pain", etc. By relating such entities with the class labels, we generate "association strength" based medical features for each question as discussed in the following.

MetaMap also provides semantic mappings for medical entities. It has a total 133 types of semantic mappings. However we only consider those medical entities for feature generation that belong to at least one of the following 16 semantic mappings, since these are relevant for our task. This is important since including entities from other semantic mappings could lead to noisy entity linking and hence poor feature generation. Types of semantic mappings we used for feature generation are: Antibiotics, Clinical Drugs, Diagnostic Procedures, Indicator-Reagent, Diagnostic Aid, Therapeutic Procedure, Drug Delivery Device, Anatomical Abnormality, Disease and Syndrome, Sign or Symptom, Family Group, Body System, Biological Region or Location, Biological Function, Body Parts, Body Space, and Age Group.

For each question $q_i \in Q$, we obtain a list l_i of medical entities belonging to any of the above semantic mapping types. Next, for each entity, we compute their Strength of Association (SoA) scores with every class label. Let K be the number of classes (in our case, $K = 7$), and let us denote the class labels by $\{c_i\}_{i=1}^K$. We define the Strength of Association (SoA) between an entity e and a class label c_i as follows.

$$SoA(e,c) = \log_2 \frac{P(e|c)}{P(e|\neg c)} = \log_2 \frac{freq(e,c) \times freq(\neg c)}{freq(e,\neg c) \times freq(c)} \qquad (1)$$

where $freq(e,c)$ is the number of questions with label c which contain the entity e. $freq(c)$ is the number of questions in class c. $freq(e,\neg c)$ is the number of times e occurs in questions in classes other than c. $freq(\neg c)$ is the number of

---

Algorithm 1. Generating Features based on SoA Scores of Medical Entities

---

Input   Entities list l_i for each $q_i \in Q$, threshold τ, Entity level SoA scores $S_e^{M \times K}$
Output   Question level SoA scores $S_q^{N \times K}$

1: Initialization: $S_q = [0]^{N \times K}$
2: **for** $i = 0$ to N **do**
3:    **for each** $entity \in l_i$ **do**
4:        $(max_1, max_2) \leftarrow$ (Maximum, Second Maximum) value in $S_e[entity]$
5:        **if** $\frac{max_1}{max_2} > \tau$ **then**
6:            $index \leftarrow argmax(S_e[entity])$
7:            $S_q[i][index] = S_q[i][index] + 1$
8: Return $S_q^{N \times K}$

---

questions with label other than c. The intuition is that the entity e is associated closely with class c if it occurs in questions labeled c much more number of times compared to occurrences in questions of another class.

Though association of individual words with the class labels is captured by TF-IDF, SoA captures the association between domain-specific multi-word medical entities and class labels. If an entity has a stronger tendency to occur in a question with a particular label than in questions with other labels, then that (entity, label) pair will have an SoA score greater than zero. Let M be the total number of unique medical entities in the dataset. Since there are K (=7) class labels, we obtain a SoA matrix $S_e^{M \times K}$. Algorithm 1 illustrates the method we use to process these entity-level SoA scores to come up question-level SoA scores $S_q^{N \times K}$ where N is the number of questions in the dataset. For each question q_i, we first retrieve the list of medical entities l_i. Then for each entity in the list we find the maximum (max_1) and the second maximum (max_2) SoA scores for the entity. We intend to capture only strong associations between entities and labels. Hence, only if the ratio of max_1 to max_2 is greater than a defined threshold τ, we let the entity e contribute to the SoA score for the question q_i. The SoA scores in S_q are the medical features fed to a fully connected layer followed by a combination of a dropout layer and a softmax layer.

3.4   Models Using Weak Supervision

First, we crawl ~100K health questions from "medhelp.org"[3] to gather a large dataset of unlabeled questions. After pre-processing the crawled questions (as discussed in Sect. 3.1), we use self-training to generate weakly supervised labels. Self-training is a popular method for weak supervision. It is an iterative method where every iteration k contains two steps: (1) based on the current labeled dataset L_{k-1}, a classifier C_k is trained, and (2) the classifier C_k is used to predict labels for instances in the current unlabeled data U_{k-1} leading to classification of more instances to generate the new U_k and L_k.

Two techniques are popularly used for robust self-training:

---

[3] http://www.medhelp.org/.

Algorithm 2. Self-training With Lookups

Input Labeled Dataset $L = \{\langle x_i, y_i \rangle\}_{i=1}^{M}$, Unlabeled Dataset $U = \{x_i\}_{i=1}^{N}$, Throttling threshold T

Output Classifier C, Augmented Labeled Dataset L.

1: Initialization: Lookup Lists $l[c] = \phi \quad \forall c \in \{c_1, \ldots, c_K\}$.
2: **while** stopping criteria is not met **do**
3:     $C \leftarrow Train\_Model(L)$
4:     Candidates $\text{Cand}[c] = \phi \quad \forall c \in \{c_1, \ldots, c_K\}$.
5:     $Pred \leftarrow Predict(U, C)$ where $Pred[u][c]$ is the probability of classifying instance $u \in U$ to
        class c.
6:     **for each** $u \in U$ **do**
7:         $r \leftarrow argmax(Pred[u])$
8:         **if** $Pred[u][r] > T$ **then**   ▷ Throttling
9:             **if** $u \in l[c_r]$ **then**
10:                 $\text{Cand}[c_r] \leftarrow \text{Cand}[c_r] \cup \{u\}$ ▷ Lookups
11:             **else**
12:                 $l[c_r] \leftarrow l[c_r] \cup \{u\}$
13:     $size \leftarrow \min(|\text{Cand}[c_1]|, \ldots, |\text{Cand}[c_K]|)$ ▷ Balancing
14:     **for each** $c \in \{c_1, \ldots, c_K\}$ **do**
15:         **for** $i \leftarrow 1$ to $size$ **do**
16:             $L \leftarrow L \cup \{\langle \text{Cand}[c][i], c \rangle\}$
17:             $U \leftarrow U - \{\text{Cand}[c][i]\}$
18: Return C, L.

- Throttling Principle: An unlabeled instance $x \in U_{k-1}$ is assigned to a class c_i (and hence to L_k) in the second step only if the classifier C_k predicts the class c_i for x with a probability greater than a threshold T.
- Balancing Principle: In order to avoid class imbalance, equal number of instances of all predicted classes are added to the labeled data for the next iteration.

One disadvantage of the traditional self-training method is that a particular unlabeled instance might be predicted to belong to a class with high confidence incorrectly just because the parameters of the model have not yet stabilized and are still being learned themselves. This could lead to propagation of errors over iterations due to corruption of training data. This is particularly critical in the first few iterations where the model has been learned using a very small labeled dataset and is therefore "weak".

In this paper, we propose a novel variant of the self-training method called "self-training with lookups" to handle this problem. Besides following the principles of throttling and balancing, it makes use of lookup lists which significantly decreases incorrect label assignments to unlabeled data compared to traditional self-training. We maintain one lookup list for each class. The proposed self-training with lookups method is illustrated in Algorithm 2. In our algorithm, if the classifier predicts label c for instance $u \in U$ with probability greater than the throttling threshold T for the first time, the instance is not directly added to the labeled set. Instead, it is added to a lookup list for the class c. However, if the instance u is labeled with the class c again in subsequent iterations (and so the lookup list for the class c already contains u), it is added to the labeled set with label as c.

On detailed analysis, we find that the proposed method provides significant reduction in noisy labels compared to the traditional self-training method, at the cost of a few more computations. In order to compare the performance of the proposed method with the traditional self-training method, we use exactly the same parameters for both. Out of 100000 unlabeled questions, our approach assigns labels to 34664 questions while the traditional self-training algorithm generates labels for 53123 samples after the same number of iterations (10). Further, we compare the quality of weakly supervised question labels generated from both the algorithms by using the generated labeled data to learn initial parameters of various deep learning models followed by fine tuning with the original labeled training data.

While the size of the labeled data generated by our method is expectedly small, the quality of labels using our method leads to better classification accuracy. Table 1 shows the comparison results. Accuracy values in each cell are the average values obtained after repeating the experiments five times. For all the models, experiments with labels generated using our approach performed better than the ones with labels generated using the traditional self-training method. Self-training with lookups outperforms traditional self-training method by ∼1% on average. Hence, for all further experiments, we use weakly supervised labels generated using the "self-training with lookups" method.

Table 1. Comparison of self-training with lookups and traditional self-training

Model name	Without lookups	With lookups
BLSTM	64.47 %	65.01%
BLSTM-A	64.75%	65.67%
H-BLSTM	68.09%	68.35%
HAN	67.10%	67.98%
TFIDF-DN	67.05%	68.96%

3.5   Overall Model Using Ensemble Learning

In recent years, ensemble based methods have been found to be successful across multiple machine learning tasks such as classification, clustering, anomaly detection, etc. For classification, it has been well studied that combining models which individually use different features can result in accuracy gains since different feature spaces result into uncorrelated errors. In our case, we combine the benefits of memorization (Deep Networks), generalization (TF-IDF based models) and domain knowledge (SoA based model) to build an ensemble model to get the final class prediction. We experiment with various ways of combining these individual methods including cascading, averaging, and weighted averaging. Due to lack of space we will list the results obtained using the top performing models only. Figure 1 shows our overall model architecture.

In supervised learning only labeled training instances are used to train the model. While in weakly supervised methods, the model is first trained using weakly supervised data and then labeled training data is used to fine tune the model. In Fig. 1, W_{sk} represents the 2-Dimensional embedding matrix of the kth sentence in a question. The embedding matrix W_{sk} is fed to a Bidirectional LSTM B_k which creates an encoding for the corresponding input sentence. Each block B_k follows the same internal structure as shown in the figure where multiple LSTM cells (L) are connected in forward and reverse fashion. Further, the encoded sentences for the question generated by the first level of BLSTM are input to another BLSTM (B_Q) which encodes the complete question. Then this question encoding is passed to the softmax layer for classification. For the SoA based deep network and the TF-IDF based deep network, as mentioned earlier, the representations are fed to a fully connected layer followed by a combination of a dropout layer and a softmax layer at the top. The class predictions P_{SOA}, P_{HBLSTM}, P_{TF-IDF} obtained from the three models are combined using weighted averaging to obtain the final prediction P_{FINAL}. For ensemble based models, we report accuracies obtained using the weighted average scheme [4]. We fine tuned the parameters $(W_{we}, W_{tfidf}, W_{soa})$ which represent the weights given to predictions generated by the models, using the original labeled dataset. Note that W_{we} denotes the weight to the word embedding based HBLSTM model, W_{tfidf} denotes the weight to the TF-IDF based model and W_{soa} denotes the weight to the SoA based model.

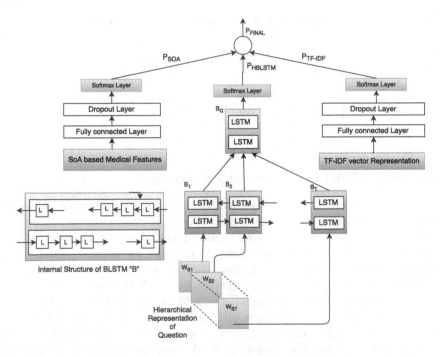

Fig. 1. Architecture of the proposed deep learning based question classifier

Results obtained using this ensemble model outperform the baselines in both supervised as well as weak supervised settings. We discuss the performance details in Sect. 4.

4   Results and Analysis

The dataset provided as part of the Healthcare Data Analytics Challenge at ICHI 2016 (see footnote 1) contains real questions posted on a health discussion forum. The training data has 8000 questions and the test data contains 3000 questions each with the question title, question text, and a category. The categories are: (1) Demographic, (2) Disease, (3) Treatment, (4) Goal-oriented, (5) Pregnancy, (6) Family support, and (7) Socializing. The dataset is balanced with almost same number of instances per class. The performance results of various models in the supervised and the weakly supervised settings are shown in Tables 2 and 3 respectively. All accuracies are computed on the test set of size 3000, provided in ICHI data analytics shared task.

4.1   Results of Supervised Models

Table 2 shows the results obtained using various supervised models. We observe that in the pure supervised environment, models built on word embedding based features or TF-IDF based features could not beat the accuracy of the ICHI 2016 winners model which is the current state-of-the-art for this task. However, the ensemble based method with weighted combination of predictions from the models trained on SoA based medical features along with the predictions from word embedding and TFIDF based models, leads to an accuracy which is significantly better.

For the case, when we used hierarchical-BLSTMs for word embeddings, we found the best weights as $W_{we} = 0.73$, $W_{tfidf} = 1.15$, $W_{SoA} = 0.66$. We observed that the SoA based model improved the overall performance of the model by 3.02%. For SoA based model, we chose the threshold $\tau = 1.35$ after tuning it between 0.5 and 1.5. When we changed the word embedding model to HANs, we found the following weights for the ensemble model $W_{we} = 0.70$, $W_{tfidf} = 1.15$, $W_{SoA} = 0.66$. In this case, we observed that the SoA based model improved the overall performance of this model by 2.47%.

4.2   Results of Weakly Supervised Models

Table 3 shows the results obtained using various weakly supervised models. Distance supervision using weak supervision based methods has been shown to improve overall performance of deep learning methods for a large variety of tasks. We obtain large amounts of labeled data using weak supervision on data obtained from "medhelp.org". Pre-training of the network parameters using such weakly supervised data improved the individual models' accuracies as shown in Table 3. This also resulted in improved accuracies for the ensemble based models.

In our study, we achieved the highest accuracy of 71.13% using the ensemble model (in the weakly supervised setting) which combines Hierarchical-BLSTM, TF-IDF based classifier, and the SoA based classifier. This also outperformed all other models in the supervised setting. It beats the current baseline model by a margin of 3.13%.

Table 2. Accuracy comparison of various supervised models

Models	Accuracy
ICHI 2016 challenge winners [18]	**68.00%**
BLSTM	62.70%
BLSTM-A	62.53%
H-BLSTM	64.03%
HAN	63.93%
SoA-DN	59.76%
TFIDF-DN	65.10%
HAN + TFIDF-DN	66.37%
H-BLSTM + TFIDF-DN	66.74%
HAN + TFIDF-DN + SoA-DN	**68.84%**
H-BLSTM + TFIDF-DN + SoA-DN	**69.76%**

Table 3. Accuracy comparison of various weakly supervised models

Models	Accuracy
H-BLSTM	**68.35%**
HAN	67.98%
TFIDF-DN	**68.96%**
HAN + TFIDF-DN + SoA-DN	**70.37%**
H-BLSTM + TFIDF-DN + SoA-DN	**71.13%**

4.3   Error Analysis

We observed that for instances which contain strong features indicative of multiple classes, sometimes the classifier fails to predict the most dominant class. For example, consider the question "Last night I got high fever. My dad consulted his doctor friend....". The most dominant category for this question is the "Disease" category but because of the terms like "My dad", "consulted", "his doctor", it got categorized as "Family". Another kind of questions where the classifier fails is when the forum question is about particular disease symptoms, and in between the question mentions possible treatment plans for that specific disease. In such cases sometimes the classifier fails to predict the correct label. Perhaps modeling the problem in a multi-label multi-class classification setting will address such problems.

5   Conclusions

In this paper, we discussed the problem of medical forum question classification. We found that using domain knowledge based features along with word embeddings provides better accuracy compared to just using word embeddings in the deep learning supervised setting. Biomedical entities can lead to substantial improvement in modeling the user generated health content. In addition to the proposed usage of medical features, we experimented with various methods to generate weakly supervised labels, and presented our new approach "self-training with lookups". Our experiments demonstrate the effectiveness of the ensemble of the three models (word embedding based sequence learning models, model with SoA based medical domain features, and TF-IDF based statistical model). As part of future work we plan to pursue two directions: (1) Explore the lookups idea further by trying combinations like varying the number of iterations in which the classifier assigns a label to an instance (currently set as 2), and varying throttling threshold over iterations. (2) Model other forms of context related to the question, for example, user who asked the question, time when the question was asked, etc.

References

1. Aronson, A.R.: Effective mapping of biomedical text to the UMLS Metathesaurus: the MetaMap program. In: AMIA, p. 17 (2001)
2. Chapelle, O., Scholkopf, B., Zien, A.: Semi-supervised learning. IEEE Trans. Neural Netw. **20**, 542–542 (2009)
3. Christopher, D.M., Prabhakar, R., Hinrich, S.: An Introduction to Information Retrieval. Cambridge University Press, Cambridge (2008). 151, 177
4. Dietterich, T.G.: Ensemble methods in machine learning. In: Kittler, J., Roli, F. (eds.) MCS 2000. LNCS, vol. 1857, pp. 1–15. Springer, Heidelberg (2000). https://doi.org/10.1007/3-540-45014-9_1
5. Guo, H., Na, X., Hou, L., Li, J.: Classifying Chinese questions related to health care posted by consumers via the internet. J. Med. Internet Res. **19**(6), e220 (2017)
6. Hochreiter, S., Schmidhuber, J.: Long short-term memory. Neural Comput. **9**, 1735–1780 (1997)
7. Le, Q., Mikolov, T.: Distributed representations of sentences and documents. In: ICML (2014)
8. Liu, F., Antieau, L.D., Yu, H.: Toward automated consumer question answering: automatically separating consumer questions from professional questions in the healthcare domain. J. Bio. Info. **44**, 1032–1038 (2011)
9. Mikolov, T., Karafiát, M., Burget, L., Cernocký, J., Khudanpur, S.: Recurrent neural network based language model. In: Interspeech (2010)
10. Mikolov, T., Sutskever, I., Chen, K., Corrado, G.S., Dean, J.: Distributed representations of words and phrases and their compositionality. In: NIPS (2013)
11. Moen, S., Ananiadou, T.S.S.: Distributional Semantics Resources for Biomedical Text Processing (2013)
12. Mrabet, Y., Kilicoglu, H., Roberts, K., Demner-Fushman, D.: Combining open-domain and biomedical knowledge for topic recognition in consumer health questions. In: AMIA, vol. 2016, p. 914 (2016)
13. Pennington, J., Socher, R., Manning, C.: Glove: global vectors for word representation. In: EMNLP, pp. 1532–1543 (2014)
14. Roberts, K., Rodriguez, L., Shooshan, S.E., Demner-Fushman, D.: Resource classification for medical questions. In: AMIA (2016)

15. Ruder, S., Ghaffari, P., Breslin, J.G.: A hierarchical model of reviews for aspect-based sentiment analysis. arXiv preprint arXiv:1609.02745 (2016)
16. Socher, R., Bengio, Y., Manning, C.D.: Deep learning for NLP (without magic). In: Tutorial Abstracts of ACL 2012, pp. 5–5. Association for Computational Linguistics (2012)
17. Tan, M., Santos, C.D., Xiang, B., Zhou, B.: LSTM-based deep learning models for non-factoid answer selection. arXiv preprint arXiv:1511.04108 (2015)
18. Verma, J., Kwon, B.C., Cheng, Y., Ghosh, S., Ng, K.: Classification of healthcare forum messages. In: ICHI (2016)
19. Yang, Z., Yang, D., Dyer, C., He, X., Smola, A.J., Hovy, E.H.: Hierarchical attention networks for document classification. In: HLT-NAACL, pp. 1480–1489 (2016)

Multi-task Learning for Extraction of Adverse Drug Reaction Mentions from Tweets

Shashank Gupta[1], Manish Gupta[1(✉)], Vasudeva Varma[1], Sachin Pawar[2], Nitin Ramrakhiyani[2], and Girish Keshav Palshikar[2]

[1] International Institute of Information Technology-Hyderabad, Hyderabad, India
shashank.gupta@research.iiit.ac.in, {manish.gupta,vv}@iiit.ac.in
[2] TCS Research, Pune, India
{sachin7.p,nitin.ramrakhiyani,gk.palshikar}@tcs.com

Abstract. Adverse drug reactions (ADRs) are one of the leading causes of mortality in health care. Current ADR surveillance systems are often associated with a substantial time lag before such events are officially published. On the other hand, online social media such as Twitter contain information about ADR events in real-time, much before any official reporting. Current state-of-the-art in ADR mention extraction uses Recurrent Neural Networks (RNN), which typically need large labeled corpora. Towards this end, we propose a multi-task learning based method which can utilize a similar auxiliary task (adverse drug event detection) to enhance the performance of the main task, i.e., ADR extraction. Furthermore, in absence of the auxiliary task dataset, we propose a novel joint multi-task learning method to automatically generate weak supervision dataset for the auxiliary task when a large pool of unlabeled tweets is available. Experiments with ∼0.48M tweets show that the proposed approach outperforms the state-of-the-art methods for the ADR mention extraction task by ∼7.2 % in terms of F1 score.

Keywords: Multi-task learning · Pharmacovigilance · Neural networks

1 Introduction

Estimates show that Adverse Drug Reactions (ADRs) are the fourth leading cause of deaths in the United States ahead of cardiac diseases, diabetes, AIDS and other fatal diseases[1]. Another study[2] conducted in the US reveals that ∼6.7% of the hospitalized patients have a serious ADR, with a fatality rate

---

M. Gupta is also a Principal Applied Scientist at Microsoft.

[1] https://ethics.harvard.edu/blog/new-prescription-drugs-major-health-risk-few-offsetting-advantages.

[2] http://bit.ly/2vaWF6e.

© Springer International Publishing AG, part of Springer Nature 2018
G. Pasi et al. (Eds.): ECIR 2018, LNCS 10772, pp. 59–71, 2018.
https://doi.org/10.1007/978-3-319-76941-7_5

of ~0.32%. Hence, it necessitates the monitoring and detection of such adverse events to minimize the potential health risks by having the relevant pharmaceutical companies issue appropriate warnings. Practically, clinical trials cannot investigate all settings in which a drug can be used, making it impractical to profile a drug's side effects before its formal approval. Typically, post-marketing drug safety surveillance (also called as pharmacovigilance) is conducted to identify ADRs after a drug's release. Such surveys rely on formal reporting systems such as Federal Drug Administration's Adverse Event Reporting System (FAERS)[3]. However, often a large fraction (~94%) of the actual ADR instances are under-reported in such systems [13]. Social media presents a plausible alternative to such systems, given its wide userbase. A recent study [10] shows that Twitter has three times more ADRs reported as compared to FAERS.

Earlier work in this direction focused on feature based pipeline followed by a sequence classifier [21]. More recent works are based on Deep Neural Networks [5]. Deep learning based methods [7,17] typically rely on the presence of a large annotated corpora, due to their large number of free parameters. Due to the high cost associated with tagging ADR mentions in a social media post and limited availability of labeled datasets, it is hard to train a deep neural network effectively for such a task. In this work, we attempt to address this problem and propose two novel multi-task learning setups which utilize similar tasks to effectively augment the rather limited existing datasets for ADR extraction.

Multi-task learning works on the basic premise that auxiliary tasks can be utilized to improve performance of the main task by exploiting the correlations between them [8]. Adverse drug event (ADE) detection is a task very similar to our original task of ADR mention extraction. The ADE detection problem deals with *detecting* an adverse drug event from a social media post. We hypothesize that due to semantic similarities between the two tasks, they can be modeled together in a joint learning setup. We propose a multi-task learning setup with ADR extraction as the main task and ADE detection as an auxiliary task which complements the learning of our main task. Furthermore, we propose a novel weakly-supervised learning based method which exploits semi-supervised learning to augment the main task (ADR extraction) dataset and also works in parallel to automatically generate auxiliary task (ADE detection) dataset.

To summarize, the main contributions of our work are: (1) We investigate the effect of adding an available auxiliary task (ADE detection) to the main task (ADR extraction) in a multi-task learning setup. (2) We propose a novel weakly-supervised and a semi-supervised learning based method to automatically generate auxiliary task dataset (ADE detection) and model it in a novel joint multi-task learning framework. (3) We perform experiments on two datasets to show the effectiveness of the proposed methods.

The remainder of the paper is organized as follows. In Sect. 2 we discuss the related work in the area of ADR extraction and Multi-task learning. In Sect. 3 we describe our proposed methods in detail. In Sect. 4 and Sect. 5, we discuss in detail our experimental results and its analysis. Finally, Sect. 6 concludes our work with a brief summary.

---

[3] http://bit.ly/2xnu7pE.

2   Related Work

In this section, we review some of the existing work in the areas of ADR extraction and Multi-task learning.

ADR Extraction: Traditional methods for ADR extraction used linguistic features such as POS tags, word embedding features and word context features along with sequence classifiers like a linear-chain CRF [21]. To avoid time consuming feature engineering, recent works use deep learning approaches [7,15,17, 20]. Cocos et al. [5] proposed a Long Short Term Memory (LSTM) based model with word embedding features to extract ADRs from Twitter posts. Stanovsky et al. [22] proposed a LSTM based model where lexical word embeddings are augmented with Knowledge-Graph based embeddings. In their model, if a word has a lexical match with a Knowledge-Graph entity (e.g., DBPedia), its corresponding lexical word embedding is replaced by embedding learned through Knowledge graph based methods [25].

Multi-task learning (MTL): Previous works in Multi-task learning have explored the use of auxiliary tasks to improve the generalization performance of a main task [2–4,8]. In the context of deep neural networks, MTL has been successfully applied in the area of Natural Language Processing [6,19] and Information Retrieval [18]. These models work on the premise that multiple related tasks share common features which allows the model to share the statistical strengths between them. Sharing statistical strengths among different tasks also acts as an implicit regularizer, allowing the model to generalize better. Due to sharing of the model between tasks, MTL also effectively acts as an implicit data augmentation method, since the same model is exposed to the training data of multiple tasks. In this work, we exploit the data augmenter role of MTL to compensate for the lack of rich training data for the ADR extraction task using a single neural network based model.

3   The Proposed Multi-task Learning Framework

In this section, we start by defining the ADR extraction and ADE detection problems. Next, we propose a multi-task learning framework for ADR extraction. Finally, we propose a joint multi-task learning framework for both the tasks.

3.1   Problem Definition

ADR Extraction: Given a social media post in the form of a word sequence $x_1...., x_n$, predict an output sequence $y_1, ..., y_n$ which indicates the presence/absence of the ADR mention, where each y_i is encoded using standard sequence labeling encoding scheme such as the IO encoding similar to that used in [5].

ADE Detection: Given a social media post in the form of a word sequence $x_1, ..., x_n$, predict a single variable y, which indicates whether there is an occurrence of an ADE in the input social media post or not. It can thus be modeled as a binary classification problem.

Algorithm 1. Multi-Task Learning for ADR Extraction

Input N: (No. of training examples / batch size) for ADR task
M: (No. of training examples / batch size) for ADE task
α: $\frac{M}{N}$

Output Model parameters: $\theta_{\text{Shared}}, \theta_{\text{ADR}}, \theta_{\text{ADE}}$

1: Initialize model parameters : $\theta_{\text{Shared}}, \theta_{\text{ADR}}, \theta_{\text{ADE}}$ randomly
2: **for** $epoch \leftarrow 1, maxEpochs$ **do**
3:     **for** $i \leftarrow 1, N$ **do**
4:         **for** $j \leftarrow 1, \alpha$ **do**
5:             X_{ADE}, Y_{ADE} = sample $(N(i-1)+j)^{th}$ batch from ADE training data
6:             L_{ADE} = ADE Loss(X_{ADE}, Y_{ADE}) from Eq. 5
7:             Compute gradients for ADE loss, and update $\theta_{\text{Shared}}, \theta_{\text{ADE}}$
8:         X_{ADR}, Y_{ADR} = sample i^{th} batch from ADR training data
9:         L_{ADR} = ADR Loss(X_{ADR}, Y_{ADR}) from Eq. 3
10:         Compute gradients for ADR loss, and update $\theta_{\text{Shared}}, \theta_{\text{ADR}}$

3.2   Multi-task Learning for ADR Extraction

Given the two tasks, ADR extraction and ADE detection, we first describe the modeling of each task individually and then discuss how to model them in a single setup.

ADR Extraction: We choose the model described in [5], which is a fully supervised bi-directional LSTM (bi-LSTM) transducer trained on a manually annotated tweet corpus with word-level ADR mention annotation. Formally, given an input word sequence $x_1, ..., x_n$, where n is the maximum sequence length, a bi-LSTM transducer [12] is employed to capture complex sequential dependencies. At each time-step t, the bi-LSTM transducer attempts to model the task as follows.

$$h_t = \text{bi-LSTM}(e_t, h_{t-1}) \tag{1}$$

where $h_t \in \mathcal{R}^{(2 \times d_h)}$, is the hidden unit representation of the bi-LSTM with d_h being the hidden unit size. Since it is a concatenation of hidden units of a forward sequence LSTM and backward sequence LSTM, its overall dimension is $2d_h$. e_t is the embedding vector corresponding to the input word x_t extracted from a pre-trained word embedding lookup table.

$$y_t = \text{softmax}(W_1 h_t + b) \tag{2}$$

where $y_t \in \mathcal{R}^{d_l}$, is the output vector at each time-step which encodes the probability distribution over the number of possible output labels (d_l) at each time-step of the sequence. $W_1 \in \mathcal{R}^{d_l * d_h}$ and $b \in \mathcal{R}^{d_l}$ are weight vectors for the affine transformation. Finally, the cross entropy loss function for the task is defined as follows.

$$L_{\text{ADR}} = -\sum_{t=1}^{n} \sum_{i=1}^{d_l} \hat{y}_{t_i} \log y_{t_i} \tag{3}$$

where $\hat{y}_t$ is the one-hot representation of the actual label at time-step t.

ADE Detection: Given an input word sequence $x_1, ..., x_n$, where n is the maximum sequence length, similar to the ADR Extraction model, a bi-directional LSTM transducer (bi-LSTM) is employed to model the sequential nature of the dataset. The LSTM transducer acts as a feature extractor in this case, which is followed by an average-pooling layer to generate a fixed-size vector representation of the input sentence followed by the classification loss function. Formally, the ADE detection model is defined as follows.

$$h_t = \text{bi-LSTM}(e_t, h_{t-1}), \quad h = \frac{1}{n} \sum_{t=1}^{n} h_t, \quad y = \text{softmax}(W_2 h + b_1) \qquad (4)$$

where h_t is similar to the one defined for the ADR task. $h \in \mathcal{R}^{(2*d_h)}$ is the average-pooled fixed size representation of the input sequence. $y \in \mathcal{R}^2$ is the output vector which encodes the probability distribution over the binary choice, with $W_2 \in \mathcal{R}^{2*(2d_h)}$ and $b_1 \in \mathcal{R}^2$, the corresponding weight vectors. Finally, the loss function for the task is the cross-entropy loss defined as follows.

$$L_{\text{ADE}} = - \sum_{i=1}^{2} \hat{y}_i \log y_i \qquad (5)$$

where $\hat{y}$ is the one-hot representation of the actual label for the input sentence.

Multi-task Learning Model: The MTL model architecture is illustrated in Fig. 1. The bi-LSTM transducer acts as the common (shared) layer between both tasks, thus receiving gradient updates from both. The network then bifurcates to task specific layers as seen in the dotted region in the figure.

The training algorithm is illustrated in Algorithm 1. To enhance the performance of the main task, we employ the following strategy for training. Since our main task of interest is ADR extraction, the number of parameter updates for this task are fixed to be N (number of training examples for ADR/batch size for ADR) for each epoch. Let M denote the ratio (number of training examples for

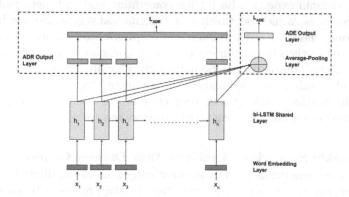

Fig. 1. Network architecture for the multi-task learning model to combine the ADR extraction and ADE detection tasks

---

Algorithm 2. Weakly Supervised Auxiliary Task Dataset Generation

---

Input U: Large collection of unlabeled tweets
   τ : threshold for self-training
   D_{ADR} : Labeled dataset for ADR task
Output   New labeled datasets D'_{ADR} and D'_{ADE}

1: Initialize model parameters, θ^0 for bi-LSTM transducer randomly.
2: $T \leftarrow D_{ADR}$
3: **while** *(stopping criteria is not met)* **do**
4:   bi-LSTM(θ^t) = finetune bi-LSTM(θ^{t-1}) minimizing L_{ADR} on T
5:   **for** $i \leftarrow 1, |U|$ **do**
6:    **if** $score(U_i) \geq \tau$ **then**
7:     $T \leftarrow T \cup U_i$
8:     $U \leftarrow U - U_i$
9:   $U \leftarrow$ re-sample large pool of unlabeled tweets
10: $D'_{ADR} \leftarrow \phi$, $D'_{ADE} \leftarrow \phi$
11: **for** $i \leftarrow 1, |U|$ **do**
12:   **if** $score(U_i) \geq \tau$ **then**
13:    $D'_{ADR} \leftarrow D'_{ADR} \cup U_i$
14:    $D'_{ADE} \leftarrow D'_{ADE} \cup \{U_i, 1\}$
15:   **else**
16:    $D'_{ADE} \leftarrow D'_{ADE} \cup \{U_i, 0\}$
17:   $U \leftarrow U - U_i$

ADE/batch size for ADE). To compensate for the likely difference in the number of training examples for the ADE task, for each parameter update of the ADR task, $\alpha = \frac{M}{N}$ parameter updates are performed for ADE.

The MTL setup can also be viewed as an iterative process where each iteration contains two steps. The first step is the detection of an adverse drug event and the second step involves its extraction. We claim that the sharing of the network between the two tasks helps in boosting performance of our main task. We validate this claim in the experiments section.

3.3   Joint Multi-task Learning

Training a good supervised model for pharmacovigilance need high quality labeled datasets, annotated by domain experts. Getting large datasets labeled by medical domain experts is both time consuming and cost-inefficient. In this section, we discuss a method which can automatically generate auxiliary task dataset in our context, in order to build a MTL pharmacovigilance system. While we discuss the method in the context of pharmacovigilance, it can be applied to other domains too, due to its generic nature. Specifically, we use semi-supervised learning and weakly-supervised learning to augment the main task dataset and generate the auxiliary task dataset respectively. We also present a joint MTL model learned using the data generated using weak-supervision.

3.3.1   Weakly Supervised Auxiliary Task Dataset Generation

Algorithm 2 outlines the method to automatically generate auxiliary task dataset (ADE detection in our case). The first stage in the process is to augment the existing training data with a larger dataset generated using semi-supervised learning. Semi-supervised learning can leverage large unlabeled dataset to assist the supervised learning model. We choose self-training [9], as the method for semi-supervised learning for this task (Line 2 to 8 of Algorithm 2), mainly because of its simplicity and effectiveness in solving various NLP and IR tasks [14, 24].

At each step of self-training, the bi-LSTM transducer is trained on the updated training dataset T (Line 4 of Algorithm 2). Note that bi-LSTM's parameters are re-used from the previous iteration. Each sample from the unlabeled example pool is scored using a scoring function computed as follows. First, the current transducer is used to decode/infer output label distribution for each word in the unlabeled sample. For each word in the output sequence, we simply choose the output label which has the maximum probability. We filter out the data sample if the transducer does not output even a single ADR label for any word in the sample. If there is at least one word labeled as ADR, we compute the score for the sample as the multiplication of the ADR probabilities for the ADR-labeled words in the sample normalized by the number of ADR words. If this confidence score of the sample is greater than some pre-defined threshold τ, the sample is added to the training data along with its output labels as generated by the transducer (Line 6 and 7).

The next stage is the generation of ADE task dataset (Line 9 to 17). The pool of unlabeled examples is re-sampled to avoid overlap with the previously used pool. Each data sample from the unlabeled pool is scored using the scoring function defined previously. If the confidence score is greater than τ, the sample is added to the ADR dataset with the decoded labels and it is also added to the ADE dataset with a label of 1 (Lines 13 and 14). Since this sample's confidence score is greater than a threshold, which indicates with high confidence that it has an ADR mention, it is safe to assume that the sentence has an ADE, thereby assigning it a label of 1. In the other case, due to the low confidence score of the sample, it is assigned a label of 0 for ADE (Line 16).

3.3.2   Joint Multi-task Learning Formulation

Algorithm 2 produces training datasets D'_{ADR} and D'_{ADE} as the output. We use these to define a joint MTL model as follows. In the dataset, for each example we have two labels, an output label sequence for ADR and a binary label for ADE. We define the joint loss function using a linear combination of the loss functions of the two tasks as $L = \lambda \cdot \mathbb{I}[y_{ADE} == 1] \cdot L_{ADR} + (1 - \lambda) \cdot L_{ADE}$ where λ controls the contribution of losses of the individual tasks in the overall joint loss. $\mathbb{I}[y_{ADE} == 1]$ is an indicator function which activates the ADR loss only when the corresponding ADE label is 1, since we do not want to back-propagate ADR loss when the corresponding ADE label is 0, which is intuitive by definition.

4   Experiments

In this section we discuss the datasets used, implementation details, experimental results and some qualitative analysis.

4.1   Datasets

The statistics of the datasets are presented in Table 1.

- We use the Twitter dataset, *Twitter ADR* described in [5]. It contains 957 tweets posted between 2007 and 2010, with mention annotations of ADR and some other medical entities. Due to Twitter's license agreement, authors released only tweet ids with their corresponding mention span annotations. At the time of collection of the original tweets using Twitter API, we were able to collect only 639 tweets.
- We use the second Twitter dataset, *TwiMed* described in [1]. It contains 1000 tweets with mention annotations of Symptoms from drug (ADR) and other mention annotations posted in 2015. Due to Twitter's license agreement, we were able to extract 663 tweets only.
- For the ADE detection task, we use the Twitter dataset *Twitter ADE* released as part of a Health application shared task[4]. The dataset consists of 13829 tweets annotated with a label of 1 or 0 indicating the presence or absence of an adverse drug event respectively.
- For the unlabeled tweets used for semi-supervised learning, we collected tweets using the keywords as drug-names and ADR lexicon publicly available[5]. This filtering step ensures that all collected tweets have at least one drug-name occurrence and one ADR phrase. The tweets were posted in 2015.

Table 1. Dataset statistics

Dataset	No. tweets	No. ADR Words	Pos. ADE	Neg. ADE
Twitter ADR	639	1,526	-	-
TwiMed	663	1,091	-	-
Twitter ADE	13,829	-	1,206	12,623
Unlabeled Tweets	4,61,522	-	-	-

4.2   Implementation Details

For implementation of the model, we use the popular python deep learning toolkits Keras[6] and Tensorflow[7].

Text Pre-processing: As part of text pre-processing, we normalized all HTML links and USER mentions to the tokens "⟨LINKS⟩" and "⟨USER⟩" respectively. We limit the vocabulary size to 40k most frequent words in case of semi-supervised learning based MTL task. We also remove all mentions of special

---

[4] https://healthlanguageprocessing.org/.
[5] http://diego.asu.edu/downloads.
[6] https://keras.io/.
[7] https://www.tensorflow.org/.

characters and emoticons from the tweet. For each method, the tweet length is padded to the maximum length from the corpus.

Hyper-parameter settings: We kept the hyper-parameter setting for the bi-LSTM transducer similar to the one reported in [5]. Word2Vec embeddings trained on a large generic tweet collection with a dimension of 400 [11] are used as input to the transducer. The hidden unit dimension (d_h) is set to 500. The number of output units (d_l) is 4. We use adam [16] as optimizer with number of epochs set to 10 for all methods. The batch-size for the ADR and ADE tasks are set to 8 and 32 respectively for the MTL method. For the semi-supervised learning method in the weak-supervision part, the batch-size for the ADR task is set to 64 with the confidence threshold value empirically set to 0.5. The stopping criteria for the self-training kicks in when the number of iterations reaches 5 or if the unlabeled tweets pool is exhausted, whichever occurs first. For the joint MTL method, the λ is empirically set to 0.8. The learning rate for all methods is set to 0.001.

4.3 Results

The results of various methods are presented in Tables 2 and 3 for the Twitter ADR and TwiMed datasets respectively. For the ADR task, to encode the output labels we use the IO encoding scheme where each word is labeled with one of the following labels: (1) I-ADR (ADR mention), (2) I-Other (mention category other than ADR), (3) O, (4) PAD (padding token). Since our entity of interest is ADR, we report the results on ADR only. An example tweet annotated with IO-encoding is as follows. "$@BLENDOS_O$ $Lamictal_O$ and_O $trileptal_O$ and_O $seroquel_O$ of_O $course_O$ the_O $seroquel_O$ I_O $take_O$ in_O $severe_O$ $situations_O$ $because_O$ $weight_I - ADR$ $gain_I - ADR$ is_O not_O $cool_O$". For performance evaluation we use approximate-matching [23], which is used popularly in biomedical entity extraction tasks [5,21]. We report the F1-score, Precision and Recall computed using approximate matching as follows.

$$\text{Precision} = \frac{\#\text{ADR approximately matched}}{\#\text{ADR spans predicted}}, \text{Recall} = \frac{\#\text{ADR approximately matched}}{\#\text{ADR spans in total}}$$

(6)

Table 2. Experimental results for Twitter ADR dataset (along with Std. Deviation)

Method	Precision	Recall	F1-score
Baseline [5]	0.7067 ± 0.057	0.7207 ± 0.074	0.7102 ± 0.049
Baseline with adam	0.7065 ± 0.058	0.7576 ± 0.083	0.7272 ± 0.051
KB-Embedding Baseline [22]	0.7171 ± 0.058	0.7713 ± 0.091	0.7397 ± 0.055
Self-training	0.6999 ± 0.047	0.8304 ± 0.039	0.7588 ± 0.039
Joint MTL (Sect. 3.3)	0.7177 ± 0.027	**0.8482 ± 0.068**	0.7770 ± 0.043
MTL (Sect. 3.2)	**0.7569 ± 0.044**	0.8386 ± 0.078	**0.7935 ± 0.045**

Table 3. Experimental results for TwiMed dataset (along with Std. Deviation)

Method	Precision	Recall	F1-score
Baseline [5]	0.6120 ± 0.116	0.5149 ± 0.099	0.5601 ± 0.100
Baseline with adam	0.6281 ± 0.094	0.5614 ± 0.110	0.5859 ± 0.079
KB-Embedding Baseline [22]	0.5960 ± 0.081	0.6144 ± 0.068	0.6042 ± 0.060
Self-training	0.5717 ± 0.056	0.7141 ± 0.082	0.6332 ± 0.057
Joint MTL (Sect. 3.3)	0.5675 ± 0.049	**0.7384 ± 0.079**	0.6401 ± 0.051
MTL (Sect. 3.2)	**0.6656 ± 0.083**	0.6380 ± 0.077	**0.6482 ± 0.065**

The F1-score is the harmonic-mean of the Precision and Recall values. All results are reported using 10-fold cross-validation along with the standard deviation across the folds.

Our baseline methods are bi-LSTM transducer [5] with traditional word embeddings and the current state-of-the-art bi-LSTM transducer which used traditional word embeddings augmented with knowledge-graph based embeddings [22].

For both the datasets, it should be noted that Cocos et al. [5] used RMSProp as an optimizer, and since we are using adam for all our methods, so for a fair comparison we also report the baseline results with adam. The corresponding results are reported in the first two rows of both the tables. It is clear that re-implementation with adam optimizer enhances the performance, which is consistent with the general consensus around adam optimizer.

The KB-embedding baseline [22] replaces word embeddings of the medical entities in the sentence with the corresponding embeddings learned from a knowledge-base. The corresponding results can be seen in row 3 of the tables. It is clear that adding KB-based embeddings enhances the performance over the baseline, due to the external knowledge added in the form of KB embeddings.

The results for our methods are presented from row 4 onwards. We first discuss the results from our joint MTL method. Since the joint MTL method involves self-training as its first step followed by the joint modeling, we also report results using self-training alone. Results from self-training are reported in row 4 in the tables. The self-training based method outperforms the KB-based method, which shows that addition of a large unlabeled corpus in the model improves the performance.

Addition of another task on top of unlabeled data and modeling it in a joint MTL setting further improves the performance. Finally, the results from the MTL method using actual ADE task dataset are presented in the last row of both the tables. It can be seen that the MTL method significantly outperforms baseline methods in terms of F1-score. These results validate our initial hypothesis that sharing two similar tasks of ADR extraction and ADE detection helps the model generalize better.

5   Qualitative Analysis

In this section, we aim to answer the following research questions.

- **Q1:** What is the effect of the auxiliary task dataset's size on the performance of the MTL method?
- **Q2:** What is the effect of size of the unlabeled corpus on the performance of self-training and joint-MTL method?
- **Q3:** What is the effect of adding more depth to the bi-LSTM transducer on the MTL method's performance?

To answer the first question, we perform MTL experiments with varying ADE dataset size. The results are presented in Fig. 2. F1-score has a clear correlation with the percentage training size for the ADE task. As the ADE dataset size is increased, the F1-score also increases monotonously. Similar trend is observed for both the datasets. It clearly indicates the importance of the auxiliary task in the MTL setting.

To answer the second question, we perform joint MTL experiments with varying unlabeled data size. Results are presented in Fig. 3. The results are fairly flat for both the datasets as the unlabeled data size increases. This clearly indicates that our joint MTL method is robust to the size of unlabeled data. It also indicates that our method works well even with a small seed set of unlabeled data-points too.

Results with varying representation capacity of bi-LSTM transducer are presented in Fig. 4. It is clear that the performance degrades as more bi-LSTM layers are stacked on top of the original model. We suspect that this might be the case due to limited manually annotated training data present.

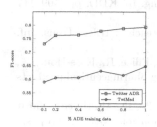

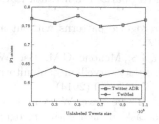

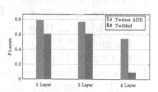

Fig. 2. Performance variation with % ADE datasize

Fig. 3. Performance variation with unlabeled datasize

Fig. 4. Performance variation with stacking bi-LSTM layers

6   Conclusions

In this paper, we proposed two multi-task learning based methods to tackle the problem of labeled data scarcity for adverse drug reaction mention extraction task. Our first method uses adverse drug event detection as an auxiliary task, and demonstrates superior results in comparison to performing the ADR extraction task independently. The second proposed method is a novel joint MTL method, which uses semi-supervised and weakly-supervised learning to automatically generate ADE detection task dataset and then uses the datasets in a novel joint-MTL setting where both tasks are simultaneously modeled. We analyzed the method on two popular ADR extraction datasets, and it demonstrates superior results as compared to the state-of-the-art methods in ADR extraction.

References

1. Alvaro, N., Miyao, Y., Collier, N.: TwiMed: Twitter and PubMed comparable corpus of drugs, diseases, symptoms, and their relations. JMIR Public Health Surveill. **3**(2) (2017)
2. Ando, R.K., Zhang, T.: A framework for learning predictive structures from multiple tasks and unlabeled data. JMLR **6**, 1817–1853 (2005)
3. Argyriou, A., Evgeniou, T., Pontil, M.: Multi-task feature learning. In: NIPS, pp. 41–48 (2006)
4. Caruana, R.: Multitask learning: a knowledge-based source of inductive bias. In: ICML, pp. 41–48 (1993)
5. Cocos, A., Fiks, A.G., Masino, A.J.: Deep learning for pharmacovigilance: recurrent neural network architectures for labeling adverse drug reactions in Twitter posts. JAMIA, p. ocw180 (2017)
6. Collobert, R., Weston, J.: A unified architecture for natural language processing: deep neural networks with multitask learning. In: ICML, pp. 160–167 (2008)
7. Collobert, R., Weston, J., Bottou, L., Karlen, M., Kavukcuoglu, K., Kuksa, P.: Natural language processing (almost) from scratch. JMLR **12**(Aug), 2493–2537 (2011)
8. Evgeniou, T., Pontil, M.: Regularized multi-task learning. In: KDD, pp. 109–117 (2004)
9. Fralick, S.: Learning to recognize patterns without a teacher. IEEE Trans. Inf. Theory **13**(1), 57–64 (1967)
10. Freifeld, C.C., Brownstein, J.S., Menone, C.M., Bao, W., Filice, R., Kass-Hout, T., Dasgupta, N.: Digital drug safety surveillance: monitoring pharmaceutical products in Twitter. Drug Saf. **37**(5), 343–350 (2014)
11. Godin, F., Vandersmissen, B., De Neve, W., Van de Walle, R.: Multimedia Lab@ ACL W-Nut NER shared task: named entity recognition for twitter microposts using distributed word representations. In: ACL-ICJNLP 2015, pp. 146–153 (2015)
12. Graves, A.: Sequence transduction with recurrent neural networks. CoRR abs/1211.3711 (2012)
13. Hazell, L., Shakir, S.A.: Under-reporting of adverse drug reactions: a systematic review. Pharmacoepidemiol. Drug Saf. **14**, S184–S185 (2005)
14. Iosifidis, V., Ntoutsi, E.: Large scale sentiment learning with limited labels. In: KDD, pp. 1823–1832. ACM (2017)
15. Kim, Y.: Convolutional neural networks for sentence classification. In: EMNLP, pp. 1746–1751 (2014)
16. Kingma, D., Ba, J.: Adam: a method for stochastic optimization. arXiv:1412.6980 (2014)

17. LeCun, Y., Bengio, Y., Hinton, G.: Deep learning. Nature **521**(7553), 436–444 (2015)
18. Liu, X., Gao, J., He, X., Deng, L., Duh, K., Wang, Y.: Representation learning using multi-task deep neural networks for semantic classification and information retrieval. In: NAACL-HLT, pp. 912–921 (2015)
19. Luong, M., Le, Q.V., Sutskever, I., Vinyals, O., Kaiser, L.: Multi-task sequence to sequence learning. CoRR abs/1511.06114 (2015)
20. Màrquez, L., Callison-Burch, C., Su, J., Pighin, D., Marton, Y. (eds.): EMNLP (2015)
21. Nikfarjam, A., Sarker, A., OConnor, K., Ginn, R., Gonzalez, G.: Pharmacovigilance from social media: mining adverse drug reaction mentions using sequence labeling with word embedding cluster features. JAMIA **22**(3), 671–681 (2015)
22. Stanovsky, G., Gruhl, D., Mendes, P.N.: Recognizing mentions of adverse drug reaction in social media using knowledge-infused recurrent models. In: EACL, pp. 142–151 (2017)
23. Tsai, R.T.H., Wu, S.H., Chou, W.C., Lin, Y.C., He, D., Hsiang, J., Sung, T.Y., Hsu, W.L.: Various criteria in the evaluation of biomedical named entity recognition. BMC Bioinform. **7**(1), 92 (2006)
24. Vieira, H.S., da Silva, A.S., Cristo, M., de Moura, E.S.: A self-training CRF method for recognizing product model mentions in web forums. In: Hanbury, A., Kazai, G., Rauber, A., Fuhr, N. (eds.) ECIR 2015. LNCS, vol. 9022, pp. 257–264. Springer, Cham (2015). https://doi.org/10.1007/978-3-319-16354-3_27
25. Wang, Z., Zhang, J., Feng, J., Chen, Z.: Knowledge graph embedding by translating on hyperplanes. In: AAAI, pp. 1112–1119 (2014)

Choices in Knowledge-Base Retrieval
for Consumer Health Search

Jimmy[1,3]($\boxtimes$), Guido Zuccon[1], and Bevan Koopman[2]

[1] Queensland University of Technology, Brisbane, Australia
jimmy@hdr.qut.edu.au, g.zuccon@qut.edu.au
[2] Australian E-Health Research Center, CSIRO, Brisbane, Australia
bevan.koopman@csiro.au
[3] University of Surabaya (UBAYA), Surabaya, Indonesia

Abstract. This paper investigates how retrieval using knowledge bases can be effectively translated to the consumer health search (CHS) domain. We posit that using knowledge bases for query reformulation may help to overcome some of the challenges in CHS. However, translating and implementing such approaches is nontrivial in CHS as it involves many design choices. We empirically evaluated the impact these different choices had on retrieval effectiveness. A state-of-the-art knowledge-base retrieval model—the Entity Query Feature Expansion model—was used to evaluate the following design choices: which knowledge base to use (specialised vs. generic), how to construct the knowledge base, how to extract entities from queries and map them to entities in the knowledge base, what part of the knowledge base to use for query expansion, and if to augment the KB search process with relevance feedback. While knowledge base retrieval has been proposed as a solution for CHS, this paper delves into the finer details of doing this effectively, highlighting both pitfalls and payoffs. It aims to provide some lessons to others in advancing the state-of-the-art in CHS.

1 Introduction and Related Work

A major challenge for users in consumer health search (CHS) is how to effectively represent complex and ambiguous information needs as a query [13,14,16]. Studies on query formulation in CHS have shown that consumers struggle to find effective query terms [14], often submitting layman and circumlocutory descriptions of symptoms instead of precise medical terms [17]. For example, people search for "skin irregularities" instead of "skin lesions" (the correct medical term for the symptom). This leads to poor retrieval effectiveness and low user satisfaction. Different approaches have been proposed to improve CHS, including query suggestion [15], learning-to-rank using syntactic, semantic or readability features [7,12], and query expansion or reformulation [8–10].

Here we focus on overcoming the CHS problem by expanding/reformulating a health query with more effective terms (e.g., less ambiguous, synonyms, etc.). Manually replacing query terms with those from medical terminologies

© Springer International Publishing AG, part of Springer Nature 2018
G. Pasi et al. (Eds.): ECIR 2018, LNCS 10772, pp. 72–85, 2018.
https://doi.org/10.1007/978-3-319-76941-7_6

(e.g., UMLS) has proven effective [8]. This shows that query reformulation in the CHS can be effective—but can it be done automatically?

In the general search domain, there have been a number of automated query reformulation approaches that link queries to entities in a knowledge base (KB) such as Wikipedia and Freebase and then used these related entities for query expansion. Bendersky et al. [1] approach involved linking the query to concepts in Wikipedia. Concepts from the query, denoted κ_Q, were weighted; the same was done for concepts in each of the documents in the corpus, denoted κ_D. The relevance score $sc(Q, D)$ between query Q and document D was calculated as a relatedness measure between κ_Q and κ_D [1]. Later, the Entity Query Feature Expansion model [2] extended this by automatically expanding queries by linking them to Wikipedia. Instead of just using entities from Wikipedia (as Bendersky et al. [1] did), the Entity Query Feature Expansion model labelled words in the user query and in each document with a set of entity mentions M_Q and M_d [2]. Each entity mention was related to KB entities $e \epsilon E$, with different relationship types. The queries were expanded by including entity aliases, categories, words, and types from Wikipedia articles. The expanded query was then matched against documents in the corpus using the query likelihood model with Dirichlet smoothing.

We posit that this Entity Query Feature Expansion model would have merit in CHS. It provides a means of mapping health queries to health entities in a health related (subset of a) KB, be this either a general KB (Wikipedia) or a specialised one (e.g., UMLS). The initial query can then be expanded based on related entities. In this paper, we investigated the use of both a specialised health KB, in line with previous work that expanded queries using, e.g., MeSH or UMLS [3,9,10], and of a general KB like Wikipedia. Our rationale for this latter choice was the observation that consumers tend to submit queries using general terms and that these are covered by Wikipedia entities. However, Wikipedia also covers many of the medical entities found in specialised medical KBs. More importantly, there are links between the general and specialised entities in Wikipedia – links that can be exploited for query expansion. Thus, we adopted the Entity Query Feature Expansion model for our empirical evaluation, determining if such a KB retrieval approach is effective for CHS.

In investigating the effectiveness of the KB retrieval approach to CHS there are a number of important design decisions. The impact of these different decisions has not been thoroughly considered when describing the proposed approach [1,2]. Therefore, in this paper we also seek to empirically evaluate the impact of a number of different choices in KB retrieval for CHS: (i) KB construction; (ii) entity mention extraction; (iii) entity mapping; (iv) source of expansion; (v) use of relevance feedback. We also determine whether the use of a specialised KB is preferred over a general one, or vice versa.

2 Expansion Model

We implemented the Entity Query Feature Expansion model for retrieval on both the Wikipedia and UMLS as the KB. For the Wikipedia KB, a single

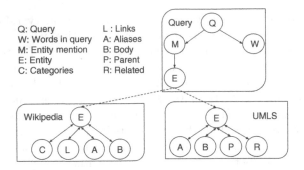

Fig. 1. Summary of expansion sources.

entity is represented by a single Wikipedia page (the page title identifies the entity). Beyond titles, Wikipedia also contains many page features useful in a retrieval scenario: entity title (E), categories (C), links (L), aliases (A), and body (B). As for the UMLS KB, a single entity is represented by the most frequently used terms for a single concept unique identifier (CUI). Features of a UMLS KB entity are aliases (A), body (B), parent concepts (P), and related concepts (R). Figure 1 shows the features we used for mapping the queries to entities in the KB and as the source of expansion terms. We formally define the query expansion model as:

$$\hat{\vartheta}_q = \sum_M \sum_f \lambda_f \vartheta_{f(EM,SE)} \tag{1}$$

where M are the entity mentions and contain uni-, bi-, and tri-gram generated from the query; f is a function used to extract the expansion terms. $\lambda_f \epsilon (0,1)$ is a weighting factor. $\vartheta_{f(EM,SE)}$ is a function to map entity mention M to the KB features EM (e.g., "Title", "Aliases", "Links", "Body", etc.) and extract expansion terms from source of expansion SE (e.g., "Title", "Aliases", etc.).

3   Choices in Knowledge Base Retrieval

3.1   Knowledge Base Construction

We investigated which entities should form the basis of our KB. The CHS focus meant that health-related entities were needed. For Wikipedia KB, we considered three choices for collecting health related pages: (WC-Type) pages with Medicine infobox[1] type[2] (e.g., "abortion method", "alternative medicine", "pandemic"); (WC-TypeLinks) pages with Medicine infobox type and with links to medical terminologies such as Mesh, UMLS, SNOMED CT, ICD; (WC-UMLS) pages

---

[1] A Wikipedia Infobox is used to summarise important aspects of an entity and its relation with other articles.

[2] http://en.wikipedia.org/wiki/Wikipedia:List_of_infoboxes#Health_and_fitness.

with title matching an UMLS entity. The last method used QuickUMLS [11] to map Wikipedia page titles to the UMLS: if the mapping was successful, we included the Wikipedia entity (page) in the KB.

For UMLS KB, we considered two choices: (UC-All) all entities and (UC-Med) entities related to four key aspects of medical decision criteria (i.e., symptoms, diagnostic test, diagnoses, and treatments) as used in [6,10]. For these choices, we included all English and non-obsolete terms.

3.2 Entity Mentions Extraction

Entity mention extraction is the process of identifying spans of text from the query that could map to some entity. It does not consider which exact entity (this is detailed in the next section). We considered three possible choices to extract entity mentions: (ME-All) include all uni-, bi- and tri-grams of the query (*default choice*); (ME-CHV) include only those uni-, bi- and tri-grams of the query that matched entities in the Consumer Health Vocabulary (CHV) [5][3]; and (ME-UMLS) include only those uni-, bi- and tri-grams of the query that matched entities in the UMLS (via QuickUMLS). These three choices were used for both the Wikipedia and UMLS KBs.

3.3 Entity Mapping

We investigated how the entity mentions from the previous section were mapped to entities in the KB. An entity mention was mapped to an entity if an exact match was found between the mention and the entity. As shown in Fig. 1, the Wikipedia entity can be represented according to six different sources; the choices considered were: (WEM-Title) titles, (WEM-Aliases) aliases, (WEM-Links) links, (WEM-Body) the entire bodies of the Wikipedia pages, (WEM-Cat) categories, (WEM-All) all the previous sources (*default choice*). For UMLS KB, the choices considered were: (UEM-Title) titles, (UEM-Aliases) aliases, (UEM-Body) the entire UMLS concept description, (UEM-Parent) parents, (UEM-Related) related entities, (UEM-All) all the previous sources (*default choice*), (UEM-QuickUmls) use QuickUMLS [11] to obtain entity mappings.

3.4 Source of Expansion

We investigated which sources in the KB were used to draw candidate terms for query expansion. We explored three choices: (SE-Title) titles associated with the entities, (SE-Aliases) aliases associated with the entities, (SE-All) both titles and aliases (*default choice*). While other information sources could be used (for example, those used for entity mapping), preliminary experiments showed that only these three choices produced meaningful results. These choices were used for both the Wikipedia and UMLS KBs.

---

[3] Only complete string matches were considered.

3.5   Relevance Feedback

We investigated the use of relevant feedback (both explicit relevance feedback (RF) and Pseudo Relevance Feedback (PRF)). We performed RF by extracting the ten most important health related words (based on tf.idf scores) from each of the top three relevant documents (relevance label greater than 0) thus resulting in a maximum of thirty expansion terms. PRF was performed by extracting the ten most important health related words from the top three ranked documents (regardless of their true relevance label). A term was considered as health related if it exactly matched a title or an alias of an entity in the target KB (either Wikipedia or UMLS).

4   Empirical Evaluation

To investigate the influence choices in KB retrieval have on query expansion for the CHS task, we empirically evaluated methods using the CLEF 2016 eHealth [18]. This collection comprises 300 query topics originating from health consumers seeking health advice online. Documents are taken from Clueweb12b-13. The collection was indexed using Elasticsearch 5.1.1, with stopping and stemming. A simple baseline was implemented using BM25F with $b = 0.75$ and $k1 = 1.2$. BM25F allows specifying boosting factors for matches occurring in different fields of the indexed web page. We considered only the title field and the body field, with boost factors 1 and 3, respectively. These were found to be the optimal weights for BM25F for this test collection in previous work [4]. This is a strong baseline as it outperforms most runs submitted to CLEF 2016.

For constructing the Wikipedia KB, we considered candidate pages from the English subset of Wikipedia (dump 1/12/2016), limited to current revisions only and without talk or user pages. Of the 17 million entries, we filtered out pages that were redirects; this resulted in a Wikipedia corpus of 9,195,439 pages. These candidate pages were then processed according to the choices available for KB construction (Sect. 3.1). Selected pages to be included in the KB were also indexed using Elasticsearch 5.1.1 with field based indexing (fields: title, links, categories, types, aliases, and body), to support the use of different fields as the source of query expansion terms (Sect. 3.4).

For constructing the UMLS KB, we indexed 3,057,234 non obsolete English terms with the following fields: title (the most frequently used term for a CUI), aliases (for all other terms used for the CUI), body (the description of a CUI), parent (title of UMLS entities with relationship type PAR), related (title of UMLS entities with relationship type RQ and RL).

Results were evaluated using nDCG@10, RBP@10 (persistence 0.5, depth 10, reporting also residuals (Res.)), in line with the CLEF 2016 collection, as users in the CHS task tend to primarily examine the first few search results. Additionally, bpref was used as a first attempt to reduce the influence of unjudged documents on evaluation (expanded queries retrieved many more unjudged documents than the baseline). In all result tables, superscripts refer to statistical significance (pairwise t-test with Bonferroni adjustment and $\alpha < 0.05$) between the result

Table 1. Influence of choices in KB construction; all queries (top) and high coverage queries (bottom).

Choice	nDCG@10	bpref	RBP@10	Res.	$\lceil exp \rceil$	$\langle e, g, l \rangle$
baseline[0]	$.2465^{1-5}$	$.1798^{1235}$	$.3263^{1-5}$	.0399		
WC-Type[1]	$.0950^{0245}$	$.1485^{04}$	$.1258^{024}$	.7071	38.99	299,55,161
WC-TypeLinks[2]	$\mathbf{.1146}^{01}$	$\mathbf{.1547}^{0}$	$\mathbf{.1532}^{01}$	.6361	43.22	300,66,157
WC-UMLS[3]	$.1090^{0}$	$.1475^{04}$	$.1439^{0}$	.6342	21.17	299,54,163
UC-All[4]	$.1256^{01}$	$\mathbf{.1653}^{13}$	$\mathbf{.1626}^{01}$	.5976	29.27	299,63,164
UC-Med[5]	$\mathbf{.1300}^{01}$	$.1558^{0}$	$.1552^{0}$	.5318	43.83	270,52,151
Choice	nDCG@10	bpref	RBP@10	Res.	$\lceil exp \rceil$	$\langle e, g, l \rangle$
baseline[0]	$.4481^{5}$	$.4700^{1345}$	.5046	.0010		
WC-Type[1]	$.4567^{5}$	$.4160^{05}$	$.4342^{34}$	.1736	3.54	13,5,6
WC-TypeLinks[2]	$\mathbf{.4816}^{45}$	$\mathbf{.4334}^{5}$	.4944	.1129	3.54	13,5,6
WC-UMLS[3]	.4602	$.4186^{05}$	$\mathbf{.6718}^{1}$	.1814	17.54	13,9,4
UC-All[4]	$\mathbf{.4285}^{25}$	$\mathbf{.3791}^{05}$	$\mathbf{.5874}^{1}$	.0143	34.46	13,8,4
UC-Med[5]	$.3542^{0124}$	$.2615^{0-4}$	.4854	.0466	46.17	12,6,5

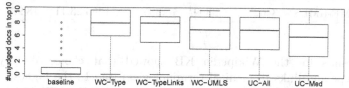

Fig. 2. Unjudged documents among the top 10 retrieved by runs in Table 1 (Top).

and the result from the choice associated with the superscript. Furthermore the average number of terms added in the expanded query ($\lceil exp \rceil$) and the number of expanded queries, queries with a gain for RBP@10 and a loss for RBP@10 were recorded as a triplet $< e, g, l >$.

Because of space limits, for each choice, we empirically evaluated the influence the choice had on retrieval effectiveness by examining each choice sequentially. We did this across both Wikipedia and UMLS KB, and drawed conclusions about which KB best supports CHS at the end. For each choice, we fixed the best setting and use this best setting for the subsequent choice. We determined the best setting firstly based on results (i.e., nDCG@10, bpref, RBP@10) for the all queries set. If no method is clearly best for this set, then we checked results from the high coverage queries set. The complete set of results is provided in an online appendix at https://github.com/ielab/ECIR2018_KnowledgeBase_CHS.

4.1   Knowledge Base Construction

The effect on retrieval of choices in KB construction is reported in Table 1 (top); results are averaged over all 300 queries in the CLEF 2016 collection.

Table 2. Influence of choices in entity mention extraction; all queries (top), high coverage queries (bottom).

Choice	nDCG@10	bpref	RBP@10	Res.	$\overline{|exp|}$	$\langle e,g,l\rangle$
baseline[0]	$.2465^{1-6}$	$.1798^{12356}$	$.3263^{1-6}$	.0399		
WME-All[1]	$\mathbf{.1146^{0}}$	$\mathbf{.1547^{0}}$	$.1532^{0}$	.6361	43.22	300,66,157
WME-CHV[2]	$.1143^{0}$	$.1487^{04}$	$\mathbf{.1573^{0}}$	.6024	36.06	285,59,155
WME-UMLS[3]	$.1031^{04}$	$.1500^{0}$	$.1426^{0}$	.6008	31.00	281,56,156
UME-All[4]	$\mathbf{.1256^{03}}$	$\mathbf{.1653^{25}}$	$\mathbf{.1626^{0}}$	.5976	29.27	299,63,164
UME-CHV[5]	$.1185^{0}$	$.1570^{04}$	$.1539^{0}$	.6080	24.85	288,48,168
UME-UMLS[6]	$.1191^{0}$	$.1640^{0}$	$.1537^{0}$	.5649	2.90	282,51,161

Choice	nDCG@10	bpref	RBP@10	Res.	$\overline{|exp|}$	$\langle e,g,l\rangle$
baseline[0]	.3218	.3388	.3647	.0042		
WME-All[1]	.3795	.3286	.4516	.1874	25.32	22,9,8
WME-CHV[2]	$\mathbf{.3907^{3}}$	**.3295**	**.4714**	.1112	28.06	16,8,6
WME-UMLS[3]	$.3606^{2}$	.3220	.4528	.0652	22.44	16,8,6
UME-All[4]	**.3503**	.3346	$\mathbf{.4162^{6}}$	.1488	28.62	21,12,7
UME-CHV[5]	.3466	**.3459**	.3992	.1574	25.71	17,9,7
UME-UMLS[6]	.3462	$.3309^{4}$	.3852	.1256	23.11	18,9,7

The results for the Wikipedia KB showed that choice WC-TypeLinks (infobox type and links to medical terminologies) lead to the highest effectiveness across most measures. However, UC-All from the UMLS KB obtained higher effectiveness for all measures. Nevertheless, the baseline performed considerably better than the KB retrieval methods.

When further analysing the results, we found that, for a large number of queries, the KB retrieval methods ranked many unjudged documents amongst the top 10; while the baseline had a much lower rate of unjudged documents amongst the top 10. Figure 2 reported the distribution of unjudged documents for each of the configurations considered. This is clearly influencing the results, as demonstrated by the large values of RBP residuals associated with the KB retrieval methods in Table 1 (compared to the residual of the baseline). Interestingly, if all unjudged documents turned out to be relevant, the RBP@10 of the KB retrieval methods would prove largely superior than that of the baseline (compare the residuals).

We then considered a subset of queries for which, on average across all runs considered for a specific choice, there were a maximum of 2 unjudged documents out of the first 10. This threshold was determined by analysing the number of unjudged documents for the baseline (the baseline does not change, irrespective of the choices), so that the threshold corresponded to 1.5 times the interquartile range above the third quartile (the upper whisker of the box-plot). Note that this produced a different subset of queries for each of the considered choices; however, the subsets had the same average "coverage" with respect to the relevance assessments. We referred to these subsets as the *high coverage queries*. This sub-

Table 3. Influence of choices in entity mapping; all queries (top), high coverage queries (bottom).

Choice	nDCG@10	bpref	RBP@10	Res.	$\overline{\lvert exp \rvert}$	$\langle e, g, l \rangle$
baseline^0	$.2465^{1-d}$	$.1798^{1-69a}$	$.3263^{1-d}$	.0399		
WEM-Title^1	$.1547^{02569ac}$	$.1602^{06-ad}$	$.1940^{025689}$	.3699	25.60	172,32,103
WEM-Aliases^2	$\mathbf{.1984^{0134679-d}}$	$\mathbf{.1689^{03689a}}$	$\mathbf{.2681^{013-79-d}}$	.2392	16.97	114,31,60
WEM-Links^3	$.1506^{02569ac}$	$.1500^{0278bcd}$	$.2067^{0269cd}$	.3130	24.23	149,22,96
WEM-Body^4	$.1427^{025689}$	$.1600^{0789ad}$	$.1826^{02589}$	.4175	71.30	204,42,121
WEM-Cat^5	$.1783^{013469-d}$	$.1624^{06789ad}$	$.2320^{0124679acd}$	.2673	25.04	107,22,70
WEM-All^6	$.1143^{0-5789bd}$	$.1487^{012578bcd}$	$.1573^{0-3589}$	.6024	36.06	285,59,155
UEM-Title^7	$.1518^{0269ac}$	$.1774^{13-69a}$	$.1801^{02589}$	.5332	16.82	287,50,160
UEM-Aliases^8	$\mathbf{.1717^{0469-d}}$	$\mathbf{.1847^{1-69-c}}$	$\mathbf{.2365^{014679-d}}$	.3633	9.96	266,75,125
UEM-Body^9	$.0734^{0-8a-d}$	$.1341^{0124578bcd}$	$.0943^{0-8a-d}$	.6772	113.14	296,35,180
UEM-Parent^a	$.1259^{0-35789b}$	$.1415^{0124578bcd}$	$.1702^{02589}$	.5616	28.25	265,44,147
UEM-Related^b	$.1463^{025689ac}$	$.1677^{3689a}$	$.1915^{0289d}$	.5154	32.53	276,62,148
UEM-All^c	$.1256^{0-35789b}$	$.1653^{36ad}$	$.1626^{023589}$	.5976	29.27	299,63,164
UEM-QuickUmls^d	$.1355^{025689}$	$.1792^{13-69ac}$	$.1563^{023589b}$	.5497	3.44	297,65,162
Choice	nDCG@10	bpref	RBP@10	Res.	$\overline{\lvert exp \rvert}$	$\langle e, g, l \rangle$
baseline^0	$.4018^{79}$	$.3886^{49}$	$.4640^{9}$	.0017		
WEM-Title^1	$.4288^{239}$	$\mathbf{.3940^{49}}$	$.4715^{9}$	.0559	18.86	7,4,3
WEM-Aliases^2	$.3789^{179}$	$.3850^{49}$	$.4593^{9}$	.0325	12.71	7,3,4
WEM-Links^3	$.3655^{179b}$	$.3469^{9b}$	$.4191^{79}$	.0619	33.56	9,3,6
WEM-Body^4	$.3554^{79b}$	$.3289^{0125789b}$	$.4070^{9b}$	.0328	101.77	13,4,9
WEM-Cat^5	$.3919^{79}$	$.3846^{49}$	$.4540^{79}$	.0017	3.50	2,0,2
WEM-All^6	$\mathbf{.4434^{9}}$	$.3711^{9}$	$\mathbf{.5412^{9}}$	.1655	24.00	15,8,6
UEM-Title^7	$\mathbf{.5051^{02-59}}$	$.3858^{49}$	$\mathbf{.6281^{359a}}$	.1612	11.10	20,11,8
UEM-Aliases^8	$.4250^{9}$	$.4001^{49}$	$.5100^{9}$	.0438	15.75	20,12,7
UEM-Body^9	$.1752^{0-8a-d}$	$.2332^{0-8a-d}$	$.1227^{0-8a-d}$	.3577	91.81	21,1,17
UEM-Parent^a	$.3800^{9}$	$.3616^{9}$	$.4351^{79}$	.2068	26.90	20,12,8
UEM-Related^b	$.4695^{349cd}$	$\mathbf{.4160^{349d}}$	$.5753^{49d}$	.0564	27.10	21,14,6
UEM-All^c	$.4114^{9b}$	$.3759^{9}$	$.5075^{9}$	.1083	31.43	21,12,7
UEM-QuickUmls^d	$.4048^{9b}$	$.3696^{9b}$	$.4615^{9b}$	.1818	27.95	21,10,9

set included 13 queries for choice 1 (Table 1, bottom). Results showed reduced residuals and reduced gaps between KB retrieval methods and the baselines; however trends in effectiveness across the considered choices for the Wikipedia KB did not change, unlike the relative effectiveness of the UMLS KB method (UC-All) that proved less effective than methods on the Wikipedia KB.

For Wikipedia, the results showed that the best setting was WC-TypeLinks. Thus, we selected WC-TypeLinks for the rest of the following analyses for Wikipedia KB; while we used UC-All for UMLS KB.

4.2 Entity Mentions Extraction

Table 2 (top: 300 queries and bottom: 22 high coverage queries) reports the results obtained when comparing choices for entity mention extraction. For

Table 4. Influence of choices in source of expansion; all queries (top), high coverage queries (bottom).

Choice	nDCG@10	bpref	RBP@10	Res.	$|exp|$	$\langle e,g,l \rangle$
baseline^0	$.2465^{2-6}$	$.1798^{1-4}$	$.3263^{2356}$	.0399		
WSE-Title^1	$\mathbf{.2425}^{2-6}$	$\mathbf{.1843}^{023}$	$\mathbf{.3230}^{2356}$	.0829	1.37	76,26,32
WSE-Aliases^2	$.1976^{01}$	$.1687^{01456}$	$.2677^{01}$	.2376	16.75	114,30,61
WSE-All^3	$.1984^{01}$	$.1689^{01456}$	$.2681^{01}$	.2392	16.97	114,31,60
USE-Title^4	$\mathbf{.2126}^{0156}$	$\mathbf{.1887}^{023}$	$\mathbf{.2996}^{56}$	.2119	2.85	235,73,98
USE-Aliases^5	$.1813^{0146}$	$.1864^{23}$	$.2449^{014}$	.3298	9.16	257,72,120
USE-All^6	$.1717^{0145}$	$.1847^{23}$	$.2365^{014}$	.3633	9.96	266,75,125
Choice	nDCG@10	bpref	RBP@10	Res.	$|exp|$	$\langle e,g,l \rangle$
baseline^0	$.2794^{1}$	$.2189^{456}$	$.3554^{1}$	.0130		
WSE-Title^1	$\mathbf{.2860}^{0}$	$\mathbf{.2211}^{45}$	$\mathbf{.3737}^{0}$	.0149	1.77	13,8,4
WSE-Aliases^2	.2734	$.2191^{4}$	.3645	.0446	1.82	28,17,10
WSE-All^3	.2754	$.2191^{4}$	.3646	.0448	11.39	28,18,9
USE-Title^4	$\mathbf{.2928}^{6}$	$\mathbf{.2400}^{0-3}$	$\mathbf{.3870}$	.0424	2.41	85,39,22
USE-Aliases^5	.2633	$.2357^{01}$	.3578	.0888	8.36	97,42,31
USE-All^6	$.2619^{4}$	$.2346^{0}$	.3544	.0999	9.11	99,43,32

Wikipedia, results showed that the choice of constructing entity mentions with uni-, bi- and tri-grams of the queries that matched CHV (WME-CHV) was overall the one that provided the highest retrieval effectiveness. While this is clear in the high coverage set, the difference between this strategy and using all grams (WME-All) for all queries set is less clear, probably due to the extent of many unjudged documents affecting some runs. We concluded that WME-CHV was the most effective choice and selected WME-CHV in the remaining analyses.

For UMLS, results showed that constructing entity mentions using all uni-, bi-, and tri-grams of the queries (UME-ALL) terms provided the highest retrieval effectiveness. Thus, we selected UME-ALL in the remaining analyses.

4.3   Entity Mapping

Table 3 (top: 300 queries and bottom: 22 queries) reports the results obtained when comparing choices for entity mapping. For both KBs, mapping entities to Aliases (WEM-Aliases and UEM-Aliases) clearly outperformed the other approaches (all queries). Results for the high coverage queries showed mixed results. Thus, we selected WEM-Aliases and UEM-Aliases for the subsequent analyses.

4.4   Source of Expansion

Table 4 (top: 300 queries and bottom: 119 queries) reports the results obtained when comparing sources of query expansion. Results clearly showed that selecting titles as source of expansion (WSE-Title and USE-Title) was the most effec-

Table 5. Influence of choices in relevance feedback; all queries (top), high coverage queries (bottom).

Choice	nDCG@10	bpref	RBP@10	Res.	$|exp|$	$\langle e,g,l \rangle$
baseline^0	$\mathbf{.2465}^{1235-9}$	$.1798^{478}$	$.3263^{23689}$	.0399		
baselineRF^1	$.2055^{024569}$	$.1777^{58}$	$\mathbf{.3412}^{23679}$	.1400	11.70	150,75,74
baselinePRF^2	$.1657^{0134578}$	$.1704^{578}$	$.2679^{01458}$	.2831	15.63	297,66,146
GUIR-3^3	$.1975^{02468}$	$\mathbf{.1803}^{8}$	$.2636^{014578}$	.2333		292,74,134
WSE-Title^4	$\mathbf{.2425}^{1235679}$	$\mathbf{.1843}^{08}$	$.3230^{235689}$	.0829	1.37	76,26,32
WSE-TitleRF^5	$.2133^{012469}$	$.1833^{1268}$	$\mathbf{.3523}^{234679}$	.1710	1.02	183,92,75
WSE-TitlePRF^6	$.1660^{0134578}$	$.1716^{578}$	$.2638^{01458}$	.2928	16.17	297,71,142
USE-Title^7	$.2126^{02469}$	$.1887^{0268}$	$.2996^{13589}$	.2119	2.85	235,73,98
USE-TitleRF^8	$\mathbf{.2245}^{02369}$	$\mathbf{.2006}^{0-79}$	$\mathbf{.3687}^{0234679}$	.2290	9.79	263,94,93
USE-TitlePRF^9	$.1784^{014578}$	$.1829^{8}$	$.2672^{014578}$	.2989	25.35	300,70,146

Choice	nDCG@10	bpref	RBP@10	Res.	$|exp|$	$\langle e,g,l \rangle$
baseline^0	$\mathbf{.2718}^{3678}$	$\mathbf{.2309}^{47}$	$.3321^{378}$	.0013		
baselineRF^1	$.2625^{68}$	$.2178^{78}$	$\mathbf{.3630}^{368}$	.0199	12.00	38,22,16
baselinePRF^2	$.2429^{6-9}$	$.2142^{789}$	$.3339^{678}$	.0662	15.66	80,25,27
GUIR-3^3	$.2363^{04789}$	$.2207^{478}$	$.2799^{0145789}$	.0875		79,22,29
WSE-Title^4	$\mathbf{.2737}^{368}$	$\mathbf{.2378}^{037}$	$.3397^{378}$	.0240	1.42	24,10,6
WSE-TitleRF^5	$.2635^{68}$	$.2193^{78}$	$\mathbf{.3669}^{368}$	.0308	9.88	48,26,15
WSE-TitlePRF^6	$.2272^{01245789}$	$.2161^{78}$	$.3131^{125789}$	.0932	15.93	80,24,28
USE-Title^7	$.2961^{0236}$	$\mathbf{.2495}^{0-6}$	$.3981^{023468}$	.0545	2.37	67,32,12
USE-TitleRF^8	$\mathbf{.3087}^{0-69}$	$.2445^{123}$	$\mathbf{.4398}^{0-79}$	.0455	11.07	72,33,12
USE-TitlePRF^9	$.2790^{2368}$	$.2323^{2}$	$.3748^{368}$	.0800	23.19	80,30,27

tive choice compared to other choices for both Wikipedia KB and UMLS KB. Therefore, we selected WSE-Title and USE-Title for the following analyses.

4.5 Relevance Feedback

Table 5 (top: 300 queries and bottom: 80 queries) reports the results obtained with and without relevance feedback. For Wikipedia, results showed that the addition of feedback produced mixed results. RF produced the best RBP@10 across all types of queries. In terms of nDCG@10 and bpref, the Wikipedia WSE-Title choice performed better without the addition of feedback. For the UMLS, results showed that RF produced the best performance for all queries set on all measures. For the high coverage queries, the USE-Title obtained better bpref without the addition of relevance feedback. The application of relevance feedback to the baseline only improved RBP@10 when using true relevance information (RF). Nevertheless, this performed worse than the KB methods.

5   Further Analysis and Discussion

In summary, we found that: (1) PRF does not improve results, independently of the KB; (2) RF instead does provide improved effectiveness, with UMLS-based best settings (USE-TitleRF) being generally better than Wikipedia-based best settings (WSE-TitleRF) for both all queries and the high coverage queries sets; (3) For the high coverage queries set (Table 5), independently of whether relevance feedback was applied, UMLS based KB best settings were more effective than Wikipedia based KB settings; for all queries set, UMLS based KB settings with RF performed better than Wikipedia based KB settings on all measures; (4) UMLS KB expanded more queries than the Wikipedia KB. This last finding is likely due to the Wikipedia KB being incomplete in that it considered only pages with health infobox and links to medical terms. Though this was the best setting, it removed many health related pages such as "headache". Further, we found that the two methods provided radically different query expansions: on average, they only had 8.9% of expansion terms in common. On top of that, we found that they retrieved different sets of documents (average overlap for the best settings without relevance feedback: 61% (55%) of the top 1,000 (10) documents). Given these differences, we suggest future work to be directed to explore the effectiveness of combining expansions from the two KBs.

To contextualise the results obtained by KB retrieval methods, in Table 5 we also report the results of the method implemented by the GUIR-3 submission to the CLEF 2016 challenge [10]. This was the best performing, comparable[4] query expansion method at CLEF 2016. The method expands queries by mapping query entities to the UMLS, and navigating the UMLS tree to gather hypernims from mapped entities as source of expansion. Post-processing is applied to the candidates to retain expansions more likely to be of benefit to retrieval. For each query, multiple expanded query variations are collected and their results aggregated using the Borda algorithm (see [10] for details). Unlike the original method, our implementation relied on BM25F rather than DFR as scoring method and QuickUMLS in place of Metamap, so as to be directly comparable with our baseline and KB retrieval methods. In Table 5 we do not report $|exp|$ for GUIR-3 as the method replaces some of the original terms with the expansion ones, thus making comparisons not trivial.

By observing the number of expansion terms added across the KB methods, we noted that the effective choices for KB query expansion tend to produce the lowest number of expansion terms (as well as expanding the smallest number of queries). While relevance feedback added a significant number of expansion terms (as well as expanding a large number of queries), PRF did so somewhat too aggressively, which may explain why RF, which is more conservative both in queries that are expanded and the extent of expansion, outperformed PRF.

Finally, we analysed the results by considering the impact of query expansion for each query. Figure 3 reported the gains/losses vs. baseline obtained by the

---

[4] ECNU-2 had the highest effectiveness, but it used Google query suggestion service to gain expansions.

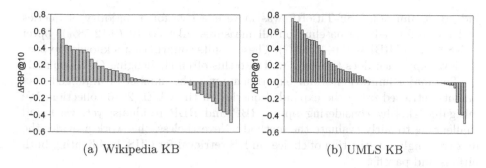

(a) Wikipedia KB   (b) UMLS KB

Fig. 3. Changes in RBP@10 between the Entity Query Feature Expansion model utilising the best settings vs. baseline. Only high coverage queries are reported.

best settings of Wikipedia KB (WSE-TitleRF) and UMLS KB (USE-TitleRF). In total, for WSE-TitleRF (USE-TitleRF), 183 (263) queries were expanded by the model (48 (72) in the high coverage set). Of these, 16 (76) showed no change in effectiveness compared to the baseline (7 (27) in the high coverage set). Of the remaining, 92 (94) showed improvements (26 (33) in the high coverage), while 75 (93) showed losses (15 (12) in the high coverage); the magnitudes of these changes are shown in the figure. These improvements (or losses) were measured using RBP@10 and thus expanded queries with low coverage are unlikely to perform as effective as expanded queries with high coverage.

6   Conclusions

In this paper, we explored the influence of different choices in knowledge base (KB) retrieval for consumer health search (CHS). Choices included KB construction, entity mentions extraction, entity mapping, source of expansion, and relevance feedback. We compared the effectiveness of a general KB (Wikipedia) and a medical specialised KB (UMLS) as the basis for query expansion. Our empirical evaluation showed that the best settings for the Wikipedia KB are: (1) index only Wikipedia pages that have health related infobox types or links to medical terminologies, (2) use uni-, bi-, and tri-grams of the original queries that matched CHV terms as entity mentions, (3) map entity mentions to Wikipedia entities based on the Aliases feature, (4) source expansion terms from the mapped Wikipedia page Title, and (5) add relevance feedback terms. As for the UMLS KB, the best settings are: (1) index all UMLS concepts, (2) use all uni-, bi-, and tri-grams of the original queries as entity mentions, (3) map entity mentions to UMLS entities based on the Aliases feature, (4) source expansion terms from the mapped UMLS Title feature, and (5) add relevance feedback terms.

Results after tuning the 5 choices showed that, overall, UMLS based KB settings were more effective than Wikipedia based ones. For all queries set, the best UMLS KB settings (USE-TitleRF) performed better than the baseline in terms of bpref (+11.56%) and RBP@10 (+12.99%). For queries with high coverage of judged documents, USE-TitleRF was more effective for a majority of queries and outperformed the baseline on all measures: nDCG@10 (+12.58%), bpref (+5.89%), and RBP@10 (+32.43%). These results confirm that a knowledge-base retrieval approach does translate well into this often challenging CHS domain.

The major limitation of our experiments was the number of unjudged documents retrieved using the expanded queries on the CLEF 2016 collection. We mitigated this by considering bpref, RBP and RBP residuals; yet, we found challenging to fairly evaluate the methods. Nevertheless, this work provides the first thorough investigation of choices in KB retrieval for CHS, highlighting both pitfalls and payoffs.

Acknowledgements. Jimmy is sponsored by the Indonesia Endowment Fund for Education (Lembaga Pengelola Dana Pendidikan/LPDP). Guido Zuccon is the recipient of an Australian Research Council DECRA Research Fellowship (DE180101579).

References

1. Bendersky, M., Metzler, D., Croft, W.: Effective query formulation with multiple information sources. In: WSDM 2012, pp. 443–452 (2012)
2. Dalton, J., Dietz, L., Allan, J.: Entity query feature expansion using knowledge base links. In: SIGIR 2014, pp. 365–374 (2014)
3. Díaz-Galiano, M., Martín-Valdivia, M., Ureña-López, L.: Query expansion with a medical ontology to improve a multimodal information retrieval system. JCBM **39**(4), 396–403 (2009)
4. Jimmy, Zuccon, G., Koopman, B.: Boosting titles does not generally improve retrieval effectiveness. In: ADCS 2016, pp. 25–32 (2016)
5. Keselman, A., Tse, T., Crowell, J., Browne, A., Ngo, L., Zeng, Q.: Relating consumer knowledge of health terms and health concepts. In: AMIA 2006 (2006)
6. Limsopatham, N., Macdonald, C., Ounis, I.: Inferring conceptual relationships to improve medical records search. In: OAIR 2013, pp. 1–8 (2013)
7. Palotti, J., Goeuriot, L., Zuccon, G., Hanbury, A.: Ranking health web pages with relevance and understandability. In: SIGIR 2016, pp. 965–968 (2016)
8. Plovnick, R., Zeng, Q.: Reformulation of consumer health queries with professional terminology: a pilot study. JMIR, **6**(3) (2004)
9. Silva, R., Lopes, C.: The effectiveness of query expansion when searching for health related content: Infolab at CLEF eHealth 2016. In: CLEF 2016 (2016)
10. Soldaini, L., Edman, W., Goharian, N.: Team GU-IRLAB at CLEF eHealth 2016: Task 3. In: CLEF (Working Notes), pp. 143–146 (2016)
11. Soldaini, L., Goharian, N.: QuickUMLS: a fast, unsupervised approach for medical concept extraction. In: SIGIR MedIR 2016, Pisa, Italy (2016)
12. Soldaini, L., Goharian, N.: Learning to rank for consumer health search: a semantic approach. In: Jose, J.M., Hauff, C., Altıngovde, I.S., Song, D., Albakour, D., Watt, S., Tait, J. (eds.) ECIR 2017. LNCS, vol. 10193, pp. 640–646. Springer, Cham (2017). https://doi.org/10.1007/978-3-319-56608-5_60
13. Toms, E., Latter, C.: How consumers search for health information. HIJ **13**(3), 223–235 (2007)
14. Zeng, Q., Kogan, S., Ash, N., Greenes, R., Boxwala, A.: Characteristics of consumer terminology for health information retrieval. JMIM **41**(4), 289–298 (2002)

15. Zeng, Q.T., Crowell, J., Plovnick, R.M., Kim, E., Ngo, L., Dibble, E.: Assisting consumer health information retrieval with query recommendations. JAMIA **13**(1), 80–90 (2006)
16. Zhang, Y.: Searching for specific health-related information in MedlinePlus: behavioral patterns and user experience. JAIST **65**(1), 53–68 (2014)
17. Zuccon, G., Koopman, B., Palotti, J.: Diagnose this if you can: on the effectiveness of search engines in finding medical self-diagnosis information. In: Hanbury, A., Kazai, G., Rauber, A., Fuhr, N. (eds.) ECIR 2015. LNCS, vol. 9022, pp. 562–567. Springer, Cham (2015). https://doi.org/10.1007/978-3-319-16354-3_62
18. Zuccon, G., Palotti, J., Goeuriot, L., Kelly, L., Lupu, M., Pecina, P., Mueller, H., Budaher, J., Deacon, A.: The IR task at the CLEF eHealth evaluation lab 2016: user-centred health information retrieval. In: CLEF 2016 (2016)

Deep Learning

An Adversarial Joint Learning Model for Low-Resource Language Semantic Textual Similarity

Junfeng Tian[1], Man Lan[1,2]([⊠]), Yuanbin Wu[1,2], Jingang Wang[3], Long Qiu[4], Sheng Li[3], Lang Jun[3], and Luo Si[3]

[1] School of Computer Science and Software Engineering,
East China Normal University, Shanghai, People's Republic of China
51151201048@stu.ecnu.edu.cn, {mlan,ybwu}@cs.ecnu.edu.cn
[2] Shanghai Key Laboratory of Multidimensional Information Processing,
Shanghai, China
[3] iDST, Alibaba Group, Hangzhou, China
{jingang.wjg,lisheng.ls,langjun.lj,luo.si}@alibaba-inc.com
[4] Onehome (Beijing) Network Technology Co. Ltd., Beijing, China
qiulong@onehome.me

Abstract. Semantic Textual Similarity (STS) of low-resource language is a challenging research problem with practical applications. Traditional solutions employ machine translation techniques to translate the low-resource languages to some resource-rich languages such as English. Hence, the final performance is highly dependent on the quality of machine translation. To decouple the machine translation dependency while still take advantage of the data in resource-rich languages, this work proposes to jointly learn the low-resource language STS task and that of a resource-rich one, which only relies on multilingual word embeddings. In particular, we project the low-resource language word embeddings into the semantic space of the resource-rich language via a translation matrix. To make the projected word embeddings resemble that of the resource-rich language, a language discriminator is introduced as an adversarial teacher. Thus the parameters of sentence similarity neural networks of two tasks can be effectively shared. The plausibility of our model is demonstrated by extensive experimental results.

Keywords: Semantic Textual Similarity · Low-resource language
Neural networks · Adversarial learning

1 Introduction

Semantic Textual Similarity (STS) is an essential basis of natural language understanding. STS aims at measuring the degree of semantic equivalence between a pair of sentences. Formally, given two sentences, STS models return a continuous valued similarity score. The study of STS can be of great benefit

© Springer International Publishing AG, part of Springer Nature 2018
G. Pasi et al. (Eds.): ECIR 2018, LNCS 10772, pp. 89–101, 2018.
https://doi.org/10.1007/978-3-319-76941-7_7

to many Natural Language Processing (NLP) and Information Retrieval (IR) applications, including Machine Translation (MT) [2,18], Question Answering [12,30], and Summarization [21,24].

Most traditional methods adopt various linguistic features with machine learning algorithms [3,27,34]. These methods require domain expert to manually design features, which is costly and time-consuming. Recent deep learning models have been widely used to build STS systems [7,14,23,28]. The advantage of deep learning models is that they can automatically learn the features, without relying on specific domain knowledge. Deep STS models benefit from word embeddings trained on large corpora, which can capture the semantic similarity information between words. The word embeddings can be conveniently acquired with unlabeled corpora, and several public pre-trained word embeddings for multiple languages are available off-the-shelf. Therefore, for multi-lingual STS, we can train an end-to-end neural network with multi-lingual word embeddings as input and their semantic similarity as output. This neural network-based model works well as long as there are enough annotated data to learn the large number of parameters, which is not a big deal for resource-rich languages.

Low-resource language STS suffers from less annotated data, and creating a suitable, large annotated dataset requires a lot of effort. Moreover, available public datasets are typically only available in the most common languages. Therefore, most previous work adopt a two-step setting to address the low-resource issue: (1) translating low-resource languages into resource-rich languages such as English; (2) predicting the similarity score using a pre-trained supervised STS system in resource-rich language. Although these two-step methods can take advantage of large-scale training data, they suffer from the high dependence on MT. What's worse, the errors in MT would propagate to the second step and lead to deteriorate overall performance.

To address the problem of MT-based methods, we propose an adversarial joint learning model for low-resource language STS. Specifically, we train two neural networks jointly, one for the low-resource language STS (*i.e.*, the main task), and the other for the resource-rich language STS (*i.e.*, the auxiliary task). The main task shares the network structure and network parameters with the auxiliary task, which has a richer supply of data. The two input word embeddings are in separate vector space, and it is difficult to learn a more general representation. Inspired by [22,33], we equip with a translation matrix in the network of the main task, and introduce an adversarial language discriminator to ensure this desired alignment. The translation matrix linearly projects word embeddings of the low-resource language into the semantic space of the resource-rich language. For the main task to successfully tap into the richer data of the auxiliary task, this word embedding projection has to be a decent alignment with the word embeddings of the resource-rich language. Fortunately, it is possible because of the similarity of the geometric arrangements in the semantic spaces of different languages, as the concepts they share are grounded in the real world.

The main contributions of our work are three-fold:

1. We propose a joint learning model to address STS task in low-resource scenarios, by taking advantage of fruitful annotated data of resource-rich languages in a multi-task setting.
2. We introduce a translation matrix to align the semantic embeddings of two languages, enabling us to evaluate the semantic similarity of low-resource language sentence pairs in another semantic space.
3. We optimize the model in an adversarial manner to ensure the neural networks to reach a translation matrix capable of accurate alignment.

2 Related Work

This paper draws on work in three general areas, which we briefly describe in this section.

Semantic Textual Similarity: Most traditional methods focus on feature engineering for the STS problem, and various useful matching features were explored. Šarić et al. [29] calculate the n-gram overlaps at the word and character level respectively to evaluate the semantic similarity of two sentences. Li et al. [13] utilize Latent Semantic Analysis (LSA) to gain statistical information from large corpora to assist the similarity judgment. Mihalcea et al. [20] estimate sentence similarity using WordNet as external resources. Sultan et al. [27] measure the sentence similarity based on word-alignments. Zhao et al. [34] integrate multiple similarity measures and adopt supervised learning algorithms to compare the contributions of different features.

With the popularity of deep learning in the IR and NLP research communities recently, multiple neural networks-based methods have been explored in STS tasks. Most of them model sentence pairs in a Siamese structure. He et al. [7] adopt CNN to capture sentence similarity from multiple perspectives, e.g., different window size of filters, different pooling types, and different distance functions. Tai et al. [28] propose a tree-structured Long Short-Term Memory (LSTM) network, which recursively constructs sentence representations based on their corresponding syntactic trees. Mueller and Thyagarajan [23] directly use a non-linear function instead of Softmax to calculate the similarity score. Liu et al. [14] build a deep fusion network with two LSTM components interacting with each other via a recursive composition mechanism. STS is related to Textual Entailment, except that the concept of semantic similarity is more specific than semantic relatedness and the latter includes concepts as antonymy and meronymy. Recently, Yanaka et al. [32] transferred semantic information from SNLI dataset to address STS task using universal sentence encoder.

Cross-lingual Word Embeddings: Cross-lingual word embeddings are the basis of our proposals. Many methods require some bilingual signals. Many works use parallel corpora, and optimize a cross-lingual constraint between embeddings [8,35], while others use bilingual dictionary, and learn a linear mapping [1,22, 26]. Recent studies employ adversarial training to learn such mapping with no bilingual supervision [33]. We also adopt the adversarial training architecture to learn the translation from low-resource languages to rich-resource languages.

Multi-task Learning: Multi-task learning (MTL) has been employed successfully in various NLP applications [4,19]. MTL is also named as joint learning, learning to learn, or learning with auxiliary tasks in some literature.[1] One advantage of MTL is the augmentation of implicit data, since a model that learns two tasks simultaneously is able to learn a more general representation. Collobert and Weston [4] propose a multi-task neural network that jointly train several relevant tasks by sharing word embeddings. Liu et al. [15] propose the multi-task neural network by modifying the recurrent neural network for text classification tasks. Lan et al. [11] use synthetic implicit data extracted from unlabeled data for implicit discourse relation recognition with multi-task setting. Liu et al. [17] propose a CNN-based multi-task learning framework to improve the performance of implicit discourse identification.

Adversarial Learning [6] has recently emerged as a practical tool to measure the equivalence between distributions and it has proven to be effective when combined with multi-task learning. Ganin and Lempitsky [5] apply adversarial training to domain adaptation, aiming at transferring the knowledge of source domain to target domain. Park and Im [25] propose a novel approach for multimodal representation learning which uses adversarial back-propagation concept. Liu et al. [16] propose an adversarial multi-task learning framework in order to learn task-independent representations. Different from these studies, our model aims to find a good enough translation matrix between tasks of different languages using adversarial training strategy.

3   Model Description

To facilitate the discussion, we first give some definitions. Given two sentences $S_1 = w_1^1, w_2^1, \cdots, w_n^1$ and $S_2 = w_1^2, w_2^2, \cdots, w_m^2$, the goal is map the sentence pair to a continues semantic similarity score $y \in \mathbb{R}$.

3.1   Single-Task Learning Model

Our single task learning model is depicted in Fig. 1. Firstly, we map each word w_i into a low-dimensional vector $\mathbf{x}_{w_i}$ using a pre-trained word vector lookup table $\mathbf{E} \in \mathbb{R}^{|V| \times d}$, where $|V|$ denotes the vocabulary size and d is the dimension of a word vector.

3.1.1   Sentence Representation

We use a BiLSTM layer to obtain sentence representation. LSTM [9] are capable of learning and remembering over long sequences of inputs. BiLSTM consists of two LSTMs that run in parallel in opposite directions, to capture information

---

[1] In this paper, MTL and joint learning are interchangeable.

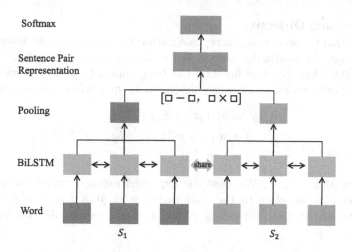

Fig. 1. An end-to-end neural network for semantic textual similarity

from both the future and the past. At time step t, we obtain hidden states $\overrightarrow{h_t}, \overleftarrow{h_t} \in \mathbb{R}^{d_h}$ from the forward LSTM and the backward LSTM, respectively. We concatenate them to obtain the intermediate state $h_t \in \mathbb{R}^{2d_h}$ in the sequence:

$$\overrightarrow{h_t} = \overrightarrow{LSTM}(x_t, \overrightarrow{h_{t-1}}) \tag{1}$$

$$\overleftarrow{h_t} = \overleftarrow{LSTM}(x_t, \overleftarrow{h_{t+1}}) \tag{2}$$

$$h_t = [\overrightarrow{h_t}, \overleftarrow{h_t}] \tag{3}$$

After that, to form a fixed-length sentence representation h_{S_i}, we perform max-pooling operation over time t to select the informative features.

$$h_S = \max_t h_t \tag{4}$$

3.1.2 Sentence Pair Representation

Given two sentence representations h_{S_1} and h_{S_2} in the pair, we obtain the sentence pair representations before predicting the similarity score. We use the following formulas from Tai et al. [28] to capture the semantic relation between a sentence pair:

$$h_+ = |h_{S_1} - h_{S_2}| \tag{5}$$

$$h_\times = h_{S_1} \odot h_{S_2} \tag{6}$$

$$h_s = [h_+, h_\times] \tag{7}$$

where the absolute value function $|-|$ and the multiplicative measure $\odot$ are both applied in an element-wise style. The absolute value function can be viewed as the distance, and the multiplicative measure can be interpreted as the comparison of the signs of the input representations. Finally, we obtain the sentence pair representation h_s by concatenating these two vectors.

3.1.3   Training Objective

After we obtain the sentence pair representation h_s, we use a multi-layer percep-tron to predict the similarity score $\hat{y}$. We minimize the objective function from Tai et al. [28]. This objective function has been shown to perform very strongly on text similarity tasks, significantly better than squared or absolute error.

$$h'_s = \sigma(W_s h_s + b_s) \tag{8}$$
$$\hat{p} = softmax(W_p h'_s + b_p) \tag{9}$$
$$\hat{y} = r^T \hat{p} \tag{10}$$

where $r^T = [0, 1, \cdots, K]$. We want the expected rating $\hat{y}$ under the predicted distribution $\hat{p}$ to be close to the gold rating $y \in [0, K]$: $\hat{y} = r^T \hat{p} \approx y$. We therefore transform the gold rating y into a sparse target distribution p that satisfies $y = r^T p$:

$$p_i = \begin{cases} y - \lfloor y \rfloor, & i = \lfloor y \rfloor + 1 \\ \lfloor y \rfloor + 1 - y, & i = \lfloor y \rfloor \\ 0 & \text{otherwise} \end{cases} \tag{11}$$

for $0 \leq i \leq K$. The goal is to minimize the distance between the predicted distribution $\hat{p}$ and the target distribution p, and KL-divergence between p and $\hat{p}$ is used as the loss function:

$$\mathcal{L} = \frac{1}{m} \sum_{k=1}^{m} \text{KL}\left(p^{(k)} \,\middle\|\, \hat{p}^{(k)}\right) \tag{12}$$

where m is the number of training pairs.

3.2   Adversarial Joint Learning Model

The single task learning model suffers from the conflict of large parameters and insufficient training data. Joint learning effectively increases the number of training samples. Here we treat STS task in low-resource language as main task and STS task in resource-rich language as auxiliary task. Figure 2 shows our joint learning model. The vanilla version of the model is a weighted combination of the two single tasks:

$$\mathcal{L} = \lambda \mathcal{L}_{main} + (1 - \lambda) \mathcal{L}_{aux} \tag{13}$$

where λ is a hyper-parameter which balances the two tasks. If $\lambda = 1$, it is a single model trained on the low-resource language task. And if $\lambda = 0$, it is trained on the the resource-rich language.

3.2.1   Translation Matrix

Although the vanilla joint learning model can take advantage of the wealthy training data in resource-rich language, it is difficult to learn a more general rep-resentation, since the two input word embeddings are in separate vector space.

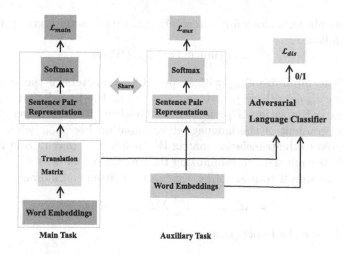

Fig. 2. The adversarial joint learning model

Fortunately, Mikolov et al. [22] observe that word embeddings trained separately on monolingual corpora have similar geometric arrangements, and the relationship between vector spaces can be captured by a linear transformation. Zhang et al. [33] propose several adversarial training methods to establish the transformation without relying on any cross-lingual supervision.

Motivated by these findings, we equip our model with a translation matrix W. Through it, we transform the word embeddings of low-resource language into the ones of resource-rich language. By optimizing the translation matrix and the two tasks simultaneously, the joint learning model can transform the low-resource language STS task into the resource-rich one, thus the shared parameters in sentence similarity neural networks can learn STS task in the resource-rich language space.

3.2.2 Adversarial Language Classifier

Inspired by adversarial networks [6], we formulate an adversarial game by introducing an adversarial language classifier. The classifier is to distinguish the language of word embeddings, while the translation matrix strives to make the transformed source word embeddings seem like the target ones, which make the classifier impossible to distinguish.

As a generic binary classifier, a standard feed-forward neural network D with one hidden layer is used. The classifier D takes the word embeddings as input, the language label as target, and its loss function is the cross-entropy loss:

$$\mathcal{L}_{adv} = \frac{1}{m} \sum_{k=1}^{m} (-\log D(y_k) - \log(1 - D(Wx_k))) \qquad (14)$$

where x_k is the low-resource language word embeddings, y_k is the rich-language word embeddings.

In training phase, we look for parameters that satisfy a min-max optimization criterion as follows:

$$\theta^* = \arg\min_W \max_D -\mathcal{L}_{adv}(W, D) \tag{15}$$

which equals to minimize $\mathcal{L}_{adv}$ with respect to parameters of the classifier D, and maximize $\mathcal{L}_{adv}$ with respect to parameters of the translation matrix W. The min-max optimization is performed by Gradient Reverse Layer [5], which reversing the gradient of the language discrimination loss $\mathcal{L}_{adv}$ when they are backpropagated to the translation matrix W. In this way, maximizing $\mathcal{L}_{adv}$ with respect to W is equivalent to minimizing the discrimination loss.

Finally, the overall training objective can be written as follows:

$$\mathcal{L} = \lambda\mathcal{L}_{main} + (1 - \lambda)\mathcal{L}_{aux} + \gamma\mathcal{L}_{adv} \tag{16}$$

where λ and γ are the hyper-parameter.

4   Experimental Setup

4.1   Datasets

SemEval 2017 STS task [3] provides multilingual annotated sentence similarity datasets. Each sentence pair is annotated with a score, ranging from 0 for no meaning overlap to 5 for meaning equivalence. Table 1 lists the statistics of the datasets[2]. We grant the limited annotated dataset (e.g., Spanish or Arabic) as low-resource language, and the large annotated English dataset as resource-rich language, to conduct two separate experiments. Due to that the datasets are from different source domains, for each language, we randomly select 80% as training set, 10% as development set, and the remaining 10% as test set.

4.2   Multilingual Embeddings

Our model relies only on language specific word embeddings, which are often easily available in low-resource language for the sake of large-scale unlabeled corpus. We use the pre-trained 300-dimensional word embeddings[3], which were obtained

Table 1. The statistics of the datasets

Language pair	# Pairs	Source
Arabic-Arabic (AR-AR)	1, 338	MSRpar, MSRvid, SMTeuroparl, SNLI (2017)
Spanish-Spanish (ES-ES)	1, 805	News, Wiki (2014, 2015), SNLI (2017)
English-English (EN-EN)	13, 592	SemEval (2012 - 2015)

---

[2] The data is available at http://alt.qcri.org/semeval2017/task1/index.php?id=data-and-tools.

[3] https://github.com/facebookresearch/fastText/blob/master/pretrained-vectors.md.

using fastText with skip-gram model. There are 7%, 12%, 18% out-of-vocabulary (OOV) words in the provided Spanish, Arabic and English embeddings, respectively.

4.3   Parameter Settings and Evaluation Metric

We initialize the word embeddings with the 300-dimensional pre-trained fastText embeddings, and don't fine-tune during training. All parameters are initialized by sampling from a uniform distribution $U(-0.1, 0.1)$. And the dimension of the hidden states in BiLSTM is set to 300. We train our model by optimizing the objective using ADAM [10] with a learning rate of 0.001 and a mini-batch size of 32. In order to avoid overfitting, we use dropout of hidden units, L2 regularization on weights, and early stopping by observing Pearson r on the development dataset. Finally, we choose the model that performs best on the development dataset for the final evaluation on the test dataset. During training, we find that the loss of the main task, the auxiliary task, and the discriminator are of the same magnitude, thus we directly set the hyper-parameter $\lambda = \gamma = 0.5$ without further tuning.

Similar to [31], we shuffle the input to make it more difficult to memorize the sequence information since the word orders are variant in different languages. We go through each sentence and shuffle the words (re-arrange all the words in a sentence) with probability 0.25, forcing the sentence representation to focus more on the identities of the words themselves and less on the order.

Evaluation Metric: On each test set, we evaluate the performance under the metric of the Pearson correlation coefficient r between gold ratings and system returned scores.

5   Results and Discussion

To evaluate our proposed model for low-resource language STS, we include two naïve but effective baselines as follows:

Lexical Overlap Similarity. As the name implies, the similarity of two sentences is calculated based on their lexical overlap ratio. Let S_i denote as the set of words in i-th sentence, and the lexical overlap similarity is defined as:

$$\text{overlap}(S_1, S_2) = 2 \cdot \left(\frac{|S_1|}{|S_1 \cap S_2|} + \frac{|S_2|}{|S_1 \cap S_2|} \right)^{-1} \tag{17}$$

Word Embedding Similarity. The similarity is calculated based on the cosine similarity of the sentences' embeddings. A sentence's embedding is the summation of the embeddings of all the words within the sentence (OOV words are omitted).

As we introduced in Sect. 3.1, we include some variants of the single-task neural network-based model for comparison.

Resource-rich (English) Only. The single-task model is trained with resource-rich language data only. The similarity score are predicted with the learned parameters directly. In other words, the training phase is fed with the resource-rich language annotated data, while the test phase is fed with the target low-resource language data.

Low-resource Language Only. The single-task model is trained with low-resource language data only. The similarity score is predicted with the learned parameters directly. The training and test phase are both fed with low-resource language data.

Machine Translation-based Method. The single-task model is trained with low-resource language data only. During test, the low-resource language input is translated into the resource-rich language via Google Translator[4].

Table 2. Pearson correlation coefficients of the comparison methods on Arabic and Spanish datasets respectively.

Methods		AR-AR	ES-ES
Baseline	Lexical Overlap Similarity	0.6850	0.6642
	Word Embedding Similarity	0.7898	0.7348
Single-Task	Resource-rich language (English) Only	0.7508	0.6722
	Low-resource Language Only	0.5504	0.5803
	Machine Translation	0.7049	0.6316
Joint Model	Vanilla	0.8301	0.6936
	+ translation matrix	0.8427	0.7436
	+ adversarial Learning	**0.8629**	**0.7630**

Table 2 presents the experimental results of all the comparison methods.

From the upper block (i.e., Rows 1-2) of Table 2, Word Embeddings Similarity outperforms Lexical Overlap similarity. This demonstrates that word embeddings from large unlabeled corpora bear more semantic information and bring benefit to similarity judgment. Nevertheless, lexical overlap-based methods consider similarity merely on strictly surface word matching.

The middle block (i.e., Rows 3-5) of Table 2 presents the experimental results of our single-task models with different training data. It's obvious that the word embedding based method is a rather strong baseline, and all the single-task model can not beat it. The single task model perform worst when it is trained with low-resource language data solely. It is reasonable because the training data for low-resource language is too limited to obtain a robust model.

The bottom block of Table 2 demonstrates the results of our joint models. All the joint models outperform the single-task models. Even the vanilla joint

---

[4] https://cloud.google.com/translate/.

model without explicit information sharing between the two tasks can improve the metrics on two test dataset. This reveals the effectiveness of our joint learning model, where the main task (i.e., low-resource language task) can benefit from the better-trained neural networks of the auxiliary task. In comparison to the word embedding-based baseline, although the vanilla joint model performs better on Arabic dataset while worse on Spanish dataset, the introduction of translation matrix can improve the joint model further to outperform the baseline. Moreover, the intact joint model with translation matrix in an adversarial learning manner can achieve the state-of-the-art performance on both datasets.

6 Conclusion

Semantic Textual Similarity (STS) for low-resource languages is a challenging task due to the lack of annotated data. Although machine translation-based methods are helpful by translating the sentences to some common languages with fruitful training data, the performance is highly dependent on the translation quality. To address this problem, we propose a joint learning model composed of two neural network-based tasks: a main task for the low-resource language STS, and an auxiliary task for a resource-rich language STS. The joint model only relies on multilingual embeddings, which are easy to obtain from large unlabeled data, getting rid of machine translation. To improve the joint model, we introduce a translation matrix that projects the low-resource language embeddings into the semantic space of the resource-rich language. During model training, a language discriminator is utilized as an adversarial teacher to make the projected word embeddings resemble that of the resource-rich language. The effectiveness of our model is proved by extensive experimental results.

Acknowledgments. We would like to thank the reviewers for their valuable comments. This work is supported by grants from Science and Technology Commission of Shanghai Municipality (15ZR1410700), the Open Project of Shanghai Key Laboratory of Trustworthy Computing (No. 07dz22304201604).

References

1. Artetxe, M., Labaka, G., Agirre, E.: Learning bilingual word embeddings with (almost) no bilingual data. In: Proceedings of ACL, pp. 451–462, July 2017
2. Béchara, H., Escartín, C.P., Orasan, C., Specia, L.: Semantic textual similarity in quality estimation. Baltic J. Mod. Comput. 4(2), 256 (2016)
3. Cer, D., Diab, M., Agirre, E., Lopez-Gazpio, I., Specia, L.: Semeval-2017 task 1: Semantic textual similarity multilingual and crosslingual focused evaluation. In: Proceedings of SemEval, pp. 1–14 (2017)
4. Collobert, R., Weston, J.: A unified architecture for natural language processing: deep neural networks with multitask learning. In: Proceedings of ICML, pp. 160–167 (2008)
5. Ganin, Y., Lempitsky, V.: Unsupervised domain adaptation by backpropagation. In: Proceedings of ICML, pp. 1180–1189 (2015)

6. Goodfellow, I., Pouget-Abadie, J., Mirza, M., Xu, B., Warde-Farley, D., Ozair, S., Courville, A., Bengio, Y.: Generative adversarial nets. In: Proceeding of NIPS, pp. 2672–2680 (2014)
7. He, H., Gimpel, K., Lin, J.J.: Multi-perspective sentence similarity modeling with convolutional neural networks. In: Proceedings of EMNLP, pp. 1576–1586 (2015)
8. Hermann, K.M., Blunsom, P.: Multilingual distributed representations without word alignment. In: Proceedings of ICLR (2014)
9. Hochreiter, S., Schmidhuber, J.: Long short-term memory. Neural Comput. 9(8), 1735–1780 (1997)
10. Kingma, D.P., Ba, J.: Adam: a method for stochastic optimization. CoRR abs/1412.6980 (2014)
11. Lan, M., Wang, J., Wu, Y., Niu, Z.Y., Wang, H.: Multi-task attention-based neural networks for implicit discourse relationship representation and identification. In: Proceedings of EMNLP, pp. 1310–1319 (2017)
12. Lan, M., Wu, G., Xiao, C., Wu, Y., Wu, J.: Building mutually beneficial relationships between question retrieval and answer ranking to improve performance of community question answering. In: Proceedings of IJCNN (2016)
13. Li, Y., McLean, D., Bandar, Z.A., O'shea, J.D., Crockett, K.: Sentence similarity based on semantic nets and corpus statistics. IEEE Trans. Knowl. Data Eng. 18(8), 1138–1150 (2006)
14. Liu, P., Qiu, X., Chen, J., Huang, X.: Deep fusion lstms for text semantic matching. In: Proceedings of ACL (2016)
15. Liu, P., Qiu, X., Huang, X.: Deep multi-task learning with shared memory for text classification. In: Proceedings of EMNLP, pp. 118–127 (2016)
16. Liu, P., Qiu, X., Huang, X.: Adversarial multi-task learning for text classification. In: Proceeding of ACL, pp. 1–10 (2017)
17. Liu, Y., Li, S., Zhang, X., Sui, Z.: Implicit discourse relation classification via multi-task neural networks. arXiv preprint arXiv:1603.02776 (2016)
18. Lo, C.k., Wu, D.: Meant: an inexpensive, high-accuracy, semi-automatic metric for evaluating translation utility via semantic frames. In: Proceedings of ACL, pp. 220–229 (2011)
19. Luong, M.T., Le, Q.V., Sutskever, I., Vinyals, O., Kaiser, L.: Multi-task sequence to sequence learning. In: Proceedings of ICLR (2016)
20. Mihalcea, R., Corley, C., Strapparava, C., et al.: Corpus-based and knowledge-based measures of text semantic similarity. In: Proceedings of AAAI (2006)
21. Mihalcea, R., Tarau, P.: Textrank: bringing order into texts. In: Proceedings of ACL (2004)
22. Mikolov, T., Le, Q.V., Sutskever, I.: Exploiting similarities among languages for machine translation. arXiv preprint arXiv:1309.4168 (2013)
23. Mueller, J., Thyagarajan, A.: Siamese recurrent architectures for learning sentence similarity. In: Proceedings of AAAI, pp. 2786–2792 (2016)
24. Nagwani, N.K., Verma, S.: A frequent term and semantic similarity based single document text summarization algorithm (0975–8887). Int. J. Comput. Appl. 17, 36–40 (2011)
25. Park, G., Im, W.: Image-text multi-modal representation learning by adversarial backpropagation. arXiv preprint arXiv:1612.08354 (2016)
26. Smith, S.L., Turban, D.H.P., Hamblin, S., Hammerla, N.Y.: Offline bilingual word vectors, orthogonal transformations and the inverted softmax. In: Proceedings of ICLR (2017)
27. Sultan, M.A., Bethard, S., Sumner, T.: Dls@cu: Sentence similarity from word alignment and semantic vector composition. In: Proceedings of SemEval (2015)

28. Tai, K.S., Socher, R., Manning, C.D.: Improved semantic representations from tree-structured long short-term memory networks. In: Proceedings of ACL (2015)

29. Šarić, F., Glavaš, G., Karan, M., Šnajder, J., Dalbelo Bašić, B.: Takelab: systems for measuring semantic text similarity. In: Proceedings of SemEval (2012)

30. Wang, B., Liu, K., Zhao, J.: Inner attention based recurrent neural networks for answer selection. In: Proceedings of ACL (2016)

31. Wieting, J., Gimpel, K.: Revisiting recurrent networks for paraphrastic sentence embeddings. In: Proceedings of ACL, pp. 2078–2088 (2017)

32. Yanaka, H., Mineshima, K., Martínez-Gómez, P., Bekki, D.: Determining semantic textual similarity using natural deduction proofs. In: Proceedings of EMNLP, pp. 681–691 (2017)

33. Zhang, M., Liu, Y., Luan, H., Sun, M.: Adversarial training for unsupervised bilingual lexicon induction. In: Proceedings of ACL, pp. 1959–1970 (2017)

34. Zhao, J., Zhu, T., Lan, M.: Ecnu: one stone two birds: ensemble of heterogenous measures for semantic relatedness and textual entailment. In: Proceedings of SemEval (2014)

35. Zou, W.Y., Socher, R., Cer, D., Manning, C.D.: Bilingual word embeddings for phrase-based machine translation. In: Proceedings of EMNLP (2013)

Reproducing a Neural Question Answering Architecture Applied to the SQuAD Benchmark Dataset: Challenges and Lessons Learned

Alexander Dür[✉], Andreas Rauber, and Peter Filzmoser

Vienna University of Technology, Vienna, Austria
e0927362@student.tuwien.ac.at, rauber@ifs.tuwien.ac.at,
p.filzmoser@tuwien.ac.at

Abstract. Reproducibility is one of the pillars of scientific research. This study attempts to reproduce the Gated Self-Matching Network, which is the basis of one of the best performing models on the SQuAD dataset. We reimplement the neural network model and highlight ambiguities in the original architectural description. We show that due to uncertainty about only two components of the neural network model and no precise description of the training process, it is not possible to reproduce the experimental results obtained by the original implementation. Finally we summarize what we learned from this reproduction process about writing precise neural network architecture descriptions, providing our implementation as a basis for future exploration.

1 Introduction

Reading comprehension tasks, which require reading a passage and then answering questions about it, are a long-standing challenge in the area of natural language processing. Progress in solving such complex problems, involving textual understanding and reasoning depend heavily on the availability of large, high-quality datasets [4]. Previous datasets in the area of reading and machine comprehension tasks are either small in size due to the high effort of human annotation or generated automatically and require only shallow reasoning [1,11].

With the release of the Stanford Question Answering Dataset, a large and high quality, human annotated dataset is available for reading comprehension based question answering [10]. The dataset was created through crowdsourcing and features a wide variety of answer types [10]. Due to its size it enables large end-to-end neural network models to be trained on the dataset [6,16,17].

Currently one of the best performing models on the dataset is a neural network model based on the Gated Self-Matching Network [16]. It achieves strong benchmark results through the use of two recently introduced neural network elements based on match-LSTM [15] and pointer networks [14]. The focus of this paper lies in the reproduction of the Gated Self-Matching Network model and

© Springer International Publishing AG, part of Springer Nature 2018
G. Pasi et al. (Eds.): ECIR 2018, LNCS 10772, pp. 102–113, 2018.
https://doi.org/10.1007/978-3-319-76941-7_8

the experiments leading to state-of-the-art benchmark results on the SQuAD dataset. Results are reported for a single model and for an ensemble model of 20 training runs. This reproduction will only focus on the single model.

We chose to reproduce this paper due to its interesting combination of recently developed neural network elements which achieves strong results on the SQuAD benchmarks. With the rising popularity of neural network models, it also becomes of increasing importance that experimental papers using complex and novel neural network architectures can still be reliably reproduced. This work is a step in highlighting aspects of neural network implementations that are required to unequivocally reimplement model architectures and reproduce experimental results.

Rojas [12] identifies three important elements in artificial neural network models:

- the structure of the nodes
- the topology of the network
- the learning algorithm used to find the weights of the network

Therefore a reproducible experimental neural network paper should convey these three aspects of the described model in such detail, that the model architecture can be implemented identically to the original model and that the model can be trained in a way, that it achieves comparable scores on the benchmark datasets used in the original paper. This requires an explicit description of the neural network structure, a detailed description of the training process, as well as access to the data, used for training and evaluation.

In the course of this study we were in contact with the authors of the original Gated Self-Matching Network. They did not provide us with any source code, but answered some of our questions per email. We summarize our communication with the authors in Sect. 4.1. When some part of the model architecture is not covered in the paper and the authors could not provide us sufficient details about their implementation, we will explain what assumptions were made in order to be able to implement a working version of the model.

This work is not meant as a criticism to the authors, but as a case study to reveal the challenges in presenting neural network architectures in an academic context. Academic papers require to describe the architecture without directly referencing large amounts of code, but still detailed enough that the implementation can be reproduced by readers of the paper to a degree, where the model performance on benchmarks is comparable.

The remainder of the paper is structured as follows: First we will give an overview of the original model's architecture and training process, followed by highlighting areas where details are unclear and how we addressed those details in our reimplementation and reproduction of the experiments. Finally we will attempt to reproduce the paper's experiments demonstrating state-of-the-art results on the development set of SQuAD and discuss the differences we observed.

2   SQuAD Task Description and Evaluation Metrics

The SQuAD dataset features 107,785 question-answer pairs [10]. The question-answer pairs refer to one of 23,215 passages from 536 documents, each extracted from one Wikipedia article. The answers are constrained to a sub-span of text from the passage. An example of such a passage and a question-answer pair can be found in Table 1.

Table 1. Example of a question-answer pair with the corresponding passage from the SQuAD dataset.

Passage: It was not until January 1518 that friends of Luther translated the 95 Theses from Latin into German and printed and widely copied them, making the controversy one of the first in history to be aided by the printing press. Within two weeks, copies of the theses had spread throughout Germany; within two months, they had spread throughout Europe.
Question: What device was one of the first to aid a controversy?
Answer: printing press

The documents in the dataset are split into a training set (80%), a development set (10%) and a hidden test set (10%) set. In this reproduction study we will only test our models on the development set. The authors of the dataset define two evaluation metrics, both of which ignore articles and punctuation:

Exact Match (EM): The percentage of answers that match any of the ground truth answers exactly.

Macro-averaged F1 Score: The predicted answer and the ground truth answer are tokenized. Then the F1 score between these two bags of tokens is calculated. When there are different ground-truth answers the one which produces the highest F1 score is chosen. To aggregate F1 scores over multiple question-answer pairs, their unweighted mean is used.

The paper to be reproduced reports an exact match score of 71.1 and an F1 score of 79.5 on the public SQuAD development set, based on the architecture and training process outlined below. When published, this was the top-performing model on the official leadership board, making it an interesting candidate to reproduce and re-use as a basis for further advances, as the original model was not made publicly available. As of writing this, a model based on the architecture described in the paper holds the place as the second best performing single model, outperformed only by the AIR-FusionNet model of the Microsoft Business AI Solutions team by 0.263 exact match score points and 0.404 F1 score points.

3 Overview of the Gated Self-Matching Network

In this section we will provide an overview of the model architecture used by the Gated Self-Matching Network as described in the original paper [16].

3.1 Architecture

This section will give the reader an understanding of the overall architecture of the neural network model, which is illustrated by Fig. 1. The inputs to the model are all m words in the question Q, $\{w_t^Q\}_{t=1}^m$, and all n words in the passage P, $\{w_t^P\}_{t=1}^n$, and the output are the two indices (p^1, p^2) of the first and last word belonging to the answer span, based on the calculations of the final output layer (h_1^a, h_2^a).

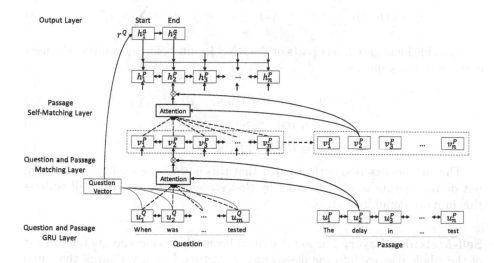

Fig. 1. Structural overview of gated self-matching network [16].

Encoding Layer. Using inputs to the model $\{w_t^Q\}_{t=1}^m$, and $\{w_t^P\}_{t=1}^n$, the encoding layer creates one context vector per word in the question and the passage.

First each word in converted to a word-level embedding, which yields two sequences of word vectors, $\{e_t^Q\}_{t=1}^m$ and $\{e_t^P\}_{t=1}^n$, for the words in the question and passage respectively. To deal with out-of-vocabulary words additionally a character-level embedding is created for each word.

The character-level embeddings, $\{c_t^Q\}_{t=1}^m$ and $\{c_t^P\}_{t=1}^n$, are created by taking the final output of a single layer of a bidirectional recurrent neural network (RNN) that processes the sequence of character level embeddings for each word.

Finally the sequences of concatenated word-level and character level embeddings $\{[e_t^Q, c_t^Q]\}_{t=1}^m$ and $\{[e_t^P, c_t^P]\}_{t=1}^n$ are each processed by bidirectional RNNs to yield the final context vectors $\{u_t^Q\}_{t=1}^m$ and $\{u_t^P\}_{t=1}^n$. Different RNNs are used for encoding question and passage. According to the implementation details section of the original paper 3 layers of bidirectional gated recurrent units (GRU) [2] are used for question and passage encoding.

Question and Passage Matching Layer. The matching layer is one of the main contributions of the original paper. It uses an RNN with attention and an additional gating mechanism to filter the input to the RNN, which generates the encoded sequence of passage words, $\{v_t^P\}_{t=1}^n$. The attention mechanism attends to the sequence of encoded question words $\{u_t^Q\}_{t=1}^m$:

$$s_j^t = v^T tanh(W_u^Q u_j^Q + W_u^P u_t^P + W_v^P v_{t-1}^P) \tag{1}$$

$$a_i^t = exp(s_i^t)/\sum_{j=1}^m s_j^t \tag{2} \qquad c_t = \sum_{i=1}^m a_i^t u_i^Q \tag{3}$$

An additional gate filters parts of the RNN input, before generating the next element of the sequence:

$$g_t = sigmoid(W_g[u_t^P, c_t]) \tag{4}$$

$$[u_t^P, c_t]^* = g_t \cdot [u_t^P, c_t] \tag{5}$$

$$v_t^P = RNN(v_{t-1}^P, [u_t^P, c_t]^*) \tag{6}$$

The authors describe in their paper that this layer is encoded bidirectionally, but do not specify what they mean by that statement exactly. We will address this in more detail in Sect. 4.2.

Self-Matching Layer. The self-matching layer tries to incorporate the context of the whole passage into the passage word vectors. It is a variant of the gated RNN presented in Sect. 3.1 that matches the sequence $\{v_t^P\}_{t=1}^n$ against itself, to produce a context aware passage representation $\{h_t^P\}_{t=1}^n$. The only modification is that the attention mechanism is independent of the RNN state:

$$s_j^t = v^T tanh(W_v^P v_j^P + W_v^{\tilde{P}} v_t^P) \tag{7}$$

Then the same gating mechanism as before is used, but the resulting final sequence is generated by a bidirectional RNN:

$$h_t^P = BiRNN(h_{t-1}^P, [v_t^P, c_t]^*) \tag{8}$$

Output Layer. The final output of the model, which consists of the first and last token index of the question span, is calculated in the output layer. To predict these two indices a pointer network as described by Vinyals et al. [14] is used. Pointer networks excel at predicting sequences of elements from some initial

sequence and can solve problems such as TSP or convex hull calculation, where
the expected answer is a sequence based on the input tokens.

The pointer network is again an RNN cell with attention:

$$s_j^t = v^T tanh(W_h^P h_j^P + W_h^a h_{t-1}^a) \tag{9}$$

$$a_i^t = exp(s_i^t)/\sum_{j=1}^{m} s_j^t \tag{10} \qquad c_t = \sum_{i=1}^{m} a_i^t h_i^P \tag{11}$$

$$h_t^a = RNN(h_{t-1}^a, c_t) \tag{12}$$

The attention weights a_i^t calculated at each time step are additionally used
to produce the next prediction output:

$$p^t = argmax(a_1^t, a_2^t, ..., a_n^t) \tag{13}$$

p^1 and p^2 are the predicted start and end position, respectively.

The initial value h_0^a of the answer pointer RNN is calculated through
attention-pooling over the sequence of encoded question word vectors $\{u_t^Q\}_{t=1}^m$:

$$s_j = v^T tanh(W_h^Q u_j^Q + b^Q) \tag{14} \qquad a_i = exp(s_i)/\sum_{j=1}^{m} s_j \tag{15}$$

$$h_0^a = \sum_{i=1}^{m} a_i u_i^Q \tag{16}$$

The network is optimized by minimizing the sum of the negative log probabilities of the ground truth answer start and end position. The predictive distributions of start and end position are given by $\{a_t^1\}_{t=1}^n$ and $\{a_t^2\}_{t=1}^n$.

3.2 Implementation and Training Details

The authors used the Stanford CoreNLP Tokenizer [8], which is a freely available
open-source tool, to split the questions and passages into a sequence of tokens.
300-dimensional case-sensitive pre-trained Glove word embeddings [9] were used
for question and passage encoding. These embeddings were fixed during training.
Words for which no word vector existed were replaced with the zero vector.

RNN cells are of the GRU [2] variant of LSTM [5]. The dimensionality of all
layers including the dimensionality of the vectors used for calculating attention
weights are set to 75.

AdaDelta [18] with an initial learning rate of 1.0 is used as optimizer for
the model [16]. The parameters ρ and ϵ are set to 0.95 and $1e^{-6}$ respectively.
Additionally the authors placed dropout [13] with a dropout-rate of 0.2 between
the layers of their model.

4   Implementation

In order to reproduce the experimental results of the original Gated Self-Matching Network [16] we need to implement the model architecture and training procedure as used in the original paper. During the implementation, we encountered several technical challenges related to unclear and ambiguous descriptions. Some of these challenges were resolved by falling back to best-practices, while for others we considered different variants, which we needed to test.

4.1   Communication with the Authors

We asked the authors to help with our reproduction efforts by providing us with additional information regarding their implementation of the neural network. In our communication with the authors we first clarified their formulation of the attention pooling for generating the initial state of the answer pointer network. In the original version of Eq. 14 Wang et al. [16] use the product of two parameters $W_v^Q \cdot V_r^Q$ instead of a single parameter b^Q, as it more closely represents their implementation. They confirmed that our formulation of the equation is equivalent to theirs.

Next we inquired about the training process used during their experiments and learned that during training, they halved the learning rate if the exact match score on the validation set failed to improve after several training batches.

Unfortunately they could not help us with further questions regarding remaining ambiguities as discussed below.

4.2   Ambiguity in the Architecture Description

Encoding Layer. The architecture of the encoding layer is very well described, except for two details. The dimensionality of the character embeddings is not provided. We assume they are of dimensionality 75, the same as all other trainable vectors.

The original paper does not mention any shared weights between the RNNs encoding question and passage words, but uses the singular when referring to the RNN(s) encoding the sequences. Other well performing models trained on the SQuAD dataset explicitly mention shared weights [17]. Therefore we will not only test the architecture without shared weights, but also a variant with shared weights. As the primary description of the encoder does not mention multiple RNN layers we will also test a variant that only uses one layer of bidirectional RNNs, in addition to configurations with 3 layers.

Matching Layers. The general architecture of the first matching layer could be implemented by us without any problems. The authors additionally mention that the question-passage matching layer is encoded bidirectionally. The most straightforward interpretation is that the layer in the description, which consists of an attention mechanism and a unidirectional RNN, is applied bidirectionally.

Other interpretations are that the output of the RNN matching question and passage is fed into 1 or 3 layers of bidirectional RNNs. Therefore three variants will be tested, two copies of the described matching layer, one in each direction and 1 and 3 layers of bidirectional RNNs processing the output of the unidirectional matching RNN.

The architecture of the self-matching layer on the other hand is clearly specified.

Output Layer. There were no problems in implementing the pointer network for the prediction of start and end position of the answer span. But the attention-pooling over the sequence of encoded question words, which generates the initial state of the pointer network has mismatching dimensions with the pointer network itself. The implementation in the original paper uses attention-pooling over the sequence of encoded question vectors $\{u_t^Q\}_{t=1}^m$ to generate the initial state of the pointer network. This sequence has dimensionality 150 as it is the output of a bidirectional GRU layer with a size of 75 per direction, but in the original paper a unidirectional GRU cell with hidden state size 75 is used for the pointer network. To make the initial state of size 75 compatible with the GRU cell we increase its hidden state size to 150.

In the original paper the authors do not mention how they handle the fact that there are sometimes multiple ground truth answers per question-answer pair during training. Ground truth answers can be any coherent span of characters from the original text, but this model can only make predictions on the token level. Ground truth answers that start or stop not between, but in the middle of tokens, cannot be properly predicted by the model described. Therefore we chose to use the ground truth answer which achieves the highest F1 score. In the case that multiple ground truth answers achieve the same F1 score we chose the first one.

4.3   Ambiguity in the Training Process

The only general issue we encountered when training the network, were that initial states for the RNNs were not mentioned. We assume they are non-trainable zero vectors.

The word embeddings are the only parameters whose initial value is specified and which are explicitly exempted from training. For the remaining parameters no initial values are specified therefore we use the Glorot Uniform Initializer [3] to generate the initial parameter values and assume that they should be trained.

In the original paper, there is no description of the training process besides the initial configuration of the optimizer. From our communications with the authors we learned that the learning rate is halved, if performance on the development set decreases after several batches of training. We assume a batch size of 32 and a validation interval of 547 batches, which is close to one fifth of an epoch and halve the learning rate if the exact match score failed to improve. We stop the optimization process once the learning rate was halved 10 times.

We provide the source code of our implementation for further investigation and re-use via Github[1].

5 Experimental Results

We measure the quality of our reproduction on the results published on the public SQuAD development set as reported by the authors in their paper. We test five different variations of the model, as the original architectural description sometimes allows for different interpretations.

Table 2. Reproduction results of experiments on the SQuAD development set.

Model	Encoding layer	Matching layer	EM-Score	F1-Score
Baseline	2 × 3 × BiGRU	bidirectional GARNN	50.1	59.0
Encoder 1	3 × BiGRU (shared)	bidirectional GARNN	57.7	67.5
Encoder 2	2 × 1 × BiGRU	bidirectional GARNN	64.1	73.2
Encoder 3	1 × BiGRU (shared)	bidirectional GARNN	**65.4**	**74.3**
Matching 1	2 × 3 × BiGRU	GARNN + 1 × BiGRU	42.4	50.9
Matching 2	2 × 3 × BiGRU	GARNN + 3 × BiGRU	34.9	43.6
Combination 1	3 × BiGRU (shared)	GARNN + 1 × BiGRU	55.6	65.7
Combination 2	3 × BiGRU (shared)	GARNN + 3 × BiGRU	58.5	68.3
Combination 3	2 × 1 × BiGRU	GARNN + 1 × BiGRU	62.9	72.3
Combination 4	2 × 1 × BiGRU	GARNN + 3 × BiGRU	62.6	72.1
Combination 5	1 × BiGRU (shared)	GARNN + 1 × BiGRU	63.0	72.4
Combination 6	1 × BiGRU (shared)	GARNN + 3 × BiGRU	61.9	72.0
Wang et al. [16]	-	-	**71.1**	**79.5**

Table 2 shows a comparison of the experimental results attained by our reimplementation compared to the original paper. None of our models can achieve performance close to the one reported for the original model.

The baseline model uses the most literal interpretation of the model description in the original paper. It uses two separate blocks of 3 layers of bidirectional GRU cells for encoding questions and passages, which do not share any weights. The matching layer consists of two instances of the gated RNN with attention, one in each direction. The performance of this model is very weak compared to the performance reported by Wang et al. [16]. The original model achieved first place on the official SQuAD leaderboard at the time of publishing, while our reproduction would only have achieved 8th place in comparison to the 10 competing entries.

The two variants of the encoding layer we tested show both much better results. Sharing the weights between the two encoding layers for questions and

---

[1] Code available at https://github.com/alexduer/squad-gated-rep.

passages, as well as reducing the size to a single layer of bidirectional GRU cells, show improved performance on the development set. A combination of the two modifications can improve the two evaluation metrics even further.

Variants of the matching layer on the other hand show decreased performance. When using a unidirectional version of the gated RNN with attention for matching question and passage and using 1 or 3 layers of GRU to bidirectionally encode the resulting sequence, the exact-match and the F1 score drops. We suspect the reason being that the increased depth of the network makes it more difficult to optimize.

Therefore we also test all three variants of the encoding layer in combination with the two variants of the matching layer. In combination with the modified encoding layers, the two matching layer variants show much better performance, but not as good as the variant used in the baseline model.

6   Discussion

The Gated Self-Matching Network by Wang et al. [16] is the basis of one of the best performing models on the SQuAD dataset. While the paper provides valuable insights into the architecture of the model, it is missing details required to implement and train the neural network model. Our best model achieves an exact-match score of 65.4 and an F1 score of 74.3, while the original implementation achieves an exact-match score of 71.1 and an F1 score of 79.5.

There are two possible sources for this discrepancy. Either our implementation of the Gated Self-Matching Network is different from the original or the training procedure used deviated too far from the one used by Wang et al. [16]. We suspect that the differences in our reproduction are largely due to differences in either the architecture of the encoding layer or the question and passage matching layer, while the choice of training procedure should have a comparably small impact.

To further reduce the gap in performance between our reproduction and the original implementation, we looked at all combinations of the different variants we tested. No variant could achieve comparable performance to the original model on the SQuAD development set. This lets us conclude that there is an important detail in the model architecture not mentioned in the original paper, that is required to successfully reproduce the benchmark results.

During our reproduction attempt we learned several lessons about describing reproducible neural network architectures.

One source of ambiguity while reproducing the Gated Self-Matching Network was that the architectural description was split between two different sections. While each component was given its own subsection, some architectural information required for reproduction is found in the implementation details section. The architecture should be explained component by component, while giving special attention to the dimensionality of all intermediate outputs. The dimensionality of each component, whether an RNN is bidirectional or not and which weights are shared is not an implementation detail, as these concepts can have different interpretations in the case of novel neural network components.

Each experimental neural network paper should contain a detailed description of the training procedure used to generate the weights. This starts with the initial parameter values, which parameters are trainable, but should also include the configuration of the optimizer and the exact procedure of adapting the learning rate. Also the batch size has to be mentioned, especially when using adaptive optimizers like AdaDelta [18] or Adam [7], as the batch size interacts with their automatic learning rate adaptation. This is essential, specifically in cases where the model itself cannot be shared publicly for whatever reasons.

7   Conclusion

The Gated Self-Matching Network by Wang et al. [16] is a large step forward in the area of using neural networks for natural language processing. It shows an effective way of matching two token sequences against one another and how a pointer network can be used to select spans of natural text.

While our reproduction attempt showed the effectiveness of these two methods on a reading comprehension based question answering task such as SQuAD, we could not fully reproduce the benchmark results reported in the original paper. The possible reasons are two ambiguous details in the model architecture and a vaguely specified training process. Our experiments showed the difficulties one encounters when trying to reproduce such state-of-the-art neural network architectures revealing the complexity of providing precise and still compact descriptions of models and training processes.

Based on the technical challenges we encountered while reproducing the Gated Self-Matching Network we recommend describing neural network architectures component by component, as complete as possible and reducing a separate implementation details section, which gives additional architectural information required for reproduction, to a minimum, as it allows for easier detection of inconsistencies in the architecture. One should also not forget that the training procedure used is almost as important as the architecture itself, as it is required to reproduce experimental results.

Besides an accurate description, one of best method to ensure such experiments are reproducible is to provide an open source implementation of the architecture, that allows anyone to run the experiments for themselves. This would further reduce the effort required to build on top of state-of-the-art systems, significantly speeding up the scientific process, while also reducing the effort for researchers to clarify questions and core aspects that could not be appropriately covered within the limits of a paper-based description of their research.

References

1. Chen, D., Bolton, J., Manning, C.D.: A thorough examination of the CNN/daily mail reading comprehension task. In: Proceedings of the 54th Annual Meeting of the Association for Computational Linguistics, pp. 2358–2367 (2016)

2. Cho, K., Van Merriënboer, B., Gulcehre, C., Bahdanau, D., Bougares, F., Schwenk, H., Bengio, Y.: Learning phrase representations using RNN encoder-decoder for statistical machine translation. In: Proceedings of the 2014 Conference on Empirical Methods in Natural Language Processing (EMNLP), pp. 1724–1734 (2014)

3. Glorot, X., Bengio, Y.: Understanding the difficulty of training deep feedforward neural networks. In: Proceedings of the Thirteenth International Conference on Artificial Intelligence and Statistics, pp. 249–256 (2010)

4. Hermann, K.M., Kocisky, T., Grefenstette, E., Espeholt, L., Kay, W., Suleyman, M., Blunsom, P.: Teaching machines to read and comprehend. In: Advances in Neural Information Processing Systems, pp. 1693–1701 (2015)

5. Hochreiter, S., Schmidhuber, J.: Long short-term memory. Neural Comput. **9**(8), 1735–1780 (1997)

6. Hu, M., Peng, Y., Qiu, X.: Mnemonic reader for machine comprehension. arXiv preprint arXiv:1705.02798 (2017)

7. Kingma, D., Ba, J.: Adam: a method for stochastic optimization. arXiv preprint arXiv:1412.6980 (2014)

8. Manning, C.D., Surdeanu, M., Bauer, J., Finkel, J.R., Bethard, S., McClosky, D.: The stanford CoreNLP natural language processing toolkit. In: ACL (System Demonstrations), pp. 55–60 (2014)

9. Pennington, J., Socher, R., Manning, C.D.: Glove: global vectors for word representation. In: Proceedings of the 2014 Conference on Empirical Methods in Natural Language Processing (EMNLP), vol. 14, pp. 1532–1543 (2014)

10. Rajpurkar, P., Zhang, J., Lopyrev, K., Liang, P.: Squad: 100,000+ questions for machine comprehension of text. In: Proceedings of the 2016 Conference on Empirical Methods in Natural Language Processing (EMNLP) (2016)

11. Richardson, M., Burges, C.J., Renshaw, E.: Mctest: a challenge dataset for the open-domain machine comprehension of text. In: Proceedings of the 2013 Conference on Empirical Methods in Natural Language Processing (EMNLP), vol. 3, p. 4 (2013)

12. Rojas, R.: Neural Networks: A Systematic Introduction. Springer, Heidelberg (2013), https://doi.org/10.1007/978-3-642-61068-4

13. Srivastava, N., Hinton, G.E., Krizhevsky, A., Sutskever, I., Salakhutdinov, R.: Dropout: a simple way to prevent neural networks from overfitting. J. Mach. Learn. Res. **15**(1), 1929–1958 (2014)

14. Vinyals, O., Fortunato, M., Jaitly, N.: Pointer networks. In: Cortes, C., Lawrence, N.D., Lee, D.D., Sugiyama, M., Garnett, R. (eds.) Advances in Neural Information Processing Systems, vol. 28, pp. 2692–2700. Curran Associates, Inc. (2015)

15. Wang, S., Jiang, J.: Learning natural language inference with LSTM. In: Proceedings of the 15th Annual Conference of the North American Chapter of the Association for Computational Linguistics: Human Language Technologies (2016)

16. Wang, W., Yang, N., Wei, F., Chang, B., Zhou, M.: Gated self-matching networks for reading comprehension and question answering. In: Proceedings of the 55th Annual Meeting of the Association for Computational Linguistics (2017)

17. Xiong, C., Zhong, V., Socher, R.: Dynamic coattention networks for question answering. arXiv preprint arXiv:1611.01604 (2016)

18. Zeiler, M.D.: Adadelta: an adaptive learning rate method. arXiv preprint arXiv:1212.5701 (2012)

Bringing Back Structure to Free Text Email Conversations with Recurrent Neural Networks

Tim Repke[✉] [ID] and Ralf Krestel

Hasso Plattner Institute, Potsdam, Germany
{tim.repke,ralf.krestel}@hpi.de

Abstract. Email communication plays an integral part of everybody's life nowadays. Especially for business emails, extracting and analysing these communication networks can reveal interesting patterns of processes and decision making within a company. Fraud detection is another application area where precise detection of communication networks is essential. In this paper we present an approach based on recurrent neural networks to untangle email threads originating from forward and reply behaviour. We further classify parts of emails into 2 or 5 zones to capture not only header and body information but also greetings and signatures. We show that our deep learning approach outperforms state-of-the-art systems based on traditional machine learning and hand-crafted rules. Besides using the well-known Enron email corpus for our experiments, we additionally created a new annotated email benchmark corpus from Apache mailing lists.

1 Introduction

Emails are an important part of day to day business communication, hence their analysis inspired research from a variety of disciplines. In Social Network Analysis, User Profiling, or Behaviour Analysis often only information contained in the well structured email protocol headers is used. However, a lot more information remains hidden in the free text body of an email, which contains additional meta-data about a discussion in the form of quoted messages that are forwarded or replied to.

In the early days of email communication, users followed clear rules, e.g. prefixing quoted text with angle brackets (>). Nowadays, due to the diversity of email programs, formatting standards, and the freedom to edit quoted text, identifying the different parts of a message body is a surprisingly challenging task. Email programs like Outlook, Thunderbird, or even online services such as Gmail, usually group emails into conversations and attempt to hide quoted parts. To this end, they try to match preceding emails by subject and sender, which fails in case the subject or quoted text was edited.

We propose a neural network based approach for extraction of the inherent structure in email text to overcome problems of error-prone rule-based

© Springer International Publishing AG, part of Springer Nature 2018
G. Pasi et al. (Eds.): ECIR 2018, LNCS 10772, pp. 114–126, 2018.
https://doi.org/10.1007/978-3-319-76941-7_9

approaches. This enables downstream tasks to work with much cleaner data and additional information by focusing on specific parts. Further we show improvements in flexibility and performance over earlier work on similar tasks.

Problem statement. Our goal is to extract the inherent structure of free text emails containing a conversation thread composed of consecutive quoted or forwarded messages. Components of an email are referred to as *zones* similar to the definition used by Lampert et al. [10]. We assume that a conversation thread is represented as a sequence of *client header* and *body blocks*. A pair of corresponding header and body is called *conversational part* or *message*.

In this context, client headers are blocks of meta-data automatically inserted by an email program, usually containing information on the sender, recipient, date, and subject of the quoted email. Generally the header indicates, whether the subsequent message body was forwarded or replied to by the text above. Bodies are the actual written messages, which on reply or forward are quoted below the newer message.

Message bodies can often be further separated into a *greeting* (such as a formal or informal address of the recipient at the beginning of the message), *authored text* (the actual message), *signoff* (closing words of the message), and a *signature* (containing contact information, advertising, or legal disclaimers). As emails with inline relies are usually copied, we consider a block of copied lines and responses as one body block.

We assume that each single line can be assigned to exactly one zone as Fig. 1 exemplary shows. In case of conflicts, the predominant or detailed type is used.

2   Related Work

Email corpora provide fascinating insights into human communication behaviour and therefore inspire research in many different areas. Datasets such as the Enron [9] or Avocado corpus [15] provide real world information about business communication and contain a mix of professional emails, personal emails, and spam. Ben Shneiderman published parts of his personal email archive for research [16]. Also popular is the 20 Newsgroups dataset [12] sampled from newsgroup postings in the early 90s, which we discard as it contains only few conversation threads. For the work at hand, we use the Enron corpus and emails we gathered from public email archives of the Apache Software Foundation[1].

A recent survey shows the diversity of email classification tasks alone [14]. Similarly interesting is the analysis of communication networks based on meta-data like sender, recipients, and time extracted from emails [1].

Models based on the written content of emails may get confused by automatically inserted text blocks or quoted messages. Thus, working with real world data requires normalisation of data prior to the problem at hand. Rauscher et al. [17] developed an approach to detect zones inside work-related emails where relevant business knowledge may be found.

---

[1] http://mail-archives.apache.org/mod_mbox/.

From: Alice		Sent: Mon, 14 May 2001 07:15 AM
To: Bob, Brian		
Subject: RE: Telephone Call with Jerry Murdock		
Body	Thank you for your help.	
Body		
Body/Signature	ISC Hotline	
Header	03/15/2001 10:32 AM	
Header		
Header	Sent by: Randi Howard	
Header	To: Jeff Skilling/Corp/Enron@ENRON	
Header	cc:	
Header	Subject: Re: My "P" Number	
Body		
Body/Greeting	Mr. Skilling:	
Body		
Body	Your P number is P00500599. For your convenience, you can also go to	
Body	http://isc.enron.com/ under Site Highlights and reset your password or	
Body	find your "P" number.	
Body/Signoff	Thanks,	
Body/Signoff		
Body/Signoff	Randi Howard	
Body/Signature	ISC HOTLINE	
Body		
Header	From: Jeff Skilling 03/15/2001 10:01 AM	
Header		
Header	To: ISC Hotline/Corp/Enron@Enron	
Header	cc:	
Header		
Header	Subject: My "P" Number	
Body		
Body	Could you please forward my "P" number. I am unable to get into the XMS	
Body	system and need this ASAP.	
Body		
Body/Signoff	Thanks for your help.	

Fig. 1. Example email with zones; consecutive blank lines reduced to one

In their work towards detecting emails containing requests for action, Lampert et al. [11] observed a relative error reduction by 40% when removing quoted sections of emails. Similar observations were made more recently predicting reply behaviour within the Avocado dataset [21].

Thread Reconstruction. Another popular area of research is the reconstruction of graphs reflecting which message responds to another. Wang et al. propose baseline approaches based on temporal relationships [20]. There are also more advanced models that use sentence-level topic features to resolve a message graph using random walks [7]. Most recently, Tien et al. [19] proposed a novel convolutional neural network over a grid built by assigning roles to extracted entities. The latent graph is derived from the configuration with the highest coherence score. In our work however, we only focus on separating conversational parts within free text messages, not the actual reconstruction of the thread.

Email Zoning with Rules and Text Alignment. We identified three approaches to email zoning: rule based, text alignment, and machine learning.

The most naïve approach is to write specific rules that match commonly used patterns in email text. Talon[2] provides a sophisticated set of patterns to

---

[2] https://github.com/mailgun/talon.

match most popular client header formats. The obvious downside is the lack of flexibility and that it's error-prone to changes.

Assuming a complete email corpus, a message in one user's outbox may be found in the inbox of other user(s). Likewise, quoted messages exist within the corpus as an original message from preceding communication. By finding overlapping text passages across the corpus, Jamison et al. managed to resolve email threads of the Enron corpus almost perfectly [6]. It has to be noted, that the claimed accuracy of almost 100% was only tested on 20 email threads.

In order to reassemble email threads, Yeh et al. considered a similar approach with a more elaborate evaluation reaching an accuracy of 98% separating email conversations into parts [22]. To do so, they rely on additional meta information in emails sent through Microsoft Outlook (thread index) and rules that match specific client headers. Thus, such an approach will not work on arbitrary emails, nor can it handle different localisation or edits by the user.

Contrary to approaches using text alignments, we don't assume a complete corpus. Our goal is to extract all information from only a single email archive or even a single email.

Machine Learning for Email Zoning. Another approach to email zoning uses machine learning with carefully designed features.

Carvalho and Cohn proposed Jangada [2], a system to remove quoted text and signature blocks from emails in the twenty newsgroup dataset [12]. They first classify emails to find those that contain quoted text or signatures and then classify each line individually using Conditional Random Fields (CRF) and sequence-aware perceptrons. Reported accuracies range from 97% to above 99%.

Other researchers applied Jangada to Hotmail emails and measured accuracies around 64% [4]. With some adaptation, they managed to extract five different zones (author text, signature, advertisement, quoted text and reply lines) with an average accuracy of up to 88%.

Lampert et al. developed the Zebra system [10] as a pre-processor to their previously mentioned work on requests for action [11]. Adversely to previous approaches, they use Support Vector Machines and therefore classify lines of an email into zones individually rather than considering a sequence of lines. For that, they describe graphic, orthographic, and lexical features to represent lines within their context reaching an average accuracy of 93% on the two-zone task and 87% on a nine-zone task. Comparing the performance by zone type, most problems are caused by signature lines (F-score around 60%), signoffs (70%) and attachments (69%). It was found, that adding contextual features didn't improve the performance [10]. Contrary to our objectives, Zebra only tries to identify the zones within the very last message within an email thread and rejects the rest as quoted text, whereas we aim to detect the zones across the entire email.

We compare results of our system described in Sect. 3 with Jangada and Zebra. We not only aim to improve upon those results, but also provide a system that is able to detect zones along the entire conversation thread contained in an email and not only the latest part. Furthermore, our system uses neural networks rendering expensive and potentially error-prone feature engineering

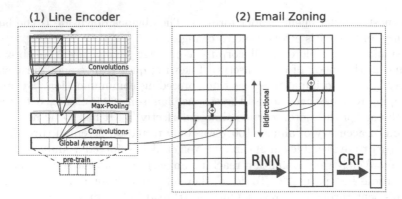

Fig. 2. Schematic model overview; Left side shows line embedding stage using the CNN approach, right side outlines email zoning model.

obsolete. This way, even very small or incomplete datasets can be utilised for downstream tasks like social network analysis, speech act recognition and other research areas using email data.

3   Segmentation of Emails

Systems for email segmentation that are discussed earlier are based on hand written rules to match common structures directly or use them as features for machine learning models. Such approaches will fail when client headers are localised, formats are changed, or quoted messages are edited by users or get corrupted.

In most cases it may seem obvious to the human eye how to segment an email into client headers and quoted text even though different or corrupted formats are used. However, even a sophisticated text parsing program will fail since client headers follow no standardised format. Usually lines start with attribute keywords such as "From:" or "Subject:", however their value may span multiple lines and use varying delimiters. This even makes it hard to detect the boundaries between header and body blocks, since one can not rely on the presence of keywords or well formed, deterministic schemas.

In this paper, we propose the *Quagga*[3] system based on neural network archi-tectures. As shown in Fig. 2, emails are processed in two stages: the line encoding and the email zoning stage. In this section we describe he how email text is rep-resented and how classifiers can be used as a reliable and robust preprocessor for a simple program to extract its inherent structure.

3.1   Representation of Email Data

In the initial stage of our system, the email text data is encoded into a low dimensional space to be used as input to the second stage as outlined on the

---

[3] The quagga is a subspecies of zebras. (https://en.wikipedia.org/wiki/Quagga).

left side in Fig. 2. The smallest fragments to be considered for email zoning are the lines in the email text. Lines are delimited by the newline character (\n), which may not necessarily be the same as wrapped lines displayed by an email program. Analysis of the annotated data shows that this granularity is sufficient for all header, body, and signature zones as was assumed by other research on similar tasks.

Each line is encoded as a sequence of one-hot vectors representing respective characters. We distinguish one hundred different case-sensitive alpha-numeric characters and basic ASCII symbols plus an out-of-scope placeholder. This is sufficient for all email corpora we looked at, where only a negligible portion of characters exceeds this set. We presume, that this could be adapted for applications with Cyrillic, Arabic or other alphabets.

Inspired by research on character-aware language models [8], we devised a recurrent and a convolutional neural network model. The recurrent model consists of a layer with varying number of gated recurrent units (GRU), where the last unit's output serves as a fixed size embedding of the line. The convolutional model uses two convolutional layers, which scan the sequence of characters in a line and are intertwined by max-pooling and global-averaging layers finally leading into a densely connected layer, where the number of neurons corresponds to the embedding size as shown on the left in Fig. 2.

In both models, the line representations are learnt in a supervised fashion. During training, a densely connected layer with softmax activation is appended so that a classifier can be trained to distinguish between lines of corresponding zone types. Optimal parameters of the topology such as the number of layers and embedding size are determined experimentally. A detailed analysis of embedding accuracy when limiting the length per line is found in Sect. 3.3.

3.2   Classification of Email Lines

A model in the first stage of our Quagga system learns line representations by classifying them into zones. That way however, the context in which a line appears is missing, resulting in less ideal performance on ambiguous or deceptive cases. Thus, we added a second stage to our system for sequence to sequence classification using a GRU-CRF model as outlined in the right part of Fig. 2, which takes a sequence of line encodings per email as input. Three of the five zone types only appear within message bodies, so we use two concatenated embeddings as input, where one is pre-trained using two- and the other with five-zone classification.

Best performance was achieved with a bidirectional GRU layer, which scans the lines from top to bottom and in reverse order and concatenates the hidden states of respective lines. In sequence to sequence classification, recurrent neural networks only consider the previous hidden state but neglect the actually predicted label sequence. We already observed small improvements by using a bidirectional layer over a unidirectional one, since each line's context reflects the previous and following lines. Like in language models [5,13], the addition of a CRF to the output shows further performance gains.

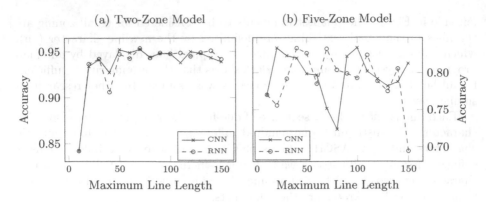

Fig. 3. Accuracy for increasing max line length using 32-dimensional embeddings

Training both parts of the system as an entire model in one pass by directly connecting the encoder output layer to the second stage model's input lead to unstable results even after pre-training the line encoder. Therefore we train the model in the second stage of our system separately from the line encoding model.

The predicted sequence of zone types can be used to extract the conversational parts of an email and also separate the message from additional content such as signatures, greetings and signoffs. Consecutive lines with the same predicted zone type are aggregated into a block. Further processing inevitably requires making some assumptions about the general structure of emails. Based on the analysis of emails in the training data, we assume that a body proceeds a header block. Furthermore, we define that in-line replies with quoted parts belong to one message, which is not problematic, since this would usually only appear in Usenet-style emails, which use indicators like repeated > at the beginning of a line and therefore don't require sophisticated processing as presented here.

Small errors in the prediction can be fixed heuristically, for example a block with a single line classified as header containing only the "Subject:" keyword likely is either a false positive or belongs to another block nearby. The introduction of such rules could reduce the initial robustness and is omitted for the evaluation in this paper. In the scope of our work, we found that using the Quagga system as a pre-processor for finding related blocks significantly improves the accuracy of parsing rules for downstream tasks like constructing communication graphs from header blocks compared to a purely rule-based parser without pre-processing.

3.3   Selection of Model Parameters

In the description of the proposed model we highlighted adjustable parameters. This includes the model for line representations in the first stage, limiting the length of each line, and finding the ideal dimension of line embeddings. We base the model's topology configuration on the analysis of related models [5,8,13].

The ideal configuration for the line encoder model is determined through grid search across mentioned parameters. We record the accuracy of the convolutional and recurrent approach for line encoding in Fig. 3. Note, that the convolutional model assumes a fixed size input, so shorter lines are zero-padded at the end. When evaluating the reported accuracy, one has to consider two things. First, this metric may not project down to the later stage of our system, and second, the line type distribution bias reduces the range of values to 0.81 (or 0.65) upwards.

We did not observe significant differences between embedding dimensions above 32, so we choose this dimensionality in favour of a less complex model. Most errors are caused by blank lines, which are usually classified as "Body". Results when training using two-zone classification are mostly stable for both models. The majority of lines in our training data are between forty and fifty characters long.

Both approaches for line representations seem to have their strengths and weaknesses. Since there is no clear winner, we continue only using the convolutional model in this work and fix the input length to 100 characters per line. We do so based on the argument, that one may want to process large corpora and prefer a faster system.

4 Experimental Setup

In this section we present an overview of the email datasets we used and discuss the sampling of emails to create an unbiased evaluation set. Further, we describe competing approaches that are used as baselines for comparison of our results. We also analyse model parameters and its robustness to changes in email text.

4.1 Dataset

We evaluate our proposed approach to email zoning on the Enron corpus [9] and emails gathered from public mail archives of the Apache Software Foundation[4] (ASF). Estival et al. [4] and Lampert et al. [10] discussed shortcomings in working with Usenet-style emails, leading us to refrain from using the twenty newsgroup dataset [12] as was done for the Jangada system. We found that more recent email threads from the ASF archives, especially those on mailinglists for users of different software projects, offer diverse formatting patterns.

Each dataset is divided into three subsets for training, validation, and testing. Emails are sampled at random from their respective original dataset and put into one of those subsets. To ensure representative results that are not biased by author or domain, sampling per subset is restricted to distinct mailboxes (Enron) or mailing lists (ASF). The ASF dataset was compiled by randomly selecting emails from the *flink-user*, *spark-user*, and *lucene-solr-user* mailing list archives.[5]

---

[4] http://mail-archives.apache.org/mod_mbox/.

[5] Annotated datasets and code can be found at https://github.com/TimRepke/Quagga.

Table 1. Annotated datasets in numbers

	Enron			ASF		
	Train	Test	Eval	Train	Test	Eval
Emails	500	200	100	350	100	50
Individual messages	1048	474	233	934	226	108
Average length of threads	3.5	3.6	3.5	3.5	3.7	3.1
Number of signatures	103	58	26	76	13	5

Table 1 shows an overview of the selected dataset and the expected number of messages to be extracted. Prior heuristic analysis of the Enron corpus estimated 60% of emails to contain conversation threads [9], which is close to our annotated data. On average an email has two parts with 20 lines per message. Only a few messages contain a signature, which on average are six lines long.

4.2   Competing Approaches

We compare our proposed model for extracting zones from emails against several other approaches. Most notably, Jangada [2] and Zebra [10] are reimplemented with slight modifications to fit the more refined problem statement. Both systems originally are intended to distinguish lines within an email, which are not part of the latest message of that thread. Clearly that deviates from our goal to extract *all* individual parts and detect zones with additional detail within those. Since the systems are supposed to detect zones within the first part of the email, their features and underlying models should in principle also work on our task.

The source code for Jangada is freely available on the author's web page[6]. We used the source code as a basis for an implementation in Python and the originally used model for sequence labelling, which is part of the MinorThird Library[7]. For the extraction of signatures, Jangada originally only considers the last ten lines of an email. In our implementation, the perceptron performs a multi-class classification along all lines of the email corresponding to zone types defined earlier. The model is trained with window-size 5 for 40 epochs.

The Zebra project web page[8] does only provide annotated data, but not the system's source code. Gossen et al. implemented[9] it for their work on classification of action items in emails [18]. We used that as a guideline for our adapted Python implementation. The SVM is trained for a maximum of 200 iterations in a one-versus-rest fashion for multi-class classification using RBF kernels.

We use a selection of features from both models as input for a recurrent neural network with two GRU layers [3], which we will refer to as *FeatureRNN*. The above models are baselines for the comparison to our proposed Quagga

---

[6] http://www.cs.cmu.edu/~vitor/software/jangada/.
[7] http://minorthird.sourceforge.net/.
[8] http://zebra.thoughtlets.org/zoning.php.
[9] https://github.com/gerhardgossen/soZebra.

Table 2. Classifying emails into zones (Precision/Recall/Accuracy)

Approach	Zones	Enron	ASF
Jangada [2]	2	0.89/0.88/0.88	0.97/0.97/0.97
Zebra [10]	2	0.66/0.25/0.25	0.88/0.18/0.18
FeatureRNN	2	0.98/0.98/0.97	0.97/0.95/0.94
Quagga	2	**0.98/0.98/0.98**	**0.98/0.98/0.98**
Jangada [2]	5	0.82/0.85/0.85	0.90/0.92/0.91
Zebra [10]	5	0.60/0.25/0.24	0.81/0.20/0.20
FeatureRNN	5	0.92/0.75/0.75	0.90/0.60/0.60
Quagga	5	**0.93/0.93/0.93**	**0.95/0.95/0.95**

system as described in Sect. 3 using a convolutional model as line encoder with fixed input sizes of 100 characters per line.

5   Results

In this section we compare Quagga to similar systems found in related work. To get a good understanding of the versatility, we not only look at the results shown in Table 2, but also consider the robustness against noise or otherwise changing data as well as how many training samples are required to get good results.

We were not able to reproduce reported accuracies of Zebra [10], which given the nature of the features and individual classification of lines disregarding their position and context in an email was expected. Jangada uses more general features and looks at a sliding window of lines and we got close to reported accuracies, especially for the ASF dataset, which is closer to the twenty newsgroup data the authors used [10]. Overall, our system shows very good performances and seamlessly adapts to other datasets without problems.

Number of Training Samples. Complex neural network based machine learning models require lots of training samples to reliably proficiently solve a given task. We limited the number of Enron emails shown to the network during training down to 10% of the Enron training set, corresponding to 50 emails and then measured the performance whilst continuing to add training samples up to all 500 emails. The model trained with the least data in this scenario only lags behind around 1% in accuracy compared to a model trained on all data in both the two- and five-zone task.

Cross Corpus Compatibility. Ideally, a system like Quagga would be trained once and work well on arbitrary emails. The Enron and ASF datasets show more differences than there are in samples in the training and testing data of one corpus. Thus we trained Quagga on Enron and tested it on ASF emails and vice versa. We observe, that by training on ASF emails, Quagga does not

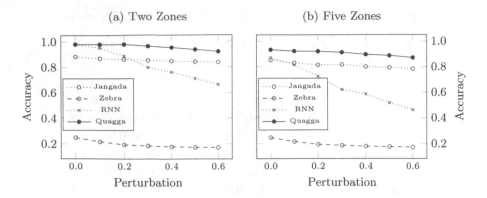

Fig. 4. Robustness against perturbation for the Enron test set

generalise as well to emails from he other corpus (Accuracy: 0.86 or 0.80, for two- or five-zones) as the other way around (Accuracy: 0.94 or 0.86). Compared to the results when training and testing using emails from the same base dataset, this results in a decrease of performance of around 4–10%.

Robustness to Noise. Our hypothesis is, that a model which learns meaningful features itself is more robust towards changes to the email text as hard coded rules responding to specific keywords or patterns. To show the flexibility of our model, we first introduce the notion of a perturbation threshold $\rho \in [0,1)$. Before passing an email to a model, a function iterates over each character and with probability ρ edits, removes, or duplicates it. Training was performed on uncorrupted data only.

The robustness of each model against increasing perturbation is shown in Fig. 4. Relatively, the drop in performance is the same the for two- and five-zone task, although at different absolute accuracies. Quagga doesn't seem to be affected up to $\rho = 0.2$ and keeps producing reliable results even at higher perturbation thresholds. Surprisingly, also Jangada is not influenced significantly by the introduction of perturbation. As opposed to Zebra and FeatureRNN, it is using more features related to small patterns or proportions of types of symbols, whereas the others depend on more complex patterns which are more error prone to change.

6   Conclusion and Future Work

In this work we presented a reliable and flexible system for finding the inherent structure of an email containing multiple conversational parts. In the first stage, the system uses a convolutional neural network to encode lines of an email which are used by a GRU-CRF to predict a sequence of zone types per line reaching accuracies of 98%. Compared to similar models, we show significant improvement and especially seamless adaptation to other datasets as well as robustness against corrupted data.

Research based on email data can largely benefit from this system by pre-processing the text and focus downstream algorithms on relevant parts of an email like client headers or the actual text cleaned of irrelevant parts.

In addition to the Quagga system, we provide a detailed annotation of a subset of the Enron corpus that can directly be used for building communication networks without further parsing including linked person aliases and, if present, contact details from email signatures as well as the new ASF corpus.

References

1. Bonchi, F., Castillo, C., Gionis, A., Jaimes, A.: Social network analysis and mining for business applications. TIST **2**(3), 22 (2011)
2. Carvalho, V., Cohen, W.: Learning to extract signature and reply lines from email. In: CEAS (2004)
3. Cho, K., van Merrienboer, B., Bahdanau, D., Bengio, Y.: On the properties of neural machine translation: encoder-decoder approaches. CoRR (2014)
4. Estival, D., Gaustad, T., Pham, S., Radford, W., Hutchinson, B.: Author profiling for English emails. In: Conference of the Pacific ACL (2007)
5. Huang, Z., Xu, W., Yu, K.: Bidirectional LSTM-CRF models for sequence tagging. CoRR (2015)
6. Jamison, E., Gurevych, I.: Headerless, quoteless, but not hopeless? Using pairwise email classification to disentangle email threads. In: RANLP (2013)
7. Joty, S., Carenini, G., Ng, R.T.: Topic segmentation and labeling in asynchronous conversations. Artif. Intell. Res. **47**, 521–573 (2013)
8. Kim, Y., Jernite, Y., Sontag, D., Rush, A.: Character-aware neural language models. CoRR (2015)
9. Klimt, B., Yang, Y.: The enron corpus: a new dataset for email classification research. In: Boulicaut, J.-F., Esposito, F., Giannotti, F., Pedreschi, D. (eds.) ECML 2004. LNCS (LNAI), vol. 3201, pp. 217–226. Springer, Heidelberg (2004). https://doi.org/10.1007/978-3-540-30115-8_22
10. Lampert, A., Dale, R., Paris, C.: Segmenting email message text into zones. In: EMNLP (2009)
11. Lampert, A., Dale, R., Paris, C.: Detecting emails containing requests for action. In: Human Language Technologies. ACL (2010)
12. Lang, K.: Newsweeder: learning to filter netnews. In: Twelfth International Conference on Machine Learning (1995)
13. Ma, X., Hovy, E.H.: End-to-end sequence labeling via bi-directional LSTM-CNNs-CRF. CoRR (2016)
14. Mujtaba, G., Shuib, L., Raj, R., Majeed, N., Al-Garadi, M.: Email classification research trends: review and open issues. IEEE Access **5**, 9044–9064 (2017)
15. Oard, D., Webber, W., Kirsch, D., Golitsynskiy, S.: Avocado Research Email Collection. Linguistic Data Consortium, Philadelphia (2015)
16. Perer, A., Shneiderman, B.: Beyond threads: identifying discussions in email archives. Technical report, MUC (2005)
17. Rauscher, F., Matta, N., Atifi, H.: Context aware knowledge zoning: traceability and business emails. In: Mercier-Laurent, E., Boulanger, D. (eds.) AI4KM 2015. IAICT, vol. 497, pp. 66–79. Springer, Cham (2016). https://doi.org/10.1007/978-3-319-55970-4_5

18. Scerri, S., Gossen, G., Davis, B., Handschuh, S.: Classifying action items for semantic email. In: LREC (2010)
19. Nguyen, D.T., Joty, S., Boussaha, B.E.A., de Rijke, M.: Thread reconstruction in conversational data using neural coherence models. In: Neu-IR (2017)
20. Wang, Y.C., Joshi, M., Cohen, W.W., Rosé, C.P.: Recovering implicit thread structure in newsgroup style conversations. In: ICWSM (2008)
21. Yang, L., Dumais, S., Bennett, P., Awadallah, A.: Characterizing and predicting enterprise email reply behavior. In: SIGIR (2017)
22. Yeh, J., Hamly, A.: Thread reassembly using similary matching. In: CEAS (2006)

A Hybrid Embedding Approach to Noisy Answer Passage Retrieval

Daniel Cohen$^{(\boxtimes)}$ and W. Bruce Croft

College of Information and Computer Sciences,
University of Massachusetts Amherst, Amherst, MA, USA
{dcohen,croft}@cs.umass.edu

Abstract. Answer passage retrieval is an increasingly important information retrieval task as queries become more precise and mobile and audio interfaces more prevalent. In this task, the goal is to retrieve a contiguous series of sentences (a passage) that concisely addresses the information need expressed in the query. Recent work with deep learning has shown the efficacy of distributed text representations for retrieving sentences or tokens for question answering. However, determining the relevancy of answer passages remains a significant challenge, specifically when there exists a lexical and semantic gap between the text representation used for training and the collection's vocabulary. In this paper, we demonstrate the flexibility of a character based approach on the task of answer passage retrieval, agnostic to the source of embeddings and with improved performance in P@1 and MRR metrics over a word based approach as the collections degrade in quality.

Keywords: Answer passage · Representation learning
Hybrid embedding

1 Introduction

A key part of an effective information retrieval (IR) system is the ability to identify the specific text relevant to a query. For some queries, the relevant text consists of documents or passages topically related, while other queries can best be answered by a few select words without the need for any additional text. The latter, known as factoid question answering (QA), has received significant attention with the rise of deep neural networks, achieving state of the art performance over term frequency based methods [5,17,19]. However, these neural models cannot be directly applied to larger bodies of text without a large degradation in performance [2].

Passage retrieval techniques have previously been developed to locate highly topically relevant passages spanning multiple sentences in documents [1,23]. Answer passage retrieval focuses on finding text passages that directly answer

© Springer International Publishing AG, part of Springer Nature 2018
G. Pasi et al. (Eds.): ECIR 2018, LNCS 10772, pp. 127–140, 2018.
https://doi.org/10.1007/978-3-319-76941-7_10

questions expressed in more precise queries. In contrast to factoid QA that identifies specific words in a sentence, each word in the passage contributes to answering the query's information need. Thus, answer passage retrieval models must be able to capture long term dependencies commonly found in language in order to determine the relevancy of a passage. This disparity can be more easily seen with the following query:

Factoid QA Queries: When did James Dean die? How high is Everest?
Answer Passage Query: What kinds of harm do cruise ships do to sea life such as coral reefs, and what is the extent of their damage?

While the factoid QA queries rely on a few keywords and a specific request for information which can be resolved with a single word or number, answer passage queries require a more elaborate answer combining multiple aspects into a unified passage. The open ended nature of answer passage retrieval makes it difficult to apply these factoid QA models directly to answer passage retrieval [3]. This difficulty is compounded by the fact that passages can have little term overlap with the query, and poses a significant obstacle for conventional approaches as they rely on transforming text into vectors using domain knowledge.

One can view the difference in vocabulary as a missing text problem, in which case past work using character n-grams [8] has shown to be an effective method, out performing equivalent word based models on the task of retrieving optical character recognition degraded text. Recent work such as the Deep Structured Semantic Model (DSSM) [10] and the Deep Relevance Matching Model [7] has bridged the gap between document and answer passage retrieval by using a similar character hashing approach or a distributed representation to model interactions. However, they do not accurately retrieve answer passages as they are unable to model the sequential long term dependencies which contribute to the relevance of a passage.

Both of these issues are exacerbated by the caveat that pre-trained embeddings do not adapt well to collections with different vocabularies or where language differs from the corpus used for the word embedding training. The lexical and semantic shift permeates the network's hidden layers, resulting in a significant loss of performance when compared to embeddings trained on the actual test collection, particularly for the answer passage retrieval task [3]. Retraining embeddings to reflect a new collection can consistently improve performance [4]; however, it is impractical to create new local embeddings at run time as recent methods [3,4,15] rely on a time consuming optimization process requiring large amounts of data.

We approach these two challenges inherent to answer passage retrieval by leveraging a long short term memory (LSTM) network to build phrase level representations of both standard word embeddings as well as with the flexibility of a character n-gram based approach as seen in the DSSM and the OCR degraded text task. We adapt the fixed window of the trigram hashing in DSSM by using varying length convolutional filters to aggregate multiple length character n-grams and then sequentially building sentence and passage embeddings using a recurrent network. This approach produces a network that (1) is robust to degradation in collection quality (2) maintains performance on high quality collections where standard character based approaches fail to perform, and (3) does not require the expensive process of retraining embeddings for each collection.

2 Related Work

Deep learning for IR related tasks excels where standard approaches have failed. These methods almost all rely on a distributed representation of the vocabulary, referred to as word embeddings. The most common method in IR is the work by Mikolov et al. [15], word2vec, where they train a small neural network to predict the context around a word. The internal hidden representation of the network when predicting the context around a word becomes the embedding for that word. This results in similar words, such as *cat, dog, pet*, to have similar hidden representations.

Wang and Nyberg [21] use these embeddings as input into their BiLSTM networks as an effective method for retrieving non-factoid answers. They use Google's pre-trained word2vec embeddings and boost the output of the network with term frequency statistics. This approach is able to outperform conventional IR approaches without the need of feature engineering.

Tan et al. [19] expand on neural retrieval for QA and create larger LSTM-CNN and CNN-LSTM networks. In this work, the initial layer is a BiLSTM layer with an attention mechanism, and feeds into a CNN. An attention mechanism allows the network to focus on information specifically relevant to both the query and answer rather than modeling the entire text independently. The final output of their networks is the cosine similarity between the question and answer embeddings. Again, this work uses google's pre-trained word2vec embeddings as the initial input. They use a similar attention mechanism, but prime the network with the query prior to processing the answer text, allowing the network to attend temporally. Santos et al. [5] investigate another attention mechanism by pooling the rows and columns of a similarity matrix and use the softmax to weight the LSTM or CNN representation of the question and answer respectively. This has been shown to outperform the method used in [19].

Cohen and Croft [3] demonstrate that updating word embeddings via back-propagation during training results in significant improvement when compared to standard embeddings. Diaz et al. [4] propose training word embedding vectors on topically-constrained corpora, instead of large topically-unconstrained corpora. These locally trained embedding vectors were shown to perform well for the query expansion task.

Due to the limitations of word embeddings with unseen vocabulary and new collections, Zhang et al. [25] demonstrate that a character level embedding fed into a deep CNN is an effective method of categorizing text. The deep CNN is able to recognize abstract text concepts and apply them to ontology classification, sentiment analysis, and text categorization. They compare their CNN to a word2vec [15] based LSTM model, and the CNN results in lower testing errors on all datasets.

Kim et al. [12] leverage this work to create a hybrid CNN-LSTM neural language model. Their network involves a single layer CNN with temporal pooling that feeds into an LSTM model to predict the next character. Their model parses each word separately as a concatenation of character embeddings as opposed to [25], which concatenates the entire passage. Again, the character based approach outperforms the word based approach using perplexity as a metric, and it is able to robustly handle words not seen during training.

In the realm of IR, Huang et al. [10] have capitalized on character level representations when creating deep structured semantic model (DSSM). They chose to represent the text as a series of character trigrams rather than the conventional word based approach. For example, the word #good# would be represented as (#go, goo, ood, od#), where # represents the start and end of a word. This reduces the dimensionality of one hot encodings from the size of the vocabulary to the number of distinct trigrams found in the collection, resulting in a 4 to 16 fold reduction. In addition, this character based approach allows for scaling up to very large vocabularies for use in realistic web searches.

There has been work in using convolutional networks to construct representations for short text; however, this has only been done on short factoid text or knowledge base (KB) question answering. In general, these models do not attempt to capture long term dependencies critical in an answer passage retrieval task and do not leverage recurrent networks for learning passage length representations. Golub and Ziadong [9] use a character CNN to capture deep representations of KB entity and predicates for factoid QA over a KB. Meng et al. [13] use a structure similar to Severyn and Moschitti [17] with the input as character embeddings, which works well on factoid QA tasks such as TREC QA, but fails to outperform traditional baselines as the passages increase in length [3].

3 Model

We propose a hybrid CNN-LSTM model that not only constructs passage level representations from word embeddings, but simultaneously builds an identical representation from a separate character representation. This hybrid approach allows the network to leverage the information contained in pretrained word embeddings while simultaneously using the character subnetwork to construct collection specific representation in its hidden layers. A simplified representation of the model is shown in Fig. 1 with the three key components illustrated. As each component plays a critical role in determining the relevance of a candidate passage, the remainder of this section explains in detail the construction and motivation for each layer's architecture within the model.

3.1 Character Embeddings

As opposed to previous work in IR with neural networks [5,14,17,19], our model's input consists of an additional sequence of characters rather than words alone.

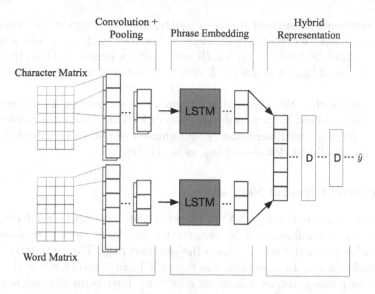

Fig. 1. A compressed representation of the Hybrid architecture.

The advantage of processing text from a character level representation is that it allows for the upper layers to learn a word representation tailored to the collection. Given a sequence of characters from a passage, we concatenate their embeddings into a $k \times l$ matrix where k is the dimension of the embedding and l is the length of the passage. The embeddings were created via the approach introduced by Mikolov et al. [15] with a skipgram window of 5 to create the character embeddings.

As in [25], the alphabet consists of 70 characters, 26 lowercase English letters, 10 digits, and 33 other characters. Characters not contained in the alphabet, including spaces, are represented as k dimensional zero vectors. Uppercase letters were removed as they did not improve performance, and k was chosen to be 20. The 33 other characters are shown below:

$$-,;.!?:'/\backslash|\_@\#\$\%\char`\^\&*\sim`+-=<>()[]\{\}$$

3.2   Embedding-Level Convolutional Layer

One can view a convolution as sliding a fixed width filter, **f**, over the sequence of character embeddings. The filter is constant as it slides over the text, and its weights are updated via backpropagation to identify specific features. This allows the model to transform the input of individual characters into words, or words into short phrases based on common, repeated patterns within the fixed width filter. In the model, the character or word sequence is converted into a passage matrix, $\mathbf{P} \in \mathbb{R}^{k \times l}$, where we convolve $\mathbf{P}$ with a filter $\mathbf{f} \in \mathbb{R}^{k \times w}$ with w as the width of the filter and has the same dimension as the embeddings. The model in this paper uses the activation function $tanh$ which allows for faster convergence compared to the standard logistic function. In order for a convolutional layer to recognize a variety of features, each layer uses a number of filters within [100,1000] for typical IR and NLP applications. Thus, the output of a convolutional layer is a matrix, $\mathbf{F} \in \mathbb{R}^{c \times k \times l}$, where c is the number of filters chosen.

After performing the convolution, temporal max-pooling is performed to select the most salient features over a portion of $\mathbf{F}$. This eliminates non-maximal values and reduces the dimensionality and number of parameters needed for the network via non-linear down-sampling as in [17,19].

3.3   Recurrent Neural Network

A recurrent neural network (RNN) is a type of neural network architecture that captures temporal information [6]. Information previously seen is represented in the network's internal state h_{t-i} from the previous step. The cumulative nature of the hidden state, $\mathbf{h}_t$ results in the network incorporating all of the inputs up to t. As passage retrieval involves capturing long term dependencies over multiple sentences, this type of network is uniquely suited to the task of answer passage retrieval.

To adequately handle the length of answer passages, we adopt a modified RNN structure, called bidirectional long short term memory (BiLSTM) networks [6]. This architecture adds additional structures to a RNN to better control the information across sequential inputs by using internal gates with their own activation functions as well as processing the text in both directions and summing the representations at each timestep. All internal activations of the BiLSTM cell were *tanh* functions.

3.4   Joint Representation

As the Hybrid model consists of two unique substructures, each processing word and character embeddings respectively, the output of the BiLSTM layers are mean pooled and concatenated across timesteps to produce a single vector, $\mathbf{v} \in \mathbb{R}^n$ which can be viewed as the embedding of the entire phrase. The combination of character and word level embeddings allows this model to leverage two unique representations, the character subnetwork is directly tailored to the collection while the word subnetwork aids in generalization.

It is then fed into three dense layers to learn the interaction between word and character phrases. The final dense layer maps to a scalar value, $\hat{y}$ in Fig. 1, that represents the relevance of the input.

3.5   Attention Mechanism

While LSTM networks are able to store internal states across sections of a sequence, they cannot capture arbitrary length dependencies that span across longer passages [6]. In order to persuade the hidden states of the model to focus on information relevant to the query contained in candidate passages, we use an attention mechanism by allowing the hidden layers of a network to compare query and document text when learning abstract representations. This reduces the information load on the network as the parameters are able to focus on modeling this interaction rather than each text individually.

With a variety of attention mechanisms available in previous work [14,16,24], we adopt a method that primes the network similar to machine translation [21] to aid in the LSTM capturing long term dependencies. Given a question-passage pair below,

$$q_1, q_2, \ldots, q_n <?> a_1, a_2, \ldots, a_n$$

The network iterates over the query until it reaches the $<?>$ token, at which point it receives a candidate answer. As discussed in Sect. 3.1, this method allows the network to imprint query specific terms and topics within the cell states that produce selective activations for related information in candidate passages. By priming the network, the recurrent layers learn to model intermediate representations of relevance rather than waiting to introduce query similarity within the final few layers [13,17,19].

4   Experiments

The three datasets used for our experiments were (1) Yahoos Webscope L4, (2) nfl6, which is a lower quality answer passage set created from Yahoos general Webscope L6[1], and (3) a web answer passage collection, called WebAP[2]. These collections were chosen to reflect the answer passage retrieval task while still possessing distinct properties. Training, validation, and testing sets were created via a 64-16-20 split. Detailed statistics for each collection are shown in Table 2.

L4 dataset: This Yahoo collection has been used previously in [3,18,21] for answer passage retrieval and is sometimes referred to as the "manner" collection and are high quality. Each question contains a noun and verb, and each answer is well formed. An example query from this collection indicative of the quality is *"How can I safely open a geode?"* All answers that were not the highest voted answer were removed for each question as multiple answers for a question could be correct. This was done to remove label noise during training and provide more accurate results during evaluation.

nfL6 dataset: Introduced in [3], this dataset consists of 87,361 questions that are best answered by a passage rather than a single sentence. Unlike L4, the questions in this dataset are more generic, such as *"Why do teachers go abroad?"* and *"Why do people steal?"*. Furthermore, answers are not as high quality due to the method of creation.

WebAP dataset: In order to investigate the performance of a character based model in a different retrieval environment, we the WebAP collection from Keikha

---

[1] https://ciir.cs.umass.edu/downloads/WebAP/.
[2] https://ciir.cs.umass.edu/downloads/nfL6.

Table 1. Architecture of the three networks evaluated. Char and Word represent the components used for processing character and word embeddings respectively in the Hybrid model; w = filter width, c = number of filters, σ = activation function, l = layer dimension.

Layer		Char	Word	BiLSTM
Conv	w	[6,7]	[1,2]	
	c	[450,525]	[600,700]	
	σ	tanh	tanh	
Conv	w	[3,4]		
	c	[225,300]		
	σ	tanh		
Conv	w	[3,4]		
	c	[225,300]		
	σ	tanh		
BiLSTM	l	[350,350]	[550,550]	[600,600]
Dense	l	500	500	500
Dense	l	300	300	300
Dense	l	1	1	1

et al. [11]. In contrast to the above collections, the queries are more open ended and can have a variety of passages that are all relevant. An example of this is seen in the query *"Describe the history of the U.S. oil industry"*. Non-relevant portions of each document are split into non-overlapping random length passages. This was done to avoid the network learning certain length passages as non-relevant. Candidate passages with a word count greater than 4000 were removed from the collection as they significantly increased the memory footprint of the models when training. This did not impact the results as they were labeled non-relevant and consistently ranked last during testing.

Table 2. Statistical description of tokens per question-answer pair in nfL6, Webscope L4, and WebAP collections after preprocessing.

Tokens	Webscope L4	nfL6	WebAP
Min	6	10	2
Max	897	722	10885
μ	91.9	50.9	61.2
σ	99.7	25.6	58.1

4.1 Baselines

We compare our Hybrid model to previous deep learning implementations and BM25. As little work has been done specifically on the answer passage retrieval task, we include additional networks used for factoid QA. We use Wang and Nyberg's [21] non-factoid BiLSTM model prior to boosting, Tan et al. [19] factoid QA CNN-LSTM model, and Severyn and Moschitti's Convolutional Deep Neural Network (CDNN) [17]. We also evaluate the individual word and char components of the Hybrid model to isolate the performance difference of word and character embeddings denoted as Word-CNN-LSTM and Char-CNN-LSTM respectively. Exact configurations of the BiLSTM, Char-CNN-LSTM, and Word-CNN-LSTM models are shown in Table 1. We also include DSSM [10] as a competitive character level baseline and DRMM [7] as a competitive neural architecture for document retrieval. All word embedding based neural models are evaluated both on embeddings training locally on the collection and pre-trained embeddings from Google's 300 dimension word2vec model[3]. Character embeddings are initialized from Wikipedia's 05-2015 data dump[4].

4.2 Evaluation

Mean reciprocal rank (MRR) and precision at 1 (P@1) are used for evaluation. Both metrics are common in IR, and reflect the small number of relevant answer passages as well as the importance on the first passage retrieved for mobile and audio search. Similar to [3,21], the test collection was created from pooling the top 10 results from a BM25 search for each question, and including the correct answer passage as the 10th answer if it is not included in the list. We perform this adjustment in order to include all queries even if BM25 fails to return a top ten result as to not bias evaluation towards models that favor term frequency features. For WebAP, the top 100 retrieved results were used for reranking in the same manner, and five fold cross-validation was performed for evaluation.

4.3 Setup and Training

Our CNN-LSTM based networks were optimized via RMSprop [20] over a binary cross entropy function. The networks were trained until the metrics over the validation set stopped improving.

5 Results and Discussion

In this section, we first evaluate the performance of the Hybrid model with respect to the baselines. In order to examine the impact of the additional character structure, we break apart the Hybrid model and evaluate the Char-CNN-LSTM and Word-CNN-LSTM subnetworks independently. Lastly, the comparative performance of local and pretrained embeddings are discussed in relation

---

[3] https://code.google.com/archive/p/word2vec/.
[4] https://dumps.wikimedia.org/enwiki/20160501/.

Table 3. Performance of networks on the three test collections. Local and Pretrained refer to the embedding types used. * denotes significance with $p < .05$ with respect to baselines using two tailed t test. † denotes same significance against subnetworks (Word/Char-CNN-LSTM)

Implementation		L4		nfl6		WebAP	
		P@1	MRR	P@1	MRR	P@1	MRR
BM25		.0738	.1412	.1312	.2660	.3000	.4120
DSSM [10]		.0805	.2477	.0905	.2576	.2150	.3127
DRMM [7]		.1416	.3291	.2844	.3350	.2558	.4064
Tan et al. [19]		.2473	.4217	.2139	.3934	.2047	.3612
CDNN [17]		.0989	.2434	.1438	.2842	.2122	.3834
BiLSTM [21]	Local	.4726	.6329	.2471	.4710	.3177	.4618
	Pretrained	.4492	.6129	.2332	.4287	.3059	.4502
Word-CNN-LSTM	Local	.4523	.6206	.3482	.5327	.2947	.4136
	Pretrained	.4514	.6190	.3406	.5236	.3159	.4411
Char-CNN-LSTM	Local	.4132	.5801	.3211	.4987	.3531	.5148
	Pretrained	.4137	.5798	.3214	.4983	**.3533**	**.5150**
Hybrid	Local	.4608†	.6241	**.3517*†**	5429*†	.3017	.4410
	Pretrained	**.4798*†**	**.6407*†**	.3516*†	**.5433*†**	.3215	.4716

to the models. The results for each of these are shown in Table 3. Of particular note is the poor performance of the traditional factoid or sentence QA models, Severyn and Moschitti's CDNN [17] and Tan et al. cosine similarity based approach [19]. While both of these models perform close to state of the art on WikiQA and TREC QA, they achieve significantly worse results on the answer passage retrieval task. While not benchmarked, Meng et al. [13] possess a similar structure to CDNN and thus, Meng et al.'s character based model would not perform well due to the shared architecture and lack of temporal structure.

Hybrid Embedding Effect. The Hybrid model outperforms the baselines on all but the WebAP collection. The close performance on L4 when compared to the BiLSTM model can be attributed to the language contained in the L4 collection. Compared to nfl6, both queries and answer passages contain significantly less slang, improper syntax, and more consistent sentence structure. The lack of improvement suggests that the convolutional layers do not provide any additional benefit when the collection consists of well formed passages. This is reinforced by the drop in performance on all recurrent word embedding based models moving from L4 to nfl6. Both the Tan et al. and BiLSTM models are most impacted by the lower quality collection. However, the character based DSSM, as well all models with a convolutional component are more robust to this degradation in quality. In particular, the Hybrid model is shown to be the most adaptable to this, achieving 0.3516 P@1 and 0.5433 MRR utilizing pretrained embeddings on the noisier nfl6 collection.

The performance over WebAP highlights the weakness of neural models. Through the lens of BM25, the baseline DSSM, DRMM, Tan et al. and CDNN models all fail to outperform the *tf.idf* baseline. Although the Hybrid model has somewhat better performance than the BiLSTM model, there is little difference between their scores across local/pretrained embeddings. As the WebAP collection only has 82 queries, with an average of 97 graded passages per query, this prevents the network from seeing a large portion of the word embedding space as one cannot increase negative sampling without performance cost if the relevant and non-relevant passages are from different distributions [22]. Thus, at testing time the model often sees new vocabulary and passages unseen during training. Just as in the nfl6 dataset, character embeddings bridge this gap by allowing almost all characters to be seen during training, and the convolutional layers allows for small morphological differences to exist in the same area of the manifold. This is reflected in the Char-CNN-LSTM component of the Hybrid model outperforming both a standard BiLSTM and the Word-CNN-LSTM on the WebAP collection.

Compositional Impact. To view the additional information gained by including character embeddings that is omitted from the word level networks, we evaluate the individual components designated as Word-CNN-LSTM and Char-CNN-LSTM consisting of only word and character embedding inputs respectively. Examining Word-CNN-LSTM's metrics suggests that the convolutional filters learned are somewhat noisy, resulting in lower performance compared to the LSTM only baseline as the attention component of the models are in the upper LSTM layers. However, the addition of the character component provides missing information to allow the Hybrid model to outperform all baselines, overcoming the reduced attention ability of the CNN-LSTM interaction. This compounding effect is present in both the L4 and nfl6 collection but does not pertain to the WebAP collection. As mentioned in the previous section, the small amount of training examples allows the Char-CNN-LSTM component to achieve the highest metrics regardless of the embedding origin. This discrepancy can be attributed to the training process, where the weights associated with the word models converge much faster than the character based network. As such, the lexical gap between training and test sets in WebAP is exacerbated by the reliance on the quickly converging word network despite the addition of the character network.

Local vs Pre-trained Embeddings. Viewing the results from an embedding initialization perspective, conventional word based models drop in performance when using pretrained embeddings on all collections. Updating word embeddings during training allows for a richer representation for the network to use in the hidden layers [3]. The collection least effected by this drift in information from pretrained to local embedding is the WebAP collection. We attribute this to the lower volume of training examples seen compared to the L4 and nfl6 datasets.

Unlike the BiLSTM network, implementing a convolutional layer as input over the word embeddings causes the upper layers to become less sensitive towards the type of embedding used. However, only the Hybrid model has the most consistent performance on the pretrained embeddings, significantly outperforming models using local embeddings. The Hybrid model allows for this robustness to embedding source by dynamically leveraging the character embeddings to bridge the gap between word embedding initializations. This is exemplified on noisier collections such as nfl6, where even the Word-CNN-LSTM model suffers when moving from local to pretrained embeddings.

Table 4. Performance of networks cross-trained on the yahoo CQA data and evaluated on the WebAP collection. BM25 score is included for reference. * denotes significance with $p < .05$ with respect to baselines using two-tailed t test.

Model	Trained	Transfer	P@1	MRR
BM25	N/A	WebAP	.3000	.4120
BiLSTM [21]	L4+nfl6	WebAP	.0941	.2116
Word-CNN-LSTM	L4+nfl6	WebAP	.1176	.2718
Char-CNN-LSTM	L4+nfl6	WebAP	.1602*	.2937*
Hybrid	L4+nfl6	WebAP	**.1836***	**.3115***

Table 5. A representation of the vocabulary overlap between training and correctly ranked candidates answers. Answers are labeled correct only if they are ranked first, reflecting the P@1 metric. Total Overlap is shared vocabulary across all questions, while μ and σ are mean and standard deviation of shared vocabulary with respect to individual questions.

P@1	Shared vocabulary	Char-CNN-LSTM	Word-CNN-LSTM	BiLSTM
Correct	Total Overlap	.42	.57	.59
	μ	.38	.42	.45
	σ	.08	.05	.06

Cross Collection Performance. As the Hybrid model was able to effectively handle the noisy nfl6, we investigated the performance of these models across collections. Specifically, we evaluated the ability of these models to generalize outside of the language distribution in which they were trained. The BiLSTM and Hybrid models, as well as the Word-CNN-LSTM and Char-CNN-LSTM subcomponents, were trained on both L4 and nfl6 collections and evaluated on the WebAP dataset. The results in Table 4 show that the Hybrid approach is again the most robust to retrieval tasks that significantly differ from the training collection. Vocabulary overlap between the training and evaluation set are shown in Table 5.

6   Conclusion and Future Work

This paper demonstrates the use of a CNN-LSTM model based on hybrid embedding for the complex task of answer passage retrieval. We leverage past work with character based models to bridge the gap between local and global embeddings performance for answer passage tasks. Additionally, we illustrate the advantage of incorporating a character embedding with temporal structure for collections that suffer from a small number of training examples or lower quality tokens.

We evaluated our models on three collections, Yahoo Webscope L4, nfl6, and WebAP. The Hybrid model outperforms all baselines while avoiding the use of local embeddings save for the case of WebAP, where only the character subnetwork has significantly greater performance. Incorporating character embeddings into the CNN-LSTM structure results in a model able to adapt especially well compared to a standard BiLSTM network. Given that the baselines perform well on the factoid QA or ad-hoc retrieval tasks, their relatively poor performance on answer passage retrieval demonstrates that this is a different task. While not looked at in this paper, a potential improvement to bridge the gap between word and character embeddings would be to use a gate mechanism when joining the word and character sub networks through a dynamic process.

Acknowledgments. This work was supported in part by the Center for Intelligent Information Retrieval, in part by NSF IIS-1160894 and in part by NSF grant #IIS-1419693. Any opinions, findings and conclusions or recommendations expressed in this material are those of the authors and do not necessarily reflect those of the sponsor.

References

1. Callan, J.P.: Passage-level evidence in document retrieval. In: ACM SIGIR 1994, pp. 302–310. Springer, New York (1994). https://doi.org/10.1007/978-1-4471-2099-5_31
2. Cohen, D., Ai, Q., Croft, W.B.: Adaptability of neural networks on varying granularity IR tasks. In: SIGIR Neu-IR Workshop, Pisa, Italy (2016)
3. Cohen, D., Croft, W.B.: End to end long short term memory networks for non-factoid question answering. In: ICTIR, Newark, DE, USA (2016)
4. Diaz, F., Mitra, B., Craswell, N.: Query expansion with locally-trained word embeddings. In: ACL, pp. 367–377. ACL (2016)
5. dos Santos, C.N., Tan, M., Xiang, B., Zhou, B.: Attentive pooling networks. CoRR, abs/1602.03609 (2016)
6. Graves, A., Jaitly, N., Mohamed, A.-R.: Hybrid speech recognition with deep bidirectional LSTM. In: ASRU, 2013, pp. 273–278 (2013)
7. Guo, J., Fan, Y., Ai, Q., Croft, W.B.: A deep relevance matching model for ad-hoc retrieval. In: CIKM 2016, pp. 55–64. ACM, New York (2016)
8. Harding, S.M., Croft, W.B., Weir, C.: Probabilistic retrieval of OCR degraded text using N-grams. In: Peters, C., Thanos, C. (eds.) ECDL 1997. LNCS, vol. 1324, pp. 345–359. Springer, Heidelberg (1997). https://doi.org/10.1007/BFb0026737
9. He, X., Golub, D.: Character-level question answering with attention. In: EMNLP, pp. 1598–1607. ACL (2016)
10. Huang, P.-S., He, X., Gao, J., Deng, L., Acero, A., Heck, L.: Learning deep structured semantic models for web search using clickthrough data. In: CIKM 2013, pp. 2333–2338. ACM (2013)
11. Keikha, M., Park, J., Croft, W.B.: Evaluating answer passages using summarization measures. In: SIGIR 2014, pp. 963–966 (2014)

12. Kim, Y., Jernite, Y., Sontag, D., Rush, A.M.: Character-aware neural language models. In: AAAI, pp. 2741–2749 (2016)
13. Meng, L., Li, Y., Liu, M., Shu, P.: Skipping word: a character-sequential representation based framework for question answering. In: CIKM 2016, pp. 1869–1872 (2016)
14. Miao, Y., Yu, L., Blunsom, P.: Neural variational inference for text processing. In: ICML, pp. 1727–1736 (2016). JMLR.org
15. Mikolov, T., Chen, K., Corrado, G., Dean, J.: Efficient estimation of word representations in vector space. In: ICLR Workshop, Scottsdale, AZ, USA (2013)
16. Seo, M.J., Kembhavi, A., Farhadi, A., Hajishirzi, H.: Bidirectional attention flow for machine comprehension. In: ICLR, Toulon, France (2017)
17. Severyn, A., Moschitti, A.: Learning to rank short text pairs with convolutional deep neural networks. In: SIGIR 2015, pp. 373–382. ACM, New York (2015)
18. Surdeanu, M., Ciaramita, M., Zaragoza, H.: Learning to rank answers on large online QA collections. In ACL:HLT, pp. 719–727 (2008)
19. Tan, M., Xiang, B., Zhou, B.: LSTM-based deep learning models for non-factoid answer selection. CoRR, abs/1511.04108 (2015)
20. Tielman, T., Hinton, G.: Lecture 6.5-rmsprop: Divide the Gradient by a Running Average of its Recent Magnitude (2012)
21. Wang, D., Nyberg, E.: A long short-term memory model for answer sentence selection in question answering. In: ACL-IJCNLP, ACL 2015 (Volume 2: Short Papers), pp. 707–712, July 26–31, 2015, Beijing, China (2015)
22. Wei, K., Iyer, R., Bilmes, J.: Submodularity in data subset selection and active learning. In: ICML, Lille, France (2015)
23. Yang, L., Ai, Q., Spina, D., Chen, R.-C., Pang, L., Croft, W.B., Guo, J., Scholer, F.: Beyond factoid QA: effective methods for non-factoid answer sentence retrieval. In: Ferro, N., Crestani, F., Moens, M.-F., Mothe, J., Silvestri, F., Di Nunzio, G.M., Hauff, C., Silvello, G. (eds.) ECIR 2016. LNCS, vol. 9626, pp. 115–128. Springer, Cham (2016). https://doi.org/10.1007/978-3-319-30671-1_9
24. Yin, W., Schütze, H., Xiang, B., Zhou, B.: ABCNN: attention-based convolutional neural network for modeling sentence pairs. In: TACL, vol. 4, pp. 259–272 (2016)
25. Zhang, X., Zhao, J., LeCun, Y.: Character-level convolutional networks for text classification. In: NIPS, NIPS 2015, pp. 649–657, Montreal, Canada (2015)

Deep Learning for Detecting Cyberbullying Across Multiple Social Media Platforms

Sweta Agrawal[1] and Amit Awekar[2]([⊠])

[1] Member of Technical Staff, Adobe Systems, Noida, India
sweagraw@adobe.com
[2] Indian Institute of Technology, Guwahati, Guwahati, India
awekar@iitg.ernet.in

Abstract. Harassment by cyberbullies is a significant phenomenon on the social media. Existing works for cyberbullying detection have at least one of the following three bottlenecks. First, they target only one particular social media platform (SMP). Second, they address just one topic of cyberbullying. Third, they rely on carefully handcrafted features of the data. We show that deep learning based models can overcome all three bottlenecks. Knowledge learned by these models on one dataset can be transferred to other datasets. We performed extensive experiments using three real-world datasets: Formspring (∼12k posts), Twitter (∼16k posts), and Wikipedia(∼100k posts). Our experiments provide several useful insights about cyberbullying detection. To the best of our knowledge, this is the first work that systematically analyzes cyberbullying detection on various topics across multiple SMPs using deep learning based models and transfer learning.

Keywords: Cyberbullying · Social media · Deep learning

1 Introduction

Cyberbullying has been defined by the National Crime Prevention Council as the use of the Internet, cell phones or other devices to send or post text or images intended to hurt or embarrass another person. Various studies have estimated that between to 10% to 40% of internet users are victims of cyberbullying [17]. Effects of cyberbullying can range from temporary anxiety to suicide [4]. Many high profile incidents have emphasized the prevalence of cyberbullying on social media. Most recently in October 2017, a Swedish model Arvida Byström was cyberbullied to the extent of receiving rape threats after she appeared in an advertisement with hairy legs[1].

Detection of cyberbullying in social media is a challenging task. Definition of what constitutes cyberbullying is quite subjective. For example, frequent use

---

[1] BBC News Article https://goo.gl/t6hQ7c.

© Springer International Publishing AG, part of Springer Nature 2018
G. Pasi et al. (Eds.): ECIR 2018, LNCS 10772, pp. 141–153, 2018.
https://doi.org/10.1007/978-3-319-76941-7_11

of swear words might be considered as bullying by the general population. However, for teen oriented social media platforms such as Formspring, this does not necessarily mean bullying (Table 2). Across multiple SMPs, cyberbullies attack victims on different topics such as race, religion, and gender. Depending on the topic of cyberbullying, vocabulary and perceived meaning of words vary significantly across SMPs. For example, in our experiments we found that for word 'fat', the most similar words as per Twitter dataset are 'female' and 'woman' (Table 8). However, other two datasets do not show such particular bias against women. This platform specific semantic similarity between words is a key aspect of cyberbullying detection across SMPs. Style of communication varies significantly across SMPs. For example, Twitter posts are short and lack anonymity. Whereas posts on Q&A oriented SMPs are long and have option of anonymity (Table 1). Fast evolving words and hashtags in social media make it difficult to detect cyberbullying using swear word list based simple filtering approaches. The option of anonymity in certain social networks also makes it harder to identify cyberbullying as profile and history of the bully might not be available.

Past works on cyberbullying detection have at least one of the following three bottlenecks. First (Bottleneck B1), they target only one particular social media platform. How these methods perform across other SMPs is unknown. Second (Bottleneck B2), they address only one topic of cyberbullying such as racism, and sexism. Depending on the topic, vocabulary and nature of cyberbullying changes. These models are not flexible in accommodating changes in the definition of cyberbullying. Third (Bottleneck B3), they rely on carefully handcrafted features such as swear word list and POS tagging. However, these handcrafted features are not robust against variations in writing style. In contrast to existing bottlenecks, this work targets three different types of social networks (Formspring: a Q&A forum, Twitter: microblogging, and Wikipedia: collaborative knowledge repository) for three topics of cyberbullying (personal attack, racism, and sexism) without doing any explicit feature engineering by developing deep learning based models along with transfer learning.

We experimented with diverse traditional machine learning models (logistic regression, support vector machine, random forest, naive Bayes) and deep neural network models (CNN, LSTM, BLSTM, BLSTM with Attention) using variety of representation methods for words (bag of character n-gram, bag of word unigram, GloVe embeddings, SSWE embeddings). Summary of our findings and research contributions is as follows.

– This is the first work that systematically analyzes cyberbullying on various topics across multiple SMPs and applies transfer learning for cyberbullying detection task.
– Presence of swear words is neither necessary nor sufficient for cyberbullying. Robust models for cyberbullying detection should not rely on such handcrafted features.
– Deep Learning based models outperform traditional Machine Learning models for cyberbullying detection task.

- Training datasets for cyberbullying detection contain only a few posts marked as a bullying. This class imbalance problem can be tackled by oversampling the rare class.
- The vocabulary of words used for cyberbullying and their interpretation varies significantly across SMPs.

2 Datasets

Please refer to Table 1 for summary of datasets used. We performed experiments using large, diverse, manually annotated, and publicly available datasets for cyberbullying detection in social media. We cover three different types of social networks: teen oriented Q&A forum (Formspring), large microblogging platform (Twitter), and collaborative knowledge repository (Wikipedia talk pages). Each dataset addresses a different topic of cyberbullying. Twitter dataset contains examples of racism and sexism. Wikipedia dataset contains examples of personal attack. However, Formspring dataset is not specifically about any single topic. All three datasets have the problem of class imbalance where posts labeled as cyberbullying are in the minority as compared to neutral posts. Variation in the number of posts across datasets also affects vocabulary size that represents the number of distinct words encountered in the dataset. We measure the size of a post in terms of the number of words in the post. For each dataset, there are only a few posts with large size. We truncate such large posts to the size of post ranked at 95 percentile in that dataset. For example, in Wikipedia dataset, the largest post has 2846 words. However, size of post ranked at 95 percentile in that dataset is only 231. Any post larger than size 231 in Wikipedia dataset will be truncated by considering only first 231 words. This truncation affects only a small minority of posts in each dataset. However, it is required for efficiently training various models in our experiments. Details of each dataset are as follows.

Table 1. Dataset statistics

Dataset	# Posts	Classes	Length @95%	Max length	Vocabulary size	Source
FormSpring	12k	2	62	1115	6058	[12]
Twitter	16k	3	26	38	5653	[16]
Wikipedia	100k	2	231	2846	55262	[18]

Formspring [12]: It was a question and answer based website where users could openly invite others to ask and answer questions. The dataset includes 12K annotated question and answer pairs. Each post is manually labeled by three workers. Among these pairs, 825 were labeled as containing cyberbullying content by at least two Amazon Mechanical turk workers.

Twitter [16]: This dataset includes 16K annotated tweets. The authors bootstrapped the corpus collection, by performing an initial manual search of common slurs and terms used pertaining to religious, sexual, gender, and ethnic minorities. Of the 16K tweets, 3117 are labeled as sexist, 1937 as racist, and the remaining are marked as neither sexist nor racist.

Wikipedia [18]: For each page in Wikipedia, a corresponding talk page maintains the history of discussion among users who participated in its editing. This data set includes over 100k labeled discussion comments from English Wikipedia's talk pages. Each comment was labeled by 10 annotators via Crowdflower on whether it contains a personal attack. There are total 13590 comments labeled as personal attack.

2.1   Use of Swear Words and Anonymity

Please refer to Table 2. We use the following short forms in this section: B = Bullying, S = Swearing, A = Anonymous. Some of the values for Twitter dataset are undefined as Twitter does not allow anonymous postings. Use of swear words has been repeatedly linked to cyberbullying. However, preliminary analysis of datasets reveals that depending on swear word usage can neither lead to high precision nor high recall for cyberbullying detection. Swear word list based methods will have low precision as $P(B|S)$ is not close to 1. In fact, for teen oriented social network Formspring, 78% of the swearing posts are non-bullying. Swear words based filtering will be irritating to the users in such SMPs where swear words are used casually. Swear word list based methods will also have a low recall as $P(S|B)$ is not close to 1. For Twitter dataset, 82% of bullying posts do not use any swear words. Such passive-aggressive cyberbullying will go undetected with swear word list based methods. Anonymity is another clue that is used for detecting cyberbullying as bully might prefer to hide its identity. Anonymity definitely leads to increased use of swear words ($P(S|A) \geq P(S)$) and cyberbullying ($P(B|A) \geq P(B)$, and $P(B|(A\&S)) \geq P(B)$). However, significant fraction of anonymous posts are non-bullying ($P(B|A)$ not close to 1) and many of bullying posts are not anonymous ($P(A|B)$ not close to 1). Further, anonymity might not be allowed by many SMPs such as Twitter.

Table 2. Swear word use and anonymity

Dataset	P(B)	P(S)	P(A)	P(B|S)	P(S|B)	P(B|A)	P(A|B)	P(S|A)	P(B| (A& S))
FormSpring	0.06	0.16	0.53	0.22	0.59	0.08	0.71	0.20	0.25
Twitter	0.31	0.13	-	0.42	0.18	-	-	-	-
Wikipedia	0.12	0.17	0.27	0.49	0.69	0.25	0.56	0.27	0.65

3   Related Work

Cyberbullying is recognized as a phenomenon at least since 2003 [13]. Use of social media exploded with launching of multiple platforms such as Wikipedia

(2001), MySpace (2003), Orkut (2004), Facebook (2004), and Twitter (2005). By 2006, researchers had pointed that cyberbullying was as serious phenomenon as offline bullying [10]. However, automatic detection of cyberbullying was addressed only since 2009 [19]. As a research topic, cyberbullying detection is a text classification problem. Most of the existing works fit in the following template: get training dataset from single SMP, engineer variety of features with certain style of cyberbullying as the target, apply a few traditional machine learning methods, and evaluate success in terms of measures such as F1 score and accuracy. These works heavily rely on handcrafted features such as use of swear words. These methods tend to have low precision for cyberbullying detection as handcrafted features are not robust against variations in bullying style across SMPs and bullying topics. Only recently, deep learning has been applied for cyberbullying detection [2]. Table 10 summarizes important related work.

4   Deep Neural Network (DNN) Based Models

We experimented with four DNN based models for cyberbullying detection: CNN, LSTM, BLSTM, and BLSTM with attention. These models are listed in the increasing complexity of their neural architecture and amount of information used by these models. Please refer to Fig. 1 for general architecture that we have used across four models. Various models differ only in the Neural Architecture layer while having identical rest of the layers. CNNs are providing state-of-the-results on extracting contextual feature for classification tasks in images, videos, audios, and text. Recently, CNNs were used for sentiment classification [7]. Long Short Term Memory networks are a special kind of RNN, capable of learning long-term dependencies. Their ability to use their internal memory to process arbitrary sequences of inputs has been found to be effective for text classification [5]. Bidirectional LSTMs [20] further increase the amount of input information available to the network by encoding information in both forward and backward direction. By using two directions, input information from both the past and future of the current time frame can be used. Attention mechanisms allow for a more direct dependence between the state of the model at different points in time. Importantly, attention mechanism lets the model learn what to attend to based on the input sentence and what it has produced so far.

The embedding layer processes a fixed size sequence of words. Each word is represented as a real-valued vector, also known as word embeddings. We have experimented with three methods for initializing word embeddings: random, GloVe [11], and SSWE [14]. During the training, model improves upon

Fig. 1. Model architecture

the initial word embeddings to learn task specific word embeddings. We have observed that these task specific word embeddings capture the SMP specific and topic specific style of cyberbullying. Using GloVe vectors over random vector initialization has been reported to improve performance for some NLP tasks. Most of the word embedding methods such as GloVe, consider only syntactic context of the word while ignoring the sentiment conveyed by the text. SSWE method overcomes this problem by incorporating the text sentiment as one of the parameters for word embedding generation. We experimented with various dimension size for word embeddings. Experimental results reported here are with dimension size as 50. There was no significant variation in results with dimension size ranging from 30 to 200.

To avoid overfitting, we used two dropout layers, one before the neural architecture layer and one after, with dropout rates of 0.25 and 0.5 respectively. Fully connected layer is a dense output layer with the number of neurons equal to the number of classes, followed by softmax layer that provides softmax activation. All our models are trained using backpropagation. The optimizer used for training is Adam and the loss function is categorical cross-entropy. Besides learning the network weights, these methods also learn task-specific word embeddings tuned towards the bullying labels (See Sect. 5.3). Our code is available at: https:// github.com/sweta20/Detecting-Cyberbullying-Across-SMPs.

5 Experiments

Existing works have heavily relied on traditional machine learning models for cyberbullying detection. However, they do not study the performance of these models across multiple SMPs. We experimented with four models: logistic regression (LR), support vector machine (SVM), random forest (RF), and naive Bayes (NB), as these are used in previous works (Table 10). We used two data representation methods: character n-gram and word unigram. Past work in the domain of detecting abusive language have showed that simple n-gram features are more powerful than linguistic and syntactic features, hand-engineered lexicons, and word and paragraph embeddings [9]. As compared to DNN models, performance of all four traditional machine learning models was significantly lower. Please refer to Table 3.

Table 3. Results for traditional ML models using F1 score

Dataset	Label	Character n-grams				Word unigrams			
		LR	SVM	RF	NB	LR	SVM	RF	NB
Formspring	Bully	0.448	0.422	0.298	0.359	0.489	0.463	0.264	0.025
Twitter	Racism	0.723	0.676	0.752	0.686	0.738	0.772	0.739	0.617
	Sexism	0.729	0.688	0.720	0.647	0.762	0.758	0.755	0.635
Wiki	Attack	0.694	0.677	0.674	0.655	0.711	0.686	0.730	0.659

All DNN models reported here were implemented using Keras. We pre-process the data, subjecting it to standard operations of removal of stop words, punctuation marks and lowercasing, before annotating it to assigning respective labels to each comment. For each trained model, we report its performance after doing five-fold cross-validation. We use following short forms.

- Datasets: F (Formspring), T (Twitter), W (Wikipedia)
- Datasets with oversampling of bullying posts: F+ (Formspring), T+ (Twitter), W+ (Wikipedia)
- Evaluation measures: P (Precision), R (Recall), F1 (F1 score)
- DNN Models: M1 (CNN), M2 (LSTM), M3 (BLSTM), M4 (BLSTM with attention)

5.1 Effect of Oversampling Bullying Instances

The training datasets had a major problem of class imbalance with posts marked as bullying in the minority. As a result, all models were biased towards labeling the posts as non-bullying. To remove this bias, we oversampled the data from bullying class thrice. That is, we replicated bullying posts thrice in the training data. This significantly improved the performance of all DNN models with major leap in all three evaluation measures. Table 4 shows the effect of oversampling for a variety of word embedding methods with BLSTM Attention as the detection model. Results for other models are similar [1]. We can notice that oversampled datasets (F+, T+, W+) have far better performance than their counterparts (F, T, W respectively). Oversampling particularly helps the smallest dataset Formspring where number of training instances for bullying class is quite small (825) as compared to other two datasets (about 5K and 13K). We also experimented with varying the replication rate for bullying posts [1]. However, we observed that for bullying posts, replication rate of three is good enough.

Table 4. Effect of oversampling bullying posts using BLSTM with attention

Dataset	Label	P			R			F1		
		Random	Glove	SSWE	Random	Glove	SSWE	Random	Glove	SSWE
F	Bully	0.52	0.56	0.63	0.40	0.49	0.38	0.44	0.51	0.47
F+	Bully	0.84	0.85	0.90	0.98	0.97	0.91	0.90	0.90	0.91
T	Racism	0.67	0.74	0.76	0.73	0.76	0.77	0.70	0.75	0.76
T+	Racism	0.94	0.90	0.90	0.98	0.95	0.96	0.96	0.93	0.93
T	Sexism	0.65	0.86	0.83	0.64	0.52	0.47	0.65	0.65	0.59
T+	Sexism	0.88	0.95	0.88	0.97	0.91	0.92	0.93	0.91	0.90
W	Attack	0.77	0.81	0.82	0.74	0.67	0.68	0.76	0.74	0.74
W+	Attack	0.81	0.86	0.87	0.91	0.89	0.86	0.88	0.88	0.87

5.2   Choice of Initial Word Embeddings and Model

Initial word embeddings decide data representation for DNN models. However during the training, DNN models modify these initial word embeddings to learn task specific word embeddings. We have experimented with three methods to initialize word embeddings. Please refer to Table 5. This table shows the effect of varying initial word embeddings for multiple DNN models across datasets. We can notice that initial word embeddings do not have a significant effect on cyberbullying detection when oversampling of bullying posts is done (rows corresponding to F+, T+, W+). In the absence of oversampling (rows corresponding to F, T W), there is a gap in performance of simplest (CNN) and most complex (BLSTM with attention) models. However, this gap goes on reducing with the increase in the size of datasets.

Table 5. Effect of choosing initial word embedding method on F1 score

Dataset	Label	Random		Glove		SSWE	
		M1	M4	M1	M4	M1	M4
F	Bully	0.30	0.44	0.34	0.51	0.34	0.47
F+	Bully	0.91	0.90	0.93	0.90	0.91	0.91
T	Racism	0.68	0.70	0.73	0.75	0.70	0.76
T+	Racism	0.90	0.96	0.95	0.93	0.93	0.93
T	Sexism	0.59	0.65	0.61	0.65	0.63	0.59
T+	Sexism	0.93	0.93	0.93	0.91	0.92	0.90
W	Attack	0.72	0.76	0.72	0.74	0.74	0.74
W+	Attack	0.83	0.88	0.89	0.88	0.88	0.87

Table 6 compares the performance of four DNN models for three evaluation measures while using SSWE as the initial word embeddings. We have noticed that most of the time LSTM performs weaker than other three models. However, performance gap in the other three models is not significant.

Table 6. Performance comparison of various DNN models

Dataset	Label	P				R				F1			
		M1	M2	M3	M4	M1	M2	M3	M4	M1	M2	M3	M4
F+	Bully	0.93	0.91	0.91	0.90	0.90	0.85	0.81	0.91	0.91	0.88	0.86	0.91
T+	Racism	0.93	0.91	0.92	0.90	0.94	0.80	0.95	0.96	0.93	0.85	0.93	0.93
	Sexism	0.92	0.84	0.88	0.88	0.92	0.93	0.94	0.92	0.92	0.88	0.92	0.90
W+	Attack	0.92	0.70	0.90	0.87	0.83	0.54	0.81	0.86	0.88	0.61	0.85	0.87

5.3   Task Specific Word Embeddings

DNN models learn word embeddings over the training data. These learned embeddings across multiple datasets show the difference in nature and style of bullying across cyberbullying topics and SMPs. Here we report results for BLSTM with attention model. Results for other models are similar. We first verify that important words for each topic of cyberbullying form clusters in the learned embeddings. To enable the visualization of grouping, we reduced dimensionality with t-SNE [8], a well-known technique for dimensionality reduction particularly well suited for visualization of high dimensional datasets. Please refer to Table 7. This table shows important clusters observed in t-SNE projection of learned word embeddings. Each cluster shows that words most relevant to a particular topic of bullying form cluster.

Table 7. Embeddings learned using DNNs

Bullying form	Observed cluster
Sexism	Kitchen, feminist, feminists, its, feminism, girl, rights, two, female, bitch, head, sexist, woman, girls, blondes, rape
Racism	Pedophile, murdered, either, israel, mohammed, slave, prophet, muslims, quran, may, islam, religion, war, pay
Personal attack	Fuck, fucking, u, little, you, shit, style, faggot, ass, off, changes, suck, see, hate, know, nigger, moron, site

We also observed changes in the meanings of the words across topics of cyberbullying. Table 8 shows most similar words for a given query word for two datasets. Twitter dataset which is heavy on sexism and racism, considers word slave as similar to targets of racism and sexism. However, Wikipedia dataset that is about personal attacks does not show such bias.

Table 8. Most similar words to the query word across platform

Query word	Similar words	
	Twitter	Wiki
Slave	Feminists, religion, jews, islam, muslims, christians	Sucks, bad, blocked, tried, cannot, can't, didn't, never
Evidence	God, opinion, eliminated, opinions, murdered, racist, raped	Interested, suggest, yes, love, good, happy, quote, note, useful
Fat	Female, woman, face, women, kids, fan, blonde, friends	Blocked, sorry, bad, used, tried, cannot, banned, never, fuck
Gay	Die, ask, fake, child, babies, females, wife, female, woman	Bad, sorry, used, blocked, tried, fuck, fucking, that's, notice, shit

5.4   Transfer Learning

We used transfer learning to check if the knowledge gained by DNN models on one dataset can be used to improve cyberbullying detection performance on other datasets. We report results where BLSTM with attention is used as the DNN model. Results for other models are similar [1]. We experimented with following three flavors of transfer learning.

Complete Transfer Learning (TL1): In this flavor, a model trained on one dataset was directly used to detect cyberbullying in other datasets without any extra training. TL1 resulted in significantly low recall indicating that three datasets have different nature of cyberbullying with low overlap (Table 9). However precision was relatively higher for TL1, indicating that DNN models are cautious in labeling a post as bully (Table 9). TL1 also helps to measure similarity in nature of cyberbullying across three datasets. We can observe that bullying nature in Formspring and Wikipedia datasets is more similar to each other than the Twitter dataset. This can be inferred from the fact that with TL1, cyberbullying detection performance for Formspring dataset is higher when base model is Wikipedia (precision = 0.51 and recall = 0.66)as compared to Twitter as the base model (precision = 0.38 and recall = 0.04). Similarly, for Wikipedia dataset, Formspring acts as a better base model than Twitter while using TL1 flavor of transfer learning. Nature of SMP might be a factor behind this similarity in nature of cyberbullying. Both Formspring and Wikipedia are task oriented social networks (Q&A and collaborative knowledge repository respectively) that allow anonymity and larger posts. Whereas communication on Twitter is short, free of anonymity and not oriented towards a particular task.

Feature Level Transfer Learning (TL2): In this flavor, a model was trained on one dataset and only learned word embeddings were transferred to another dataset for training a new model. As compared to TL1, recall score improved

Table 9. Comparison of transfer learning methods using precision

Metric	Test	Train								
		F+			T+			W+		
		TL1	TL2	TL3	TL1	TL2	TL3	TL1	TL2	TL3
Precision	F	-	-	-	0.38	0.90	0.88	0.51	0.92	0.85
	T	0.83	0.88	0.90	-	-	-	0.72	0.91	0.90
	W	0.82	0.92	0.91	0.68	0.90	0.91	-	-	-
Recall	F	-	-	-	0.04	0.98	0.98	0.66	0.98	0.99
	T	0.01	0.99	0.99	-	-	-	0.17	0.98	0.99
	W	0.21	0.96	0.96	0.05	0.97	0.96	-	-	-
F1-score	F	-	-	-	0.07	0.95	0.93	0.58	0.95	0.92
	T	0.03	0.93	0.94	-	-	-	0.28	0.94	0.94
	W	0.35	0.94	0.94	0.10	0.94	0.94	-	-	-

Table 10. Summary of related work

Paper	Year	Target SMPs	Data features	Model used	Cyberbullying topics	Bottlenecks	Dataset size	Metric used	Metric value
[6]	2011	Youtube	TF-IDF, list of swear words etc.	Naive bayes, SVM, J48, JRip	Sexuality, Race and culture, Intelligence	B1, B3	4500	Accuracy	Sexuality(80.20%) Intelligence(70.39%) Race(68.30%)
[12]	2011	Formspring	Number of "bad" words (NUM), density of "bad" words (NORM)	SMO, IBK, J48, JRip	Bully/ Not bully	B1, B2, B3	3915	Accuracy	78.5%
[3]	2015	Yahoo	Distributed representations of comments(paragraph2vec)	Continuous BOW (CBOW)	Hate Speech/ Clean	B1, B2	~951k	AUC	0.80
[15]	2015	Ask.fm	Word unigram and bigram bags-of-words, character trigram bag-of-words, sentiment lexicon features(comment2vec)	SVM	Threat, Insult, Defense, Sexual Talk, Defamation, Encouragements and Swear	B1, B3	~91k	F1-score	55.39
[9]	2016	Yahoo	Word and character N-grams, Linguistic features, Syntactic and Distributional Semantics	Vowpal Wabbits regression model	Abusive/ Non Abusive	B1, B2, B3	Finance (~759k), News (~1390k)	AUC	0.90
[2]	2017	Twitter	TF-IDF values, BoWV over Global Vectors, task-specific embeddings	FastText, CNNs, LSTMs, GBDT	Sexism, Racism, None	B1	16k	F1-score	0.93
[18]	2017	Wikipedia	char-n grams, word n-grams	LR, MLP	Personal Attack	B1, B2	100k	AUC	96.59

dramatically with TL2 (Table 9). Improvement in precision was also significant (Table 9). These improvements indicate that learned word embeddings are an essential part of knowledge transfer across datasets for cyberbullying detection.

Model Level Transfer Learning (TL3): In this flavor, a model was trained on one dataset and learned word embeddings, as well as network weights, were transferred to another dataset for training a new model. TL3 does not result in any significant improvement over TL2. This lack of improvement indicates that transfer of network weights is not essential for cyberbullying detection and learned word embeddings is the key knowledge gained by the DNN models.

DNN based models coupled with transfer learning beat the best-known results for all three datasets. Previous best F1 scores for Wikipedia [18] and Twitter [2] datasets were 0.68 and 0.93 respectively. We achieve F1 scores of 0.94 for both these datasets using BLSTM with attention and feature level transfer learning (Table 9). For Formspring dataset, authors have not reported F1 score. Their method has accuracy score of 78.5% [12]. We achieve F1 score of 0.95 with accuracy score of 98% for the same dataset.

5.5   Conclusion and Future Work

We have shown that DNN models can be used for cyberbullying detection on various topics across multiple SMPs using three datasets and four DNN models. These models coupled with transfer learning beat state of the art results for all three datasets. These models can be further improved with extra data such as information about the profile and social graph of users. Most of the current datasets do not provide any information about the severity of bullying. If such fine-grained information is made available, then cyberbullying detection models can be further improved to take a variety of actions depending on the perceived seriousness of the posts.

References

1. More experimental results. https://goo.gl/BBFxYH
2. Badjatiya, P., Gupta, S., Gupta, M., Varma, V.: Deep learning for hate speech detection in tweets. In: WWW, pp. 759–760 (2017)
3. Djuric, N., Zhou, J., Morris, R., Grbovic, M., Radosavljevic, V., Bhamidipati, N.: Hate speech detection with comment embeddings. In: WWW, pp. 29–30 (2015)
4. Hinduja, S., Patchin, J.W.: Bullying, cyberbullying, and suicide. Arch. Suicide Res. **14**(3), 206–221 (2010)
5. Johnson, R., Zhang, T.: Supervised and semi-supervised text categorization using LSTM for region embeddings. In: ICML, pp. 526–534 (2016)
6. Karthik, D., Roi, R., Henry, L.: Modeling the detection of textual cyberbullying. In: Workshop on the Social Mobile Web, ICWSM (2011)
7. Kim, Y.: Convolutional neural networks for sentence classification. In: EMNLP, pp. 1746–1751 (2014)
8. van der Maaten, L., Hinton, G.: Visualizing data using t-SNE. J. Mach. Learn. Res. **9**(Nov), 2579–2605 (2008)

9. Nobata, C., Tetreault, J., Thomas, A., Mehdad, Y., Chang, Y.: Abusive language detection in online user content. In: WWW, pp. 145–153 (2016)
10. Patchin, J.W., Hinduja, S.: Bullies move beyond the schoolyard: a preliminary look at cyberbullying. Youth Violence Juvenile Justice 4(2), 148–169 (2006)
11. Pennington, J., Socher, R., Manning, C.D.: Glove: global vectors for word representation. In: EMNLP, pp. 1532–1543 (2014)
12. Reynolds, K., Kontostathis, A., Edwards, L.: Using machine learning to detect cyberbullying. In: ICMLA, pp. 241–244 (2011)
13. Servance, R.L.: Cyberbullying, cyber-harassment, and the conflict between schools and the first amendment. Wis. Law Rev. 12–13 (2003)
14. Tang, D., Wei, F., Yang, N., Zhou, M., Liu, T., Qin, B.: Learning sentiment-specific word embedding for twitter sentiment classification. In: ACL, pp. 1555–1565 (2014)
15. Van Hee, C., Lefever, E., Verhoeven, B., Mennes, J., Desmet, B., De Pauw, G., Daelemans, W., Hoste, V.: Automatic detection and prevention of cyberbullying. In: International Conference Human and Social Analytics, pp. 13–18 (2015)
16. Waseem, Z., Hovy, D.: Hateful symbols or hateful people? Predictive features for hate speech detection on twitter. In: NAACL SRW, pp. 88–93 (2016)
17. Whittaker, E., Kowalski, R.M.: Cyberbullying via social media. J. Sch. Violence 14(1), 11–29 (2015)
18. Wulczyn, E., Thain, N., Dixon, L.: Ex machina: personal attacks seen at scale. In: WWW, pp. 1391–1399 (2017)
19. Yin, D., Xue, Z., Hong, L., Davison, B.D., Kontostathis, A., Edwards, L.: Detection of harassment on web 2.0. In: The Workshop on Content Analysis in the WEB 2.0, WWW, pp. 1–7 (2009)
20. Zhou, P., Qi, Z., Zheng, S., Xu, J., Bao, H., Xu, B.: Text classification improved by integrating bidirectional LSTM with two-dimensional max pooling. In: COLING, pp. 3485–3495 (2016)

Affective Neural Response Generation

Nabiha Asghar[1](✉)(iD), Pascal Poupart[1], Jesse Hoey[1], Xin Jiang[2], and Lili Mou[1]

[1] Cheriton School of Computer Science, University of Waterloo, Waterloo, Canada
{nasghar,ppoupart,jhoey}@cs.uwaterloo.ca,doublepower.mou@gmail.com
[2] Noah's Ark Lab, Huawei Technologies, Sha Tin, Hong Kong
xin.jiang@huawei.com

Abstract. Existing neural conversational models process natural language primarily on a lexico-syntactic level, thereby ignoring one of the most crucial components of human-to-human dialogue: its affective content. We take a step in this direction by proposing three novel ways to incorporate affective/emotional aspects into long short term memory (LSTM) encoder-decoder neural conversation models: (1) affective word embeddings, which are cognitively engineered, (2) affect-based objective functions that augment the standard cross-entropy loss, and (3) affectively diverse beam search for decoding. Experiments show that these techniques improve the open-domain conversational prowess of encoder-decoder networks by enabling them to produce more natural and emotionally rich responses.

Keywords: Dialogue systems · Human computer interaction
Natural language processing · Affective computing

1 Introduction

Human-computer dialogue systems have wide applications ranging from restaurant booking [24] to emotional virtual agents [13]. In a neural network-based dialogue system, discrete words are mapped to real-valued vectors, known as *embeddings*, capturing abstract meanings of words [14]; then an encoder-decoder framework—with long short term memory (LSTM)-based recurrent neural networks (RNNs)—generates a response conditioned on one or several previous utterances. Recent advances in this direction have demonstrated its efficacy for both task-oriented [24] and open-domain dialogue generation [11].

While most of the existing neural conversation models generate syntactically well-formed responses, they are prone to being short, dull, or vague. Latest efforts to address these issues include diverse decoding [22], diversity-promoting objective functions [10], human-in-the-loop reinforcement/active learning [1,11] and content-introducing approaches [15]. However, one shortcoming of these existing open-domain neural conversation models is the lack of *affect* modeling of natural language. These models, when trained over large dialogue datasets, do not capture the emotional states of the two humans interacting

G. Pasi et al. (Eds.): ECIR 2018, LNCS 10772, pp. 154–166, 2018.
https://doi.org/10.1007/978-3-319-76941-7_12

in the textual conversation, which are typically manifested through the choice of words or phrases. For instance, the attention mechanism in a sequence-to-sequence (Seq2Seq) model can learn syntactic alignment of words within the generated sequences [2]. Also, neural word embedding models like Word2Vec learn word vectors by context, and can preserve low-level word semantics (e.g., "king" – "male" ≈ "queen" – "woman"). However, emotional aspects are not explicitly captured by existing methods.

Our goal is to alleviate this issue in open-domain neural dialogue models by augmenting them with affective intelligence. We do this in three ways.

1. We embed words in a 3D affective space by retrieving word-level affective ratings from a cognitively engineered affective dictionary [23], where affectively similar constructs are close to one other. In this way, the ensuing neural model is aware of words' emotional features.
2. We augment the standard cross-entropy loss with affective objectives, so that our neural models are taught to generate more emotional utterances.
3. We inject affective diversity into the responses generated by the decoder through *affectively diverse* beam search algorithms, and thus our model actively searches for affective responses during decoding.

We also show that these emotional aspects can be combined to further improve the quality of generated responses in an open-domain dialogue system. Overall, in information-retrieval tasks like question-answering, our proposed models can help retain the users by interacting in a more human way.

2 Related Work

Affectively cognizant virtual agents are attracting interest both in the academia [13] and the industry,[1] due to their ability to provide emotional companionship to humans. Past research has mostly focused on developing hand-crafted speech and text-based features to incorporate emotions in retrieval-based or slot-based spoken dialogue systems [3,18]. Our work is related to two very recent studies:

- Affect Language Model [6, Affect-LM] is an LSTM-RNN language model which leverages the LIWC [17] text analysis program for affective feature extraction through keyword spotting. It considers binary affective features, namely *positive emotion, angry, sad, anxious*, and *negative emotion*. Our work differs from Affect-LM in that we consider affective dialogue systems instead of merely language models.
- Emotional Chatting Machine [26, ECM] is a Seq2Seq model. It takes as input a prompt and the desired emotion of the response, and produces a response. It has 8 emotion categories, namely *anger, disgust, fear, happiness, like, sadness, surprise*, and *other*. Our approach does not require the input of desired emotion as in ECM, which is unrealistic in applications. Instead, we intrinsically model emotion by affective word embeddings as input, as well as objective functions and inference criterion based on these embeddings.

---

[1] https://www.ald.softbankrobotics.com/en/robots/pepper.

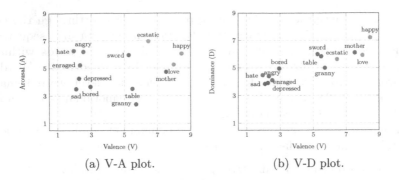

(a) V-A plot.   (b) V-D plot.

Fig. 1. Relationship between several adjectives, nouns, and verbs on 3-D VAD scale.

3   Background

In NLP, **word embeddings** map words (or tokens) to real-valued vectors of fixed dimensionality. Typically, they are learned from the co-occurrence statistics of words in large natural language corpora, and the learned embedding vector space has such a property that words sharing similar syntactic and semantic context are close to each other. However, it is known that co-occurrence statistics are insufficient to capture sentiment/emotional features, because words different in sentiment often share context (e.g., "a *good* book" vs. "a *bad* book").

A **sequence-to-sequence (Seq2Seq) model** is an encoder-decoder neural framework that maps a variable length input sequence to a variable length output sequence [21]. It consists of an encoder and a decoder, both of which are RNNs (typically with LSTM units). The encoder network sequentially accepts the embedding of each word in the input sequence, and encodes the input sentence as a vector. The decoder network takes the vector as input and sequentially generates an output sequence. Given a message-response pair (X, Y), where $X = x_1, \cdots, x_m$ and $Y = y_1, \cdots, y_n$ are sequences of words, Seq2Seq models (parametrized by θ) are typically trained with cross entropy loss (XENT):

$$L_{\text{XENT}}(\theta) = -\log p(Y|X) = -\sum_{i=1}^{n} \log p(y_i|y_1, \cdots, y_{i-1}, X), \qquad (1)$$

4   The Proposed Affective Approaches

In this section, we propose affective neural response generation, which augments traditional neural conversation models with emotional cognizance. We leverage a cognitively engineered dictionary to propose three strategies for affective response generation, namely affective word embeddings as input, affective training objectives, and affectively diverse beam search. As will be shown later, these affective strategies can be combined to further improve Seq2Seq dialogue systems.

4.1   Affective Word Embeddings

As said, traditional word embeddings trained with co-occurrence statistics are insufficient to capture affect aspects. We propose to augment traditional word embeddings with a 3D affective space by using an external cognitively-engineered affective dictionary [23].[2] The dictionary we use consists of 13,915 lemmatized English words, each of which is rated on three traditionally accepted continuous and real-valued dimensions of emotion: Valence (V, the pleasantness of a stimulus), Arousal (A, the intensity of emotion produced by a stimulus), and Dominance (D, the degree of power exerted by a stimulus). Sociologists hypothesize that the VAD space captures almost 70% of the variance in affective meanings of concepts [16]. VAD ratings have been previously used in sentiment analysis and empathetic tutors, among other affective computing applications [8,19]. To the best of our knowledge, we are the first to introduce VAD to dialogue systems.

The scale of each dimension in the VAD space is from 1 to 9, where a higher value corresponds to higher valence, arousal, or dominance. Thus, V $\simeq 1, 5$ and 9 corresponds to a word being very negative (*pedophile*), neutral (*tablecloth*) and very positive (*happiness*), respectively. Similarly, A $\simeq 1, 5$ and 9 corresponds to a word having very low (*dull*), moderate (*watchdog*), and very high (*insanity*) emotional intensity, respectively. Finally, D $\simeq 1, 5$ and 9 corresponds to a word that is very powerless (*dementia*), neutral (*waterfall*) and very powerful (*paradise*), respectively. The VAD ratings of each word were collected through a survey in [23] over 1800 participants. We directly take them as the 3-dimensional word-level affective embeddings. Some examples of words (including nouns, adjectives, and verbs) and their corresponding VAD values are depicted in Fig. 1.

For words missing in this dictionary, such as stop words and proper nouns, we set the VAD vector to be the neutral vector $\boldsymbol{\eta} = [5, 1, 5]$, because these words are neutral in pleasantness (V) and power (D), and evoke no arousal (A). Formally, we define "*word to affective vector*" (W2AV) as:

$$\texttt{W2AV}(w) = \begin{cases} \text{VAD}(l(w)), & \text{if } l(w) \in dict \\ \boldsymbol{\eta} = [5, 1, 5], & \text{otherwise} \end{cases} \tag{2}$$

where $l(w)$ is the lemmatization of the word w. In this way, words depicting similar emotions are close together in the affective space, and affectively dissimilar words are far apart from each other. Thus W2AV is suitable for neural processing.

The simplest approach to utilize W2AV, perhaps, is to feed it to a Seq2Seq model as input. Concretely, we concatenate the W2AV embeddings of each word with its traditional word embeddings, the resulting vector being the input to both the encoder and the decoder.

4.2   Affective Loss Functions

Equipped with affective vectors, we further design affective training loss functions to explicitly train an affect-aware Seq2Seq conversation model. The philosophy

---

[2] Available for free at http://crr.ugent.be/archives/1003.

of manipulating loss function is similar to [10], but we focus on affective aspects (instead of diversity in general). We have several heuristics as follows.

Minimizing Affective Dissonance. We start with the simplest approach: maintaining affective consistency between prompts and responses. This heuristic arises from the observation that typical open-domain textual conversations between two humans consist of messages and responses that, in addition to being affectively loaded, are affectively similar to each other. For instance, a friendly message typically elicits a friendly response and provocation usually results in anger or contempt. Assuming that the general affective tone of a conversation does not fluctuate too suddenly and too frequently, we emulate human-human interactions in our model by minimizing the *dissonance* between the prompts and the responses, i.e. the Euclidean distance between their affective embeddings. This objective allows the model to generate responses that are emotionally aligned with the prompts. Thus, at time step i, the loss is computed by

$$L_{\mathrm{DMIN}}^i(\theta) = -(1-\lambda)\log p(y_i|y_1,\cdots,y_{i-1},X) + \lambda\,\hat{p}(y_i)\left\|\sum_{j=1}^{|X|}\frac{\mathrm{W2AV}(x_j)}{|X|} - \sum_{k=1}^{i}\frac{\mathrm{W2AV}(y_k)}{i}\right\|_2 \quad (3)$$

where $\|\cdot\|_2$ denotes ℓ_2-norm. The first term is the standard XENT loss as in Eq. 1. The sum $\sum_j \frac{\mathrm{W2AV}(x_j)}{|X|}$ is the average affect vector of the source sentence, whereas $\sum_k \frac{\mathrm{W2AV}(y_k)}{i}$ is the average affect vector of the target sub-sentence generated up to the current time step i.

In other words, we penalize the distance between the average affective embeddings of the source and the target sentences. Notice that this affect distance is not learnable and that selecting a single predicted word makes the model indifferentiable. Therefore, we relax hard prediction of a word by its predicted probability $\hat{p}(y_i)$. λ is a hyperparameter balancing the two factors.

Maximizing Affective Dissonance. Admittedly, minimizing the affective dissonance does not always make sense while we model a conversation. An over-friendly message from a stranger may elicit anger or disgust from the recipient. Furthermore, responses that are *not* too affectively aligned with the prompts may be perceived as more interesting, by virtue of being less predictable. Thus, we design an objective function L_{DMAX} that *maximizes* the dissonance by flipping the sign in the second term in Eq. 3. (Details are not repeated here.)

Maximizing Affective Content. Our third heuristic encourages Seq2Seq to generate affective content, but does not specify the polarity of sentiment. This explores the hypothesis that most of the casual human responses are not dull or emotionally neutral. Concretely, we maximize the affective content of the model's responses, so that it avoids generating generic responses like *"yes," "no," "I don't know,"* and *"I'm not sure."* That is, at the time step i, the loss function is

$$L_{\mathrm{AC}}^i(\theta) = -(1-\lambda)\log p(y_i|y_1,\cdots,y_{i-1},X) - \lambda\,\hat{p}(y_i)\left\|\mathrm{W2AV}(y_i) - \boldsymbol{\eta}\right\|_2 \quad (4)$$

The second term is a regularizer that discourages non-affective words. We penalize the distance between y_i's affective embedding and the affectively neutral vector $\boldsymbol{\eta} = [5, 1, 5]$, so the model pro-actively chooses emotionally rich words.

4.3   Affectively Diverse Decoding

In this subsection, we propose affectively diverse decoding that incorporates affect into the decoding process of neural response generation.

Traditionally, beam search (BS) has been used for decoding in Seq2Seq models because it provides a tractable approximation of searching an exponentially large solution space. However, in the context of open-domain dialogue generation, BS is known to produce nearly identical samples like *"This is great!"* and *"This is so great!"*. Diverse beam search (DBS) [22] is a recently proposed variant of BS that explicitly considers diversity during decoding; it has been shown to outperform BS and other diverse decoding techniques in many NLP tasks.

Below, we describe BS, DBS, and our proposed affective variants of DBS.

Beam Search (BS) and Diverse Beam Search (DBS). BS maintains top-B most likely (sub)sequences, where B is known as the *beam size*. At each time step t, the top-B subsequences at time step $t - 1$ are augmented with all possible actions available; then the top-B most likely branches are retained at time t, and the rest are pruned. Let V be the set of vocabulary tokens, X be the input sequence, $\mathbf{y}_{i,[t-1]}$ be the ith beam stored at time $t - 1$, and $Y_{[t-1]} = \{\mathbf{y}_{1,[t-1]}, \cdots, \mathbf{y}_{B,[t-1]}\}$ be the set of beams stored by BS at time $t - 1$. Then at time t, the BS objective is

$$Y_{[t]} = y_{1..t}^{1^*}, \cdots, y_{1..t}^{B^*} = \underset{\substack{\mathbf{y}_{1,[t]}, \cdots, \mathbf{y}_{B,[t]} \\ \in Y_{[t-1]} \times V}}{\arg\max} \sum_{b=1}^{B} \sum_{i=1}^{t} \log p(y_{b,i}|\mathbf{y}_{b,[i-1]}, X) \quad (5)$$

subject to $\mathbf{y}_{i,[t]} \neq \mathbf{y}_{j,[t]}$, where $Y_{[t-1]} \times V$ is the set of all possible extensions based on the beams stored at time $t - 1$. DBS aims to overcome the diversity problem in BS by incorporating diversity among candidate outputs. It divides the top-B beams into G groups (each group containing $B' = G/B$ beams) and adds to traditional BS (Eq. 5) a dissimilarity term $\Delta(Y_{[t]}^1, \cdots, Y_{[t]}^{g-1})[y_t]$ which measures the dissimilarity between group g and previous groups $1, \cdots, g - 1$ if token y_t is selected to extend any beam in group g. This is given by

$$Y_{[t]}^g = \underset{\substack{\mathbf{y}_{1,[t]}^g, \cdots, \mathbf{y}_{B',[t]}^g \\ \in Y_{[t-1]}^g \times V}}{\arg\max} \sum_{b=1}^{B'} \sum_{i=1}^{t} \log p(y_{b,i}^g|\mathbf{y}_{b,[i-1]}^g, X) + \lambda_g \Delta(Y_{[t]}^1, \cdots, Y_{[t]}^{g-1})[y_{b,t}^g] \quad (6)$$

subject to $\mathbf{y}_{i,[t]}^g \neq \mathbf{y}_{j,[t]}^g$, where $\lambda_g \geq 0$ is a hyperparameter controlling the diversity strength. Intuitively, DBS modifies the probability in BS by adding a dissimilar term between a particular sample (i.e., $y_{b,1}^g \cdots y_{b,t}^g$) and samples in other groups (i.e., $Y_{[t]}^1, \cdots, Y_{[t]}^{g-1}$). We refer readers to [22] for the details of DBS. Here, we focus on the dissimilarity metric that can incorporate affective aspects into the decoding phase.

Affectively Diverse Beam Search (ADBS). The dissimilarity metric for DBS can take many forms as used in [22]: Hamming diversity that penalizes tokens based on the number of times they are selected in the previous groups, n-gram diversity that discourages repetition of n-grams between groups, and neural-embedding diversity that penalizes words with similar embeddings across groups. Among these, the neural-embedding diversity metric is the most relevant to us. When used with Word2Vec embeddings, this metric discourages semantically similar words (e.g., synonyms) to be selected across different groups.

To decode affectively diverse samples, we propose to inject affective dissimilarity across the beam groups based on affective word embeddings. This can be done either at the word level or sentence level. We formalize these notions below.

- *Word-Level Diversity for ADBS (WL-ADBS).* We define the word-level affect dissimilarity metric Δ_W to be

$$\Delta_W(Y_{[t]}^1, \cdots, Y_{[t]}^{g-1})[y_{b,t}^g] = -\sum_{j=1}^{g-1}\sum_{c=1}^{B'} \mathrm{sim}\left(\mathtt{W2AV}(y_{b,t}^g), \mathtt{W2AV}(y_{c,t}^j)\right) \quad (7)$$

where $\mathrm{sim}(\cdot)$ denotes a similarity measure between two vectors. In our experiments, we use the cosine similarity function. $y_{b,t}^g$ denotes the token under consideration at the current time step t for beam b in group g, and $y_{c,t}^j$ denotes the token chosen for beam c in a previous group j at time t.

Intuitively, this metric computes the cosine similarity of group g's beam b with all the beams generated in groups $1, \cdots, g-1$. The metric operates at the word level, ensuring that the word affect at time t is diversified across groups.

- *Sentence-Level Diversity for ADBS (SL-ADBS).* The word-level metric Δ_W in Eq. 7 does not take into account the overall sentence affect for each group. We propose an alternative sentence-level affect diversity metric, given by

$$\Delta_S(Y_{[t]}^1, \cdots, Y_{[t]}^{g-1})[y_{b,t}^g] = -\sum_{j=1}^{g-1}\sum_{c=1}^{B'} \mathrm{sim}\left(\Psi(\mathbf{y}_{b,[t]}^g), \Psi(\mathbf{y}_{c,[t]}^j)\right) \quad (8)$$

$$\text{where} \qquad \Psi(\mathbf{y}_{i,[t]}^k) = \sum_{w\in\mathbf{y}_{i,[t]}^k} \mathtt{W2AV}(w) \quad (9)$$

Here, $\mathbf{y}_{i,[t]}^k$ for $k \leq g$ is the ith beam in the kth group stored at time t; $\mathbf{y}_{b,[t]}^g$ is the concatenation of $\mathbf{y}_{b,[t-1]}^g$ and $y_{b,t}^g$. Intuitively, this metric computes the *cumulative dissimilarity* (given by the function $\Psi(\cdot)$) between the current beam and all the previously generated beams in other groups. This bag-of-affective-words approach is simple but works well in practice, as will be shown later.

Table 1. The effect of affective word embeddings as input.

Model	Syntactic coherence	Natural	Emotional approp.
Word emb. (baseline)	1.48	0.69	0.41
Word+Affective emb.	**1.71** ↑	**1.05** ↑	**1.01** ↑

Table 2. The effect of affective loss functions.

Model	Syntactic coherence	Naturalness	Emotional approp
$L_{\mathtt{XENT}}$ (baseline)	1.48	0.69	0.41
$L_{\mathtt{DMIN}}$	**1.75** ↑	0.83 ↑	0.56 ↓
$L_{\mathtt{DMAX}}$	1.74 ↑	0.85 ↑	0.58 ↑
$L_{\mathtt{AC}}$	1.71 ↑	**0.95** ↑	**0.71** ↑

It should be also noticed that several other beam search-based diverse decoding techniques have been proposed in recent years, including DivMBest [7] and MMI objective [10]. All of them use the notion of a *diversity term* within BS; therefore our affect-injecting technique can be used with these algorithms.

5   Experiments

We evaluated our approach on the Cornell Movie Dialogs Corpus [4], which contains ~300k utterance-response pairs. All our model variants used a single-layer LSTM encoder and a single-layer LSTM decoder, each layer containing 1024 cells. We set the vocabulary size to 12,000 and used Adam [9] optimizer.

For the baseline $L_{\mathtt{XENT}}$ loss, we used 1024-D Word2Vec embeddings as input and trained the Seq2Seq model for 50 epochs using Eq. 1. For the affective embeddings as input, we used 1027-D vectors, each a concatenation of 1024-D Word2Vec and 3-D W2AV embeddings. Training was done for 50 epochs. For each of the affective loss functions ($L_{\mathtt{AC}}$, $L_{\mathtt{DMIN}}$, and $L_{\mathtt{DMAX}}$), we trained the model using $L_{\mathtt{XENT}}$ loss for 40 epochs, followed by 10 epochs using the affective loss functions. The ADBS metrics Δ_W and Δ_S were deployed at test time with $G = B$.

5.1   Results

Recent work employs both automated metrics (e.g., BLEU, ROUGE, and METEOR) and human judgments to evaluate dialogue systems. While automated metrics enable high-throughput evaluation, they have weak or no correlation with human judgments [12]. It is also unclear how to evaluate affective aspects by automated metrics. Therefore, in this work, we recruited 3–5 human judges to evaluate our models, following several previous studies [15,20].

To evaluate the quality of the generated responses, we used 5 workers to evaluate 100 test samples for each model variant in terms of *syntactic coherence*

Table 3. Effect of affectively diverse decoding. H-DBS refers to Hamming-based DBS used in [22]. WL-ADBS and SL-ADBS are the proposed word-level and sentence-level affectively diverse beam search, respectively.

Model	Syntactic diversity	Affective diversity	# Emotionally approp. responses
BS (baseline)	1.23	0.87	0.89
H-DBS	1.47 ↑	0.79 ↓	0.78 ↓
WL-ADBS	**1.51** ↑	1.25 ↑	1.30 ↑
SL-ADBS	1.45 ↑	**1.31** ↑	**1.33** ↑

Table 4. Combining different affective strategies.

Model	Syntactic coherence	Naturalness	Emotional approp.
Traditional Seq2Seq (baseline)	1.48	0.69	0.41
Seq2Seq+Affective embeddings	1.71 ↑	1.05 ↑	1.01 ↑
Seq2Seq+Affective emb. & Loss	**1.76** ↓	1.03 ↓	1.07 ↑
Seq2Seq+Affective emb. & Loss & Decoding	1.69 ↓	**1.09** ↑	**1.10** ↓

(Does the response make grammatical sense?), *naturalness* (Could the response have been plausibly produced by a human?) and *emotional appropriateness* (Is the response emotionally suitable for the prompt?). For each axis, the judges were asked to assign each response an integer score of 0 (bad), 1 (satisfactory), or 2 (good). The scores were then averaged for each axis (Tables 1 and 2). We evaluated the inter-annotator consistency by Fleiss' κ score [5], and obtained a κ score of 0.447, interpreted as "moderate agreement" among the judges.[3] We also computed the statistical significance of the results using one-tailed Wilcoxon's Signed Rank Test [25] with significance level set to 0.05. This is indicated in Tables 1 and 2 through arrows: a down-arrow indicates that the model performed equally well as the baseline, and an up-arrow indicates that the model performed significantly better than the baseline.

The evaluation of diversity was conducted separately (Table 3). In this experiment, an annotator was presented with top-three decoded responses and was asked to judge *syntactic diversity* (How syntactically diverse are the five responses?) and *emotional diversity* (How affectively diverse are the five responses?). The rating scale was 0, 1, 2, and 3 with labels bad, satisfactory, good, and very good, respectively. The annotator was also asked to state the

---

[3] https://en.wikipedia.org/wiki/Fleiss%27_kappa

number of beams that were emotionally appropriate to the prompt. The scores obtained for each question were averaged. We used three annotators in this experiment (fewer than the previous one), as it required more annotations (3 responses for every test sample). The Fleiss' κ score for this protocol was 0.471, signifying "moderate agreement" between the judges. As before, Wilcoxon's Signed Rank significance test was used to compare each model with the baseline (vanilla BS).

Next, we evaluate the performance of 3 affective strategies individually, namely affective word embeddings as input, affective loss functions, and affectively diverse decoding. We then show how these strategies can be integrated.

Experiment #1: Affective word embeddings as input. Table 1 compares Seq2Seq open-domain dialogue systems with and without the affective word embeddings. We see that the cognitively engineered affective embeddings, even with 3 additional features, improve the baseline Seq2Seq model significantly (as indicated by the up-arrows). The improvement is consistent in terms of all 3 evaluation aspects, and its effect is the most significant in emotional appropriateness. The results show that traditional word embeddings learned end-to-end during training are not sufficient to capture emotional aspects; using additional knowledge makes the model more aware of affect.

Experiment #2: Affective loss functions. We compare in Table 2 the proposed loss functions—namely minimizing affective dissonance (DMIN), maximizing affective dissonance (DMAX), and maximizing affective content (AC)—with traditional cross-entropy loss (XENT).

Table 5. Examples of the responses generated by the baseline and affective models.

Message	Model	Response
I wish I was a billionaire	Traditional Seq2Seq	Um, you know...
	$L_{\texttt{DMIN}}$	But you work for the casino?
	$L_{\texttt{DMAX}}$	You are personally responsible
	$L_{\texttt{AC}}$	It's suicide!
	$L_{\texttt{XENT}}+$ Aff. emb.	Don't trouble yourself about it
	$L_{\texttt{AC}}+$Aff. emb.	Would it matter in the end
	$L_{\texttt{AC}}+$Aff. emb.+SL-ADBS	Would it change you?
I hate you	Traditional Seq2Seq	It was supposed to be
	$L_{\texttt{DMIN}}$	Aww c all guys do...
	$L_{\texttt{DMAX}}$	Keeps fairies away!
	$L_{\texttt{AC}}$	I haven't done anything crazy
	$L_{\texttt{XENT}}+$Aff. emb.	Don't say that!
	$L_{\texttt{AC}}+$Aff. emb.	I still love you!
	$L_{\texttt{AC}}+$Aff. emb. + SL-ADBS	I don't want to fight you

As shown in Table 2, DMIN and DMAX yield similar results, both outperforming XENT. Moreover, AC generally outperforms DMIN and DMAX in terms of naturalness and appropriateness. The results imply that forcing the affect vector in either direction (towards or against the previous utterance) helps the model, but its performance is worse than AC. The mediocre performance of L_{DMIN} and L_{DMAX} could be explained by the fact that the relationship between a prompt and a response is not always as simple as minimum or maximum affective dissonance. It is usually much more subtle; therefore it makes more sense to model this relationship through established sociological interaction theories like the Affect Control Theory [8]. By contrast, the AC loss function encourages affective content without specifying the affect direction; it works well in practice and significantly out-performs the baseline XENT loss on all three axes.

Considering both Tables 1 and 2, we further notice that the affective loss function alone is not as effective as affective embeddings. This makes sense because the loss function does not explicitly provide additional knowledge to the neural network, but word embeddings do. However, as will be seen in Experiment #4, these affective aspects can be directly combined. Another interesting observation is the improved syntactic coherence of the affect-based models; we hypothesize that these models replace grammatically incorrect words with affectively suitable options that turn out to be more grammatically sound.

Experiment #3: Affectively Diverse Decoding. We now evaluate our affectively diverse decoding methods. Since evaluating diversity requires multiple decoded utterances for a test sample, we adopted a different evaluation setting as described before. Table 3 compares both word-level and sentence-level affectively diverse BS (WL-ADBS and SL-ADBS, respectively) with the original BS and Hamming-based DBS used in [22]. We see that WL-ADBS and SL-ADBS beat the baselines BS and Hamming-based DBS by a statistically significant margin on affective diversity as well as number of emotionally appropriate responses. SL-ADBS is slightly better than WL-ADBS as expected, since it takes into account the cumulative affect of sentences as opposed to individual words.

Experiment #4: Putting them all together. We show in Table 4 how the affective word embeddings, loss functions, and decoding methods perform when they are combined. Here, we chose the best variants in the previous individual tests: the loss function maximizing affective content (L_{AC}) and the sentence level diversity measure (SL-ADBS). In this table, the statistical significance arrows denote the comparison of each row with the previous row, rather than with the baseline. As shown, the performance of our model generally increases when we gradually add new components to it, though some of the incremental improvements are statistically insignificant.

Note that our setting is different from ECM [26], the only other known emotion-based neural dialogue system to the best of our knowledge. ECM requires a desired affect category as input, which is unrealistic in applications. It also differs from our experimental setting (and our research goal), making direct comparison infeasible. However, our proposed affective approaches can be potentially integrated to ECM.

Finally, we present several sample outputs of all models in Table 5 to give readers a taste of how the responses differ.

6 Conclusion

In this work, we advance the development of affectively cognizant neural encoder-decoder dialogue systems by three affective strategies. We embed linguistic concepts in an affective space with a cognitively engineered dictionary, propose several affect-based heuristic objective functions, and introduce affectively diverse decoding methods. In information retrieval tasks such as question-answering and dialogue systems, these techniques can help retain the users by interacting with them in a more empathetic and human way.

References

1. Asghar, N., Poupart, P., Jiang, X., Li, H.: Deep active learning for dialogue generation. In: Proceedings of Joint Conference on Lexical and Computational Semantics, pp. 78–83 (2017)
2. Bahdanau, D., Cho, K., Bengio, Y.: Neural machine translation by jointly learning to align and translate. In: ICLR (2015)
3. Callejas, Z., Griol, D., López-Cózar, R.: Predicting user mental states in spoken dialogue systems. EURASIP J. Adv. Signal Process. **2011**(1), 6 (2011)
4. Danescu-Niculescu-Mizil, C., Lee, L.: Chameleons in imagined conversations: a new approach to understanding coordination of linguistic style in dialogs. In: Proceedings Workshop on Cognitive Modeling and Computational Linguistics, pp. 76–87 (2011)
5. Fleiss, J.L.: Measuring nominal scale agreement among many raters. Psychol. Bull. **76**(5), 378–382 (1971)
6. Ghosh, S., Chollet, M., Laksana, E., Morency, L.P., Scherer, S.: Affect-LM: a neural language model for customizable affective text generation. In: ACL (2017)
7. Gimpel, K., Batra, D., Dyer, C., Shakhnarovich, G., Tech, V.: A systematic exploration of diversity in machine translation. In: EMNLP, pp. 1100–1111 (2013)
8. Hoey, J., Schröder, T., Alhothali, A.: Affect control processes: intelligent affective interaction using a partially observable markov decision process. Artif. Intell. **230**, 134–172 (2016)
9. Kingma, D., Ba, J.: Adam: a method for stochastic optimization. In: ICLR (2015)
10. Li, J., Galley, M., Brockett, C., Gao, J., Dolan, B.: A diversity-promoting objective function for neural conversation models. In: NAACL-HLT, pp. 110–119 (2016)
11. Li, J., Monroe, W., Ritter, A., Jurafsky, D.: Deep reinforcement learning for dialogue generation. In: EMNLP, pp. 1192–1202 (2016)
12. Liu, C.W., Lowe, R., Serban, I., Noseworthy, M., Charlin, L., Pineau, J.: How not to evaluate your dialogue system: an empirical study of unsupervised evaluation metrics for dialogue response generation. In: EMNLP, pp. 2122–2132 (2016)
13. Malhotra, A., Yu, L., Schröder, T., Hoey, J.: An exploratory study into the use of an emotionally aware cognitive assistant. In: AAAI Workshop: Artificial Intelligence Applied to Assistive Technologies and Smart Environments (2015)
14. Mikolov, T., Sutskever, I., Chen, K., Corrado, G.S., Dean, J.: Distributed representations of words and phrases and their compositionality. In: NIPS (2013)

15. Mou, L., Song, Y., Yan, R., Li, G., Zhang, L., Jin, Z.: Sequence to backward and forward sequences: a content-introducing approach to generative short-text conversation. In: COLING, pp. 3349–3358 (2016)
16. Osgood, C.E.: The nature and measurement of meaning. Psychol. Bull. **49**(3), 197–237 (1952)
17. Pennebaker, J.W., Francis, M.E., Booth, R.J.: Linguistic Inquiry and Word Count. Erlbaum Publishers, Mahwah (2001)
18. Pittermann, J., Pittermann, A., Minker, W.: Emotion recognition and adaptation in spoken dialogue systems. Int. J. Speech Technol. **13**(1), 49–60 (2010)
19. Robison, J., McQuiggan, S., Lester, J.: Evaluating the consequences of affective feedback in intelligent tutoring systems. In: Proceedings of International Conference on Affective Computing and Intelligent Interaction, pp. 1–6 (2009)
20. Shang, L., Lu, Z., Li, H.: Neural responding machine for short-text conversation. In: ACL, pp. 1577–1586 (2015)
21. Sutskever, I., Vinyals, O., Le, Q.V.: Sequence to sequence learning with neural networks. In: NIPS, pp. 3104–3112 (2014)
22. Vijayakumar, A.K., Cogswell, M., Selvaraju, R.R., Sun, Q., Lee, S., Crandall, D., Batra, D.: Diverse beam search: decoding diverse solutions from neural sequence models. arXiv preprint arXiv:1610.02424 (2016)
23. Warriner, A.B., Kuperman, V., Brysbaert, M.: Norms of valence, arousal, and dominance for 13,915 English lemmas. Behav. Res. Methods **45**(4), 1191–1207 (2013)
24. Wen, T.H., Gasic, M., Mrkšić, N., Su, P.H., Vandyke, D., Young, S.: Semantically conditioned LSTM-based natural language generation for spoken dialogue systems. In: EMNLP, pp. 1711–1721 (2015)
25. Wilcoxon, F.: Individual comparisons by ranking methods. Biom. Bull. **1**(6), 80–83 (1945)
26. Zhou, H., Huang, M., Zhang, T., Zhu, X., Liu, B.: Emotional chatting machine: emotional conversation generation with internal and external memory. arXiv preprint arXiv:1704.01074 (2017)

Web2Text: Deep Structured Boilerplate Removal

Thijs Vogels, Octavian-Eugen Ganea, and Carsten Eickhoff[(✉)]

Department of Computer Science, ETH Zurich, Zürich, Switzerland
t.vogels@me.com, octavian.ganea@inf.ethz.ch, c.eickhoff@acm.org

Abstract. Web pages are a valuable source of information for many natural language processing and information retrieval tasks. Extracting the main content from those documents is essential for the performance of derived applications. To address this issue, we introduce a novel model that performs sequence labeling to collectively classify all text blocks in an HTML page as either boilerplate or main content. Our method uses a hidden Markov model on top of potentials derived from DOM tree features using convolutional neural networks. The proposed method sets a new state-of-the-art performance for boilerplate removal on the CleanEval benchmark. As a component of information retrieval pipelines, it improves retrieval performance on the ClueWeb12 collection.

1 Introduction

Modern methods in natural language processing and information retrieval are heavily dependent on large collections of text. The World Wide Web is an inexhaustible source of content for such applications. However, a common problem is that Web pages include not only main content, but also ads, hyperlink lists, navigation, previews of other articles, banners, *etc.* This boilerplate/template content has often been shown to have negative effects on the performance of derived applications [15,24].

The task of separating main text in a Web page from the remaining content is known in the literature as "boilerplate removal", "Web page segmentation" or "content extraction". Established popular methods for this problem use rule-based or machine learning algorithms. The most successful approaches first perform a splitting of an input Web page into text blocks, followed by a binary labeling of each block as either main content or boilerplate.

In this paper, we propose a hidden Markov model on top of neural potentials for the task of boilerplate removal. We leverage the representational power of convolutional neural networks (CNNs) to learn unary and pairwise potentials over blocks in a page-based on complex non-linear combinations of DOM-based traditional features. At prediction time, we find the most likely block labeling by maximizing the joint probability of a label sequence using the Viterbi algorithm [23]. The effectiveness of our method is demonstrated on standard benchmarking datasets.

© Springer International Publishing AG, part of Springer Nature 2018
G. Pasi et al. (Eds.): ECIR 2018, LNCS 10772, pp. 167–179, 2018.
https://doi.org/10.1007/978-3-319-76941-7_13

The remainder of this document is structured as follows. Section 2 gives an overview of related work. Section 3 formally defines the main-content extraction problem, introduces the block segmentation procedure and details our model. Section 4 empirically demonstrates the merit of our method on several benchmark datasets for content extraction and document retrieval.

2   Related Work

Early approaches to HTML boilerplate removal use a range of heuristics and rule-based methods. Finn *et al.* [7] design an effective system called *Body Text Extractor* (BTE). It relies on the observation that the main content contains longer paragraphs of uninterrupted text, where HTML tags occur less frequently compared to the rest of the Web page. Looking at the cumulative distribution of tags as a function of the position in the document, Finn *et al.* identify a flat region in the middle of this distribution graph to be the main content of the page. While simple, their algorithm has two drawbacks: (1) it only makes use of the location of HTML tags and not of their structure, thus losing potentially valuable information, and (2) it can only identify one continuous stretch of main content which is unrealistic for a considerable percentage of modern Web pages.

To address these issues, several other algorithms have been designed to operate on DOM trees, thus leveraging the semantics of the HTML structure [6,11,19]. The problem with these early methods is that they make intensive use of the fact that pages used to be partitioned into sections by `<table>` tags, which is no longer a valid assumption.

In the next line of work, the DOM structure is used to jointly process multiple pages from the same domain, relying on their structural similarities. This approach was pioneered by Yi *et al.* [24] and was improved by various others [22]. These methods are very suitable for detecting template content that is present in all pages of a website, but have poor performance on websites that consist of a single Web page only. In this paper we focus on single-page content extraction without exploiting the context of other pages from the same site.

Gottron *et al.* [10] propose *Document Slope Curves* and *Content Code Blurring* methods that are able to identify multiple disconnected content regions. The latter method parses the HTML source code as a vector of 1's, representing pieces of text, and 0's, representing tags. This vector is then smoothed iteratively, such that eventually it finds active regions where text dominates (content) and inactive regions where tags dominate (boilerplate). This idea of smoothing was extended to also deal with the DOM structure [4,21]. Chakrabarti *et al.* [3] assign a likelihood of being content to each leaf of the DOM tree while using isotonic smoothing to combine the likelihoods of neighbors with the same parents. In a similar direction, Sun *et al.* [21] use both the tag/text ratio and DOM tree information to propagate *DensitySums* through the tree.

Machine learning methods offer a convenient way to combine various indicators of "contentness", automatically weighting hand-crafted features according to their relative importance. The FIASCO system by Bauer *et al.* [2] uses Support

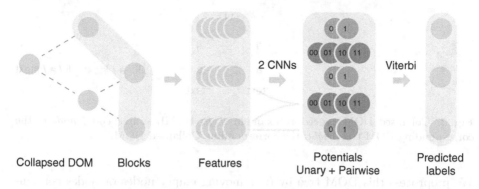

Fig. 1. The Web2Text pipeline. The leaves of the Collapsed DOM tree of a Web page form an ordered sequence of blocks to be labeled. For each block, we extract a number of DOM tree-based features. Two separate convolutional networks operating on this sequence of features yield two respective sets of potentials: unary potentials for each block and pairwise potentials for each pair of neighboring blocks. These define a hidden Markov model. Using the Viterbi algorithm, we find an optimal labeling that maximizes the total sequence probability as predicted by the neural networks.

Vector Machines (SVM) to classify an HTML page as a sequence of blocks that are generated through a DOM-based segmentation of the page and are represented by linguistic, structural and visual features. Similar works of Kohlschütter *et al.* [17] also employ SVMs to independently classify blocks. Spousta et al. [20] extend this approach by reformulating the classification problem as a case of sequence labeling where all blocks are jointly tagged. They use conditional random fields to take advantage of correlations between the labels of neighboring content blocks. This method was the most successful in the CleanEval competition [1].

In this paper, we propose an effective set of block features that capture information from adjacent neighbors in the DOM tree. Additionally, we employ a deep learning framework to automatically learn non-linear features combinations, giving the model an advantage over traditional linear approaches. Finally, we jointly optimize the labels for the whole Web page according to local potentials predicted by the neural networks.

3   Web2Text

Boilerplate removal is the problem of labeling sections of the text of a Web page as *main content* or *boilerplate* (anything else) [1]. In the following, we discuss the various steps of our method. The complete pipeline is also illustrated in Fig. 1.

3.1   Preprocessing

We expect raw Web page input to be written in (X)HTML markup. Each document is parsed as a Document Object Model tree (DOM tree) using Jsoup [12].

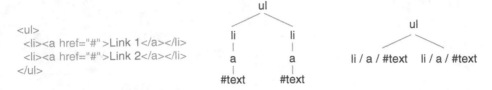

```
<ul>
  <li><a href="#">Link 1</a></li>
  <li><a href="#">Link 2</a></li>
</ul>
```

Fig. 2. Collapsed DOM procedure example. *Left*: HTML source code, *middle*: the corresponding DOM tree, *right*: the corresponding Collapsed DOM.

We preprocess this DOM tree by (i) removing empty nodes or nodes containing only whitespace, (ii) removing nodes that do not have any content we can extract: *e.g.* `<br>`, `<checkbox>`, `<head>`, `<hr>`, `<iframe>`, `<img>`, `<input>`.

We make use of the parent and grandparent DOM tree relations. In a raw DOM tree, however, these relationships are not always meaningful. Figure 2 shows a typical fragment of a DOM tree where two neighboring nodes share the same semantic parent (`<ul>`) but not the same DOM parent. To improve the expressiveness of tree based features (such as "the number of children of a node's parent"), we recursively merge single child parent nodes with their respective child. We call the resulting tree-structure the *Collapsed DOM* (CDOM).

3.2   Block Segmentation

Our content extraction algorithm is based on sequence labeling. A Web page is treated as a sequence of blocks that are labeled *main content* or *boilerplate*. There are multiple ways to split a Web page into blocks, the most popular currently used being (i) Lines in the HTML file, (ii) DOM leaves, (iii) Block-level DOM leaves. We opt for using the most flexible *DOM leaves* strategy, described as follows. Sections on a page that require different labels are usually separated by at least one HTML tag. Therefore, it is safe to consider DOM leaves (`#text` nodes) as the blocks of our sequence. A potential disadvantage of this approach is that a hyperlink in a text paragraph can receive a different label than its neighboring text. Under this scheme, an empirical evaluation of Web2Text shows no cases where parts of a textual paragraph are wrongly labeled as *boilerplate*, while the rest are marked as *main content*.

3.3   Feature Extraction

Features are properties of a node that may be indicative of it being content or boilerplate. Such features can be based on the node's text, CDOM structure or a combination thereof. We distinguish between block features and edge features.

Block features capture information on each block of text on a page. They are statistics collected based on the block's CDOM node, parent node, grandparent node and the root of the CDOM tree. In total, we collect 128 features for each text block, *e.g.* "the node is a `<p>` element", "average word length", "relative position in the source code", "the parent node's text contains an email address", "ratio of stopwords in the whole page", *etc.* We clip and standardize all non-binary features to be approximately Gaussian with zero mean and unit variance across the training set. For a full overview of all 128 features, please refer to Appendix A.

Edge features capture information on each pair of neighboring text blocks. We collect 25 features for each such pair. Define the *tree distance* of two nodes as the sum of the number of hops from both nodes to their first common ancesor. The first edge features we use are binary features corresponding to a tree distance of 2, 3, 4 and > 4. Another feature signifies if there is a *line break* between the nodes in an unstyled HTML page. Finally, we collect features b70–b89 from Appendix A for the *common ancestor* CDOM node of the two text blocks.

3.4   CNN Unary and Pairwise Potentials

We assign unary potentials to each text block to be labeled and pairwise potentials to each pair of neighboring text blocks. In our case, potentials are probabilities as explained below. The unary potentials $p_i(l_i = 1)$, $p_i(l_i = 0)$ are the probabilities that the label l_i of a text block i is content or boilerplate, respectively. The two potentials sum to one. The pairwise potentials $p_{i,i+1}(l_i = 1, l_{i+1} = 1)$, $p_{i,i+1}(l_i = 1, l_{i+1} = 0)$, $p_{i,i+1}(l_i = 0, l_{i+1} = 1)$ and $p_{i,i+1}(l_i = 0, l_{i+1} = 0)$ are the transition probabilities of the labels of a pair of neighboring text blocks. These pairwise potentials also sum to one for each text block pair.

The two sets of potentials are modeled using CNNs with 5 layers, ReLU non-linearity between layers, filter sizes of $(50, 50, 50, 10, 2)$ for the unary network and of $(50, 50, 50, 10, 4)$ for the pairwise network. All filters have a stride of 1 and kernel sizes $(1, 1, 3, 3, 3)$ respectively. The unary CNN receives a sequence of block features corresponding to the sequence of text blocks to be labeled and outputs unary potentials for each block. The pairwise CNN receives a sequence of edge features corresponding to the sequence of edges to be labeled and outputs the pairwise potentials for each block. We use zero padding to make sure that each layer produces a sequence of the same size as its input sequence. The outputs for the unary network are sequences of 2 values per block that are normalized using softmax. The outputs for the pairwise network are sequences of 4 values per block-pair that are normalized in the same way. Thus, the output for the block i depends indirectly on a range of blocks around it. We employ dropout regularization with rate 0.2 and L_2 weight decay with rate 10^{-4}.

For the unary potentials, we minimize the cross-entropy

$$\theta^*_{\text{unary}} = \operatorname{argmin}_{\theta_{\text{unary}}} - \sum_{i=0}^{n} \log p_i(l_i = l_i^* \mid \theta_{\text{unary}}), \tag{1}$$

where l_i^* is the true label of block i, θ_{unary} are the parameters of the unary network and n is the index of the last text block in the sequence.

For the pairwise network, we minimize the cross-entropy

$$\theta^*_{\text{pairwise}} = \text{argmin}_{\theta_{\text{pairwise}}} - \sum_{i=0}^{n-1} \log p_{i(i+1)}(l_i = l^*_i, \ l_{i+1} = l^*_{i+1} \mid \theta_{\text{pairwise}}), \quad (2)$$

where θ_{pairwise} are the parameters of the pairwise network.

3.5   Inference

The joint prediction of the most likely sequence of labels given an input Web page works as follows. We denote the sequence of text blocks on the page as $(b_0, b_1, \ldots, b_n)$ and write the probability of a corresponding labeling $(\ell_0, \ell_1, \ldots, \ell_n) \in \{0, 1\}^n$ being the correct one as

$$p(\ell_0, \ldots, \ell_n) = \left(\prod_{i=0}^{n} p_i(\ell_i) \right) \left(\prod_{i=0}^{n-1} p_{i(i+1)}(\ell_i, \ell_{i+1}) \right)^{\lambda}, \quad (3)$$

where λ is an interpolation factor between the unary and pairwise terms. We use $\lambda = 0.1$ in our experiments. This expression describes a hidden Markov model and it is maximized using the Viterbi algorithm [23] to find the optimal labeling given the predicted CNN potentials.

4   Experiments

Our experiments are grouped in two stages. We begin by assessing Web2Text's performance at boilerplate removal on a high-quality manually annotated corpus of Web pages. In a second step, we turn towards a much larger collection and investigate how improved content extraction results in superior information retrieval quality. Both experiments highlight the benefits of Web2Text over state-of-the-art alternatives.

4.1   Training Data

CleanEval 2007 [1] is the largest publicly available dataset for this task. It contains 188 text blocks per Web page on average. It consists of an original split of development (60 pages) and test (676 pages) sets. We divide the development set into a training set (55 pages) and a test set (5 pages). Since our model has more than 10,000 parameters, it is likely that the original training set is too small for our method. Thus, we did a second split of the CleanEval as follows: training (531 pages), validation (58 pages) and test (148 pages).

Automatic Block Labeling. To our knowledge, the existing corpora (including CleanEval) for boilerplate detection pose an additional difficulty. These datasets consist only of pairs of Web pages and corresponding cleaned text (manually extracted). As a consequence, the alignment between the source text and cleaned text, as well as block labeling, have to be recovered. Some methods (*e.g.* [20]) rely on expensive manual block annotations. One of our contributions is the following automatic recovery procedure of the aligned (block, label) pairs from the original (Web page, clean text) pairs. This allows us to leverage more training data compared to previous methods.

We first linearly scan the cleaned text of a Web page using windows of 10 consecutive characters. Each such snippet is checked for uniqueness in the original Web page (after spaces trimming). If such a unique match is found, then it can be used to divide both the cleaned text and the original Web page in two parts on which the same matching method can be applied recursively in a *divide-et-impera* fashion. After all unique snippets are processed, we use dynamic programming to align the remaining splitted parts of the clean text with the corresponding splitted parts of the original Web page blocks. In the end, in the rare case that the content of a block is only partially matched with the cleaned text, we mark it as *content* iff at least 2/3 of its text is aligned.

4.2 Training Details

The unary and pairwise potential-predicting networks are trained separately with the Adam optimizer [14] and a learning rate of 10^{-3} for 5000 iterations. Each iteration processes a mini-batch of 128 9-text-block long Web page excerpts. We perform early stopping, observing no improvements after this number of steps. We then pick the model corresponding to the lowest error on the validation set.

4.3 Baselines

We compare Web2Text to a range of methods described in the literature or deployed in popular libraries. BTE [7] and Unfluff [8] are heuristic methods. [16,17] is a popular machine learning system that offers various content extraction settings[1] which we used in our experiments (see Table 1). CRF [20] achieves one of the best results on CleanEval. This machine learning model trains a Conditional Random Field on top of block features in order to perform block classification. However, as explained in Sect. 4.1, CRF relies on a different Web page block splitting and on expensive manual block annotations. As a consequence, we were not able to re-train it and thus only used their out-of-the-box model pre-trained on the original CleanEval split. For a fair comparison, we also train on the original CleanEval split, but note below that our neural network has many more parameters and will suffer from using so few training instances.

---

[1] We were not able to find code for re-training this system.

Table 1. Boilerplate removal results on the CleanEval dataset. We use two different splits of this dataset, the original split (55p, 5p, 676p) and our split (531p, 58p, 148p). It is confirmed that our method benefits from bigger training sets.

Method	Original test (676 pages)				Our test (148 pages)			
	Acc.	Precision	Recall	F_1	Acc.	Precision	Recall	F_1
CRF [20] original train 55p + 5p	0.82	0.87	0.81	0.84	0.82	0.88	0.81	0.84
BTE [7]	0.79	0.79	**0.89**	0.83	0.75	0.76	0.84	0.80
default-ext [16]	0.80	0.89	0.75	0.81	0.79	0.89	0.74	0.81
article-ext [16]	0.72	0.91	0.59	0.71	0.67	0.89	0.50	0.64
largest-ext [16]	0.60	**0.93**	0.36	0.52	0.59	**0.93**	0.33	0.48
Unfluff [8]	0.71	0.90	0.57	0.70	0.68	0.90	0.51	0.65
Web2Text original train 55p, val 5p	**0.84**	0.88	0.85	**0.86**				
Web2Text our train 531p, val 58p					**0.86**	0.87	**0.90**	**0.88**

Model Sizes. The CRF model [20] contains 9,705 parameters. In comparison, our unary CNN network contains 17,960 parameters, while the pairwise CNN contains 12,870 parameters, the total number of parameters for the joint structured model being 30,830. This explains why the original train set is too small for our model.

4.4   Content Extraction Results

Table 1 shows the results of this experiment. All the metrics are block based, where all blocks are weighted equally. We note that Web2Text obtains state-of-the-art accuracy, recall and F1 scores compared to popular baselines including previous CleanEval winners. Note that these numbers are obtained by evaluating each method using the same block segmentation procedure, namely the DOM leaves strategy described in Sect. 3.2. We additionally note that, compared to using Web2Text only with the unary CNN, the gains of the hidden Markov model are marginal in this experiment.

Running Times. Web2Text takes 54 ms per Web page on average; 35 ms for DOM parsing and feature extraction, and 19 ms for the neural network forward pass and Viterbi algorithm. These measurements were done on a Macbook with a 2.8 GHz Intel Core i5 processor.

4.5   Impact on Retrieval Performance

Besides the previously presented intrinsic evaluation of text extraction accuracy, we are interested in the performance gains that other derived tasks experience

Table 2. The effect of boilerplate removal on *ad hoc* retrieval performance.

Collection	Ret. model	Method	P@10	R@10	F_1@10	MAP	nDCG
CW12-A	QL	Raw content	0.316	0.056	0.095	0.137	0.459
CW12-A	QL	CRF [20]	0.342*	0.068*	0.113*	0.147*	0.543*
CW12-A	QL	BTE [7]	0.301	0.048	0.083	0.128	0.435
CW12-A	QL	default-ext [16]	0.318	0.055	0.094	0.138	0.462
CW12-A	QL	article-ext [16]	0.298	0.049	0.084	0.126	0.433
CW12-A	QL	largest-ext [16]	0.279	0.044	0.076	0.112	0.417
CW12-A	QL	Unfluff [8]	0.304	0.051	0.087	0.128	0.428
CW12-A	QL	Web2Text	**0.361***†	**0.079***†	**0.130***†	**0.154***†	**0.578***†
CW12-A	RM	Raw content	0.278	0.048	0.082	0.121	0.439
CW12-A	RM	CRF [20]	0.301*	0.057*	0.096*	0.138*	0.487*
CW12-A	RM	BTE [7]	0.262	0.041	0.071	0.110	0.409
CW12-A	RM	default-ext [16]	0.277	0.048	0.082	0.123	0.442
CW12-A	RM	article-ext [16]	0.260	0.039	0.068	0.109	0.411
CW12-A	RM	largest-ext [16]	0.248	0.032	0.057	0.097	0.401
CW12-A	RM	Unfluff [8]	0.264	0.041	0.071	0.111	0.407
CW12-A	RM	Web2Text	**0.325***†	**0.069***†	**0.114***†	**0.145***†	**0.525***†
CW12-B	QL	Raw content	0.210	0.025	0.045	0.037	0.134
CW12-B	QL	CRF [20]	0.241*	0.031*	0.055*	0.048*	0.165*
CW12-B	QL	BTE [7]	0.193	0.019	0.035	0.030	0.121
CW12-B	QL	default-ext [16]	0.212	0.026	0.046	0.038	0.132
CW12-B	QL	article-ext [16]	0.199	0.017	0.031	0.031	0.120
CW12-B	QL	largest-ext [16]	0.178	0.015	0.028	0.024	0.107
CW12-B	QL	Unfluff [8]	0.195	0.020	0.036	0.029	0.121
CW12-B	QL	Web2Text	**0.266***†	**0.038***†	**0.067***†	**0.055***†	**0.181***†
CW12-B	RM	Raw content	0.172	0.021	0.037	0.030	0.122
CW12-B	RM	CRF [20]	0.198*	0.028*	0.049*	0.041*	0.143*
CW12-B	RM	BTE [7]	0.158	0.015	0.027	0.022	0.111
CW12-B	RM	default-ext [16]	0.170	0.020	0.036	0.029	0.124
CW12-B	RM	article-ext [16]	0.156	0.015	0.027	0.019	0.109
CW12-B	RM	largest-ext [16]	0.145	0.013	0.024	0.015	0.095
CW12-B	RM	Unfluff [8]	0.159	0.016	0.029	0.021	0.112
CW12-B	RM	Web2Text	**0.213***†	**0.032***†	**0.056***†	**0.046***†	**0.165***†

when operating on the output of boilerplate removal systems of varying quality. To this end, our extrinsic evaluation studies the task of *ad hoc* document retrieval. Search engines that index high-quality output of text extraction systems should be better able to answer a given user-formulated query than systems indexing raw HTML or naïvely cleaned content. Our experiments are based on the well-known ClueWeb12 collection of Web pages.[2] It is organized in two

---

[2] http://lemurproject.org/clueweb12/.

well-defined document sets, the full CW12-A corpus of 733M organic Web documents (27.3 TB of uncompressed text) as well as the smaller, randomly sampled subset CW12-B of 52M documents (1.95 TB of uncompressed text). The collection is indexed using the Indri search engine and retrieval runs are conducted using two state-of-the-art probabilistic retrieval models, the query likelihood model [13] (QL) as well as a relevance-based language model [18] (RM). Our 50 test queries alongside their relevance judgments originate from the 2013 edition of the TREC Web Track [5].

Table 2 highlights the quality of each combination of retrieval model and collection when indexing either raw or cleaned Web content. Within each combination, statistical significance of performance differences between raw and cleaned HTML content is denoted by an asterisk. Models that significantly outperform all other text extraction methods are indicated by †. We can note that, in general, retrieval systems indexing CW12-A deliver stronger results than those operating only on the CW12-B subset. Due to the random sampling process, many potentially relevant documents are missing from this smaller collection. Similarly, across all comparable settings, the query likelihood model (QL) performs significantly better than the relevance model (RM). As hypothesized earlier, text extraction can influence the quality of subsequent document retrieval. We note that low-recall methods (BTE, article-ext, largest-ext, Unfluff) cause losses in retrieval performance, as relevant pieces of content are incorrectly removed as boilerplate. At the same time, the most accurate models (CRF, Web2Text) were able to introduce improvements across all metrics. Web2Text, in particular, outperformed all baselines at significance level 0.05. We note that, for this experiment, Web2Text was trained on our CleanEval split as explained in Sect. 4.1.

5   Conclusion

This paper presents Web2Text[3], a novel algorithm for main content extraction from Web pages. The method combines the virtues of popular sequence labeling approaches such as CRFs [9] with deep learning methods that leverage the DOM structure as a source of information. Our experimental evaluation on CleanEval benchmarking data shows significant performance gains over all state-of-the-art methods. In a second set of experiments, we demonstrate how highly accurate boilerplate removal can significantly increase the performance of derived tasks such as *ad hoc* retrieval.

A   A List of Block Features

See Table 3.

---

[3] Our source code is publicly available: https://github.com/dalab/web2text.

Table 3. List of all block features used. 1/0 indicated a binary feature: 1 if true, 0 if false. Non-binary features are normalized to have zero mean and unit variance.

ID	Name	Description
b1	has duplicate	1/0: is there another node with the same text?
b2	has 10 duplicates	1/0: are there at least 10 other nodes with the same text?
b3	r same class path	ratio of nodes on the page with the same class path (e.g. body>div>a.link>b)
b4	has word	1/0: there is at least one word in the text block
b5	log(n words)	log(number of words) (clipped between 0 and 3.5)
b6	avg. word length	average word length (clipped between 3 and 15)
b7	has stopword	1/0: block contains a stopword
b8	stopword ratio	ratio of words the are in our stopword list
b9	log(n characters)	log(number of characters) (clipped between 2.5 and 5.5)
b10	log(punctuation ratio)	log(ratio of characters $\in \{, , ?, ;, :, !\}$ to the total) (clipped between -4 and -2.5)
b11	has numeric	1/0: the node contains numeric characters
b12	numeric ratio	ratio of numeric characters to the total character count
b13	log(avg sentence length)	log(average sentence length) (clipped between 2 and 5)
b14	ends with punctuation	1/0: the node ends with a character $\in \{, , ?, ;, :, !\}$
b15	ends with question mark	1/0: the node ends with a question mark
b16	contains copyright	1/0: the node contains a copyright symbol
b17	contains email	1/0: the node contains an email address
b18	contains url	1/0: the node contains a URL
b19	contains year	1/0: the node contains a word consisting of 4 digits
b20	ratio words with capital	ratio of words starting with a capital letter
b21	ratio words with capital^2	b25 squared
b22	ratio words with capital^3	b25 to the power 3
b23	contains punctuation	node contains a character $\in \{, , ?, ;, :, !\}$
b24	n punctuation	number of characters $\in \{, , ?, ;, :, !\}$
b25	has multiple sentences	1/0: there are more than 1 sentences in the text
b26	relative position	relative position of the start of this block in the source code
b27	relative position^2	17 squared
b28	has parent	1/0: the CDOM leaf has a parent node
b29	p body percentage	ratio of the source code characters that is within the parent CDOM node
b30	p link density	ratio of characters within ¡a¿ elements to total character count
b31–b47	parent features	b6–b22, but for the parent CDOM node
b48	p contains form element	1/0: the parent CDOM node contains a form element
b49–b69	parent tag features	encoding of the parent CDOM node's HTML tags as 1's and 0's
		(td, div, p, tr, table, body, ul, span, li, blockquote, b, small, a, ol, ul, i, form, dl, strong, pre)
b69	has grandparent	1/0: the node has a grandparent CDOM node
b70–b89	grandparent features	b29–b48, but for the grandparent CDOM node
b90–b109	root features	b29–b48, but for the root CDOM node (body)
b110–b128	tag features	encoding of the CDOM node's HTML tags as 1's and 0's (a, p, td, b, li, span, I, tr, div, strong, em, h3, h2, table, h4, small, sup, h1, blockquote)

References

1. Baroni, M., Chantree, F., Kilgarriff, A., Sharoff, S.: CleanEval: a competition for cleaning web pages. In: LREC (2008)
2. Bauer, D., Degen, J., Deng, X., Herger, P., Gasthaus, J., Giesbrecht, E., Jansen, L., Kalina, C., Kräger, T., Märtin, R., Schmidt, M., Scholler, S., Steger, J., Stemle, E., Evert, S.: FIASCO: filtering the internet by automatic subtree classification, Osnabruck. In: Building and Exploring Web Corpora: Proceedings of the 3rd Web as Corpus Workshop, Incorporating CleanEval, vol. 4, pp. 111–121 (2007)
3. Chakrabarti, D., Kumar, R., Punera, K.: Page-level template detection via isotonic smoothing. In: Proceedings of the 16th International Conference on World Wide Web, pp. 61–70. ACM (2007)
4. Chakrabarti, D., Kumar, R., Punera, K.: A graph-theoretic approach to webpage segmentation. In Proceedings of the 17th International Conference on World Wide Web, pp. 377–386. ACM (2008)
5. Collins-Thompson, K., Bennett, P., Diaz, F., Clarke, C., Voorhees, E.: Overview of the TREC 2013 web track. In: Proceedings of the 22nd Text Retrieval Conference (TREC 2013) (2013)
6. Debnath, S., Mitra, P., Pal, N., Giles, C.L.: Automatic identification of informative sections of web pages. IEEE Trans. Knowl. Data Eng. **17**(9), 1233–1246 (2005)
7. Finn, A., Kushmerick, N., Smyth, B.: Content classification for digital libraries. Unrefereed, Fact or fiction (2001)
8. Geitgey, A.: Unfluff - an automatic web page content extractor for node.js! (2014)
9. Gibson, J., Wellner, B., Lubar, S.: Adaptive web-page content identification. In: Proceedings of the 9th Annual ACM International Workshop on Web Information and Data Management, pp. 105–112. ACM (2007)
10. Gottron, T.: Content code blurring: a new approach to content extraction. In: 19th International Workshop on Database and Expert Systems Application, DEXA 2008, pp. 29–33. IEEE (2008)
11. Gupta, S., Kaiser, G., Neistadt, D., Grimm, P.: DOM-based content extraction of HTML documents. In: Proceedings of the 12th International Conference on World Wide Web, pp. 207–214. ACM (2003)
12. Hedley, J.: Jsoup HTML parser (2009)
13. Jin, R., Hauptmann, A.G., Zhai, C.: Language model for information retrieval. In: Proceedings of the 25th Annual International ACM SIGIR Conference on Research and Development in Information Retrieval, pp. 42–48. ACM (2002)
14. Kingma, D., Ba, J.: Adam: a method for stochastic optimization. arXiv preprint arXiv:1412.6980 (2014)
15. Kohlschütter, C.: A densitometric analysis of web template content. In: Proceedings of the 18th International Conference on World Wide Web, pp. 1165–1166. ACM (2009)
16. Kohlschütter, C., et al.: Boilerpipe - boilerplate removal and fulltext extraction from HTML pages. Google Code (2010)
17. Kohlschütter, C., Fankhauser, P., Nejdl, W.: Boilerplate detection using shallow text features. In: Proceedings of the Third ACM International Conference on Web Search and Data Mining, pp. 441–450. ACM (2010)
18. Lavrenko, V., Croft, W.B.: Relevance based language models. In: Proceedings of the 24th Annual International ACM SIGIR Conference on Research and Development in Information Retrieval, pp. 120–127. ACM (2001)

19. Lin, S.-H., Ho, J.-M.: Discovering informative content blocks from web documents. In: Proceedings of the Eighth ACM SIGKDD International Conference on Knowledge Discovery and Data Mining, pp. 588–593. ACM (2002)
20. Spousta, M., Marek, M., Pecina, P.: Victor: the web-page cleaning tool. In: 4th Web as Corpus Workshop (WAC4)-Can We Beat Google, pp. 12–17 (2008)
21. Sun, F., Song, D., Liao, L.: DOM based content extraction via text density. In: Proceedings of the 34th International ACM SIGIR Conference on Research and Development in Information Retrieval, pp. 245–254. ACM (2011)
22. Vieira, K., Da Silva, A.S., Pinto, N., De Moura, E.S., Cavalcanti, J., Freire, J.: A fast and robust method for web page template detection and removal. In: Proceedings of the 15th ACM International Conference on Information and Knowledge Management, pp. 258–267. ACM (2006)
23. Viterbi, A.J.: Error bounds for convolutional codes and an asymptotically optimum decoding algorithm. In: The Foundations of the Digital Wireless World: Selected Works of AJ Viterbi, pp. 41–50. World Scientific (2010)
24. Yi, L., Liu, B., Li, X.: Eliminating noisy information in web pages for data mining. In: Proceedings of the Ninth ACM SIGKDD International Conference on Knowledge Discovery and Data Mining, pp. 296–305. ACM (2003)

Attention-Based Neural Text Segmentation

Pinkesh Badjatiya[(✉)], Litton J. Kurisinkel, Manish Gupta,
and Vasudeva Varma

IIIT-H, Hyderabad, India
{pinkesh.badjatiya,litton.jkurisinkel}@research.iiit.ac.in,
{manish.gupta,vv}@iiit.ac.in

Abstract. Text segmentation plays an important role in various Natural Language Processing (NLP) tasks like summarization, context understanding, document indexing and document noise removal. Previous methods for this task require manual feature engineering, huge memory requirements and large execution times. To the best of our knowledge, this paper is the first one to present a novel supervised neural approach for text segmentation. Specifically, we propose an attention-based bidirectional LSTM model where sentence embeddings are learned using CNNs and the segments are predicted based on contextual information. This model can automatically handle variable sized context information. Compared to the existing competitive baselines, the proposed model shows a performance improvement of ∼7% in WinDiff score on three benchmark datasets.

1 Introduction

The task of text segmentation is defined as the process of segmenting a chunk of text into meaningful sections based on their topical continuity. Text segmentation is one of the fundamental NLP problems which finds its use in many tasks like summarization [14], passage extraction [16], discourse analysis [26], Question-Answering [16], context understanding, document noise removal, etc. Fine grained segmentation of a document into multiple sections provides a better understanding about the document structure which can also be used to generate better document representations, which in turn could benefit other natural language applications. Complexity of text segmentation varies with the type of text and writing styles – informational, conversational, narrative, descriptive, etc. In some cases, context is a very important signal for the task, while in other cases, dependence on context may be minimal. Also, complex topic shifts in the text and use of abstract cue phrases in the sentences make the task challenging.

Multiple supervised and unsupervised methods have been already proposed to tackle some of these challenges. Many unsupervised methods are heuristic and ad hoc in nature, need huge memory, have large execution times, and

Manish Gupta—A Principal Applied Scientist at Microsoft.

do not generalize well across multiple text types. Supervised methods require labeled data and often the performance of such systems comes at the cost of hand-crafted highly tuned feature engineering. None of these methods can automatically tune the degree of dependence on the context. Sequence-to-sequence models like Recurrent Neural Networks (RNNs) and Long Term-Short Memory (LSTMs) can model sequences effectively by controlling information flow across time. Such models can in general help capture long range dependencies (context) but they work well with short sequences. They can be enhanced by giving varying attention weights to sentences in the context, where the weight denotes the relative importance of a context sentence for segmentation. Attention thus allows us to learn the focus points from the context. To the best of our knowledge, this is the first work to explore the use of attention-based neural mechanism for text segmentation. We propose a novel Attention-based CNN-BiLSTM model that learns to represent the context of the sentence by learning the attention weights. The proposed model does not require any manually designed features, is domain independent and scalable. The proposed neural model architecture is illustrated in Fig. 1. We compare the proposed method with competitive baselines on three benchmark datasets.

In Sect. 2, we review the existing work on text segmentation. Section 3 describes the proposed Neural model with Attention-based approach. Section 4 compares the performance of various methods on benchmark datasets. In Sect. 5, we analyze the results and conclude with a brief summary in Sect. 6.

2   Related Work

Unsupervised methods for text segmentation include lexical cohesion [4,8], statistical modeling [1,25], affinity propagation based clustering [11,22], and topic modeling [4,13,18,20]. Topic modeling approaches include PLDA [18] (captures the amount of topic distribution that a paragraph shares with its predecessor), SITS [15] (chains a set of Hierarchical Dirichlet Process LDAs), TSM [2] (integrates point-wise boundary sampling with topic modeling), and [3] (ordering-based probabilistic topic models to incorporate the ordering irregularity into the probabilistic approach). These methods are globally informed, i.e., they consider the whole document when generating the most probable segment boundaries. However, huge memory requirements and large execution times make these methods unpractical for use in real applications.

Various classifiers like decision trees [6,24] and probabilistic models [1,7,19,25] have been proposed for supervised text segmentation. Popular features include lexical (like lexical similarities [8]), conversational (acoustic indicators, long pauses, shifts in speaking rates, higher maximum accent peak, cue phrases, silences, overlaps, speaker change [5]) and knowledge-based features [10]. Supervised methods require labeled data and hand-crafted highly tuned feature engineering. Also, they are locally informed and often fail to capture the overall global topic structure of the document.

Some previous studies, although scarce and somewhat preliminary, have explored neural approaches for domain-specific text segmentation. Sheikh et al. [23] proposed a method for segmentation in transcripts using RNNs, Wang et al. [28] attempt to learn a coherence function using the partial ordering relations, Wang et al. [27] use BiLSTM-CNN to model the task as a simple binary classification task for Chinese. In this paper we explore the use of attention-based deep neural architecture for the task of automatic linear text segmentation, which provides a good trade off between the locally informed and globally informed behavior by varying the amount of context information used.

3   The Proposed Method

In this section, we start by presenting the formal problem definition. Further we discuss steps related to the data preparation and pre-processing. Finally, we present our neural model architecture.

3.1   Problem Definition

We model the text segmentation problem as a binary classification problem. Given a document, we define the problem with respect to the ith sentence in the document, as follows.

Given: A sentence s_i with its K sized *left-context* $\{s_{i-K}, \ldots, s_{i-1}\}$ (i.e., K sentences before s_i) and K sized *right-context* $\{s_{i+1}, \ldots, s_{i+K}\}$ (i.e., K sentences after s_i). Here K is the context size.
Predict: Whether the sentence s_i denotes the beginning of a new text segment.

In this paper, we propose a neural framework to tackle this problem. Using a neural framework, we aim at using the context for learning distinctive features for sentences that mark the beginning of the segment. The architecture of the proposed model is illustrated in Fig. 1.

3.2   Data Preparation

In this section, we discuss two main steps in data preparation: pre-processing and custom batch creation to incorporate neighboring context.

Data Pre-processing. We fix the length of sentences to L words and truncate/pad as required to achieve appropriate fixed length embedding of the sentences. To represent words, we use the 300D word2vec[1] embeddings which are trained on Google News dataset containing ~100B words with a vocabulary size of ~3M words.

Let V represent the vocabulary, and let d be the word embedding size. Let $E^{V \times d}$ be the embedding matrix whose each row represents the embedding of

---

[1] https://code.google.com/archive/p/word2vec/.

a particular word in the model vocabulary. Let η_i be a matrix whose j^{th} row corresponds to the one-hot representation of the jth word of the sentence s_i. Thus, η_i has L rows and V columns. Given the word embedding matrix E, we obtain the sentence embedding matrix $e(s_i)$ for sentence s_i as $e(s_i)^{L \times d} = \eta_i^{L \times V} \times E^{V \times d}$.

While creating η_i for a sentence s_i, we perform basic text cleaning steps like skipping the punctuations and stop words. For all the missing words in the word2vec vocabulary we perform WordNet-based lemmatization and use the lemmatized word instead. If the embedding is still missing from the vocabulary then we replace it with a special token $\langle UNK \rangle$.

Custom Batch Creation. We wish to exploit the context around a sentence to decide whether the sentence indicates a segment boundary. For a sentence s_i, let lc_i and rc_i be the one-hot representations of the *left-context* and *right-context* respectively. Both lc_i and rc_i therefore contain $K \times L$ words each. Their embeddings $e(lc_i)$ and $e(rc_i)$ can then be computed as $e(lc_i)^{K \times L \times d} = lc_i^{K \times L \times V} \times E^{V \times d}$ and $e(rc_i)^{K \times L \times d} = rc_i^{K \times L \times V} \times E^{V \times d}$ respectively. Note that we also refer to the sentence s_i as the *mid-sentence*. We perform padding as required to obtain a fixed length representation of size $K \times L \times d$ for both the contexts. The input to the model is a batch of samples with the ith sample, S_i, defined as the concatenation of the embeddings of the *left-context, mid-sentence* and the *right-context* as follows.

$$S_i = [e(lc_i)^{K \times L \times d}, e(s_i)^{L \times d}, e(rc_i)^{K \times L \times d}] \tag{1}$$

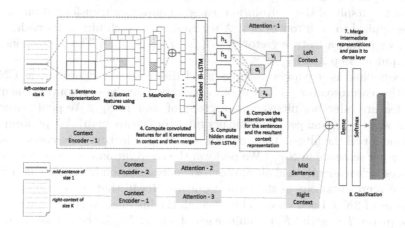

Fig. 1. Architecture diagram for the proposed model

The context size K should be such that the covered neighborhood information is enough to make conclusive decision about the current sentence being a segment boundary or not. Higher values of K provide the model with extra unnecessary context along with increase in the number of parameters. Lower values of K reduces the model complexity, but also restricts the model's ability to capture relations across near sentences only. K can be tuned using validation data. We study sensitivity of results to variation in K in Sect. 5.

3.3   The Proposed Neural Model

We now discuss in detail the proposed neural network model as illustrated in Fig. 1. The data pre-processing and sentence embedding discussed in the previous sub-section provides us instances of the form S_i, which consist of *left-context*, *right-context* and the *mid-sentence* word-embedding representations. This corresponds to the output at Step 1 in Fig. 1.

CNN Transformations. We leverage the widely used CNN architecture to obtain rich feature representations for each sentence in the *left-context, mid-sentence* as well as the *right-context*. Recall that the embedding S_i has $2K + 1$ rows each having $L \times d$ dimensions. Let us denote the jth such embedding matrix in S_i as $S_{(i,j)}^{L \times d}$. On each such $S_{(i,j)}$, we perform 1D Convolution operations with z number of filters. Let us denote the lth filter with a set of weights $\{\omega_l^{h \times d}, b_l\}$. Such a filter with height h can be applied on the input $S_{(i,j)}$ to obtain feature maps as follows.

$$f_{kl} = \phi(\omega_l^{h \times d} \cdot S_{(i,j)}[k - \frac{h}{2} : k + \frac{h}{2}]^{h \times d} + b_l) \tag{2}$$

Note that the convolution operations on text data involve filters with width same as input dimensionality (d). Thus, a filter has dimensions $d \times h$. Here, f_{kl} denotes the result of the convolution using a non-linear transformation ϕ. The filter is applied to each row of $S_{i,j}$. After applying z such filters, for each row k of $S_{i,j}$, we obtain a feature vector $f_k = \{f_{k1}, f_{k2}, \ldots, f_{kz}\}$. This corresponds to the output at Step 2 of Fig. 1.

Max-pooling is a popular sample-based discretization operation in CNNs. Given the feature vector f_k, max pooling operation involves computing the maximum feature value per filter across a group of rows in $S_{i,j}$. We pool across all the L rows in $S_{(i,j)}$ and get one value per filter (or feature map). We perform this operation for all the filters. Thus, overall, we obtain a feature rich representation of size z per sentence in $S_{(i,j)}$. We perform the convolution operation for all the sentences independently and obtain context representation by concatenating the sentence representations in the same sequence. Recall that S_i contained representations of $2K + 1$ sentences. Thus, overall the instance S_i is now represented by a sequence TS_i with $2K + 1$ units each of size z. TS_i is the output at Step 4 of Fig. 1.

We use shared filters for the *left-context* and *right-context* because: (1) It reduces the number of trainable parameters drastically making it easier for the model to train. (2) The representation vectors generated for both the *right-context* and *left-context* have the same semantics and lie in the similar vector space.

Stacked BiLSTMs with Attention. The problem could have been modeled as a sequence to sequence label generation task where each training sample is a whole document. But this model would be difficult to generalize for variable document length. Also, LSTMs have been shown to work well for shorter sequences. Hence, we first used CNNs to generate sentence embeddings and then use BiLSTM network on a smaller sequence that consists of only the main sentence and its neighbors.

To obtain a unified rich feature representation, we use Attention Bidirectional Long-Short Term Memory Network (Attention-BiLSTM) on top of this sequence TS_i. LSTMs [9] have been shown to model sequences better than vanilla Recurrent Neural Networks (RNNs) for various NLP tasks. LSTMs keep memories to capture long range dependencies. These memory cells allow error messages to flow at different strengths depending on the inputs. LSTMs have the ability to control the flow of information that flows to the memory cell state by using structures called *gates*. The reader is referred to [9] for details about LSTMs. To obtain a unified context representation which has rich feature set, we use a bidirectional LSTM (BiLSTM). The resultant embeddings are the concatenation of the two embeddings obtained through a forward pass LSTM and a reverse pass LSTM, capturing information from both the directions.

As shown in Fig. 1, we model the sequence of size K for both the left and the right context parts of S_i using separate LSTM networks each having K such memory cells, intuition being, two sentences at equal distance from the middle sentence might not have similar effect on the sentence being a segment boundary.

Traditional sequence encoder architecture which uses stacked BiLSTMs, forces the encoder to capture the information in a single fixed length representation. This also has a drawback as the hidden state at h_t is dependent on h_{t-1} across consecutive layers in the stack, thus the hidden state h_0 will have a significant effect on the future states. To overcome this, we use two vertically stacked BiLSTMs followed by soft attention [29]. Attention allows the model to give more importance to certain set of sentences in the context while ignoring the others, effectively learning the focus points to better predict if a sentence forms a segment boundary. We introduce an attention vector, z_s and use it to measure the relative importance of the sentences in the context as follows. Let $H_i^{K \times sz}$ denote the output of the last BiLSTM layer. Here sz is the size of the BiLSTM output corresponding to a sentence in the context. As shown in Fig. 1, $H_i = \{h_1, \ldots, h_K\}$.

$$e_i^{K \times 1} = H_i^{K \times sz} \times W^{sz \times 1} + b_i^{K \times 1} \tag{3}$$

$$a_i = \exp(\tanh(e_i^T z_s)), \quad \alpha_i = \frac{a_i}{\sum_p a_p}, \quad v_i = \sum_{j=1}^{K} \alpha_j h_j \tag{4}$$

The resultant context embedding, v_i, which is jointly learned during the training process captures the essential information from the context. We compute this context embedding for both the *left-context* and *right-context*. The merged vector corresponds to the output at Step 7 of Fig. 1.

Dense Fully Connected Layer with Softmax. The resultant embeddings obtained from a shared encoder for the *left-context* and *right-context* and the embedding for *mid-sentence* obtained from separate but similar encoder are passed to a dense fully connected layer followed by an output.

$$P(y|s_i) = \text{softmax}(W_h \times v_i + b_i), \quad y_{pred} = \arg\max P(y|s_i) \qquad (5)$$

For classification we have a softmax layer over the output vectors. Finally, we take arg max over the predicted probability distribution to generate predictions.

4   Experiments

In this section, we discuss datasets, metrics and parameter settings for our experiments. Source code and datasets are available at https://github.com/pinkeshbadjatiya/neuralTextSegmentation.

4.1   Datasets

Text segmentation is quite a subjective task making evaluation of the text segmentation systems challenging. Hence, we use standard benchmark datasets for evaluation. Figure 2 shows summary of statistics about the datasets.

1. **Clinical** [12]: Consists of a set of 227 chapters from a medical textbook. Each chapter is marked into sections indicated by the author which forms the segmentation boundaries. It contains a total of 1136 sections.
2. **Fiction** [11]: Consists of a collection of 85 fiction books downloaded from Project Gutenberg. Segmentation boundaries are the chapter breaks in each of the books.
3. **Wikipedia:** Consists of randomly selected set of 300 documents having an average segment size of 26. These documents widely fall under the narrative category. Each document is divided into sections as marked in the original XML dump of the website. We use these section markers to create a segmentation boundary.

		Clinical	Fiction	Wikipedia
#Documents		227	85	300
#Sentences or #Samples		31868	27551	58071
Segment Length	Mean	35.72	24.15	25.97
	Std Dev	29.37	18.24	9.98

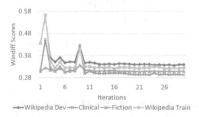

Fig. 2. Statistics of various datasets used for performance evaluation

Fig. 3. Variation of windiff scores wrt iterations for various datasets

4.2   Metrics

We evaluate the performance of the model with respect to two metrics: Pk [1] and WinDiff [17]. Both the metrics use a sliding window of fixed size w over the document and compare the hypothesized segments with the reference ones. The window size, k, is generally set to half the gold-standard segmentation length [1]. Pk is the probability that two segments drawn from a document are incorrectly identified as belonging to the same segment. Windiff moves a sliding window across the text and counts the number of times the hypothesized and reference segment boundaries are different within the window. Counts are scaled to obtain probability values. Both Pk and Windiff thus lie between 0 and 1 and an algorithm that assigns all boundaries correctly receives a score of 0. WinDiff is considered a better measure than Pk as the Pk metric suffers from multiple issues as described in [17]. Pevzner et al. [17] proposed WinDiff as an update to the Pk metric. *Both the metrics are a loss measure. The lower the score, the better.*

4.3   Model Parameters and Training

For the purpose of training, we randomly select a set of pages from Wikipedia. We use section splits as our segment splits for training the model and skip the section headers, considering only the section content for training. We perform a 80-20 training-development split to obtain the training and development datasets. Figure 3 shows variation of WinDiff scores on the Wikipedia development dataset and other datasets as training progresses. The figure suggests that training converges well after 30 iterations, hence we fix number of iterations as 30. We trained our model on ~270 documents from a sample of the Wikipedia corpus, creating ~49k training sentences/samples. The average segment size of the whole Wikipedia corpus is around 9 sentences, while the test datasets have higher segment sizes (Fig. 2). We filter the documents that have average segment size less than 20 which results in training set having average segment length of 25. We train our model in batches of size 40. We set the context size, K, to 10 for all the experiments mentioned in Table 1.

The training dataset class distribution is heavily skewed with about 92% samples belonging to class 0. Hence, we use weighted-binary-cross-entropy as our loss function to penalize the classifier more heavily on mis-classification of a segment boundary. The loss function is defined as $loss = -\frac{1}{N}\sum_{i=1}^{N}(t\log(o) + \frac{f_1}{f_0}(1 - t)\log(1 - o)))$ where t and o are the target and the predicted outputs respectively. f_0 and f_1 are the frequencies of class 0 and class 1 respectively.

We use 'AdaDelta' [33] as the optimizer and use dropouts of 0.2–0.3 for input and recurrent gates in the recurrent layers. We also use dropouts of 0.3 after the dense fully connected layers to prevent over fitting on the training dataset. We use filters of sizes $\{2, 3, 4, 5\}$ with 200 filters for each of the sizes. The recurrent layers have 600 neurons.

4.4 Comparison with Other Baseline Methods

We compare the performance of our proposed model against various competitive baselines, four basic neural models, and three BiLSTM model variants. Each of those models help us understand contributions of the various components of the proposed model. Table 1 shows the summary of the results obtained using various models on all three benchmark datasets. In the following, we describe the baseline systems in brief.

1. **U&I** [25]: It is a probabilistic framework based on maximizing the compactness of the language models induced for each segment using ideas similar to the noisy channel and minimum description length methods.
2. **MinCut** [12]: This method treats text segmentation as a graph-partitioning task aiming to optimize the normalized-cut criterion. It simultaneously optimizes the total similarity within each segment and dissimilarity across segments.
3. **BayesSeg** [4]: This method models the words in each topic segment as draws from a multinomial language model associated with the segment. Segmentation is obtained by maximizing the observation likelihood in such a model.
4. **APS** [11]: Affinity Propagation for Segmentation receives a set of pairwise similarities between data points and produces segment boundaries and segment centers. Data points which best describe all other data points within the segment are considered segment centers. APS iteratively passes messages in a cyclic factor graph, until convergence.
5. **PLDA** [18]: PLDA is a generative model that uses Bayesian inference to simultaneously address the problems of topic segmentation and topic identification. It chains a set of LDAs by assuming a Markov structure on topic distributions.
6. **TSM** [2]: Structured Topic Model is a hierarchical Bayesian model for unsupervised topic segmentation. It uses an MCMC inference to split/merge segment(s).

For all these baseline algorithms, we use the publicly available source codes. We also fine tune the parameters, using the scripts provided by the authors, for our experiment on the Wikipedia dataset to get the optimal set of parameters. We could not perform some experiments where source codes were not available publicly. We mark those instances with NA. We also compare the performance on "Random" baseline where we place the segment boundaries randomly in the text. Part A of Table 1 shows the observed results from our experiments on these baselines.

To understand the contribution of various components in our proposed model we compare the performance of other neural models as well without using any context information. We discuss these neural models in brief below. Part B of Table 1 presents the performance for four such neural architectures.

1. **Perceptron:** We encode the sentence using mean of the word2vec representations of the corresponding words and then learn a 5-layered perceptron.

2. **LSTM:** We represent each sentence using a sequence of words and then learn a combined dense representation for each sentence in the vector space using the word2vec embeddings for words.

3. **Stacked-LSTM:** This model is similar to LSTM, except that it provides more flexibility at the cost of more trainable parameters.

4. **CNN:** We use the CNN based sentence representations obtained by convolving multiple variable length filters with the word embeddings to obtain rich feature representations for each sentence.

We also compare the performance of our proposed model with other BiLSTM based neural models. Each of these models obtain specific sentence representations which are then passed to a BiLSTM architecture to obtain context representations. Part C of Table 1 presents the performance for these neural architectures besides the proposed method, CNN+Attn-BiLSTM.

1. **MeanBoW-BiLSTM:** We use mean of all the word vectors to obtain a sentence representation, which along with its neighboring context, is passed to a stacked BiLSTM encoder architecture to obtain the context representations.

2. **TFIDF MeanBoW-BiLSTM:** To obtain a sentence embedding, we compute weighted mean of the word2vec word embeddings where TF-IDF (Term Frequency-Inverse Document Frequency) scores are used as weights. TF-IDF scores capture the relative relevance of a particular word in the sentence.

3. **CNN-BiLSTM:** We use the CNN based sentence representations obtained by convolving multiple variable length filters with the word embeddings to obtain rich feature representations for each sentence. There is no attention layer in this method.

Table 1. Accuracy comparison of the proposed approach with competitive baselines. **Lower values are better.** Experiments marked with **NA** could not be performed due to non-availability of publicly available source codes. Some of the cell values have been directly taken from respective papers, if they mentioned them for the same (method, dataset) pair.

	Model	Clinical		Fiction		Wikipedia	
		WinDiff	Pk	WinDiff	Pk	WinDiff	Pk
Part A: Competitive baselines	U&I [25]	0.376	0.370	0.459	0.459	0.368	0.368
	MinCut [12]	0.382	0.368	0.405	0.371	0.389	0.364
	BayesSeg [4]	0.353	0.339	0.337	**0.278**	0.390	0.359
	APS [11]	0.399	0.396	0.480	0.451	0.380	0.392
	PLDA [18]	0.373	0.324	0.430	0.361	NA	NA
	TSM [2]	0.345	**0.306**	0.408	0.325	NA	NA
	Random	0.459	0.441	0.510	0.475	0.486	0.480
Part B: Neural models without context	Perceptron	0.338	0.357	0.336	0.335	0.421	0.415
	Stacked-LSTMs	0.381	0.393	0.329	0.394	0.437	0.420
	LSTMs	0.486	0.471	0.366	0.417	0.508	0.455
	CNN	0.309	0.329	0.314	0.386	0.363	0.380
Part C: Context based neural models	MeanBoW + BiLSTM	0.349	0.365	0.319	0.389	0.405	0.398
	TF-IDF MeanBoW + BiLSTM	0.345	0.366	0.328	0.382	0.382	0.392
	CNN + BiLSTM	0.334	0.316	0.331	0.324	0.378	**0.328**
	CNN + Attn-BiLSTM	**0.294**	0.318	**0.308**	0.378	**0.315**	0.344

5   Analysis of Results

In this section we analyze the results of our experiments and compare their performance with the existing non-neural as well as neural baselines. We also briefly discuss about the choice of metrics in Subsect. 5.4.

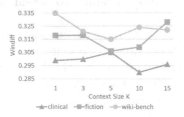
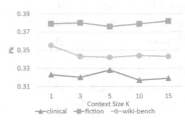

Fig. 4. Variation of WinDiff scores on various datasets with varying context size K

Fig. 5. Variation of Pk scores on various datasets with varying context size K

5.1   Comparison with Baseline Models

Table 1 compares the performance of the proposed Attention-based model on various benchmark datasets with other existing models. We observe that the use of Attention-based supervised models provide a performance improvement over other methods across all the datasets on the Windiff metric, and compares well with the best method on the Pk metric.

The proposed model has additional benefits with respect to runtime performance. Once the training is finished, during the prediction phase, our model takes on an average 0.09 s on a batch of 40 sentences on GeForce GTX 1060 GPU which is much faster than other methods in Part A of Table 1 which take time to the order of minutes to days during prediction phase as most of the computation takes place during that time.

5.2   Comparison with Neural Models

It is important to note that without context the neural models (Part B) sometimes perform worse than the baseline models (Part A). All of the variants of LSTMs in Part C are better than the context-unaware LSTM model in Part B. In Part C, we note that the use of TFIDF-weighted-Mean BoW word2vec embeddings for representing sentences (TFIDF MeanBoW + BiLSTM) only provides a slight improvement. We conclude that for the task of text segmentation, TF-IDF features do not add additional information as compared to the word-embeddings.

Our experiments with the CNN model provides us with good results compared to other model variants overall. Results in Part B show improved performance even without using any context information with the use of CNN for obtaining sentence representations. Our experiments with CNNs along with BiLSTM do not show much improvement in the results on the Windiff metric, though they show significant improvement on the Pk metric results across all the datasets encouraging us to use BiLSTM as part of our proposed model. The use of Attention further improves the results by a significant margin across all the datasets.

We also observe improved performance with the use of Attention on the neighboring context. Use of soft Attention [29] allows the model to focus on certain regions more than others in order to generate better context representations. Using the attention layer shows an increase in Windiff performance by 4%, 2.3% and 6.3% on Clinical, Fiction and Wikipedia datasets respectively over the CNN+BiLSTM model. Overall, the proposed model shows an improvement of 5.1%, 10% and 6.5% on Windiff metric on Clinical, Fiction and Wikipedia datasets respectively, and improvement of 3.1% on Pk metric on the Wikipedia dataset compared to the existing competitive baselines.

5.3 Varying Context Size K

We also experiment with the variation in context size K on all the datasets and report the results in Figs. 4 and 5. We train our model on the same training dataset for 30 epochs and report the results with variable context sizes. We observe a decreasing trend (recall low Windiff scores are good) in the WinDiff scores as Context size increases which gradually starts increasing as context size grows. All the three datasets follow this trend for the Windiff metric, while only the Fiction and Wikipedia datasets have shown similar results for Pk metric. For Clinical dataset, we lose a lot of domain specific information while converting words to vectors using the word-embeddings as word2vec is not very rich in domain specific information. This leads to poor results on the Windiff metric with low context-size. As context size grows, it gathers enough information to correctly classify the segment boundaries.

5.4 Implication of Windiff and Pk Metric Results

Since the models are trained on the Wikipedia Dataset, it is focused towards presenting an evenly spread out segmentation of the paragraph, as learned from the Wikipedia documents, which often have a uniform section distribution. The Clinical and Fiction datasets both have quite a significant number of segments with less number of sentences (as evident from the very high standard deviation reported in Fig. 2) resulting in less number of predicted segment boundaries. Assuming a window of size λ, each miss of segmentation boundary will produce a false negative. Each such false negative will receive a total of λ penalties. Since the model often results in less number of segmentations, it often receives higher number of penalties than expected, resulting in higher Pk scores. Hence, as also seen in Table 1, baseline neural models consistently perform poorly on the Pk metric. But the Pk scores for neural models on Wikipedia dataset are comparable to the state-of-the-art methods.

Choice of WinDiff over Pk: Pk metric suffers from multiple issues due to which it does not provide a good measure of the hypothesized segmentations. These issues are covered in detail by Pevzner et al. [17], motivating us to use the WinDiff metric as our primary measure for comparing performance with the existing models and baselines. We still report the results in both the metrics as they might provide beneficial information to the readers.

6   Conclusions

In this paper, we studied the problem of text segmentation from a neural perspective. We presented a model which first learns rich features for every sentence using Convolutional Neural Networks followed by sequential learning using temporal data. Finally, we also learn focus on various sentences in the context using the attention layer. We performed extensive experiments to compare against well-established non-neural baselines, as well as against recent neural models. Experimenting with three different datasets, we empirically proved that our proposed model provides lowest Windiff loss with very little supervision, and with low execution times.

In the future, we plan to test this model on non-English datasets, especially morphologically rich languages where huge datasets are not available for training the model and most of the contextual information is captured at the word level rather than at the sentence level.

References

1. Beeferman, D., Berger, A., Lafferty, J.: Statistical models for text segmentation. Mach. Learn. **34**(1), 177–210 (1999)
2. Du, L., Buntine, W.L., Johnson, M.: Topic Segmentation with a structured topic model. In: HLT-NAACL, pp. 190–200 (2013)
3. Du, L., Pate, J.K., Johnson, M.: Topic segmentation in an ordering-based topic model. In: AAAI, pp. 2232–2238 (2015)
4. Eisenstein, J., Barzilay, R.: Bayesian unsupervised topic segmentation. In: EMNLP, pp. 334–343. ACL (2008)
5. Galley, M., McKeown, K., Fosler-Lussier, E., Jing, H.: Discourse segmentation of multi-party conversation. In: ACL, pp. 562–569 (2003)
6. Grosz, B., Hirschberg, J.: Some intonational characteristics of discourse structure. In: Proceedings of the Second International Conference on Spoken Language Processing (1992)
7. Hajime, M., Takeo, H., Manabu, O.: Text segmentation with multiple surface linguistic cues. In: ACL, pp. 881–885 (1998)
8. Hearst, M.A.: TextTiling: segmenting text into multi-paragraph subtopic passages. Comput. Linguist. **23**(1), 33–64 (1997)
9. Hochreiter, S., Schmidhuber, J.: Long short-term memory. Neural Comput. **9**(8), 1735–1780 (1997)
10. Joty, S., Carenini, G., Murray, G., Ng, R.T.: Supervised topic segmentation of email conversations. In: ICWSM (2011)
11. Kazantseva, A., Szpakowicz, S.: Linear text segmentation using affinity propagation. In: EMNLP, pp. 284–293. ACL (2011)

12. Malioutov, I., Barzilay, R.: Minimum cut model for spoken lecture segmentation. In: ACL, pp. 25–32 (2006)
13. Misra, H., Yvon, F., Jose, J.M., Cappe, O.: Text segmentation via topic modeling: an analytical study. In: CIKM, pp. 1553–1556. ACM (2009)
14. Mitrat, M., Singhal, A., Buckley, C.: Automatic text summarization by paragraph extraction. In: Intelligent Scalable Text Summarization (1997)
15. Nguyen, V.A., Boyd-Graber, J., Resnik, P.: SITS: a hierarchical nonparametric model using speaker identity for topic segmentation in multiparty conversations. In: COLING, pp. 78–87 (2012)
16. Oh, H.J., Myaeng, S.H., Jang, M.G.: Semantic passage segmentation based on sentence topics for question answering. Inf. Sci. **177**(18), 3696–3717 (2007)
17. Pevzner, L., Hearst, M.A.: A critique and improvement of an evaluation metric for text segmentation. Comput. Linguist. **28**(1), 19–36 (2002)
18. Purver, M., Griffiths, T.L., Körding, K.P., Tenenbaum, J.B.: Unsupervised topic modelling for multi-party spoken discourse. In: ACL, pp. 17–24 (2006)
19. Reynar, J.C.: Statistical models for topic segmentation. In: ACL, pp. 357–364 (1999)
20. Riedl, M., Biemann, C.: TopicTiling: a text segmentation algorithm based on LDA. In: ACL Student Research Workshop, pp. 37–42 (2012)
21. Rush, A.M., Chopra, S., Weston, J.: A neural attention model for abstractive sentence summarization. In: EMNLP, pp. 379–389. ACL (2011)
22. Sakahara, M., Okada, S., Nitta, K.: Domain-independent unsupervised text segmentation for data management. In: ICDMW, pp. 481–487 (2014)
23. Sheikh, I., Fohr, D., Illina, I.: Topic segmentation in ASR transcripts using bidirectional RNNs for change detection. In: IEEE Automatic Speech Recognition and Understanding Workshop (2017)
24. Tür, G., Hakkani-Tür, D., Stolcke, A., Shriberg, E.: Integrating prosodic and lexical cues for automatic topic segmentation. Comput. Linguist. **27**(1), 31–57 (2001)
25. Utiyama, M., Isahara, H.: A statistical model for domain-independent text segmentation. In: ACL, pp. 499–506 (2001)
26. Dijk, T.A.V.: Episodes as units of discourse analysis. In: Analyzing Discourse: Text and Talk, pp. 177–195 (1982)
27. Wang, L., Li, S., Xiao, X., Lyu, Y.: Topic segmentation of web documents with automatic cue phrase identification and BLSTM-CNN. In: Lin, C.-Y., Xue, N., Zhao, D., Huang, X., Feng, Y. (eds.) ICCPOL/NLPCC -2016. LNCS (LNAI), vol. 10102, pp. 177–188. Springer, Cham (2016). https://doi.org/10.1007/978-3-319-50496-4_15
28. Wang, L., Li, S., Lyu, Y., Wang, H.: Learning to rank semantic coherence for topic segmentation. In: EMNLP, pp. 1340–1344. ACL (2017)
29. Xu, K., Ba, J., Kiros, R., Cho, K., Courville, A., Salakhudinov, R., Zemel, R., Bengio, Y.: Show, attend and tell: neural image caption generation with visual attention. In: ICML, pp. 2048–2057 (2015)
30. Yang, Z., Yang, D., Dyer, C., He, X., Smola, A., Hovy, E.: Hierarchical attention networks for document classification. In: HLT-NAACL, pp. 1480–1489 (2016)
31. Yin, W., Schütze, H., Xiang, B., Zhou, B.: ABCNN: attention-based convolutional neural network for modeling sentence pairs. In: ACL, pp. 259–272 (2016)
32. Yu, J., Xiao, X., Xie, L., Chng, E.S.: Topic embedding of sentences for story segmentation. In: APSIPA ASC (2017)
33. Zeiler, M.D.: ADADELTA: an adaptive learning rate method. arXiv preprint arXiv:1212.5701 (2012)

Evaluations and User Behavior

Modelling Randomness in Relevance Judgments and Evaluation Measures

Marco Ferrante[1], Nicola Ferro[2]([✉]), and Silvia Pontarollo[1]

[1] Department of Mathematics, University of Padua, Padua, Italy
{ferrante,spontaro}@math.unipd.it
[2] Department of Information Engineering, University of Padua, Padua, Italy
ferro@dei.unipd.it

Abstract. We propose a general stochastic approach which defines relevance as a set of binomial random variables where the expectation p of each variable indicates the quantity of relevance for each relevance grade. This represents the first step in the direction of modelling evaluation measures as a transformation of random variables, turning them into random evaluation measures. We show that a consequence of this new approach is to remove the distinction between binary and multi-graded measures and, at the same time, to deal with incomplete information, providing a single unified framework for all these different aspects. We experiment on TREC collections to show how these new random measures correlate to existing ones and which desirable properties, such as robustness to pool downsampling and discriminative power, they have.

1 Introduction

Relevance judgements are at the core of *Information Retrieval (IR)* evaluation since they determine and inform all the subsequent scoring and comparison of IR systems. For this reason, over the years, a lot of effort has been put in their creation and in ensuring their quality, see e.g. [7,20], also in a crowd-sourcing context [1].

We know that relevance assessment is a not deterministic process, as witnessed by different studies on inter-assessor agreement [16,17] and as exploited by algorithms to merge relevance labels in a crowd-sourcing context [5]. However, once relevance judgments have been created – either by traditional assessors or with sophisticated algorithms merging labels from crowd-assessors – we seem to forget their intrinsic randomness and we consider them as if they were deterministic: for example, evaluation measures just handle the relevance judgment associated with a document as exact.

In this paper, we move a step forward to account for the intrinsic randomness in relevance judgements and we frame them into a general stochastic approach where the judgement assigned to a document is a binomial random variable whose expectation p indicates the quantity of relevance assigned to that document.

G. Pasi et al. (Eds.): ECIR 2018, LNCS 10772, pp. 197–209, 2018.
https://doi.org/10.1007/978-3-319-76941-7_15

We show how to apply the proposed framework to the definition of *random evaluation measures*, i.e. IR evaluation measures able to incorporate the inherent randomness in relevance judgements, and how this new approach not only eliminates the distinction between binary and multi-graded evaluation measures but also deals with incomplete information, by providing us with a single unifying vision which can be coherently applied to all the IR evaluation measures.

We apply our framework to two widely known measures, namely *Average Precision (AP)* and *Rank-Biased Precision (RBP)* [10], in order to show the generality of the proposed solution. We also conduct a systematic experimentation using TREC collections which shows that these new random evaluation measures are a coherent extension of their non-random counterparts and that they have many desirable properties in terms of robustness to incomplete information and sensitivity in discriminating among systems.

The paper is organized as follows: Sect. 2 discusses some related works; Sect. 3 introduces our stochastic framework; Sect. 4 reports the evaluation of the proposed approach; and, Sect. 5 draws some conclusions and outlooks possible future works.

2 Related Work

To the best of our knowledge, this paper represents one of the first attempts to explicitly model relevance judgements as a stochastic process with the specific goal to introduce random IR evaluation measures, benefiting from a single unifying view on multi-graded measures and incomplete information.

One of the closest areas is dealing with incomplete information in relevance judgments, i.e. how to account for unjudged documents [2,11,14,18]. All these works differ from our approach in that they focus on unjudged documents in the pool and how to reliably estimate a proportion of relevant documents for them. On the contrary, we model each single relevance judgement as a binomial random variable and we derive a general stochastic framework where evaluation measures account for randomness in the assessment of each retrieved document, both judged and unjudged documents in the pool.

Finally, when it comes to multi-graded judgements, either we have evaluation measures which are natively multi-graded, such as *Normalized Discounted Cumulated Gain (nDCG)* [6] and *Expected Reciprocal Rank (ERR)* [3], or extensions from the binary to the multi-graded case, such as *Graded Average Precision (GAP)* [12]. However, all these cases treat relevance judgements as deterministic and the extensions from the binary to the multi-graded case are typically ad-hoc, i.e., they work only for a specific measure, while our approach is general and can be seamlessly applied to any IR evaluation measure.

3 Proposed Stochastic Model

3.1 Random Relevance

We stem from the notation proposed by [4] for defining the basic concepts of topics, documents, ground-truth, run and judged run and we extend it to account for random relevance instead of deterministic one.

Let us consider a set of **documents** D and a set of **topics** T. Let $(REL, \preceq)$ be a totally ordered set of **relevance degrees**, where we assume the existence of a minimum that we call the **non-relevant** relevance degree $\mathbf{nr} = \min(REL)$. We assume that REL is a finite set. Moreover, given $m \in \mathbb{N}$ such that $|REL| = m+1$, we denote its strictly ordered elements as $rel_0 \prec \cdots \prec rel_m$, where $rel_0 = \mathbf{nr}$.

For each pair $(t, d) \in T \times D$, the **ground-truth** GT is a map which assigns a relevance degree $rel \in REL$ to a document d with respect to a topic t. The **recall base** is the map RB from T into $\mathbb{N}$ defined as the total number of relevant documents for a given topic $t \mapsto RB_t = \left| \{ d \in D : GT(t, d) \succ \mathbf{nr} \} \right|$.

Given a run $r_t = (d_1, \ldots, d_N)$ of length N, let $\hat{r}_t[i]$ be the relevance assigned to the document d_i for the topic t, i.e. $\hat{r}_t[i] = GT(t, d_i)$.

Given a positive integer N, the length of the run, we define the **set of retrieved documents** as $D(N) = \{ (d_1, \ldots, d_N) : d_i \in D, d_i \neq d_j \text{ for any } i \neq j \}$, i.e. the ranked list of retrieved documents without duplicates, and the **universe set of retrieved documents** as $\mathcal{D} := \bigcup_{N=1}^{|D|} D(N)$.

The already existing binary (when $m = 1$) and multi-graded ($m > 1$) evaluation measures usually map each relevance degree into an integer number. For example, if $REL = \{ \mathbf{nr}, \mathbf{r} \}$, then AP assigns the value 0 to every non-relevant document while 1 is used for the relevant ones. Similarly, if $REL = \{ \mathbf{nr}, \mathbf{pr}, \mathbf{r}, \mathbf{hr} \}$, nDCG [6] assigns an integer number to each relevance degree, e.g. 0, 5, 10 and 15, consistently with the ordering among the relevance degrees. If it is very natural to assign 0 to a non-relevant document and 1 to a relevant one, being this latter value just any possible positive number different from zero that simply indicates the "presence" of some relevance, the situation is not so clear in the case of multi-graded relevance. For example, if 5 is the value assigned to a partially relevant (**pr**) document and 10 is the one for a relevant document (**r**), this does not necessary mean that relevant documents are twice as relevant as partially relevant ones, even though their contribution to some measures, e.g. nDCG, is actually doubled.

Could there exist a right or at least a common way to assign integers to different degrees of relevance? For example, [9] proposed magnitude estimation as a way to let users to estimate relevance on their own scale and raised the question whether a single view of relevance is actually appropriate to describe a population of users? The answer to this question is not easy and in the present paper, to account for a population of users, we consider the relevance of each document as a **random** number chosen between $\{0, 1\}$, where again 0 means completely "non-relevant" and 1 means "fully relevant".

Therefore, we describe the relevance of a document via a **binomial random variable** $B(1, p)$ with parameters 1 and p, where p roughly defines the *quantity* *of relevance* of that document. Recall that such a binomial random variable is a function from Ω, i.e. a suitable sample space, into $\{0,1\}$ and it is equal to 1 with probability p and 0 with probability $1-p$.

In accordance with this construction, we redefine the ground-truth as follows: for each pair $(t, d_i) \in T \times D$, the **random Ground-truth** RGT, also called random relevance, is a binomial random variable of parameters $(1, p_{t,d_i})$, where p_{t,d_i} is the parameter associated to the document d_i with respect to a topic t. $p_{t,d_i} = 0$ corresponds to a document completely not relevant and $p_{t,d_i} = 1$ to a fully relevant document. For simplicity, in the sequel we will write $p_{t,i}$ instead of p_{t,d_i}. Moreover, we replace the deterministic recall base RB_t defined before with $\widehat{RB}_t$, the expected total relevance present in D, i.e. $\widehat{RB}_t = \sum_{d \in D} \mathbb{E}[RGT(t,d)]$ whose true value will be most of the times just estimated.

Let $\mathcal{R}$ be the set $\bigcup_{N=1}^{|D|} \{0,1\}^N$; a **random judged run** is the function $\hat{r}_t$ from $\Omega \times T \times D$ into $\mathcal{R}$, which assigns a random relevance to each retrieved document in the ranked list

$$(\omega, t, r_t) \mapsto \hat{r}_t(\omega) = \big(RGT(t, d_1)(\omega), \ldots, RGT(t, d_N)(\omega)\big) .$$

3.2   Random Evaluation Measures

Generally speaking, a *random evaluation measure* is an application

$$M : \Omega \times T \times D \to \mathbb{R}_+$$

obtained by the composition of the random judged run with the map

$$\mu : \mathcal{R} \to \mathbb{R}_+$$

giving $M = \mu\big(RGT(t, d_1)(\omega), \ldots, RGT(t, d_N)(\omega)\big)$.

To show how to apply the proposed approach, we provide the definition of the random version of two well-known evaluation measures, namely RBP and AP.

Random Rank Biased Precision (RRBP) of parameter $q \in (0,1)$ is defined as

$$RBP[\hat{r}_t(\omega)] = (1-q) \sum_{n=1}^{N} q^{n-1} \hat{r}_t[n](\omega) .$$

where q denotes the persistence of the user in scanning the results list.

Random Average Precision (RAP) is defined as

$$AP[\hat{r}_t(\omega)] = \frac{1}{\widehat{RB}_t} \sum_{n=1}^{N} \left(\frac{1}{n} \sum_{m=1}^{n} \hat{r}_t[m](\omega) \right) \hat{r}_t[n](\omega) .$$

To compare different systems, we need to define an ordering among runs of documents. Since the relevance is now stochastic, the ordering of the systems has to be defined in terms of the laws of the random relevances of the documents retrieved in the runs.

Definition 1. *Given a topic* t, *two runs of documents* r_t *and* s_t *and a random evaluation measure* $M(\cdot, t)(\omega)$, *we define a weak order on* $\mathcal{R}$ *as*

$$r_t \preceq s_t \quad \Leftrightarrow \quad \mathbb{E}[M(r_t, t)(\omega)] \leq \mathbb{E}[M(s_t, t)(\omega)] \ .$$

Therefore let us now take into account the expectations of the random versions of RBP and AP. We assume the random relevances of different documents to be independent random variables.

We define the **expected Rank Biased Precision (eRRBP)** as the expectation of RBP when computed over runs with random relevance degrees. Since RBP is a linear combination of independent random variables, the computation of its mean is quite simple, giving rise to the following expression:

$$\mathbb{E}[RBP[\hat{r}_t(\omega)]] = (1 - q) \sum_{n=1}^{N} q^{n-1} p_{t,n} \ . \tag{1}$$

Similarly, **expected Random Average Precision (eRAP)** is the expectation of AP, whose computation is slightly more complicated, since we here have the sum of partial sums of the same random variables. The mean is:

$$\mathbb{E}[AP[\hat{r}_t(\omega)]] = \frac{1}{\widehat{RB}_t} \sum_{n=1}^{N} \frac{1}{n} \left(1 + \sum_{m=1}^{n-1} p_{t,m} \right) p_{t,n} \ , \tag{2}$$

where we have made use of the fact that all the moments of a $B(1, p_{t,k})$ random variable are equal to $p_{t,k}$.

Summing up, the proposed random measures decouple the problem of determining the presence of relevance from that of indicating the amount of relevance. Indeed, the former is represented by the output of the binomial random variables, either 0 in case of absence of relevance or 1 in case of presence of relevance; the latter is represented instead by the parameter p of the binomial random variables, which accounts for the amount of relevance. In this way, the same mechanism for indicating the presence and amount of relevance is used for both the binary and multi-graded case, thus eliminating the distinction between them. Furthermore, these random measures allow us to "seed" some relevance also for the not relevant documents by setting the parameter p slightly greater than 0 in that case. This is especially useful in the case of unjudged documents and incomplete information, since it allows us to somehow capture what we might call the "dark relevance" present in the document's universe. For these reasons, we can say that the proposed random measures are able to seamlessly describe both multi-graded and incomplete information.

4   Experiments

We focus on the following existing evaluation measures to compare ours against: nDCG [6] and ERR [3] as examples of natively multi-graded evaluation measures; GAP [12] as an example of extension of AP to graded judgments; and

Graded Rank-Biased Precision (gRBP) [15] as an example of use of RBP [10] with graded judgements; *Binary Preference (bpref)* [2] and *Inferred Average Precision (infAP)* [18] as examples of binary measures for incomplete information.

We used the following collections: *TREC Terabyte track* T14 using the GOV2 collection with 50 topics, deep pools at depth 100, and graded relevance judgments – i.e., not relevant, relevant and highly relevant; 58 runs were submitted, retrieving 1,000 documents for each topic; *TREC Web track* T21 using the ClueWeb09 collection with 50 topics, shallow pools using depths 20 and 30, and graded relevance judgments – i.e., junk, not relevant, relevant, highly relevant, key and nav; we considered junk and not relevant as a single not relevant level and key and nav as a single key level; 27 runs were submitted, retrieving 10,000 documents for each topic.

For nDCG we use a log base $b = 10$ and gains 0, 5, 10, and 15 for not relevant, relevant, highly relevant, and key documents, respectively. For ERR we instead use 0, 1, 2 and 3 as gains. For RBP we set the persistence parameter q to 0.8, which works well for both deep and shallow pools as pointed out by [10].

Although our approach provides a very fine-grained level of detail in defining the random relevance up to each (topic, document) pair, e.g. by using magnitude estimation techniques [9], in the following evaluation we let the parameter $p_{t,d}$ to be fixed for each relevance degree to a value p_k, independently from the document at hand, since this is the way in which all the IR measures we compare against work and this is the information available in the pools of the used collections. We can view each p_k as how much an assessor is confident that every given document with relevance degree equal to rel_k is actually relevant. The different values of the parameters p_k are reported in the caption of the figures which display the experimental results later on.

To ease the reproducibility of the experiments, the code for running them is available at: https://bitbucket.org/frrncl/ecir2018.

4.1   RQ1: Relation to Other Evaluation Measures

Figure 1 report the outcomes of the correlation analysis on T14 and T21, respectively, using both Kendall's τ correlation [8] and τ_{ap} correlation [19]. Each row represents an alternative configuration of the parameters ranging from hard to lenient in the sense that, for example in the case of three relevance degrees, GAP with threshold probabilities [0.00, 1.00] corresponds to AP when you perform a hard mapping to binary relevance, i.e. only the top relevance degree is considered relevant; on the other hand, GAP with threshold probabilities [1.00, 0.00] corresponds to AP when you perform a lenient mapping to binary relevance, i.e. every relevance degree above not relevant is considered relevant.

For each set of parameters (hard, medium, lenient), we explore two options for eRAP and eRRBP. Option 1 makes eRAP to behave as close as possible to GAP by constraining the eRAP probabilities à la GAP: for example, in the case of three relevance degrees if the GAP threshold probabilities are $[g_1, g_2]$ we

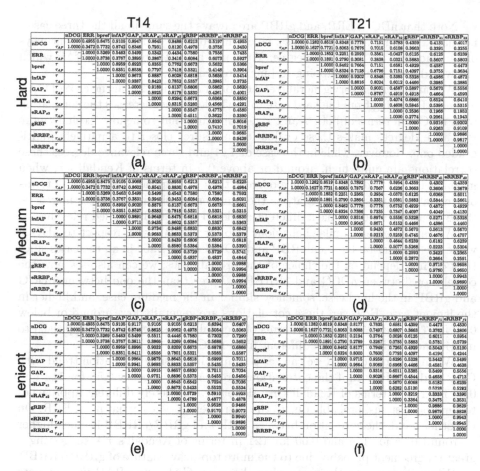

Fig. 1. Correlation analysis on T14 (first column) and T21 (second column) for different sets of parameters. The first row of subfigures reports the *hard case*: (a) GAP with threshold probabilities $a = [0.20, 0.80]$, eRAP end eRRBP with probabilities $a1 = [0.00, 0.20, 1.00]$ and $a2 = [0.05, 0.20, 0.95]$ on T14; (b) GAP with threshold probabilities $b = [0.10, 0.40, 0.50]$, eRAP end eRRBP with probabilities $b1 = [0.00, 0.10, 0.50, 1.00]$ and $b2 = [0.05, 0.10, 0.50, 0.95]$ on T21. The second row of subfigures reports the *medium case*: (c) GAP with threshold probabilities $c = [0.50, 0.50]$, eRAP end eRRBP with probabilities $c1 = [0.00, 0.50, 1.00]$ and $c2 = [0.05, 0.50, 0.95]$ on T14; (d) GAP with threshold probabilities $d = [0.20, 0.40, 0.40]$, eRAP end eRRBP with probabilities $d1 = [0.00, 0.20, 0.60, 1.00]$ and $d2 = [0.05, 0.20, 0.60, 0.95]$ on T21. The third row of subfigures reports the *lenient case*: (e) GAP with threshold probabilities $e = [0.70, 0.30]$, eRAP end eRRBP with probabilities $e1 = [0.00, 0.70, 1.00]$ and $e2 = [0.05, 0.70, 0.95]$ on T14; (f) GAP with threshold probabilities $f = [0.40, 0.40, 0.20]$, eRAP end eRRBP with probabilities $f1 = [0.00, 0.40, 0.80, 1.00]$ and $f2 = C2 = [0.05, 0.40, 0.80, 0.95]$ on T21.

constraint the eRAP probabilities to $[0, g_1, g_1 + g_2]$. Option 2 lets eRAP to behave in its intended way of use with more freedom in the choice of the probabilities, still being in the hard, medium or lenient cases.

Note that nDCG, ERR, and gRBP are always the same in all the three cases, i.e. hard, medium and lenient case, since they do not depend on different ways of thresholding the relevance degree. The same holds for bpref and infAP which always adopt a lenient mapping to binary relevance.

We can observe a very general trend on both T14 and T21: in the case of GAP and eRAP moving from the hard case to the lenient case increases the correlation with nDCG in terms of both τ and τ_{ap}, indicating that somehow "seeing" more relevance degrees makes these evaluation measures closer to a natively multi-graded one. However, while in the case of T14 and three relevance degrees these correlations are quite high, around or above 0.9 in terms of Kendall's τ, when it comes to T21 and four relevance degrees they are typically below 0.8 in terms of Kendall's τ; this suggest that, as the number of relevance degrees increases, these measures tend to take a different angle on what multi-graded relevance is.

We can observe a similar behaviour also for eRRBP with respect to nDCG, while the correlation between gRBP and nDCG is the same in all the cases since they do not depend on the choice of the probabilities; we can also see how the correlation between nDCG and eRRBP is lower than the one between nDCG and gRBP in the hard case, somehow similar in the medium case, and higher in the lenient case. However, these correlations are generally low, around or below 0.60 for T14 and 0.45 for T21 in terms of Kendall's τ, suggesting an approach to multi-graded relevance quite different from nDCG.

When it comes to ERR we can see a similar increasing correlation trend with GAP and eRAP, even if the correlations are very low – below 0.55 for T14 and 0.35 for T21 in terms of Kendall's τ – denoting completely different approaches to ranking systems, probably due to the extremely top-heavy nature of ERR. On the other hand, the correlation between ERR and gRBP/eRRBP is much higher – around 0.75 for T14 and 0.60 for T21 in terms of Kendall's τ – suggesting a greater agreement probably due to the more top-heavy nature of gRBP/eRRBP.

When it comes to the comparison between GAP and eRAP we can note that eRAP with probabilities constrained à la GAP (option 1) is very highly correlated with GAP with both τ and τ_{ap} almost always well above 0.9 on both T14 and T21; moreover, this correlation tends to increase moving from the hard to the lenient cases.

If we consider GAP and eRAP used in its intended way (option 2), we can observe high correlations in the case of T14 above 0.8 for both τ and τ_{ap} while they drop below 0.65 in the case of T21, indicating that the more the number of relevance degrees the more GAP and eRAP depart from each other; moreover, as in the previous cases, we can note an increasing trend as we pass from the hard to the lenient cases.

The comparison between gRBP and eRRBP shows that there is not much difference between using option 1 and 2, since both are very highly correlated with gRBP, being τ and τ_{ap} above 0.9 on both T14 and T21. This hints that the intrinsic structure of the RBP somehow "prevails" on the way in which you make it multi-graded and, as a result, all the different ways to make it multi-graded turn out to be very correlated.

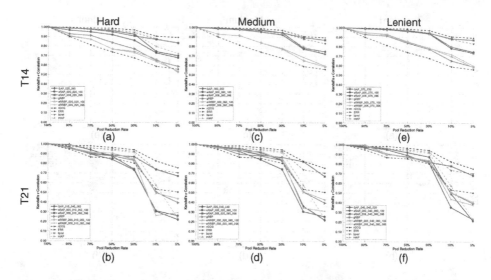

Fig. 2. Robustness to pool downsampling for both T14 (first row) and T21 (second row). The probabilities for the hard, medium, and lenient cases are the same as in Fig. 1 for T14 and T21.

When it comes to the comparison with measures for incomplete information, i.e., bpref and infAP, we can observe that there is an increasing correlation trend passing from the hard to the lenient case for both eRAP and eRRBP on both T14 and T21; this makes sense since both bpref and infAP adopt a lenient approach for mapping multi-graded judgements to binary ones. In particular, in the case of infAP and eRAP, we can observe quite high τ correlations, from 0.88 onwards on T14 and from 0.83 onwards on T21 when eRAP is constrained à la GAP (option 1) where slightly lower correlations on T21 are due to the more multi-graded nature of this track. When we allow for more degrees of freedom in eRAP (option 2), correlations get lower, in the range 0.80-0.86 on T14 and 0.53-0.63 on T21, still being coherent with infAP, but more affected by multi-graded judgments.

Overall, the correlation analysis shows how introducing the idea of random relevance and turning evaluation measures into random evaluation measures allows us to seamlessly manage both binary and multi-graded judgements, keeping a coherent vision with respect to both binary and multi-graded measures. Moreover, the same approach provide us also with an unifying view with respect to addressing incomplete information, as also investigated in the next section.

4.2   RQ2: Properties of the Evaluation Measures

Figure 2 shows the robustness of the evaluation measures to pool downsampling for both T14 and T21 in the hard, medium, and lenient cases considered before. We downsampled pools as in [2] and we computed the Kendall's τ correlation of each measure with respect to its version on the full pool as an indicator of

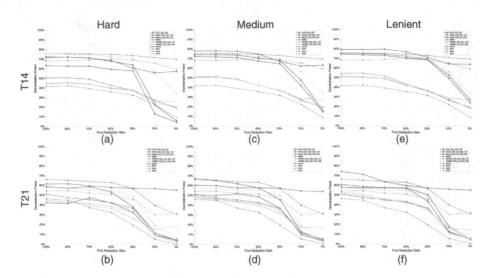

Fig. 3. Discriminative power at pool samples for both T14 (first row) and T21 (second row). The probabilities for the hard, medium, and lenient cases are the same as in Fig. 1 for T14 and T21.

how robust a measure is to pool downsampling. Consider that, downsampling the pools, in the case of T14 we are passing from a deep to a shallow pool while for T21 we are passing from a shallow to an extremely shallow pool. We can see that, consistently with previous findings in the literature, nDCG, bpref, and infAP are among the most robust measures to pool downsampling while ERR is more sensitive to it.

In the case of T14 (first row of Fig. 2) we can observe two separate clusters for all the cases (hard, medium, lenient): one for GAP and eRAP and the other for gRBP and eRRBP where the former is more robust to pool downsampling than the latter. In the case of GAP/eRAP we can see that the more we pass from the hard to the lenient case the closer is their behaviour to the nDCG, bpref, and infAP ones even if GAP has stable performances up to 30% reduction rate while eRAP used in its intended way (option 2, continuous blue line with squares) outperforms it and remains stable also for the more extreme reduction rates, basically behaving like nDCG and bpref.

In the case of T21 (second row of Fig. 2), we do not have anymore two separate clusters but all the measures look very similar up to the 30% reduction rate, being nDCG and bpref always the top performing ones. However, for the 10% and 5% reduction rates all the measures have a sudden drop, even below the level of ERR, with the sole exceptions of nDCG, bpref, and eRAP used in its intended way (option 2, continuous blue line with squares) which remain stable and perform very similarly.

Figure 3 shows the *(Discriminative Power (DP)* [13] achieved at different pool samples for both T14 and T21 in the hard, medium, and lenient cases.

If we consider the full pools, we can see that, consistently with previous findings in the literature, nDCG (DP = 75.44% on T14 and DP = 66.10% on T21) together with bpref (DP = 68.18% on T14 and DP = 54.13% on T21) and infAP (DP = 75.38% on T14 and DP = 57.26% on T21) are among the most discriminative measures while ERR (DP = 41.38% on T14 and DP = 46.15% on T21) is one of the least discriminative ones, due to its strongly top-heavy nature. As a general trend, we can see how the DP improves passing from the hard to the lenient case for both eRRBP and eRAP. In particular, we can see how gRBP (DP = 50.57% on T14 and DP = 51.00% on T21) and eRRBP (DP = 44.00%–55.00% on T14 and DP = 43.00%–53.00% on T21), due to their more top-heavy nature, behave somehow similarly to ERR and tend to be less discriminative than the other measures. On the other hand, GAP (DP = 72.00%–75.00% on T14 and DP = 58.00%–60.00% on T21) and eRAP (DP = 63.00%–79.00% on T14 and DP = 42.00%–74.00% on T21) behave closer to nDCG, bpref and infAP. In particular, on both T14 and T21, eRAP used in its intended way (option 2, continuous blue line with squares) outperforms even nDCG, bpref, and infAP in both the medium case (DP = 77.92% on T14 and DP = 67.24% on T21) and in the lenient case (DP = 79.19% on T14 and DP = 74.07% on T21).

If we consider the different down-sampled pools, we can observe that the above mentioned trends are roughly respected and the discriminative power is stable enough up to the 50% reduction rate. On the other hand, for higher pool reduction rates, there is typically a drop in the performance with the exception of nDCG, bpref, and eRAP used in its intended way (option 2, continuous blue line with squares). These three latter measures behave quite similarly on T14 while eRAP used in its intended way (option 2) substantially outperforms both nDCG and bpref on T21 in the hard, medium, and lenient cases.

Overall, this analysis suggests that the proposed random measures maintain (or even improve) desirable properties in terms of robustness to incomplete information and discriminative power but providing a single unified vision which account for binary and multi-graded judgements as well as for incomplete information.

5   Conclusions and Future Work

In this paper, we have proposed a general stochastic approach for modelling relevance as a random binomial variable. Besides modelling the intrinsic randomness present in the relevance assessment process and the different viewpoints of a user population, the random relevance allows us to turn evaluation measures into random evaluation measures and to provide a single unifying vision between binary and multi-graded measures as well as on the robustness to incomplete information.

A systematic experimentation on TREC collections has shown how these new random measures relate and differ from existing measures and that they have desirable properties in term of robustness to pool downsampling and ability of discriminating among different systems. Overall, this suggested that the proposed new stochastic approach provides the aforementioned benefit at no cost or even improving the properties of the random measures derived from it.

Future work will concern the investigation of further applications of the random relevance and the random evaluation measures. In the crowd-sourcing context, random relevance could be exploited as an alternative way, e.g. to majority voting, to merge pools produced by multiple crowd-assessors, since the relevance of a document could be expressed via a binomial random variable of parameter $(1, p)$, where p is determined accordingly to the different assessments given by the assessors.

References

1. Alonso, O., Mizzaro, S.: Using crowdsourcing for TREC relevance assessment. IPM **48**(6), 1053–1066 (2012)
2. Buckley, C., Voorhees, E.M.: Retrieval evaluation with incomplete information. SIGIR **2004**, 25–32 (2004)
3. Chapelle, O., Metzler, D., Zhang, Y., Grinspan, P.: Expected reciprocal rank for graded relevance. CIKM **2009**, 621–630 (2009)
4. Ferrante, M., Ferro, N., Maistro, M.: Towards a formal framework for utility-oriented measurements of retrieval effectiveness. ICTIR **2015**, 21–30 (2015)
5. Hosseini, M., Cox, I.J., Milić-Frayling, N., Kazai, G., Vinay, V.: On aggregating labels from multiple crowd workers to infer relevance of documents. In: Baeza-Yates, R., de Vries, A.P., Zaragoza, H., Cambazoglu, B.B., Murdock, V., Lempel, R., Silvestri, F. (eds.) ECIR 2012. LNCS, vol. 7224, pp. 182–194. Springer, Heidelberg (2012). https://doi.org/10.1007/978-3-642-28997-2_16
6. Järvelin, K., Kekäläinen, J.: Cumulated gain-based evaluation of IR techniques. TOIS **20**(4), 422–446 (2002)
7. Kazai, G., Craswell, N., Yilmaz, E., Tahaghoghi, S.S.M.: An analysis of systematic judging errors in information retrieval. CIKM **2012**, 105–114 (2012)
8. Kendall, M.G.: Rank Correlation Methods. Griffin, Oxford (1948)
9. Maddalena, E., Mizzaro, S., Scholer, F., Turpin, A.: On crowdsourcing relevance magnitudes for information retrieval evaluation. TOIS **35**(3), 19:1–19:32 (2017)
10. Moffat, A., Zobel, J.: Rank-biased precision for measurement of retrieval effectiveness. TOIS **27**(1), 201–227 (2008)
11. Park, L.A.F.: Uncertainty in rank-biased precision. ADCS **2016**, 73–76 (2016)
12. Robertson, S.E., Kanoulas, E., Yilmaz, E.: Extending average precision to graded relevance judgments. SIGIR **2010**, 603–610 (2010)
13. Sakai, T.: Evaluating evaluation metrics based on the bootstrap. SIGIR **2006**, 525–532 (2006)
14. Sakai, T.: Alternatives to Bpref. SIGIR **2007**, 71–78 (2007)
15. Sakai, T., Kando, N.: On information retrieval metrics designed for evaluation with incomplete relevance assessments. Inf. Retriev. **11**(5), 447–470 (2008)
16. Voorhees, E.M.: Variations in relevance judgments and the measurement of retrieval effectiveness. SIGIR **1998**, 315–323 (1998)
17. Webber, W., Chandar, P., Carterette, B.A.: Alternative assessor disagreement and retrieval depth. CIKM **2012**, 125–134 (2012)

18. Yilmaz, E., Aslam, J.A.: Estimating average precision with incomplete and imperfect judgments. CIKM **2006**, 102–111 (2006)
19. Yilmaz, E., Aslam, J.A., Robertson, S.E.: A new rank correlation coefficient for information retrieval. SIGIR **2008**, 587–594 (2008)
20. Zobel, J.: How reliable are the results of large-scale information retrieval experiments. SIGIR **1998**, 307–314 (1998)

Information Scent, Searching and Stopping

Modelling SERP Level Stopping Behaviour

David Maxwell[1(✉)] and Leif Azzopardi[2]

[1] School of Computing Science, University of Glasgow, Glasgow, Scotland
d.maxwell.1@research.gla.ac.uk
[2] Computer and Information Sciences, University of Strathclyde, Glasgow, Scotland
leif.azzopardi@strath.ac.uk

Abstract. Current models and measures of the *Interactive Information Retrieval (IIR)* process typically assume that a searcher will always examine the first snippet in a given *Search Engine Results Page (SERP)*, and then with some probability or cutoff, he or she will stop examining snippets and/or documents in the ranked list (snippet level stopping). Prior work has however shown that searchers will form an initial impression of the SERP, and will often abandon a page without clicking on or inspecting in detail any snippets or documents. That is, the *information scent* affects their decision to continue. In this work, we examine whether considering the information scent of a page leads to better predictions of stopping behaviour. In a simulated analysis, grounded with data from a prior user study, we show that introducing a SERP level stopping strategy can improve the performance attained by simulated users, resulting in an increase in gain across most snippet level stopping strategies. When compared to actual search and stopping behaviour, incorporating SERP level stopping offers a closer approximation than without. These findings show that models and measures that naïvely assume snippets and documents in a ranked list are actually examined in detail are less accurate, and that modelling SERP level stopping is required to create more realistic models of the search process.

1 Introduction

Interactive Information Retrieval (IIR) is a complex, non-trivial process in which during a search session, searchers may issue multiple queries and examine a varying number of snippets and documents per query [12]. One particularly important part of this process is knowing *when to stop* [25]. Stop too early, and you could miss useful information; stop too late, and you could be wasting valuable time examining non-relevant material. Research into examining stopping behaviour has been until recently relatively sparse, with a series of studies finding that people stop based upon their intuition, or what is simply *"good enough"* [38]. Formally, stopping behaviour has typically been considered at two levels: *(i)* the query (or snippet) level; and *(ii)* the session level. As such,

© Springer International Publishing AG, part of Springer Nature 2018
G. Pasi et al. (Eds.): ECIR 2018, LNCS 10772, pp. 210–222, 2018.
https://doi.org/10.1007/978-3-319-76941-7_16

researchers have attempted to quantify the sense of *"enough"* at both levels by proposing a series of *stopping rules* and heuristics that attempt to encode this intuition (e.g. [3,6,15,25]). Models of stopping behaviour have also been encoded with measures used to evaluate the quality of ranked lists, and within simulations of interaction. However, the majority of work in this area currently assumes that a searcher will always examine the first snippet, and will either examine to a fixed depth, or stop based upon some probability on continuing. Yet the *Search Engine Results Page (SERP)* provides various cues which searchers use to decide when to stop, or even whether to begin examining the SERP in detail at all. Thus, current stopping models tend to be agnostic of the *information scent* [5,30], which has been previously shown to affect a searcher's (stopping) behaviours [4,37]. This scent can be used to determine whether a given SERP *smells good enough* to *enter* and examine individual snippets within the SERP in more detail, as per the *Patch Model* in *Foraging Theory* [32].

To this end, this paper: *(i)* introduces a new SERP level decision point in an established interaction model, allowing searchers following the model to obtain an initial impression (or *'overview'*) of the SERP before deciding to enter or abandon it; and *(ii)* enumerates a series of simple SERP level stopping strategies, implementing the new decision point in several ways. These strategies are grounded using analysis from a prior user study [21] examining information scent. We report on a large-scale simulation, allowing us to address our two main research questions. Does incorporating a SERP level stopping decision point, motivated by information scent, lead to: **RQ1** higher overall performance, and **RQ2** better approximations of searcher stopping behaviour?

2 Background

A user is said to *abandon* a SERP when he or she fails to click on any of the results returned for the given query [7,10]. This may be for a variety of reasons, the primary reason being user satisfaction (or lack of) [10]. Satisfaction from simply examining snippets may lead to *good abandonment* [16,37]. Alternatively, if the presented SERP looks poor, dissatisfaction occurs. This phenomenon has been shown to lead to differences in information seeking behaviour, which have been analysed and subsequently modelled [14]. We consider in this study *Information Foraging Theory (IFT)* [30] as a means for attempting to model such a process, where a user abandons a SERP through dissatisfaction with the presented SERP – good abandonment in this study is not considered.

IFT is primarily composed of three models: the *Information Scent model*, the *Information Patch model*, and the *Information Diet model* [30]. Of particular relevance to this work are the related studies on *scent* and *patches* (as discussed in Sect. 3). Pirolli and Card [30] argue that information seekers are like animals foraging in the wild, and as such will follow a scent to find food. Similarly, information seekers follow *proximal cues* provided by hypertext links, titles, snippets and thumbnails to help locate relevant information [5,26,28–30]. In the context of news search, cues were examined by Sundar et al. [33]. Cues

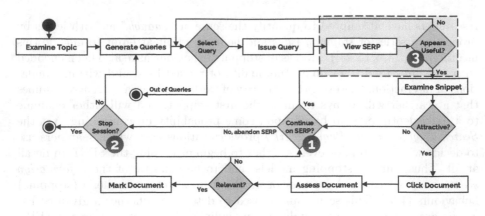

Fig. 1. The updated *Complex Searcher Model (CSM)* with the key decisions (shown in grey) and actions (shown in white). Stopping level decision points are numbered *1–3*. The new SERP level stopping decision point is highlighted in the dashed box.

such as an article's source were shown to have a powerful effect on the perception of said article. If cues can provide a rationale as to what leads to a promising scent trail, it follows that scent also provides a rationale as to when a searcher will stop examining a set of results [30, 36, 37]. The distribution of relevant search items also matters: a searcher may continue to forage to greater depths if the SERP appears to contain many relevant items [37]. A similar trend was also observed by Card et al. [4], who found that when navigating through webpages, searchers were more likely to leave when the information scent began to decline.

Examining searcher behaviours when considering scent has been examined by several researchers (e.g. [8, 21, 27, 33, 37]). Wu et al. [37] conducted a user study where the scent of the first SERP was manipulated. They created low, medium and high scent SERPs by changing the number and distribution of relevant items on the page. Subjects interacting with SERPs with a higher scent examined more content and clicked to greater depths, while subjects on low scent pages examined less, and were more likely to abandon the page altogether. A study by Ong et al. [27] replicated the same experimental setup as used by Wu et al. [37], but for both desktop and mobile environments, where similar findings were observed – subjects using the desktop interface however tended to perform better. Maxwell et al. [21] conducted a user study where information scent was varied by manipulating the length of snippets (changing proximal cues) as opposed to manipulating performance of SERPs as done before [27, 37]. It was found that as result snippets increased in length from title only to title plus four line summaries, subjects examined fewer snippets – and were more likely to click on documents, but with lower accuracy. Taken together, these studies suggest that the information scent does indeed influence stopping behaviour. In this study, we operationalise scent as the performance of a SERP (as done in [27, 37]), examining how scent affects search, stopping and overall performance.

3   Updating the *Complex Searcher Model*

We propose the introduction of a SERP level decision point within the *Complex Searcher Model (CSM)*. The CSM combines several frameworks previously proposed in the literature [2,18,20,34] that model the search process for the simulation and evaluation of user behaviour and performance. The process models ad-hoc topic retrieval based tasks, where the searcher has to identify documents that are relevant to a given information need. Represented as a flow diagram, the CSM is illustrated in Fig. 1.[1] The key stopping decision points in the CSM are highlighted in the figure as: *(1)* the *snippet level* decision point (referred to as the *query level* in the literature); *(2) session level* stopping; and *(3)* the proposed *SERP level* stopping decision point.

The new decision point considers the SERP as a whole, and the impression that the searcher obtains from the cues and information visible to them within the browser's viewport. Searchers could, for example, *skim* the SERP, examining the titles and/or URLs of visible results, before determining – through the process of *information triage* [17] – whether enough relevant content is present to examine them in more detail. By considering the SERP as a whole, this provides a way to model abandonment within the search process, rather than assuming that a searcher will assess the first snippet specifically. This therefore marks a departure from assumptions encoded within many *Information Retrieval (IR)* models and measures, such as *P@k*, *RBP* [24], and *INST* [1,23,31]. The motivation for including this additional decision point stems from empirical research (i.e. query abandonment [9]) and theory. In Foraging Theory, as mentioned in Sect. 2, a forager, when presented with a patch, will survey said patch to assess its potential gain before making a decision as to whether it would be worth their while entering it [32]. For example, McNair [22] showed that foragers assess patches – and even select different strategies – based upon this initial patch impression. When considering a SERP as being analogous to a patch, we posit that, given the opportunity to judge a SERP for potential usefulness, a searcher will be able to save time by abandoning SERPs that appear to offer poor yields, and thus search more efficiently. In this work, we will compare the stopping behaviour and overall performance with and without the SERP level stopping decision point across established snippet level stopping strategies, along with an examination as to which approach best approximates actual searcher behaviour.

4   Experimental Method

To address **RQ1** and **RQ2**, we first conducted a large scale simulation to assess the performance when different SERP level stopping strategies are employed (***performance runs***). Then we conducted a simulated analysis replaying actual user queries to determine which SERP level stopping strategy offers the best approximation to actual searcher stopping behaviour (***comparison runs***).

---

[1] For a more detailed description of the flow of interaction and various processes represented within the CSM, refer to Maxwell et al. [18].

Table 1. Interaction costs and probabilities (as observed from the user study by Maxwell et al. [21]) that are used to ground our simulated analysis. Refer to Sect. 4.2 for an explanation of each of the probabilities listed.

Time required to...	Seconds
...issue a query	9.42
...examine a SERP	3.93
...examine a snippet	2.35
...examine a document	17.19
...mark a document	1.26
Session Time	**360**

(a) Interaction costs

	Probability	Avg.	Savvy	Naïve
SERP	$P(E\|LS)$	0.34	0.00	0.74
SERP	$P(E\|HS)$	0.77	0.80	0.82
Snip.	$P(C\|R)$		0.35	
Snip.	$P(C\|N)$		0.25	
Doc.	$P(M\|R)$		0.67	
Doc.	$P(M\|N)$		0.58	

(b) Interaction probabilities

4.1 Corpus, Topic and System

For this study, we used the *TREC AQUAINT* newswire collection, complete with the *TREC 2005 Robust Track* topic set. The set consists of a total of 50 topics, all of which were used for our performance runs. The AQUAINT collection was indexed with the *Whoosh IR toolkit*[2] (version 2.7.4), where stopwords were removed with Porter stemming applied. The retrieval model used for all simulations was BM25 ($b = 0.75$). The simulation framework *SimIIR*[3] was used, where we added the proposed SERP level component to the framework.

4.2 User Study, Subjects, Costs and Probabilities

Log interaction data was obtained from a within-subjects user study by Maxwell et al. [21], using the same collection and retrieval model as above. In the study, 53 subjects undertook ad-hoc topic retrieval using the same configuration of search engine and corpus as described above. Subjects were asked to identify (mark) as many relevant documents as they could over four topics, with each subject allocated a total of 10 min per topic[4]. For each topic, the search system was configured to present *query biased snippets* [35] of different lengths. For this study however, we consider only one of those interfaces – where two fragments were presented. This decision was taken: *(i)* to simplify the reporting of our results; and *(ii)* because the interfaces all yielded similar interaction probabilities. Two snippet fragments

---

[2] Whoosh is available on *PyPi* at https://pypi.python.org/pypi/Whoosh/.

[3] SimIIR is available at https://github.com/leifos/simiir.

[4] Despite the allocation of 10 min per topic, only the first six minutes (360 s) of interaction data were considered in the results of Maxwell et al. [21]. As such, we use this as our simulated search session time limit. Refer to Maxwell et al. [21] for the rationale behind this decision.

(roughly equivalent to two lines of surrogate text) is considered to provide a good tradeoff between length and examination cost [11].

Given the log data for the interface, we were able to estimate the interaction probabilities and costs to ground our simulations for this study. Table 1(b) presents: the probability of clicking on a result summary, given it is TREC relevant or not ($P(C|R)$ and $P(C|N)$, respectively); and the probability of marking a document relevant, given it has been clicked on and is TREC relevant or not ($P(M|R)$ and $P(M|N)$, respectively). The table also includes the probabilities of *examining* a SERP yielding a *high* information scent (good results), represented by $P(E|HS)$ – with the converse for *low* information scent (poor results) defined as $P(E|LS)$. To compute the latter two probabilities, we first categorised queries issued by each subject according to scent, such that if $P@10 = 0.0$, the scent level would be considered to be low. This definition follows from work by Wu et al. [37], who state that a page that returns little or no relevant content can be considered to offer a low information scent. We then counted the number of SERPs that recorded no clicks as abandoned SERPs (as per Hassan and White [9]), and divided this value by the number of queries issued. From Table 1(b), we observe that the probability of continuing after observing a SERP of high scent ($P(E|HS)$) is greater than the probability of continuing to examine a low scent SERP ($P(E|LS)$). This provides evidence that searchers do indeed attempt to avoid low quality SERPs. In this study, we used three sets of SERP interaction probabilities to examine the effect of information scent on search behaviour. These are subsequently detailed in Sect. 4.3.

4.3   User Simulations Setup

To run a given simulation, we instantiated each component of the CSM (as illustrated in Fig. 1). Since we employed various interaction probabilities, the stochastic components using these were each trialled ten times. Each run's pseudo-random number generator was seeded to ensure reproducible results, with the same seed used across SERP conditions. This allowed us to perform a pairwise comparison of performance. Considering 50 topics, each component, and the numerous parameter settings trialled, a total of approximately 356, 000 runs were executed to produce the required results. Each of the different simulation components and their instantiated configurations are described below.

Query Generation. Keskustalo et al. [13] proposed a series of *idealised, prototypical* approaches to generating queries, as identified from a user study. In particular, strategy **QS3** identified by the authors offered reasonably good performance. The strategy generates queries of three terms in length.[5] Two *pivot terms* were selected, with an additional third term added. However, users have been shown to steadily build up their queries as they acquire more information, first issuing short queries then increasing their length [13]. To this end, we

---

[5] Human subjects issued queries of 3.31 terms on average. This means that the three term queries generated by **QS3** can be considered as a reasonable approximation.

created a modified querying strategy, taking the pivot terms, issuing these as individual queries first, and then combining them as the pivot. This approach: *(i)* makes the querying strategy more realistic [13]; and *(ii)* allows us to test the robustness of both the SERP level and snippet level stopping strategies when faced with both good and poor performing queries.[6]

SERP Decision Making. The new SERP level decision point allows for a searcher to begin examining individual snippets for attractiveness, or abandon the SERP completely. To examine this component in detail, we report on three different implementations.

- **Always** *(No SERP Judgements – Always Examine):* With this strategy, a user will always *enter* the SERP and examine a number of snippets, determined by the snippet level stopping strategy. This is the current state of the art that we consider as our baseline approach.
- **Perfect** *(Perfect SERP Judgements):* Here, a simulated user will only begin to examine a SERP in detail if $P@k > 0$ (the patch threshold). If $P@k = 0$, the user will abandon the SERP and proceed to the next action as dictated by the CSM. This is the upper bound for our simulations – analogous to, as an example, the *ideal user* of Hagen et al. [8].
- **Stochastic SERP Judgements:** This strategy uses a stochastic element to determine whether the simulated user should enter the SERP or not. Like above, the $P@k$ of the SERP is computed. If the SERP is of high scent, $P(E|HS)$ is used to determine whether the user enters the SERP. If the SERP is considered to be of low scent, $P(E|LS)$ is used to determine the likelihood of abandonment. The three sets of probabilities we considered in this study are detailed below.
  - **Average:** $P(E|HS)$ and $P(E|LS)$ are estimated over all users.
  - **Savvy:** $P(E|HS)$ and $P(E|LS)$ are estimated based on the top 15 users with the lowest $P(E|LS)$.
  - **Naïve:** $P(E|HS)$ and $P(E|LS)$ are estimated based on the top 15 users with the highest $P(E|LS)$.

Table 1(b) shows the probabilities used for **Average**, **Savvy** and **Naïve**. The information scent of the SERP was estimated using the associated TREC QRELs for the given topic and based on the top seven snippets returned ($k = 7$). This value was selected as the interface in the user study displayed, on average, seven snippets in the browser's viewport. We also considered additional ways to estimate the scent of page, such as considering the uniqueness of relevant documents within a SERP (i.e. stop if lots of *purple links* – visited links – were on the SERP page). However, the findings were similar to those reported here – and so were not included due to space constraints.

---

[6] For example, a robust snippet level stopping strategy would ideally stop early in the ranked list for poor performing queries, and later for good queries – good queries will return more relevant documents in the ranked list of results.

Snippet and Document Decision Making. As done in prior simulations [2, 19], the decision to click on a snippet – and the subsequent decision to mark a document relevant – are based upon interaction probabilities. The clicking $(P(C))$ and marking $(P(M))$ probabilities used here are reported in Table 1(b).

Snippet Level Stopping Strategies. If the scent of a result page appears to be good enough to examine in more detail, a simulated user will employ the use of one of the following established SERP level stopping strategies.

- **SS1 (Fixed Depth):** A simulated user will stop examining the ranked list once they have observed x_1 snippets, regardless of their relevance.
- **SS2 (Adaptive):** A simulated user will stop once they have observed x_2 non-relevant snippets on the provided SERP.
- **SS3 (Adaptive):** A simulated user will stop once they have observed x_3 non-relevant snippets in a row (contiguously).

All three strategies have been used in prior simulations [18,20]. **SS1** can be considered to be the *de facto* approach used by many models and measures we use in the field (e.g. *P@k*). The adaptive strategies **SS2** and **SS3** that consider a user's tolerance to non-relevance are based upon the *frustration point* and *disgust* rules, proposed by Cooper [6] and Kraft and Lee [15] respectively. These strategies were selected as they had previously been shown to provide good approximations of actual searcher behaviours [20]. We explored a range of values for x_1, x_2 and x_3, trialling 1–10 in steps of 1, and 15–30 in steps of 5.

4.4   User Comparisons

Comparison runs used the same configurations as described previously, save for the querying strategy. Here, we *replayed* each of the queries issued by the real-world subjects from the associated user study.[7] Each query was issued over the different configurations, allowing us to then calculate the simulated click depths per query. We then compared the actual click depths against the simulated click depths for each query, calculating the *Mean Squared Error (MSE)* between the two. For this analysis, we used a total of 175 user queries.

5   Results

Here, the performance of simulated users is reported both in terms of *Cumulative Gain (CG)* and the click depth reached per query *(D/Q)*. CG is measured by summing the TREC QRELs judgement scores of all documents marked as relevant over the course of a search session.

---

[7] This also meant that for our comparison runs, only the four topics selected by Maxwell et al. [21] were trialled, rather than the full set of 50 topics from the TREC 2005 Robust Track as used in our performance runs.

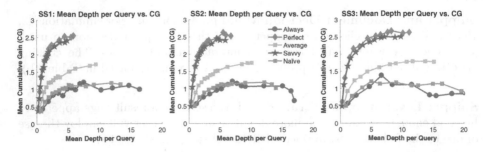

Fig. 2. Plots illustrating the results of our performance runs over each SERP stopping strategy and snippet level stopping strategies **SS1** (L), **SS2** (C) and **SS3** (R).

RQ1 *(Examining Performance).* Figure 2 shows three plots, each of which illustrates the maximum levels of CG attained by simulated users at varying D/Q values. These are shown over the different SERP level stopping strategies: **Always** (baseline); **Perfect**; **Average**; **Savvy**; and **Naïve**, over the three snippet level stopping strategies **SS1** (left), **SS2** (centre) and **SS3** (right). From the plots, we can immediately see that compared to our baseline approach **Always**, **Perfect** attained a much higher level of CG (e.g. 2.55 for **Perfect** at a D/Q of 5.77 vs. 1.2 for **Always** at a D/Q of 7.39 over **SS1**). Turning to our three stochastic variants, the **Savvy** searcher always abandoned a low scent SERP and examined a high scent SERP about 80% of the time. This led to a general trend similar to that of **Perfect**, yet with a slightly lower maximum level of CG (e.g. 2.41 at a D/Q of 4.78 over **SS1**). This is in line with intuition, as 20% of the time, the **Savvy** user would have abandoned a high scent SERP, accounting for the slightly lower levels of performance. On the other hand, the **Naïve** searcher followed a similar trend to our baseline approach, **Always**. This is again in line with expectations, as the probabilities used by **Naïve** led to a high probability of examining high *and* low scent SERPs. In turn, this led to an inefficient search strategy (like **Always**) – one where searchers would by and large waste time examining low scent SERPs. The final **Average** searcher however fell between the extremes of **Savvy** and **Naïve**, and attained a maximum CG of 1.72 at a D/Q of 9.32 over **SS1**. Similar trends as discussed previously can be seen across all three snippet level stopping strategies **SS1**, **SS2**, and **SS3** – although for **SS3**, simulated users on average examined to slightly greater depths per query. Overall, the highest CG was attained by **Perfect** over **SS3**, with the lowest CG of 1.18 reached by **Naïve** – baseline **Always** was close with a CG of 1.2. Overall, the **Savvy**, **Average** and **Naïve** searchers tended to outperform the **Always** baseline, and suggests that performance improvements can be made to varying degrees depending upon how well the searcher can identify good quality SERPs. Interestingly, searchers need not be **Perfect**, with **Average** searchers still performing much better than **Always**. These findings show that including the SERP level decision point does indeed lead to improvements in performance.

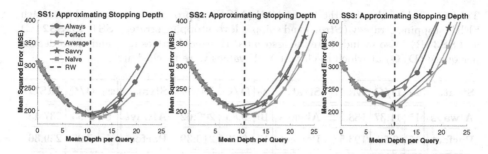

Fig. 3. Plots illustrating the results of our comparison runs over each SERP stopping strategy and snippet level stopping strategies **SS1** (L), **SS2** (C) and **SS3** (R). Also included for comparison is the mean click depth exhibited by the real world subjects.

RQ2 *(Approximating Stopping Depths)*. To determine whether including the SERP decision point could lead to better approximations of stopping behaviour, we calculated and plotted the MSE for each SERP and snippet level stopping strategy (see Fig. 3). Again, **SS1** is shown on the left, with **SS2** in the centre and **SS3** on the right. Also included in each of the plots – denoted by the black dashed line – is the actual mean click depth that the 53 subjects of the user study examined to – a depth computed across all issued queries as 10.65. From the plots, we can immediately observe that the lowest (and therefore best) MSE values were found to be close to the real mean click depth for both **SS1** and **SS2**, but the approximations offered by **SS3** were slightly further away, with the best approximation for **SS3** yielding a D/Q of 10.09. The best MSE approximations – and the corresponding x_n threshold and D/Q that it was attained at – can be seen in Tables 2(a), (b) and (c) for **SS1**, **SS2** and **SS3** respectively. Closer examination of the tables show that the best approximation over **SS1** was achieved at a D/Q of 10.90 ($x_1 = 20$) for **Naïve**, with a D/Q of 10.79 ($x_2 = 15$) for **Average**. Indeed, the stochastic users gave the best approximations over all three snippet level stopping strategies. This finding is intuitive as nobody from the user study correctly identified high and low scent SERPs 100% of the time, making **Perfect** an unrealistic strategy to use. Interestingly, stopping behaviour was best approximated by **Average** searchers.

Closer inspection of the results for **SS3** shows that this snippet level stopping strategy consistently yielded higher (and thus poorer) MSE values, although D/Q approximations remained close to the actual mean click depth – at least for the stochastic users, **Average**, **Perfect** and **Naïve**. This finding is interesting because the same strategy yielded the best approximations for searcher behaviour in previous work [20], and suggests the strategy may not be robust when applied in other contexts. Overall, the actual stopping behaviour of searchers *is* better approximated when incorporating a SERP level decision point.

Table 2. Tables showing the lowest MSE (**MSE**) approximations attained over each SERP stopping strategy (*Strat.*) and snippet level stopping strategy **SS1** (L), **SS2** (C) and **SS3** (R). Also included are the associated threshold values (x_n) and mean depths per query (**D/Q**) at which the lowest MSE values were reached at.

Strat.	x_1	D/Q	MSE	Strat.	x_2	D/Q	MSE	Strat.	x_3	D/Q	MSE
Always	15	11.37	188.12	**Always**	10	9.85	200.62	**Always**	5	8.55	237.45
Perfect	15	9.59	193.57	**Perfect**	10	8.28	210.39	**Perfect**	5	7.03	239.36
Avg.	**20**	**10.90**	**184.66**	**Avg.**	**15**	**10.79**	**190.91**	**Avg.**	**6**	**10.09**	**208.23**
Savvy	20	11.05	192.53	**Savvy**	15	10.90	199.32	**Savvy**	6	10.22	214.67
Naïve	20	12.68	185.36	**Naïve**	15	12.58	191.47	**Naïve**	6	11.85	213.71
(a) SS1				(b) SS2				(c) SS3			

6   Discussion and Future Work

In this paper, we have considered how information scent affects search and stopping behaviour, and have encoded this within the *Complex Searcher Model (CSM)* to provide a more realistic model of the search process. This was operationalised by the inclusion of a new SERP level decision point, where the *scent* of a SERP is attained from an initial impression of the page. This information is then used to decide if individual snippets should be examined, or whether to simply abandon the SERP. We found that the inclusion of this additional decision point can lead to more effective searching, but only if the searcher is able to discern between SERPs of high and low scent. Our study shows that **Savvy** users can easily avoid poor quality SERPs, while **Naïve** users find it hard to recognise the quality of SERPs. This suggests that work should be directed towards improving how SERPs are rendered to increase how well people can identify good SERPs from the bad, as well as research into what cues searchers look for in a good SERP. Furthermore, we found that including the SERP level decision point led to more accurate modelling of actual stopping behaviour. This represents a major shift in modelling interaction – and has ramifications for how IR systems are measured, which typically assumes people examine ranked lists. These results suggest that future work needs to be directed towards measures that consider abandonment, and should also include how the sequence and quality of queries affects interactions taking place with ranked lists.

Acknowledgements. Our thanks to Horaţiu Bota and Alastair Maxwell for their feedback – including Horaţiu's helpful comments on our results. We would also like to thank the anonymous reviewers for their comments and feedback. Finally, the lead author is funded by the UK Government though the EPSRC, grant number 1367507.

References

1. Bailey, P., Moffat, A., Scholer, F., Thomas, P.: User variability and IR system evaluation. In: Proceedings of 38th ACM SIGIR, pp. 625–634 (2015)
2. Baskaya, F., Keskustalo, H., Järvelin, K.: Modeling behavioral factors in interactive information retrieval. In: Proceedings of 22nd ACM CIKM, pp. 2297–2302 (2013)
3. Browne, G., Pitts, M., Wetherbe, J.: Stopping rule use during web-based search. In: Proceedings of HICSS-38, p. 271b (2005)
4. Card, S., Pirolli, P., Van Der Wege, M., Morrison, J., Reeder, R., Schraedley, P., Boshart, J.: Information scent as a driver of web behavior graphs: results of a protocol analysis method for web usability. In: Proceedings of 19th ACM CHI, pp. 498–505 (2001)
5. Chi, E., Pirolli, P., Chen, K., Pitkow, J.: Using information scent to model user information needs and actions and the web. In: Proceedings of 19th ACM CHI, pp. 490–497 (2001)
6. Cooper, W.: On selecting a measure of retrieval effectiveness part II. Implementation of the philosophy. J. Am. Soc. Info. Sci. **24**(6), 413–424 (1973)
7. Diriye, A., White, R., Buscher, G., Dumais, S.: Leaving so soon? Understanding and predicting web search abandonment rationales. In: Proceedings of 21st ACM CIKM, pp. 1025–1034 (2012)
8. Hagen, M., Michel, M., Stein, B.: Simulating ideal and average users. In: Proceedings of 12th AIRS, pp. 138–154 (2016)
9. Hassan, A., White, R.: Personalized models of search satisfaction. In: Proceedings of 22nd ACM CIKM, pp. 2009–2018 (2013)
10. Hassan, A., Shi, X., Craswell, N., Ramsey, B.: Beyond clicks: query reformulation as a predictor of search satisfaction. In: Proceedings of 22nd CIKM, pp. 2019–2028 (2013)
11. Hearst, M.: Search user interfaces. Cambridge University Press, Cambridge (2009)
12. Ingwersen, P., Järvelin, K.: The Turn: Integration of Information Seeking and Retrieval in Context. Springer, Dordrecht (2005). https://doi.org/10.1007/1-4020-3851-8
13. Keskustalo, H., Järvelin, K., Pirkola, A., Sharma, T., Lykke, M.: Test collection-based IR evaluation needs extension toward sessions – a case of extremely short queries. In: Proceedings of 5th AIRS, pp. 63–74 (2009)
14. Kiseleva, J., Kamps, J., Nikulin, V., Makarov, N.: Behavioral dynamics from the SERP's perspective: what are failed SERPs and how to fix them? In: Proceedings of 24th ACM CIKM, pp. 1561–1570 (2015)
15. Kraft, D., Lee, T.: Stopping rules and their effect on expected search length. IPM **15**(1), 47–58 (1979)
16. Loumakis, F., Stumpf, S., Grayson, D.: This image smells good: effects of image information scent in search engine results pages. In: Proceedings of 20th ACM CIKM, pp. 475–484 (2011)
17. Marshall, C.C., Shipman III, F.M.: Spatial hypertext and the practice of information triage. In: Proceedings of 8th ACM HYPERTEXT, pp. 124–133 (1997)
18. Maxwell, D., Azzopardi, L.: Agents, simulated users and humans: an analysis of performance and behaviour. In: Proceedings of 25th ACM CIKM, pp. 731–740 (2016)
19. Maxwell, D., Azzopardi, L., Järvelin, K., Keskustalo, H.: An initial investigation into fixed and adaptive stopping strategies. In: Proceedings of 38th ACM SIGIR, pp. 903–906 (2015)

20. Maxwell, D., Azzopardi, L., Järvelin, K., Keskustalo, H.: Searching and stopping: an analysis of stopping rules and strategies. In: Proceedings of 24th ACM CIKM, pp. 313–322 (2015)
21. Maxwell, D., Azzopardi, L., Moshfeghi, Y.: A study of snippet length and informativeness: behaviour, performance and UX. In: Proceedings of 40th ACM SIGIR (2017)
22. McNair, J.N.: Optimal giving-up times and the marginal value theorem. Am. Nat. **119**(4), 511–529 (1982)
23. Moffat, A., Bailey, P., Scholer, F., Thomas, P.: INST: An adaptive metric for IR evaluation. In: Proceedings of 20th ADCS, pp. 5:1–5:4 (2015)
24. Moffat, A., Zobel, J.: Rank-biased precision for measurement of retrieval effectiveness. ACM Trans. Info. Syst. **27**(1), 2:1–2:27 (2008)
25. Nickles, K.: Judgment-based and reasoning-based stopping rules in decision making under uncertainty. Ph.D. thesis, University of Minnesota (1995)
26. Olston, C., Chi, E.: ScentTrails: integrating browsing and searching on the web. ACM Trans. Comput. Hum. Interact. **10**(3), 177–197 (2003)
27. Ong, K., Järvelin, K., Sanderson, M., Scholer, F.: Using information scent to understand mobile and desktop web search behavior. In: Proceedings of 40th ACM SIGIR (2017)
28. Pirolli, P.: Information Foraging Theory: Adaptive Interaction with Information, 1st edn. Oxford University Press, New York (2007)
29. Pirolli, P., Card, S.: Informayion foraging in information access environments. In: Proceedings of 13th ACM SIGCHI, pp. 51–58 (1995)
30. Pirolli, P., Card, S.: Information foraging. Psychol. Rev. **106**, 643–675 (1999)
31. Smucker, M., Clarke, C.: Modeling optimal switching behavior. In: Proceedings of 1st ACM CHIIR, pp. 317–320 (2016)
32. Stephens, D., Krebs, J.: Foraging Theory. Princeton University Press, Princeton (1986)
33. Sundar, S., Knobloch-Westerwick, S., Hastall, M.: News cues: information scent and cognitive heuristics. J. Am. Soc. Inf. Sci. Technol. **58**(3), 366–378 (2007)
34. Thomas, P., Moffat, A., Bailey, P., Scholer, F.: Modeling decision points in user search behavior. In: Proceedings of 5th IIiX, pp. 239–242 (2014)
35. Tombros, A., Sanderson, M.: Advantages of query biased summaries in information retrieval. In: Proceedings of 21st ACM SIGIR, pp. 2–10 (1998)
36. Wu, W.: How far will you go? Using need for closure and information scent to model search stopping behavior. In: Proceedings of 4th IIiX, pp. 328–328 (2012)
37. Wu, W., Kelly, D., Sud, A.: Using information scent and need for cognition to understand online search behavior. In: Proceedings of 37th ACM SIGIR, pp. 557–566 (2014)
38. Zach, L.: When is "enough" enough? Modeling the information-seeking and stopping behavior of senior arts administrators. J. Am. Soc. Info. Sci. Tech. **56**(1), 23–35 (2005)

Investigating Result Usefulness in Mobile Search

Jiaxin Mao[1(✉)], Yiqun Liu[1], Noriko Kando[2], Cheng Luo[1], Min Zhang[1], and Shaoping Ma[1]

[1] Tsinghua University, Beijing, China
maojiaxin@gmail.com , yiqunliu@tsinghua.edu.cn
[2] National Institute of Informatics, Tokyo, Japan

Abstract. The existing evaluation approaches for search engines usually measure and estimate the utility or usefulness of search results by either the explicit relevance annotations from external assessors or implicit behavior signals from users. Because the mobile search is different from the desktop search in terms of the search tasks and the presentation styles of SERPs, whether the approaches originated from the desktop settings are still valid in the mobile scenario needs further investigation. To address this problem, we conduct a laboratory user study to record users' search behaviors and collect their usefulness feedbacks for search results when using mobile devices. By analyzing the collected data, we investigate and characterize how the relevance, as well as the ranking position and presentation style of a result, affects its user-perceived usefulness level. A moderating effect of presentation style on the correlation between relevance and usefulness as well as a position bias affecting the usefulness in the initial viewport are identified. By correlating result-level usefulness feedbacks and relevance annotations with query-level satisfaction, we confirm the findings that usefulness feedbacks can better reflect user satisfaction than relevance annotations in mobile search. We also study the relationship between users' usefulness feedbacks and their implicit search behavior, showing that the viewport features can be used to estimate usefulness when click signals are absent. Our study highlights the difference between desktop and mobile search and sheds light on developing a more user-centric evaluation method for mobile search.

Keywords: Mobile search · Evaluation · User behavior analysis

1 Introduction

With the rapid growth of mobile search, the evaluation of the mobile search engine is becoming an important research topic in Information Retrieval. Previous research has shown that the mobile search is different from desktop search in several aspects including search intents [23], user interfaces (e.g. a much smaller screen), and users' search behavior patterns including querying [13,22], SERP scanning [14], and relevance assessment [24]. Recently, to further reduce user's

© Springer International Publishing AG, part of Springer Nature 2018
G. Pasi et al. (Eds.): ECIR 2018, LNCS 10772, pp. 223–236, 2018.
https://doi.org/10.1007/978-3-319-76941-7_17

interaction cost on mobile devices, a larger number of search results that aim to satisfy users without requiring them to click, such as knowledge graphs [16] and direct answers [26], are incorporated into mobile SERPs, making the mobile search results become even more diverse. Due to these differences, whether the existing *system-oriented* and *user-oriented* evaluation methods that were developed for desktop search are as effective in mobile search needs further investigation.

For the system-oriented evaluation, the Cranfield-like evaluation paradigm [2] is widely used. In this paradigm, we measure the effectiveness of search systems by computing some *evaluation metrics* such as MAP and NDCG [10], based on a set of *relevance judgments*. While Verma and Yilmaz [24] found that the relevance judgments on desktop and mobile are different, some recent studies [11, 15,20] in desktop search have spotted a gap between the relevance annotations from assessors and the *usefulness* [4] feedbacks from users and showed that usefulness feedback has a stronger correlation with user satisfaction. However, to what extent the relevance annotation can reflect the result-level user-perceived usefulness and be adopted to estimate the query-level user satisfaction in mobile search has not been extensively studied yet.

In the user-oriented evaluation, user's click [12] and post-click dwell time [5] on landing pages have been widely used as implicit feedbacks to measure the user satisfaction in Web search. However, to reduce user's interaction cost, modern mobile search engines often present search results in the form of information cards [16,26], which aim to meet user' information needs on SERPs, without requiring further clicks. Therefore, the click-based online evaluation methods may not be as reliable in mobile search. To address this problem, some recent studies (e.g. [16,17]) proposed to utilize the viewport[1] changes on mobile devices to capture user's viewing behavior and estimate their attention in mobile search. These studies suggested that user's viewing behavior captured by viewport changes is valuable in measuring user satisfaction in mobile search. But to the best of our knowledge, no existing research has systematically investigated the relationship between user's viewing behavior and their explicit usefulness feedbacks *per result* in mobile search.

To fill these two research gaps, we conducted a laboratory user study to address the following research questions:

- **RQ1:** What factors may affect the user-perceived usefulness of a search result in mobile search?
- **RQ2:** How do the user-perceived usefulness of search results correlate with the query-level user satisfaction in mobile search?
- **RQ3:** How can we use search behavior features to estimate user-perceived usefulness in mobile search?

The laboratory user study enables us to get explicit feedbacks from users, record rich behaviors, and control the undesired variabilities. In particular, we use the explicit result-level usefulness feedbacks and query-level satisfaction

---

[1] The region on the display screen for viewing the content of Web pages.

feedbacks from the participants to measure the user-perceived usefulness of a search result and the query-level user satisfaction in this study (See Section Data Collection for more details).

2   Data Collection

The data was collected through a laboratory user study with 43 participants.

Search Tasks. 20 search tasks were adopted in the user study. Each search task is defined by a query selected from the query log of a commercial mobile search engine. Among these 20 search tasks, 13 are informational, 6 are transactional, and only 1 of them is navigational (i.e. finding the official website of a university). The informational tasks covers a variety of topics such as QA, news, healthcare etc. and the transactional tasks are about finding specific videos, images, and mobile games. The authors further created a background story according to each sampled query to reduce the potential ambiguity of a single query. For each search task, we crawled four SERPs from four popular mobile search engines on one day of October, 2016, using the corresponding query. Because the search tasks cover different topics, the search results on these SERPs cover a variety of vertical types such as Image, Video, News, QA, and Knowledge Graph. In this way, we collected $20 \times 4 = 80$ SERPs for 20 search tasks from 4 different mobile search engines.

Participants. We hired 43 undergraduate students (20 females and 23 males, aged from 19 to 23) from our university as participants via emails and online social networks. In the pre-experiment questionnaire, most participants reported that they were familiar with search engines (Mean = 5.68 in a 7-point Likert scale from *not familiar at all* to *very familiar*) and smart phones (Mean = 5.79 in a 7-point Likert scale), which indicates that they had adequate search expertise to complete the mobile search tasks.

Apparatus. We implemented a Web-based experimental system to host the crawled SERPs and used an Android smartphone which has a 5-inch touch screen with a resolution of 1280×720. Using the `WebView` widget provided in Android SDK, we developed an experimental mobile browser which can record rich user interaction logs including the content of visited pages, scrolling, touch gestures, clicks, and switchings between pages.

The widths of the crawled SERPs is equal to the width of the viewport and zooming was not allowed. Therefore, all the scrolling actions in the collected log are in vertical directions. Depending on the heights of search results, the initial viewport usually contains the first 2–4 results of each SERP.

User Study Procedure. Each participant were required to complete 20 search tasks. For each search task, only one of the four SERPs from four different search engines would be shown to a participant. To balance the sources of the SERPs, we divided the 20 search tasks into 4 groups (task 1–5 as the first group, 6–10 as the second group, and etc.) and used a Latin square of size 4 to assign the SERPs from one of the four search engines to each group. In this way, we created 4 different settings for assigning the sources of SERPs to search tasks. The participants were assigned to the four settings in a balanced way, therefore, each SERP was shown to 10 or 11 participants. To control the order effects of search tasks, we also rotated the 20 search tasks using a Latin square of size 20.

For each search task, we first showed the task description (i.e. a query and the background story) to the participant. Then the participant was required to search with the query and complete the search task using the experimental mobile browser. The instruction given to the participants was the following:

> *"Assuming you have the information need described in the background story, please search with this query in our system as you usually do with a mobile search engine."*

Because we only crawled the first SERP for each search task, query reformulations and paginations were not allowed.

After completing the search task, the participant would give usefulness feedback for the search results and satisfaction feedback for the query. Unlike previous studies in desktop settings [20] that only require the participant to give usefulness feedbacks for the *clicked* results, we asked the participants to select all the results they had *examined* on the mobile SERPs and give usefulness feedbacks for all these *examined* results. The collected result-level usefulness feedback in our study should reflect whether the presented snippet on the SERP was useful or not if the result was not clicked, or whether both the content on the SERP *and* the content on the landing page were useful if the result was clicked. We use the 4-level graded usefulness feedback and corresponding feedback instruction ($U \in \{1, 2, 3, 4\}$) that were adopted by Mao et al. [20].

In this way, we collected participants' self-reported examination feedbacks (E) and their explicit usefulness feedbacks (U) for the examined results simultaneously. E is a binary variable and $E = 1$ means the participant reported that she examined the result. We further assume that the unexamined results did not contribute to the completion of the search task, therefore, their usefulness feedbacks U were set to 1: not useful at all.

For query-level satisfaction (SAT), a 5-level graded scale [18] was used and the instruction was:

> *"Are you satisfied with your search experience with the query and search results? 1: not satisfied at all - 5: very satisfied".*

Data Annotation. To investigate what factors affect the user-perceived usefulness of mobile search results, we further hired professional assessors to assess *relevance* and *click necessity* [19] for all the search results. Because previous research [24] shows that the relevance annotation will be affected by the device used in the annotation process, we required the assessors to make annotations on the same smartphone that was used in the user study.

We used a typical 4-level graded relevance following the TREC criteria [25]. Each search result was annotated by three assessors. The Fleiss' κ of relevance annotation is 0.388, which demonstrates a fair agreement between the assessors.

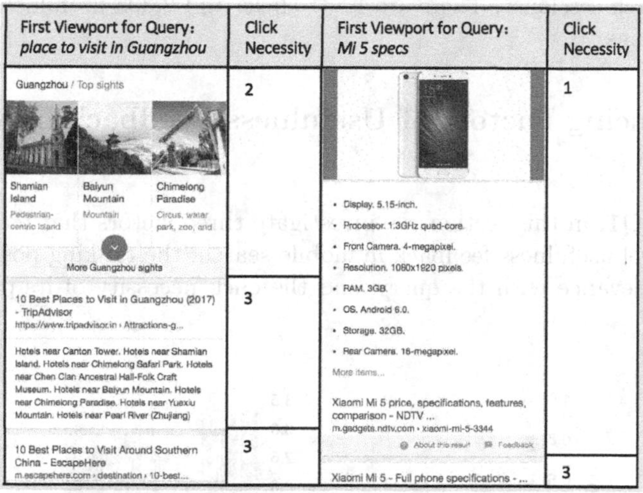

First Viewport for Query: *place to visit in Guangzhou*	Click Necessity	First Viewport for Query: *Mi 5 specs*	Click Necessity
Guangzhou / Top sights — Shamian Island (Pedestrian-centric Island), Baiyun Mountain (Mountain), Chimelong Paradise (Circus, water park, zoo, and More Guangzhou sights	2	Display. 5.15-inch. · Processor. 1.3GHz quad-core. · Front Camera. 4-megapixel. · Resolution. 1080x1920 pixels. · RAM. 3GB. · OS. Android 6.0. · Storage. 32GB. · Rear Camera. 16-megapixel. More items... Xiaomi Mi 5 price, specifications, features, comparison - NDTV ... m.gadgets.ndtv.com · xiaomi-mi-5-3344	1
10 Best Places to Visit in Guangzhou (2017) - TripAdvisor https://www.tripadvisor.in › Attractions-g... Hotels near Canton Tower. Hotels near Shamian Island. Hotels near Chimelong Safari Park. Hotels near Chen Clan Ancestral Hall-Folk Craft Museum. Hotels near Baiyun Mountain. Hotels near Chimelong Paradise. Hotels near Yuexiu Mountain. Hotels near Pearl River (Zhujiang)	3	About this result Feedback	
10 Best Places to Visit Around Southern China - EscapeHere m.escapehere.com › destination › 10-best...	3	Xiaomi Mi 5 - Full phone specifications - ...	3

Fig. 1. Examples of results with different click necessity (1: Not Necessary; 2: Possibly Necessary; 3: Definitely Necessary).

Different types of vertical results are federated into the SERP of mobile search engines. The contents and the presentation styles of these heterogeneous results may have an effect on how the user interacts with them. Traditionally, such effect is investigated by categorizing the results into different vertical types, such as knowledge graph [16], direct answer [26] as well as weather, travel, finance, and etc. [8]. However, the search results in our dataset were crawled from four different search engines and have many different presentation styles. It is tricky to develop a taxonomy to cover all the presentation styles in our dataset properly. Therefore, in this study, we adopted the *click necessity* measure and the corresponding annotation procedure proposed by Luo et al. [19] to investigate the effect on usefulness brought by the abundant information presented in the snippets of heterogeneous mobile search results.

Similar to relevance annotation, each result was annotated by three assessors. The Fleiss' κ of the click necessity annotation is 0.475, which reaches a moderate agreement level and shows that the click necessity can be annotated reliably by external assessors. We show some examples of results with different click necessity scores in Fig. 1. 137 (17.6%) of the unique results were annotated as

"1: not necessary", 136 (17.4%) as "2: possibly necessary", and 507 (65.0%) as "3: definitely necessary". Over half of results were annotated as "3: definitely necessary" because organic results constitute a major proportion of the search results.

Collected Data. After a throughly inspection of the collected dataset, we removed 3 informational search tasks because of the malfunctioning of the experimental apparatus, especially the search behavior logging function. We collected 731 valid search sessions[2]. There are 1,831 clicks and 2,305 usefulness feedbacks in these sessions.

3   Influencing Factors of Usefulness Feedback in Mobile Search

Regarding **RQ1**, in this section, we investigate three factors that may influence the result-level usefulness feedback in mobile search: the ranking position of the result, the relevance with the query, and the click necessity of its presentation style.

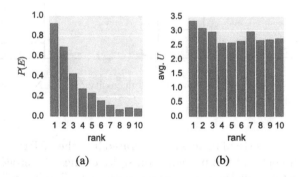

(a)   (b)

Fig. 2. The effects of rank on user's (a) examination and (b) usefulness feedbacks.

Effect of Ranking Positions. Previous research showed that in desktop setting, the examination of search results is affected by the *position bias* [6]. Higher-ranked results tend to received more user attention and larger probabilities of examination. Because examination is a prerequisite for usefulness, the rank of a result may also influence its usefulness feedback.

We first show the effect of ranking positions on participants' self-reported examination in Fig. 2a. The results confirm that the probability of examination $P(E)$ is decreasing with the rank of results. While 92.0% of the results in the 1st position were examined by participants, only 22.9% of the results in the 5th position were examined.

---

[2] The dataset will be open to public for research purpose after the double-blind review process.

We then show the average usefulness feedbacks for the examined results in different ranks in Fig. 2b. From this figure, we find that: (1) the top-3 results have significantly higher usefulness feedbacks than other results (independent t-test, two-tailed, $p < 0.001$). (2) the usefulness feedbacks of the top-3 results decrease with the rank significantly (one-way ANOVA test, $F(2, 1503) = 22.24$, $p < 0.001$). (3) There is no significant difference in the usefulness feedbacks of the results from 4th to 10th positions[3] (one-way ANOVA test, $F(6, 729) = 1.69$, $p = 0.12$).

These observations indicate that the rank of results affects not only user's examination behavior but also their usefulness judgments on examined results. It is also interesting to see that the users treat the results in the initial viewport (i.e. the results in top 3 positions) differently than the other results. They rate the examined results in top 3 positions as more useful than the other examined results. While the position bias affects their usefulness feedbacks in the initial viewport, the position effect seems to become less important when users scroll downwards to examine the results in the 4th to 10th positions.

Effect of Relevance. To investigate the relationship between relevance and usefulness in mobile search, we compute the Pearson's r and Cohen's Weighted κ [3] between the relevance annotations from assessors and usefulness feedbacks from participants. For all the displayed results, there are only a weak linear correlation ($r = 0.29$) and a slight agreement ($\kappa = 0.11$) between relevance and usefulness. But if we only consider the examined results (i.e. the results that have usefulness feedbacks from the participants), a moderate linear correlation ($r = 0.50$) and a fair agreement ($\kappa = 0.33$) are detected. The reason for this apparent difference is that many relevant results were not examined by the participants because of the position bias on examination shown in Fig. 2a. We also note that the correlation between the relevance annotations and usefulness feedbacks of examined results in our study is stronger than the correlation in desktop search reported by Mao et al. [20] ($r = 0.332, \kappa = 0.209$). Compared to their study, the search tasks is more specific and query reformulation is not allowed in our study. Therefore, it is easier for the relevance assessors to guess user's information needs and make relevance judgments that can better reflect the user-perceived usefulness.

Effect of Click Necessity. We are also interested in understanding how the click necessity of results affect the user-perceived usefulness.

While the Pearson's r between the 3-level click necessity annotation and 4-level usefulness feedback of the examined results is not significantly different from 0 ($r = 0.034$, $p = 0.10$), we hypothesize that the click necessity has interaction effects with relevance on user-perceived usefulness and these effects may differ for the clicked results and unclicked results. Therefore, we conduct two

---

[3] We omitted the results ranked below the 10th position here because some SERPs only contains 10 results.

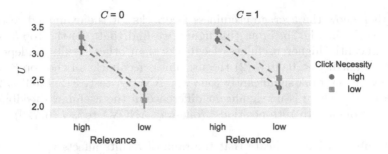

Fig. 3. Effects of click necessity and relevance on the usefulness feedbacks of unclicked results ($C = 0$) and clicked results ($C = 1$).

2×2 two-way ANOVA tests, that regard both relevance and click necessity as factors, for both clicked and unclicked results. Binary labels for relevance and click necessity are generated based on the 4-level and 3-level graded annotations. Because 50.4% of results are highly relevant ($R = 4$), we regard the results with $R = 4$ as results with *high relevance* and other results as with *low relevance*. For click necessity, we group the results with annotations "1: not necessary" and "2: possibly necessary" as results with *low click necessity* ($n = 273$, 35.0%) and the results with annotation "3: definitely necessary" as results with *high click necessity* ($n = 507$, 65.0%).

The interaction effects are shown in Fig. 3. From the left part of the figure, we observe that the click necessity and relevance of unclicked results has an interaction effect on usefulness feedback. The ANOVA test shows that the interaction effect is statistically significant ($F(1, 811) = 9.62, p = 0.002$). Presenting highly relevant information directly on the SERP can bring more usefulness even when the result is not clicked. From the right part of the figure, we find no interaction effect of relevance and click necessity for the clicked documents ($F(1, 1486) = 0.03$, $p = 0.85$). However, the clicked results with low click necessity is significantly more useful than the clicked results with high click necessity ($F(1, 811) = 13.14$, $p < 0.001$). The low-click-necessity results usually come from high-quality sources, such as online encyclopedia and online Q&A sites. Therefore, were they clicked, the usefulness feedbacks of them may be higher than the organic results with high click necessity.

4   Usefulness vs. Satisfaction in Mobile Search

Regarding **RQ2**, we examine the relationship between result-level usefulness and query-level satisfaction by correlating some metrics based on participants' usefulness feedbacks with their satisfaction feedbacks. We use the same metrics based on relevance annotations as baselines. The results measured in Pearson's r are shown in Table 1.

We first compute the Discounted Cumulative Gain (DCG) [10] truncated at different positions using relevance annotations and usefulness feedbacks.

Table 1. Pearson's r between satisfaction feedbacks and metrics based on relevance annotations and usefulness feedbacks. The darker and lighter shadings indicate the correlation is significant at $p < 0.01$ and 0.05.* (or **) indicates the difference is significant at $p < 0.05$ ($p < 0.01$), comparing to the same metric based on relevance annotation R.

	Relevance (R)	Usefulness (U)
$DCG@1$	0.147	0.350**
$DCG@3$	**0.192**	**0.381**
$DCG@5$	0.172	0.320**
$DCG@10$	0.122	0.282**
CG_C	-0.087	0.014
MAX_C	-0.062	0.114**
AVG_C	0.030	0.206**
CG_E	-0.057	0.072*
MAX_E	0.088	0.541**
AVG_E	**0.252**	**0.548**

We assume the unexamined results (i.e. the results without usefulness feedbacks from the participant) are "not useful at all ($U = 1$)". From the upper part of Table 1, we observe that: (1) the correlation between usefulness feedbacks and satisfaction is stronger than that between relevance annotations and satisfaction, which is similar to the findings in desktop settings [20]. (2) For both relevance annotations and usefulness feedbacks, $DCG@3$ is the best among the DCGs truncated at different positions in terms of the correlation with satisfaction feedbacks, which indicates the results in the initial viewport play an important role in determining the user experience in mobile search.

We also compute some online metrics and correlate them with satisfaction feedbacks. Because users may acquire useful information without clicking the results in mobile search, we take all the examined results into consideration. The following online metrics are adopted:

- CG_C/CG_E: the sum of all result-level judgments of clicked/examined results.
- MAX_C/MAX_E: the maximum of result-level judgments of clicked/examined results.
- AVG_C/AVG_E: the average result-level judgments of clicked/examined results.

The correlations between these online metrics and users' query-level satisfaction are shown in the lower part of Table 1. It is observed the examined results, especially MAX_E and AVG_E, have stronger correlations with the query-level satisfaction than those based on only the clicked results, which confirms our hypothesis that the unclicked but examined results also contribute to the mobile search experience. We also see that, while the online metrics based on usefulness feedbacks can better reflect satisfaction, the online metrics based on relevance annotations perform poorly in estimating query-level satisfaction in mobile search.

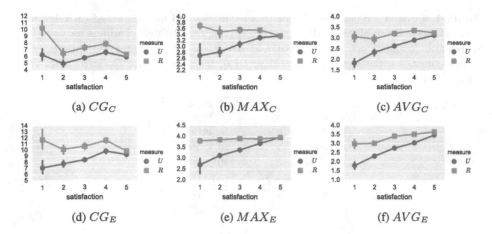

Fig. 4. The average examination-based online metrics of the search sessions with different satisfaction level (SAT).

We further investigate the reason for this apparent difference between usefulness and relevance judgments by showing the average values of the online metrics based on them for sessions with different satisfaction level in Fig. 4. An increasing gap between the relevance-based online metrics and the usefulness-based ones is spotted in these figures as the decrease of satisfaction level. These gaps suggest the relevance annotations systematically overestimate the utility of the results in unsatisfied search sessions. This finding in mobile search confirms Mao et al. [20]'s finding in desktop search that relevance is not sufficient for usefulness and satisfaction.

It is also interesting to find non-monotonous relationships between CG_C, CG_E, and satisfaction (SAT). In the sessions with moderate satisfaction level ($SAT = 2$–4), CG_C and CG_E based on relevance and usefulness are positively correlated with SAT. However, when the session is highly unsatisfactory ($SAT = 1$), the user will compensate for the low result quality by clicking on more results, which results in a higher CG_C of both result-level measures and a higher CG_E of relevance annotation. The users will be satisfied if they can find enough useful information with minimum effort, therefore, the average CG_C and average CG_E of extremely satisfied sessions ($SAT = 5$) are lower than those of the sessions with $SAT = 4$.

5   Usefulness vs. User Behavior in Mobile Search

Addressing **RQ3** will help us to infer user's experience during search using the behavior signals that can be logged passively. Recent studies have proposed to use the viewport time [16] to estimate the attention and satisfaction of users when click signals are absent or at least inaccurate in mobile search. So in this section, we first investigate the relationship between the viewport time and explicit usefulness feedback from the user.

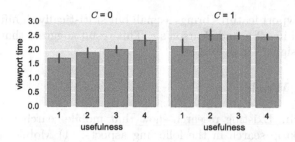

Fig. 5. Average viewport time of the clicked and unclicked results.

Figure 5 shows the average viewport time on the snippets of clicked and unclicked results. It is interesting to see from the left part of the figure that there is a weak but statistically significant positive correlation ($F(3, 811) = 2.71$, $p = 0.044$, Pearson's $r = 0.10$) between the viewport time and usefulness feedback for the unclicked results, while from the right part that there is no such correlation for the clicked results ($F(3, 1386) = 0.89$, $p = 0.44$). This results suggest that in mobile search, the viewport time can be a useful signal in estimating the usefulness of unclicked results but are not so helpful in inferring the usefulness of the clicked results.

To further test whether the viewport time can help in estimating user-perceived usefulness in mobile search, we build regression models to predict the 4-level usefulness feedbacks of the examined results ($n = 2,305$). We compute four viewport features for each result: (1) viewport time of the snippet (viewport time); (2) viewport time divided by the session time (viewport time%); (3) viewport time divided by the area of the snippet (viewport time per pixel); (4) the number of snippets that have been covered by the viewport (#snippet in viewport). We combine the viewport features with existing click and dwell time based features (e.g. the features used in [20]) to train the Gradient Boosted Regression Tree (GBRT) to predict user-perceived usefulness and test whether the viewport features can improve the prediction performance via 10-fold cross-validations. We randomly shuffle the dataset for three times and apply the 10-fold cross-validation for each shuffled dataset, which generates $3 \times 10 = 30$ test folds.

Table 2. The results of usefulness prediction for examined results ($n = 2,305$).

	MSE	MAE	Pearson's r
Click&Dwell time	0.865	0.748	0.374
Click&Dwell time + Viewport	0.849**	0.747	0.394**

The average performance of the models on the 30 test folds, measured in the Mean Squared Error (MSE), Mean Absolute Error (MAE), and Pearson's r between the predicted values and true values of the usefulness feedbacks, are shown in Table 2. By comparing the performance of these two models, we find that adding viewport features brings a small but statistically significant improvement, measured in MSE and Pearson's r. This result suggests that the viewport time is a valid signal for usefulness.

6   Related Work

Mobile Search. Existing research show that mobile search is different from traditional desktop search in the following aspects: (1) Mobile search is often conducted in different contexts compared to desktop search: people are more likely to search for news, location-based information and etc. with "fragmented attention" [9]. This observation is also confirmed by analysis on commercial search engines' query logs [23]. (2) The screen space of mobile devices is much smaller than a particular desktop display. Thus mobile users have to incur more effort to read the same amount of information. This will also impact users' behavior pattern and experience [14,16,17,19,21]. (3) Since modern mobile devices are often equipped with a touch screen, users usually interact with SERPs with Multiple Touch Interactions [7], which provides a new opportunity to model users' search processes in a finer grain. These differences between mobile and desktop environment motivate the study of the evaluation of mobile search engines.

Usefulness as an Evaluation Criteria for Search. Search evaluation sits at the center of IR studies. While the *system-oriented* evaluation methods aim to build reusable test collections to evaluate the effectiveness of search systems, the *user-oriented* evaluation methods try to measure user's experience during the information seeking process.

Since Belkin et al. [1] and Cole et al. [4] has proposed usefulness as a criteria for the evaluation of interactive information retrieval, some recent effort has been put into filling up the gap between relevance judgment from assessors and usefulness feedback from searchers [11,15,20]. They found that usefulness feedback has a stronger correlation with user satisfaction. Although these studies have already gained much success in modeling user satisfaction, the effectiveness of usefulness in the context of mobile search has not been extensively investigated.

7   Discussions and Conclusions

To summarize, via a carefully designed user study, we collect users' search behaviors along with their explicit usefulness feedbacks for both the clicked and unclicked results in mobile search. Using the collected data, we investigate the relationships between usefulness feedbacks, ranking positions, relevance annotations, and click necessity annotations to address **RQ1**. We find that the ranking positions have an effect on the usefulness feedbacks of the results in initial viewports. While a moderate linear correlation is found between usefulness feedbacks and relevance annotations, we find that the presentation style of results, reflected by their click necessity, is a moderating factor of the relationship between relevance and usefulness. Regarding **RQ2**, we correlate the result-level measures with query-level user satisfaction and confirm in mobile environment that the usefulness feedbacks have a stronger correlation with user satisfaction than the relevance annotations, showing a potential limitation of system-oriented evaluation in estimating actual user satisfaction. Regarding **RQ3**, we use ANOVA tests and regression models to examine the relationships between usefulness feedbacks and user's search behaviors, especially the viewport time of snippets on mobile SERPs. The results suggest that the viewport time can be a useful feature in estimating usefulness in mobile search. Because the usefulness feedback can: (1) better reflect user satisfaction in mobile search; (2) be estimated by search behavior features, it is promising to be adopt in the user-oriented evaluation of mobile search engines.

References

1. Belkin, N.J., Cole, M., Liu, J.: A model for evaluation of interactive information retrieval. In: Proceedings of the SIGIR 2009 Workshop on the Future of IR Evaluation, pp. 7–8 (2009)
2. Cleverdon, C., Mills, J., Keen, M.: Aslib cranfield research project: factors determining the performance of indexing systems (1966)
3. Cohen, J.: Weighted kappa: nominal scale agreement provision for scaled disagreement or partial credit. Psychol. Bull. **70**(4), 213 (1968)
4. Cole, M., Liu, J., Belkin, N., Bierig, R., Gwizdka, J., Liu, C., Zhang, J., Zhang, X.: Usefulness as the criterion for evaluation of interactive information retrieval. In: Proceedings of the HCIR, pp. 1–4 (2009)
5. Fox, S., Karnawat, K., Mydland, M., Dumais, S., White, T.: Evaluating implicit measures to improve web search. ACM TOIS **23**(2), 147–168 (2005)
6. Granka, L.A., Joachims, T., Gay, G.: Eye-tracking analysis of user behavior in WWW search. In: SIGIR 2004, pp. 478–479. ACM (2004)
7. Guo, Q., Jin, H., Lagun, D., Yuan, S., Agichtein, E.: Mining touch interaction data on mobile devices to predict web search result relevance. In: SIGIR 2013, pp. 153–162. ACM (2013)
8. Guo, Q., Song, Y.: Large-scale analysis of viewing behavior: towards measuring satisfaction with mobile proactive systems. In: CIKM 2016, pp. 579–588. ACM (2016)
9. Harvey, M., Pointon, M.: Searching on the go: the effects of fragmented attention on mobile web search tasks. In: SIGIR 2017, pp. 155–164. ACM (2017)
10. Järvelin, K., Kekäläinen, J.: Cumulated gain-based evaluation of IR techniques. ACM Trans. Inf. Syst. (TOIS) **20**(4), 422–446 (2002)
11. Jiang, J., He, D., Kelly, D., Allan, J.: Understanding ephemeral state of relevance. In: CHIIR 2017, pp. 137–146. ACM (2017)
12. Joachims, T.: Optimizing search engines using clickthrough data. In: KDD 2002, pp. 133–142. ACM (2002)
13. Kamvar, M., Baluja, S.: A large scale study of wireless search behavior: Google mobile search. In: SIGCHI 2006, pp. 701–709. ACM (2006)
14. Kim, J., Thomas, P., Sankaranarayana, R., Gedeon, T., Yoon, H.J.: Eye-tracking analysis of user behavior and performance in web search on large and small screens. JASIST **66**(3), 526–544 (2015)
15. Kim, J.Y., Teevan, J., Craswell, N.: Explicit in situ user feedback for web search results. In: SIGIR 2016, pp. 829–832. ACM (2016)

16. Lagun, D., Hsieh, C.H., Webster, D., Navalpakkam, V.: Towards better measurement of attention and satisfaction in mobile search. In: SIGIR 2014, pp. 113–122. ACM (2014)
17. Lagun, D., McMahon, D., Navalpakkam, V.: Understanding mobile searcher attention with rich ad formats. In: CIKM 2016, pp. 599–608. ACM (2016)
18. Liu, Y., Chen, Y., Tang, J., Sun, J., Zhang, M., Ma, S., Zhu, X.: Different users, different opinions: predicting search satisfaction with mouse movement information. In: SIGIR 2015, pp. 493–502. ACM (2015)
19. Luo, C., Liu, Y., Sakai, T., Zhang, F., Zhang, M., Ma, S.: Evaluating mobile search with height-biased gain. In: SIGIR 2017. ACM (2017)
20. Mao, J., Liu, Y., Zhou, K., Nie, J.Y., Song, J., Zhang, M., Ma, S., Sun, J., Luo, H.: When does relevance mean usefulness and user satisfaction in web search? In: SIGIR 2016, pp. 463–472. ACM (2016)
21. Ong, K., Järvelin, K., Sanderson, M., Scholer, F.: Using information scent to understand mobile and desktop web search behavior. In: SIGIR 2017, pp. 295–304. ACM (2017)
22. Shokouhi, M., Jones, R., Ozertem, U., Raghunathan, K., Diaz, F.: Mobile query reformulations. In: SIGIR 2014, pp. 1011–1014. ACM (2014)
23. Song, Y., Ma, H., Wang, H., Wang, K.: Exploring and exploiting user search behavior on mobile and tablet devices to improve search relevance. In: WWW 2013, pp. 1201–1212. ACM (2013)
24. Verma, M., Yilmaz, E.: Characterizing relevance on mobile and desktop. In: Ferro, N., Crestani, F., Moens, M.-F., Mothe, J., Silvestri, F., Di Nunzio, G.M., Hauff, C., Silvello, G. (eds.) ECIR 2016. LNCS, vol. 9626, pp. 212–223. Springer, Cham (2016). https://doi.org/10.1007/978-3-319-30671-1_16
25. Voorhees, E.M., Harman, D.K., et al.: TREC: Experiment and Evaluation in Information Retrieval, vol. 1. MIT press Cambridge, Cambridge (2005)
26. Williams, K., Kiseleva, J., Crook, A.C., Zitouni, I., Awadallah, A.H., Khabsa, M.: Detecting good abandonment in mobile search. In: WWW 2016, pp. 495–505 (2016)

A Comparative Study of Native and Non-native Information Seeking Behaviours

David Brazier(✉) and Morgan Harvey

Northumbria University, Newcastle upon Tyne, UK
{d.brazier,morgan.harvey}@northumbria.ac.uk

Abstract. The proliferation of web-based technologies has led most national governments to begin transitioning to a so called "e-service," where provision is made through purely digital means. Despite their obvious benefits for most users, these on-line systems present barriers of access to certain groups in society. In this study we consider the information behaviour of English as a second language (ESL) and native English speaking participants as they conduct search tasks designed to reflect actual information seeking situations in a UK governmental context. Results show that the ESL users rely more on query assistance, delve deeper into the Search Engine Results Page (SERP) and obtain better performance the longer they read documents. This was not the case for the natives, despite spending the most time reading documents. There are some similarities in their information seeking behaviours as both groups submit similar length queries, and are equally proficient in identifying when a failed query did not meet their information need. This proficiency was not reflected in their performance in some tasks, with both groups unable to consistently predict when they had not performed well. The results of this work have potentially profound repercussions for how e-government services are provided and how users are assisted in their use of these.

1 Introduction

With the global-scale proliferation of web-based technologies and the subsequent uptake of electronic services (so called "e-services"), the number of non-English language users on the web is, unsurprisingly, rising also. Despite the fact that recent figures suggest that only slightly over a quarter of all Internet users are English native speakers [21], relatively little research effort is put into improving the quality of non-English web search [15]. Research has found that, despite the increasing number of users who speak English as a second language (ESL), or do not speak English at all, the extent and quality of content in other languages often does not meet the needs of said users [3]. In addition to this, even when there is sufficient content available, there are a considerable number of mostly unresolved complexities and issues of monolingual search in non-English languages [15,22].

© Springer International Publishing AG, part of Springer Nature 2018
G. Pasi et al. (Eds.): ECIR 2018, LNCS 10772, pp. 237–248, 2018.
https://doi.org/10.1007/978-3-319-76941-7_18

Although there are numerous works on Cross Language Information Retrieval (CLIR) [18] and translation services for ESL users reading English language content [9], adoption of these technologies is certainly not universal. As such, a large number of users often still need to seek information by searching in the English language, regardless of whether it is their native language or not. This issue is made more serious by the policies of most national governments, the UK's included, to begin transitioning their services from a "traditional" face-to-face and paper-based paradigm to "e-services," where provision is made through purely digital means [11]. For those in society, however, who are not adept in the use of such technologies, or are not able to readily make sense of the important information delivered through them, this raises concerns around the barriers that may be erected and the risk this poses of segregating users, especially those in vulnerable groups [13], such as refugees and migrants [16].

Before any transition to such a self-service, e-government model, all attempts must be made to try and to assist those most at risk of being segregated and to understand any issues they may have in accessing and using these services. It is with this in mind that this paper seeks to identify the current information seeking behaviours of ESL users when performing e-government-related tasks, to ascertain where and why issues arise during this process and how their behaviour differs from those of native English speakers when performing the same tasks under the same conditions.

2   Related Work

In recent years, researchers have investigated the issues users can face when attempting to access and comprehend important information sources, e-services in particular. Lloyd et al. [16] found that refugees trying to access e-government services experience information poverty due to social exclusion of the participants as a result of barriers e-services can erect. Vinson suggests that such information poverty can lead to serious negative outcomes, including "limited support networks, [an] inability to access the labour market, alienation from society and poorer educational outcomes" [23].

Numerous consider governmental e-services, the public's engagement with such services and barriers to their use [1,7], as well as e-government use within the field of information retrieval [12]. With the notable exception of work by Scantlebury on e-health information seeking [20], a large portion of this research is in a governmental context outside of the UK. This is surprising, given the UK government's drive for e-governance, in line with other governments worldwide, which culminated in the "Digital by Default" campaign [11]. Aham and Li [1] investigated user engagement with governmental digital services and found that one of the most influential factors was the content and, more specifically how long documents were and how complex the use of language within the documents was. Burroughs' [7] work aimed to overcome barriers to citizens' ability to access e-services in South Africa and concluded that awareness of, and sensitivity to, the user's native language are crucial variables in how well such a service is used by those who "do not speak a 'world language' (such as English)".

Savolainen [19] discusses the socio-cultural barriers of information seeking, of which institutional and user language barriers are just some. He posits that these aspects have been considered in a number of contexts, by a number of researchers, but there still remains work to be done on the extent to which these barriers are hindering, delaying or preventing information access, as well as the possibilities of offering alternative routes to information. This raises questions about users whose native language is not English, and the barriers they face if governmental services are solely accessible on-line. Brazier and Harvey [5] studied the search behaviours and performance of ESL users when given search tasks that new immigrants to a country might need to perform and found that, while most users were very confident of their English language searching abilities, they did not perform very well.

Some fairly recent work has compared search behaviour and performance of native and ESL speakers. Chu et al. [8,9] suggest that users searching using a second language require significantly more time, submit more query reformulations and view/assess a greater number of websites and those with only an intermediate grasp of the language struggle with query reformulation. Bogers et al. [4] considered the problem of searching for books and found, somewhat in contrast, that ESL users spend more time on task than native speakers, but that there is very little difference between natives and ESL users in relation to the number of queries, query length, or depth of result inspection. They surmised this could be as a result of their users' experience in searching for books in English and having acceptable foreign language skills.

In this work we integrate elements of the literature mentioned to specifically investigate the search behaviours and performance of both native and ESL searchers on contextually-relevant tasks, taken unadulterated from the work of Brazier and Harvey [5,6]. In doing so, we can gain a better understanding of how e-services should be developed and provisioned such that they are of benefit to all users, regardless of whether or not English is their mother tongue. In addition to this we can also learn more about the differences (and similarities) between how native and ESL users use English-language search engines.

3   Methodology

3.1   Procedure

The study utilised a mixed methods approach, gathering query log information, manually extracted from screen and video recordings, to gain a rich insight into user information seeking behaviour. To compliment this data, semi-structured focus group discussions were conducted after each experiment to elicit self-reported behaviours and anecdotal evidence, which we explore using thematic analysis. Study sessions for each participatory group were conducted separately, with a total of nine sessions: four for ESL and five for English native speakers. Each session began with participants filling in a demographic questionnaire, which collected information on their area of study; age; gender; nationality; language(s) spoken and proficiency; IT use; search engine use in English and their native tongue; search engine competency and preference and their own UK governmental service experience. The participants performed four contextually (to UK government) relevant search tasks [6]. Using the Chrome browser, each participant was instructed to use Google to perform each task, but were not limited to the search results page.

Tasks were a maximum of 10 min, although participants were provided the opportunity to end the task early if they felt they had a sufficient number of documents to complete the task. Participants were given up to 5 additional minutes to read the task and complete pre- and post-questionnaires, allowing the experiment to take no more than one hour in total. Post-study discussions then ensued with time-scales dictated by the discourse, ranging from 25 to 55 min. Tasks were distributed to participants using a Latin square design to account for task fatigue and potential learning effects.

For each task, participants were asked to read the scenario, then fill in a pre-task questionnaire [10] to gauge their domain knowledge, interest in the topic and the perceived difficulty of the task using a five-point Likert scale. Participants then began their search for relevant documents/sources, bookmarking any deemed of relevance as they went. At the end of each task the participant was also required to complete a post-task questionnaire, as seen in Table 1.

Table 1. Post-task questions

Q1	I was given enough information to complete the task
Q2	It was clear what was being asked
Q3	The task was relevant to me
Q4	The task was easy to understand
Q5	I was engaged in the task
Q6	I performed the task to the best of my ability
Q7	I found the task difficult
Q8	I'm confident the content I found satisfied the task
Q9	I am confident about the search query terms I used
Q10	I'm confident I identified relevant websites
Q11	I'm confident in my ability to read the website content
Q12	I am confident in my ability to understand the content of the websites I visited
Q13	I am confident the search task was completed

Participants for the study were recruited via face-to-face inquiry, university mailing lists and poster advertising. Interested parties registered their interest on the callforparticipants.com website, where participants were able to indicate whether or not they were native speakers. Once recruited, sessions were organised based on availability of the participant, venue and technology as aforementioned. Each was remunerated for their participation with a £10 Amazon voucher.

3.2   Measures and Metrics

Using Morae Manager each recorded session was manually tagged to calculate several measures and metrics. Total task time was defined as the period between when users clicked start task and end task; number of queries was the total number queries submitted by participants, including suggested queries; length of query is the total number of terms; number of assisted terms are the number of query terms entered through the assistance functionality; length of time querying is the time between a click on the search field and the time a query is submitted; time on SERP is the time between SERP load and when the participant navigates away, either by a result click or switching tab; link position is dependent on the listing number of the SERP link clicked assuming there are 10 links per SERP page; times bookmarked are the total number of documents bookmarked during that click-through session; the number of times in-site search and in-site link click are the total number per click-through session.

To determine relevance, all bookmarks were assessed by two native English-speaking IR researchers [14] using a voting strategy - any bookmarks not given the same score were discussed and a single score was agreed - and given scores between 1 and 4, where 1 is not relevant, 2 is tangentially relevant, 3 is partially relevant and 4 is relevant. Query classification is after Chu et al. [8] and determined by the same researchers.

Unless otherwise stated, the non-parametric Wilcoxon signed rank test is utilised to determine statistical significance.

4   Results

4.1   Participants

Initially there were thirty participants recruited, however, one native user was removed as they failed to bookmark any documents, opting instead to write notes (not URLs) about their interactions. During initial data analysis, it was identified that two of the native participants, who had acknowledged they were (non UK) native English speakers, actually registered on their pre-study demographic questionnaire as only being fluent in English and spoke Hindi and Hausa natively. As a result, these participants have been grouped with the ESL users, resulting in 12 native and 17 ESL participants (N = 29), all of whom were postgraduate students conducting a PhD project at a large UK university.

Non-natives were from countries across Africa (18%), Asia (59%) and Europe (24%) with a total of 18 languages spoken natively, and 27 languages in total up to a competent level. 82% self-assessed as being fluent in the English language, with 18% competent. 41% of the ESL participants were female with an average age of 28 $(SD = 4.619)$ and 59% were male with an average age of 31.5 $(SD = 3.440)$. All use IT daily, with 94% using a search engine in English daily, and 6% every few days. 83% of English-natives were British born, with 8% African and 8% Caribbean. 42% of the native participants were female with an average age of 37.4 $(SD = 10.229)$ and 58% were male with an average age of

27 ($SD = 2.268$). All use IT daily, with 94% using a search engine daily, and 6% every few days. 88% of ESL users and 100% native were confident or very confident in formulating queries, identifying relevant search results and information on website in English. The majority of both groups had used UK government e-services previously (ESL 59% Native 75%), 18% (ESL) and 17% (native) hadn't, and 23% (ESL) and 8% (native) weren't sure.

4.2 Tasks

Differences in task relevance were statistically significant (W = 2059.5, p-value = 0.015), with relevance highest among the ESL users (see Table 2), while natives generally found the tasks less relevant. It is unsurprising that relevance of the tasks for natives are lower than those of ESL users considering the method in which the tasks were formulated [5]. However, it is interesting to note that, despite there being no native English speaker participation in the topic selection, no one topic was deemed completely irrelevant, with the *housing* task of most and the *digital by default* task of least relevance to both groups. When discussed post-task, the task descriptions were determined believable and realistic, although somewhat vague and general at times i.e. the *health* task.

Table 2. Task relevance for both groups

Topic	Non-native		Native	
	Mean	Median	Mean	Median
1	3.529	4	2.583	2
2	3.588	4	3.000	3
3	3.294	3	2.667	3
4	2.471	3	2.083	2

The native participants spent more time on task overall (541.25 to 551.09 s), although not significantly so (W = 1335.5, p-value = 0.1359). This is contrary to research by Chu [8], who found the opposite to be true, with quite disproportionate average time differences between natives and ESL users.

4.3 Relevance and Document Classification

Although this study focusses on e-governance, participants were not limited to relying solely on governmental sources, and were actively encouraged by the researcher to "bookmark whichever sources were deemed of most use".

In total the ESL group bookmarked 459 (27 per participant) bookmarks and the natives bookmarked 249 (21 per participant). 55.6% of the ESL bookmarked URLs were governmental and 51.4% for the natives with these no more relevant than the non-governmental ones (see Table 3). This does appear to be quite topic-dependent as government sources were more relevant in topic 1 for both groups and topic 2 for the ESL users. Non-governmental documents were higher scoring in topic 2 for natives, topic 3 for both and topic 4 for both. This is explained by post-task discussion comments on topics 1 and 4, where participants found governmental information was of most use and highly informative in visa applications, whereas for topic 4 it was revealed that the governmental documents, although official and informative, did not best match the task as they did not consider practical application of the information and were mostly policy documents.

Table 3. Bookmark relevance

	Non-native		Native	
	Count	Relevance	Count	Relevance
Gov	255	3.02	128	3.02
Non-gov	204	3.01	121	3.25

Table 4. Native and Non-native bookmark relevance

	Non-native	Native
Relevant	195	111
Partially	91	63
Tangentially	159	72
Non-relevant	14	3

4.4 Performance

The native group bookmarked fewer documents per task on average (5.213, compared to 6.647) but performed marginally better, in terms of average precision, than the ESL users overall - 0.69 compared to 0.623 (see Table 4) - although not significantly so (W = 1487.5, p-value = 0.525).

Table 5. Performance by task

Task	Non-native			Native		
	Count	Avg. Prec	Gov	Count	Avg. Prec	Gov
1	11	0.885	0.756	9	0.863	0.740
2	8	0.649	0.288	3	0.821	0.268
3	3	0.586	0.606	3	0.576	0.525
4	4	0.339	0.542	3	0.508	0.459

When broken down by task (see Table 5) both groups performed better in task 1 with the ESL users, surprisingly, performing best, which could be explained through the design of the visa section of the gov.uk website. For users able to find this site, there is a wizard which guides them through the process systematically, thereby ensuring relevant documents are accessed on each click. In other tasks there was no such functionality present, either in governmental or non-governmental documents. It must be noted that estate and letting agents' websites (accessed as part of Task 2 on housing) do contain filtering functionality, which may explain marked differences in both performance and number of bookmarked documents in this task.

Despite both groups relying on similar proportion of non-governmental documents, and although the ESL users bookmarked a larger number of documents, their performance is lower. Performance for task 4 is interesting, in that both groups have similar bookmarked documents and both rely almost equally on governmental and non-governmental sources, and yet perform worst here, the ESL users markedly so. Reasons for such poor performance have been touched on in Sect. 4.3, with users struggling to balance contextual relevance with (governmental) document trustworthiness and, therefore, reliability. It is curious that despite acknowledging the lack of contextual relevance in some policy documents, there was still a large proportion of users who bookmarked said documents.

As shown in Figs. 1 and 2, in terms of post-task perception, users felt that they had enough information were engaged, that tasks were clear and weren't difficult, and that they were confident in the content they identified and that the tasks were complete (refer to Table 1 in Sect. 3.1 for question descriptions). In 3 of the 4 tasks for ESL users and 2 of the 4 tasks for natives, between 35% and 66% of documents bookmarked were not relevant.

The mostly positive nature of their post-task review is in stark contrast to their actual performance, which was identified before for ESL users [6].

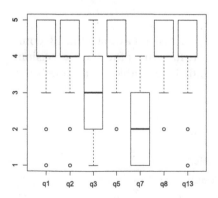
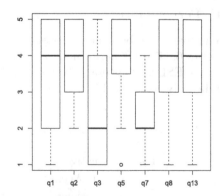

Fig. 1. Non-native task 4 post response.   **Fig. 2.** Native task 4 post response.

4.5   Behaviours

Querying

Natives submitted more queries yet spent less time querying (4 queries per task taking 8 s per query, compared to 3 queries with 9 s per query for ESL users), appearing to contradict the study by Bogers et al. [4], which found ESL users to query much more. Both the Bogers et al. study and this one found query length to be equal. Use of query assistance was significantly different between the groups (W = 109390, p-value $\ll$ 0.01): 6% of all ESL users query terms were provided by or amended through Google's assistive functionality, but only 5% of the natives' terms. Some users were particularly heavy users of this feature, as there was a range between users of 0 to 75 terms for ESL users and 0 to 40 terms for the natives. There were very few instances of misspelling from both groups, which may be accounted for by the education and language fluency levels of the participants [4], although ESL users did make the majority of errors (16 compared to 5). The experimental conditions may have influenced participant behaviour as one native user (A1) acknowledged that they were aware of the recording of the study and made a conscious effort to spell correctly, whereas in a more relaxed setting they would often rely on assistance. This was echoed by native participant B1, who explained that assistance would be used (in other settings) to complete queries to save time. A comparison of queries found that there were no differences in the distribution of queries submitted across both groups, with new queries and reformulations (66.43% for ESL users and 68.91%) making up the majority of submitted queries, despite being contrary to the initial study [8], this has been identified previously [5].

Search Results and Reading

Non-natives looked significantly deeper (W = 117350, p-value $\ll$ 0.01) into search results than natives with an average depth of 9 (see Fig. 3), while the natives averaged a depth of 3. As such it is of little surprise that ESL users spent more time on the SERP (31.11 s) than natives (29.10 s). When discussing governmental links on the SERP, it was noted by several participants that they had to actively search for governmental links (specifically gov.uk links), as they often did not occupy the top positions of the SERP. This may explain why the ESL users both search deeper and longer than the native users, who bookmarked fewer governmental documents (Table 3). It is worth noting that although not statistically significant, approximately a quarter of all queries submit resulted in zero SERP link clicks, also known as a failed query, for both the native and ESL groups. This is a reasonable indicator that they are equally proficient in identifying when a query or SERP link did not meet their information need. Although this could be explained by the level of education and English language proficiency of the participants.

Natives were found to spend more time reading documents than ESL users and significantly so (W = 90662, p-value $\ll$ 0.01), as shown clearly in Fig. 4. This is somewhat surprising, as it could be assumed that those less familiar with the language are more likely to read the documents in more depth and take more time to do so [14], however this was not the case. It may be that natives are willing to spend more time reading the documents as it is less effort for

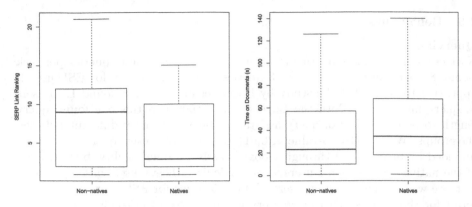

Fig. 3. SERP depth.   **Fig. 4.** Time spent reading documents.

them to do so. Once outlier C is removed due to their unique search behaviours, time spent reading documents significantly predicts performance for ESL users (adjusted R-squared: 0.6818, p-value $\ll 0.01$) and for every 1 additional second of time spent on the document, the expected performance (in terms of precision) increases by 0.004.

Search Strategy
A number of users in both groups utilised the shortcut find method (ctrl+F) to look for keywords on the current page, rather than using the in-site search functionality. In post discussion reasons for such strategies were explained due to the trust and observable success from utilising web search engines, in this case Google, rather than the in-site search facilities. This is further displayed by the usage of in-site search by both groups (mean = 0.031 for natives compared to 0.110 for natives). These behaviours have been identified previously by Nielsen [17] and the concern is that in the time since this article, the situation has not changed. This is, perhaps, in part due to the trust placed in the results presented by major search engines and the lack of trust in bespoke search or unbranded systems. The UK Government's Digital Service have plans to update and improve the in-site search function, possibly to address this [2], however, as these behaviours appear not to be specific to any content or source, there is some way to go for users to reap the full potential of the in-site search function.

5   Limitations

An obvious limitation of this study is the experimental conditions influencing participant behaviours, something acknowledged by some of the native users, and must be considered a factor for others behaviours also. Although such a controlled study does bring benefits, future work could utilise a more hands-off approach. Educational background and number of participants is also a consideration. Although no generalisable hypothesis can be drawn from this limited user representation, the results allow us some insights into the search behaviours of both ESL and native English users and their e-government services information interactions. Relevance assessment is also a limitation, considering the effect of language on interpretation of information (from both a researcher and user perspective), and must be considered in future studies.

6 Conclusions

This study expanded on previous work in multilingual IR from an information seeking behaviour perspective by examining the ways in which ESL users approach a number of important search tasks in comparison to native English users. The study has identified some marked and statistically significant differences between the groups, with ESL users using more query assistance (auto-correct), delving deeper into the SERP and spending longer in doing so. Additionally, the longer they spent reading documents, the higher their performance, which was not the case for the natives, despite spending the most time reading documents. Nevertheless, there are also some similarities in their information seeking behaviours as both groups submit similar length queries and are equally proficient in identifying when a failed query did not meet their information need. This proficiency was not reflected in their performance in some tasks, with both groups unable to consistently predict when they had not performed particularly well. Relevance of the bookmarked documents, in this case, was found to be subject to the contextual and practical application of the information, and the official and trustworthy (yet not contextually-relevant) nature of governmental documents, which could go some way to explaining poorer performance among both groups. These results are somewhat alarming as it is reasonable to assume that as users' educational levels, (English) language proficiency and/or information literacy lower in comparison to those of the study participants, their own performance would in turn diminish. In light of a solely e-government system, this raises significant concerns about users and the information they rely on to make judgements that can have real world implications. One way of mitigating such concerns is to consider the use of wizards. Performance was high among both groups when this system design was implemented, and in post discussion, there was positive sentiment (from both groups) towards such a tool as they provide a clear and structured platform to information. Future works in this area plan to include a more expansive and representative population of ESL users.

References

1. Aham-Anyanwu, N., Li, H.: E-public engagement: formulating a citizen content engagement model. In: 25th European Conference on Information Systems (ECIS), 5–10 June 2017, pp. 753–770 (2017)
2. Allum, J.: How we're making gov.uk work harder for users (2017). https://gds.blog.gov.uk/2017/02/27/how-were-making-gov-uk-work-harder-for-users/. Accessed 19 Dec 2017
3. Berendt, B., Kralisch, A.: A user-centric approach to identifying best deployment strategies for language tools: the impact of content and access language on web user behaviour and attitudes. Inf. Retrieval **12**(3), 380–399 (2009)

4. Bogers, T., Gäde, M., Hall, M., Skov, M.: Analyzing the influence of language proficiency on interactive book search behavior. In: iSchools (2016)
5. Brazier, D., Harvey, M.: E-government and the digital divide: a study of english-as-a-second-language users' information behaviour. In: Jose, J.M., Hauff, C., Altıngovde, I.S., Song, D., Albakour, D., Watt, S., Tait, J. (eds.) ECIR 2017. LNCS, vol. 10193, pp. 266–277. Springer, Cham (2017). https://doi.org/10.1007/978-3-319-56608-5_21
6. Brazier, D., Harvey, M.: Strangers in a strange land: a study of second language speakers searching for e-services. In: CHIIR, pp. 281–284. ACM (2017)
7. Burroughs, J.M.: What users want: assessing government information preferences to drive information services. Gov. Inf. Q. **26**(1), 203–218 (2009)
8. Chu, P., Jozsa, E., Komlodi, A., Hercegfi, K.: An exploratory study on search behavior in different languages. In: IIiX, pp. 318–321. ACM (2012)
9. Chu, P., Komlodi, A.: Transearch: a multilingual search user interface accommodating user interaction and preference. In: 2017 CHI Conference Extended Abstracts, pp. 2466–2472. ACM (2017)
10. Edwards, A., Kelly, D.: How does interest in a work task impact search behavior and engagement? In: CHIIR, pp. 249–252. ACM, March 2016
11. Freeguard, G., Andrews, E., Devine, D., Munro, R., Randall, J.: Whitehall monitor 2015 (2015). http://www.instituteforgovernment.org.uk/publications/whitehall-monitor-2015. Accessed 19 Dec 2017
12. Freund, L.: A cross-domain analysis of task and genre effects on perceptions of usefulness. Inf. Process. Manage. **49**(5), 1108–1121 (2013)
13. Helbig, N., Gil-García, J.R., Ferro, E.: Understanding the complexity of electronic government: implications from the digital divide literature. Gov. Info. Q. **26**(1), 89–97 (2009)
14. Józsa, E., Köles, M., Komlódi, A., Hercegfi, K., Chu, P.: Evaluation of search quality differences and the impact of personality styles in native and foreign language searching tasks. In: IIiX 2012, pp. 310–313. ACM (2012)
15. Lazarinis, F., Vilares, J., Tait, J., Efthimiadis, E.N.: Current research issues and trends in non-english web searching. Inf. Retrieval **12**(3), 230–250 (2009)
16. Lloyd, A., Kennan, M.A., Thompson, K.M., Qayyum, A.: Connecting with new information landscapes: information literacy practices of refugees. J. Doc. **69**(1), 121–144 (2013)
17. Nielsen, J.: Information foraging: why google makes people leave your site faster (2003). https://www.nngroup.com/articles/information-scent/. Accessed 19 Dec 2017
18. Peters, C., Braschler, M., Clough, P.: Multilingual Information Retrieval: From Research to Practice, 1st edn. Springer, Heidelberg (2012). https://doi.org/10.1007/978-3-642-23008-0
19. Savolainen, R.: Approaches to socio-cultural barriers to information seeking. Libr. Inf. Sci. Res. **38**(1), 52–59 (2016)
20. Scantlebury, A., Booth, A., Hanley, B.: Experiences, practices and barriers to accessing health information: a qualitative study. Int. J. Med. Inform. **103**, 103–108 (2017)
21. Internet World Stats, June 2017. http://www.internetworldstats.com/stats7.htm
22. Steichen, B., Freund, L.: Supporting the modern polyglot: a comparison of multilingual search interfaces. In: CHI, pp. 3483–3492. ACM (2015)
23. Vinson, T.: Social inclusion: the origins, meaning, definitions and economic implications of the concept of inclusion/exculsion (2009)

Representation

Indiscriminateness in Representation Spaces of Terms and Documents

Vincent Claveau[⊠]

Univ. Rennes, CNRS, IRISA, Campus de Beaulieu, Rennes, France
vincent.claveau@irisa.fr

Abstract. Examining the properties of representation spaces for documents or words in Information Retrieval (IR) – typically $\mathbb{R}^n$ with n large – brings precious insights to help the retrieval process. Recently, several authors have studied the real dimensionality of the datasets, called intrinsic dimensionality, in specific parts of these spaces [14]. They have shown that this dimensionality is chiefly tied with the notion of indiscriminateness among neighbors of a query point in the vector space. In this paper, we propose to revisit this notion in the specific case of IR. More precisely, we show how to estimate indiscriminateness from IR similarities in order to use it in representation spaces used for documents and words [7,18]. We show that indiscriminateness may be used to characterize difficult queries; moreover we show that this notion, applied to word embeddings, can help to choose terms to use for query expansion.

Keywords: Intrinsic dimensionality · Indiscriminability
RSV scores · Distributional thesauri · Query expansion

1 Introduction

Examining the properties of representation spaces for documents or words in Information Retrieval (IR) – typically $\mathbb{R}^n$ with n large– brings precious insights to help the retrieval process. It is well-known that the dimensionality of the representation space is not the same as the dimensionality of the data. In the usual vector space model used in IR, the dimensionality of the representation space is the number of different words in the document collection, yet it often possible to represent the same documents in a space with much less dimensions. This fact is at the heart of techniques like *Latent Semantic Indexing* or *Latent Dirichlet Allocation* which reduce the dimensionality of the original (very sparse) vector space to a much smaller (and dense) space.

In this paper, we focus on the intrinsic dimensionality of the data, not from a global perspective (as for LSI or LDA), but more locally on portions of the space. For that purpose, we rely on the work of [2,14] which permit to define and estimate the local intrinsic dimensionality of the data (see Sect. 2). They showed that it can be used to measure the indiscriminateness of neighbors of a query, and thus to indirectly assess the potential quality of the answers to this query.

© Springer International Publishing AG, part of Springer Nature 2018
G. Pasi et al. (Eds.): ECIR 2018, LNCS 10772, pp. 251–262, 2018.
https://doi.org/10.1007/978-3-319-76941-7_19

Since indiscriminateness depends on how the neighborhood is defined, and thus on the distance metric used, it is necessary to adapt it to RSV (*Relevance Status Value*) if one wants to use it in IR (Sect. 3). Then, we show in Sect. 4 how it can be used to analyze the representation space of documents in IR, for instance to detect difficult queries.

2   Related Work

Characterizing the intrinsic dimensionality of data sets have been studied in different ways. For instance, embedding techniques or projection techniques build spaces with lower dimensionality in which data points are projected under certain conditions of discriminateness, like *Principal Component Analysis* (PCA), *Latent Semantic Indexing* (LSI), *Latent Dirichlet Allocation* (LDA) [8,11] or *manifold learning* [21,22,26]. The intrinsic dimensionality of the whole data set is then the one of this new space obtained through projection.

Recently, [14] proposed a generalized expansion measure defining the local intrinsic dimension by examining how many points are met around a query point within a certain distance, and how it evolves when the distance augments. More formally, consider two balls centered in x_1 and x_2 with radius ϵ_1 et ϵ_2 in $\mathbb{R}^m$. The ratio between the volumes of these balls can be expressed as:

$$\frac{volume(B(x, \epsilon_1))}{volume(B(x, \epsilon_2))} = \left(\frac{\epsilon_1}{\epsilon_2}\right)^m$$

From that, one can define the dimensionality:

$$m = \frac{\ln(volume(B(x, \epsilon_1))) - \ln(volume(B(x, \epsilon_2)))}{\ln \epsilon_1 - \ln \epsilon_2}$$

The idea at the heart of the intrinsic dimension measure is to divert the previous equation by replacing the number of points in the volume instead of the volume itself [14, Sect. IV B for justification]. Let us note $|B(x, \epsilon)|$ the number of points in $B(x, \epsilon)$; we have:

$$\hat{m} = \frac{\ln |(B(x, \epsilon_1))| - \ln |B(x, \epsilon_2)|}{\ln \epsilon_1 - \ln \epsilon_2}$$

The dimensionality is now the one of the data, not the one of the representation space. It is worth noting that this estimate is local to a point x (when considering $x = x_1 = x_2 = x$).

This intrinsic dimensionality model has been used in several ways for analyzing and building indexing structures for similarity search [4,12,13], and for anomaly detection [27]. Let us also cite the work of [3]; it does not rely on intrinsic dimensionality but the authors also exploit statistics on distance distribution in a similar context than ours in Sect. 4.

3   Use for Information Retrieval

The interest of intrinsic dimensionality for IR is its capacity to characterize the neighborhood of a query based on the documents surrounding it in the representation space. If this intrinsic dimensionality is very high, it means that a slight distance variation may completely change the set of documents that are considered as the closest to the query. Therefore, a high intrinsic dimensionality implies a high indiscriminateness of the documents around the query [14]. This is this exact property that we want to exploit here, provided that we can adapt it to the particular case of the similarity (Relevance Status Value, RSV) functions used in IR.

3.1   Limits

The intrinsic dimensionality definition previously given is set for a space in which the metric used is a distance, typically a L2 distance (Euclidean distance). It is used to define the set of points contained in the balls with different diameters (eg. for a ball centered in x with radius $r > 0$, the points d_i considered are those with $L2(x, d_i) \leq r$). Yet, in IR, L2 is rarely used as RSV in the vector space model; instead, cosine has been widely used. As it has been shown [14], the intrinsic dimension definition can be used with such angular distances. The approach is the same as before: for a given vector, we compare the number of vectors in its neighborhood, that is, with angles lesser than ϵ_1 and ϵ_2.

Most of the common and modern RSV function can be written:

$$RSV(q, d) = \sum_{t \in q} w_q(t) \cdot w_d(t)$$

with $w_q(t)$ the weight of term t in query q and $w_d(t)$ the weight in document d, as illustrated in Table 1 (from [16]).

With the following notations:

$c(t, d)$ number of occurrences of term t in document d
$c(t, q)$ number of occurrences du term t in query q
N number of documents in the collection
$df(t)$ number of documents containing term t
$dl(d)$ length of document d
$avdl$ average length of documents
$c(t, C)$ number of occurrences of term t in collection C
$p(t|C)$ probability of term t for a language model of the collection.

The RSV functions can be seen as simple scalar product between document vector d and query vector q, which we note $\langle q, d \rangle$. It differs from cosine in that it does not impose a L2 normalization of the vectors (Figs. 1 and 2).

The absence of normalization is an important issue since it makes impossible to use the same principle as before to compute the intrinsic dimension. Indeed, for a query q, close documents (in terms of scalar product) may be at any L2 distance. More formally, for two thresholds values for the scalar product ϵ_1 and ϵ_2 ($\epsilon_1 \geq \epsilon_2$), the part of space containing points d_i such that $\epsilon_1 \geq \langle d_i, q \rangle \geq \epsilon_2$ is infinite, as illustrated in Fig. 3.

Table 1. Weighting functions of terms in the query and the document for different state-of-the-art IR models: BM25+ [16,20], Divergence From Randomness PL2 [1,9], Language modeling with Dirichlet smoothing Dir [28], Pivoted Normalization Piv [23]

model	weighting
BM25+ $w_d(t)$	$\left(\frac{(k1+1)c(t,d)}{k1(1-b+b\cdot dl(d)/avdl)+c(t,d)} + \delta\right) \cdot \log\frac{N+1}{df(t)}$
BM25+ $w_q(t)$	$\frac{(k3+1)c(t,q)}{k3+c(t,q)}$
	with $k1, k_3, b$ and δ fixed parameters
PL2 $w_d(t)$	$\frac{tfn(t,d)\cdot\log_2(tfn(t,d)\cdot\lambda_t)+\log_2 e\cdot(1/\lambda_t-tfn(t,d))+0.5\log_2(2\pi\cdot tfn(t,d))}{tfn(t,d)+1}$
PL2 $w_q(t)$	$c(t,q)$
	with $tfn(t,d) = c(t,d)\cdot\log_2\left(1+c\cdot\frac{avdl}{dl(d)}\right)$
	$c > 0$ a search parameter and $\lambda_t = \frac{N}{c(t,C)}$
Dir $w_d(t)$	$\log\left(\frac{\mu}{dl(d)+\mu} + \frac{c(t,d)}{(dl(d)+\mu)p(t|C)}\right)$
Dir $w_q(t)$	$c(t,q)$
	$\mu > 0$ a smoothing parameter
Piv $w_d(t)$	$\frac{1+log(1+log(c(t,d)))}{1-s+s\cdot dl(d)/avdl} \cdot \log\frac{N+1}{df(t)}$ si $c(t,d) > 0$ and 0 else
Piv $w_q(t)$	$c(t,q)$
	with s a fixed parameter

3.2 Estimate with the Power Law Exponent

In spite of the limit caused by the scalar product form of most RSV, we still aim at characterizing the intrinsic dimensionality, or at least the indiscriminateness, locally in the space. In previous work, [2,15] has shown that the intrinsic dimensionality could be estimated from the repartition of the L2 distances between a query and the other points. In line with this, we propose to characterize indiscriminateness by examining the evolution of the number of neighbors (in our

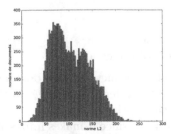

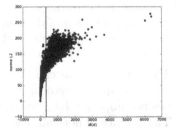

Fig. 1. Distribution of the L2 norms of documents in Tipster collection under BM25+ (modified version of BM25 proposed by [16])

Fig. 2. L2 norms of documents in Tipster collection according to their length $dl(d)$ with BM25+; horizontal line is the average length ($avdl$)

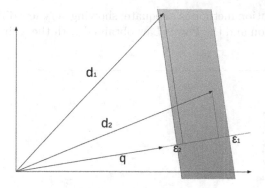

Fig. 3. In gray: portion of space defined by the set of points whose scalar products with a normed vector q lie between ϵ_1 and ϵ_2

case, documents considered as close to the query) according to the RSV. This evolution can be interpreted as the repartition function of a random variable X which represents the RSV score between a given query and a document. More precisely, since we are only interested in the local behavior of X, that if to the closest documents, we only examine the repartition function for the highest RSV scores. The hypothesis we make is that the distribution of RSV can be locally modeled as a Power Law, that is:

$$f(x) = \lambda x^{-\alpha} \text{ with } \lambda \text{ a constant and } \alpha > 1 \tag{1}$$

This is the exponent α which is characteristic of the indiscriminateness of the data. Formally, α cannot be linked to the intrinsic dimensionality as defined by [14] due to the problem raised by the use of scalar product as previously explained. In the IR case, we have n observations x_i, that is n RSV values of a query with its n closest neighbors (n highest RSV scores). It is thus possible to estimate α from these x_i. Among the various methods in the literature, the estimation based on log-likelihood has been shown to be the less biased [5]. Let the x_i be all the observations greater than a threshold x_{min}; α is then estimated as:

$$\hat{\alpha} = 1 + n \cdot \left(\sum_{i=1}^{n} \ln \frac{x_i}{x_{min}} \right)^{-1} \tag{2}$$

In the experiments reported below, we consider $n = 100$ observations x_i to estimate α, that is the 100 highest RSV scores (x_{min} is the 101st highest RSV score).

Figure 4 shows the RSV score repartition as an histogram for a given query, and the corresponding Power Law whose exponent α is estimated as previously explained. Two facts are worth noting: first, the hypothesis we make about the distribution following a Power Law seems reasonable, since the histogram is typical for this distribution (the same observation holds for every query tested), and second, the estimation method is adequate, showing very few differences between the real distribution and the Power Law obtained with the estimated α.

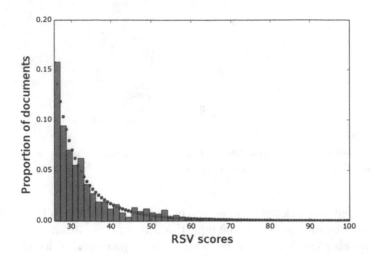

Fig. 4. Example RSV values repartition (red histogram) and the corresponding Power Law (blue) obtained with log-likelihood estimate of α from the RSV values (Color figure online)

4   Experiments in the Document Space

In this section, we study how the indiscriminateness index α, as defined above, can be used in a standard IR framework. We show how α can be used to characterize the documents close to a query, either within the vector space model or in other spaces where the RSV can nonetheless be seen scalar products in Euclidean spaces.

4.1   Data and Evaluation Scores

Two IR collections are used in our experiments: Tipster and OHSUMED [10]. Tipster contains more than 170 000 documents and 50 queries; it was used in TREC-2. The queries are composed of several parts, including the query itself and a narrative detailing the relevance criteria); in the experiments reported below, only the actual query part is used. OHSUMED contains 350,000 bibliographical notices from Medline and 106 queries from the TREC-9 filtering task. Performance are assessed with standard scores: Precision at different threshold (P@x), R-precision (R-prec), *Mean Average Precision* (MAP).

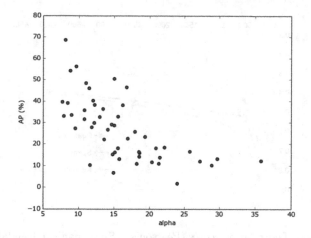

Fig. 5. Performance (AP) of queries from Tipster according to their index α with a BM25+ model

4.2   Detecting Difficult Queries

The distribution of documents around a query, or more precisely, the distribution of distances (or RSV) between a query and its closest documents can help characterize the difficulty of the query. In order to assess that, we look for a correlation between the index α (cf. Eq. 2) around a query and the Average Precision (AP) of this query. This is illustrated in Fig. 5 on the Tipster collection, with BM25+ [16] as RSV.

Table 2. Correlations (and their associated p-values) between AP and index α on Tipster with a BM25+ model

Coefficient	Value	p-value
Pearson r	-0.7150	$5.43e^{-09}$
Spearman ρ	-0.7753	$3.82e^{-11}$
Kendall τ	-0.5755	$3.69e^{-09}$

Table 3. Correlations (and their associated p-values) between AP and index α on OHSUMED with a Dirichlet LM ($\mu = 1000$)

Coefficient	Value	p-value
Pearson r	-0.4919	$9.85e^{-08}$
Spearman ρ	-0.6141	$3.26e^{-12}$
Kendall τ	-0.4494	$1.14e^{-11}$

The set of points exhibit the expected dependency between index α and the performance. In Table 2, we indicate the Pearson, Spearman and Kendall correlations (and their p-values) between the list of queries ordered by AP and the list of queries ordered by α on Tipster with a BM25+ model. The same information is given in Table 3 for OHSUMED with a Dirichlet LM (μ is set to 1000 for which it maximizes the MAP). The inverse correlation clearly appears:

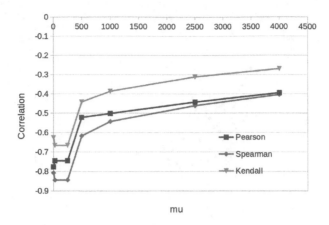

Fig. 6. Evolution of correlation scores (Pearson's r, Spearman's ρ, Kendall's τ) according to Dirichlet smoothing paramater μ.

a low retrieval performance for query is related to a high indiscriminateness around this query.

Same observations also hold with other RSV functions on both collections, but some model parameters have a high impact on the results. For instance, we can observe the influence of smoothing on indiscriminateness within languages models. When smoothing is heavy (high μ in Dirichlet smoothing for example), documents tends to have similar weights for every query term, and thus similar RSV score, which thus makes the documents more difficult to discriminate (high α index). When μ is low, (inverse) correlation is high and tends to diminish when μ gets higher. It can be verified in Fig. 6, where the correlation between AP and α are given for several μ.

5 Query Expansion

Projecting words in continuous representation spaces, such as vector spaces, has been widely studied recently. In these spaces, it is possible to find semantic proximity between words with the help vector-based distances. It makes it possible to build semantic lexicons in which each word is associated with its closest neighbors. In order to do so, these embedding techniques rely on the distributional hypothesis: close words share close contexts. By comparing contexts of words (words occurring before and after), these techniques infer proximities between the words and/or a vector representation of the words. In this field, WORD2VEC [18] is very well-known: words are represented as vectors with the help of a neural approach. In recent work, [7] have shown that IR techniques could also be used to build such vector representation for words. In this later work, the authors use similarities like Okapi-BM25 between contexts of two words to build distributional thesauri. Finally, a word is represented by its similarity scores to every other word. This approach, called spectral, has yielded good results in numerous tasks [6].

Semantic lexicons can be used for query expansion: the closest semantic neighbors of query words are added to the query. It has been used as a way to evaluate the quality of the lexicons. In this section, we examine indiscriminateness, as previously defined, in representation spaces generated by WORD2VEC or the spectral approach. Since they also rely on similarity based on RSV, we use the same estimation technique of index α to characterize the local properties of the word spaces. In particular, we examine how α can be used in the query expansion task.

5.1 Framework

We adopt the following experimental framework. The IR collection is Tipster (see previous section). The distributional lexicons are the spectral one developed by [7] and a Word2Vec model trained on the GoogleNews (freely available at https://code.google.com/p/word2vec/). The IR system used is Indri [17,24]. It is known to offer state-of-the-art performance, and moreover this probabilistic systems implements a combination of language modeling [19] and inference networks [25] which makes it possible to use operators AND OR... In the experiments report below, standard settings are used (Dirichlet smoothing parameter $\mu = 2500$). Thanks to its query language, this system allows us to easily expand the query with the semantic neighbors found in the distributional lexicon: the operator '#syn' aggregates counts of words considered as synonyms. In order to limit the effect of inflection (plural/singular) on the results, both plural and singular forms of the words are added to the query (original words of the query or words added from the lexicon). Performance is evaluated with the standard IR scores, by comparing results with and without expansion.

5.2 Experiments

Many uses for the α index of the words in the semantic lexicon are possible. Here, we report the results of two experiments where the α indexes of the words are used to filter expansions, with two different settings. In the first setting (Filter 1 hereafter) we compute α for each word of the original query, and we only add semantic neighbors for words of the original queries having α lower than a certain threshold. In the experiment, the threshold is fixed as the average α of the words of all the queries. In the second setting (Filter 2), we first filter words with Filter 1, and moreover, only semantic neighbors with α below a certain threshold are used to expand the queries.

Results with the spectral lexicon are given in Table 4, and those for WORD2VEC are in Table 5. Statistical significance (Wilcoxon with $p = 0.05$) are given: expansion results are compared with non-expanded version; expansion + filter 1 or 2 are compared with expansion (with no filtering). Non significant results are in italics.

Table 4. Relative performance gain (%) on Tipster with query expansion with and without filtering; spectral lexicon

No expansion	MAP	R-Prec	P@5	P@10	P@50	P@100
	21.78	30.93	92.80	89.40	79.60	70.48
With expansion	+13.80	+9.58	+2.16	+4.03	+5.58	+8.26
With expansion + Filter 1	+16.22	+10.78	+3.02	+4.47	+9.20	+12.51
With expansion + Filter 1 & 2	+22.83	+13.00	+2.56	+6.31	+14.10	+21.39

Table 5. Relative performance gain (%) on Tipster with query expansion with and without filtering; Word2Vec

No expansion	MAP	R-Prec	P@5	P@10	P@50	P@100
	21.78	30.93	92.80	89.40	79.60	70.48
With expansion	+13.52	+9.50	+2.59	+3.36	+8.29	+9.99
With expansion + Filter 1	+15.73	+9.27	+2.22	+4.96	+9.63	+14.41
With expansion + Filter 2	+20.76	+13.63	+3.88	+5.82	+10.15	+14.27

The good results of expanding query (without filtering) with lexical resources are already known. Yet, it is worth noting that they are slightly made better, but not significantly, by filtering the words to be expanded (Filter 1) with the α index. When filtering also the words to add to the query (Filter 2), the gains are significantly better. These results holds for both lexical resources. In practice, a close examination of the expanded queries shows that words whose α are above the maximum threshold are indeed polysemic or general ones: *choice, term, use, way, young...*

6   Concluding Remarks

In this article, we have shown how to adapt the notion of intrinsic dimensionality [14] to RSV similarities used in IR. In the follow up of [2], we have defined the α index to characterize indiscriminateness among neighbors of any point in the representation space. In a standard IR setting, we can exhibit the link between this index computed for any query and the performance of to be expected for this query. We have also applied this approach to word representation spaces, as generated by embedding techniques. We have shown how to improve query expansion by filtering the words to add to the query based on their indiscriminateness.

This work opens many research avenues. From a theoretical point of view, defining intrinsic dimensionality in the case of similarities used in IR, instead of distances studied by [14], raises questions. In this article, we have used the α index with the hypothesis that the RSV scores follow a Power Law distribution. Although this is experimentally verified, the precise link between intrinsic dimensionality and α should be formally investigated. From an applicative point of view, one can finds many use for this indiscriminateness notion. For instance, it may allow to propose LSI or LDA representation by adapted to the local complexity. Another pragmatic use would be to help formulate a query during an online search: when typing a word making the query α to raise, a user could be asked to precise or reformulate the query, which may improve the results as seen in Sect. 4.

References

1. Amati, G., Rijsbergen, C.J.V.: Probabilistic models of information retrieval based on measuring the divergence from randomness. ACM Trans. Inf. Syst. **20**, 357–389 (2002)
2. Amsaleg, L., Oussama, C., Furon, T., Girard, S., Houle, M.E., Kawarabayashi, K.I.: Estimating local intrinsic dimensionality. In: 21st Conference on Knowledge Discovery and Data Mining, KDD 2015, Sidney, Australia, August 2015. https://hal.inria.fr/hal-01159217
3. Bellogín, A., de Vries, A.P.: Understanding similarity metrics in neighbour-based recommender systems. In: Proceedings of the 2013 Conference on the Theory of Information Retrieval, ICTIR 2013, pp. 13:48–13:55. ACM, New York (2013). http://doi.acm.org/10.1145/2499178.2499186
4. Beygelzimer, A., Kakade, S., Langford, J.: Cover trees for nearest neighbors. In: Proceedings of International Conference on Machine Learning (ICML), pp. 97–104 (2006)
5. Clauset, A., Shalizi, C.R., Newman, M.E.J.: Power-law distributions in empirical data. SIAM Rev. **51**(4), 661–703 (2009)
6. Claveau, V., Kijak, E.: Direct vs. indirect evaluation of distributional thesauri. In: Proceedings of COLING 2016, the 26th International Conference on Computational Linguistics: Technical Papers, pp. 1837–1848. The COLING 2016 Organizing Committee, Osaka, December 2016. http://aclweb.org/anthology/C16-1173
7. Claveau, V., Kijak, E., Ferret, O.: Improving distributional thesauri by exploring the graph of neighbors. In: International Conference on Computational Linguistics, COLING 2014, Dublin, August 2014. https://hal.archives-ouvertes.fr/hal-01027545
8. Deerwester, S., Dumais, S.T., Furnas, G.W., Landauer, T.K., Harshman, R.: Indexing by latent semantic analysis. J. Am. Soc. Inf. Sci. **41**, 391 (1990)
9. Fang, H., Tao, T., Zhai, C.: Diagnostic evaluation of information retrieval models. ACM Trans. Inf. Syst. **29**, 7 (2011)
10. Hersh, W., Buckley, C., Leone, T.J., Hickam, D.: OHSUMED: an interactive retrieval evaluation and new large test collection for research. In: Proceedings of the 17th Annual International ACM SIGIR Conference on Research and Development in Information Retrieval, SIGIR 1994, pp. 192–201. Springer-Verlag, New York Inc., New York (1994). http://dl.acm.org/citation.cfm?id=188490.188557
11. Hoffman, M., Bach, F.R., Blei, D.M.: Online learning for latent dirichlet allocation. In: Lafferty, J., Williams, C., Shawe-Taylor, J., Zemel, R., Culotta, A. (eds.) Advances in Neural Information Processing Systems, vol. 23, pp. 856–864. Curran Associates, Inc. (2010). http://papers.nips.cc/paper/3902-online-learning-for-latent-dirichlet-allocation.pdf

12. Houle, M.E., Ma, X., Nett, M., Oria, V.: Dimensional testing for multi-step similarity search. In: Proceedings of the 12th IEEE International Conference on Data Mining (ICDM), pp. 299–308 (2012)
13. Houle, M.E., Nett, M.: Rank cover trees for nearest neighbor search. In: Brisaboa, N., Pedreira, O., Zezula, P. (eds.) SISAP 2013. LNCS, vol. 8199, pp. 16–29. Springer, Heidelberg (2013). https://doi.org/10.1007/978-3-642-41062-8_3
14. Houle, M., Kashima, H., Nett, M.: Generalized expansion dimension. In: Proceedings of the 12th IEEE International Conference on Data Mining Workshops (ICDMW), pp. 587–594 (2012)
15. Levina, E., Bickel, P.J.: Maximum likelihood estimation of intrinsic dimension. In: Advances in Neural Information Processing Systems (NIPS) (2004)
16. Lv, Y., Zhai, C.: Lower-bounding term frequency normalization. In: Proceedings of the 20th ACM International Conference on Information and Knowledge Managementm, CIKM 2011, pp. 7–16. ACM, New York (2011). http://doi.acm.org/10.1145/2063576.2063584
17. Metzler, D., Croft, W.: Combining the language model and inference network approaches to retrieval. Inf. Process. Manag. **40**(5), 735–750 (2004). Special Issue on Bayesian Networks and Information Retrieval
18. Mikolov, T., Yih, W.t., Zweig, G.: Linguistic regularities in continuous space word representations. In: 2013 Conference of the North American Chapter of the Association for Computational Linguistics: Human Language Technologies (NAACL HLT 2013), Atlanta, Georgia, pp. 746–751 (2013)
19. Ponte, J.M., Croft, W.B.: A language modeling approach to information retrieval. In: Proceedings of the 21st Annual international ACM SIGIR Conference on Research and Development in information Retrieval (SIGIR 1998), pp. 275–281 (1998)
20. Robertson, S.E., Walker, S., Hancock-Beaulieu, M.: Okapi at TREC-7: automatic ad hoc, filtering, VLC and interactive track. In: Proceedings of the 7th Text Retrieval Conference, TREC-7, pp. 199–210 (1998)
21. Roweis, S.T., Saul, L.K.: Nonlinear dimensionality reduction by locally linear embedding. Science **290**(5500), 2323–2326 (2000)
22. Scholkopf, B., Smola, A.J., Muller, K.R.: Nonlinear component analysis as a kernel eigenvalue problem. Neural Comput. **10**(5), 1299–1319 (1998)
23. Singhal, A.: Modern information retrieval: a brief overview. Bull. IEEE Comput. Soc. Tech. Committee Data Eng. **24**, 35–43 (2001)
24. Strohman, T., Metzler, D., Turtle, H., Croft, W.: Indri: a language-model based search engine for complex queries (extended version). Technical report CIIR (2005)
25. Turtle, H., Croft, W.: Evaluation of an inference network-based retrieval model. ACM Trans. Inf. Syst. **9**(3), 187–222 (1991)
26. Venna, J., Kaski, S.: Local multidimensional scaling. Neural Netw. **19**, 889–899 (2006)
27. de Vries, T., Chawla, S., Houle, M.E.: Density-preserving projections for large-scale local anomaly detection. Knowl. Inf. Syst. **32**(1), 25–52 (2012)
28. Zhai, C., Lafferty, J.D.: A study of smoothing methods for language models applied to ad hoc information retrieval. In: Proceedings of the SIGIR Conference, pp. 334–342 (2001)

On the Reproducibility
and Generalisation of the Linear
Transformation of Word Embeddings

Xiao Yang$^{(\boxtimes)}$, Iadh Ounis, Richard McCreadie, Craig Macdonald,
and Anjie Fang

University of Glasgow, Glasgow, UK
{xiao.yang,iadh.ounis,richard.mccreadie,craig.macdonald}@glasgow.ac.uk,
a.fang.1@research.gla.ac.uk

Abstract. Linear transformation is a way to learn a linear relationship between two word embeddings, such that words in the two different embedding spaces can be semantically related. In this paper, we examine the reproducibility and generalisation of the linear transformation of word embeddings. Linear transformation is particularly useful when translating word embedding models in different languages, since it can capture the semantic relationships between two models. We first reproduce two linear transformation approaches, a recent one using orthogonal transformation and the original one using simple matrix transformation. Previous findings on a machine translation task are re-examined, validating that linear transformation is indeed an effective way to transform word embedding models in different languages. In particular, we show that the orthogonal transformation can better relate the different embedding models. Following the verification of previous findings, we then study the generalisation of linear transformation in a multi-language Twitter election classification task. We observe that the orthogonal transformation outperforms the matrix transformation. In particular, it significantly outperforms the random classifier by at least 10% under the F1 metric across English and Spanish datasets. In addition, we also provide best practices when using linear transformation for multi-language Twitter election classification.

Keywords: Embedding · Linear transformation · Twitter classification

1 Introduction

Word embeddings are particularly useful as text representations, since semantically (rather than textually) similar words can be found using similarity metrics (e.g. *cosine* similarity) [1]. Therefore, there is an increasing interest in using multilingual word embeddings to capture semantic similarities among different languages. For example, recent works have learned multilingual word embeddings using monolingual text corpora along with a parallel corpus of aligned words

© Springer International Publishing AG, part of Springer Nature 2018
G. Pasi et al. (Eds.): ECIR 2018, LNCS 10772, pp. 263–275, 2018.
https://doi.org/10.1007/978-3-319-76941-7_20

and/or sentences [2–4]. Based on the observation that similar words in different languages have similar geometric arrangements in word embedding spaces, Mikolov et al. [1] showed that multilingual word embeddings can be obtained "offline" using a *linear transformation*. Despite the simplicity of linear transformation, it has been shown to be effective for machine translation, i.e. when aiming to translate words from a source language to another language. Using a large scale training dictionary of more than 10^{10} English and Spanish word pairs, a linear transformation approach achieved 0.53 precision@1 [1].

Furthermore, recent enhancements have been proposed to make linear transformation more effective, namely: by retrieving translation pairs [5]; or learning a linear transformation matrix based on orthogonal transformation (e.g. by leveraging canonical correlation analysis (CCA) [6,7] or singular value decomposition (SVD) [8–10]). Given the current research interest in the use of word embeddings in various tasks, such as information retrieval [11–13] and text classification [14–16], the reproduction, validation, and generalisation of findings from the literature of linear transformation are important for extending that research for multilingual scenarios. As such, in this paper, we examine the reproducibility and generalisation of linear transformation of word embeddings in different languages.

We begin by reproducing two previous linear transformation approaches:

1. Matrix transformation (denoted MT) proposed by Mikolov et al. [1]
2. Orthogonal transformation that uses SVD (denoted OT) [9].

We choose these two approaches because, to the best of our knowledge, MT is the first attempt to address linear transformation of word embedding, while OT is a recent approach that claims to provide better performance over previous approaches. Over a simple machine translation task using our own word-aligned translation corpus of English and Spanish words, we validate the consistency and performance of linear transformation. We also evaluate the generalisation of linear transformation by applying it to a multi-language Twitter election classification task that classifies each tweet as "election-related" or "other". This task aims to adapt or transfer an existing classifier trained on a Twitter election dataset in English to that of Spanish and vice versa. This is particularly useful in monitoring emerging topics during the lead-up to an election, where well-designed training/test collections are not available. Our results on 3 Twitter election datasets (in two different languages) show that linear transformation is generalisable to the multi-language Twitter election classification task.

The remainder of this paper is organised as follows: We first describe the linear transformation approaches in Sect. 2. We report our experimental setup in Sect. 3, describing the datasets we used, classifier and the evaluation process. In Sect. 4, we present the results of a simple machine translation task, validating the reproducibility of linear transformation. In Sect. 5, we study the generalisation of linear transformation and present results for multi-language Twitter election classification. Finally, Sect. 6 summarises our conclusions.

2 Linear Transformation

Linear transformation approaches, for example the matrix transformation (MT) approach [1] and the orthogonal transformation approach that uses SVD (OT) [9], allow the transfer of pre-trained monolingual embedding models "offline" using aligned words in two languages. In particular, experiment using the orthogonal transformation approach (OT) demonstrate that a linear mapping between embedding spaces should be orthogonal to achieve enhanced performance [8,9]. In the rest of this section, we detail the implementation of the 2 approaches we used.

MT Approach [1]: In this approach, a list of word pairs $\{x_i, y_i\}_{i=1}^n$ is generated by using *Google Translate*, where y_i is the translation of x_i. Word x_i in source language is extracted from a background text corpora (e.g. comprised of Google News articles). As such, words x_i and y_i have the same meaning but in two different languages. Then, a linear matrix W is trained by using gradient descent to minimise the squared reconstruction error, as shown in:

$$\min_W \sum_{i=1}^n |y_i - Wx_i|^2 \tag{1}$$

After the training process, one word vector can be projected to a vector in another space by applying $y_i' = Wx_i$. To find a similar word in another space, one can simply use cosine similarity to find the translation of x_i, whose vector is the closest to y_i'.

OT Approach [9]: Smith et al. [9] provided an enhanced version of MT based on orthogonal transformation. When matrix W maps one embedding space A to the embedding space B, W^T should be able to map the embedding space B back to embedding space A, i.e. we have $y \sim Wx$ and $x \sim W^Ty$. This means that the transformation matrix W is supposed to be an orthogonal matrix O with $O^TO = I$, where I is the identity matrix. Therefore, using this orthogonal matrix O, one can obtain a word similarity matrix $S = YOX^T$, where $S_{i,j} = |y_i||Ox_j|cos(\theta_{i,j}) = cos(\theta_{i,j})$ if X and Y are normalised. Note that matrix S contains the similarity of any word pairs from the embedding spaces of the two languages. Similarly, an orthogonal matrix is trained by maximizing the similarity of the ground truth word pairs $\{x_i, y_i\}_{i=1}^n$. This process is shown in the following equation:

$$\max_O \sum_{i=1}^n y_i^T Ox_i, \text{where } O^TO = I \tag{2}$$

To implement the training process, vectors of words in $\{x_i, y_i\}_{i=1}^n$ (denoted as $\{X_D, Y_D\}$) are first retrieved from their embedding spaces, respectively. Next, a singular value decomposition (SVD) is applied following $M = Y_D^T X_D = U\Sigma V^T$, where U and V are made up of the orthonormal vectors and Σ contains singular values. The optimised similarity matrix can be obtained as follows:

$$S = YUV^TX^T, \text{where } S_{i,j} = y_i^T UV^T x_j = (U^Ty_i) \cdot (V^Tx_j) \tag{3}$$

Therefore, both embedding spaces can be mapped into a single space by applying V^T to X and U^T to Y. In this paper, we use the MT implementation of Dinu et al. [5], which solves Eq. (1) using least squares[1]. For OT, we directly use the codes from Smith et al. [9], which is publicly available[2].

3   Experimental Setup

In this section, we briefly describe the word embedding models, as well as provide details about the evaluation datasets and metrics used in our experiments.

3.1   Word Embeddings

For the purpose of reproducibility, we use pre-trained and publicly available word embedding models instead of training our own models. The publicly available word embedding models were trained using *fastText*[3] from Wikipedia corpora since *fastText* has proved to be both effective and efficient [17]. We only choose the English and Spanish embedding models from 294 available languages[4], as these are the languages of our Twitter election datasets. In particular, these embedding models have 300 dimensions and are obtained using the skip-gram model with default *fastText* settings.

3.2   Translation Corpus

In order to learn and test the transformation matrix of MT and OT, a translation corpus is required to provide word-level alignment of the two languages. We also use the translation corpus to reproduce the translation task in [1,9]. Each word alignment is a pair of a Spanish word and its translation in English. We extract the most common 50k words from a Spanish Wikipedia snapshot dated 02/10/2015, excluding stopwords (e.g. "un", "es", "yo" and etc.). Afterwards, their corresponding English translations are obtained using the *Google Translate* service. Due to the nature of languages, the English translations may contain multiple words (e.g. "lanzado" is translated as "thrown out"). Indeed, 3,817 such translations are not considered as word-level alignments in this paper. In addition, *Google Translate* fails to translate 14,504 words extracted from the Spanish Wikipedia snapshot (e.g. "lobería", "porrón" and "ciénega"). These cases are removed from the translated corpus. We choose Wikipedia as the source to obtain translation pairs since the publicly available word embedding models from *fastText* are trained on Wikipedia corpora. As such, we can minimise occurrences of the out-of-vocabulary (OOV) problem when training and testing the linear transformation matrix. The Spanish Wikipedia snapshot (dated 02/10/2015) we use contains $1.15M+$ documents and about $436K$ unique words excluding stopwords. The final translation corpus consists of 29,907 Spanish-English word pairs.

---

[1] clic.cimec.unitn.it/~georgiana.dinu/down/.

[2] github.com/Babylonpartners/fastText_multilingual.

[3] github.com/facebookresearch/fastText.

[4] github.com/facebookresearch/fastText/blob/master/pretrained-vectors.md.

3.3   Twitter Election Datasets

To evaluate the generalisation of linear transformation on a multi-language Twitter election classification task, we use 3 Twitter election datasets[5] in this paper.

Venezuela Election. We target the 2015 Venezuela parliamentary election, which was held on the 6th December 2015 to elect the 164 deputies and 3 indigenous representatives of the National Assembly.

Philippines Election. We target the 2016 Philippines general election, which was held on the 9th May 2016 for the executive and legislative branches of all levels of government. The Philippine presidential and vice-presidential elections of 2016 were held as part of the general election, and are covered in this dataset.

Ghana Election. We target the 2016 Ghana general election, which was held on the 7th December 2016 to elect a president and members of parliament.

Before collecting Twitter posts about each election, we had political science experts selected a number of keywords (e.g. PHVote and GHElection) and Twitter user accounts (e.g. election candidates and news media) by browsing the Twitter posts related to a given election. We then use the Twitter Streaming API to collect Twitter posts that contain either one of the specified keywords or that are posted by one of the selected Twitter user accounts. In addition, we only collect Twitter posts that were published during the period of one month before and after the election date since this period potentially covers more relevant pre- and post election topics.

Since millions of tweets were collected for each target election, we adopt the classical TREC-style pooling methodology [18] that will be described later. This allows human assessors to identify election-related tweets without having to judge all of the tweets. Moreover, we allow our political science experts to suggest queries (keywords related to the election e.g. election, vote buying and supporters clash), and use the Terrier IR platform [19] to rank the retrieved tweets of each query per day. When ranking tweets, we use the DFReeKLIM [20] weighting model, which is designed for the effective retrieval of short documents like tweets. Finally, only the top-ranked 7 tweets for each query per day are added to the *pool* of tweets to be assessed because this gives a tweet collection of approximately 4k – 5k tweets, which allows our human annotators to finish the annotation job in a short time. Each sampled tweet is labelled as: "election-related" or "other" by our 5 political science experts. The final label of a tweet is then determined by a majority vote. Overall, for Venezuela, Philippines and Ghana datasets, we found moderate agreements of 52%, 68% and 71% respectively between all assessors using Cohen's *kappa*. The general statistics of our datasets such as the dominant language and the number of tweets in each category are shown in Table 1. The datasets cover two languages: English and Spanish, which are used to evaluate the performance of linear transformation in the multi-language Twitter election classification task.

---

[5] dx.doi.org/10.5525/gla.researchdata.564.

Table 1. Statistics of the Twitter election classification datasets.

Election	Language	Election-related	Other	Total
Venezuela	Spanish	2,273	3,474 (60%)	5,747
Philippines	English	1,755	2,408 (58%)	4,163
Ghana	English	1,254	1,999 (61%)	3,253

Using the generated election datasets, we consider two settings in this paper: Train a classifier on an English election dataset A and test the classifier on a Spanish election dataset B (denoted A $\Rightarrow$ B), and vice versa (denoted B $\Rightarrow$ A). We also split our election datasets into different subsets for each setting. For example, for A $\Rightarrow$ B, 60% of instances are randomly sampled from dataset A as D_s to train classifiers and the remaining 40% in dataset A as validation set D_s^v. 90% of instances from dataset B are sampled as the out-of-sample D_t^o that is used as the test set to evaluate the performance in another election; the remaining 10% (D_t^v) in the dataset B is used to track the performance of linear transformation during the training of the classifiers.

3.4 Classifier

In order to study the generalisation of linear transformation on the multi-language Twitter election classification task and evaluate its performance, we need to learn a text classifier on the training dataset in one language and apply it to a test dataset in another language. A variety of learning algorithms are available for such a task, such as random forest and support vector machines (SVM). However, one of the most recent and effective algorithms is based upon *Convolutional Neural Networks* (CNN) [15,16]. CNN classifiers have shown their effectiveness for Twitter classification tasks, such as sentiment analysis [21,22]. In addition, CNN can work with word embeddings by simply stacking the word vectors. Through the convolution operations, local indicators that are important for the classification task can be learned from the labelled dataset by sliding filters over the vector features. Therefore, in this paper, we use CNN classifiers with word embeddings to evaluate the classification performance of linear transformation. In particular, we train CNN classifiers on a training dataset D_s and then test it on a test dataset D_t^o that is in another language. The words in the test datasets are transformed into the embedding space we used to train the CNN classifiers. When transforming a word from a source language to the target language, the transformation matrix W is applied to its word vector x_i. As such, the transformed vector $y_i' = W x_i$ is used as part of the vector representations of a tweet. Such representations can then be used by the CNN classifier to classify "election-related" tweets in the unseen test dataset. Furthermore, a regularisation technique, namely dropout, is also applied to the CNN to only keep a neuron active with some probability p during training [15]. To evaluate the effectiveness of linear transformation, MT and OT are compared with a random baseline that makes predictions randomly according to the distribution of election-related tweets in the training datasets.

3.5 Training, Hyper-parameters and Metrics

To evaluate the performance in the translation task, for consistency, we use the same metrics that are used by Mikolov et al. [1] and Smith et al. [9], namely: precision@1 (P@1), precision@5 (P@5) and precision@10 (P@10). These three metrics evaluate how many words in the test translation corpus have the correct translations in the retrieved top k translations ranked by the cosine similarity. For the multi-language Twitter election classification task, we report the precision, recall and F1 score. We set up classifiers with filter size $m = 1$ and dropout rate $p = 0.5$. We pad short tweets to the length of the longest tweet using a special token, which are initialised as zero vectors.

Table 2. Translation results using *pseudo-dictionary* with various dictionary sizes. Best scores are highlighted in bold.

Training Set size	Algorithm	English to Spanish			Spanish to English		
		P@1	P@5	P@10	P@1	P@5	P@10
~5k	MT	0.014	0.040	0.062	0.021	0.051	0.072
	OT	0.264	0.471	0.537	0.407	0.567	0.611
~250k	MT	0.010	0.041	0.064	0.033	0.062	0.083
	OT	0.345	0.562	0.625	0.505	0.651	0.685
~608k	MT	0.010	0.039	0.062	0.032	0.063	0.084
	OT	**0.348**	**0.566**	**0.628**	**0.509**	**0.654**	**0.689**

4 Reproducibility – Linear Transformation Performance

This section reports our attempts to reproduce the results presented in the recent linear transformation paper [9] that uses SVD based orthogonal transformations. In this paper, we reproduce the results of the simple machine translation task, which attempts to retrieve the correct translation of a given word in a source language. The word in a source language is transformed into the target language using the linear transformation mentioned in Sect. 2. Then the transformed vector is used to retrieve the closest word in the target language by cosine similarity. In previous work, linear transformation has been evaluated in the translation task using an English-Italian translation corpus [5,8,9]. Thus, we use a different translation corpus of Spanish-English to validate whether the previous findings can be reproduced. We sample 1,000 Spanish-English translation pairs from our translation corpus as the translation test set, while using the rest as the translation training set. To reproduce previous findings in [9], we also include a *pseudo-dictionary* as another translation training set, which consists of identical character strings shared by both Spanish and English word embedding models. 608,772 such identical words appear in both embedding models, including loanwords from the two languages such as "TV", "IBM" and "fanatica". However, such identical word pairs in two languages do not necessarily have

Table 3. Translation results using our *translation corpus* with various dictionary sizes. Best scores are highlighted in bold.

Training Set size	Algorithm	English to Spanish			Spanish to English		
		P@1	P@5	P@10	P@1	P@5	P@10
~5k	MT	0.446	0.652	0.712	0.590	0.752	**0.793**
	OT	0.464	0.669	0.726	0.604	0.756	0.789
~15k	MT	0.430	0.628	0.698	0.577	0.746	0.783
	OT	0.466	**0.678**	**0.732**	0.615	**0.760**	0.789
~25k	MT	0.442	0.624	0.681	0.568	0.733	0.777
	OT	**0.469**	0.675	0.729	**0.616**	0.758	0.785

the same meaning, e.g. "once" is written identically in English and Spanish but has different meanings in each language. Moreover, in addition to experiments originally performed in [9], we also vary the size of the translation training set to evaluate the performance in different sizes.

In Table 2, we first present the translation performance using the pseudo-dictionary. We train the orthogonal transformation matrix W and transform the source embedding space to the target embedding space. Afterwards, we predict translations of words in the source embedding by a nearest neighbour retrieval as detailed by Mikolov et al. [1].

In particular, we vary the size of the pseudo-dictionary to validate the effect of the dictionary size. We randomly sample 5k, 250k and 608k pairs without replacement from the entire pseudo-dictionary to train the transformation matrix W. At the end, the learned transformation matrix is applied and evaluated on the test set of our translation corpus. From Table 2, we see clearly that OT outperforms MT in both translations from English to Spanish and from Spanish to English. On our test dictionary set, OT achieved best P@1 of 0.348 and 0.509 for English to Spanish and Spanish to English respectively. However, MT only achieved 0.01 and 0.032 respectively. This validates the previous finding of Smith et al. [9], and shows the advantage of orthogonal transformations over the original linear transformation proposed by Mikolov et al. [1]. In addition to the experiments in [9], by increasing the size of training dictionary, our results show that both OT and MT can slightly improve their performance, however the improvement is minimal when the size is greater than 250k.

In Table 3, we show the translation performance using the training dictionary from our translation corpus. Similar to the experiment on the pseudo-dictionary, we vary the size of the training dictionary. However, we split the training dictionary based on the word frequency in the Spanish Wikipedia snapshot we used. Compared with the results of using a pseudo-dictionary, it is unsurprising that the performance is much better for both OT and MT since the quality of aligned dictionary is better than the pseudo-dictionary. This also validates the results in [9] that an accurate translation dictionary is important for learning an effective translation matrix. In particular, MT shows comparable performance with OT on P@5 and P@10. However, in our additional expriments, we show that

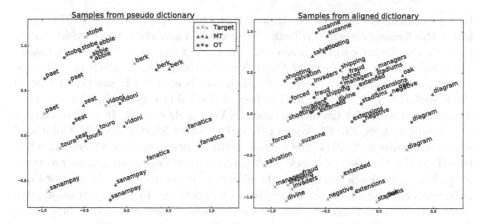

Fig. 1. Two dimensional PCA projections of words sampled from pseudo and aligned dictionaries. The target language is Spanish (translated in this figure). English words are transformed to the Spanish word embedding using MT and OT.

increasing the size of the training dictionary does not lead to an improvement on P@5 and P@10 after 15k for both OT and MT.

To provide insights on the difference between MT and OT, we show the two dimensional PCA projections of sampled words in Fig. 1. Using samples from the pseudo-dictionary, we show that source words (in English) transformed by OT are generally closer to the corresponding target words (in Spanish). The samples from the aligned dictionary show that OT has better performance since the transformed words are closer to the target language than MT. Overall, we reproduced results and findings of the previous work [9], which shows that OT is more effective in the translation tasks, even when trained on a pseudo-dictionary. Furthermore, we observe that by increasing the size of the training set, improvements in translation performance rapidly diminish.

5   Generalisation – Multi-language Twitter Classification

By applying linear transformation to a multi-language classification task, we report the generalisation of linear transformation. The obtained results are shown in Tables 4 and 5 where the first column shows the algorithms, and other columns show the classification performance by scenario, for example "Philippines ⇒ Venezuela" shows the results of training a classifier on the Philippines dataset (D_s) and testing on the Venezuela dataset (D_t^o). When testing the classifier on a test dataset, we use linear transformation to translate the corresponding word embedding vectors into the embedding space we used to train the classifiers.

As shown in Table 4, the classifier trained using OT indeed outperforms MT in the "Venezuela ⇒ Philippines" scenario. However, by training classifiers on the Philippines dataset and testing on the Venezuela dataset, MT shows a slightly better performance under all of the metrics tested. In particular, only OT outperforms the Random classifier in both of the two classification tasks, which shows that OT can better capture linear transformation between the two languages. We note that, by transforming Spanish embeddings into English embeddings, better performance can be achieved. Such an observation is similar to that of the translation task we examined in Sect. 4. In Table 5, we evaluate the performance between the Ghana dataset and the Venezuela dataset. In both of the two classification tasks, the F1 score of OT outperforms Random and MT. Compared with the results in Table 4, the overall performance drops for MT and OT when tested on the Venezuela dataset. Many factors may lead to such a performance drop. For example, depending on the election period and candidates, word distributions can vary in different elections. In addition, by simply translating a word into another language, it neglects the word order in different languages. Another factor is that the size of the Philippines dataset is larger than the Ghana dataset, therefore it may have better overlap with the Venezuela dataset on election topics. These factors can all affect the classification performance, which shows the complexity of the multi-language classification when compared with the simpler translation task examined previously in Sect. 4. In addition, we observe that MT performs poorly performance when applied in the "Venezuela ⇒ Philippines" and "Venezuela ⇒ Ghana" scenarios. Overall, only OT achieved significant improvements over the random classifier on all the aforementioned classification tasks, which yields p-value <0.05 using McNemar's test.

Table 4. Classification results using transformed embedding models. † indicates significant improvement over random classifier.

Algorithm	Philippines ⇒ Venezuela			Venezuela ⇒ Philippines		
	Precision	Recall	F1	Precision	Recall	F1
Random	0.399	0.418	0.409	0.417	**0.379**	0.398
MT	**0.612**	**0.785**	**0.688†**	**0.956**	0.195	0.324
OT	0.608	0.783	0.684†	0.949	0.322	**0.481†**

Table 5. Classification results using transformed embedding models. † indicates significant improvement over random classifier.

Algorithm	Ghana ⇒ Venezuela			Venezuela ⇒ Ghana		
	Precision	Recall	F1	Precision	Recall	F1
Random	0.399	**0.418**	0.409	0.397	0.390	0.394
MT	0.788	0.270	0.402	**0.916**	0.116	0.206
OT	**0.793**	0.387	**0.520†**	0.890	**0.450**	**0.598†**

When training classifiers, we track the performance of the linear transformation models when training each classifier. The performances over the training steps are shown in Fig. 2 where we track the F1 scores of classifiers on the validation set D_s^v of the training dataset (shown as "Validation"), on the subset D_t^v of the test dataset (in another language) using two approaches OT and MT. As shown in Fig. 2, the performances of both OT and MT improve at the beginning of the training process until the performance of classifiers converges on the classification training dataset. However, the performance of MT tends to drop when continuing to train the classifiers. In contrast, OT is more stable and able to retain the attained performance along the entire training process. In particular, in the task "Venezuela $\Rightarrow$ Ghana", OT tends to improve the performance continuously while the performance of MT decreases dramatically. The diverging behaviours of MT and OT in Fig. 2 shows that OT can map the relationship of two embedding spaces better than that of using MT. Therefore, when training classifiers, OT is less sensitive to the new batches of training instances. Additionally, our results show that, as a best practice, stopping the training process of classifiers earlier when the classification performance converges can potentially help MT avoid a decline in performance by 10% to 20% F1.

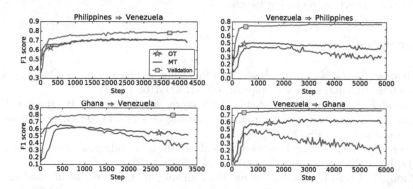

Fig. 2. Performance of linear transformation along the training process of classifiers. Square and star signs are used to distinguish the lines of Validation, OT and MT.

6   Conclusions

In this paper, we have reproduced, validated, and generalised findings of linear transformation of word embeddings from the literature. We evaluated linear transformation approaches on two different tasks (e.g. machine translation and multi-language Twitter election classification), making further observations from our experiments. In conclusion, we have confirmed that the orthogonal transformation using SVD [9] indeed outperforms the original approach proposed by Mikolov et al. [1] in Sect. 4. In particular, the orthogonal transformation can still learn a reasonable transformation matrix only using a pseudo-dictionary that contains words shared by the embedding models. Moreover, we show that by increasing the size of the training dictionary set, further gains in translation performance rapidly diminish. Furthermore, in Sect. 5, we apply linear transformation approaches to a multi-language Twitter election classification task, which is a more complex task than the translation task commonly examined in the literature. We observe that again the orthogonal transformation is more effective in all the classification scenarios than the original approach. Moreover, its performance is significantly better than a random classifier with at least 10% improvement in F1 score, thus we show the effectiveness of linear transformation without any prior knowledge from the test dataset (in another language). We also showed that a best practice is to halt the training process the classifier when convergence is reached, as this can potentially avoid a performance drop off. Finally, given that the performance of linear transformation varies on different datasets, we conclude that future work should investigate what are the factors that affect the translation and classification performance and how to leverage on these factors to improve linear transformation.

Acknowledgements. This paper was supported by a grant from the Economic and Social Research Council, (ES/L016435/1). The authors would like to thank the assessors for their efforts in reviewing tweets.

References

1. Mikolov, T., Le, Q.V., Sutskever, I.: Exploiting similarities among languages for machine translation. arXiv preprint arXiv:1309.4168 (2013)
2. Chandar, S., Lauly, S., Larochelle, H., Khapra, M., Ravindran, B., Raykar, V.C., Saha, A.: An autoencoder approach to learning bilingual word representations. In: Proceedings of NIPS (2014)
3. Eger, S., Hoenen, A.: Language classification from bilingual word embedding graphs. arXiv preprint arXiv:1607.05014 (2016)
4. Zhou, H., Chen, L., Shi, F., Huang, D.: Learning bilingual sentiment word embeddings for cross-language sentiment classification. In: Proceedings of ACL (2015)
5. Dinu, G., Lazaridou, A., Baroni, M.: Improving zero-shot learning by mitigating the hubness problem. arXiv preprint arXiv:1412.6568 (2014)
6. Ammar, W., Mulcaire, G., Tsvetkov, Y., Lample, G., Dyer, C., Smith, N.A.: Massively multilingual word embeddings. arXiv preprint arXiv:1602.01925 (2016)
7. Faruqui, M., Dyer, C.: Improving vector space word representations using multilingual correlation. In: Proceedings of EACL (2014)
8. Artetxe, M., Labaka, G., Agirre, E.: Learning principled bilingual mappings of word embeddings while preserving monolingual invariance. In: Proceedings of EMNLP (2016)
9. Smith, S.L., Turban, D.H.P., Hamblin, S., Hammerla, N.Y.: Offline bilingual word vectors, orthogonal transformations and the inverted softmax. In: Proceedings of ICLR (2017)
10. Xing, C., Wang, D., Liu, C., Lin, Y.: Normalized word embedding and orthogonal transform for bilingual word translation. In: Proceedings of HLT-NAACL (2015)
11. Mitra, B., Nalisnick, E., Craswell, N., Caruana, R.: A dual embedding space model for document ranking. arXiv preprint arXiv:1602.01137 (2016)
12. Moran, S., McCreadie, R., Macdonald, C., Ounis, I.: Enhancing first story detection using word embeddings. In: Proceedings of ACM SIGIR (2016)

13. Fang, A., Macdonald, C., Ounis, I., Habel, P., Yang, X.: Exploring time-sensitive variational Bayesian inference LDA for social media data. In: Jose, J.M., Hauff, C., Altıngovde, I.S., Song, D., Albakour, D., Watt, S., Tait, J. (eds.) ECIR 2017. LNCS, vol. 10193, pp. 252–265. Springer, Cham (2017). https://doi.org/10.1007/978-3-319-56608-5_20
14. Yang, X., Macdonald, C., Ounis, I.: Using word embeddings in Twitter election classification. In: Proceedings of Neu-IR Workshop at SIGIR (2016)
15. Kim, Y.: Convolutional neural networks for sentence classification. In: Proceedings of EMNLP (2014)
16. Severyn, A., Nicosia, M., Barlacchi, G., Moschitti, A.: Distributional neural networks for automatic resolution of crossword puzzles. In: Proceedings of IJCNLP (2015)
17. Bojanowski, P., Grave, E., Joulin, A., Mikolov, T.: Enriching word vectors with subword information. arXiv preprint arXiv:1607.04606 (2016)
18. Voorhees, E.M., Harman, D.K.: TREC: Experiment and Evaluation in IR. MIT Press, Cambridge (2005)
19. Macdonald, C., McCreadie, R., Santos, R.L., Ounis, I.: From puppy to maturity: experiences in developing Terrier. In: Proceedings of OSIR Workshop at SIGIR (2012)
20. Amati, G., Amodeo, G., Bianchi, M., Marcone, G., Bordoni, F.U., Gaibisso, C., Gambosi, G., Celi, A., Di Nicola, C., Flammini, M.: FUB, IASI-CNR, UNIVAQ at TREC 2011 Microblog track. In: Proceedings of TREC (2011)
21. Severyn, A., Moschitti, A.: UNITN: Training deep convolutional neural network for Twitter sentiment classification. In: Proceedings of SemEval (2015)
22. Tang, D., Wei, F., Yang, N., Zhou, M., Liu, T., Qin, B.: Learning sentiment-specific word embedding for Twitter sentiment classification. In: Proceedings of ACL (2014)

Discriminative Path-Based Knowledge Graph Embedding for Precise Link Prediction

Maoyuan Zhang[1], Qi Wang[1(✉)], Wukui Xu[2], Wei Li[1], and Shuyuan Sun[1]

[1] School of Computer, Central China Normal University,
Wuhan, People's Republic of China
nlpwq@mails.ccnu.edu.cn
[2] Intelligent and Distributed Computing Laboratory,
Huazhong University of Science and Technology,
Wuhan, People's Republic of China

Abstract. Representation learning of knowledge graph aims to transform both the entities and relations into continuous low-dimensional vector space. Though there have been a variety of models for knowledge graph embedding, most existing latent-based models merely explain triples via latent features, while supplementary rich inference patterns hidden in the observed graph features have not been fully employed. For this reason, in this paper we propose the discriminative path-based embedding model (DPTransE) which jointly learns from the latent features and graph features. Our model builds interactions between these two features, and uses the graph features as the crucial prior to offer precise and discriminative embedding. Experimental results demonstrate that our method outperforms other baselines on the task of link prediction and entity classification.

Keywords: Knowledge representation · Knowledge graph
Distributed representation

1 Introduction

Knowledge Graphs (KGs) encoding structured information of entities and relations have become crucial resources to support many intelligent applications, such as question answering and web search [17]. A typical KG usually explains knowledge as relational data and represents it as a triple denoted as (h, r, t), which is consist of an entity pair and the relation.

In the latent feature models, relations between entities can be derived from interactions of their latent features, which can be inferred automatically from data [13]. The translation-based methods, e.g., TransE [1] encodes KGs including both entities and relations into a continuous low-dimensional vector space, and the triple will be plausible if the embeddings of its entities and relation are close according to some distance measure.

© Springer International Publishing AG, part of Springer Nature 2018
G. Pasi et al. (Eds.): ECIR 2018, LNCS 10772, pp. 276–288, 2018.
https://doi.org/10.1007/978-3-319-76941-7_21

Despite their success in modelling relational fact, TransE and its variants merely translate the global latent feature representations via relation-specific offset [1], so most existing methods solely learn from latent features. Accordingly, these common methods are not sufficient to provide precise semantic embedding for some scenarios. For example, a specific relation may cover multiple semantics, such as the relation HasPart which is related to the semantic of component and location. As one more example, to illustrate more clearly, as showed in Fig. 1, the entity pair (Gary Rydstrom, Minority Report) in Freebase is connected by more than one relation. This phenomenon is quite common in KGs, but previous translation-based models simply learn the embeddings from global triples and adopt the principle of $t - h = r$, which causes the embeddings of relation "$/*$ $/films\_crewed./film/film\_crew\_gig/film$" and "$/award/award\_nominee/*$ $/nominated\_for$"[1] are quiet similar and ambiguous, thus the previous models made more errors.

It is known that there are also substantial supplementary semantic information contained in local graph patterns, such as the neighbors of entities and short paths in the graph. For instance, the triple (Barack Obama, MarryTo, Michelle Obama) can be well explained from the existence path $BarackObama \xrightarrow{parentOf} MaliaObama \xleftarrow{parentOf} MichelleObama$, which indicates a common child. The graph feature models employing the information are well-suited for modelling the local graph patterns and can be treated as a significant prior, and in fact it can be most effective for capturing the information for entity pairs [18], which can enhance the discrimination of triples and make relations more interpretable. Therefore, combining the strengths of latent-based and graph-based model is a promising method to improve the embedding performance, since these two models focus on different aspects of relational data.

Our proposed model builds interactions between the latent features and graph features. Graph features are employed to provide more precise and discriminative semantic embedding. The proposed semantic relevance hypothesis plays a key role in modelling graph features, which assumes the direct relation can be roughly reconstructed by a liner combination of the multi-step relation paths, and these paths could improve the discriminative ability among multiple relations. Namely, it allows different multi-step relation paths to play different roles for different direct relations. For example, when we query about the relation between the entity pair (Gary Rydstrom, Minority Report), there are two possible options: "film crewmember" and "award nomination". It is hard to distinguish if only learning from the latent features. However, the multi-step relation

---

[1] The original relations are

(1) $/film/film\_crewmember/films\_crewed./film/film\_crew\_gig/film$,

(2) $/award/award\_nominee/award\_nominations./award/award\_nomination/nom$ $inated\_for$, here we use a wildcard $*$ to reduce occupation without ambiguous expression.

path $\langle award\_winner \ /* \ /award, \ /* \ /nominees. \ /*/nominated\_for\rangle^2$ helps to refine the semantic "award nomination" in a more precise way by tuning weights of paths according to the priori which can be learned from the graph features. It's also notable that learning from latent features is always the main procedure and graph-based features must interact with triples for more distinct embeddings. Only in this way, the supplementary of semantic effects could make more sense.

To summarize, our contributions are as follows: DPTransE combines the strengths of latent features and graph-based features, and models the strong interactions between them by treating the graph features as crucial prior to incorporate the supplementary semantic information for precise embedding.

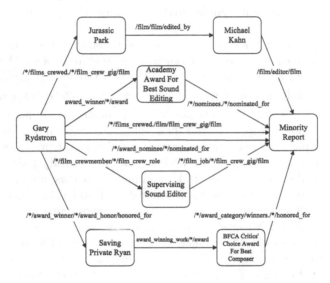

Fig. 1. An example in Freebase, using a wildcard $*$ to reduce occupation without ambiguous expression

2   Related Work

The methods proposed recently usually explain the triples via the interactions of the latent features which are obtained from the entities and their relations. All these models infer these latent features from data automatically, and the main differences between these models are the score functions $f_r(h, t)$ [24]. Here, we briefly list some related work.

TransE [1] proposed a paradigm which interpreted the relation r as a translations from head entity h to tail entity t, and the score function $f_r(h, t) =$

---

^{2} The original relation path is $/award/award\_winner/awards\_won./award/award\_$
$honor/award -> /award/award\_category/nominees./award/award\_nomination/$
$nominated\_for$.

$\|h + r - t\|$ was used to measure the compatibility of (h, r, t). Although TransE performs well in modeling the one-to-one relations, there are many issues when facing more complex relationships. In order to solve the problem of TransE, many variants (**TransH** [20], **TransR** [11]) were proposed. Both TransH and TransR aimed to model the distinctive representations of entities for different relations. TransH firstly projected the head/tail entity vector (h/t) into a relation-specific hyper-plane, where the relation vector was l_r and the entity vector was (l_{h_r}/l_{t_r}), then measured the score using the function $f_r(h, t) = \|l_{h_r} + l_r - l_{t_r}\|_{L_1/L_2}$. Similarly, given a specific relation r, TransR addressed this issue by projecting entities from entity space to a relation space using a transform matrix M_r, the score function was correspondingly defined as $f_r(h, t) = \|hM_r + l_r - tM_r\|_{L_1/L_2}$. Similar researches include **TransG** [22], **TransD** [6], **TransM** [7]. **SSE** [4] tried to discover geometric structure of the embedding space, and enforced the embedding space to be semantically smooth with the entities' semantic information. **KG2E** [5] involved the probabilistic embedding method to characterize the uncertainty of knowledge base and employed the KL-divergence for scoring triples. **HolE** [14] concentrated on learning compositional operator instead of adding two entity embeddings element-wise, which can capture rich interactions but still remains efficient to compute.

The related models mentioned above only considered the triples' interactions from the perspective of latent features, which ignored many observed graph features. There are more complex models using the path informations. **CVSM** [12] employed RNN-structure to compose a sequence of relation embeddings taking the advantages of path information. Although it performs well in some specific tasks, it is unsuitable for generalized entity and relation prediction task. **RTransE** [3] adopted the pre-selected paths called quadruples $< h, r_1, r_2, t >$ to solve the aforementioned flaw caused by CVSM, where the h, r_i, t are the embeddings learned by TransE. The model is further complicated. **PTransE** [10] is another path-based model where the path-constraint resource algorithm (PCRA) was used to measure the confidence level of a relation path. Although it has achieved much success, there is still one shortcoming, that is the weak-correlation issue, demonstrating current model could hardly characterize the strong correlation between the latent features and graph features.

3 Methodology

We have entity set E and relation set R. The facts observed in KG are stored as a set of triples $O = < h_i, r_k, t_j >$ where $h_i, t_j \in E$ and their relation $r_k \in R$, our model tries to encode both the entities and relations in $\mathbb{R}^k$ with expect to offer compatible and precise representations.

3.1 PTransE and DPTransE

For each triple (h, r, t), PTransE [10] follows the basic frame of TransE and considers the multiple-step relation paths simultaneously, the score function is correspondingly defined as

$$f_\mathbf{r}(\mathbf{h}, \mathbf{t}) = \| \mathbf{h} + \mathbf{r} - \mathbf{t} \| + \| \mathbf{h} + \mathbf{P} - \mathbf{t} \| \tag{1}$$

where $P(h,t) = \{p_1, p_2, \ldots, p_n\}$ is the set of multiple relation paths, and each path connects the given entity-pair (h,t). Here a path $p_i = <r_1, r_2, \ldots, r_l>$ is a sequence of relations. Confidence of each path is pre-fixed by PCRA which is based on the resource amount, and regardless the distinction of relations connecting the entity-pair. The pipeline method ignores the mutual promotion between graph features and the latent features. Therefore, it performs a weak correlation between the two features.

Different from PTransE, jointly learning strategy is adopted by DPTransE. It combines two models: the graph-based feature model and latent-based feature model. The former exploits the statistical properties of entity and relation as features based on the entity similar hypothesis, and explains triples from a local perspective. The latter uses the local graph patterns based on semantic relevance hypothesis as crucial prior. The combined model integrates the characteristic provided by two totally different models, where the path clustering refines the confidence level of path features and solves the data sparse problem.

3.2 Latent Feature Model

Since multi-step relations and intermediate entities bring more interactive information, constructing strong correlations between the multi-step relations and the direct relations is very necessary. Here, we propose the semantic relevance hypothesis, described as follows.

Semantic relevance hypothesis. The semantic of the direct relation can be approximately reconstructed by a linear combination of the multi-step relations given the same entity-pair, i.e., $r_k = \alpha_k P(h,t)$ where the $\alpha_k = (\alpha_{k1}, \alpha_{k2}, \ldots, \alpha_{km})$, m is the number of paths, and $\alpha_{ki} = (\theta_1, \theta_2, \ldots, \theta_n)$. For each $p_i \in P(h,t)$, $p_i = (r_1, r_2, \ldots, r_n)$, and n is the number of connected relations of path p_i. α_{ki} is the contribution of the path p_i to relation r_k given the triple (h, r_k, t).

It is acknowledged that the loss of semantic information based on basic latent feature model is intrinsic. Therefore, we have to minimize intrinsic loss together with the semantic loss according to the above hypothesis.

Formally, the score function is defined as follows:

$$f_e(h, r_k, t) = \| h + r_k - t \| + \frac{\lambda_1}{N} \| \alpha_k P(h,t) - r_k \| \tag{2}$$

where $N = \sum_{p_j \in P(h, r_k, t)} \prod_{\theta \in \alpha_{kj}} \theta$ is the normalization factor, and λ_1 is used to balance the two parts. A small score means the triple is well translated according to the latent features, and the multi-step paths provide more explicit semantic explanations simultaneously.

3.3 Graph Feature Model

In this section, graph features are used as the significant prior information for promoting latent-specific learning. Specifically, in order to measure the reliability of each path, a PRA-style method is adopted, i.e., a classifier f_{r_G} is trained for a group of relations which is different from the original PRA [8] which uses paths as features to predict the relations between the entity pair. Each entity pair (h_i, t_i) is described as (x_i, y_i), where x_i is the graph features vector, y_i is a boolean with a positive label if $(h_i, r_k, t_i) \in O$ or a negative value otherwise. Our goal is to learn the classifier such that $f_{r_G}(x_i) \simeq y_i$, and make every $r_k \in r_G$ meet. The workflow is: (1) to relation cluster, (2) generate and select the features, (3) feature computation and (4) train an individual classifier for every relation group [19].

In the phase of relation cluster, we empirically use K-Means to make the similar relations to be partitioned into one group. Specifically, we directly use the embeddings of relation learned from PTransE as input with the intuition that the multi-step relations can reflect the similarity of direct relations, and the path information is well employed in PTransE. In the phase of feature generation, path types Π_G which represent the features of relation group r_G are found by performing random walks on the KG, elements are instances of relation group r_G of interest. Note that the paths are considerable because of the intermediate entities, we filter the paths which lengths exceed 3 steps and simply pick those by frequency. In the phase of feature computation, values of each path features are computed based on entity similar hypothesis, i.e., relations connecting more similar entities are supposed to be of higher existence, and similarity is defined in light of resource allocation index [23].

$$S(h_i, t_i) = \sum_{z \in \Gamma(h_i) \bigcap \Gamma(t_i)} \frac{1}{d_z} \tag{3}$$

where z are intermediate entities of each path given the entity-pair (h_i, t_i), d_z is the degree of entity z. And then we train an individual classifier for each relation group given the training data associated with $T_G = (x_{iG}, y_{iG})_{i=1}^M$ by assuming that the each classifier has a linear form:

$$f_G(x) := \omega_G^T x$$

The loss function is defined from a logistic regression perspective:

$$L(\omega_G) = \sum_{i=1}^M \log(1 + \exp(-y_{iG} f_G(x_{iG}))) + \lambda_2 \|\omega_G\|_2^2 \tag{4}$$

3.4 Combined Model

The proposed path clustering could make the two models bring the biggest advantage into full play. Its key idea is that paths sharing more common relations are equally more reliable, and hence should be coupled [19]. Specifically, the new set Π_k is defined as the union of paths which are multi-step relations of r_k and the relation group r_G which r_k belongs to. Path of each source is consistently represented as the sequence of relations. Let π_{kp_i} represent the element of set Π_k, i.e., $\pi_{kp_i} = (r_1, r_2, \cdots, r_L)$.

Given the Π_k, each cluster contains a single path $\pi_{kp_i} \in \Pi_k$ at the beginning. Then the iteratively merge is performed. That is to say, if the overlap is above the predefined threshold δ, merge π_{kp_i} and π_{kp_j} into a new path cluster π_{kp}. Once the path clusters are merged, the path π_{kp} and the $\omega_G(\pi_{kp_{i/j}})$ will be updated to $\pi_{kp} = \pi_{kp_i} \cup \pi_{kp_j}$, the $\omega_G(\pi_{kp}) = \omega_G(\pi_{kp_i}) + \omega_G(\pi_{kp_j})$, $\omega_G(\pi_{kp_{i/j}}) = \omega_G(\pi_{kp})$, the overlap between two clusters is defined as:

$$O(\pi_{kp_i}, \pi_{kp_j}) = \frac{|\pi_{kp_i} \cap \pi_{kp_j}|}{min(|\pi_{kp_i}|, |\pi_{kp_j}|)} \tag{5}$$

The path clustering operation not only let the similar paths share weights but also solve the weights sparsity and path missing problem that the path p_j may not contained in Π_k mainly because of relation cluster and the random walks. The contribution of each path given the triple (h, r_k, t) in the latent model is further stated as follows, where $Pr(p_j)$ is the frequency obtained from all entity pairs which are connected directly by r_k.

$$\theta_j = Pr(p_j) \times \omega_G(\pi_{kp_j}) \tag{6}$$

3.5 Objective and Training

We formalize the optimization objective of DPTransE as

$$L = \sum_{(h,r,t)\in O} \sum_{(h',r',t')\in O'} \left[\gamma + f_e(h, r, t) - f_e(h', r', t')\right]_+ \tag{7}$$

where $[\cdot]_+ = max(0, x)$ returns the maximum between 0 and x, $\gamma > 0$ is a margin hyperparameter, and the corrupted triple set with respect to (h, r, t) is defined as follows, with one of three components replaced randomly.

$$O' = \{(h', r, t)\} \cup \{(h, r, t')\} \cup \{(h, r', t)\} \tag{8}$$

Optimization. For optimization, we use stochastic gradient descent (SGD) in mini-batches to iteratively update embeddings of entities and relations with additional constrains, i.e., $\forall h, r, t$, we set

$$\|h\|_2 \leq 1, \|r\|_2 \leq 1, \|t\|_2 \leq 1. \tag{9}$$

Implementation Details. The graph-based methods always add an inverse version for each relation to enhance the graph connectively [8,9], so in the training phase of graph feature model, we add triple (t, r^{-1}, h) for each observed triple (h, r, t). Furthermore, we follow the [15,18] to generate the negative instances, and ensure that the negative instances are true-negative ones.

Table 1. The statistic of datasets.

Dataset	#Ent	#Rel	#Train	#Valid	#Test
WN11	38,696	11	112,581	2,609	10,544
WN18	40,943	18	141,442	5,000	5,000
FB13	75,043	13	316,232	5,908	23,733
FB15K	14,951	1,345	483,142	50,000	59,071

4 Experiment

In this work, we empirically evaluate the proposed model and other baselines for two task: link prediction [1,2] and triple classification [11,16,20].

4.1 Data Sets

In this paper, we use the datasets which are commonly used in previous methods. FB15K and FB13 are extracted from Freebase [25] which is a large collective knowledge graph of general facts. WN18 and WN11 are adopted from WordNet [26] which is a lexical database of English language. The statistics of datasets are shown in Table 1.

4.2 Link Prediction

The task of link prediction concerns about knowledge graph completion: the embedding model tries to predict the triple (h, r, t), when one of h, r, t is missing. The WN18 and FB15K are two benchmark datasets for this task.

Evaluation Protocol. For evaluation, we use the same protocol as in TransE and its variants: for each test triple (h, r, t), we replace the head entity h (or the tail entity t) with every entity e existing in the knowledge graph. Then, a similarity score of the triple is calculated according to the score function $f_e(h, r, t)$.

Table 2. Evaluation results on link prediction

Datasets	WN18				FB15K			
Metric	Mean Rank		Hist@10(%)		Mean Rank		Hist@10(%)	
	Raw	Filter	Raw	Filter	Raw	Filter	Raw	Filter
TransE	263	251	75.4	89.2	243	125	34.9	47.1
TransH	401	388	73.0	82.3	212	87	45.7	64.4
TransR	238	225	79.8	92.0	198	77	48.2	68.7
PTransE	N/A	N/A	N/A	N/A	207	58	51.4	84.6
KG2E	362	348	80.5	93.2	**183**	69	47.5	71.5
DPTransE(2-step)	**220**	202	**80.7**	**94.7**	191	**51**	**58.1**	**88.5**
DPTransE(3-step)	233	**193**	80.1	90.3	209	77	57.8	84.1

By ranking these triples with their scores in ascending order, we get the original triple. Based on these ranking lists, we use two evaluation metrics: 1) the average rank of valid entities (denoted as Mean Rank) and 2) the proportion of testing triple whose rank is not larger than N (denoted as Hist@N), Hist@10 is applied for common reasoning, and Hist@1 is interested in precise embedding performance. Considering the fact that a corrupted triple for (h, r, t) may be exist in the KG, such a prediction should also be deemed correct. To eliminate the factor, we filter out the corrupted triples already existed in training, validation and testing sets. The former evaluation setting is named as "Raw" and the latter setting is set as "Filter". Notebaly, relations in KG can be categorized into four main classes, i.e., 1-to-1, 1-to-N, N-to-1, N-to-N [1], in terms of the cardinalities of the head entity h and tail entity t.

Implementation. As the datasets are the same, we directly report the experimental results of several baselines from the literature [21]. For all datasets, the optimal configure is determined by monitoring the mean rank on the validation set. The best configuration are as follows. We also consider the path with at most 2-steps and 3-steps, and all of the hyperparameters are same for 2-steps and 3-steps path model. For WN18, the number of relations is 18, and each type of relation is regard as a class. We select the path clustering threshold $\delta = 0.5$, the balance factor $\lambda_1 = 0.6$, the regularization parameter $\lambda_2 = 0.5$, the embedding dimension $k = 50$, learning rate for SGD $\alpha = 0.01$, the margin $\gamma = 4$ and take L1 as dissimilarity. For FB15K, the number of relation categories $K = 25$. We select the path clustering threshold $\delta = 0.5$, the balance factor $\lambda_1 = 0.8$, the regularization parameter $\lambda_2 = 0.5$, the embedding dimension $k = 50$, learning rate for SGD $\alpha = 0.01$, the margin $\gamma = 1$ and taking L1 as dissimilarity.

Results. Evaluation results on WN18 and FB15K are reported with respected to different types of relations in Tables 2 and 3. We observe that:

1. DPTransE outperforms all the baselines in all the sub-tasks, demonstrating the effectiveness of our model. It indicates that strengths of the latent features

Table 3. Evaluation results **(Hist@10)** on FB15K by mapping properties of relations(%)

Tasks	Predicting head entities				Predicting tail entities			
Relation Category	1-1	1-N	N-1	N-N	1-1	1-N	N-1	N-N
TransE	43.7	65.7	18.2	47.2	43.7	19.7	66.7	50.0
TransH	66.8	87.6	28.7	64.5	65.5	39.8	83.3	67.2
TransR	78.8	89.2	34.1	69.2	79.2	37.4	90.4	72.1
PTransE	90.1	92.0	58.7	86.1	90.1	70.7	87.5	88.7
KG2E	92.3	93.7	**66.0**	69.6	92.6	67.9	**94.4**	73.4
DPTransE(2-step)	**92.5**	**95.0**	58.0	86.6	**93.5**	**71.1**	93.9	88.2
DPTransE(3-step)	92.2	93.3	54.8	**88.4**	93.2	69.4	93.1	**90.0**

and graph-based features are complementary for representation learning of KG.

2. PTransE only considers the supplement of relation path, and the interaction between latent feature and graph feature is unsatisfactory. By focusing on the strong correlation and weighting the multi-step relation paths for different direct relations discriminatively, DPTransE outperforms it. Notice that most 2-step and 3-step achieve comparable results, it may be unnecessary to exploit the graph features that depend on longer paths.

3. From Table 2, we observe that: the improvement of FB15K is greater than that in WN18. One reason may be that the density of FB15K is greater than that in WN18, more path information is incorporated, which makes the semantic relevance hypothesis to take effect.

Table 4. Evaluation results (**Hist@1**)on FB15K by mapping properties of relations(%)

Tasks	Predicting head entities				Predicting tail entities			
Relation Category	1-1	1-N	N-1	N-N	1-1	1-N	N-1	N-N
TransE	35.4	50.7	8.6	18.1	34.5	10.6	56.1	20.3
TransH	35.3	48.7	8.4	16.9	35.5	10.4	57.5	19.3
TransR	29.5	42.8	6.1	14.5	28.0	7.7	44.1	16.2
PTransE	57.5	83.0	46.2	60.3	58.1	**58.4**	73.9	61.7
KG2E	62.3	73.9	39.4	30.4	62.3	33.9	76.2	33.8
DPTransE(2-step)	**66.4**	**88.9**	**46.7**	**64.2**	**64.8**	54.3	**80.6**	66.8
DPTransE(3-step)	62.5	81.5	38.5	61.7	64.3	48.0	75.7	**70.3**

Table 5. Hits@10 of KG2E and DPTransE on some examples of **1-to-N**[a], **N-to-1**[b], **N-to-N**[c], and **reflexivex**[d] relations.

Relation	Predict head		Predict tail	
	KG2E_KL	DPTransE	KG2E_KL	DPTransE
football_position/players[a]	100	100	100	100
production_company/films[a]	97.3	**98.9**	29.4	**45.2**
director/film[a]	93.4	**95.6**	85.8	**88.1**
discase/treatments[b]	66.6	**70.1**	100	100
person/place_of_birth[b]	34.1	**36.2**	84.6	**87.8**
film/production_companies[b]	44.2	**46.8**	97.8	96.2
field_of_study/students/majoring[c]	86.8	**92.1**	81.1	**86.3**
award_winner/awards_won[c]	**88.4**	87.8	89.2	**92.5**
sports_position/players[c]	100	100	100	100
person/sibling[d]	89.5	**93.6**	**94.7**	90
person/spose[d]	**77.8**	74.3	85.2	**86.4**

4. From Table 3, we observe that: on all types of relations, DPTransE consistently achieves significant improvements compared with TransE and its variants. As to the metric hist@1, the simple relation 1-to-1 improves relatively by 87.6% by DPTransE (2-step) than TransE, while DPTransE achieves only slight improvements compared with PTrasnE when predicting head entities in N-to-1 relations and predicting tail entities in 1-to-N relations. But for N-to-N relation, the performance of DPTransE is much better than TransE and all its variants, possibly may because we adopt the relation cluster strategy to divide the relations into different sub-types (Table 4). We list the results of some typical 1-to-1, 1-to-N, N-to-1, N-to-N relations on metric @Hist10 in Table 5, and the comparison results are directly copy from [5]. We can speculate that the relation cluster strategy can effectively enhance the performance of link prediction.

4.3   Triple Classification

Triple classification aims to judge weather the given triple (h, r, t) is correct or not. In this task, we use two datasets FB13 and WN11 for evaluation, and these datasets have already built with negative triples.

Evaluation Protocol. For each triple (h, r, t), if the score function $f_e(h, r, t)$ is below the specific threshold σ_r, then the triple will be classified as positive. Otherwise, it will be classified as negative. The thresholds σ_r are determined on the validation set by maximizing the classification accuracy.

Implementation. As all methods used the same datasets, we directly compare our models with the same baselines reported in [5]. The optimal configurations of DPTransE (2-step & 3-step) are as follows. For WN11, we select the number of relation categories $K = 11$, path clustering threshold $\delta = 0.5$, the balance factor $\lambda_1 = 0.5$, the regularization parameter $\lambda_2 = 1$, the embedding dimension $k = 100$, learning rate for SGD $\alpha = 0.001$, the margin $\gamma = 10$. For FB13, we select the number of relation categories $K = 13$, path clustering threshold $\delta = 0.5$, the balance factor $\lambda_1 = 0.8$, the regularization parameter $\lambda_2 = 0.6$, the embedding dimension $k = 100$, learning rate for SGD $\alpha = 0.001$, the margin $\gamma = 4$. Each type of relation is regard as a class for both datasets.

Table 6. Evaluation results of triple classification

Methods	WN11	FB13	AVG
TransE	75.9	81.5	78.7
TransH	78.8	83.3	81.1
TransR	85.9	82.5	84.2
KG2E	85.4	85.3	85.4
DPTransE(2-step)	**88.2**	**87.3**	**87.7**
DPTransE(3-step)	87.1	86.0	86.6

Results. The accuracy of triple classification on the two datasets is shown in Table 6. We observe that:

1. On the whole, DPTransE (2-step) yields the best performance. This further illustrates our model could promote the representation learning. DPTransE (2-step) performs better than DPTransE (3-step), which is consistent with the results of link prediction.
2. More specifically, on WN11, the accuracy of classification improves from 75.9% of TransE to 88.2% of DPTransE (2-step), while on FB13, the accuracy of classification improves from 81.5% of TransE to 87.3% of DPTransE(2-step). The reason is that, except the additional path information, the neighbor attributes are incorporated simultaneously, which refines the learning process of graph-based model.

5   Conclusion

In this paper, we propose DPTrsnsE, a new method for KG representation, where graph-based features promote the discriminative embedding learning. The combination of the latent features and graph features could bring the biggest advantage into full play by characterizing the strong correlation between direct relation and multi-step relations. The graph-based features have much effect on discovering the semantic relevance and offering more explicit and interpretable embeddings for both entities and relations. We evaluate DPTransE on link prediction and triple classification. Experimental results show that DPTransE achieves substantial improvements against the TransE and other baselines.

Acknowledgments. We thank all the anonymous reviewers for their detailed and insightful comments on this paper. This work was supported by Humanity and Social Science Youth Foundation of Ministry of Education of China (No.15YJC870029) and Research Planning Project of National Language Committee (No. YB135-40).

References

1. Bordes, A., Usunier, N., Garcia-Duran, A., Weston, J., Yakhnenko, O.: Translating embeddings for modelling multi-relational data. In: Proceedings of NIPS, pp. 2787–2795 (2013)
2. Bordes, A., Weston, J., Collobert, R., Bengio, Y.: Learning structured embeddings of knowledge bases. In: Proceedings of AAAI (2011)
3. García-Durán, A., Bordes, A., Usunier, N.: Composing relationships with translations. In: Proceedings of EMNLP, pp. 286–290 (2015)
4. Guo, S., Wang, Q., Wang, B., Wang, L., Guo, L.: Semantically smooth knowledge graph embedding. In: Proceedings of ACL, pp. 84–94 (2015)
5. He, S., Liu, K., Ji, G., Zhao, J.: Learning to represent knowledge graphs with Gaussian embedding. In: Proceedings of CIKM, pp. 623–632 (2015)
6. Ji, G., He, S., Xu, L., Liu, K., Zhao, J.: Knowledge graph embedding via dynamic mapping matrix. In: Proceedings of ACL, pp. 687–696 (2015)

7. Fan, M., Zhou, Q., Chang, E., Zheng, T.F.: Transition-based knowledge graph embedding with relational mapping properties. In: Proceedings of PACLIC, pp. 328–337 (2014)
8. Lao, N., Cohen, W.W.: Relational retrieval using a combination of path-constrained random walks. Mach. Learn. **81**, 53–67 (2010)
9. Lao, N., Mitchell, T., Cohen, W.W.: Random walk inference and learning in a large scale knowledge base. In: Proceedings of EMNLP, pp. 529–539 (2011)
10. Lin, Y., Liu, Z., Luan, H., Sun, M., Rao, S., Liu, S.: Modeling relation paths for representation learning of knowledge bases. In: Proceedings of EMNLP, pp. 705–714 (2011)
11. Lin, Y., Liu, Z., Sun, M., Liu, Y., Zhu, X.: Learning entity and relation embeddings for knowledge graph completion. In: Proceedings of AAAI, pp. 2181–2187 (2015)
12. Neelakantan, A.: Compositional vector space models for knowledge base inference. In: Proceedings of AAAI, pp. 1–16 (2015)
13. Nickel, M., Murphy, K., Tresp, V., Gabrilovich, E.: A review of relational machine learning for knowledge graphs. In: Proceedings of the IEEE, vol. 104, pp. 11–33 (2015)
14. Nickel, M., Rosasco, L., Poggio, T.: Holographic embeddings of knowledge graphs Nickel. In: Proceedings of AAAI, pp. 1955–1961 (2016)
15. Shi, B., Weninger, T.: Fact checking in large knowledge graphs - a discriminative predicate path mining approach. Knowl.-Based Syst. **104**, 123–133 (2015)
16. Socher, R., Chen, D., Manning, C.D., Ng, A.Y.: Reasoning with neural tensor networks for knowledge base completion. In: Proceedings of International Conference on Intelligent Control & Information Processing, pp. 464–469 (2015)
17. Szumlanski, S., Gomez, F.: Automatically acquiring a semantic network of related concepts. In: Proceedings of CIKM, pp. 19–28 (2015)
18. Toutanova, K., Chen, D.: Observed versus latent features for knowledge base and text inference. In: The Workshop on Continuous Vector Space Models and their Compositionality (2015)
19. Wang, Q., Liu, J., Luo, Y., Wang, B., Lin, C.Y.: Knowledge base completion via coupled path ranking. In: Proceedings of ACL, pp. 1308–1318 (2016)
20. Wang, Z., Zhang, J., Feng, J., and Chen, Z.: Knowledge graph embedding by translating on hyperplanes. In: Proceedings of AAAI (2014)
21. Xiao, H., Huang, M., Zhu, X.: From one point to a manifold: knowledge graph embedding for precise link prediction. In: Proceedings of IJCAI, pp. 1315–1321 (2016)
22. Xiao, H., Huang, M., Zhu, X.: TransG: a generative model for knowledge graph embedding. In: Proceedings of ACL, pp. 2316–2325 (2016)
23. Zhou, T., Lü, L., Zhang, Y.C.: Predicting missing links via local information. Eur. Phys. J. B **71**, 623–630 (2009)
24. Cai, H., Zheng, V.W., Chang, C.C.: A comprehensive survey of graph embedding: problems, techniques and applications. In: IEEE Transactions on Knowledge and Data Engineering (2017)
25. Bollacker, K., Evans, C., Paritosh, P., Sturge, T., Taylor, J.: Freebase: a collaboratively created graph database for structuring human knowledge. In: Proceedings of SIGMOD, pp. 1247–1250 (2008)
26. Mille, G.A.: Wordnet: a lexical database for English. In: Communications of the ACM, vol. 38(11), pp. 39–41 (1995)

Spherical Paragraph Model

Ruqing Zhang$^{(\boxtimes)}$, Jiafeng Guo, Yanyan Lan, Jun Xu, and Xueqi Cheng

CAS Key Lab of Network Data Science and Technology,
Institute of Computing Technology, University of Chinese Academy of Sciences,
Beijing, China
zhangruqing@software.ict.ac.cn,
{guojiafeng,lanyanyan,junxu,cxq}@ict.ac.cn

Abstract. Representing texts as fixed-length vectors is central to many language processing tasks. Most traditional methods build text representations based on the simple Bag-of-Words (BoW) representation, which loses the rich semantic relations between words. Recent advances in natural language processing have shown that semantically meaningful representations of words can be efficiently acquired by distributed models, making it possible to build text representations based on a better foundation called the Bag-of-Word-Embedding (BoWE) representation. However, existing text representation methods using BoWE often lack sound probabilistic foundations or cannot well capture the semantic relatedness encoded in word vectors. To address these problems, we introduce the Spherical Paragraph Model (SPM), a probabilistic generative model based on BoWE, for text representation. SPM has good probabilistic interpretability and can fully leverage the rich semantics of words, the word co-occurrence information as well as the corpus-wide information to help the representation learning of texts. Experimental results on topical classification and sentiment analysis demonstrate that SPM can achieve new state-of-the-art performances on several benchmark datasets.

1 Introduction

A central question to many language understanding problems is how to capture the essential meaning of a text in a machine-understandable format (*e.g.*, fixed-length vector representation). Most traditional methods either directly use the Bag-of-Words (BoW) representation [1], or built upon BoW using matrix factorization [2,3] or probabilistic topical models [4,5]. However, by using BoW as the foundation, rich semantic relatedness between words is lost. The text representation thus is learned purely based on the word-by-text co-occurrence information. However, humans understand a piece of text not solely based on its content (*i.e.*, the word occurrences), but also her background knowledge (*e.g.*, semantics of the words). Recent advances in the Natural Language Processing (NLP) community have shown that semantics of the words or more formally the distances between the words can be effectively revealed by distributed word representations [6], also referred to as "word embeddings" or "word vectors". Therefore, a natural idea is that one can build text representations based on

© Springer International Publishing AG, part of Springer Nature 2018
G. Pasi et al. (Eds.): ECIR 2018, LNCS 10772, pp. 289–302, 2018.
https://doi.org/10.1007/978-3-319-76941-7_22

a better foundation, namely the Bag-of-Word-Embeddings (BoWE) representation, by replacing distinct words with word vectors learned a priori with rich semantic relatedness encoded.

There have been some recent attempts to use BoWE for text representations. Perhaps the simplest way is to represent the text as a weighted average of all its word vectors [7]. Besides, [8] aggregated the word vectors into a text-level representation under the Fisher Kernel framework. Another well-known approach is the Paragraph Vector (PV) [9], which jointly learns the word and text representations as a direct optimization problem. There are several clear drawbacks with existing methods: (1) Existing methods often lack sound probabilistic foundations, making them heuristic or weak in interpretability; (2) All the methods assume the independency between texts, limiting their ability to leverage the corpus-wide information to help the representation learning of each piece of text. This limitation is analogous to that of Probabilistic Latent Semantic Indexing (PLSI) [4] in topic modeling, which has been addressed by Latent Dirichlet Allocation (LDA) [5]; (3) Simple weighted sum or aggregation using fisher kernel cannot well capture the semantic relatedness encoded in word vectors, which is typically revealed by the distance (or similarity) between word vectors.

To address these problems, we introduce a novel Spherical Paragraph Model (SPM), which learns text representations through modeling the generation of the corpus based on BoWE representations. Specifically, each piece of text is first represented as a bag of ℓ_2-normalized word vectors. Note that by normalization, the cosine similarity between word vectors are equal to the dot product between them, and all the word vectors lie on a unit hypersphere. We then assume the following generation process of the whole corpus. A text vector is first sampled from a corpus-wide prior distribution, and a word vector is then sampled from a text-level distribution given the text vector. The von Mises-Fisher (vMF) distribution [10] is employed for both corpus-wide and text-level distributions, which arises naturally for data distributed on the unit hypersphere and model the directional relation (i.e., dot product) between vectors. The text representations can then be inferred by maximizing the likelihood of the generation of the whole corpus. We develop a variational EM algorithm to learn the SPM efficiently.

Compared with previous methods, SPM enjoys the following merits: (1) By modeling the generation process of the whole corpus based on BoWE, SPM can fully leverage the rich semantics of words, the word-by-text co-occurrences information as well as the corpus-wide information to help the representation learning of texts; (2) By employing the vMF distribution, SPM can well capture the semantic relatedness encoded in words vectors (i.e., cosine similarity between word vectors); (3) SPM has good probabilistic interpretability as traditional topic models (e.g., LDA), while allows unlimited hidden topics (i.e., word clusters) as neural embedding models (e.g., PV) by eliminating the topic layer.

We evaluated the effectiveness of our SPM by comparing with existing text presentation methods based on several benchmark datasets. The empirical results demonstrate that our model can achieve new state-of-the-art performances on several topical classification and sentiment analysis tasks.

2 Related Work

In this section, we briefly review researches related to our work, including existing text representation methods and text models using the vMF distribution.

2.1 Existing Models for Texts

The most common fixed-length representation is Bag-of-Words (BoW) [1]. For example, in the popular TF-IDF scheme [11], each text is represented by *tfidf* values of selected feature-words. However, the BoW representation often suffers from data sparsity and high dimension. Meanwhile, by viewing each word as a distinct feature dimension, the BoW representation has very little sense about the semantics of the words.

To address this shortcoming, several dimensionality reduction methods have been proposed based on BoW, including matrix factorization methods such as LSI [2] and NMF [3], and probabilistic topical models such as PLSI [4] and LDA [5]. The key idea of LSI is to map texts to a vector space of reduced dimensionality (the latent semantic space), based on a Singular Value Decomposition over the term-document co-occurrence matrix. NMF is distinguished from the other methods by its non-negativity constraints, which leads to a parts-based representation because they allow only additive, not subtractive combinations. In PLSI, each word is generated from a single topic, and different words in a document may be generated from different topics. LDA is proposed by introducing a complete generative process over the documents, and demonstrated as a state-of-the-art document representation method. However, as built upon the BoW, all these methods do not leverage the rich semantics of the words, and learn text representations purely based on the word-by-text co-occurrence information.

Recent developments in distributed word representations have succeeded in capturing semantic regularities in language. Specifically, neural embedding models, *e.g.*, Word2Vec model [6] and Glove model [12], learn word vectors (also called word embeddings) efficiently from very large text corpus. The learned word vectors can reveal the semantic relatedness between words and perform word analogy tasks successfully.

With rich semantics encoded in word vectors, a natural question is how to obtain the text representation based on word vectors. A simple approach is to use a weighted average [8] or sum of all the word vectors. Besides, Fisher Vector (FV) [8] transforms the variable-cardinality word vectors into a fixed-length text representation based on the Fisher kernel framework [13]. However, these methods often lack sound probabilistic foundations. Meanwhile, simple weighted sum or aggregation using fisher kernel cannot well capture the semantic relatedness encoded in word vectors, which is typically revealed by the distance (or similarity) between word vectors. Later, Paragraph Vector (PV) which has two different model architectures (*i.e.*, PV-DM and PV-DBOW) [9] is introduced to jointly learn the word and text representations. Although these models seem to work well in practice, there is a strong independence assumption between texts in these methods, limiting their ability to leverage the corpus-wide information to help the representation learning of each piece of text.

Besides these unsupervised representation learning methods, there have been many supervised deep models which directly learn text representations for the prediction tasks. Recursive Neural Network [14] has been proven to be efficient in terms of constructing sentence representations. Recurrent Neural Network [15] can be viewed as an extremely deep neural network with weight sharing across time. Convolution Neural Network [16] can fairly determine discriminative phrases in a text with a max-pooling layer. However, these deep models are usually task dependent and time-consuming in training due to the complex model structures.

2.2 vMF in Text Models

The von Mises-Fisher distribution is known in the literature on directional statistics [17], and suitable for data distributed on the unit hypersphere. Here we first review the vMF distribution.

A d-dimensional unit random vector x (i.e., $x \in \mathbb{R}^K$ and $||x|| = 1$) is said to have K-variate von Mises-Fisher distribution if its probability density function is given by,

$$f(x|\mu, \kappa) = c_K(\kappa)e^{\kappa \mu^T x},$$

where $||\mu|| = 1$, $\kappa \geq 0$ and $K \geq 2$. The normalizing constant $c_K(\kappa)$ is given by,

$$c_K(\kappa) = \frac{\kappa^{K/2-1}}{(2\pi)^{K/2}I_{K/2-1}(\kappa)},$$

where $I_r(\cdot)$ represents the modified Bessel function of the first kind and order r. The density $f(x|\mu, \kappa)$ is parameterized by the mean direction μ, and the concentration parameter κ. The concentration parameter κ characterizes how strongly the unit vectors drawn from the distribution are concentrated on the mean direction μ. The vMF distribution has properties analogous to those of the multivariate Gaussian distribution for data in $\mathbb{R}^K$, parameterized by cosine similarity rather than Euclidean distance. Evidence suggests that this type of directional measure (i.e., cosine similarity) is often superior to Euclidean distance in high dimensions [18,19].

The vMF distribution has been applied in text representations based on BoW in literature. For example, [10] introduced the mixture of von Mises-Fisher distributions (movMF) that serves as a generative model for directional text data. The movMF model treats each normalized text vector (i.e., normalized *tf* or *tf-idf* vector) as drawn from one of the M vMF distributions centered on one cluster mean, selected by a mixing distribution. The cluster assignment variable for instance x_i is denoted by $z_i \in \{1, 2, \ldots, M\}$. The probabilistic generative process is given by,

$$z_i \sim \text{Categorical}(.|\pi), \quad x_i \sim \text{vMF}(.|\mu_{z_i}, \kappa),$$

where parameters $\Theta = \{\pi, \mu, \kappa\}$ are treated as fixed unknown constants and $Z = \{z_i\}_{i=1}^{M}$ are treated as a latent variables. Later, [20] introduced the Spherical Admixture Model (SAM), a Bayesian admixture model of normalized vectors on $\mathbb{S}^{K-1}$.

All these vMF-based methods treat the text as a single object (*i.e.*, a normalized feature vector), and successfully integrate a directional measure of similarity into a probabilistic setting for text modeling. However, the foundations of these methods are still BoW, which means that they cannot leverage the rich semantic relatedness between the words for text representation. Unlike these methods, we use vMF to capture the semantic relatedness encoded in word vectors revealed by cosine similarity, and build text representations based on a better BoWE foundation.

3   Spherical Paragraph Model

In this section, we describe our proposed SPM in detail, including the notations, the model definition, the inference and parameter estimation algorithms. Besides, we also provide some discussions on SPM as compared with existing advanced text representation methods.

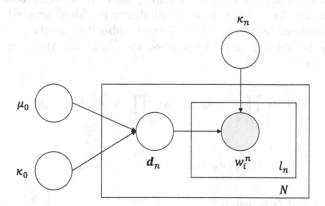

Fig. 1. A graphical model representation of Spherical Paragraph Model (SPM). (The boxes are "plates" representing replicates; a shaded node is an observed variable; an unshaded node is a hidden variable.)

3.1   Notation

Before presenting our model, we first introduce the notations used in this paper. Let $D = \{d_1, \ldots, d_N\}$ denote a corpus of N texts, where each text $d_n = (w_1^n, w_2^n, \ldots, w_{l_n}^n)$ and $n \in 1, 2, \ldots, N$ is an l_n-length word sequence over the word vocabulary V of size M. Let c_n denote all the words in text d_n. Each text $d \in D$ and each word $w \in V$ is associated with a vector $d \in \mathbb{R}^K$ and $w \in \mathbb{R}^K$, respectively, where K denotes the embedding dimensionality.

3.2 Model Definition

SPM is a probabilistic generative model over a text corpus based on BoWE. Specifically, each piece of text is first represented as a bag of ℓ_2-normalized word vectors. Note that by normalization, the cosine similarity between word vectors is equal to the dot product between them, and all the word vectors lie on a unit hypersphere. SPM then assumes the following generative process of the corpus:

For each text $d_n \in \boldsymbol{D}, n = 1, 2, \ldots, N$:
(a) Draw paragraph vector $\boldsymbol{d}_n \sim \text{vMF}(\mu_0, \kappa_0)$
(b) For each word $w_i^n \in d_n, i = 1, 2, \ldots, l_n$:
  Draw word vector $\boldsymbol{w}_i^n \sim \text{vMF}(\boldsymbol{d}_n, \kappa_n)$

where μ_0 is the corpus mean direction, κ_0 controls the concentration of text vectors around μ_0, and κ_n controls the concentration of word vectors around the text vector $\boldsymbol{d}_n$. Figure 1 provides the graphical model of the SPM.

As we can see from the above generative process, in SPM the text vectors in a corpus are determined by the corpus-wide prior distribution over the unit hypersphere, as well as the word vectors contained in the text. By using the vMF distribution, all the relations between these vectors are modeled by the dot product, which is equal to the cosine similarity measure between them (due to the ℓ_2-normalization). As we known, cosine similarity is widely adopted in revealing semantic relatedness in previous neural word embedding methods [6,21].

Based on the above generative process, we can obtain the joint probability of the whole corpus as follows,

$$p(\boldsymbol{D}) = \prod_{n=1}^{N} \int p(d_n|\mu_0, \kappa_0) \prod_{w_i^n \in d_n} p(w_i^n|d_n, \kappa_n) dd_n,$$

where:

$$P(w_i^n; d_n; \kappa_n) = e^{\kappa_n \boldsymbol{d}_n^{\mathrm{T}} \boldsymbol{w}_i^n} c_K(\kappa_n).$$

3.3 Variational Inference

The key inferential problem that we need to solve in order to use SPM is that of computing the posterior distribution of the hidden text vector given its word vectors and the corpus prior:

$$p(d_n|c_n, \mu_0, \kappa_0, \kappa_n) = \frac{p(d_n, c_n|\mu_0, \kappa_0, \kappa_n)}{p(c_n|\mu_0, \kappa_0, \kappa_n)}.$$

Unfortunately, this distribution is intractable to compute in general. Thus we develop an efficient variational inference algorithm to perform approximate inference in SPM.

The basic idea of convexity-based variational inference is to make use of Jensen's inequality [22] to obtain an adjustable lower bound on the log likelihood.

We approximate the posterior by introducing an distinct vMF distribution for each document,

$$q(d_n) \sim \text{vMF}(.|\mu'_n, \kappa'_n).$$

Here, μ'_n, κ'_n are the free variational parameters. To approximate the posterior distribution of the latent variables, the mean-field approach finds the optimal parameters of the fully factorizable $q(i.e., q(d_n))$ by maximizing the Evidence Lower Bound (ELBO),

$$\begin{aligned}
\mathcal{L} &= E_q[\log P(\mathcal{D})] - \mathcal{H}(q) \\
&= E_q[\log P(\boldsymbol{D}, \boldsymbol{V}|\mu_0, \kappa_0, \kappa_n)] - E_q[\log q(\boldsymbol{D})] \\
&= E_q[\log P(\boldsymbol{D}|\mu_0, \kappa_0)] + E_q[\log P(\boldsymbol{V}|\boldsymbol{D}, \kappa_n)] - E_q[\log q(\boldsymbol{D})].
\end{aligned}$$

Note that the expectations in this expression are taken over the variational distribution q. The posterior expectation of text vector $\boldsymbol{d}_n$ is given by,

$$E_q[\boldsymbol{d}_n] = \mu'_n \left(\frac{I_{d/2}(\kappa'_n)}{I_{d/2-1}(\kappa'_n)} \right),$$

where $\frac{I_{d/2}(\kappa'_n)}{I_{d/2-1}(\kappa'_n)}$ is a ratio of Bessel functions [23] that differ in their order by just one.

Thus the optimizing values of the variational parameters μ'_n and κ'_n are found by minimizing the KL divergence between the variational distribution q and the true posterior $p(d_n|c_n, \mu_0, \kappa_0, \kappa_n)$. Optimizing the ELBO with respect to μ'_n and κ'_n, we have

$$\kappa'_n = ||\kappa_0 \mu_0 + \sum_{i=1}^{l_n} \kappa_n \boldsymbol{w}_i^n||,$$

$$\mu'_n = \frac{\kappa_0 \mu_0 + \sum_{i=1}^{l_n} \kappa_n \boldsymbol{w}_i^n}{||\kappa_0 \mu_0 + \sum_{i=1}^{l_n} \kappa_n \boldsymbol{w}_i^n||} = \frac{\kappa_0 \mu_0 + \sum_{i=1}^{l_n} \kappa_n \boldsymbol{w}_i^n}{\kappa'_n}.$$

3.4 Parameter Estimation

We use an empirical Bayes method for parameter estimation in our SPM model. As described above, variational inference provides us with a tractable lower bound on the log likelihood. We can thus find approximate empirical Bayes estimates via an alternating variational EM procedure that maximizes the lower bound with respect to the variational parameters μ'_n and κ'_n. Then, for fixed values of the variational parameters, we maximize the lower bound with respect to the model parameters μ_0, κ_0 and κ_n. The variational EM algorithm is as follows:

- (E-step) For each text, find the optimizing values of the variational parameters μ'_n, κ'_n, as described in the previous Sect. 3.3.
- (M-step) Maximize the lower bound with respect to the parameters μ_0, κ_0 and κ_n.

These two steps are repeated until the lower bound on the log likelihood converges.

The M-step update for μ_0, κ_0 are given by,

$$\mu_0 = \frac{\sum_{n=1}^N E_q[\boldsymbol{d}_n]}{\| \sum_{n=1}^N E_q[\boldsymbol{d}_n] \|},$$

$$\kappa_0 = \frac{\bar{r}K - \bar{r}^3}{1 - \bar{r}^2} \text{ where } \bar{r} = \frac{\| \sum_{n=1}^N E_q[\boldsymbol{d}_n] \|}{N}.$$

The M-step update for κ_n is given by,

$$\kappa_n = \frac{\bar{r}K - \bar{r}^3}{1 - \bar{r}^2}, \text{ where } \bar{r} = \frac{E_q[\boldsymbol{d}_n] \sum_{i=1}^{l_n} \boldsymbol{w}_i^{n\mathrm{T}}}{l_n}.$$

3.5   Model Discussion

SPM is a probabilistic generative model based on BoWE for text representation. As it bridges two well-known branches in text representation methods, namely the probabilistic generative models and neural embedding models, here we compare SPM with these two types of methods to show its benefits.

Probabilistic generative models, also called probabilistic topic models (*e.g.*, PLSI and LDA), are advanced text modeling approaches. By assuming a generative process of the texts under a probabilistic framework, these methods usually have sound theoretical foundation and good model interpretability. However, there are two major problems in traditional topic models: (1) As built upon the BoW representation, traditional topic methods do not leverage the rich semantic relatedness of the words, and learn the text representations purely based on the word-by-text co-occurrence information; (2) There is an explicit topic layer in these models to guide the word clustering. The topic number is usually heuristically defined *a prior* which may lead to non-optimal word clustering. As we can see, SPM enjoys the merits of good interpretability as a probabilistic generative model. Meanwhile, SPM can avoid the arbitrary definition of topic numbers by eliminating the topic layer, while allows unlimited hidden topics (*i.e.*, word clusters) learned by any prior neural word embedding models based on very large corpus.

As compared with neural embedding models, here we take the state-of-the-art PV model as an example. The PV model can also be viewed as a probabilistic model based on its prediction definition. However, from the probabilistic view, PV is not a full Bayesian model and suffers a similar problem as PLSI that it provides no model on text vectors. Therefore, texts from the same corpus are assumed to be independent from each other and no corpus-wide constraint is employed in text modeling. Moreover, it is unclear how to infer the representations for texts outside of the training set with the learned model. Although PV makes itself as an optimization problem so that one can learn representations for new texts anyway, it loses the sound probabilistic foundation in that way. In contrary, SPM solves this problem by defining a complete Bayesian model. In this way, it can not only leverage corpus-wide information to help constrain the text vectors, but also infer the representations of unseen texts based on the learned model, at the expense of the usage of an approximate variational method.

4   Experiments

In this section, we conduct experiments to verify the effectiveness of SPM based on two text classification tasks.

4.1   Baselines

- **Bag-of-Words**. The Bag-of-Words model (BoW) [1] represents each text as a bag of words using *tf* as the weighting scheme. We select top 5,000 words according to *tf* scores as discriminative features.
- **LSI** and **LDA**. LSI [2] maps both texts and words to lower-dimensional representations using SVD decomposition. In LDA [5], each word within a text is modeled as a finite mixture over an set of topics. We use the vanilla LSI and LDA in the gensim library[1] with topic number set as 50.
- **movMF** and **SAM**. movMF[2] [10] is the mixture of von-Mises Fisher clustering with soft assignments. SAM[3] [20] is a class of topic models that represent data using directional distributions on the unit hypersphere. The topic numbers are both 50.
- **cBow**. We use average pooling to compose a text vector from a set of word vectors [6], where the dimension of text vectors is set as 50.
- **PV**. Paragraph Vector [9] is an unsupervised model to learn distributed representations of words and texts. We implement PV-DBOW and PV-DM model initialized with 50-dimension word embeddings due to the original code has not been released.
- **skip-thought** and **FastSent**. skip-thought[4] [24] encodes a sentence to predict sentences around it using 2400-dimension vector representation. FastSent[5] [25] is a simple additive sentence model designed to exploit the same signal, but at much lower computational expense under 100 dimension.

4.2   Setup

We perform experiments on two text classification tasks: topical classification and sentiment analysis. We utilize 50-dimension word embeddings trained on Wikipedia with word2vec [6]. The corpus in total has 3,035,070 articles and about 1 billion tokens. The vocabulary size is about 400,000. The vectors are post-processed to have unit ℓ_2-norm. In our model, text vectors are randomly

---

[1] http://radimrehurek.com/gensim/.
[2] https://github.com/mrouvier/movMF.
[3] https://github.com/austinwaters/py-sam.
[4] https://github.com/ryankiros/skip-thoughts.
[5] https://github.com/fh295/SentenceRepresentation.

Table 1. Classification accuracies (%) of different models on topical classification and sentiment analysis. Best scores are shown in bold.

Model	Topical classification				Sentiment analysis	
	Different	Similar	Same	Reuters	Subj	MR
BoW	91.4	81.8	75.6	**95.4**	89.5	74.3
LSI	85.2	80.1	68.2	93.1	85.4	64.2
LDA	73.3	67.5	56.7	89.6	72.7	58.2
movMF	71.4	64.5	59.4	87.1	67.6	53.4
SAM	88.6	81.2	70.5	88.2	74.2	61.8
cBow	91.6	81.6	75.9	91.8	90.8	74.4
PV-DBOW	91.4	80.2	**76.2**	89.6	90.1	73.9
PV-DM	91.5	80.8	76.1	90.4	90.4	74.4
FastSent	89.6	80.1	61.5	89.4	88.7	70.8
Uni-skip	86.4	77.8	59.2	77.4	92.1	**75.5**
SPM	**91.8**	**82.0**	70.0	93.2	**92.5**	75.0

initialized with values uniformly distributed in the range of $[-0.5, +0.5]$ with 50-dimension and then ℓ_2-normalized, κ_0 is intialized as 1500 and κ_n are randomly initialized with values uniformly distributed in the range of $[1000, 1500]$. Through our experiments, we use support vector machines (SVM)[6] as the classifier. Preprocessing steps were applied to all datasets: words were lowercased, non-English characters and stop words occurrence in the training set are removed. If explicit split of train/test is not provided, we use 10-fold cross-validation instead.

4.3   Topical Classification

We used two standard topical classification corpora: the 20Newsgroups[7] and the Reuters corpus[8]. The 20Newsgroups contains about 20,000 newsgroup documents harvested from 20 different Usenet newsgroups, with about 1,000 documents from each newsgroup. Following [26], three subsets of 20News are used for evaluation: (1) **news-20-different** consists of three newsgroups that cover different topics (*rec.sport.baseball*, *sci.space* and *alt.atheism*); (2) **news-20-similar** consists of three newsgroups on the more similar topics (*rec.sport.baseball*, *talk.politics.guns* and *talk.politics.misc*); (3) **news-20-same** consists of three newsgroups on the highly related topics (*comp.os.ms-windows.misc*, *comp.windows.x* and *comp.graphics*). The Reuters contains 10,788 documents, where each document is assigned to one or more categories. Documents appearing in two or more categories were removed and we selected the largest 10 categories, leaving 8,025 documents in total.

---

[6] http://www.csie.ntu.edu.tw/~cjlin/libsvm/.
[7] http://qwone.com/~jason/20Newsgroups/.
[8] http://www.nltk.org/book/ch02.html.

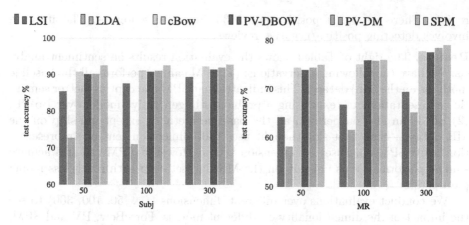

Fig. 2. Sentiment classification accuracies on two datasets. Results are grouped by dimension.

Results. The left of Table 1 shows the evaluation results on topical classification. We have the following observations: (1) The BoW representation, although simple, can achieve surprising accuracy using much larger dimensionality (*i.e.*, 5,000 dimension). Meanwhile, our SPM, using only 50-dimension text vector, can achieve slightly better or comparable performance as BoW. (2) As compared with the text representation methods built upon BoW (*i.e.*, LSI, LDA, movMF and SAM), SPM can outperform these methods almost. The results indicate that learning text representations over BoWE can in general achieve better performances than that over BoW by involving rich semantics between words. (3) Comparing with the three BoWE based representation methods, namely cBow, PV-DBOW and PV-DM, we find our SPM can outperform them on three out of four datasets. Recall that in cBow, PV-DBOW and PV-DM, texts in a corpus are actually assumed to be independent from each other. These results indicate that by modeling texts under a sound probabilistic generative framework, SPM can well leverage the corpus-wide information to help improve the text representation. (4) Compared with FastSent and uni-skip, SPM can outperform both of them over the four datasets. It seems that FastSent and uni-skip, which were proposed for short texts (*i.e.*, sentences) modeling originally, cannot work well on long texts.

4.4 Sentiment Analysis

We run the sentiment classification experiments on two publicly available datasets: (1) **Subj**, Subjectivity dataset [27][9] which contains 5,000 subjective instances and 5,000 objective instances. The task is to classify a sentence as being subjective or objective; (2) **MR**, Movie reviews [28] with one sentence per

---

[9] http://www.cs.cornell.edu/people/pabo/movie-review-data/.

review. There are 5,331 positive and 5,331 negative sentences. Classification involves detecting positive/negative reviews.

Results. The right of Table 1 shows the evaluation results on sentiment analysis. We have the following observations: (1) SPM can outperform all the baseline methods on the Subj dataset. This indicates that SPM can capture better semantic representations of texts using a probabilistic generative model over BoWE. (2) SPM can also outperform all the baseline methods except uni-skip on the MR dataset. Note that skip-thought uses 2400-dimension sentence representation while SPM only uses 50-dimension vector. However, SPM can still achieve similar performance as uni-skip on the MR dataset even with much less model parameters.

We conduct evaluations over different dimensions (*i.e.*, 50, 100, 300) to see the impact of the dimensionality on different models. For cBow, PV and SPM, we utilize 50, 100 and 300 dimensional word embeddings trained on Wikipedia using word2vec. For LSI and LDA, we set the topic numbers as 50, 100 and 300 for comparison. Figure 2 shows the results on the two different datasets. As we can see, with the increase of the dimensionality, all the models can improve their performance while SPM can consistently outperform all the other baselines. Moreover, we can find that the SPM model under dimensionality 100 can already beat the uni-skip under dimensionality 2400 (76.0% vs 75.5%).

5   Conclusion

In this paper, we propose the SPM, a novel generative model based on BoWE for text modeling. The SPM is a full Bayesian framework which models the generation of both the text vectors and word vectors, where the vMF distribution is employed to capture the directional relations between these vectors. SPM has good probabilistic interpretability and can fully leverage the rich semantics of words, the word co-occurrence information as well as the corpus-wide information to help the representation learning. The experimental results demonstrate that SPM can achieve new state-of-the-art performances on several topical classification and sentiment analysis tasks.

For the future work, we would like to explore the possibility to jointly learn word and text vectors in SPM. One idea is to leverage the word vectors learned from other large corpus as the initialization, and fine-tune them on the training data. Moreover, word order information is often critical in capturing the meaning of texts. We would also try to accommodate n-grams in the generative process to enhance the model ability. We may also test SPM on other text processing tasks to verify its generalization ability.

Acknowlegements. This work was funded by the 973 Program of China under Grant No. 2014CB340401, the National Natural Science Foundation of China (NSFC) under Grants No. 61232010, 61433014, 61425016, 61472401, 61203298 and 61722211, the Youth Innovation Promotion Association CAS under Grants No. 20144310 and 2016102, and the National Key R&D Program of China under Grants No. 2016QY02D0405.

References

1. Harris, Z.S.: Distributional structure. Word **10**(2–3), 146–162 (1954)
2. Deerwester, S., Dumais, S.T., Furnas, G.W., Landauer, T.K., Harshman, R.: Indexing by latent semantic analysis. J. Am. Soc. Inf. Sci. **41**(6), 391 (1990)
3. Lee, D.D., Seung, H.S.: Learning the parts of objects by non-negative matrix factorization. Nature **401**(6755), 788–791 (1999)
4. Hofmann, T.: Probabilistic latent semantic indexing. In: SIGIR, pp. 50–57. ACM (1999)
5. Blei, D.M., Ng, A.Y., Jordan, M.I.: Latent dirichlet allocation. J. Mach. Learn. Res. **3**(Jan), 993–1022 (2003)
6. Mikolov, T., Chen, K., Corrado, G., Dean, J.: Efficient estimation of word representations in vector space. arXiv preprint arXiv:1301.3781 (2013)
7. Vulic, I., Moens, M.F.: Cross-lingual semantic similarity of words as the similarity of their semantic word responses. In: NAACL-HLT, pp. 106–116. ACL (2013)
8. Clinchant, S., Perronnin, F.: Aggregating continuous word embeddings for information retrieval. In: Proceedings of the Workshop on Continuous Vector Space Models and Their Compositionality, pp. 100–109 (2013)
9. Le, Q.V., Mikolov, T.: Distributed representations of sentences and documents. ICML **14**, 1188–1196 (2014)
10. Banerjee, A., Dhillon, I.S., Ghosh, J., Sra, S.: Clustering on the unit hypersphere using von mises-fisher distributions. J. Mach. Learn. Res. **6**(Sep), 1345–1382 (2005)
11. Salton, G., McGill, M.J.: Introduction to Modern Information Retrieval (1986)
12. Pennington, J., Socher, R., Manning, C.D.: Glove: global vectors for word representation. EMNLP **14**, 1532–1543 (2014)
13. Jaakkola, T.S., Haussler, D., et al.: Exploiting generative models in discriminative classifiers. In: NIPS, pp. 487–493 (1999)
14. Socher, R., Perelygin, A., Wu, J.Y., Chuang, J., Manning, C.D., Ng, A.Y., Potts, C.: Recursive deep models for semantic compositionality over a sentiment treebank. In: EMNLP, vol. 1631, p. 1642. Citeseer (2013)
15. Sutskever, I., Martens, J., Hinton, G.E.: Generating text with recurrent neural networks. In: ICML-11, pp. 1017–1024 (2011)
16. Kim, Y.: Convolutional neural networks for sentence classification. In: EMNLP, pp. 1746–1751 (2014)
17. Mardia, K.V., Jupp, P.E.: Directional Statistics, vol. 494. Wiley, New York (2009)
18. Manning, C.D., Schütze, H., et al.: Foundations of statistical natural language processing, vol. 999. MIT Press, Cambridge (1999)
19. Zhong, S., Ghosh, J.: Generative model-based document clustering: a comparative study. Knowl. Inf. Syst. **8**(3), 374–384 (2005)
20. Reisinger, J., Waters, A., Silverthorn, B., Mooney, R.J.: Spherical topic models. In: ICML-10, pp. 903–910 (2010)
21. Mikolov, T., Sutskever, I., Chen, K., Corrado, G.S., Dean, J.: Distributed representations of words and phrases and their compositionality. In: NIPS, pp. 3111–3119 (2013)
22. Jordan, M.I., Ghahramani, Z., Jaakkola, T.S., Saul, L.K.: An introduction to variational methods for graphical models. Mach. Learn. **37**(2), 183–233 (1999)
23. Watson, G.N.: A Treatise on the Theory of Bessel Functions. Cambridge University Press, Cambridge (1995)
24. Kiros, R., Zhu, Y., Salakhutdinov, R.R., Zemel, R., Urtasun, R., Torralba, A., Fidler, S.: Skip-thought vectors. In: NIPS, pp. 3294–3302 (2015)

25. Hill, F., Cho, K., Korhonen, A.: Learning distributed representations of sentences from unlabelled data. In: NAACL-HLT (2016)
26. Banerjee, A., Basu, S.: Topic models over text streams: a study of batch and online unsupervised learning. In: SDM. vol. 7, pp. 437–442. SIAM (2007)
27. Pang, B., Lee, L.: A sentimental education: sentiment analysis using subjectivity summarization based on minimum cuts. In: ACL, p. 271. ACL (2004)
28. Pang, B., Lee, L.: Seeing stars: exploiting class relationships for sentiment categorization with respect to rating scales. In: ACL, pp. 115–124. ACL (2005)

Aggregating Neural Word Embeddings for Document Representation

Ruqing Zhang[1,2](✉), Jiafeng Guo[1,2], Yanyan Lan[1,2], Jun Xu[1,2],
and Xueqi Cheng[1,2]

[1] University of Chinese Academy of Sciences, Beijing, China
[2] CAS Key Lab of Network Data Science and Technology,
Institute of Computing Technology, Beijing, China
zhangruqing@software.ict.ac.cn,
{guojiafeng,lanyanyan,junxu,cxq}@ict.ac.cn

Abstract. Recent advances in natural language processing (NLP) have shown that semantically meaningful representations of words can be efficiently acquired by distributed models. In such a case, a text document can be viewed as a bag-of-word-embeddings (BoWE), and the remaining question is how to obtain a fixed-length vector representation of the document for efficient document process. Beyond those heuristic aggregation methods, recent work has shown that one can leverage the Fisher kernel (FK) framework to generate document representations based on BoWE in a principled way. In this work, words are embedded into a Euclidean space by latent semantic indexing (LSI), and a Gaussian Mixture Model (GMM) is employed as the generative model for nonlinear FK-based aggregation. In this work, we propose an alternate FK-based aggregation method for document representation based on neural word embeddings. As we know, neural embedding models have been proven significantly better performance in word representations than LSI, where semantic relations between neural word embeddings are typically measured by cosine similarity rather than Euclidean distance. Therefore, we introduce a mixture of Von Mises-Fisher distributions (moVMF) as the generative model of neural word embeddings, and derive a new FK-based aggregation method for document representation based on BoWE. We report document classification, clustering and retrieval experiments and demonstrate that our model can produce state-of-the-art performance as compared with existing baseline methods.

1 Introduction

Representing text documents as fixed-length vectors is central to many language processing tasks. Perhaps the most popular fixed-length vector representation for documents is the bag-of-words (BoW) representation [1], where each word is viewed as a distinct feature dimension based on strong independent assumption. Most traditional methods either directly use the BoW representation (e.g., tf-idf vector), or are built upon BoW (e.g., matrix factorization [2,3] and probabilistic

© Springer International Publishing AG, part of Springer Nature 2018
G. Pasi et al. (Eds.): ECIR 2018, LNCS 10772, pp. 303–315, 2018.
https://doi.org/10.1007/978-3-319-76941-7_23

topical models [4,5]). Apparently, by using BoW as the foundation, rich semantic relatedness between words is lost. The document representation thus is obtained purely based on the word-by-document co-occurrence information.

Recent developments in distributed word representations [6,7] have succeeded in revealing rich linguistic regularities between words. Specifically, by mapping each word into a continuous vector space, both syntactic and semantic relatedness between words can be captured using simple algebra over word vectors. Therefore, a natural idea is that one can build document representations based on a better foundation, namely the Bag-of-Word-Embeddings (BoWE) representation, by replacing distinct words with word vectors learned a priori with rich semantic relatedness encoded. The follow-up question is how to obtain a fixed-length vector representation of document based on BoWE for efficient document processing.

There have been several heuristic ways to obtain the document vector based on word embeddings, e.g., by using the average or weighted sum of all the word vectors contained in a document [8]. Another well-known approach is the Paragraph Vector (PV) [9] method, which jointly learns the word and document vectors through some prediction task. A common problem of all these methods is that they assume that the document vector lies in the same semantic space as words vectors. However, this may not be a necessary condition in practice since documents usually convey much richer semantics than individual words.

Recent work [10] has shown that one can use the Fisher kernel (FK) framework [11] as a flexible and principled way to generate document representations based on BoWE. It consists in non-linearly mapping the word embeddings into a higher-dimensional space and in aggregating them into a document representation. Specifically, in the FK-based aggregation, words are embedded into a Euclidean space by latent semantic indexing (LSI), and a Gaussian Mixture Model (GMM) is employed as the generative model of the word embeddings. The gradients of the GMM parameters are then used to generate the document representation. This FK-based aggregation method is highly efficient (i.e., simple adding operation to generate a new document representation), and has shown its superiority in several document clustering and retrieval tasks.

However, recent advances have shown that neural word embedding models (e.g., word2vec [6]) can produce significantly better performance in word representations than LSI. Such neural word embeddings can be efficiently acquired from large text corpus. Therefore, a natural question is whether we could leverage neural word embeddings for better document representation under the FK framework. Unfortunately, directly using the existing FK-based aggregation method [10] over neural word embeddings may not be appropriate. The major reason is that the generative model (i.e., GMM) in [10] is employed to capture the Euclidean distances between word embeddings from LSI, while semantic relations between neural word embeddings (e.g., Glove and word2vec) are typically measured by cosine similarity. Therefore, we propose an alternate FK-based aggregation method for document representation based on neural word embeddings. As we known, the von Mises-Fisher (vMF) distribution is well-suited to model directional data distributed on the unit hypersphere and capture the directional relations (i.e., cosine similarity) between vectors. Therefore, we introduce a Mixture of von Mises-Fisher distributions (moVMF) [12] as the generative model of neural word embeddings, and derive a new aggregation algorithm based moVMF model under the FK framework. We evaluated the effectiveness of our model by comparing with existing document representation methods. The empirical results demonstrate that our model can achieve new state-of-the-art performances on several document classification, clustering and retrieval tasks.

2   Related Work

We provide a short review of the works on those topics which are most related to our work: Bag-of-Words, Bag-of-Word-Embeddings, vMF and Fisher Vector.

- **Bag-of-Words.** The most common fixed-length representation is Bag-of-Words (BoW) [1]. For example, in the popular TF-IDF scheme, each document is represented by *tfidf* values of a set of selected feature-words. Besides, several dimensionality reduction methods have been proposed based on BoW, including matrix factorization methods such as LSI [2] and NMF [3], and probabilistic topical models such as PLSA [4] and LDA [5]. LDA, the generative counterpart of PLSA, has played a major role in the development of probabilistic models for textual data. As a result, it has been extended or refined in a countless studies [13,14]. Besides, several studies reported that LDA does not generally outperform LSI in IR or sentiment analysis tasks [15,16]. To further tackle the prediction task, Supervised LDA [17] is developed by jointly modeling the documents and the labels.
- **Bag-of-Word-Embeddings.** Recent advances in the natural language processing (NLP) community have shown that semantics of words or more formally the distances between words can be effectively revealed by distributed word representations. Specifically, neural embedding models, *e.g.*, Word2Vec [6] and Glove [7], learn word vectors efficiently from very large text corpus. Word embeddings are useful because they encode both syntactic and semantic information of words into continuous vectors and similar words are close in vector space. With rich semantics encoded in word vectors, there have been many methods [8,9,18–20] built upon Bag-of-Word-Embedding (BoWE) for document representations.
- **vMF in topic models.** The vMF distribution has been used to model directional data by placing points on a unit sphere and is known in the literature on directional statistics [21]. [12] proposed an admixture model (moVMF) that uses vMF to model the document corpus based on normalized word frequency vectors. [22] used vMF as the observational distribution of each word and used a Hierarchical Dirichlet Process (HDP) [23], a Bayesian nonparametric variant of Latent Dirichlet Allocation (LDA), to automatically infer the number of topics.
- **Fisher Kernel.** Fisher kernel is a generic framework introduced in [11] for classification purposes to combine the strengths of the generative and discriminative worlds. The idea is to characterize a signal with a gradient vector

derived from a probability density function (pdf) which models the generation process of the signal. This representation can then be used as input to a discriminative classifier. This framework has been successfully applied to computer vision [24,25] and text analysis [10]. The gradient representation of the Fisher kernel has a major advantage over the histogram of occurrences of the BoW: for the same vocabulary size, it is much larger. Hence, there is no need to use costly kernels to (implicitly) project these very high-dimensional gradient vectors into a still higher dimensional space.

3   Model

In this section, we describe our proposed FK framework in detail, including the generation process of words with continuous mixture models and the FK-based aggregation. The proposed procedure is as follows:

Learning phase: Given an unlabeled training set of documents:

- Learn the neural word embedding in a low-dimensional space, e.g., by word2vec. After this operation, each word w is then represented by a vector E_w of size d.
- Fit a probabilistic model, i.e., a mixture of Von Mises-Fisher model (moVMF), on these neural word embeddings. The detailed description of moVMF is shown in the following Probabilistic modeling Section.

Document representation: Given a document whose BoW representation is $\{w_1, \ldots, w_T\}$:

- Transform the BoW representation into the BoWE representation:

$$\{w_1, \ldots, w_T\} \rightarrow \{E_{w_1}, \ldots, E_{w_T}\}$$

- Aggregate the neural word embeddings E_{w_t} using the Fisher Kernel framework. We detail the framework in the following Fisher kernel aggregation Section.

3.1   Probabilistic Modeling

We use the mixture of Von Mises-Fisher distributions (moVMF) as the generative model of neural word embeddings. Here we describe the vMF distribution and moVMF model in detail.

The von Mises-Fisher distribution is known in the literature on directional statistics, and suitable for data distributed on the unit hypersphere. A d-dimensional unit random vector x (i.e., $x \in \mathbb{R}^d$ and $||x|| = 1$) is said to have d-variate von Mises-Fisher distribution if its probability density function is given by,

$$f(x|\mu, \kappa) = c_d(\kappa) e^{\kappa \mu^\mathrm{T} x}, \tag{1}$$

where $||\mu|| = 1$, $\kappa \geq 0$ and $d \geq 2$. The normalizing constant $c_d(\kappa)$ is given by,

$$c_d(\kappa) = \frac{\kappa^{d/2-1}}{(2\pi)^{d/2} I_{d/2-1}(\kappa)}, \qquad (2)$$

where $I_r(\cdot)$ represents the modified Bessel function of the first kind and order r. The density $f(x|\mu, \kappa)$ is parameterized by the mean direction μ, and the concentration parameter κ. The concentration parameter κ characterizes how strongly the unit vectors drawn from the distribution are concentrated on the mean direction μ. Larger values of κ imply stronger concentration about the mean direction.

Later, [12] introduce the mixture of von Mises-Fisher distributions (moVMF) that serves as a generative admixture model for directional data. Let $f_i(x|\theta_i)$ denote a vMF distribution with parameter $\theta_i = (\mu_i, \kappa_i)$ for $1 \leq i \leq N$. Then a mixture of these N vMF distributions has a density given by

$$f(x|\Theta) = \sum_{i=1}^{N} \alpha_i f_i(x|\theta_i), \qquad (3)$$

where parameters $\Theta = \{\alpha_1, \ldots, \alpha_N, \theta_1, \ldots, \theta_N\}$ and the α_i are non-negative and sum to one. To sample a point from this mixture density we choose the i-th vMF randomly with probability α_i, and then sample a point on $\mathbb{S}^{d-1}$ ($\mathbb{S}^{d-1}$ denotes the $(d-1)$-dimensional sphere embedded in $\mathbb{R}^d$) following $f_i(x|\theta_i)$. To train the model, we can use the familiar EM algorithm, to efficiently iterate between estimating the most likely conditional distribution of $\{\alpha_1, \ldots, \alpha_N\}$ in the E-step and optimizing $\{\theta_1, \ldots, \theta_N\}$ to maximize the likelihood in the M-step. The moVMF generalizes clustering methods parameterized by cosine distance and it successfully integrates a directional measure of similarity into a probabilistic setting.

3.2 Fisher Kernel Aggregation

In this work, we describe a given document, $X = \{x_t, t = 1 \ldots T\}$, as a set of d-dimensional neural word embeddings whose generation process can be modeled by the probability density function (pdf) of moVMF. Evidence suggests that this type of directional measure (i.e., cosine similarity) is often superior to Euclidean distance in high dimensions [26]. In this moVMF, each vMF distribution p_i can be viewed as a visual word and N is the vocabulary size. We denote $\lambda = \{w_i, \mu_i, \kappa_i, i = 1 \ldots N\}$, where $\{w_i, \mu_i, \kappa_i\}$ are respectively the mixture weight, mean vector and concentration of i-th vMF.

In practice, the moVMF is estimated offline with a set of neural word embeddings learned *a prior* from a large training set of documents. The parameters Θ are estimated through the optimization of a Maximum Likelihood (ML) criterion using the Expectation-Maximization (EM) algorithm.

Since the partial derivatives with respect to mixture weights α_Θ and concentration parameters κ_Θ carry little additional information, we only focus on the partial derivatives with respect to the mean parameters μ_Θ. Given μ_Θ, X can be described by the gradient vector:

$$G_\Theta^X = \nabla_{\mu_\Theta}^T \log f(X|\Theta). \tag{4}$$

Intuitively, it describes in which direction the parameters Θ of the model should be modified so that the model μ_Θ better fits the data. Assuming that the word embeddings x_t in X are iid, we have:

$$G_\Theta^X = \sum_{t=1}^{T} \nabla_{\mu_\Theta} \log f(x_t|\Theta). \tag{5}$$

In the following, $\gamma_t(i)$ denotes the occupancy probabiltity, i.e. the probability for observation x_t to be generated by the i-th vMF. Bayes formula gives:

$$\gamma_t(i) = p(i|x_t, \Theta) = \frac{\alpha_i f_i(x|\theta_i)}{\sum_{j=1}^{N} \alpha_j f_j(x|\theta_j)}. \tag{6}$$

Simple mathematical derivation with respect to μ_i has:

$$G_{\mu_i}^X = \sum_{t=1}^{T} \gamma_t(i)\kappa_i x_t. \tag{7}$$

To normalize the dynamic range of different dimensions of gradient vectors, it is important to normalize the vectors. As in [11], the Fisher information matrix (FIM) F_Θ of μ_Θ is suggested for this purpose:

$$F_\Theta = E_{x \sim \mu_\Theta}[\nabla_\Theta \log f(x|\Theta)\nabla_\Theta \log f(x|\Theta)']. \tag{8}$$

As F_Θ is symmetric and positive definite, it has a Cholesky decomposition. Then, [11] proposed to measure the similarity between two samples X and Y:

$$K(X, Y) = G_\Theta^{X'} F_\Theta^{-1} G_\Theta^Y. \tag{9}$$

Then $K(X, Y)$ can be rewritten as a dot-product between normalized vectors $\mathcal{G}_\Theta$ with:

$$\mathcal{G}_\Theta^X = F_\Theta^{-1/2} G_\Theta^X, \tag{10}$$

where $\mathcal{G}_\Theta^X$ is referred to as the *Fisher Vector* (FV) of X [27].

Let f_{μ_i} denote the diagonal approximation of FM which corresponds respectively to μ_i. According to Eq. 8, we can get

$$f_{\mu_i} = \int_X f(X|\Theta)[\sum_{t=1}^{T} \gamma_t(i)\kappa_i x_t]^2 dX. \tag{11}$$

Using the diagonal approximation of the FIM, we finally obtain the following formula for the gradient with respect to μ_i:

$$\mathcal{G}_i^X = f_{\mu_i}^{-1/2} G_{\mu_i}^X = \sum_{t=1}^T \frac{\gamma_t(i)x_t d}{w_i \kappa_i ||\mu_i||}. \tag{12}$$

The FV $\mathcal{G}_\Theta^X$ is the concatenation of the $\mathcal{G}_i^X, \forall i$, and is therefore N×d dimensional, where d is the dimensionality of the continuous word embeddings and N is the number of vMFs.

4   Experiments

In this section, we conduct experiments to verify the effectiveness of our model over document classification, clustering and retrieval tasks.

4.1   Baselines

- **Bag-of-word.** The Bag-of-Words model (BoW) [1] represents each document as a bag of words using *tf-idf* [28] as the weighting scheme. We select top 5,000 words according to *tf-idf* scores and use the vanilla TFIDF in the gensim library[1].
- **LSI.** LSI [2] maps both documents and words to lower-dimensional representations in a so-called latent semantic space using singular value decomposition (SVD) decomposition. We use the vanilla LSI in the gensim library with topic number set as 50.
- **LDA.** In LDA [5], each word within a document is modeled as a finite mixture over an set of topics. We use the vanilla LDA in the gensim library with topic number set as 50.
- **cBow.** Continuous Bag-of-Words model [6]. We use average pooling to compose a document vector from a set of word vectors.
- **PV.** Paragraph Vector [9] is an unsupervised model to learn distributed representations of words and documents. We implement PV-DBOW and PV-DM model by ourselves since no original code is available.
- **FV-GMM.** Fisher Kernel based on Gaussian mixture model (GMM) [10] is used for document representation from word embeddings. It treats documents as bags-of-embedded-words (BoEW) and to learn probabilistic mixture models once words were embedded in a Euclidean space.

We refer to our FK-based aggregation method as **FV-moVMF**.

---

[1] http://radimrehurek.com/gensim/.

4.2   Setup

We used two datasets for classificaiton, one for clustering and one for information retrieval. Preprocessing steps were applied to all the datasets: words were lowercased, non-English characters and stop words were removed. All the neural word embeddings used in the above methods were trained on the corresponding document collections in each task under 50-dimension by word2vec[2]. For FK-based aggregation methods, the number of mixture components were set as 15 since we observed ignorable performance differences with larger value. In previous work, FV-GMM [10] obtained the word embeddings by LSI. For comparison, we also tried FV-GMM based on neural word embeddings.

We refer to these two types of aggregation methods as FV-GMM$_{LSI}$ and FV-GMM$_{Neu}$, respectively. Similarly, we also have two versions of FV-moVMF, namely FV-moVMF$_{LSI}$ and FV-moVMF$_{Neu}$.

4.3   Classification

We run the classification experiments on two publicly available datasets:

- **Subj,** Subjectivity dataset [29][3] which contains $5,000$ subjective instances (snippets) and $5,000$ objective instances (snippets). The task is to classify a sentence as being subjective or objective;
- **MR,** Movie reviews [30] with one sentence per review. There are $5,331$ positive sentences and $5,331$ negative sentences. Classification involves detecting positive/negative reviews.

We use 10-fold cross-validation and Logistic Regression as the classifier.

Table 1 shows the evaluation results on the two datasets. The results show that learning text representations over BoWE (e.g., cBow, PV-DBOW, PV-DM) can in general achieve better performances than that over BoW (e.g., BoW, LSI and LDA) by involving richer semantics between words. For the FV models, the consistent improvements of neural embedding based methods over LSI based methods (i.e., FV-moVMF$_{Neu}$ and FV-GMM$_{Neu}$ vs FV-moVMF$_{LSI}$ and FV-GMM$_{LSI}$) verify the effectiveness of neural embeddings in capturing word semantics. Furthermore, each version of FV-moVMFs works better than FV-GMMs (e.g., FV-moVMF$_{Neu}$ vs FV-GMM$_{Neu}$), indicating that moVMF is a better statistical model for neural word embeddings than GMMs. Finally, FV-moVMF$_{Neu}$ can outperform all the baselines on the two datasets, demonstrating the effectiveness of our approach.

4.4   Clustering

We used one well-known and publicly available dataset: the 20 Newsgroups[4], for clustering. The 20Newsgroups contains about $20,000$ newsgroup documents

---

[2] https://code.google.com/p/word2vec/.

[3] http://www.cs.cornell.edu/people/pabo/movie-review-data/.

[4] http://qwone.com/~jason/20Newsgroups/.

Table 1. Classification accuracies (%) of different models. Best scores are bold. Two-tailed t-tests demonstrate the improvements of our model to all the baseline models are statistically significant ([‡] indicates p-value < 0.05).

Model	Subj	MR
BoW	89.5	74.3
LSI	85.4	64.2
LDA	72.7	58.2
cBow	90.9	74.8
PV-DBOW	90.1	73.9
PV-DM	90.4	74.4
FV-GMM$_{LSI}$	87.8	68.5
FV-GMM$_{Neu}$	90.3	72.6
FV-moVMF$_{LSI}$	88.6	71.5
FV-moVMF$_{Neu}$	**91.8**[‡]	**75.7**[‡]

harvested from 20 different Usenet newsgroups, with about $1,000$ documents from each newsgroup. We compared k-means over all the methods and use two standard evaluation metrics[5] to assess the quality of the clusters, namely the Adjusted Rand Index (ARI) [31] and Normalized Mutual Information (NMI) [32]. These measures compare the clusters with respect to the partition induced by the category information. For all the clustering methods, the number of clusters is set to the true number of classes of the collections.

Table 2. Clustering experiments of different models (in %). Best scores are bold. Two-tailed t-tests demonstrate the improvements of our model to all the baseline models are statistically significant ([‡] indicates p-value < 0.05).

Model	20News	
	ARI	NMI
BoW	9.2	29.7
LSI	33.2	43.1
LDA	30.8	47.2
cBow	38.6	53.8
PV-DBOW	42.6	56.6
PV-DM	42.9	56.8
FV-GMM$_{LSI}$	38.5	46.7
FV-GMM$_{Neu}$	42.3	54.6
FV-moVMF$_{LSI}$	37.9	51.9
FV-moVMF$_{Neu}$	**44.1**[‡]	**57.8**[‡]

---

[5] http://scikit-learn.org/stable/.

From Table 2, we can observe similar performance trending of different methods as that on the classification tasks. Moreover, the PV methods show better performances than FV-GMM$_{Neu}$. It indicates that dot product employed by PV works better than Euclidean distance used in FV-GMM$_{Neu}$. Finally, our FV-moVMF$_{Neu}$ outperforms all the other baseline models, showing the power of FK framework for document representation with the appropriate generative distribution.

4.5   Document Retrieval

We use one TREC collection: Robust04[6], for the document retrieval task. The topics of Robust04 are collected from TREC Robust Track 2004. It has approximately 500, 000 documents and the vocabulary size is about 600, 000. The retrieval experiments described in this section are implemented using the Galago Search Engine[7]. We use the standard cosine similarity to produce the relevance scores between documents and the query based on different models. For evaluation, the top-ranked 1, 000 documents are compared using the mean average precision (MAP) and precision at rank 20 (P@20). We also compare with the traditional retrieval model, namely BM25 [33], and linearly combine the normalized scores of BM25 and the other models :

$$score(d, Q) = \lambda score_{BM25}(d, Q) + (1 - \lambda)score_{model}(d, Q), \qquad (13)$$

where (d, Q) is the document-query pair and λ is the interpolation parameter. In our experiments, we select λ as 0.8 based on the development set.

Table 3. Retrieval experiments of different models (in %). Best scores are bold.

Model	Robust04	
	MAP	P@20
BM25	24.1	33.7
LSI	3.4	3.9
LDA	4.7	5.6
cBow	7.2	11.1
FV-GMM$_{Neu}$	9.8	12.4
FV-moVMF$_{Neu}$	11.2	13.9
BM25+LSI	25.3	36.6
BM25+LDA	25.4	36.3
BM25+cBow	25.3	36.5
BM25+FV-GMM$_{Neu}$	25.4	36.3
BM25+FV-moVMF$_{Neu}$	**25.6**	**36.7**

---

[6] http://trec.nist.gov/.
[7] http://www.lemurproject.org/galago.php.

From Table 3 we can see that, simple cosine similarity between documents and query based on different representation models cannot work well in the retrieval task since many exact matching singles are lost in this way. When combined with BM25 method, improved performance can be obtained as semantic relatedness between document and query is captured. Moreover, our proposed FV-moVMF$_{Neu}$ can bring the largest improvement among all the combinations, indicating that our model offers a better similarity with latent representations.

5   Conclusion

In this paper we introduced an alternate FK framework for document representations based on BoWE. Our new FK-based aggregation method builds upon neural word embeddings by employing a moVMF distribution as the generative model. The experimental results demonstrate that our model can achieve new state-of-the-art performances on several document processing tasks.

Nevertheless, there is still room to improve our model in the future. For example, we could like to learn the parameters of moVMF together with the FV framework, instead of estimating offline. Moreover, it is interesting to validate the effectiveness of using other word embedding techniques like Glove [7] and other statistical models for Bag-of-Word-Embeddings.

Acknowlegements. This work was funded by the 973 Program of China under Grant No. 2014CB340401, the National Natural Science Foundation of China (NSFC) under Grants No. 61232010, 61433014, 61425016, 61472401, 61203298 and 61722211, the Youth Innovation Promotion Association CAS under Grants No. 20144310 and 2016102, and the National Key R&D Program of China under Grants No. 2016QY-02D0405.

References

1. Harris, Z.S.: Distributional structure. Word **10**(2–3), 146–162 (1954)
2. Deerwester, S., Dumais, S.T., Furnas, G.W., Landauer, T.K., Harshman, R.: Indexing by latent semantic analysis. J. Am. Soc. Inf. Sci. **41**(6), 391 (1990)
3. Lee, D.D., Seung, H.S.: Learning the parts of objects by non-negative matrix factorization. Nature **401**(6755), 788–791 (1999)
4. Hofmann, T.: Probabilistic latent semantic indexing. In: Proceedings of the 22nd Annual International ACM SIGIR, pp. 50–57. ACM (1999)
5. Blei, D.M., Ng, A.Y., Jordan, M.I.: Latent dirichlet allocation. J. Mach. Learn. Res. **3**, 993–1022 (2003)
6. Mikolov, T., Chen, K., Corrado, G., Dean, J.: Efficient estimation of word representations in vector space. arXiv preprint arXiv:1301.3781 (2013)
7. Pennington, J., Socher, R., Manning, C.D.: Glove: global vectors for word representation. In: Proceedings of the 2014 Conference on Empirical Methods in Natural Language Processing (EMNLP), vol. 14, pp. 1532–1543 (2014)
8. Vulic, I., Moens, M.F.: Cross-lingual semantic similarity of words as the similarity of their semantic word responses. In: NAACL-HLT 2013, pp. 106–116. ACL (2013)

9. Le, Q.V., Mikolov, T.: Distributed representations of sentences and documents. In: ICML, vol. 14, pp. 1188–1196 (2014)
10. Clinchant, S., Perronnin, F.: Aggregating continuous word embeddings for information retrieval. In: Proceedings of the Workshop on Continuous Vector Space Models and their Compositionality, pp. 100–109 (2013)
11. Jaakkola, T., Haussler, D.: Exploiting generative models in discriminative classifiers. In: NIPS, pp. 487–493 (1999)
12. Banerjee, A., Dhillon, I.S., Ghosh, J., Sra, S.: Clustering on the unit hypersphere using von mises-fisher distributions. J. Mach. Learn. Res. **6**, 1345–1382 (2005)
13. Eisenstein, J., Ahmed, A., Xing, E.P.: Sparse additive generative models of text (2011)
14. Weng, J., Lim, E.P., Jiang, J., He, Q.: Twitterrank: finding topic-sensitive influential twitterers. In: Proceedings of the Third ACM International Conference on Web Search and Data Mining, pp. 261–270. ACM (2010)
15. Wang, Q., Xu, J., Li, H., Craswell, N.: Regularized latent semantic indexing. In: Proceedings of the 34th International ACM SIGIR Conference on Research and Development in Information Retrieval, pp. 685–694. ACM (2011)
16. Maas, A.L., Daly, R.E., Pham, P.T., Huang, D., Ng, A.Y., Potts, C.: Learning word vectors for sentiment analysis. In: Proceedings of the 49th Annual Meeting of the Association for Computational Linguistics: Human Language Technologies, vol. 1, pp. 142–150. Association for Computational Linguistics (2011)
17. David, M., Blei, J.D.: Supervised topic models. In: Proceedings of Advances in Neural Information Processing Systems (2007)
18. Socher, R., Perelygin, A., Wu, J.Y., Chuang, J., Manning, C.D., Ng, A.Y., Potts, C.: Recursive deep models for semantic compositionality over a sentiment treebank. In: EMNLP, vol. 1631, Citeseer, p. 1642 (2013)
19. Sutskever, I., Martens, J., Hinton, G.E.: Generating text with recurrent neural networks. In: ICML, vol. 11, pp. 1017–1024 (2011)
20. Zhao, H., Lu, Z., Poupart, P.: Self-adaptive hierarchical sentence model. In: IJCAI, pp. 4069–4076 (2015)
21. Fisher, R.: Dispersion on a sphere. In: Proceedings of the Royal Society of London A: Mathematical, Physical and Engineering Sciences, vol. 217, pp. 295–305. The Royal Society (1953)
22. Batmanghelich, K., Saeedi, A., Narasimhan, K., Gershman, S.: Nonparametric spherical topic modeling with word embeddings. arXiv preprint arXiv:1604.00126 (2016)
23. Teh, Y.W., Jordan, M.I., Beal, M.J., Blei, D.M.: Sharing clusters among related groups: hierarchical dirichlet processes. In: NIPS, pp. 1385–1392 (2005)
24. Perronnin, F., Dance, C.: Fisher kernels on visual vocabularies for image categorization. In: CVPR, pp. 1–8. IEEE (2007)
25. Bressan, M., Cifarelli, C., Perronnin, F.: An analysis of the relationship between painters based on their work. In: ICIP, 113–116. IEEE (2008)
26. Mikolov, T., Sutskever, I., Chen, K., Corrado, G.S., Dean, J.: Distributed representations of words and phrases and their compositionality. In: NIPS, pp. 3111–3119 (2013)
27. Perronnin, F., Liu, Y., Sánchez, J., Poirier, H.: Large-scale image retrieval with compressed fisher vectors. In: CVPR, pp. 3384–3391. IEEE (2010)
28. Salton, G., McGill, M.: Introduction to Modern Information Retrieval. McGraw-Hill, New York (1983)

29. Pang, B., Lee, L.: A sentimental education: sentiment analysis using subjectivity summarization based on minimum cuts. In: Proceedings of the 42nd Annual Meeting on Association for Computational Linguistics, p. 271. Association for Computational Linguistics (2004)
30. Pang, B., Lee, L.: Seeing stars: exploiting class relationships for sentiment categorization with respect to rating scales. In: Proceedings of the 43rd Annual Meeting on Association for Computational Linguistics, p. 115–124. Association for Computational Linguistics (2005)
31. Hubert, L., Arabie, P.: Comparing partitions. J. Classif. **2**(1), 193–218 (1985)
32. Estévez, P.A., Tesmer, M., Perez, C.A., Zurada, J.M.: Normalized mutual information feature selection. IEEE Trans. Neural Netw. **20**(2), 189–201 (2009)
33. Robertson, S.E., Walker, S.: Some simple effective approximations to the 2-poisson model for probabilistic weighted retrieval. In: SIGIR, pp. 232–241. Springer, New York (1994). https://doi.org/10.1007/978-1-4471-2099-5_24

Unsupervised Sentiment Analysis of Twitter Posts Using Density Matrix Representation

Yazhou Zhang[1] , Dawei Song[1,2]([✉]), Xiang Li[1], and Peng Zhang[1]

[1] School of Computer Science and Technology, Tianjin University, Tianjin, China
{yzhou_zhang,dwsong,xlee,pz}@tju.edu.cn,
[2] School of Computing and Communications, The Open University,
Milton Keynes, UK
dawei.song@open.ac.uk

Abstract. Nowadays, a series of pioneering studies provide the evidence that quantum probability theory can be applied in information retrieval as a mathematical framework, such as Quantum Language Model (QLM) and its variants. In these studies, the density matrix, which is defined on the quantum probabilistic space, is used to represent query and document. However, these studies are only designed for information retrieval tasks, which are unable to model sentiment information. In this paper, we investigate the feasibility of quantum probability theory for twitter sentiment analysis, and propose a density matrix based unsupervised sentiment analysis approach. The main idea is to artificially create two sentiment dictionaries, generate density matrices of documents and dictionaries using an extended QLM, then employ the quantum relative entropy to judge the similarity between density matrices of documents and dictionaries. Extensive experiments are conducted on two widely used twitter datasets, which are the Obama-McCain Debate (OMD) dataset and Sentiment Strength Twitter Dataset (SS-Tweet). The experimental results show that our approach significantly outperforms a number of baselines, demonstrating the effectiveness of the proposed density matrix based sentiment analysis approach.

Keywords: Sentiment analysis · Quantum Language Model
Density matrix

1 Introduction

Along with the rapid development of WWW and social networking services, a growing number of people express their opinions on Twitter by publishing blogs and posting comments. Millions of pages of social data are posted on internet every day, which has grown to be one of the most important sources for people to acquire information and make decision. The importance of discovering the sentiment orientation of user generated tweets has been recognized in a wide

© Springer International Publishing AG, part of Springer Nature 2018
G. Pasi et al. (Eds.): ECIR 2018, LNCS 10772, pp. 316–329, 2018.
https://doi.org/10.1007/978-3-319-76941-7_24

range of application domains, e.g., to help manufacturers predict the attitudes of consumers toward their products, and to help political association understand the general public opinion. Sentiment analysis of Twitter is of great theoretical and practical significance, and has attracted more and more attention from both academia and industry [1, 2].

Text representation is a critical component of Twitter sentiment analysis [3]. Nowadays, vector based technologies are the mainstream approaches, such as n-gram features [4], one-hot representation [5] and word embeddings [6], etc. For instance, Pang and Lee [7] first applied unigrams and bigrams features and machine learning methods to identify the overall sentiment of movie reviews. Yin and Jin [8] used word embedding and SVM to perform document sentiment classification. Giatsoglou et al. [9] combined bag of words representation and average emotion representation to construct a hybrid vector for sentiment analysis. Pham and Le designed a multiple layer architecture of knowledge representation for representing each input text [10]. These lines of work mainly focus on encoding the semantic information via vector to improve the performance of sentiment classification, which are proved effective in sentiment analysis.

Recently, there are emerging a series of quantum theory (QT) based studies in information retrieval, which provide the evidence that quantum probability theory can be applied as a mathematical framework [11, 12]. The representatives are the Quantum Language Model (QLM) and its variants [13, 14]. In these studies, the density matrix, which is defined on the quantum probabilistic space, is used to represent query and document, respectively. They utilize an EM-based training algorithm to estimate density matrix by maximizing a likelihood function. The effectiveness of QLM and its variants are demonstrated from both theoretical and experimental perspectives.

Despite QLM and its variants success in information retrieval, they are only designed for information retrieval tasks, which are not suitable for sentiment analysis. QLM is unable to model the sentiment information. However, in QT, the density matrix is a kind of probability distribution, corresponding to the probabilities of all events. Compared with vector-based representation, density matrix can encode more semantic information and distribution information. Hence, using quantum probability theory to manipulate probability and vector space is still a compelling idea for developing a novel sentiment representation model.

In this paper, we investigate the feasibility of quantum theory for twitter sentiment analysis, and propose a density matrix based unsupervised sentiment analysis approach. Specifically, we first create two sentiment dictionaries, one of which is comprised of positive sentiment words and the other is made of negative sentiment words. Second, we learn the density matrix representation of documents and sentiment dictionaries, using a globally convergence based QLM (GQLM). Last, we employ the quantum relative entropy to evaluate the similarity between density matrices of documents and dictionaries. The sentiment recognition results are computed on the ground of the similarity results. Extensive experiments are conducted on two widely used twitter datasets, which are the Obama-McCain Debate (OMD) dataset and Sentiment Strength Twitter

(SS-Tweet) Dataset. The experimental results demonstrate the effectiveness of our proposed density matrix based approach in comparison with the state of the art algorithms.

2   Preliminary of Quantum Probability Theory

Quantum probability theory, which is developed by von Neumann, is an quantum generalization of the classical probability theory. It aims at interpreting the mathematical foundations of quantum theory.

2.1   Events and Projectors

In quantum probability theory, the quantum probability space is naturally encapsulated in an infinite Hilbert space, noted as $\mathbb{H}$.

As we all know that event in classical probability theory corresponds to a subset of the sample space, which is represented by a random variable $\mathbf{X}$. However, in quantum probability theory, event is defined to be a subspace of Hilbert space, which is represented by any orthogonal projector Π. With the Dirac's notation, a quantum state vector $\boldsymbol{u} \in \mathbb{H}$ and its transpose $\boldsymbol{u}^T$ can be expressed as a ket $|u\rangle$ and a bra $\langle u|$. The inner product between two quantum state $|u\rangle$, $|v\rangle$ is represented as $\langle u|v\rangle$. For a unit state vector $|u\rangle$ of $\mathbb{H}$, $\|\boldsymbol{u}\|_2 = 1$, the projector on the direction u can be written as $\Pi = |u\rangle\langle u|$, which is a elementary event of quantum probability space. Assuming that $e_1, e_2,...,e_n$ are n events, $\Pi_{e_1}, \Pi_{e_2},...,\Pi_{e_n}$ are corresponding projectors, then $\prod_{i=1}^{n} \Pi_{e_i}$ represents "simultaneous occurrence of n events".

2.2   Density Matrix

The density matrix is an excellent representation of the state of a system. The density matrix is a quantum generalization of probability distribution of position and state in classical theory, and combines information about both quantum state and classical uncertainty into a single construction [14].

The density matrix ρ for a quantum state $|u_i\rangle$ can be defined as:

$$\rho = \sum_i \phi_i |u_i\rangle\langle u_i| \tag{1}$$

where ϕ_i is the probability for the qubit to be in state $|u_i\rangle$, sufficing that $\phi_i \geq 0$ and $\sum \phi_i = 1$.

The density matrix ρ has the following properties: (a) ρ is Hermetian, $\rho_{ij} = \rho_{ji}^*$; (b) ρ is positive definite, $\rho \geq 0$; (c) the diagonal elements of the density matrix are probability values of the same distribution and the sum of the diagonal elements is one, namely, $tr(\rho) = 1$; (d) $tr(\rho^2) = 1$ for pure state, and $tr(\rho^2) < 1$ for mixed state

Based on the Gleason's Theorem, the result of the probability that event $|u\rangle\langle u|$ belongs to the system ρ is as follows:

$$\mu_\rho\left(|u\rangle\langle u|\right) = tr\left(\rho|u\rangle\langle u|\right) = \langle u|\rho|u\rangle \tag{2}$$

where tr denotes trace, $\mu_\rho \in [0,1]$, behaves like a probability measurement.

3   Methodology

In this section, we first create two sentiment dictionaries, one of which is comprised of positive sentiment words and the other is made of negative sentiment words. Second, we develop a globally convergence based QLM (GQLM) to train density matrices of documents and dictionaries. Last, we analyze the polarity of the text by means of the quantum relative entropy.

3.1   Creating Sentiment Dictionaries

Typically, the authors usually express their feelings via a variety of sentiment words on Twitter. For instance, they would like to use a few positive sentiment words, such as "happy", "amazing", "awesome", etc. to imply that they feel good. They choose some negative sentiment words, such as "bored", "sad", "hate", etc. to express uncomfortable feelings. Therefore, we believe that a tweet implies a positive sentiment when it is more strongly associated with positive sentiment words and a negative sentiment when it is more strongly associated with negative sentiment words.

Motivated by this, we artificially create two sentiment dictionaries based on a few seed words and training documents, which intends to represent two basic types of sentiments, positive and negative. For simplicity, We would like to call them as positive sentiment dictionary (PSD) and negative sentiment dictionary (NSD). Here are our procedures of creating sentiment dictionaries: first, we manually choose seven opposing pairs of words (which are positive/negative, good/bad, love/hate, excellent/poor, amazing/shit, nice/terrible, awesome/crap) as the seed words. Therefore, the original PSD = (positive, good, love, excellent, amazing, nice, awesome), and the original NSD = (negative, bad, hate, poor, shit, terrible, crap).

Second, since adjectives and adverbs have been proved to be good indicators of personal sentiments [15], we extract all adjectives and adverbs from the training documents using a Hidden Markov Models (HMM) part-of-speech tagger, and treat these extracted words as the candidate sentiment words. The HMM part-of-speech tagger involves an assumption that picking the most likely tag t_i for each word w_i in a given sentence, and finally choosing the tag sequence t_{seq} that maximizes: $P\left(word|tag\right) \cdot P\left(tag|previous\, n\, tags\right)$, which can be formulated as:

$$t_{seq} = argmax\, P\left(t_{seq}|w_{seq}\right) \approx argmax \prod_{i=1}^{n} P\left(w_i|t_i\right) P\left(t_i|t_{i-1}\right) \tag{3}$$

Third, we calculate the strength of semantic association between each candidate word and seed words in expanding PSD and NSD, using the PMI-IR algorithm [16]. The sentiment score of a word is calculated as follows:

$$Score\,(word) = \sum_{seeds \in PSD} PMI\,(word, seeds) - \sum_{seeds \in NSD} PMI\,(word, seeds) \quad (4)$$

where the candidate word is positive if its score is greater than zero and negative if its score is less than zero. Figure 1 shows the distribution of all candidate sentiment words of the SS-Tweet dataset. In this paper, we would like to select the candidate words with strongly positive and negative sentiment, whose score is greater than (or less than) two. Last, the PSD contains 142 positive sentiment words, and the NSD contains 72 negative sentiment words. Table 1 introduces a few positive and negative sentiment words from our sentiment dictionaries for illustration.

3.2 A Globally Convergence Based Quantum Language Model (GQLM)

Aiming at an effective representation learning model, we base our computational framework on the Quantum Language Model (QLM). In QLM, both single terms and compound dependencies are modeled as projectors in a probability space. Each document and query are represented as a sequence of projectors. Then QLM utilizes a $R\rho R$ training algorithm to estimate density matrix of each document and query.

However, the $R\rho R$ algorithm has exposed the convergence problem [12,14]. To address this problem, we employ a globally convergent algorithm to guarantee convergence, which has been strictly proved in [17]. Based on this, we develop a globally convergence based QLM (GQLM) to train density matrices of each document and dictionary.

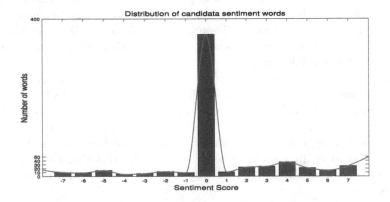

Fig. 1. Distribution of all candidate sentiment words.

Table 1. An illustration of PSD and NSD.

Dictionary	Sentiment words	Score	Sentiment words	Score
PSD	best	4.66	excellent	11.01
	healthy	4.55	glad	6.12
	amazing	6.12	favorite	4.71
	beautiful	3.86	awesome	8.84
	thankful	10.38	smart	2.76
	juicy	3.92	thankful	10.38
NSD	anxious	−7.70	fake	−4.95
	dirty	−5.55	bloody	−6.16
	sad	−8.04	weird	−3.46
	scary	−6.13	not	−2.03
	dumb	−5.95	bad	−7.28
	offensively	−7.14	terrible	−11.04

In GQLM, all textual words can be seen as elementary events, and events in QT are defined to be subspaces, which are represented by any orthogonal projectors Π. The projectors Π are used to estimate density matrices ρ of the document, corresponding to the probabilities of all events. Theoretically, compared with vector-based representation, density matrices can more efficiently encode the semantic dependencies and their probabilistic distribution information.

Specifically, for each word w_i, supposing $|w_i\rangle$ is its corresponding one-hot vector. The projector Π_{word} for a single word w_i is below as in Eq. (5).

$$\Pi_i = |w_i\rangle\langle w_i| \tag{5}$$

After defining projectors for each word, we can represent a document with a sequence of projectors, $\mathcal{P}_U = \{\Pi_1, \Pi_2, ..., \Pi_n\}$, where n is the number of terms in the document. Then we use the Maximum Likelihood estimation to train density matrices ρ of documents through maximizing the product of probability for each quantum elementary event (single term). The likelihood function is the probability of getting the observed events given density matrices:

$$\zeta(\rho) \propto \prod tr(\mathcal{P}_U\rho) = \prod_i tr(\Pi_i\rho) \tag{6}$$

Since the log function is monotonic, the objective function $F(\rho)$ is formulated as:

$$F(\rho) \equiv \max_{\rho} \sum_i \log(tr(\Pi_i\rho)) \tag{7}$$

Different from the original QLM, the GQLM uses the globally convergence algorithm to estimate the maximum likelihood value. The globally convergent algorithm extends the RρR algorithm, where the ascent direction of likelihood is determined by two ascent directions controlled by the step size t. The globally convergent algorithm is able to find a value which ensures a sufficient improvement in the likelihood function.

Based on the gradient of the objective function $F(\rho)$, this algorithm defines that $\nabla F(\rho) = \sum_i \frac{f_i}{tr(\Pi_i \rho)} \Pi_i$, where f_i is the term frequency. It also determines a definition that a direction D^k is an ascent direction at the kth iteration if $tr(\nabla F(\rho^k) D^k) > 0$, and this definition ensures the progressively increasing of the function.

The search direction D^k is gradient related, which is a combination of the direction $\bar{D}^k$ and $\tilde{D}^k$, where $\bar{D}^k$ and $\tilde{D}^k$ are also ascent directions for any ρ^k controlled by t. Using the Armijo condition and a backtracking procedure, the search direction D^k at the kth iteration is given by:

$$D^k = \frac{2}{q(t_k)} \bar{D}^k + \frac{t_k tr(\nabla F(\rho^k) \rho^k \nabla F(\rho^k))}{q(t_k)} \tilde{D}^k \tag{8}$$

where $q(t_k)$, $\bar{D}^k$, $\tilde{D}^k$ are also defined in the paper [17]. From Eq.(8), we can deduce that as $t_k \to 0$, D^k is close to $\bar{D}^k$, whereas $t_k \to \infty$, D^k goes to the direction $\tilde{D}^k$. In this paper, $t_k \in [0, 1]$, $q(t_k) \geq 1$. To show the robustness of this algorithm, we random initialize a diagonal matrix ρ^0 while it satisfies $\rho^0 > 0$ and $tr(\rho^0) = 1$.

At the k-th iteration, after generating an ascent direction D^k, this method tries to update the iteration:

$$\rho^{k+1} = \rho^k + t_k D^k \tag{9}$$

This process will stop when the number of iterations is greater than 17. Because we observe that the value of objective function keeps unchanged after 17 iterations. The complete procedure of estimating density matrix is described in Algorithm 1.

After training density matrices, the Dirichlet smoothing method is applied to smooth the density matrices. If ρ_{doc} is a document GQLM, its smoothed version is obtained by interpolation with the collection GQLM ρ_{col}:

$$\rho_d = (1 - \gamma) \rho_{doc} + \gamma \rho_{col} \tag{10}$$

where $\gamma \in [0, 1]$ controls the amount of smoothing.

3.3   Quantum Relative Entropy (QRE)

We introduce quantum relative entropy to calculate the similarity between density matrices of documents and dictionaries. As a generalization of classical relative entropy, quantum relative entropy can measure of distinguishability between two quantum states. Closely, it is a kind of "distance" measure between density matrices [18]. Quantum relative entropy has been widely applied in quantum information theory.

Algorithm 1. Algorithm of estimating density matrix

Require: Each word vector w_i, the initial density matrix ρ^0 and each document d
Ensure: Density matrix of each document ρ
1: $\mathcal{P}_U \Leftarrow \phi$; // $\mathcal{P}_U$ is the projector sequence
2: **for** each $d \in D$ **do**
3:    **for** each $w \in d$ **do**
4:      **for** $i = 1$; $i \leq \#(w,d)$; $i++$ **do**
5:        $\Pi_i = |w_i\rangle\langle w_i|$;
6:        $\mathcal{P}_U \Leftarrow \mathcal{P}_U \bigoplus \Pi_i$;
7:      **end for**
8:    **end for**
9: **end for**
10: **for** each $\mathcal{P}_U$ of d **do**
11:    Maximize $F(\rho) \equiv \sum_i \log(tr(\Pi_i\rho))$;
12:    **for** $k = 1$; $F(\rho^{k+1}) - F(\rho^k) \leq \epsilon = 10^{-5}$; $k++$ **do**
13:      $\rho^{k+1} = \rho^k + t_k D^k$;
14:    **end for**
15: **end for**
16: **return** ρ

In this paper, given two density matrices ρ_{text} and ρ_{dict} for a twitter text and a dictionary respectively, quantum relative entropy can be defined as:

$$
\begin{aligned}
S(\rho_{text}||\rho_{dict}) &= tr(\rho_{text}log\rho_{text}) - tr(\rho_{text}log\rho_{dict}) \\
&= tr(\rho_{text}(log\rho_{text} - log\rho_{dict}))
\end{aligned}
\tag{11}
$$

where $S(\rho_{text}||\rho_{dict}) \geq 0$, with equality if and only if $\rho_{text} = \rho_{dict}$.

4 Empirical Evaluation

4.1 Experimental Settings

Our main research questions are: (1) Is the quantum probability theory feasible for sentiment analysis? (2) Does the GQLM guarantee the global convergence? (3) How effective is our density matrix based approach in twitter sentiment analysis?

To address (1), we run our approach on a binary sentiment classification task. We choose two widely used twitter datasets with different topic diversity, and the task for every model is to predict whether a tweet belongs to positive or negative class. To answer (2), we first evaluate the performance of GQLM on two datasets and compare its performance to the original QLM. Moreover, we conduct the convergence analysis of GQLM and QLM. To answer (3), we first compare the performance of GQLM on two datasets to a number of baselines, and analyze the reason to study how successful our approach is on text representation.

Dataset, pre-processing, evaluation metrics: Currently, benchmark datasets for sentiment analysis are very thin on the ground. As what other researchers did in [19,20], we choose two widely used sentiment datasets: the Obama-McCain Debate (OMD) dataset [21], and Sentiment Strength Twitter (SS-Tweet) Dataset [22]. The OMD dataset contains 509 positive and 741 negative tweets, crawled during the first U.S. presidential TV debate. The SS-Tweet dataset consists of 4242 tweets manually labelled with their positive and negative sentiment scores. We extract each positive tweet only if its positive sentiment score divided by negative sentiment score is greater than 1.5. We extract each negative tweet using the similar method. Last, we obtain 2113 tweets. Table 2 introduces the details of the two twitter datasets.

Table 2. Two twitter datasets used for our experiments.

Dataset	Num of positive	Num of negative	Total
OMD	509	741	1250
SS-Tweet	1252	861	2113

For pre-processing tweets, we remove stopwords and punctuations included in the standard stopword list from Python's NLTK package.

For choosing evaluation metrics, since our approach is an unsupervised sentiment analaysis method, we adopt **Precision, Recall, F1 score, Accuracy** instead of **ROC** curves as the evaluation metrics to evaluate the classification performance of each method.

Baselines and parameters settings: We compare the following models: (i) **Bag of words model (Bow)**, the classical representation model is used to represent tweets, and a Random Forest (RF) classifier (in which the number of tree is 100) is trained. To limit the size of the feature vectors, we use the 1000 most frequent words; (ii) **Doc2vector**: we use doc2vector [23] technique to represent tweets, and apply RF to identify emotions. The dimension of each document vector is 100; (iii) **PMI-IR**: Turney's unsupervised learning algorithm [15] for classifying sentiment. We realize this algorithm as a baseline; (iv) **SentiStrength**: SentiStrength [24] assigns to each tweet three sentiment strengths: a negative strength between -1 to -5, a positive strength between $+1$ to $+5$, and a neutral strength with 0; (v) **QLM**, the original quantum language model which is regarded as the baseline model. To limit the size of vocabulary, we use the 1000 most frequent words.

Furthermore, in order to validate the effectiveness of quantum relative entropy, we compare a common similarity measure method: Standardized Euclidean Distance (SED).

4.2   Results on OMD Dataset

Table 3 reports the results on OMD dataset. We can notice that bag of words model achieves the best accuracy result among all baselines. It shows that bag

Table 3. Results on OMD dataset. Best results are highlighted in boldface. Numbers in parentheses indicate relative improvement over the QLM model.

Algorithm	OMD dataset			
	Precision	Recall	F1	Accuracy
Bow	0.5151	0.3761	0.4347	0.6237
Doc2vector	0.4782	0.3407	0.3979	0.6103
PMI-IR	0.4693	0.5088	0.4883	0.5969
SentiStrength	0.4926	**0.8938**	**0.6352**	0.6110
QLM	0.5962	0.5313	0.5619	0.6140
GQLM+SED	0.5590	0.5497	0.5543	0.6030
GQLM+QRE	**0.8136**	0.5089	0.6261	**0.6298**
	(+36.46%)	(−4.21%)	(+11.42%)	(+2.57%)

of words model is able to extract semantic information by building the vocabulary. Doc2vector algorithm gets a bit lower results than bag of words model. This implies that the 100-dimensional vector is not enough to encode semantic information compared with 1000-dimensional vocabulary vector. However, Doc2vector algorithm outperforms the PMI-IR algorithm and achieves competitive accuracy result against SentiStrength algorithm, which are two unsupervised sentiment analysis methods. The PMI-IR method performs very poorly, gets the worst results among all baselines, as we expected. Because its sentiment detection rules are a bit simple. SentiStrength algorithm produces lower precision and accuracy results than bag of words model, even though existing a fact that SentiStrength algorithm gets the best recall and F1 results. These results indicate only creating simple sentiment detection rules or relying on low-level textual features is not enough to deal with twitter sentiment analysis task. It is necessary to develop better text representation model for unsupervised sentiment analysis methods. Through constructing term projectors based on word vector and training density matrices, QLM produces a modest improvement over Doc2vector, PMI-IR, SentiStrength algorithms. This demonstrates that applying quantum probability theory is highly feasible for sentiment analysis task.

Compared with the above baselines, GQLM+SED gets poor results while GQLM+QRE outperforms all baselines, gets the best precision and accuracy results. These results show that: (a) the performance obtained by our method illustrates the benefits of using density matrices, which are probability distributions of words. (b) Quantum relative entropy performs better than the traditional euclidean distance when quantifying the similarity between two density matrices.

4.3 Results on SS-Tweet Dataset

Since the SS-Tweet twitter dataset is popular and has been used to evaluate various supervised and unsupervised learning methods, the results on SS-Tweet dataset will be more persuasive.

Table 4 shows the experimental results on SS-Tweet dataset. From the table, we can observe that Bag of words model performs very well, which gets best accuracy result among all baselines. It indicates that bag of words model can effective represent each tweet by counting occurrences of words. Doc2vector algorithm achieves higher recall and F1 results but lower accuracy result than bag of words model. Because Doc2vector algorithm may depend on the dimension of word embeddings. Doc2vector algorithm still outperforms PMI-IR and SentiStrength algorithms, both of which are unsupervised sentiment analysis methods. As two classical lexicon based methods, PMI-IR and SentiStrength algorithms perform very poorly, get the worst results among all baselines. This indicates that supervised learning algorithms tend to exceed unsupervised learning algorithms when unsupervised algorithms only create simple sentiment detection rules, such as simple adding the positive scores and the negative scores. QLM gets better results than Doc2vector, PMI-IR, SentiStrength algorithms. This may be because quantum projectors help to model mid-level term features, and density matrix contains more semantic information. Compared with the above baselines, GQLM+SED posts competitive results against the Bow model, which demonstrates the effectiveness of the globally convergent algorithm. Last, GQLM+QRE outperforms all baselines significantly. The performance obtained by our method illustrates the benefits of using density matrices, which are probability distributions of words.

Table 4. Results on SS-Tweet dataset. Best results are highlighted in boldface. Numbers in parentheses indicate relative improvement over the QLM model.

Algorithm	SS-Tweet			
	Precision	Recall	F1	Accuracy
Bow	0.5919	0.6932	0.6386	0.5920
Doc2vector	0.5916	**0.7091**	**0.6450**	0.5636
PMI-IR	0.6617	0.3585	0.4651	0.5118
SentiStrength	0.5875	0.5640	0.5755	0.5183
QLM	0.6078	0.5276	0.5648	0.5796
GQLM+SED	0.5729	0.5477	0.5600	0.5917
GQLM+QRE	**0.6777**	0.5267	0.5927	**0.6188**
	(+11.50%)	(−0.17%)	(+4.94%)	(+6.76%)

4.4   Convergence Analysis

In order to demonstrate the effectiveness of the globally convergent algorithm that is used in our GQLM, we select two illustrative examples, as shown in Fig. 2. Overall, we can notice that the objective functions of GQLM converge very fast at the first five iterations on both datasets. Then, the convergence speed is slowing down between five and fifteen iterations. Finally, the values of objective function keep unchanged after 17 iterations. The objective functions obtain their maximum values using the globally convergent algorithm, which are -26.56 and -28.68. However, we can observe that the objective function of QLM does not converge at each iteration, using the $R\rho R$ algorithm. The objective function only has two certain values. Furthermore, at any number of iterations, the objective function of GQLM stays significantly above the objective function of QLM. These results indicate that our GQLM could guarantee the global convergence, then train a density matrix of high quality.

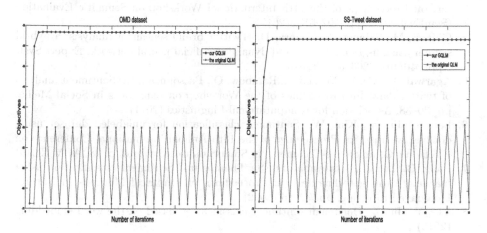

Fig. 2. Plots of the objective function of GQLM and QLM against the number of iterations for OMD and SS-Tweet dataset (left and right).

5   Conclusion

Sentiment analysis is an interesting and challenging task, which has been paid increasing attention. In this paper, we propose a density matrix based approach, which is able to analyze the sentiment information of twitter data. Specifically, we first create two sentiment dictionaries based on a few seed words and training documents. Then, words of twitter document and dictionary are modeled as a sequence of projectors, which represent the elementary events in the probabilistic space. The globally convergent algorithm is employed to train density matrices. Last, the sentiment recognition on two widely used twitter datasets is performed through the quantum relative entropy. The experimental results show that our proposed approach outperforms a number of state-of-art baselines. This demonstrates that the effectiveness of our approach, and applying quantum probability theory is highly feasible for developing sentiment analysis models.

Acknowledgements. This work is supported in part by the Chinese National Program on Key Basic Research Project (973 Program, grant No. 2014CB744604), Natural Science Foundation of China (grant No. U1636203, 61272265, 61402324), and the

European Union's Horizon 2020 research and innovation programme under the Marie Skłodowska-Curie grant agreement No 721321.

References

1. Giachanou, A., Crestani, F.: Like it or not: a survey of twitter sentiment analysis methods. ACM Comput. Surv. (CSUR) **49**(2), 28 (2016)
2. Pang, B., Lee, L.: Opinion mining and sentiment analysis. Found. Trends Inf. Retr. **2**(1–2), 1–135 (2008)
3. Rosenthal, S., Farra, N., Nakov, P.: Semeval-2017 task 4: sentiment analysis in twitter. In: Proceedings of the 11th International Workshop on Semantic Evaluation (SemEval-2017), pp. 502–518. (2017)
4. Ghiassi, M., Skinner, J., Zimbra, D.: Twitter brand sentiment analysis: a hybrid system using n-gram analysis and dynamic artificial neural network. Expert Syst. Appl. **40**(16), 6266–6282 (2013)
5. Agarwal, A., Xie, B., Vovsha, I., Rambow, O., Passonneau, R.: Sentiment analysis of twitter data. In: Proceedings of the Workshop on Languages in Social Media, pp. 30–38. Association for Computational Linguistics (2011)
6. Lee, S., Jin, X., Kim, W.: Sentiment classification for unlabeled dataset using doc2vec with jst. In: Proceedings of the 18th Annual International Conference on Electronic Commerce: e-Commerce in Smart connected World, p. 28. ACM (2016)
7. Pang, B., Lee, L., Vaithyanathan, S.: Thumbs up? Sentiment classification using machine learning techniques. In: Proceedings of Empirical Methods in Natural Language Processing, pp. 79–86 (2002)
8. Yin, Y., Jin, Z.: Document sentiment classification based on the word embedding (2015)
9. Giatsoglou, M., Vozalis, M.G., Diamantaras, K., Vakali, A., Sarigiannidis, G., Chatzisavvas, K.C.: Sentiment analysis leveraging emotions and word embeddings. Expert Syst. Appl. **69**, 214–224 (2017)
10. Pham, D.H., Le, A.C.: Learning multiple layers of knowledge representation for aspect based sentiment analysis. Data Knowl. Eng. (2017)
11. Van Rijsbergen, C.J.: The Geometry of Information Retrieval. Cambridge University Press, Cambridge (2004)
12. Li, Q., Li, J., Zhang, P., Song, D.: Modeling multi-query retrieval tasks using density matrix transformation. In: Proceedings of the 38th International ACM SIGIR Conference on Research and Development in Information Retrieval, pp. 871–874. ACM (2015)
13. Sordoni, A., Nie, J.Y., Bengio, Y.: Modeling term dependencies with quantum language models for IR. In: Proceedings of the 36th International ACM SIGIR Conference on Research and Development in Information Retrieval, pp. 653–662. ACM (2013)
14. Li, J., Zhang, P., Song, D., Hou, Y.: An adaptive contextual quantum language model. Phys. A **456**, 51–67 (2016)
15. Turney, P.D.: Thumbs up or thumbs down?: Semantic orientation applied to unsupervised classification of reviews. In: Proceedings of the 40th Annual Meeting on Association for Computational Linguistics, pp. 417–424. Association for Computational Linguistics (2002)
16. Turney, P.D.: Mining the web for synonyms: PMI-IR versus LSA on TOEFL. In: De Raedt, L., Flach, P. (eds.) ECML 2001. LNCS (LNAI), vol. 2167, pp. 491–502. Springer, Heidelberg (2001). https://doi.org/10.1007/3-540-44795-4_42

17. Goncalves, D.S., Gomes-Ruggiero, M.A., Lavor, C.: Global convergence of diluted iterations in maximum-likelihood quantum tomography. arXiv preprint arXiv:1306.3057 (2013)
18. Schumacher, B., Westmoreland, M.D.: Relative entropy in quantum information theory. Contemp. Math. **305**, 265–290 (2002)
19. Pak, A., Paroubek, P.: Twitter as a corpus for sentiment analysis and opinion mining. In: LREc, vol. 10 (2010)
20. Kouloumpis, E., Wilson, T., Moore, J.D.: Twitter sentiment analysis: the good the bad and the omg!. ICWSM **11**(164), 538–541 (2011)
21. Diakopoulos, N.A., Shamma, D.A.: Characterizing debate performance via aggregated twitter sentiment. In: Proceedings of the SIGCHI Conference on Human Factors in Computing Systems, pp. 1195–1198. ACM (2010)
22. Thelwall, M., Buckley, K., Paltoglou, G.: Sentiment strength detection for the social web. J. Assoc. Inf. Sci. Technol. **63**(1), 163–173 (2012)
23. Le, Q.V., Mikolov, T.: Distributed representations of sentences and documents. Int. Conf. Mach. Learn. **4**(2), 1188 (2014)
24. Thelwall, M., Buckley, K., Paltoglou, G., Cai, D., Kappas, A.: Sentiment strength detection in short informal text. J. Assoc. Inf. Sci. Technol. **61**(12), 2544–2558 (2010)

Recommendation

Explicit Modelling of the Implicit Short Term User Preferences for Music Recommendation

Kartik Gupta[✉], Noveen Sachdeva, and Vikram Pudi

IIIT-H, Hyderabad, India
{kartik.gupta,noveen.sachdeva}@research.iiit.ac.in,
vikram@iiit.ac.in

Abstract. Recommender systems are a key component of music sharing platforms, which suggest musical recordings a user might like. People often have implicit preferences while listening to music, though these preferences might not always be the same while they listen to music at different times. For example, a user might be interested in listening to songs of only a particular artist at some time, and the same user might be interested in the top-rated songs of a genre at another time. In this paper we try to explicitly model the short term preferences of the user with the help of Last.fm tags of the songs the user has listened to. With a session defined as a period of activity surrounded by periods of inactivity, we introduce the concept of a subsession, which is that part of the session wherein the preference of the user does not change much. We assume the user preference might change within a session and a session might have multiple subsessions. We use our modelling of the user preferences to generate recommendations for the next song the user might listen to. Experiments on the user listening histories taken from Last.fm indicate that this approach beats the present methodologies in predicting the next recording a user might listen to.

Keywords: Recommendation systems · User modelling

1 Introduction

Today with an increase in the use of digital platforms like Last.fm [9] for music sharing which have large databases, it has become a common practice to use recommendation systems to suggest musical recordings (songs) a user might listen to next. The next section describes the various types of techniques used for music recommendation such as collaborative filtering [2,22], content based systems [1,4,6], sequence based systems [3,13,16,24]. In addition to this, hybrid systems [5,18,23] are discussed which combine two or more techniques to enhance the performance of the recommender systems.

We feel the above systems are not fully able to capture the short-term preferences of the users in terms of features and the features which a user might give

© Springer International Publishing AG, part of Springer Nature 2018
G. Pasi et al. (Eds.): ECIR 2018, LNCS 10772, pp. 333–344, 2018.
https://doi.org/10.1007/978-3-319-76941-7_25

importance to might be much more in number than what is considered by these systems. For example, if a user wants to check out all the songs of a particular artist then content-based systems using the acoustic features to recommend songs might not perform very well but the systems which consider the metadata of the song might do good. To fully capture the preference of the users in terms of the attributes of the songs, we use the Last.fm [9] tags of each song which can be retrieved using their public API. These tags are able to describe a large variety of features of the song which may or may not be derivable from the audio of the song. For example tags like 'Guitar Solo' might be derivable from the audio of the song but there are tags like 'heartbreak' which might not be derivable from the audio.

Often users have breaks in their listening activity and the periods of activity on music sharing platforms surrounded by periods of inactivity are called sessions. Often people might have different preferences during different sessions and recommendation based on the sessions where the user preference was different might be rendered useless. There are recommender systems which take into account the sessions of a user and recommend music based on the sessions which had the same songs as the current ongoing session [3,15]. In our work, we try to model the user preferences based on the tags of the songs that the user listens to. We introduce the concept of subsession which we define as the part of a session during which the preferences of the user remains constant. We then recommend a set of possible next songs user might listen to using based on the current subsession. The preference of the user is modelled by the Last.fm tags of the songs user has listened to in a subsession. Hence in this paper, we make the following contributions:

- We propose a model for capturing the short term preferences of the user by introducing the notion of a subsession.
- A simple recommendation system is built and tested using our model and we show that it indeed performs better than the current state of the art.
- We also test our model in predicting the immediate next set of preferences that the user might have and present the results for the same.

2   Related Works

We here present the previous works done to model the user preference for recommendation systems:

2.1   Collaborative Filtering

Collaborative filtering is a very widely used approach and recommends items to users based on the items selected by other similar users. It does not take into account the short-term preference of the user while recommending an item to the user. Similar is the item level collaborative filtering [2], wherein two items selected by the same user are considered to be similar.

Matrix factorization [22] brings about an improvement in collaborative filtering by finding out the latent user and item vectors and using them to find the probability if the user will like that item or not. Further improvements have been made to matrix factorization such as BPR [17] which is optimized for the task of ranking items for each user.

Session-based collaborative filtering [15] works by dividing the user activity into sessions and considering only songs of the active session to find similar sessions and recommend items to the user. This was further improved by another system by also taking into account the temporal context of the session and also apply topic modelling at the session level giving the role of a document to the session and songs acting as words [14].

In [12] the authors have tried to model the context of the user with the help of inputs from the user device such as the location, motion, calender etc. and apply collaborative filtering on the context level. Their approach assumes an implicit correlation between the external environment and the songs the user might like at that particular time. However, we feel this might not always be the case. Sometimes a user might want to listen to something totally different than what their external environment represents.

2.2   Content Based Reccomendation

Content-based filtering is gaining popularity in recommender systems. In these systems, the items are recommended on the basis of their content [1,4]. The content of a song usually considered are the audio features of the song such as the lyrics, instruments and the tempo of the song. Some of them might also use the metadata of the song such as the artist info or the tags associated with the song. These may or may not be machine derivable. If the content of a song is similar to the ones the user likes, then that song is recommended to the user. However, in most of the current systems, the parameters on which the interests of the user are calculated are limited and hence might not able to fully capture the interests of the user. For example, there are systems which recommend songs based on the melody of the song and hence assuming that melody is one of the features which a user might give importance to, however, this might not always be the case [7].

Liang [6] uses a very different approach and have trained the neural network to generate a latent factor vector for each song which is highly correlated to semantic tags and then used collaborative filtering to recommend songs to the user, however the neural network is trained for a small number of tags (561) which again might not be able to capture the complete preference of the user. Also, they recommend the songs based on the overall interest of the user rather than the short term preferences which our approach takes into account.

2.3   Sequence Based Recommendation

These recommender systems have started to gain popularity just recently and perform better than most of the non sequence based techniques such as collaborative filtering. Such systems base their recommendation on the sequence of items the user has already selected. Recommendation was first modeled as a sequence prediction problem by Brafman [13]. Initially, Markov chains were used to predict the recommendation and some of the improvements such as having a personal Markov chain model for each user [16]. Later on, with the popularity of neural networks, recurrent networks were successfully applied to next item prediction problem [24]. Sequential recommender systems have been extensively combined with other types of modelling to give out better results and we present some of them under the next subsection.

2.4 Hybrid Recommender Systems

There have been systems which have employed the techniques of content based as well as collaborative filtering based recommender systems. One such system was proposed by Yoshii [23] wherein they take into account the rating of the item as well as the content features. The content features are modelled using the polyphonic timbres in the song our approach is indeed a hybrid approach but in our case instead of taking into account the content of not only one item but the features of the songs recently heard by the user which might be important to the user at that point of time.

In [3] the authors combine the concept of sessions and sequence-based recommender systems. They apply sequence-based approach on the sessions rather than entire user history. We take this one step further and say that not even all items of the session are important and only last few items are important while making the prediction of the next item, and the number of few items on which the next item might depend is not constant.

In [8] Hariri applies topic modelling to the sequence of the songs heard by the user and tries to predict the topic of the next song the user might be interested in. They learn the transitions between different topics from the playlists. Though this approach seems similar to what we propose in this paper but is indeed different. Our model allows explicit specification of the point in the history to which the current preferences of the user correspond to and hence allowing us to model the next song prediction problem as an information retrieval problem. Also, our model does not take into account the sequence but the group of items which were heard together while the user preference did not change much and we claim that its the items of the group which matters but not the sequence in which the items occurred.

3   Method

3.1   User Preference and Subsession

In this section, we present definitions of the terms used and formalization of the problem statement that we tackle in our paper.

Definition 1. Session: *A session is a period of activity surrounded by periods of inactivity of at least x minutes, where x is the given threshold.*

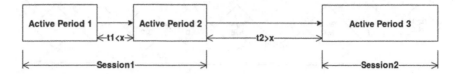

Fig. 1. Division of user activity into sessions

There is no constraint on the duration of a session, however the gap between the end and start of two different sessions should at least be x minutes for them to be considered as different sessions.

In Fig. 1 the user activity is divided into three periods. The time difference between the end of period 1 and start of period 2, t_1 is less than x minutes, and hence are considered to be the part of a single session, The time difference between the end of period 2 and the start of period 3, t_2 is more than x minutes and hence this marks a session boundary putting period 3 in a different session.

Definition 2. Preference Tag Set: *Considering each song as a transaction and each tag of that song as an item, the set of items with a minimum support p, in q consecutive transactions, is called as the Preference Tag Set and is denoted by Tags(q).*

Note that in the above definition *item* means a tag of a song, however in rest of the paper it means a song itself. We have presented the above definition for an easy understanding of the reader. In the above definition we model the preference of the user over the q consecutive songs listened by the user. The tags present in the preference tag set can be assumed to be related to the features of the items which matter most to the user. The tags which are associated to some of the items among the q items but not present in the preference tag set are assumed to be given less importance by the user and hence we can skip them while deciding the next items to recommend.

Definition 3. Subsession Seed: *A window w of l consecutive items in a user session with a preference tag set Tags(w), for a given fixed l.*

Definition 4. Subsession: *A window m ending with a given subsession seed w, where Tags(w) = Tags(m) and m and w occurring in the same session.*

Note that for all practical purposes, two consecutive subsessions with same preference tag sets would be merged together.

We describe the subsession mining process with the help of example given in Fig. 2. In the example, we keep the minimum support for a tag to be in the preference tag set to be 0.51 and the length of subsession seed as 5. A value of 0.51 as minimum support was chosen so that the tags associated to only the majority of songs can make it to the preference tag set. In the figure if we see the subsession 1, the seed will be the last 5 items for which the preference tag set will be $[A, B]$ as only these correspond to more than 51% of the items.

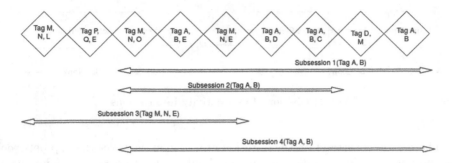

Fig. 2. Division of user activity into subsessions

Only the tags A and B occur more than or equal to 3 times and hence make it to the preference tag set of that subsession seed. To find the subsession with the seed of last 5 items, we increase the window to contain two more items towards the left(less recently heard songs) as the preference tag set for those 7 items and the preference tag set of the seed remain the same. We cannot expand it any further as the preference tag then changes. Note that a subsession seed always ends the subsession i.e.: A subsession will never include items which occurred after the subsession seed in the user history. For *Subsession*2 the seed is the 5 items before the last two items. If we try to increase the window towards less recently heard songs we see the preference tag set changes hence this subsession is equal to its subsession seed. Now since the preference tag set of *Subsession*1 and *Subsession*2 we combine them into *Subsession*4. If we consider *Subsession*3, the preference tag set $[M, N, E]$ for that is different and cannot be combined with *Subsession*4 and hence is kept separate. As is clear, we allow for overlapping subsessions.

This way all the subsession seeds are considered with a hop size of 2 and subsessions corresponding to them are found.

Below we present the problem statements we tackle in this paper.

Predicting Next Song: *Given the q most recently listened songs by the user, predict the top k candidate songs which the user is most likely to listen to after these q songs.*

Once we find an appropriate session and subsession we can recommend items which the user might like after the given items.

Predicting Next User Preference: *Given the m most recent subsessions of the user, predict the preference tag set the user can have during the next subsession*

3.2   Using Subsessions to Recommend Music

As we have defined subsessions in the previous section we here explain how we use subsession to find the top k candidates songs which the user might listen to next. Given q most recently heard songs heard by the user, we model the active subsession. Then we find the similar subsessions in the training data to the active subsession. The recommendation uses a n similar subsessions of the active subsession. While mining each subsession we also store the user from whose log the subsession is mined and we define that as the *Owner* of the subsession.

There are two matrices in our approach. One is the subsession-song matrix and the other one is the subsession-tag matrix. One of the axes of these matrices are the subsessions and the other axis are songs and tags respectively. The entries of these matrices are often called as ratings. The columns of these matrices tell what all songs were present in a subsession and what tags were present in the preference tag set of the subsession. The ratings in the subsession-song matrix are 1 if the song was present in the subsession and 0 otherwise. Similarly the ratings in the subsession-tag matrix are 1 if the tag was present in the preference tag set in that subsession and 0 if otherwise. Below we present the similarity measure Tanimoto coefficient [20] to find the similarity between two subsession-item vectors S_i and S_j

$$sim(S_i, S_j)_{item} = \frac{\sum_{k=1}^{t} r_{ik} \cdot r_{jk}}{||S_i|| + ||S_j|| - \sum_{k=1}^{t} r_{ik} \cdot r_{jk}} \tag{1}$$

where item corresponds to songs or tags and t is the total number of items, and r_{ik} denotes the rating of item k in subsession i

$$||S_i|| = \sum_{k=1}^{t} r_{ik} \cdot r_{ik} \tag{2}$$

We use the Tanimoto coefficient instead of the generally used cosine similarity because we wanted the total number of items which are present in only one set (either S_i or S_j) to decrease the similarity value which is not taken care of in the cosine similarity. Cosine similarity only accounts for the number of similar items in the two sets. For example: If two sets have 5 items each and 3 items are common, the cosine similarity would give us the similarity value 0.6 whereas our measure would give us a value of 0.428 which tells us that there is a large number of items which are not common in both the sets.

To calculate the similarity between any two given subsessions we find a weighted sum of similarity between subsession-tag interaction vector of the two subsessions, similarity between subsession-song interaction vector of the two subsessions and Kronecker delta function having $Owner_i$ and $Owner_j$ as inputs.

$$Sim(S_i, S_j) = w_1 * sim(S_i, S_j)_{tags} + w_2 * sim(S_i, S_j)_{songs} + w_3 * \delta_{Owner_i, Owner_j} \tag{3}$$

where

$$\delta_{a,b} = \{ 1, \text{ if } a = b, \ 0, \text{ if } a \neq b. \tag{4}$$

The reason we use the songs contained in a subsessions to calculate similarity is that for the same value of the preference tag set there might be a very large number of subsessions and hence the existence of common songs in two subsessions is used to increase the rank of subsession having common songs with the active subsession.

Once we have the n nearest neighbour subsessions we need to find the top k songs which occur in these subsessions. The score $Score_i^A$ for item i is then calculated to predict if it is likely to occur in the given active subsession A.

$$Score_i^A = \sum_{j=0}^{n} r_{ji}.Sim(S_A, S_j) \tag{5}$$

Based on the scores for each item obtained from the above equation, we choose top k candidates and recommend them to the user.

4   Experiments

In this paper, we have tackled two tasks. Section 4.1 presents some statistics about the dataset, and Sect. 4.2 talks about the mining of subsessions from the dataset. Section 4.3 presents the methodology and results for the next preference set prediction task. Section 4.4 onwards we present the baseline models and the results for our approach for the next song prediction task.

4.1   Dataset

The dataset was a subset taken from the Last.fm dataset [10]. The original dataset consisted of the entire listening history of about 1000 users until May, 5th 2009. Each log in the dataset consisted of user id, song name, artist name and time stamp. We performed experiments on a subset consisting of 6 month histories of all the users. The original dataset did not include tags for each song but we could retrieve it using the Last.fm public API. We only considered the users who had some activity in the period considered and the number of such users was 759. Below we present some dataset statistics (Table 1).

Table 1. Dataset and subsession statistics

Description	Value	Description	Value
Total Logs	3553321	Total Subsessions	498800
Total Users	759	Subsessions of len 5	294410
Total Sessions	110410	Subsessions of len 7	111694
Total Unique Songs	386046	Subsessions of len 9	43052
Average Songs Per Session	32.18	Subsessions of len 11 to 19	42471
Average logs per user	4681.58	Subsessions longer than 19	7173

4.2   Sessions and Subsessions

The entire log for each of the user was divided into sessions with the difference between the start and end of two sessions $x = 120$ min. The minimum length for a session was kept as 5 songs. If we found a session which had a length less than 5 songs we discarded it since it won't have any valid subsessions. Once we had all the sessions for each user they were chronologically sorted and 70% of the first occurring sessions were put in the train data and the last 30% of the sessions put into the test data. For the models which don't require the history to be divided into sessions, the session boundaries were discarded while training and testing.

For each session in the training set, all the valid subsessions were found keeping the length of subsession seeds to 5 and min support to 0.51 for preference tag set. To predict the next song, we find the n most similar subsessions to the active subsessions and provide k highest ranked songs to the user. For our experiments we found $n = 50$ and the weights for the similarity measure between two subsessions $w_1 = 0.25$, $w_2 = 0.5$, $w_3 = 0.25$ giving the best results.

4.3 Next Preference Set Prediction

As per our knowledge, this kind of task has never been taken up in the past and our modelling of the user preference allows us to be the first ones to do this. In this problem we treat each subsession as a basket of tags and given the latest sequence of baskets from the user history, this problem reduces to the next basket prediction task. We use the neural network model described in [19]. The number of subsessions while predicting the preference tag for the next subsession was kept as 5. For each tag, a distributed representation [21] was computed based on the tags appearing together in a preference tag set of subsessions in the training dataset. The neural network implemented predicts the tags which might occur in the preference tag set of the next subsession. We report $precision@k$ and $recallk@$ for this task.

Metric	$k = 5$	$k = 10$	$k = 15$	$k = 20$	$k = 25$
Precision (%)	76.67 ± 2.45	65.71 ± 2.14	57.45 ± 2.24	50.89 ± 1.98	45.69 ± 1.83
Recall (%)	28.11 ± 1.16	43.82 ± 1.57	54.11 ± 1.59	61.42 ± 2.08	66.87 ± 2.13

A high precision for low values of k shows that most of the tags predicted by the model were indeed present in the actual preference tag set and a high recall for higher values of k show that we were indeed able to predict of the tags of the preference tag set successfully.

4.4 Baseline Models

We use the following baseline models to compare with our model. For POP, BPR-MF, RNN we use the implementation provided by the authors of [11]. BPR-MF and RNNs were run 5 times each and we report the mean values for them. We refer to our model as SBRS (Subsession Based Recommender System).

1. POP: Users are recommended the most popular items in the training set.
2. BPR-MF: It is a matrix factorization based state of the art model which ranks items differently for each user [17]. The original implementation was used from MyMediaLite, using all the default values except the number of features which we kept 100 for our experiments as they gave the best results. The user histories in the training data were used to build the user profiles.
3. Session Based Collaborative Filtering (SSCF): In this method, the similar sessions are found in the training data and songs listened to in those sessions are ranked based on the number of similar sessions they occurred in and are then recommended to the user. For our experiments, we find the similar sessions based on the last 5 songs heard by the user and the number of similar sessions to consider were kept to 100 as they gave the best results.
4. RNN: In this method, the sequence of items occurring together are fed to a recurrent neural network and the target being the next item in the sequence. All the sequences in the training data are used to learn the model and to get the next recommendation, all the items heard by the user until that point are fed to the network. The Categorical Cross Entropy loss function was used with the number of hidden units 100, learning rate of 0.1. The implementation we used uses mini match stochastic gradient descent method and we kept the batch size to 20.

We built an *oracle system* which magically knows the most listened songs for each user from the user history in the test data and recommends them to the user each time. Note that it always recommends the same items to the user. This is an upper bound on the performance of the systems which always provide a constant recommendation to the user as most of the matrix factorization based methods do. The systems which take into account the recently listened songs by the user to generate recommendations can beat this oracle. In addition to the above baselines, we also compare our model with this oracle.

4.5   Testing

We iterate through entire user histories of the user in the test data to predict the next song given all the songs heard by the user till then in the test history. We report the Hit Ratio $HR@k$ [15] by varying k from 10 to 100 (Table 2).

Table 2. Results

Model	k = 10	k = 20	k = 30	k = 40	k = 50	k = 60	k = 70	k = 80	k = 90	k = 100
POP	0.85	0.97	1.24	1.69	2.14	2.31	2.58	2.83	2.98	3.08
BPR-MF	7.34	8.13	8.56	8.98	9.27	9.36	9.72	9.87	10.03	10.16
Oracle	13.22	19.48	23.95	27.56	30.48	**32.79**	**34.93**	**36.80**	**38.55**	**40.07**
SSCF	13.69	17.12	19.66	21.30	22.34	23.22	23.93	24.46	24.88	25.24
RNN	14.42	16.26	16.74	17.09	17.38	18.02	18.88	19.10	19.76	20.43
SBRS	**19.15**	**26.14**	**28.83**	**30.35**	**31.40**	32.19	32.81	33.32	33.77	34.16

5   Results and Conclusions

As is clear from the results table our model outperforms the baseline models by a huge margin. It even outperforms the oracle up to $k = 50$ which we think is an achievement and as is expected, the models which take limited user history into account while making recommendations perform better as they implicitly do consider the short term preference of the user, though only our model beats the oracle. The strength of our approach is that it models the recommendation task as a problem of retrieving same instances of user preferences in the training data and recommend items to the user based on those past instances. The results for the next user preference prediction task are encouraging enough to apply this kind of modelling to other domains as well to understand user behaviour and preferences.

This indeed looks like a promising and simple approach and can be applied to a diverse set of fields and tasks such as trends in user preferences and recommendation in other multimedia domains.

Acknowledgements. We would like to thank Alastair Porter for his valuable feedback on the initial drafts of this paper and improving the quality of this paper.

References

1. Oord, A.V.D., Dieleman, S., Schrauwen, B.: Deep content-based music recommendation. In: NIPS, pp. 2643–2651 (2013)
2. Sarwar, B., Karypis, G., Konstan, J., Riedl, J.: Item-based collaborative filtering recommendation algorithms. In: Proceedings of the 10th International Conference on World Wide Web, WWW 2001, New York, pp. 285–295 (2001)
3. Hidasi, B., Karatzoglou, A., Baltrunas, L., Tikk, D.: Session-based recommendations with recurrent neural networks. In: Proceedings of ICLR 2016 (2016)
4. McFee, B., Barrington, L., Lanckriet, G.: Learning content similarity for music recommendation. IEEE Trans. Audio Speech Lang. Process. **20**(8), 2207–2218 (2012)
5. Claypool, M., Gokhale, A., Miranda, T., Murnikov, P., Netes, D., Sartin, M.: Combining content-based and collaborative filters in an online newspaper. In: SIGIR 1999 Workshop on Recommender Systems: Algorithms and Evaluation, Berkeley, CA (1999)
6. Liang, D., Zhan, M., Ellis, D.P.: Content-aware collaborative music recommendation using pre-trained neural networks. In: Proceedings of the 16th International Society for Music Information Retrieval Conference, ISMIR 2015, Malaga, Spain, 26–30 October 2015
7. Kuo, F.F., Shan, M.K.: A personalized music filtering system based on melody style classification. In: 2002 IEEE International Conference on Data Mining, Proceedings, pp. 649–652 (2002)
8. Hariri, N., Mobasher, B., Burke, R.: Context-aware music recommendation based on latenttopic sequential patterns. In: Sixth ACM Conference on Recommender Systems, Dublin, pp. 131–138 (2012)
9. Last.fm: http://www.last.fm. Accessed 23 Apr 2017
10. Last.fm-dataset-1k-users. http://www.dtic.upf.edu/ocelma/MusicRecommendation Dataset/lastfm-1K.html. Accessed 23 Apr 2017

11. Devooght, R., Bersini, H.: Long and short-term recommendations with recurrent neural networks. In: Proceedings of the 25th Conference on User Modeling, Adaptation and Personalization, pp. 13–21 (2017)

12. Pagare, R., Naser, I., Pingale, V., Wathap, N.: Enhancing collaborative filtering in music recommender system by using context based approach

13. Brafman, R.I., Heckerman, D., Shani, G.: Recommendation as a stochastic sequential decision problem. In: ICAPS, pp. 164–173 (2003)

14. Dias, R., Fonseca, M.J.: Improving music recommendation in session-based collaborative filtering by using temporal context. In: International Conference on Tools with Artificial Intelligence, pp. 783–788. IEEE (2013)

15. Park, S.E., Lee, S., Lee, S.G.: Session-based collaborative filtering for predicting the next song. In: 2011 First ACIS/JNU International Conference on Computers, Networks, Systems and Industrial Engineering, pp. 353–358 (2011)

16. Rendle, S., Freudenthaler, C., Schmidt-Thieme, L.: Factorizing personalized Markov chains for next-basket recommendation. In: WWW, pp. 811–820. ACM (2010)

17. Rendle, S., Freudenthaler, C., Gantner, Z., Schmidt-Thieme, L.: BPR: Bayesian personalized ranking from implicit feedback. In: Proceedings of the Twenty-Fifth Conference on Uncertainty in Artificial Intelligence, pp. 452–461. AUAI Press (2009)

18. Suglia, A., Greco, C., Musto, C., de Gemmis, M., Lops, P., Semeraro, G.: A deep architecture for content-based recommendations exploiting recurrent neural networks. In: Proceedings of the 25th Conference on User Modeling, Adaptation and Personalization, pp. 202–211. ACM (2017)

19. Wan, S., Lan, Y., Wang, P., Guo, J., Xu, J., Cheng, X.: Next basket recommendation with neural networks. In: Poster Proceedings of RecSys 2015 (2015)

20. Tanimoto, T.: An elementary mathematical theory of classification and prediction. In: Internal IBM Technical report (1958)

21. Bengio, Y., Ducharme, R., Vincent, P., Jauvin, C.: A neural probabilistic language model. JMLR 3, 1137–1155 (2003)

22. Hu, Y., Koren, Y., Volinsky, C.: Collaborative filtering for implicit feedback datasets. In: ICDM, pp. 263–272 (2008)

23. Yoshii, K., Goto, M., Komatani, K., Ogata, T., Okuno, H.G.: Hybrid collaborative and content-based music recommendation using probabilistic model with latent user preferences. In: Proceedings of the International Conference on Music Information Retrieval (2006)

24. Zhang, Y., Dai, H., Xu, C., Feng, J., Wang, T., Bian, J., Wang, B., Liu, T.: Sequential click prediction for sponsored search with recurrent neural networks. In: AAAI, pp. 1369–1375 (2014)

Benefits of Using Symmetric Loss
in Recommender Systems

Gaurav Singh[1]($\boxtimes$) and Sandra Mitrović[2]

[1] UCL, London, UK
gaurav.singh.15@ucl.ac.uk
[2] KU Leuven, Leuven, Belgium
sandra.mitrovic@kuleuven.be

Abstract. The majority of online users do not engage actively with
what they are offered: they mostly use few items, give feedback on even
fewer. Additionally, in many cases, the only feedback available about
the item is positive feedback. These issues are well-known in the area
of personalized recommendation and there have been many attempts to
develop recommendation algorithms based on data consisting of only
positive feedback. Most such state-of-the-art recommendation methods
use convex loss functions, and either interpret non-interactivity with an
item as negative feedback or ignore such entries altogether, none of which
in principal reflects the reality. In this work, we provide reasons to motivate the usage of a non-convex loss in implicit feedback scenario to deal
with unlabelled data, and devise an algorithm to minimize the proposed
loss in collaborative setting. We analyse the effects of the proposed loss
both qualitatively and quantitatively on a benchmark public dataset.

1 Introduction

Recommendation systems have been an active area of research for more than a
decade. There have been a number of attempts - both recently and in the past -
to increase the accuracy and efficiency of recommendation systems in online web
services. The main challenge in achieving this goal resides in the fact that, in most
web applications, majority of users consume a very small number of items and/or
services and very few loyal users consume a large number of items. Therefore,
we have very little information about most of the users and, in addition, the
implicit feedback is restricted to positive class. It basically means that we only
get to know the items that the user liked based on implicit user actions. We
do not know anything about the items the user never interacted with. It would
obviously be much more suitable for our purpose to have negative feedback from
the user as well, since negative feedback can help us use the current state-of-the-
art algorithms that utilise user dislikes in addition to likes. However, it is not
always possible to collect negative feedback.

G. Singh was an intern at Yahoo at the time of the work.

© Springer International Publishing AG, part of Springer Nature 2018
G. Pasi et al. (Eds.): ECIR 2018, LNCS 10772, pp. 345–356, 2018.
https://doi.org/10.1007/978-3-319-76941-7_26

One of the widely used methods in recommendation systems is based on matrix completion techniques, and one of the approaches widely used for matrix completion is SVD [3]. In cases where both content and collaborative information is available, collective matrix factorization techniques perform better [15]. These methods work really well in situations where there is an abundance of positive and negative feedback, but as described above, in many situations it is not possible to obtain anything more than implicit positive feedback. In the past, some works discussed collaborative filtering in implicit feedback scenario [8,12].

Recently, some work has been done to solve the problems associated with implicit positive feedback and improve upon the area under curve (AUC). One such and also very popular method is BPR [13]. The drawback of the above mentioned recommendation methods is in assuming that the unknown feedback must be negative or lesser in comparison to liked items, expecting that low-rank approximation of the original sparse user-item matrix using a convex loss function will lead to predicting correct feedback. But, many of the unlabelled items could potentially be positive, and a *convex loss* function that is not bounded from above will bias the decision boundary into predicting such unlabelled items as negative.

In this paper, we take a semi-supervised approach motivated by positive-unlabelled classification to identify positive and negative feedback in a solely positive feedback data. In du Plessis *et al.* [5], they analyse classification for positive-unlabelled datasets and show that a non-convex bounded and symmetric[1] loss function like ramp loss is better than a convex loss function (e.g. hinge loss) for classifying in the absence of negative labels. It is due to the presence of symmetry in the ramp loss function that the expected loss decreases [5]. If a user prefers item j_1 compared to item j_2, then we can assume the presence of a decision boundary between the two items for a given user. Therefore, classification between different documents for a general classifier is analogous to classifying different items for a given user. The above mentioned decision boundaries can easily get biased by the presence of large number of unlabelled items. Some of these items are positive, and therefore, should ideally not be considered negative during model optimization, as shown in [5]. Hence, we believe that collaborative filtering from implicit feedback could benefit from a symmetric loss function.

Our contributions in this work can be summarized as follows:

- We provide reasons to motivate the use of a non-convex loss function well suited to implicit feedback scenario;
- We devise an algorithm to minimize the proposed non-convex loss function in a collaborative setting;
- We analyse the results to show that the proposed loss indeed leads to improved performance.

---

[1] Symmetric is defined as: $f(x) + f(-x) = const.$, like in [5].

2   Model

Our initial hypothesis is that, for certain number of users, we are given the information regarding their preference for different items. In other words, we assume that we are given an user-item matrix $X \in \mathbb{R}^{n \times d}$, where n is the number of users and d is the number of items. For example, in an on-line news service provider, we might assume that we record which user clicked on a particular news article. A standard way of performing collaborative filtering is matrix factorization, where the aim is to approximate original matrix with the product of two matrices of lower ranks, and optimizing an appropriate loss function.

One of the most popular convex loss function is the hinge loss. It leads to good generalization performance in classification, and has been extensively used in SVMs. It has also been used to perform max-margin matrix factorization [16] in the past. But, it was shown in du Plessis et al. [5] that a convex loss function is not suitable for positive-unlabelled classification. In du Plessis et al., the expected loss for positive-unlabelled classification is defined using a hinge loss as:

$$J_{PU-H}(g) = \underbrace{\pi \mathbb{E}_1[\ell_H(g(X))] + (1 - \pi)\mathbb{E}_{-1}[\ell_H(-g(X))]}_{\text{ordinary error term}}$$
$$+ \underbrace{\pi \mathbb{E}_1[\ell_H(g(X)) + \ell_H(-g(X))]}_{\text{superfluous penalty term}} - \pi$$

where $g(X) \in \mathbb{R}$ is continuous decision function, π is unknown class prior, $\mathbb{E}_y$ is expectation over class y, ℓ_H is the hinge loss and $J_{PU-H}(g)$ is the expected loss. If the loss is not symmetric, which means that if $\ell_H(g(X)) + \ell_H(-g(X)) \neq 1$ then $\pi \mathbb{E}_1[\ell_H(g(X)) + \ell_H(-g(X))] - \pi \neq 0$, and that adds a superfluous penalty term to the loss. It should be noted that no convex loss function would be symmetric, and therefore a non-convex loss is required. In case of ramp loss, $\ell_R(g(X)) + \ell_R(-g(X)) = 1$, where $\ell_R(z) = \frac{1}{2} \max(0, \min(2, 1 - z))$. Therefore, replacing the hinge loss (ℓ_H) with non-convex ramp loss (ℓ_R) leads to the canceling of the superfluous penalty term.

Motivated by the above analysis, we perform matrix factorization $X \approx UV^T$, but instead of defining loss as squared error ($\|X - UV^T\|^2$) or hinge loss, we define it as a ramp function computed per each user i and item j by:

$$\mathcal{R}_{ij} = \mathcal{R}(x_{ij}; \theta_{ij}) = \mathcal{H}_1(x_{ij}; \theta_{ij}) - \mathcal{H}_r(x_{ij}; \theta_{ij}) \tag{1}$$

where x_{ij} is the feedback stored in X matrix for user i and item j, $\theta_{ij} \overset{\text{def}}{=\!=} (u_{i.}, v_{j.})$, $u_{i.}$ is the row vector of matrix U corresponding to user i, $v_{j.}$ is the row vector of matrix V corresponding to item j, $\mathcal{H}_r$ is hinge loss function with the hinge located at r,

$$\mathcal{H}_r(x_{ij}; \theta_{ij}) = \max(0, r - (u_{i.} v_{j.}^T) x_{ij}) \tag{2}$$

such that r can take any negative value and $\mathcal{H}_1 = \mathcal{H}_{r|r=1}$. It is essential to include regularization on all learned parameters to avoid over-fitting, which leads to the following modified hinge loss:

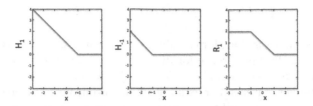

Fig. 1. Ramp loss (right) represented as a subtraction of hinge loss $\mathcal{H}_1$ (left) and $\mathcal{H}_{-1}$ (middle).

$$\mathcal{L}(x_{ij}; \theta_{ij}) = \mathcal{H}_r(x_{ij}; \theta_{ij}) + \lambda \left(\|u_{i.}\|^2 + \|v_{j.}\|^2 \right) \tag{3}$$

We use the above loss in the algorithm (see Algorithm 1) to ultimately minimize the ramp loss. We do that as minimizing ramp loss is not straightforward due to the gradient of $\mathcal{R}_{ij}$ vanishing whenever $\mathcal{R}_{ij} \geq 1 - r$ (see Fig. 1). Hence, we devise a detailed algorithm for the optimization task. The algorithm proceeds by progressively eliminating all the instances that have a $\mathcal{R}_{ij} \geq 1 - r$ (see line 9 and 14 of Algorithm 1). Due to its flatness, the ramp loss (i.e. upper bounded value) minimizes the effect of unlabelled instances by bounding the error contributed by an instance. It means that even though all unrated items are considered -1 during the max-margin matrix factorization, it is obvious that some of those unrated items would have been rated favourably given that the user noticed them. Therefore, it would be futile to classify those items as -1 during model optimization, and such outliers should be skipped during learning. We demonstrate that our algorithm can iteratively identify and eliminate such outliers using the ramp loss.

We use sample bootstrapping as mentioned in [13] to deal with large number of user-item pairs and alternatively sample $x_{ij} \in I_i^+$ (positive items for user i) and $x_{ij} \in I_i \backslash I_i^+$ from the data.

2.1 Algorithm Description

We begin by getting samples of user-item pairs (line 4), and set the learning rate α (line 6). We start iterating (line 7) and keep doing so until we do not have a significant number of items to be added to the skip list. We iterate over all the user-item pairs (line 8), and if the pair is not contained in the skip list ($\mathcal{T}$) then we perform gradient descent using the pair in lines 10–11. After iterating over all the user-item pairs in $\mathcal{S}$, we add all the pairs that contribute a loss $\mathcal{R}_{ij} \geq 1 - r$ to $\mathcal{T}$. We repeat the process till we do not have significant number of items to be added to $\mathcal{T}$ (measured in terms of δ) in an iteration. The algorithm gets terminated once we can not find any more user-item pairs to be added to $\mathcal{T}$. The algorithm achieves the optimization by removing elements that contribute more than the upper bound of the ramp loss.

Algorithm 1. Minimizing Ramp Loss

1: **procedure** LEARN U AND V
2: $\quad$ $\mathcal{T} = \{\}$
3: $\quad$ $L_0 = 0$
4: $\quad$ $\mathcal{S} = \text{getSamples}()$
5: $\quad$ $k = 1$
6: $\quad$ Learn Rate $(\alpha) = 0.05$
7: $\quad$ **while** $\delta > \epsilon$ **do**
8: $\quad\quad$ **for** $\forall x_{ij} \in S$ **do**
9: $\quad\quad\quad$ **if** $x_{ij} \notin \mathcal{T}$ **then**   #Skip instances that have $\mathcal{R}_{ij} \geq 1 - r$
10: $\quad\quad\quad\quad$ $u_{i.} \leftarrow u_{i.} - \alpha \frac{\partial \mathcal{L}(x_{ij}; \theta_{ij})}{\partial u_i}$ (Eq. 3)
11: $\quad\quad\quad\quad$ $v_{j.} \leftarrow v_{j.} - \alpha \frac{\partial \mathcal{L}(x_{ij}; \theta_{ij})}{\partial v_j}$ (Eq. 3)
12: $\quad\quad\quad$ **end if**
13: $\quad\quad$ **end for**
14: $\quad\quad$ **for** $\forall x_{i,j} \in \mathcal{S}$ **do**
15: $\quad\quad\quad$ **if** $x_{ij}(u_{i.} v_{j.}^T) < r$ **then**
16: $\quad\quad\quad\quad$ $\mathcal{T} := \mathcal{T} \cup x_{ij}$ #Record instances that have $\mathcal{R}_{ij} \geq 1 - r$
17: $\quad\quad\quad$ **end if**
18: $\quad\quad$ **end for**
19: $\quad\quad$ $L_k = |\mathcal{T}|$
20: $\quad\quad$ $\delta = \frac{L_k - L_{k-1}}{L_k}$
21: $\quad\quad$ $k = k + 1$
22: $\quad$ **end while**
23: **end procedure**

---

3 Experimentation

3.1 Datasets

We collected two datasets, one each from Yahoo news and Yahoo video portal. In addition, we use the benchmark MovieLens dataset [7]. As expected, the user engagement on both the collected datasets exhibits a long tail distribution (see Fig. 2).

– **News Dataset:** The news dataset consists of the set of the news articles displayed to the users along with their feedback. For the purpose of this study, we only use as content the headline's features of the article, as this is the piece of information viewed by the user on the front page leading her/him to click on the article or to skip it. For each article, we also record all the users that clicked on it during the three-month period. These users have clicked on total of around 48,527 news articles during three months. From these three months data, we randomly selected 120,000 users (original distribution), out of those 120K, we remove 50% of the entries (replaced with -1) for 20K users to create a test set. All the news items clicked by the user are rated as $+1$ (assuming that the user liked the headline, and clicked as a result), and all the other items are rated as -1.

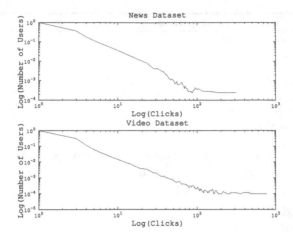

Fig. 2. Log of number of users versus log of click count for news dataset (up) and video dataset (down). The y-axis is normalized in the figure. It shows the power law distribution of the number of users that engage actively with the online service.

- **Video Dataset:** The video dataset is collected in similar fashion consisting of videos viewed by users and features of the videos. We collect as features the title and description of the video that a user reads before clicking it. A click on the video refers to an intentional play of the video. For these videos, we record all the users that clicked on it during a period of three months. These users have clicked on a total of around 116,926 videos during 3 month period. Similarly to previous dataset, $20K$ users are used to create a test set and all the videos clicked by the user are rated as +1 (and others as −1).
- **MovieLens Dataset:** We used the benchmark MovieLens 1M dataset that consists of 1M ratings on 3952 movies by 6040 users. We replace all the ratings ≥ 4 with 1, and ≤ 3 with −1 to create an implicit feedback dataset [11]. We randomly select 1000 users and uniformly sample 50% of ratings to be replaced with −1 for testing. We then evaluate the predicted ratings against the ground truth of ratings for these users.

3.2   Evaluation Metrics

We use four different well-known evaluation measures: Mean Average Precision (MAP), Normalized Discounted Cumulative Gain (NDCG) [1], Misclassification Rate (MCR) and (ROC) AUC.

3.3   Evaluation Settings

We tune all the regularization parameters on an independent dataset. We tune the parameters for the optimum value of NDCG. We create the test set using a subset of users, and replacing 50% of their ratings positive ratings with −1 (all unrated items are labelled −1). We evaluate all the methods over identical train and test sets to ensure a fair comparison.

Table 1. Results for different methods in recommendation task. The statistically significant winner is shown in bold using a t-test with p-value < 0.05. The values of MAP are lower for video and news dataset as the number of items are extremely large compared to movielens dataset.

	News			Video			MovieLens		
	Ramp	BPR	Hinge	Ramp	BPR	Hinge	Ramp	BPR	Hinge
MAP	**4.379e−4**	4.295e−4	4.247e−4	1.384e−3	**1.407e−3**	1.383e−3	0.0356	0.0359	0.0353
NDCG	**0.1741**	0.1702	0.1699	0.3719	0.3732	0.3716	**0.5429**	0.5340	0.5315
MCR	**0.1088**	0.3292	0.1111	**0.1413**	0.3526	0.1491	**0.0648**	0.4420	0.0688
AUC	0.8656	0.8082	0.8636	**0.8913**	0.7964	0.8869	0.7435	**0.7674**	0.7579

3.4   Results

We present and compare performances of mentioned evaluation measures using three different loss functions: ramp loss, Bayesian Pairwise Ranking (BPR) [13] and hinge loss [4,16]. We can see that under purely collaborative setting, the proposed approach leads to 2–5% improvement in the different metrics compared to hinge loss (see Table 1). It performs better in terms of NDCG, MAP and AUC compared to hinge loss in two of the three datasets. These results show that the flatness of the loss function in the form of ramp leads to better performance. Surprisingly, the approach manages to perform much better than BPR in terms of AUC on two datasets, even though BPR remains competitive on all other metrics. It should be noted that BPR essentially optimizes for AUC, while the ramp and hinge loss optimize for misclassification. Inspite of that we remain competitive over AUC as well. We should note that the proposed approach does not require an arbitrary value of k for recommending top-k items to a user, since all the items that are contrary to users' likes are predicted as negative. It can avoid recommending disliked items to users just because the system has run out of good items to recommend. The recommendation system does not have to recommend items with negative predicted value. Additionally, it can be observed that all the approaches are stable across different number of latent dimensions k, $50 \leq k \leq 500$ (see Fig. 3).

We empirically observed that the proposed algorithm takes approximately 2-3x the time required for minimizing hinge loss.

4   Analysis

We analysed the impact of applying ramp loss on MovieLens dataset. Among the three datasets considered, we opted specifically for this one for two reasons. First, MovieLens is a publicly available dataset, which makes our results easily verifiable. Second, this dataset contains additional information (genre) for each item (movie) which together with quantitative, enables us also providing an additional qualitative analysis of obtained results. As already mentioned, we randomly sample 1000 users and 50% of their 1 ratings which we then replaced

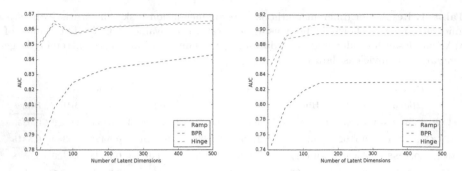

Fig. 3. AUC vs. k (Latent Dimensions) for News (top) and Video (bottom) dataset. We observe that the results are stable over a wide range. We obtained similar results for MovieLens dataset, and do not present the results to avoid redundancy.

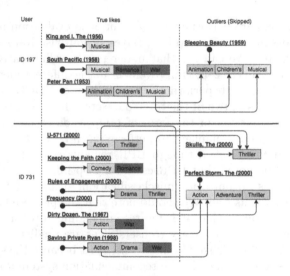

Fig. 4. Graphical illustration of true likes and outliers identified by our algorithm, for two different users. Outliers are the ones which we have masked as unrated (-1) in the test dataset, but which algorithm detected as those that should be liked ($+1$) based on user's previous preferences.

with -1. In other words, we mask positive ratings (likes) generating a set of outliers and we then test whether these outliers will be skipped by our method (as they preferably should be).

Quantitative analysis. It is important to mention that not only the total number of possible ratings in MovieLens dataset is huge (above 24 million), but also, only a minority of these ratings are positive (that is, do not equal to -1). This is due to the fact that 1) dataset initially consisted of only 1 million non-zero entries, 2) we replaced all ratings ≥ 4 with $+1$ and ≤ 3 with -1 and, 3) we created a set of outliers for testing purposes by masking 50% of random entries of 1000 randomly selected users. More precisely, we artificially created 47624 outliers, hence only 1.21% of total ratings. Our aim was to verify whether our algorithm will (and if so, to which extent) recognize these outliers. Our method skipped a total of 53102 ratings, which means that the percentage skipped was 1.34%. As explained before, element is skipped if it contributes too much to the loss, which might indicate that it is an outlier. The total number of outliers skipped by our algorithm, based on ramp loss, was 9282. Therefore, the recall obtained is 19.49%. Although we cannot calculate exact precision value due to the fact that the given data represent only partial view of reality, we can calculate lower bounds on this measure. Hence, we can claim that precision obtained is at least 17.47% (since there could be more outliers than what we added by artificially removing ratings from the dataset). It is worth emphasizing that despite the fact that the total size of dataset is huge and that, on the contrary, compared to it, both outliers and skipped elements sets are fairly small, the amount of overlap between these two sets is very impressive (almost 20%). Based on this, we can claim that ramp loss indeed helps in implicitly eliminating outliers in an implicit feedback data.

Qualitative analysis. Due to the fact that MovieLens dataset contains information on each movie name and genre, we are in a position to analyse the skipped items of our algorithm as compared to ground truth, from a qualitative perspective as well. It is worth mentioning that one movie in MovieLens dataset could be (and usually is) categorised in several genres (which we will further refer to as genre combination, unless movie is classified into only one genre, in which case we will refer to it as genre only). An item is skipped (i.e. recognised as an outlier) by the algorithm, if the algorithm, based on other user preferences, identified that the item rating was masked (set to -1) and that instead of being unrated, should have been positively rated. Identifying outliers is crucial since treating such instances as unrated (-1) would cause grief to the learned model and lead to bad decision boundary. For illustration purposes, we present results for two different users (see Fig. 4). Justifiability of algorithm's choice on outliers can be assessed based on the matches between genres/genre combinations of liked and skipped movies. For example, in the case of user #197, skipped item is of the same genre combination as one of the movies that the user rated positively. In the case of user #731, we can see that suggested outliers indeed go in line with user's mostly liked genres (and genre combinations), as user obviously likes action and thriller movies.

5   Related Works

Recommender systems have been an area of active research for over a decade. In cases where there is availability of both content and collaborative information, collective matrix factorization (CMF) techniques have shown to perform better [2,15]. A collective matrix factorization technique uses a given item-feature relation to factorize the user-item relation and vice-versa, it can be interpreted as hybrid of both content and collaborative filtering. It has been shown that CMF performs better than purely collaborative or content based approaches for recommendation [15]. More recently, [2] presented a convex formulation of the collective matrix factorization approach.

Unfortunately, in many situations we only have collaborative information in the form of implicit user feedback. Implicit feedback refers to the feedback obtained from implicit user actions like, viewing a video/clicking on a news article. It would be beneficial to obtain explicit positive and negative feedback from the user, but in many situations that is not possible. In fact, in some scenarios the only feedback obtained from the user is in the form of implicit actions. On top of that, the obtained feedback is restricted to be positive due to the nature of user actions. Hence, it is referred to as the problem of recommendation using implicit feedback. Additionally, we assume the absence of feature representation for the items during this work. We assume the presence of only collaborative information regarding the use of different videos/movies/news by each users. Hence, we discuss related works focusing on the problem of collaborative filtering using implicit feedback.

One of the approaches widely used for recommendations is based on matrix completion using singular value decomposition (SVD). It is based on decomposition of a given matrix $X \approx U\Sigma V$ where $X \in \mathbb{R}^{n \times m}$, $U \in \mathbb{R}^{n \times k}$, $V \in \mathbb{R}^{k \times m}$ and Σ is the diagonal matrix of singular values, usually $k \ll min(n, m)$. Recently, in [3], the authors use SVD for matrix completion. The orthogonality constraints can lead to overfitting of latent space in some cases. For example, topical distribution of documents does not need to be orthogonal as a document can belong to different topics. In the recent past, a number of publications have focused their attention on low rank matrix factorization based on NMF [9,17]. [6,10] discuss online versions of NMF for large scale and streaming data.

In the past, there have been works that focused specifically on the problem of collaborative filtering in implicit feedback scenario. One of the first works to discuss the problem in its current form was [8]. It presented the problem in the form of factorizing a given matrix using a squared loss over the actual and predicted value of an entry. It weighs each entry with a confidence parameter in the loss function based on the learned belief in the correctness of the rating. Initially, it gives low confidence to the unseen items, and higher confidence to the items liked by the user. Pan *et al.* [12] deal with the problem of collaborative filtering in implicit feedback as one of One-Class Collaborative Filtering. They rightly argue that since the implicit feedback is extremely sparse, it resembles the class imbalance problem in binary classification. They propose using sampling techniques to deal with class imbalance problem as in the case of imbalanced one-class classification. They also suggest using weighted optimization to give lower weights to unknown or missing entries. We should mention that the idea of learning confidence in different entries, and weighing those entries with that confidence resembles our idea of a bounded symmetric loss. But, it is not motivated by theoretical reasoning, and involves parameters that can vary over large range.

Rendle *et al.* [13] deal with the problem as bayesian pairwise ranking task. It defines an arbitrary probability of item i being more preferred than item j using a sigmoid function. It maximizes the log-likelihood for the data given the model, and learns pairwise ranking between the items. The log-likelihood is a convex function, and therefore can again lead to biased decision boundaries. In Schnabel *et al.* [14], the authors proposed a new evaluation measure for computing top-N recommendations from implicit positive feedback and a measure for optimizing the given evaluation metric. Schnabel *et al.* [14] presented the idea for max-margin formulation for matrix factorization as in the case of classification. A hinge loss works really well in case of binary classification task, and therefore has been used with success in matrix factorization. The loss generalises well to test data due to the presence of a margin between two classes, but in the presence of unlabelled data the convex loss might lead to biased decision boundary.

6 Reproducibility

The values of r (see Eq. (1)) that gave best results for news, video and MovieLens datasets were -1.0, -0.5 and -1.0 respectively. The codes used in the experiment are available at https://github.com/gauravsc/ecir2018.

7 Conclusion

The proposed approach draws parallels between positive-unlabelled classification and collaborative recommendations using implicit feedback. We motivate the use of a bounded non-convex loss function and devise an algorithm that minimizes the proposed non-convex loss in a collaborative setting. We present results that shows the effectiveness of the method. The suggested approach can be considered an implicit instance selection method that filters out points that could lead to wrong decision boundary. We perform extensive qualitative and quantitative analysis of the experimental results, and show the effectiveness of the proposed method in removing instances that can cause grief to the learned model. Our future work will focus on implicit instance selection methods for recommender systems. There has been little attention to instance selection in the context of collaborative filtering and collective matrix factorization, and we believe that it can lead to more accurate recommendation algorithms.

Acknowledgement. GS would like to acknowledge support from the FRE Program at Yahoo!

References

1. Baeza-Yates, R., Ribeiro-Neto, B., et al.: Modern Information Retrieval, vol. 463. ACM press, New York (1999)
2. Bouchard, G., Yin, D., Guo, S.: Convex collective matrix factorization. In: AIS-TATS, vol. 31, pp. 144–152 (2013)

3. Cai, J.-F., Candès, E.J., Shen, Z.: A singular value thresholding algorithm for matrix completion. SIAM J. Optim. **20**(4), 1956–1982 (2010)
4. DeCoste, D.: Collaborative prediction using ensembles of maximum margin matrix factorizations. In: Proceedings of the 23rd International Conference on Machine Learning, pp. 249–256. ACM (2006)
5. du Plessis, M.C., Niu, G., Sugiyama, M.: Analysis of learning from positive and unlabeled data. In: Advances in Neural Information Processing Systems, pp. 703–711 (2014)
6. Guan, N., Tao, D., Luo, Z., Yuan, B.: Online nonnegative matrix factorization with robust stochastic approximation. IEEE Trans. Neural Netw. Learn. Syst. **23**(7), 1087–1099 (2012)
7. Harper, F.M., Konstan, J.A.: The movielens datasets: history and context. ACM Trans. Interact. Intell. Syst. **5**(4), 19:1–19:19 (2015)
8. Hu, Y., Koren, Y., Volinsky, C.: Collaborative filtering for implicit feedback datasets. In: 2008 Eighth IEEE International Conference on Data Mining, pp. 263–272. IEEE (2008)
9. Lee, D.D., Seung, H.S.: Algorithms for non-negative matrix factorization. In: Advances in Neural Information Processing Systems, pp. 556–562 (2000)
10. Lefevre, A., Bach, F., Févotte, C.: Online algorithms for nonnegative matrix factorization with the Itakura-Saito divergence. In: 2011 IEEE Workshop on Applications of Signal Processing to Audio and Acoustics (WASPAA), pp. 313–316. IEEE (2011)
11. Lim, D., McAuley, J., Lanckriet, G.: Top-n recommendation with missing implicit feedback. In: Proceedings of the 9th ACM Conference on Recommender Systems, pp. 309–312. ACM (2015)
12. Pan, R., Zhou, Y., Cao, B., Liu, N.N., Lukose, R., Scholz, M., Yang, Q.: One-class collaborative filtering. In: 2008 Eighth IEEE International Conference on Data Mining, pp. 502–511. IEEE (2008)
13. Rendle, S., Freudenthaler, C., Gantner, Z., Schmidt-Thieme, L.: BPR: Bayesian personalized ranking from implicit feedback. In Proceedings of the Twenty-Fifth Conference on Uncertainty in Artificial Intelligence, pp. 452–461. AUAI Press (2009)
14. Schnabel, T., Swaminathan, A., Singh, A., Chandak, N., Joachims, T.: Recommendations as treatments: debiasing learning and evaluation. arXiv preprint arXiv:1602.05352 (2016)
15. Singh, A.P., Gordon, G.J.: Relational learning via collective matrix factorization. In: Proceedings of the 14th ACM SIGKDD International Conference on Knowledge Discovery and Data Mining, KDD 2008, pp. 650–658. ACM, New York (2008)
16. Srebro, N., Rennie, J., Jaakkola, T.S.: Maximum-margin matrix factorization. In: Advances in Neural Information Processing Systems, pp. 1329–1336 (2004)
17. Xu, W., Liu, X., Gong, Y.: Document clustering based on non-negative matrix factorization. In: Proceedings of the 26th Annual International ACM SIGIR Conference on Research and Development in Information Retrieval, pp. 267–273. ACM (2003)

Time-Aware Novelty Metrics
for Recommender Systems

Pablo Sánchez(✉) and Alejandro Bellogín

Universidad Autónoma de Madrid, Madrid, Spain
pablo.sanchezp@estudiante.uam.es, alejandro.bellogin@uam.es

Abstract. Time-aware recommender systems is an active research area where the temporal dimension is considered to improve the effectiveness of the recommendations. Even though performance evaluation is dominated by accuracy-related metrics – such as precision or NDCG –, other properties of the recommended items like their novelty and diversity have attracted attention in recent years, where several metrics have been defined with this goal in mind. However, it is unclear how suitable these metrics are to measure novelty or diversity in temporal contexts. In this paper, we propose a formulation to capture the time-aware novelty (or *freshness*) of the recommendation lists, according to different time models of the items. Hence, we provide a measure to account for how much a system is promoting fresh items in its recommendations. We show that time-aware recommenders tend to provide more fresh items, although this is not always the case, depending on statistical biases and patterns inherent to the data. Our results, nonetheless, indicate that the proposed formulation can be used to extend the knowledge about what items are being suggested by any recommendation technique aiming to exploit temporal contexts.

1 Introduction

Recommender Systems (RS) have become necessary applications in a large number of companies that offer personalized content to users. The ability to not only get useful and relevant recommendations, but also to offer novel and diverse items, can increase the number of users of a system and generate more benefits for companies. However, even though most recommenders are optimized to make accurate recommendations, nowadays it is recognized that other evaluation dimensions besides accuracy should be considered to properly model the user needs and understand why she wants a recommendation [1]; serendipity, novelty, and diversity, among others, are some of the criteria that are starting to get attention in the RS community beyond accuracy metrics [2].

At the same time, recommendation techniques that consider at some point the context of the user – either social, geographical, or temporal – are becoming more and more popular. Time-aware RS, in particular, allow to consider a type of context that is very easy to capture and provides very useful information to the system, since it allows to discriminate at different levels: moment of the day,

© Springer International Publishing AG, part of Springer Nature 2018
G. Pasi et al. (Eds.): ECIR 2018, LNCS 10772, pp. 357–370, 2018.
https://doi.org/10.1007/978-3-319-76941-7_27

day of the week, seasonality, etc. [3]. This type of systems have gained even more attraction in the past years since the Netflix prize, where modeling the temporal bias of users and items when rating movies was decisive to improve the performance of the algorithms [4].

However, the evaluation of RS has overlooked this important dimension, and most of the work has focused on how different evaluation methodologies (how the data partitioning should be made and which items should be considered when generating the rankings) should be applied to the recommendation data [3], leaving aside the definition of evaluation metrics specifically tailored to the problem of time-aware recommendation.

Therefore, in this paper we address the problem of formulating a time-aware evaluation metric based on the concept of novelty. To achieve this, we first extend the novelty and diversity framework presented in [5] so it can work with temporal data; we then propose different measurements to capture how novel a recommendation list is with respect to some item time model (in other terms, how *fresh* the recommended items are by considering what makes an item *new* or *old* according to a specific item time model). Finally, we validate such measurements using well-known recommendation algorithms – both time-aware and time-agnostic – on three real-world datasets.

2 Background

2.1 Recommender Systems

One of the earliest and most popular collaborative filtering approaches is the neighborhood-based recommender, either user-based (UB) or item-based (IB). These approaches are normally represented as an aggregation function of the ratings from the k most similar users or items [6]. On the other hand, matrix factorization (MF) algorithms, or model-based techniques, are often the preferred methods due to their superior accuracy in some domains and datasets [7]. These models try to explain the ratings by characterizing both users and items as k latent factors derived from the original rating matrix.

As an extension of these classical algorithms, context-based RS aim to exploit additional signals such as the time of the day, the user's social connections, the user mood, or whether the user is alone or in a group [3]. Among these signals, the temporal dimension of the interaction – e.g., a rating – is the easiest to capture and exploit. Because of this, several approaches have been proposed in terms of algorithms, as in [8] where the author exploits temporal biases and incorporates them into an MF algorithm, or in [9] where a temporal decay weight is introduced in the IB formulation so recent information is deemed more important.

2.2 Novelty Metrics for Recommender Systems

Originally, RS were evaluated using error metrics such as MAE or RMSE, largely due to the Netflix prize [4], where the objective was to decrease the RMSE error a 10% with respect to the Netflix algorithm. However, this type of evaluation method has now become outdated because it does not matches the user experience [1]. As a consequence, Information Retrieval (IR) metrics such as precision or recall are used to measure the effectiveness of the rankings generated by recommendation algorithms.

Even though these evaluation metrics are closer to the user experience, by optimizing only this type of metrics we ignore other concepts important for the interaction with the system such as the discovery of new or surprising items [10]. Because of this, different ways to define novelty and diversity have emerged in the last years. In particular, in [5] the authors generalized novelty and diversity metrics as follows:

$$m(R_u \mid \theta) = C \sum_{i_n \in R_u} \text{disc}(n)p(rel \mid i_n, u)\text{nov}(i_n \mid \theta), \qquad (1)$$

where R_u denotes the items recommended to user u, θ stands for a generic contextual variable – e.g., the user profile or the ranking R_u-, C is a normalizing constant, $\text{disc}(n)$ represents a discount function, the term $p(rel \mid i_n, u)$ introduces relevance in the definition of the metric, and $\text{nov}(i \mid \theta)$ is an item novelty model. For instance, by taking $\text{nov}(i \mid \theta) = 1 - p(seen \mid i)$ (the complement of the probability that the item was seen, i.e., its popularity) this formulation leads to the Expected Popularity Complement (EPC) novelty metric.

3   Time-Aware Novelty Metrics

Our proposed time-aware novelty metrics extend the traditional novelty metrics for RS by integrating the time dimension of user-item interactions with the system. In this section, we explain how we integrate these metrics within the framework presented in [5] (Sect. 3.2) and the different time models derived for items (Sect. 3.1).

3.1   Modeling Time Profiles for Items

In order to extend novelty metrics to consider temporal aspects of the items, we first need to represent and model their temporal dimension. From now on, we shall call these item time models as item temporal representations or item profiles.

Our goal is, for each item i, to produce an item profile $\langle t_1(i), \cdots, t_n(i) \rangle$ according to the time model t. In general, this representation will use the metadata available in the system about the item, such as its creation date, or its release time, modification time, or inclusion in the system catalog (which is not necessarily the same as the creation time, as in music databases). In fact, an item profile could include all these different times in the same representation, since an aggregation function will be later used to summarize it into a single value (see next subsection). An example of this item profiling is used in [11], where the authors exploit the release dates of music songs.

Nonetheless, when dealing with collaborative datasets, another source of information becomes available. It is possible to define the temporal profile of an item as the instants when a user interacted with it in the system. Even though the most typical interaction type in the literature are ratings, this definition is easily extended to any other interaction between users and items, such as comments, reviews, clicks, purchases, or even impressions outside of the system (such as system mentions in social networks). In this way, the item profile would represent any interaction patterns across these different information sources.

Obviously, these time profiles will model the item in a different way. Whereas the metadata-based item representations are based on some objective, static information (such as the release date), an interaction-based representation produces dynamic profiles, which will change depending on when the profiling is performed. At the same time, these latter representations are probably more subjective in the sense that they depend on how users interact with the system, but, because of this, they allow for profiles more tailored to how the community is actually using a specific item in the system.

We can draw a parallellism with how documents are treated when building temporal query profiles. According to [12], documents in IR are annotated with a timestamp corresponding to the date the document was published; this would correspond to the first group of time models presented, those based on metadata. However, we could also annotate the documents according to when they are accessed – or retrieved – as a response to a query; this document profiling strategy would correspond to the second group of time models, those based on interactions.

3.2 Measuring Time-Aware Novelty Models for Items

As we introduced in Sect. 2.2, the framework presented in [5] allows to formalize and generalize many different diversity and novelty metrics under a common formulation, where item rank and relevance are seamlessly plugged in the model and, hence, considered in the final measurements.

The core idea in this framework is how to define the item novelty model $\text{nov}(i \mid \theta)$, where θ stands for a generic contextual variable. Section 2.2 shows how a novelty defined as the complement of popularity is formulated (EPC), where θ is actually ignored; moreover, when $\text{nov}(i \mid \theta) = \min_{j \in \theta} d(i, j)$ a distance-based measure – where θ represents the items recommended to user u, R_u – is modeled, similar to the intra-list diversity metric (ILD) [13].

In order to model a time-aware novelty metric, we propose to encode the time model of the items in the θ variable as follows:

$$\theta_t = \{\theta_t(i)\} = \{(i, \langle t_1(i), \cdots, t_n(i)\rangle)\}. \tag{2}$$

Here, $t_j(i)$ represents the j-th component of the temporal representation or profile of item i by a specific time model t, as it was described in Sect. 3.1. We propose to compute the item novelty model based on different statistics of each item temporal profile. We leave the analysis of actual time series measures similar to what is done for queries [12] – and a study of their applicability – for future work, since, in most cases, the item profiles are very sparse.

In this context, the most basic model would account the item novelty based on the first appearance of that item in the system (**FIN**, from *First Item Novelty*), that is, $\mathrm{nov}^F(i \mid \theta_t) \propto \min_{j \in \theta_t(i)} \theta_t(i)$; in this sense, an item is novel when looking at the system timeline (by default, starting with the first logged interaction, although this initial timestamp could be configured to have a later value) and onwards. On the other hand, we may also model the item novelty with respect to the point of view of the evaluation split, where an item is more novel if it is closer to the end of the training split, hence, closer to the test split (**LIN**, *Last Item Novelty*). Note that the LIN model does not necessarily simplify to the complement of FIN, since its actual value will depend on the time model being used. Additionally, we can also model the item novelty by computing the average (**AIN**) or median (**MIN**) of the item profile.

These item novelty models cannot be integrated like that in Eq. (1), since the aforementioned framework is based on probability models, and hence, these quantities need to be, at least, normalized to produce valid values of the novelty metric. At the moment, the range of the item novelty models is the same as the range of the item time models, which, in general, return timestamps. A simple normalization scheme would divide the output of the item novelty model by the maximum possible value of a specific time model, another possibility would be applying a min-max normalization. Other more complex schemes could be used, providing different relative comparisons between the normalized values, but we found that the min-max normalization produced good-enough results. Therefore, we can formalize the item novelty model as follows:

$$\mathrm{nov}^{f,n}(i \mid \theta_t) = n(f(\theta_t(i)), \theta_t), \tag{3}$$

where f is one of the previous functions LIN, FIN, AIN, or MIN, and n is a normalization function, either the simple normalization function defined as $n(x, s) = x/\max s$, or the min-max normalization formulated as $n(x, s) = (x - \min s)/(\max s - \min s)$.

In summary, we propose a family of time-aware item novelty models that depend on a specific item time model, an aggregation function that summarizes the item temporal profile into a single number, and a normalization function that allows a fair comparison among novelty values for different items. Then, these item novelty models would lead to time-aware novelty metrics when integrated with item rank and relevance models within the framework presented in Eq. (1).

4   Experimental Analysis

4.1   Experimental Settings

Datasets. The experiments have been performed using three datasets where temporal information is realistic: MovieLens20M (ML), MovieTweetings (MT), and Epinions (Ep). The first two datasets cover the movie domain – one contains the

Table 1. Statistics of the datasets used in the experiments.

Dataset	Users	Items	Ratings	Density	Scale	Date range
Ep (2-core)	22,556	15,196	75,533	0.022%	[1,5]	Jan 2001–Nov 2013
ML	138,493	26,744	20,000,263	0.540%	[0.5,5]	Jan 1995–Mar 2015
MT (5-core)	15,411	8,443	518,558	0.398%	[0,10]	Feb 2013–Apr 2017

ratings provided within the MovieLens[1] recommendation system and the other collects the IMDb[2] ratings made public by users on Twitter[3] as explained in [14]–, while the other dataset contains ratings to different types of products on Epinions[4]. In MT and Ep we have performed a so-called k-core subset, such that each of the remaining users and items have at least k ratings each. We used $k = 5$ for MT and $k = 2$ for Ep, where repetitions were also ignored before the k-core (removing the older interactions). Table 1 summarizes some statistics of these datasets. It should be noted that, since MT is a dataset that is constantly updated, we selected a specific snapshot of 600K ratings (before computing the 5-core) that can be obtained in this url[5]. Moreover, the Ep dataset is made available by the authors from [15]. Finally, ML is available in the GroupLens website.

Evaluation methodology. We have performed a temporal split in each dataset where 80% of the (oldest) ratings have been included in the training set and the rest in the test set. For each user in the test set, every item in the training set is considered by the recommender except those items the user has already rated (i.e., we follow the TrainItems methodology [3]). Furthermore, we use a relevance threshold during the computation of the evaluation metrics so that those items whose ratings are higher or equal than this threshold are considered as relevant. More specifically, for Ep and ML a threshold of 5 is considered, whereas for MT this threshold is 9. The evaluation metrics reported are precision (P) and normalized discounted cumulative gain (NDCG), as implemented by the RankSys library[6], both of them at a cutoff of 5. We also report the ratio of users that receive at least one recommendation (USC, from user space coverage) [2].

Algorithms. We report two non-personalized algorithms: one that produces random recommendations ("Rnd") and another based on item popularity ("Pop"). We also report two neighborhood-based RS: a user-based approach ("UB") and an item-based approach ("IB"). A competitive matrix factorization approach [7] is also included ("HKV") in the comparison. All of these methods can be found in the RankSys framework.

---

[1] https://movielens.org.
[2] http://www.imdb.com.
[3] https://twitter.com.
[4] http://www.epinions.com.
[5] SiDooms/MovieTweetings (7a1fae8d9f).
[6] http://ranksys.org.

We complement this pool of algorithms with one technique based on implicit information instead of explicit – ratings – ("BPR") [16] implemented in MyMediaLite [17], and two methods that exploit the temporal information in the dataset, to actually analyze whether the proposed metrics are sensitive to this behavior. The first one introduces an exponential time decay weight in a user-based nearest-neighbor algorithm ("TD"), mirroring the method proposed in [9] but for users instead of items; the second one combines Markov chains with similarity measures to capture the sequential behavior of the user ("Fossil") [15].

Furthermore, we also include two non-personalized recommenders that may evidence some biases in the dataset: one that recommends the items according to how they were included in the system (increasing item id, "IdAsc") or the other way around (decreasing item id, "IdDec"). And lastly, to have an idea of how far from the optimal performance (and which algorithms are closer to such skyline) we are, we have also implemented a recommender that directly returns the test set of each user ("SkyPerf"), and another that returns items according to their last interaction (optimizing, hence, the LIN novelty metric, named "SkyFresh").

For the sake of reproducibility, the optimal parameters for these recommenders, together with the source code of the proposed metrics, are available in the following Bitbucket repository: PabloSanchezP/TimeAwareNoveltyMetrics.

4.2 Performance Results

Tables 2, 3 and 4 show the performance results of the recommenders presented before on the three datasets described. Here we use the four variations of the time-aware novelty metric defined in Sect. 3.2 with and without a relevance component; moreover, the discount component ($\mathrm{disc}(n)$ in Eq. (1)) is ignored to simplify the results. Furthermore, in these results we use interaction-based item profiles – based on ratings – since they can be applied to all the datasets we are testing; metadata-based item profiles are presented and discussed later.

The first thing we notice in our results are the low values of accuracy-related metrics (P and NDCG). This is a well-known result in the RS literature, and it is mostly due to a high sparsity in the data, especially in the groundtruth [18].

Table 2. Performance comparison in the Epinions dataset. Best value per column is in bold, the second best value is marked with a ‡, and the third one with a †.

Algorithm	P	NDCG	USC	No relevance				Relevance			
				FIN	LIN	AIN	MIN	FIN	LIN	AIN	MIN
Rnd	0.0000	0.0001	100.0	0.3812	0.6391	0.4901	0.4753	0.0000	0.0000	0.0000	0.0000
IdAsc	0.0000	0.0000	100.0‡	0.2357	0.5083	0.3599	0.3401	0.0000	0.0000	0.0000	0.0000
IdDec	0.0000	0.0001	100.0†	0.3851	0.5790	0.4766	0.4728	0.0000	0.0000	0.0000	0.0000
Pop	0.0009‡	0.0012†	100.0	0.0788	0.7936	0.2670	0.2152	0.0003	0.0009‡	0.0006‡	0.0005‡
IB	0.0002	0.0005	49.7	0.4567†	0.6705	0.5505	0.5411	0.0001	0.0001	0.0001	0.0001
UB	0.0004	0.0007	49.7	0.3325	0.7625	0.4871	0.4601	0.0001	0.0004	0.0003	0.0003
TD	0.0004	0.0008	49.7	0.6000‡	0.9150‡	0.7365	0.7238	0.0003†	0.0004	0.0003	0.0003
HKV	0.0006	0.0018‡	50.6	0.2445	0.8808†	0.4366	0.3977	0.0002	0.0006	0.0004	0.0004
BPR	0.0007†	0.0011	50.6	0.1964	0.7917	0.3705	0.3362	0.0004‡	0.0007†	0.0005†	0.0005†
Fossil	0.0002	0.0004	31.1	0.2821	0.7806	0.4527	0.4200	0.0001	0.0001	0.0001	0.0001
SkyPerf	**0.1337**	**0.4441**	66.5	**0.6170**	0.8695	0.7286‡	0.7197‡	**0.2397**	**0.3416**	**0.2845**	**0.2807**
SkyFresh	0.0000	0.0000	100.0	0.4557	**0.9999**	0.6588†	0.5976†	0.0000	0.0000	0.0000	0.0000

Table 3. Performance comparison in the MovieLens dataset.

Algorithm	P	NDCG	USC	No relevance				Relevance			
				FIN	LIN	AIN	MIN	FIN	LIN	AIN	MIN
Rnd	0.0009	0.0010	100.0	0.5573	0.9834	0.6993	0.6711	0.0004	0.0009	0.0006	0.0006
IdAsc	0.0099	0.0162	100.0‡	0.0716	0.9991	0.3550	0.2437	0.0007	0.0099	0.0044	0.0038
IdDec	0.0000	0.0000	100.0†	**0.9995**	0.9995	**0.9995**	**0.9995**	0.0000	0.0000	0.0000	0.0000
Pop	0.1027‡	0.1110‡	100.0	0.0781	0.9999†	0.4361	0.3772	0.0080	0.1027‡	0.0455‡	0.0397‡
IB	0.0347	0.0414	17.8	0.3111	0.9998	0.6145	0.5878	0.0107	0.0347	0.0218	0.0212
UB	0.0498†	0.0618†	17.8	0.2431	0.9999	0.5835	0.5594	0.0117	0.0498†	0.0289	0.0277
TD	0.0420	0.0520	17.8	0.6108‡	0.9999†	0.7838‡	0.7710‡	0.0258†	0.0420	0.0331†	0.0327†
HKV	0.0498	0.0611	17.8	0.3068	0.9998	0.6122	0.5885	0.0143†	0.0498	0.0303	0.0292
BPR	0.0443	0.0532	17.8	0.3274	0.9998	0.6202	0.5958	0.0124	0.0443	0.0267	0.0256
Fossil	0.0297	0.0348	17.8	0.3085	0.9999	0.6116	0.5816	0.0086	0.0297	0.0181	0.0174
SkyPerf	**0.7094**	**0.8396**	99.7	0.6069†	0.9993	0.7764†	0.7618†	**0.4359**	**0.7116**	**0.5565**	**0.5472**
SkyFresh	0.0027	0.0027	100.0	0.4999	**1.0000**	0.7236	0.7026	0.0016	0.0027	0.0021	0.0020

Table 4. Performance comparison in the MovieTweetings dataset.

Algorithm	P	NDCG	USC	No relevance				Relevance			
				FIN	LIN	AIN	MIN	FIN	LIN	AIN	MIN
Rnd	0.0002	0.0003	100.0	0.1693	0.8473	0.4435	0.4086	0.0001	0.0002	0.0002	0.0002
IdAsc	0.0004	0.0003	100.0‡	0.1729	0.8873	0.5485	0.5938	0.0000	0.0004	0.0002	0.0002
IdDec	0.0005	0.0004	100.0†	**0.9628**	0.9800	**0.9688**	**0.9669**	0.0005	0.0005	0.0005	0.0005
Pop	0.0028	0.0023	100.0	0.1499	0.9921	0.2534	0.2074	0.0007	0.0028	0.0009	0.0008
IB	0.0082	0.0093	78.5	0.4753	0.9848	0.6002	0.5744	0.0055	0.0081	0.0065	0.0064
UB	0.0104	0.0120	78.5	0.4902	0.9951	0.5937	0.5657	0.0075	0.0104	0.0084	0.0082
TD	0.0264‡	0.0337‡	78.5	0.8487‡	0.9988‡	0.9298‡	0.9282‡	0.0239‡	0.0264‡	0.0253‡	0.0253‡
HKV	0.0150†	0.0190†	78.5	0.4131	0.9939	0.5935	0.5621	0.0093	0.0149†	0.0115†	0.0113†
BPR	0.0100	0.0121	78.5	0.4524	0.9895	0.5984	0.5690	0.0072	0.0100	0.0083	0.0082
Fossil	0.0129	0.0154	75.2	0.5186	0.9951†	0.6294	0.6051	0.0102†	0.0129	0.0111	0.0110
SkyPerf	**0.3468**	**0.5374**	81.6	0.4262	0.9686	0.6514	0.6289	**0.1528**	**0.4485**	**0.2775**	**0.2657**
SkyFresh	0.0037	0.0041	100.0	0.6715†	**1.0000**	0.8072†	0.7924†	0.0033	0.0037	0.0035	0.0035

Nonetheless, we should note that the evaluation methodology used in this paper is even more restrictive in this aspect, because the split is temporal, and, thus, there could exist users and items that appear only on training or on test. This has a strong impact on coverage (USC), especially for the CF algorithms.

From these results, however, we can infer that there is a strong popularity bias in two of the datasets (Ep and ML), since Pop is among the best recommendation techniques in terms of accuracy metrics. This is attributed to the temporal dynamics of the dataset, as evidenced in Fig. 1, where the most popular items continue increasing their number of interactions after the training-test splitting point, hence, they are relevant for many users in test.

Regarding the results about our proposed novelty metrics, we shall focus on those without the relevance component, since those values reflect more closely the formulations presented in Sect. 3, whereas those including relevance are distorted by the actual accuracy obtained by the algorithms. We first note that the two skyline algorithms (SkyPerf and SkyFresh) obtain the performance we expect: whereas SkyPerf is optimal in terms of accuracy and relevant novelty, SkyFresh is optimal in terms of the LIN metric. Sometimes, this method also obtains good results for other metrics, mostly because the novelty metrics are not independent

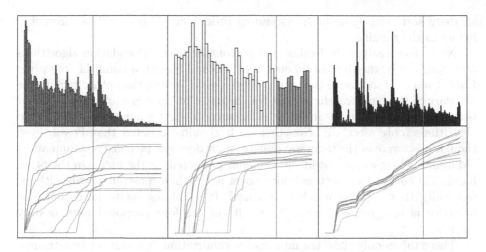

Fig. 1. Comparison of rating distributions for Ep (left), MT (center), and ML (right). Top row shows the amount of ratings created throughout time, bottom row shows the cumulative rating distribution of the top-10 most popular items. The vertical line indicates the training-test splitting point.

from each other. In this context, it might be surprising that SkyPerf does not obtain perfect values for every accuracy metric; this is because we are computing cutoff versions of the metrics, which together with the fact that we are using a relevance threshold, may leave some users without any relevant item in its pool; hence, in those cases, the metric would average 0's (because of those "empty" users in test) with other values up to 1. In any case, the reported results are the maximum achievable results, which helps us on finding the gap between observed and maximum accuracy.

We also observe that Rnd usually achieves high values of the novelty metrics (without relevance). In fact, this type of algorithm is, to some extent, showing the *a priori* probability of retrieving a novel item according to FIN, LIN, and so on. Because these values depend on the dataset and its inherent distributions (see Fig. 1), it makes sense the metrics return very different values for this technique. This result also evidences that such a simple technique should always be reported to properly assess the effect of inner biases in the data.

Related to the data biases, IdAsc and IdDec algorithms show that such simple techniques may produce large improvements in terms of novelty, but not due to an algorithmic rationale but because of how the data is built. In this way, we observe that in ML and MT, IdDec achieves very high values of our novelty metrics. This is because the ids of the items are provided in these datasets (this is not the case for Ep) and they were created sequentially, hence, by knowing that an id has a larger id than another, we know that the former is newer (more *temporally novel* or fresh) than the latter. Even though such an algorithm is not technically interesting, this issue may have a profound impact on how ties are broken when producing a ranking: depending on the dataset, if the tie-breaking strategy sorts the items in an ascending order, older items will be presented before in the ranking.

Now, if we analyze the freshness of personalized recommendation algorithms and, specifically, the time-aware ones (i.e., TD and Fossil), we find the following. First, the Fossil technique does not achieve better results than other techniques unaware of the temporal dimension; for instance, in Ep it is worse than UB but better than BPR, whereas in ML it is the second worst among the personalized algorithms, only after UB. Nonetheless, it should be noted that Fossil, even though it is aware of the temporal dimension, it does not penalize recommending old items as long as they fit in the sequence predicted for the user. On the other hand, TD consistently returns more novel items than most of the algorithms, especially than UB, on which it is based. It is interesting to note that this behavior is in agreement with the results of the four proposed item novelty models.

These tables only show the min-max normalization due to space constraints. We present these results because this normalization has better properties than the simple normalization mentioned in Sect. 3.2. In particular, we have observed it allows for more fine-grained discrimination of the reported values; for instance, LIN in EP for UB and HKV is, as seen in Table 2, 0.7625 and 0.8808, whereas when the simple normalization technique is applied this metric produces values as close as 0.9595 and 0.9797. This effect is more noticeable in some recommenders – e.g., Pop – when its recommended items present their initial interactions located near the beginning of the system timeline, which would produce a very small value with the min-max normalization, however, this value is still very large with the simple normalization; in fact, the obtained values in this situation (again, for E_p) with FIN are 0.0788 vs. 0.8430, respectively.

In summary, based on these results we conclude that the proposed metrics allow to discover whether some recommenders promote more temporal novel items than others; however, depending on the item novelty model some differences arise. For example, LIN always produces very high values, which does not help to discriminate between the recommenders. This might be also attributed to the popularity bias observed in Fig. 1, since this means that it is very easy for almost any algorithm to return novel items interacted recently, due to the observed bursting events near the splitting point, which would create – somewhat artificially – very popular fresh items. Regarding the FIN model, we believe it might not be very useful in datasets where several items appear at the very beginning, since it would not discriminate those cases, just like the LIN model. On the other hand, the two models that aggregate all the interactions received by an item (AIN and MIN) are more robust to these situations, and, in particular, the one based on the median (MIN) is expected to be more robust to outliers.

Finally, since the previous analysis was only done using the interaction-based item profiles, let us analyze now the metadata-based item profiles. Table 5 shows a comparison between a metadata-based item profile where the release date of the product (in our case, movies from ML and MT, because this information is

Table 5. Performance comparison in the MovieLens dataset (left) and MovieTweetings (right) when using metadata-based time profiles.

Algorithm	No relevance		Relevance		Algorithm	No relevance		Relevance	
	Y-*IN	R-FIN	Y-*IN	R-FIN		Y-*IN	R-FIN	Y-*IN	R-FIN
Rnd	0.7707	0.5573	0.0008	0.0004	Rnd	0.8764	0.1693	0.0002	0.0001
IdAsc	0.8387	0.0716	0.0083	0.0007	IdAsc	0.2264	0.1729	0.0001	0.0000
IdDec	0.7581	**0.9995**	0.0000	0.0000	IdDec	**0.9907**	**0.9628**	0.0005	0.0005
Pop	0.8227	0.0781	0.0848‡	0.0080	Pop	0.9693	0.1499	0.0027	0.0007
IB	0.8242	0.3111	0.0282	0.0107	IB	0.9629	0.4753	0.0079	0.0055
UB	0.8164	0.2431	0.0405†	0.0117	UB	0.9745†	0.4902	0.0102	0.0075
TD	**0.8822**	0.6108‡	0.0372	0.0258‡	TD	0.9817‡	0.8487‡	0.0260‡	0.0239‡
HKV	0.8102	0.3068	0.0397	0.0143†	HKV	0.9494	0.4131	0.0141†	0.0093
BPR	0.8354	0.3274	0.0364	0.0124	BPR	0.9621	0.4524	0.0097	0.0072
Fossil	0.8445†	0.3085	0.0251	0.0086	Fossil	0.9741	0.5186	0.0127	0.0102†
SkyPerf	0.8602‡	0.6069†	**0.6126**	**0.4359**	SkyPerf	0.9184	0.4262	**0.4122**	**0.1528**
SkyFresh	0.6305	0.4999	0.0018	0.0016	SkyFresh	0.9689	0.6715†	0.0036	0.0033

not available in Ep) is used as the time model. Since the release date (years) of an item is unique, any of the item novelty models would produce the same result, that is why we denote it in the table as Y-*IN. Additionally, we compare these results against the FIN model using ratings as interactions (R-FIN), because this model is conceptually the closest one to Y-*IN: a time representation using the release date would indicate the first possible interaction with such item.

Based on these results, we observe a different situation depending on the dataset. Indeed, whereas in ML the recommender that returns more novel items regarding their release date is TD, in MT is IdDec, an almost identical situation to what we found for the interaction-based profiles. The main rationale behind this is, as noted by the authors of the MT dataset in [14], this dataset is more dynamic and mostly contains new movies; this may explain why looking at the id provides both information about the interaction age but also about the item age in terms of its release date. It should be noted that, in contrast to the interaction-based profile models, we have not evaluated a recommendation technique that exploits the release date of the items, so there was no expected ranking of methods according to Y-*IN. In any case, it is interesting to observe that TD – a time-aware recommender including a time-decay factor – achieves very good results when item profiles are based on metadata, exactly what it showed for interaction-based time models.

5 Related Work

The concept of freshness has been widely used in IR, since most search engines aim at returning relevant but fresh (novel) documents. The authors in [19] propose a scoring method that combines freshness and relevance, and a similar idea is explored but using latent factors and learning-to-rank approaches in [20] based on click logs. This concept has received more attention in the field of Music IR [21], specifically to retrieve those songs a user may have forgotten or new releases from liked artists. In [22], freshness is defined as the strength of strangeness (of a specific song to a user) by applying the Forgotten Curve, which is modeled as a (negative) exponential function.

In recommendation, freshness or temporal novelty has usually referred to the capacity of the system to return new recommendations each time the user interacts with it [23]. In this way, in [24] the authors define temporal diversity and novelty metrics based on the overlap between previous recommendation lists. It is worth mentioning that, as described in [5], these temporal diversity and novelty metrics can be instantiated within the framework presented in Sect. 2.2, and, hence, they also fit the model presented herein. Finally, the only formulation we have found that is similar to one of the proposed time-aware novelty models was presented in [11] in the context of music recommendation. In that work, the authors define freshness as the average of the release date of the recommended songs, which is equivalent to our AIN model using a metadata-based profile. However, in that paper no time-aware RS were analyzed, so no relationships between the metric and the learning algorithm could be derived, as we have presented here.

6   Conclusion and Future Work

The temporal dimension has been mostly neglected when measuring novelty and diversity in RS, only few works mention it but in a context where the user receives repeating recommendations from the system. In this paper, we have introduced this important dimension in the definition of a family of novelty models that allows us to measure if a recommendation technique is prone to return fresh items or not. Our results show that, while the proposed metrics work as expected, they also open the possibility to be affected by some biases in the data that are not necessarily considered when measuring accuracy-based metrics – such as temporal bursts of activity and relationships between item age and their ids.

Our approach could open up new possibilities towards producing more time-aware novel recommendations, for instance, by reranking optimized rankings for accuracy based on the proposed item freshness models, as it is done for general diversity and novelty measures [10,24]. Furthermore, the proposed time-aware item novelty models and representations could be applied to streaming RS, where the temporal model of the recommender is not necessarily the same as the one for the metric: while the metric could be recomputed every day, the recommender could be trained every week, for example. In this way, a more fine-grained analysis may be derived so, for instance, the optimal period to train the recommender can be calculated, or a day-by-day sensitivity could be explored for different recommenders. It is worth mentioning that, although this is also possible to achieve with offline data, the conclusions will not be as significant because most of the datasets are very sparse, hence, it is necessary to use real, online data.

Acknowledgments. This research was supported by the Spanish Ministry of Economy, Industry and Competitiveness (TIN2016-80630-P).

References

1. McNee, S.M., Riedl, J., Konstan, J.A.: Being accurate is not enough: how accuracy metrics have hurt recommender systems. In: CHI, pp. 1097–1101. ACM (2006)
2. Gunawardana, A., Shani, G.: Evaluating recommender systems. In: Ricci, F., Rokach, L., Shapira, B. (eds.) Recommender Systems Handbook, pp. 265–308. Springer, Boston (2015). https://doi.org/10.1007/978-1-4899-7637-6_8
3. Campos, P.G., Díez, F., Cantador, I.: Time-aware recommender systems: a comprehensive survey and analysis of existing evaluation protocols. UMUAI **24**(1–2), 67–119 (2014)
4. Bell, R.M., Koren, Y.: Lessons from the netflix prize challenge. SIGKDD Explor. **9**(2), 75–79 (2007)
5. Vargas, S., Castells, P.: Rank and relevance in novelty and diversity metrics for recommender systems. In: RecSys, pp. 109–116. ACM (2011)
6. Ning, X., Desrosiers, C., Karypis, G.: A comprehensive survey of neighborhood-based recommendation methods. In: Ricci, F., Rokach, L., Shapira, B. (eds.) Recommender Systems Handbook, pp. 37–76. Springer, Boston (2015). https://doi.org/10.1007/978-1-4899-7637-6_2
7. Hu, Y., Koren, Y., Volinsky, C.: Collaborative filtering for implicit feedback datasets. In: ICDM, pp. 263–272. IEEE Computer Society (2008)
8. Koren, Y.: Collaborative filtering with temporal dynamics. CACM **53**(4), 89–97 (2010)
9. Ding, Y., Li, X.: Time weight collaborative filtering. In: CIKM, pp. 485–492. ACM (2005)
10. Castells, P., Hurley, N.J., Vargas, S.: Novelty and diversity in recommender systems. In: Ricci, F., Rokach, L., Shapira, B. (eds.) Recommender Systems Handbook, pp. 881–918. Springer, Boston (2015). https://doi.org/10.1007/978-1-4899-7637-6_26
11. Chou, S., Yang, Y., Lin, Y.: Evaluating music recommendation in a real-world setting: on data splitting and evaluation metrics. In: ICME, pp. 1–6. IEEE Computer Society (2015)
12. Jones, R., Diaz, F.: Temporal profiles of queries. ACM TOIS **25**(3), 14 (2007)
13. Ziegler, C., McNee, S.M., Konstan, J.A., Lausen, G.: Improving recommendation lists through topic diversification. In: WWW, pp. 22–32. ACM (2005)
14. Dooms, S., Bellogín, A., Pessemier, T.D., Martens, L.: A framework for dataset benchmarking and its application to a new movie rating dataset. ACM TIST **7**(3), 41:1–41:28 (2016)
15. He, R., McAuley, J.: Fusing similarity models with Markov chains for sparse sequential recommendation. In: ICDM, pp. 191–200. IEEE (2016)
16. Rendle, S., Freudenthaler, C., Gantner, Z., Schmidt-Thieme, L.: BPR: Bayesian personalized ranking from implicit feedback. In: UAI, pp. 452–461. AUAI Press (2009)
17. Gantner, Z., Rendle, S., Freudenthaler, C., Schmidt-Thieme, L.: MyMediaLite: a free recommender system library. In: RecSys, pp. 305–308. ACM (2011)
18. Bellogín, A., Castells, P., Cantador, I.: Statistical biases in information retrieval metrics for recommender systems. Inf. Retr. J. **20**(6), 606–634 (2017)
19. Sato, N., Uehara, M., Sakai, Y.: A case study on freshness based scoring for fresh information retrieval. In: ISCIT, vol. 1, pp. 210–215 (2004)
20. Wang, H., Dong, A., Li, L., Chang, Y., Gabrilovich, E.: Joint relevance and freshness learning from clickthroughs for news search. In: WWW, pp. 579–588. ACM (2012)

21. Knees, P., Schedl, M.: Music Similarity and Retrieval - An Introduction to Audio-and Web-based Strategies. The Information Retrieval Series, vol. 36, 1st edn. Springer, Heidelberg (2016). https://doi.org/10.1007/978-3-662-49722-7
22. Hu, Y., Ogihara, M.: Nextone player: a music recommendation system based on user behavior. In: ISMIR, pp. 103–108. University of Miami (2011)
23. McNee, S.M., Riedl, J., Konstan, J.A.: Making recommendations better: an analytic model for human-recommender interaction. In: CHI, pp. 1103–1108. ACM (2006)
24. Lathia, N., Hailes, S., Capra, L., Amatriain, X.: Temporal diversity in recommender systems. In: SIGIR, pp. 210–217. ACM (2010)

Employing Document Embeddings to Solve the "New Catalog" Problem in User Targeting, and Provide Explanations to the Users

Ludovico Boratto[1]([✉]), Salvatore Carta[2], Gianni Fenu[2], and Luca Piras[2]

[1] Data Science and Big Data Analytics, EURECAT,
Carrer de Bilbao, 72 (Edifici A), 08005 Barcelona, Spain
ludovico.boratto@acm.org
[2] Dipartimento di Matematica e Informatica, Università di Cagliari,
Via Ospedale 72, 09124 Cagliari, Italy
{salvatore,fenu,lucapiras}@unica.it

Abstract. In the current digital era, items that were consumed in a physical form are now available in online platforms that allow users to stream or buy them. However, not all of the items are available in digital form. When the companies that run these platforms acquire the rights to add a new catalog of items, the problem that arises is to identify who, among the customers, should be advertised with this new addition. Indeed, although the items may have existed for a long time, the preferences of the users for these items are not available. In this paper, we propose an approach that selects a set of users to target, to advertise a new catalog. In order to do so, we consider the textual description of these items and employ document embeddings (i.e., vector representations of a document) to model both the new catalog and the users. We also propose an approach to generate an explanation list to a user, represented by the top-n artists she evaluated that are most similar to the one of the new catalog. Experimental results show the effectiveness of both our targeting approach and of the explanation lists.

Keywords: User targeting · Document embeddings · Explanation

1 Introduction

Nowadays, it is common to have content that was usually consumed in physical formats (e.g., DVDs and magazines) in digital form. To supply this content, platforms that provide streaming services (Netflix, Spotify), sell copies of the digital items (iTunes), or do both (Amazon) have been developed. However, it is common not to find in these platforms everything that is available in a physical format[1]. When a digital platform adds new items after many years since its

---

[1] http://www.indiewire.com/2013/07/netflix-explains-why-it-doesnt-always-have-that-film-or-tv-show-you-really-want-to-see-video-166368/.

© Springer International Publishing AG, part of Springer Nature 2018
G. Pasi et al. (Eds.): ECIR 2018, LNCS 10772, pp. 371–382, 2018.
https://doi.org/10.1007/978-3-319-76941-7_28

original release, these additions become events that are highly advertised. The most famous example of such a scenario in the recent years is the addition of the whole discography of the Beatles on Spotify[2].

The problem that arises in this scenario is to identify the customers that can become the target of this new catalog and should receive personal notifications (e.g., via email), to advertise the new arrivals and stimulate them to buy or stream these new items. In this paper, we refer to this as the *new catalog problem*. Indeed, even if the items of the catalog have existed for a long time, no information about the preferences of the users for these items is available. This problem is similar to the "new item" problem handled by recommender systems (i.e., how to recommend to the users items with no or little amounts of ratings) [9]. However, our scenario is more complex with respect to dealing with a single item. Indeed, our problem setting has to characterize the multiple items that belong to the catalog and build a unique model that can be compared with the preferences of the users, in order to identify who should be targeted.

Recently, the vector representation of the words in a corpus (*word embedding*) has been largely employed. Word embeddings are built by considering as input the corpus, which leads to the building of a vocabulary and to the learning of the vector representation of the words. The state of the art is represented by *word2vec* [13], in which the vector representation of a word is learned thanks to a neural network. A variant, known as *doc2vec*, creates a vector for a whole document (*document embedding*) [10]. It is worth noting that other forms of embeddings exist, such as *Latent Dirichlet Allocation (LDA)* [1], *Latent Semantic Analysis (LSA)* [6], *Non-negative Matrix Factorization (NMF)* [5], and the so-called *manifold methods* (i.e., *Multidimensional scaling (MDS)* [8], *Locally Linear Embeddings (LLE)* [16], and *t-distributed Stochastic Neighbor Embeddings (t-NSE)* [11]). However, Manifold methods are usually employed to visualize high-dimensional data [4], and word2vec is known to outperform both LDA [3,4,12] and LSA [3,12,15,17].

In this paper, we tackle the problem of selecting the set of users to target, given a new catalog of items, by employing document embeddings generated with *doc2vec*. More specifically, we consider the content of the items in the new catalog in order to build a model of the catalog. We do the same for the items a user evaluated, in order to create a user model. The model of the catalog and that of a user are compared with a similarity metric, to evaluate if they match above a given threshold, and if the user should be part of the target or not. The choice to create a unique model both for the catalog and the user was made since our approach is meant to operate in large-scale online platforms; therefore, given the details of the items in the catalog, it should be able to reach the users to target in effective and efficient ways. An approach that compares the items in the catalog and those evaluated by a user one by one would not be appropriate for real-world scenarios like this.

To facilitate the understanding of our proposal and stay consistent with the previous example of the Beatles, we consider a new catalog as associated with a new artist (or director, if we switch from the music to the movie context). Since

---

[2] https://news.spotify.com/us/2015/12/23/finally/.

a targeted user might not be familiar with the artist she gets recommended, our approach provides a form of *explanation* of why the user was targeted. An explanation is a list of the top-n artists (directors) evaluated by the user that are most similar to the artist of the new catalog. It is important to stress that the targeting approach does not consider the artists a user evaluated, but considers her preferences as a whole in a model. By providing an explanation that explores the preferences of the users in terms of other similar artists, the list also serves as a validation of our model-based approach; indeed, if the artists evaluated by a user are highly similar to the one she gets recommended, our models are able to characterize both the catalog and the user.

The contributions of the paper are now summarized:

- we employ document embeddings to model both the characteristics of a new catalog of items and the preferences of the users;
- we present an approach that targets users in presence of a new catalog;
- we develop an approach to provide to each user in the target an explanation of why she was recommended the new catalog;
- we perform experiments on a real-world dataset that show the effectiveness of both the targeting and the explanation lists.

The rest of the paper is structured as follows: Sect. 2 presents our proposal, Sect. 3 the experimental framework, while Sect. 4 contains the conclusions.

2   User Targeting with Explanations

Our approach works in five steps:

1. **Document Embedding Extraction**: processing of the description of the items, to extract the document embeddings;
2. **New Catalog Modeling**: creation of a model built by considering the embeddings of the items in the new catalog;
3. **User Modeling**: creation of a model built by considering the embeddings of the items evaluated by a user;
4. **User Targeting**: selection of the users whose model is similar to that of the new catalog;
5. **Explanation List Generation**: selection of the top-n most similar artists to that associated to the new catalog.

In the following, we will describe in detail how each step works.

2.1   Document Embedding Extraction

To generate a vector representation of an item, doc2vec builds a model, by considering the item descriptions. A set of parameters is required as input:

- the layer size (Z) is the number of dimensions in the resulting vectors;
- *window* is the maximum skip length between words;

- *sample* sets the threshold for the occurrence of the words (those that appear with higher frequency in the training data are randomly down-sampled);
- *hs* indicates if Hierarchical Softmax is used in the modeling (1) or not (0);
- *negative* represents the number of negative examples;
- *threads* indicates the number of threads to use;
- *iter* indicates the number of training iterations;
- *min_count* sets the minimum number of times in which a word has to appear in the corpus, in order for it to be considered;
- *alpha* sets the starting learning rate;
- *cbow* indicates if the Continuos Bag Of Words (CBOW) model (1) or if the skip-gram model (0) should be employed.

The values that these parameters took in our study are reported in the experimental framework, to ensure the repeatability of the experiments.

The output is a set of vectors, named E, which contain the representation of each item (both in the set I of items evaluated by the users and in the set $\hat{I}$ of items in the new catalog) as an embedding in Z dimensions. Each element of the vector contains the relevance of the dimension for that item.

2.2   New Catalog Modeling

For each item of the new catalog, $\hat{i}_m \in \hat{I}$, we consider its document embedding $E(\hat{d}_m)$. We create a model of the new catalog (m_c) as the sum of the embeddings of the items in it (additive compositionality property) [14]:

$$m_c = \sum_{\hat{i}_m \in \hat{I}} E(\hat{d}_m) \tag{1}$$

In other words, if an item $\hat{i}_m$ belongs to the new catalog, we add to the catalog model m_c the relevance of each dimension that represents the item. By doing this for all the items $\hat{i}_m \in \hat{I}$, the output is a vector, in which each element contains the relevance of the corresponding dimension for the catalog.

2.3   User Modeling

For each user $u \in U$, this step considers the set of items I_u she evaluated, and builds a user model m_u that describes how each dimension characterizes the user profile. In order to build the user model m_u, we consider the document embedding $E(d_m)$ of each item $i_m \in I_u$ and perform the following operation:

$$m_u = \sum_{i_m \in I_u} E(d_m) \tag{2}$$

If a user expressed a preference for item i_m, we add the relevance of each dimension of the item to the user model m_u. The output is a vector, in which each element contains the relevance of the corresponding dimension for the user.

2.4   User Targeting

This step compares the output of the two previous steps (i.e., the model of the new catalog m_c, and the user models m_u), in order to infer which users might be interested in the catalog. The key concept behind this step is that *we do not consider the items evaluated by a user or that belong to the catalog anymore*. Since both the user and the catalog models are vectors, the best form of comparison is to calculate the angle between them, through the cosine similarity. A user u is targeted for a new catalog of items c if their similarity is higher than or equal to a threshold ϕ, as shown in Eq. 3.

$$cos(m_c, m_u) \geq \phi \tag{3}$$

The output of this step is a set $T \subseteq U$ of users to target.

It is straightforward to notice that we are trading precision against recall, when varying the threshold ϕ. Indeed, by modeling the catalog and the user with a unique vector, our approach might not capture specific characteristics of the user and of the catalog, so it might detect more users than those that will actually follow the recommendation (i.e., buy/stream an item in the catalog). However, since the purpose is to advertise the items and since we are operating in an online scenario (so there are no additional costs if more emails are sent), we opted for a matching-based approach, which should be able to detect the users interested in the catalog.

2.5   Explanation List Generation

For each user $u \in T$, this step generates a list of the top-n artists she evaluated and that are the most similar to the one associated to the new catalog. We model each artist a evaluated by a user similarly to the approach we used to model the catalog. Let f_k be the feature of the item associated to the artist. An artist model m_a is built as follows:

$$m_a = \sum_{i_m \in I} E(d_m), \ s.t. \ f_k(i_m) = a \tag{4}$$

In other words, we sum all the embeddings of the items that have a as an artist, and generate a vector that has an element for each dimension, whose value is the relevance of that dimension for that artist. In order to understand if the artist evaluated by a user is similar to that of the new catalog, we employ the cosine between the two models $(cos(m_c, m_a))$. The similarity values obtained for each artist are ranked in descending order, and the top-n are selected.

We assume that this list should serve well as an explanation of the recommendation because, on the one hand, the user who receives it is familiar with all the artists included in it and, on the other hand, these authors are semantically related to the recommended artist.

3   Experimental Framework

3.1   Experimental Setup

The framework was developed on a machine with Intel Core i7 2630QM Processor and 8GB of RAM, using the Python language with the support of the interface to word2vec and doc2vec[3].

The doc2vec parameters are set as follows: $Z = 100$, $window = 5$, $sample = 1e - 3$, $hs = 0$, $negative = 0$, $threads = 12$, $iter = 5$, $min\_count = 5$, $alpha = 0.025$, $cbow = 1$.

In order to validate the use of document embeddings to model the users and the catalogs, we performed a comparison of the results obtained by our approach with those obtained employing three state-of-the-art embeddings' solutions, i.e., *Latent Dirichlet Allocation (LDA)*, *Latent Semantic Analysis (LSA)*, and *Nonnegative Matrix Factorization (NMF)*. As stated in the Introduction, Manifold methods are usually employed for data visualization; therefore, when creating vectors of large size like the ones we created with doc2vec, these methods are unsuitable, since very long execution times are needed. In conclusion, these methods cannot be used for comparison with our approach.

In order to build the embeddings for the approaches we compare with, we used the term frequency (TF) to run *LDA*, and term frequency-inverse document frequency (TF-IDF) for *LSA* and *NMF*. The choice to use TF for *LDA* was made since the probabilistic generative model creates documents by sampling a topic for each word and a word from the sampled topic. Therefore, a vector of frequencies is needed. Regarding the parameters necessary to run these three algorithms, they have been tuned as follows:

– **LDA**. The parameter evaluation has been performed using the perplexity as a metric to evaluate the learning decay (κ), the learning offset (τ_0), and the batch size (S), since the model that returns the lowest perplexity is the one with the best parameter setting [7]. The estimation has returned the following values: $\kappa = 0.5$, $\tau_0 = 1024$, and $S = 64$. The number of topics we considered is 100, which leads to vectors of the same size as the ones generated by doc2vec, to allow a fair comparison between the approaches.
– **LSA**. No parameter had to be set. Also in this case, the vector size is 100.
– **NMF**. To create the model, we used NNDSVD (Nonnegative Double Singular Value Decomposition), which is the most performing NMF strategy in case of sparse data (as our TF-IDF matrix) [2].

3.2   Dataset

The used dataset is Yahoo! Webscope Movie dataset (R4)[4], which contains the preferences expressed by the Yahoo! Movies community on a scale from 1 to 5. The data is composed by 7,642 users, 11,915 items, and 211,231 ratings. In

---

[3] https://github.com/danielfrg/word2vec.
[4] http://webscope.sandbox.yahoo.com.

addition to this, the dataset provides a list of features for every movie, including a brief textual description in English and the name of the director. In this scenario, a catalog is defined as the whole set of movies made by the same director.

3.3   Metrics

To evaluate the accuracy of the generated targets, we measured the *precision*, *recall*, and *F1 score*, as follows:

$$precision = \frac{|\{relevant\ users\} \cap \{retrieved\ users\}}{\{retrieved\ users\}}$$

$$recall = \frac{|\{relevant\ users\} \cap \{retrieved\ users\}}{\{relevant\ users\}}$$

$$F1 = 2 \cdot \frac{precision \cdot recall}{precision + recall}$$

3.4   Experimental Strategy

The accuracy of our proposal has been analyzed under different conditions, with a leave-one-out approach, in which each set of experiments did not consider a specific director in the dataset (whose items formed the set $\hat{I}$) and the user models were trained considering the items of the rest of the directors (set I). Following this strategy, the following three sets of experiments are performed:

1. **Parameter setting.** In order to decide if a user can be targeted with a new catalog, a parameter ϕ sets the minimum similarity between the user model and the catalog. In this set of experiments we will set the parameter for our approach and for the approaches we compare with, by analyzing the accuracy of the targeting and the size of the targets.
2. **Approaches comparison.** After the threshold that allows each approach to include a user in a target has been set, we can compare the different approaches in terms of accuracy and size of the generated targets.
3. **Explanation list effectiveness evaluation.** For each user and catalog, we produce an explanation list of 10 elements (i.e., the 10 directors most similar to those that a user evaluated). For each catalog, we obtain a list of directors that in their Wikipedia page have a link to the director of the catalog[5]. If at least one director is in the explanation list, we consider that list as effective for the user, since the catalog is strictly connected to one or more of the directors in the list. In this set of experiments, we measured the percentage of effective lists for the true positive and the true negative users.

---

[5] This is done by developing a parser to the "What links here" function of Wikipedia (https://en.wikipedia.org/wiki/Help:What_links_here). As previously mentioned, out of all the entities that might link to a director (e.g., actors, movies, etc.), we selected only the pages associated to other directors.

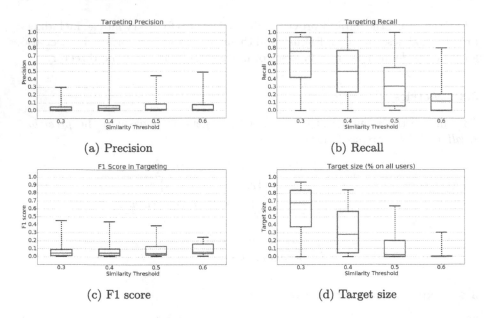

(a) Precision

(b) Recall

(c) F1 score

(d) Target size

Fig. 1. Accuracy and target size of our approach for different thresholds.

3.5   Experimental Results

Parameter Setting. The boxplots in Fig. 1 show the precision (Fig. 1a), recall (Fig. 1b), F1 score (Fig. 1c), and size of the targets (Fig. 1d) of our approach for different values of the threshold ϕ. As expected, since our approach compares two models in order to detect the users to target, the precision values are low (i.e., we detect a lot of false positive, which are users who did not evaluate the director associated to the new catalog). However, the recall values are high; indeed, even if we focus on the lowest threshold value in Fig. 1b, we can see that for 50% of the directors, the recall is close to 0.8, and for 25% of them it is close to 1. This means that our approach accurately detects almost all the users interested in the director. The low precision also has an impact on the F1 scores, which also mostly report low values. The fact that we have an increasing precision and a lowering recall for higher values of ϕ can be easily explained considering that higher threshold values mean lower amounts of detected users, so the number of false positive users lowers. As we previously mentioned, the fact that the recall values are good but the precision ones are low is related to the fact that in our targets we include many more users with respect to those that have actually evaluated the director. In order to inspect on this phenomenon, in Fig. 1d we present a boxplot for each threshold value, which indicates the distribution of the size of the detected targets. Indeed, the detected groups are large and, in the $\phi = 0.3$ case, 50% of them have a size that involves more than half of the user set.

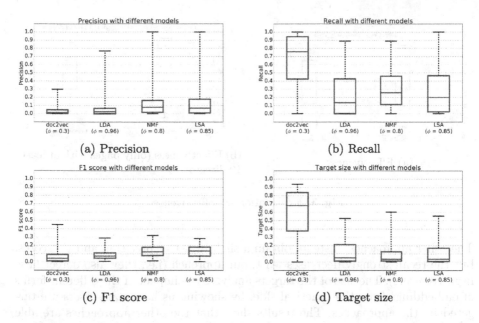

(a) Precision

(b) Recall

(c) F1 score

(d) Target size

Fig. 2. Comparison of the approaches in terms of accuracy and target size.

Given the results of these analyses, the value of ϕ chosen for our approach is 0.3. Indeed, the F1 scores reach the highest peak (even though it is the one with the lowest median), the recall reports the highest values, and the size of the targets is much higher with respect to the other threshold values (this would allow to reach more users).

Due to space constraints, we cannot report the analysis performed to set the ϕ parameter for the other forms of embeddings. The chosen values are: 0.96 for LDA, 0.8 for NMF, and 0.85 for LSA. It is clear that the values of the threshold are much higher for the other forms of embeddings; this means that if lower values had been chosen, almost all the user set would be targeted. In other words, the other forms of embeddings are unable, in this context, to model the users and the catalogs ensuring that the preferences of the users can be inferred and the differences between the users are detected by the approach.

Approaches Comparison. The boxplots in Fig. 2 compare the precision (Fig. 2a), recall (Fig. 2b), F1 scores (Fig. 2c) and size of the targets (Fig. 2d) of our approach (which employs doc2vec) and of the other three state-of-the-art approaches. Even though our approach obtains the lowest values of precision and F1 score (i.e., we target users that did not consider the director associated to the new catalog), our approach has by far the highest recall. This means that we target almost all the users that actually evaluated the director, plus additional users for which we do not have any information (again, the effectiveness of the explanation lists will tell us if these users have been properly targeted).

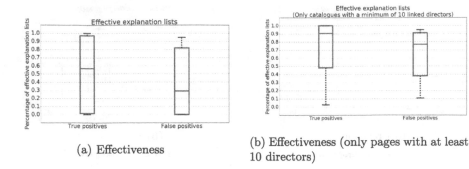

(a) Effectiveness

(b) Effectiveness (only pages with at least 10 directors)

Fig. 3. Explanation list effectiveness.

The other approaches, instead, obtain a slightly better precision and F1 scores, but the recall strongly decreases w.r.t. our approach (i.e., they also detect false negative users). The size of the targets analyzes the impact of the different forms of embeddings from the practical side, by showing us how many users are targeted by the approaches. The results show that the other approaches are able to target only small amounts of users, so this would affect the advertising campaign, since they would reach only a limited part of the user base; moreover, as the accuracy values show, these users would not be targeted accurately (both precision and recall are low). In conclusion, we can say that our approach outperforms the other approaches from most of the metrics we analyzed. Indeed, even though the others are able to obtain higher precision values (i.e., they detect less false positive users w.r.t. our approach), this leads to small groups of users to target and to low recall values.

Explanation List Effectiveness Evaluation. After the accuracy analysis, we evaluate for how many users the explanation list is effective. Figure 3 shows the results obtained by our approach[6]; more specifically, Fig. 3a shows the results considering the Wikipedia pages of all the directors in the dataset, while Fig. 3b shows the results considering only those pages that were linked by at least 10 pages of other directors (i.e., we consider a "fair" approach, in which all the directors in an explanation list could actually be linked to the recommended director). As the results show, when the Wikipedia pages of all the directors are considered, 60% or more of the explanation lists provided to true positive users are effective for half of the catalogs, and for 25% of them almost all the explanation lists are effective. Regarding the false positive users, at least 30% of the lists provided to these users are effective for half of the catalogs, and for 25% of them between 80% and 100% of the lists are effective. When considering only directors whose Wikipedia page was linked by at least 10 other directors, the

---

[6] We also measured the effectiveness of the explanation lists generated by employing the other forms of embeddings. However, the percentage of effective lists was very low. We omitted these results due to space constraints.

results show a great improvement. Indeed, for half of the catalogs, 90% or more of the explanation lists provided to true positive users are effective, while when we give explanations to false positive users, around 80% or more lists are effective for half of the catalogs. It is important to emphasize the fact that the good results of the explanation phase in the false positive case give us a confirmation of the effectiveness of the targeting algorithm, in spite of the low precision and F1 scores. Indeed, not only the true positive, but also the false positive users are properly targeted, since most of them evaluated directors linked to that of the new catalog (so they appear in some way related to it).

4 Conclusions and Future Work

In this paper, we tackled the problem of selecting a set of users to target when a new catalog of items becomes available in a digital platform. We proposed an approach that, thanks to document embeddings built by analyzing the description of the items, creates a model of both the new catalog and the users. Then, to help a user understand why she was targeted and to improve the effectiveness of our recommendation, we generate an explanation list that contains the top-n artists most similar to the one associated with the new catalog. Results show the effectiveness of our approach at recommending a new catalog to the users. Future work will refine this approach by focusing on specific features that characterize both the catalog and the users, in order to narrow the size of the detected segments while still generating an accurate targeting.

Acknowledgment. This work is partially funded by Regione Sardegna under project NOMAD (Next generation Open Mobile Apps Development), through PIA - Pacchetti Integrati di Agevolazione "Industria Artigianato e Servizi" (annualità 2013).

References

1. Blei, D.M., Ng, A.Y., Jordan, M.I.: Latent dirichlet allocation. J. Mach. Learn. Res. **3**, 993–1022 (2003)
2. Boutsidis, C., Gallopoulos, E.: SVD based initialization: a head start for nonnegative matrix factorization. Pattern Recogn. **41**(4), 1350–1362 (2008)
3. Campr, M., Ježek, K.: Comparing semantic models for evaluating automatic document summarization. In: Král, P., Matoušek, V. (eds.) TSD 2015. LNCS (LNAI), vol. 9302, pp. 252–260. Springer, Cham (2015). https://doi.org/10.1007/978-3-319-24033-6_29
4. Christou, D.: Feature extraction using latent dirichlet allocation and neural networks: a case study on movie synopses. CoRR abs/1604.01272 (2016)
5. Dhillon, I.S., Sra, S.: Generalized nonnegative matrix approximations with bregman divergences. In: Advances in Neural Information Processing Systems, NIPS 2005, 5–8 December, 2005, Vancouver, British Columbia, Canada, vol. 18, pp. 283–290 (2005)
6. Dumais, S.T.: Latent semantic analysis. ARIST **38**(1), 188–230 (2004)

7. Hoffman, M.D., Blei, D.M., Bach, F.R.: Online learning for latent dirichlet allocation. In: Advances in Neural Information Processing Systems 23: 24th Annual Conference on Neural Information Processing Systems 2010, pp. 856–864. Curran Associates, Inc. (2010)

8. Kruskal, J.B., Wish, M.: Multidimensional Scaling. Sage Publications, Beverly Hills (1978)

9. Lam, X.N., Vu, T., Le, T.D., Duong, A.D.: Addressing cold-start problem in recommendation systems. In: Proceedings of the 2nd International Conference on Ubiquitous Information Management and Communication, ICUIMC 2008, New York, pp. 208–211. ACM (2008)

10. Le, Q.V., Mikolov, T.: Distributed representations of sentences and documents. In: Proceedings of the 31th International Conference on Machine Learning, ICML 2014. JMLR Proceedings, vol. 32, pp. 1188–1196 (2014). JMLR.org

11. van der Maaten, L., Hinton, G.E.: Visualizing high-dimensional data using t-SNE. J. Mach. Learn. Res. **9**, 2579–2605 (2008)

12. Mikolov, T., Chen, K., Corrado, G., Dean, J.: Efficient estimation of word representations in vector space. CoRR abs/1301.3781 (2013)

13. Mikolov, T., Le, Q.V., Sutskever, I.: Exploiting similarities among languages for machine translation. CoRR abs/1309.4168 (2013)

14. Mikolov, T., Sutskever, I., Chen, K., Corrado, G.S., Dean, J.: Distributed representations of words and phrases and their compositionality. In: Proceedings of the 27th Annual Conference on Neural Information Processing Systems 2013, pp. 3111–3119 (2013)

15. Mikolov, T., Yih, W., Zweig, G.: Linguistic regularities in continuous space word representations. In: Human Language Technologies: Conference of the North American Chapter of the Association of Computational Linguistics, Proceedings, pp. 746–751. The Association for Computational Linguistics (2013)

16. Roweis, S.T., Saul, L.K.: Nonlinear dimensionality reduction by locally linear embedding. Science **290**, 2323–2326 (2000)

17. Zhila, A., Yih, W., Meek, C., Zweig, G., Mikolov, T.: Combining heterogeneous models for measuring relational similarity. In: Human Language Technologies: Conference of the North American Chapter of the Association of Computational Linguistics, pp. 1000–1009. The Association for Computational Linguistics (2013)

Retrieval

Statistical Stemmers:
A Reproducibility Study

Gianmaria Silvello[✉], Riccardo Bucco, Giulio Busato, Giacomo Fornari,
Andrea Langeli, Alberto Purpura, Giacomo Rocco, Alessandro Tezza,
and Maristella Agosti

Department of Information Engineering, University of Padua, Padua, Italy
{gianmaria.silvello,riccardo.bucco,giulio.busato,giacomo.fornari,
andrea.langeli,alberto.purpura,giacomo.rocco,alessandro.tezza,
maristella.agosti}@unipd.it

Abstract. Statistical stemmers are important components of Information Retrieval (IR) systems, especially for text search over languages with few linguistic resources. In recent years, research on stemmers produced relevant results, especially in 2011 when three language-independent stemmers were published in relevant venues. In this paper, we describe our efforts for reproducing these three stemmers. We also share the code as open-source and an extended version of Terrier system integrating the developed stemmers.

1 Introduction

The research on stemmers has focused for a long time on the English language and successively on a subset of other, mostly European, languages. Hence, for English several highly-effective rule-based stemmers such as Porter [11], Krovetz [5] and Lovins [6] are commonly available. For highly studied languages such as German, French, Italian and Spanish the effective rule-based stemmers implemented by Snowball[1] are typically employed by IR systems. On the other hand, for languages as the Slavic or Asian ones, there are few linguistic resources and rule-based stemmers are less effective, if available at all. In these cases, statistical stemmers can play a key role, since being language-independent they can be employed without any prior knowledge of the language at hand. Despite their relevance, statistical stemmers are not commonly taken into account in baseline IR systems or considered in longitudinal studies in IR even when non-English corpora are considered [2,4].

In recent years, we have witnessed a new interest in research on statistical stemmers with a spike in 2011 when three new stemming algorithms were proposed by the same core subset of authors – i.e., Jiaul H. Paik and Swapan K. Parui – in relevant IR venues, namely: "A fast corpus-based stemmer" (FCB) published in the ACM TALIP [10], "A novel corpus-based stemming algorithm

---

[1] http://snowballstem.org/.

© Springer International Publishing AG, part of Springer Nature 2018
G. Pasi et al. (Eds.): ECIR 2018, LNCS 10772, pp. 385–397, 2018.
https://doi.org/10.1007/978-3-319-76941-7_29

using co-occurrence statistics" (SNS) presented at SIGIR [9] and "GRAS: An effective and efficient stemming algorithm for information retrieval" published in the ACM TOIS [8]. More recently, these works have been reconsidered and discussed in a comprehensive survey about text stemming [13].

We decided to reproduce these three papers with the aim of making the stemmers they propose readily available to the research community such that they can be easily included in baseline systems and longitudinal studies. To this end, we also share an extended version of the Terrier system [7] where these stemmers have been integrated and are ready to be used in a typical IR experimental setting.

The paper is organized as follows: Sect. 2 introduces the stemmers and their main characteristics; Sect. 3 presents the main issues we faced when implementing the stemming algorithms; Sect. 4 reports the experimental results and the differences with the reference papers; finally, Sect. 5 discusses the pros and cons of the reproduced papers and what can be learned for the future.

2   Overview of the Statistical Stemmers

FCB. FCB is a statistical stemmer that relies on the frequency of suffixes of the terms in a language. FCB associates a frequency to each suffix appearing in a corpus of documents equal to the number of words that end with it. If the frequency of a suffix exceeds a certain threshold α, then that suffix is called a *potential* suffix. The algorithm then starts to group the terms in the collection into k-equivalence classes according to their prefix. It iteratively groups the terms with a prefix of length k or longer in common, at each step decreasing the minimum common prefix length until a pre-determined lower threshold. As suggested in the reference paper we set 5 as a starting prefix length and 2 as lower threshold. After the terms have been grouped into equivalence classes, the longest common prefix of each class is evaluated as a candidate stem for the class. To do this, FCB considers the size of the subsets of elements in the class that contains only terms whose suffixes, induced by the candidate stem, all belong to the potential suffixes set. Finally, we compute the ratio of the size of the largest of the aforementioned subsets – i.e., *potential-class* – and the size of the class to be evaluated. If this ratio exceeds a certain threshold δ, the longest prefix of that class is considered a valid stem for all of the terms. Otherwise, a better stem for the class is chosen amongst the values in it and the evaluation process is repeated by iteratively extracting all the subsets of terms that have a common valid stem in the class.

SNS. The goal of SNS is to group words that are morphologically related by computing their co-occurrence in the corpus. The starting hypothesis is that a document relevant to a given topic will probably contain many words relevant to the topic itself. SNS is composed of three main steps: (i) the computation of the co-occurrence strength of word pairs; (ii) the re-calculation of the strengths; and (iii) the clustering of the words. Once the co-occurrence strengths (CO) are computed, they are represented in a weighted graph where words, say $w1$

and $w2$, are nodes connected by an edge if at least one of these two conditions holds: (i) $CO(w1, w2) > 0$; and (ii) if "they have a common prefix of a given length (l_1) along with the suffixes (after truncating the longest common prefix) which are the residues (the ends after removal of longest common prefix) of more than one pair of co-occurring words with long common prefix (length larger than 5 which is the second static parameter, l_2)" [9].

The second step performs the re-calculation of co-occurrence strength between word pairs. The strength assigned to a word pair $(w1, w2)$ is proportional to the number of other words in the corpus that co-occur with both $w1$ and $w2$. Afterwards, the co-occurring words are clustered together. To do this SNS identifies the strong edges; an edge (u, v) is defined as "strong" if two conditions hold: (i) (u, v) has the highest weight w.r.t all the edges insisting on u; and, (ii) (u, v) has the highest weight w.r.t all the edges insisting on v. Lastly, the non-strong edges are removed from the graph and the remaining connected components of the graph represent the morphologically related groups.

GRAS. GRAS is a stemmer conceived for highly inflectional languages (e.g. Hungarian) where words are formed from the root by a process of suffixation. Hence, the role of suffixes is central for this algorithm. GRAS can be described as a sequence of five main steps.

In the first step, GRAS identifies the word partitions sharing a l-long prefix; l is set to be the average word length for the given language. The second step determines the common suffixes of the words sharing a prefix and it checks if there exist other word pairs with a common prefix followed by the same identified suffix. Two suffixes are considered a candidate pair if they are shared "frequently enough" by word pairs in the lexicon; the suffix frequency threshold is defined by a parameter α. In the third step, GRAS creates a graph where the identified words are mapped to nodes which are connected by an edge if the words are morphologically related – i.e., they share a non-empty prefix and the suffix pairs that remain after the removal of the common prefix are candidate pairs identified in the second phase. In the fourth phase, *pivot* nodes – words with a large number of edges — are identified. In the fifth step, equivalence classes of words are created. A word is put in the same class as the pivot to which it is connected if it has a *cohesion* of at least δ. The cohesion value determines the likelihood that two words – the candidate word and the pivot – are morphologically related.

3 Realization of the Stemmers

All the described stemmers have been implemented in Java and integrated into Terrier v4.1.[2] The input of the stemmer is composed of the lexicon and the inverted index created by Terrier and the output is a text file (i.e., the `lookup table`) containing the words in the lexicon and their stems. We extended Terrier v4.1 in order to use the statistical stemmers we realized.

---

[2] http://github.com/giansilv/statisticalStemmers/.

FCB. Since FCB relies on the frequency of suffixes in a given corpus, we developed a suffix extraction algorithm. In the reference paper there is no description of the suffix extraction process, therefore we decided to extract all the suffixes in a given corpus, without considering the inclusion relations between them. For example, if we assume that for English the suffix "ing" is a possible suffix, we extract and compute the frequency of the suffixes "g", "ng" and "ing". Afterwards, we employed the "frequency-based filtering strategy" described in the reference paper to select the most frequent suffixes.

With regards to the implementation of the core of the stemming algorithm, we made an assumption about the evaluation of the k-equivalence classes. A *potential-class* is defined in the reference paper as the largest subset of words with a common prefix R ending with frequent suffixes induced by R. However, we realized two versions of the algorithm: FCB v.1 and FCB v.2. FCB v.1 considers the prefix R to compute the suffixes, whereas FCB v.2 ignores it. This is a crucial part of the algorithm and more details about how it has been realized would have been important for reproducibility purposes. FCB v.1 considers the strings composed of all the characters that follow the common prefix R as suffixes of the terms in a k-equivalence class and then compares them against the set of frequent suffixes extracted from the collection prior to the execution of the algorithm. FCB v.2 on the other hand considers the whole terms in a k-equivalence class and qualifies them as ending with a frequent suffix if they end with any of the frequent suffixes mentioned before. Therefore, in this case, we allow the presence of a few characters between R and the beginning of the suffix. We noticed a great improvement in the performances of the algorithm by using the latter approach, especially for languages with greater inflection such as Hungarian.

Finally, the reference paper does not describe how to deal with singleton classes; singleton classes have as longest common prefix the whole term that belongs to the class, therefore, in this case, the induced suffix is always empty. This implies that many terms with a unique prefix, but ending with a frequent suffix are not stemmed.

SNS. Our implementation of SNS is composed of six main steps. *Step 1.* We extract the data from the lexicon and the inverted index and contextually we discard the terms whose first character is a digit or that are shorter than l_1 ($l_1 = 3$ in the reference paper) in order to avoid useless computations. *Step 2.* We compute the co-occurrence between two words if they have a common prefix with length greater than or equal to l_1; if their strength is not zero, then we check if their common prefix is greater than or equal to l_2 ($l_2 = 5$ in the reference paper) and we store this information to be used in later phases. *Step 3.* We create a weighted graph where words are nodes and the edges are weighted by their co-occurrence strength. *Step 4.* We update the edge strength by re-calculating the co-occurrence of terms. *Step 5.* We remove the non-strong edges. *Step 6.* We find the connected components of the graph. Each stem is generated by finding the longest common prefix amongst the connected words; this is an assumption we made since in the reference paper this phase is not described and the proposed algorithm stops right after the creation of word clusters.

GRAS. In the first step of GRAS implementation, we process the lexicon by creating partitions of words sharing a common prefix of length l. This step reduces the size of the lexicon at hand and reduces the running time of the algorithm. In the second step, for each common prefix class, we individuate the α-frequent suffixes and we store them along with their frequencies. In the third step, we build the graph (one for each common prefix class) and we calculate the cohesion for each pair of connected words.

As we discuss below, in the reference paper there are a few moot points that may create some problems from the reproducibility viewpoint. First-of-all, the l parameter is set to be "the average word length for the language concerned", but no further details are given. We chose to calculate the average length of the words in the lexicon at hand; we would have liked to have the actual l parameter employed in the reference paper since this parameter has a high influence on the stemmer and a small difference here can be sizable performance-wise.

4   Reproduction of the Results

The results for FCB, SNS and GRAS have been reproduced for the CLEF and TREC collections employed in the reference papers; more details are reported below. CLEF collections have been downloaded from DIRECT[3] and TREC collections from the TREC Website.[4] In the reference papers also the Marathi and Bengali test collections from FIRE are employed, but these collections are not currently available in the FIRE Website.[5] Nevertheless, the CLEF and TREC collections form a solid comparative testbed for assessing our reproducibility effort. All the experiments were conducted using Terrier v4.1 (with the IFB2 model) as a baseline system. All the stemmers were tested after a stopwords removal phase; for the English language we adopted the stoplist provided by Terrier and for the other languages those provided by Jacques Savoy.[6] In the following we show the results obtained by the original algorithms and the reproduced one and we report their absolute differences.

FCB. The algorithm was evaluated on the CLEF 2006–2007 collection for the Hungarian language (98 topics) and on the Wall Street Journal sub-corpus of the TIPSTER collection (topics 1–200). The experiments are divided into two sets, one where the queries are composed of the topic title (T) and the other where the queries are composed of title and description (TD). The version of Terrier employed in the reference paper is not specified. The experiments were performed by removing the stopwords and numbers from the lexicon before the execution of the stemming algorithm. We tested both the versions of the algorithm we realized for different values of δ. In the following tables, we report the results achieved with the best tested δ value.

---

[3] http://direct.dei.unipd.it/.

[4] http://trec.nist.gov/.

[5] http://fire.irsi.res.in/fire/static/data/.

[6] http://members.unine.ch/jacques.savoy/clef/.

Table 1. English WSJ TREC collection. Original and reproduced (FCB v.1) results for $\delta = 0.6$, the reference paper reports only 3 digits after the decimal point. Differences greater than 0.0100 are reported in bold.

		Original			Reproduced			Difference		
		MAP	RPrec	P@10	MAP	RPrec	P@10	MAP	RPrec	P@10
T	No Stem	0.225	0.267	0.399	0.2250	0.2674	0.3990	0.000	0.000	0.000
	FCB	0.258	0.289	0.437	0.2399	0.2791	0.4020	**−0.018**	**−0.010**	**−0.035**
	Porter	0.261	0.296	0.432	0.2621	0.2971	0.4362	+0.001	+0.001	+0.004
TD	No Stem	0.272	0.312	0.477	0.2722	0.3125	0.4765	0.000	+0.000	0.000
	FCB	0.295	0.331	0.493	0.2811	0.3181	0.4715	**−0.014**	**−0.013**	**−0.0215**
	Porter	0.294	0.325	0.477	0.2958	0.3262	0.4800	+0.002	+0.001	+0.0030

In Table 1 we report the results obtained for English. In this case, we employed FCB v.1 and even though there are sizable differences between our implementation and the original stemmer, our results are consistent with the original ones since FCB is better than no stemmer and worse than Porter; this is reasonable for a language with good linguistic resources.

Table 2. CLEF 2006–2007 Hungarian collection. Original and reproduced (FCB v.2) results for $\delta = 0.5$, the reference paper reports only 3 digits after the decimal point. Differences greater than 0.0100 are reported in bold.

		Original			Reproduced			Difference		
		MAP	RPrec	P@10	MAP	RPrec	P@10	MAP	RPrec	P@10
T	No stem	0.185	0.199	0.258	0.1830	0.1956	0.2547	−0.0020	−0.0034	−0.0033
	FCB	0.293	0.315	0.353	0.2863	0.2942	0.3284	−0.0067	**−0.0208**	**−0.0246**
	RB	0.267	0.280	0.343	0.2610	0.2737	0.3245	−0.0060	−0.0063	**−0.0185**
TD	No stem	0.239	0.252	0.314	0.2375	0.2528	0.3133	−0.0015	+0.0008	−0.0007
	FCB	0.341	0.352	0.390	0.3355	0.3263	0.3949	−0.0055	**−0.0257**	+0.0049
	RB	0.335	0.340	0.389	0.3347	0.3358	0.4102	−0.0020	**−0.0532**	**+0.0212**

In Table 2, we report the results obtained for Hungarian by employing FCB v.2, which turns out to be quite stable to the variations of δ. We can see that in this case the difference between the original algorithm and FCB v.2 is smaller than for English. There can be a sizable difference between our Terrier setup and the one of the reference paper, since also for the no stemmer and the rule-based stemmer (RB) cases we register a difference comparable to the one we get for FCB.

In order to enable the comparison of performances between FCB and the other stemmers we reproduced, we tested FCB also on the Bulgarian and Czech CLEF collections that are considered by both SNS and GRAS; the results are reported in Table 3.

Table 3. FCB v.2 results for the Bulgarian (2006–2007) and Czech (2007) CLEF collections with $\delta = 0.5$.

		Bulgarian			Czech		
		MAP	RPrec	P@10	MAP	RPrec	P@10
T	No Stem	0.165	0.191	0.220	0.220	0.248	0.244
	FCB	0.239	0.270	0.293	0.306	0.306	0.314
TD	No Stem	0.203	0.230	0.257	0.238	0.261	0.268
	FCB	0.276	0.301	0.333	0.338	0.318	0.348

SNS. SNS was evaluated on three CLEF collections – i.e. the 2006–2007 CLEF Bulgarian and Hungarian collections and the CLEF 2007 Czech collection – and one TREC collection – i.e. the corpus is the TIPSTER Disk 4&5 minus the congressional record and the federal register and the TREC6, TREC7 and TREC8 topics. The queries used for the experiments are composed of the title and description fields of the topics. The results in the reference paper were obtained using Terrier, the version of which is not specified.

In Table 4, we report the results for Bulgarian for a system employing no stemmer, a rule-based stemmer and SNS. The rule-based stemmer employed in the reference paper for Bulgarian is not further specified. There are at least three rule-based stemmers that can be used: an aggressive one, a light one using transliterated terms and one which is the same as the light stemmer except that it processes documents in Cyrillic. The closest performance value to the original one is obtained with the third stemmer and it is the one we report in the result table.

Table 4. CLEF 2006–2007 Bulgarian test collection. Original and reproduced results for the SNS stemmer and difference between the reproduced stemmer and the original one. Differences greater than 0.0100 are reported in bold.

	Original			Reproduced			Difference		
	NO	RB	SNS	NO	RB	SNS	NO	RB	SNS
MAP	0.2166	0.2794	0.3256	0.2038	0.2786	0.2980	**−0.0128**	−0.0008	**−0.0276**
RPrec	0.2293	0.2930	0.3289	0.2291	0.3033	0.3253	−0.0002	**+0.0103**	−0.0036
P@10	0.2570	0.3270	0.3520	0.2580	0.3410	0.3540	+0.0010	**+0.0140**	+0.0020

From Table 4 we can see that there is a sizable difference in terms of MAP between the systems not employing any stemmer; this difference is increased when we consider the MAP values for the systems employing SNS. As we can see for RPrec and P@10 the difference is quite small and it does not affect the reproduced results in an appreciable way. Another difference can be seen for RPrec and P@10 when the rule-based stemmer is employed.

In general, we see that SNS improves the baseline systems (no stemmer and rule-based stemmer) both in the reference paper and in the reproduced version, even though in the reproduced case the improvement is less marked, especially in terms of MAP. The problem with the reproducibility of this stemmer for Bulgarian seems to be related to the starting difference between the baseline systems (i.e. no stemmer) rather than to the specific implementation of SNS. This fact can be further assessed by considering the results for the other test collections.

For Czech, the authors claim to use the stemmer defined in [3], which actually presents two rule-based stemmers, one light and one aggressive. We tested both these stemmers and we found that the light one was used in the reference paper.

Table 5. CLEF 2007 Czech test collection. Original and reproduced results for the SNS stemmer and difference between the reproduced stemmer and the original one. Differences greater than 0.0100 are reported in bold.

	Original			Reproduced			Difference		
	NO	RB	SNS	NO	RB	SNS	NO	RB	SNS
MAP	0.2381	0.3409	0.3624	0.2382	0.3405	0.3569	+0.0001	−0.0004	−0.0055
RPrec	0.2611	0.3456	0.3441	0.2611	0.3456	0.3449	0.0000	0.0000	−0.0008
P@10	0.2680	0.3480	0.3700	0.2680	0.3480	0.3640	0.0000	0.0000	−0.0060

In Table 5 we can see that the results obtained by using no stemmer and the rule-based stemmer are reproduced with a very marginal error and thus we can state that the initial condition for the experiment with SNS is the same as in the reference paper. Also the results obtained with SNS are very close to the original ones with marginal differences for all the measures.

For Hungarian, we have a problem with the rule-based stemmer since the authors have specified that they use the stemmer defined in [12], where both a light and an aggressive stemmer are defined. We ran the experiments for these stemmers and we report the results obtained with the light stemmer that are closer to those in the reference paper.

Table 6. CLEF 2006–2007 Hungarian test collection. Original and reproduced results for the SNS stemmer and difference between the reproduced stemmer and the original one. Differences greater than 0.0100 are reported in bold.

	Original			Reproduced			Difference		
	NO	RB	SNS	NO	RB	SNS	NO	RB	SNS
MAP	0.2386	0.3132	0.3588	0.2375	0.3369	0.3583	−0.0011	**+0.0237**	−0.0005
RPrec	0.2518	0.3117	0.3585	0.2528	0.3459	0.3556	+0.0010	**+0.0342**	−0.0029
P@10	0.3143	0.3990	0.4224	0.3133	0.4153	0.4163	−0.0010	**+0.0163**	−0.0061

In Table 6 we report the original and the reproduced results for Hungarian, where we can see that the major differences concern the rule-based stemmer. Possibly, the authors employed a different rule-based stemmer than one of those defined in [12]. Nevertheless, the reproduced results for SNS (as well as for the no stemmer system) are close to the original ones and this allows us to state that SNS has been correctly reproduced for Hungarian. We can also see that SNS is still slightly superior to the rule-based stemmer we employed, especially in terms of MAP and P@10, even though this difference is less marked than the one reported in the reference paper.

Table 7. TREC 06-07-08 Ad-Hoc test collection. Original and reproduced results for the SNS stemmer and difference between the reproduced stemmer and the original one. Differences greater than 0.0100 are reported in bold.

	Original			Reproduced			Difference		
	NO	RB	SNS	NO	RB	SNS	NO	RB	SNS
MAP	0.2290	0.2599	0.2582	0.2289	0.2596	0.2319	−0.0001	−0.0003	**−0.0263**
RPrec	0.2733	0.3008	0.3001	0.2736	0.3013	0.2722	+0.0003	+0.0005	**−0.0279**
P@10	0.4327	0.4833	0.4727	0.4320	0.4827	0.4267	−0.0007	−0.0006	**−0.0460**

In Table 7 we report the results for the TREC English collection. As in the reference paper, the Porter stemmer performs the best amongst the other approaches and the reproducibility of the baseline strategies (no stemmer and rule-based stemmer) are successfully reproduced. Unfortunately, SNS for TREC presents rather different results from the original paper with a consistent decrease of performances. Despite this drop in performances, SNS still introduced a small improvement in terms of MAP with respect to the no stemmer system even though this is not comparable to the results obtained in the reference paper. These results are quite surprising especially if we consider that for the non-English collections the SNS behavior has been reproduced quite accurately.

GRAS. GRAS was evaluated on four CLEF collections – i.e. the 2006–2007 CLEF Bulgarian and Hungarian collections, the 2007 CLEF Czech collection and the 2005–2006 CLEF French collection – and one TREC collection – i.e. the corpus is the TIPSTER Disk 4&5 minus the congressional record and the federal register and the TREC6, TREC7 and TREC8 topics. The queries are formed of the title and description fields of the topics. The reference paper adopted Terrier, the version of which is not reported.

In the reference paper 87, 281 documents in the Bulgarian corpus are reported, but the number of documents in this corpus is 69, 195 [1], thus we register a conspicuous difference of 18, 086 documents that turn out to produce a lexicon which is 9% smaller than the one of the reference paper. In Table 8 we reported the number of documents and indexed tokens reported by the reference paper and the values we obtained for the same corpora; we can see that the number of documents is the same for all the corpora except the Bulgarian,

Table 8. Number of documents and unique words in the considered corpora.

	Original					Reproduced				
	EN	FR	HU	BG	CZ	EN	FR	HU	BG	CZ
Docs	472,525	177,452	49,530	87,281	81,735	472,525	177,452	49,530	69,281	81,735
Words	522,381	303,349	528,315	320,673	457,164	502,280	325,292	534,813	292,077	457,149

where the number of unique words differs quite a bit. The greatest differences are recorded for the French where the reference paper reported 7.23% more unique words than we counted and the Bulgarian with 8.92% more unique words. For the English, we have a difference of 3.84%, 1.23% for the Hungarian and less than 0.01% for the Czech language.

GRAS employs a parameter l which is set to be "the average word length for the language concerned"; no further details are given. We chose the values reported in Table 9 by calculating the average word length weighted by their frequency in the considered lexicons. The α and δ parameters have been set up to $\alpha = 4$ and $\delta = 0.8$ as specified in the reference paper.

Table 9. The average word length weighted by their frequency for the considered languages.

	Czech	Bulgarian	English	French	Hungarian
l	6	7	7	7	8

For the reproduction of the experimental results, we focused on the GRAS stemmer and in this case we do not report the results obtained for the no stemmer and the rule-based stemmer cases since they are the same as those reported for the FCB and SNS cases reported above. In the GRAS paper there are no further details about the rule-based stemmers employed thus, even after our reproducibility attempts, we remain uncertain about what stemmer was used for the Hungarian collection.

In Table 10 we report the results we obtained compared to those in the reference paper. By focusing on the MAP values, we can see that we have a sizable difference, between the original and the reproduced stemmer, only for Bulgarian and Hungarian. For Bulgarian, this difference is almost certainly related to the number of documents considered and, consequently, to the size of the lexicon. For Hungarian, the difference between the originally used corpus and the one we employed is quite contained, but as Hungarian is a highly inflected language, it may be more sensitive to the differences related to the lexicon used for training purposes. Another possible reason for these differences may be the selected l parameter, the value of which was not reported in the original paper.

Table 10. Experimental results obtained by reproducing the GRAS stemmer. Differences greater than 0.0100 are reported in bold.

		MAP	R-Prec	P@5	P@10	Rel Ret
BG	Original	0.3260	0.3340	0.4240	0.3550	2110
	Reproduced	0.3410	0.3580	0.4730	0.3720	2043
	Diff	**+0.0150**	**+0.0240**	**+0.0490**	**+0.0170**	−67
CZ	Original	0.3660	0.3600	0.4480	0.3760	689
	Reproduced	0.3630	0.3580	0.4460	0.3720	690
	Diff	−0.0030	−0.0020	−0.0020	−0.0040	+1
EN	Original	0.2700	0.3090	0.5430	0.4790	7873
	Reproduced	0.2749	0.3128	0.5492	0.4859	7904
	Diff	+0.0049	+0.0038	+0.0062	+0.0069	+31
FR	Original	0.3870	0.3980	0.5330	0.4910	4078
	Reproduced	0.3867	0.3886	0.5495	0.4838	4115
	Diff	−0.0003	−0.0094	**+0.0165**	−0.0072	+37
HU	Original	0.3510	0.3600	0.4740	0.4220	1924
	Reproduced	0.3319	0.3467	0.4701	0.4104	1846
	Diff	**−0.0191**	**−0.0133**	−0.0039	**−0.0116**	−78

5   Discussion

We considered three statistical stemmers – i.e., FCB, SNS and GRAS – proposed by the same subset of core authors in 2011 and presented in relevant IR venues. In some cases, the reproduction of the results reported in the reference papers has been challenging also for the baseline systems where no stemmer or a standard rule-based stemmer were employed.

The considered papers have some pros and cons when it comes to their reproducibility. (i) They employed a standard open-source system as Terrier for the experiments, but they do not report the version used. The use of Terrier limits the number of uncontrolled variables in an experimental setting, but from version to version some key features change and they may impact the reproducibility. (ii) They used standard test collections. This is a good practice of the IR community that enables the comparability of results by guaranteeing that the same corpus, topics and qrels are used; nevertheless, for some collections – e.g., Bulgarian for the GRAS stemmer – the corpus size used in the reference papers does not match the one reported in the CLEF documentation [1]. Moreover, for FCB only the WSJ sub-corpus was employed; this choice requires modification of the official qrels used in the TREC ad-hoc tracks with the possibility of introducing mistakes. (iii) The pseudo-code of the key algorithms is given for two algorithms: SNS and GRAS; this is a good practice because it reduces the ambiguities intrinsic with text descriptions of algorithms.

In general, the considered stemmers have been reproduced quite successfully given that, for at least one test collection per stemmer, we obtained MAP values whose difference with the one reported in the reference papers is less than 0.01.

The differences with the reference papers involve the no stemmer and the rule-based stemmer cases and, in most cases, when we failed to reproduce the baseline cases we also found sizable differences with the statistical stemmers – e.g., SNS and GRAS for Bulgarian. In these cases, the difference is possibly due to the base setting of Terrier rather than to the specific implementation of the statistical stemmer at hand.

Finally, we tested the three stemmers on three common test collections (i.e. Bulgarian, Czech and Hungarian) and we see that GRAS outperforms SNS and FCB for Bulgarian and Czech, whereas SNS outperforms the other two for Hungarian. As expected, for English none of the statistical stemmers outperforms the Porter stemmer, whereas they outperform the rule-based stemmer for the other considered languages, proving their suitability for languages with few linguistic resources.

References

1. Di Nunzio, G.M., Ferro, N., Mandl, T., Peters, C.: CLEF 2007: ad hoc track overview. In: Peters, C., et al. (eds.) CLEF 2007. LNCS, vol. 5152, pp. 13–32. Springer, Heidelberg (2008). https://doi.org/10.1007/978-3-540-85760-0_2
2. Dietz, F., Petras, V.: A component-level analysis of an academic search test collection. In: Jones, G.J.F., et al. (eds.) CLEF 2017. LNCS, vol. 10456, pp. 29–42. Springer, Cham (2017). https://doi.org/10.1007/978-3-319-65813-1_3
3. Dolamic, L., Savoy, J.: Indexing and stemming approaches for the Czech language author links open overlay panel. Inf. Proces. Manage. **45**(6), 714–720 (2009)
4. Ferro, N., Silvello, G.: CLEF 15th birthday: what can we learn from ad hoc retrieval? In: Kanoulas, E., Lupu, M., Clough, P., Sanderson, M., Hall, M., Hanbury, A., Toms, E. (eds.) CLEF 2014. LNCS, vol. 8685, pp. 31–43. Springer, Cham (2014). https://doi.org/10.1007/978-3-319-11382-1_4
5. Krovetz, R.: Viewing morphology as an inference process. In: Proceedings of 16th Annual International ACM SIGIR Conference on Research and Development in Information Retrieval (SIGIR 1993), pp. 191–202. ACM Press (1993)
6. Lovins, J.B.: Development of a Stemming algorithm. Mech. Transl. Comput. Linguist. **11**(1/2), 22–31 (1968)
7. Macdonald, C., McCreadie, R., Santos, R.L.T., Ounis, I.: From puppy to maturity: experiences in developing terrier. In: Proceedings of OSIR at SIGIR, pp. 60–63 (2012)
8. Paik, J.H., Mitra, M., Parui, S.K., Järvelin, K.: GRAS: an effective and efficient stemming algorithm for information retrieval. ACM Trans. Inf. Syst. **29**(4), 19 (2011)
9. Paik, J.H., Pal, D., Parui, S.K.: A novel corpus-based stemming algorithm using co-occurrence statistics. In: Proceedings of 34th Annual International ACM SIGIR Conference on Research and Development in Information Retrieval (SIGIR 2011), pp. 863–872. ACM Press (2011)
10. Paik, J.H., Parui, S.K.: A fast corpus-based stemmer. ACM Trans. Asian Lang. Inf. Process. **10**(2), 1–16 (2011)

11. Porter, M.F.: An algorithm for suffix stripping. Program **14**(3), 130–137 (1980)
12. Savoy, J.: Searching strategies for the Hungarian language. Inf. Process. Manage. **44**(1), 310–324 (2008)
13. Singh, J., Gupta, V.: Text stemming: approaches, applications, and challenges. ACM Comput. Surv. (CSUR) **49**(3), 45:1–45:46 (2016)

Cross-Lingual Document Retrieval Using Regularized Wasserstein Distance

Georgios Balikas[1(✉)], Charlotte Laclau[1], Ievgen Redko[2],
and Massih-Reza Amini[1]

[1] Univ. Grenoble Alpes, CNRS, Grenoble INP, LIG, Grenoble, France
geompalik@hotmail.com,
{charlotte.laclau,massih-reza.amini}@univ-grenoble-alpes.fr
[2] Univ. Lyon, INSA-Lyon, Univ. Claude Bernard Lyon 1, UJM-Saint Etienne,
CNRS, Inserm, CREATIS UMR 5220, U1206, F69XXX, Lyon, France
ievgen.redko@creatis.insa-lyon.fr

Abstract. Many information retrieval algorithms rely on the notion of
a good distance that allows to efficiently compare objects of different
nature. Recently, a new promising metric called Word Mover's Distance
was proposed to measure the divergence between text passages. In this
paper, we demonstrate that this metric can be extended to incorporate
term-weighting schemes and provide more accurate and computationally efficient matching between documents using entropic regularization.
We evaluate the benefits of both extensions in the task of cross-lingual
document retrieval (CLDR). Our experimental results on eight CLDR
problems suggest that the proposed methods achieve remarkable
improvements in terms of Mean Reciprocal Rank compared to several
baselines.

1 Introduction

Estimating distances between text passages is in the core of information retrieval
applications such as document retrieval, summarization and question answering.
Recently, in [11], Kusner et al. proposed the Word Mover's Distance (WMD),
a novel distance metric for text data. WMD is directly derived from the optimal transport (OT) theory [10,14] and is, in fact, an implementation of the
Wasserstein distance (also known as Earth Mover's distance) for textual data.
For WMD, a source and a target text span are expressed by high-dimensional
probability densities through the bag-of-words representation. Given the two
densities, OT aims to find the map (or transport plan) that minimizes the
total transportation cost given a ground metric for transferring the first density to the second. The ground metric for text data can be estimated using word
embeddings [11].

One interesting feature of the Wasserstein distance is that it defines a proper
metric on a space of probability measures. This distance presents several advantages when compared to other statistical distance measures, such as, for instance,

© Springer International Publishing AG, part of Springer Nature 2018
G. Pasi et al. (Eds.): ECIR 2018, LNCS 10772, pp. 398–410, 2018.
https://doi.org/10.1007/978-3-319-76941-7_30

the f- and the Jensen-Shannon divergences: (1) it is parametrized by the ground metric that offers the flexibility in adapting it to various data types; (2) it is known to be a very efficient metric due to its ability of taking into account the geometry of the data through the pairwise distances between the distributions' points. For all these reasons, the Wasserstein distance is of increasing interest to the machine learning community for various applications like: computer vision [18], domain adaptation [6], and clustering [12].

In this paper, our goal is to show how information retrieval (IR) applications can benefit from the Wasserstein distance. We demonstrate that for text applications, the Wasserstein distance can naturally incorporate different weighing schemes that are particularly efficient in IR applications, such as the inverse document frequency. This presents an important advantage compared to uniform weighting considered in the previous works on the subject. Further, we propose to use the regularized version of OT [7], which relies on entropic regularization allowing to obtain smoother, and therefore more stable results, and to solve the OT problem using the efficient Sinkhorn-Knopp matrix algorithm. From the application's perspective, we evaluate the use of Wasserstein distances in the task of Cross-Lingual Document Retrieval (CLDR) where given a query document (e.g., English Wikipedia entry for "Dog") one needs to retrieve its corresponding document in another language (e.g., French entry for "Chien"). In this specific context we propose a novel strategy to handle out-of-vocabulary words based on morphological similarity.

The rest of this paper is organized as follows. In Sect. 2, we briefly present the OT problem and its entropic regularized version. Section 3 presents the proposed approach and investigates different scenarios with respect to (w.r.t.) the weighting schemes, the regularization and the word embeddings. Empirical evaluations, conducted in eight cross-lingual settings, are presented in Sect. 4 and demonstrate that our approach is substantially more efficient than other strong baselines in terms of Mean Reciprocal Rank. The last section concludes the paper with a discussion of future research perspectives.

2   Preliminary Knowledge

In this section we introduce the OT problem [10] as well as its entropic regularized version that will be later used to calculate the regularized Wasserstein distance.

2.1   Optimal Transport

OT theory, originally introduced in [14] to study the problem of resource allocation, provides a powerful geometrical tool for comparing probability distributions.

In a more formal way, given access to two sets of points $X_S = \{x_i^S \in \mathbb{R}^d\}_{i=1}^{N_S}$ and $X_T = \{x_i^T \in \mathbb{R}^d\}_{i=1}^{N_T}$, we construct two discrete empirical probability distributions as follows

$$\hat{\mu}_S = \sum_{i=1}^{N_S} p_i^S \delta_{\boldsymbol{x}_i^S} \text{ and } \hat{\mu}_T = \sum_{i=1}^{N_T} p_i^T \delta_{\boldsymbol{x}_i^T},$$

where p_i^S and p_i^T are probabilities associated to $\boldsymbol{x}_i^S$ and $\boldsymbol{x}_i^T$, respectively and $\delta_{\boldsymbol{x}}$ is a Dirac measure that can be interpreted as an indicator function taking value 1 at the position of $\boldsymbol{x}$ and 0 elsewhere. For these two distributions, the Monge-Kantorovich problem consists in finding a probabilistic coupling γ defined as a joint probability measure over $X_S \times X_T$ with marginals $\hat{\mu}_S$ and $\hat{\mu}_T$ that minimizes the cost of transport with respect to some metric $l : X_s \times X_t \to \mathbb{R}^+$:

$$\min_{\gamma \in \Pi(\hat{\mu}_S, \hat{\mu}_T)} \langle A, \gamma \rangle_F$$

where $\langle \cdot, \cdot \rangle_F$ is the Frobenius dot product, $\Pi(\hat{\mu}_S, \hat{\mu}_T) = \{\gamma \in \mathbb{R}_+^{N_S \times N_T} | \gamma \mathbf{1} = \boldsymbol{p}^S, \gamma^T \mathbf{1} = \boldsymbol{p}^T\}$ is a set of doubly stochastic matrices and D is a dissimilarity matrix, i.e., $A_{ij} = l(\boldsymbol{x}_i^S, \boldsymbol{x}_j^T)$, defining the energy needed to move a probability mass from $\boldsymbol{x}_i^S$ to $\boldsymbol{x}_j^T$. This problem admits a unique solution γ^* and defines a metric on the space of probability measures (called the Wasserstein distance) as follows:

$$W(\hat{\mu}_S, \hat{\mu}_T) = \min_{\gamma \in \Pi(\hat{\mu}_S, \hat{\mu}_T)} \langle A, \gamma \rangle_F.$$

The success of algorithms based on this distance is also due to [7] who introduced an entropic regularized version of optimal transport that can be optimized efficiently using matrix scaling algorithm. We present this regularization below.

2.2 Entropic Regularization

The idea of using entropic regularization has recently found its application to the optimal transportation problem [7] through the following objective function:

$$\min_{\gamma \in \Pi(\hat{\mu}_S, \hat{\mu}_T)} \langle A, \gamma \rangle_F - \frac{1}{\lambda} E(\gamma).$$

The second term $E(\gamma) = -\sum_{i,j}^{N_S, N_T} \gamma_{i,j} \log(\gamma_{i,j})$ in this equation allows to obtain smoother and more numerically stable solutions compared to the original case and converges to it at the exponential rate [3]. The intuition behind it is that entropic regularization allows to transport the mass from one distribution to another more or less uniformly depending on the regularization parameter λ. Furthermore, it allows to solve the optimal transportation problem efficiently using Sinkhorn-Knopp matrix scaling algorithm [16].

3 Word Mover's Distance for CLDR

In this section, we explain the main underlying idea of our approach and show how the regularized optimal transport can be used in the cross-lingual information retrieval. We start with the formalization of our problem.

3.1 Problem Setup

For our task, we assume access to two document collections $\mathcal{C}^{\ell_1} = \{d_1^{\ell_1}, \ldots, d_N^{\ell_1}\}$ and $\mathcal{C}^{\ell_2} = \{d_1^{\ell_2}, \ldots, d_M^{\ell_2}\}$, where $d_n^{\ell_1}$ (resp. $d_m^{\ell_2}$) is the n-th (resp. m-th) document written in language ℓ_1 (resp. ℓ_2). Let the vocabulary size of the two languages be denoted as V^{ℓ_1} and V^{ℓ_2}. For the rest of the development, we assume to have access to dictionaries of embeddings E^{ℓ_1}, E^{ℓ_2} where words from ℓ_1 and ℓ_2 are projected into a *shared* vector space of dimension D, hence $E^{\ell_1} \in \mathbb{R}^{V^{\ell_1} \times D}, E^{\ell_2} \in \mathbb{R}^{V^{\ell_2} \times D}$ and $E_k^{\ell_1}, E_j^{\ell_2}$ denote the embeddings of words k, j. As learning the bilingual embeddings is not the focus of this paper, any of the previously proposed methods can be used e.g., [19,24]. A document consists of words and is represented using the Vector Space Model with frequencies. Hence, $\forall n, m : d_n^{\ell_1} \in \mathbb{R}^{V^{\ell_1}}, d_m^{\ell_2} \in \mathbb{R}^{V^{\ell_2}}$; the value $d_{nj}^{\ell_1}$ (resp. $d_{mk}^{\ell_2}$) then represents the frequency of word j (resp. k) in $d_n^{\ell_1} \in \mathbb{R}^{V^{\ell_1}}$ (resp. $d_m^{\ell_2} \in \mathbb{R}^{V^{\ell_2}}$). Calculating the distance of words in the embedding's space is naturally achieved using the Euclidean distance with lower values meaning that words are similar between them. For the rest, we denote by $A(j, k) = \|E_j^{\ell_1} - E_k^{\ell_2}\|_2$ the Euclidean distance between the words k and j in the embedding's space. Our goal is to estimate the distance of $d_n^{\ell_1}, d_m^{\ell_2}$, that are written in two languages, while taking advantage of the expressiveness of word embeddings and the Wasserstein distance.

3.2 Proposed Method

In order to use the Wasserstein distance on documents, we consider that the documents $d_n^{\ell_1}$ and $d_m^{\ell_2}$ from different languages are both modeled as empirical probability distributions, i.e.

$$d_n^{\ell_1} = \sum_{j=1}^{V^{\ell_1}} w_{nj} \delta_{d_{nj}^{\ell_1}} \text{ and } d_m^{\ell_2} = \sum_{k=1}^{V^{\ell_2}} w_{mk} \delta_{d_{mk}^{\ell_2}},$$

where w_{nj} and w_{mk} are probabilities associated with words j and k in $d_n^{\ell_1}$ and $d_m^{\ell_2}$, respectively. In order to increase the efficiency of optimal transport between these documents, it would be desirable to incorporate a proper weighting scheme that reflects the relative frequencies of different words appearing in a given text corpus. To this end, we use the following weighting schemes:

- *term frequency* (*tf*), that represents a document using the frequency of its word occurrences. This schema was initially proposed in [11] and corresponds to the case where $w_{nj} = d_{nj}$ and $w_{mk} = d_{mk}$.
- *term frequency-inverse document frequency* (*idf*), where the term frequencies are multiplied by the words' inverse document frequencies. In a collection of N documents, the document frequency $df(j)$ is the number of documents in the collection containing the word j. A word's inverse document frequency penalizes words that occur in many documents. As commonly done, we use a smoothed version of *idf*. Hence, we consider $w_{nj} = d_{nj} \times \log \frac{N+1}{df(j)+1}$ and $w_{mk} = d_{mk} \times \log \frac{M+1}{df(k)+1}$.

Furthermore, we use the Euclidean distance between the word embeddings of the two documents [11] as a ground metric in order to construct the matrix A. Now, we seek solving the following optimization problem:

$$\min_{\gamma \in \Pi\left(d_n^{\ell_1}, d_m^{\ell_2}\right)} \langle A, \gamma \rangle_F. \tag{1}$$

Given the solution γ^* of this problem, we can calculate the Wasserstein distance between documents as

$$W(d_n^{\ell_1}, d_m^{\ell_2}) = \langle A, \gamma^* \rangle_F = \mathrm{tr}(A\gamma^*).$$

As transforming the words of $d_n^{\ell_1}$ to $d_m^{\ell_2}$ comes with the cost $A(j, k)$, the optimization problem of Eq. (1) translates to the minimization of the associated cumulative cost of transforming all the words. The value of the minimal cost is the distance between the documents. Intuitively, the more similar the words between the documents are, the lower will be the costs associated to the solution of the optimization problem, which, in turn, means smaller document distances. For example, given "the cat sits on the mat" and its French translation "le chat est assis sur le tapis", the weights after stopwords filtering of "cat", "sits", "mat", and "chat", "assis", "tapis" will be 1/3. Given high-quality embeddings, solving Eq. (1) will converge to the one-to-one transformations "cat-chat", "sits-assis" and "mat-tapis", with very low cumulative cost as the paired words are similar.

This one-to-one matching, however, can be less efficient when documents with larger vocabularies are used. In this case, every word can be potentially associated with, not a single, but several words representing its synonyms or terms often used in the same context. Furthermore, the problem of Eq. (1) is a special case of the Earth Mover's distance [17] and presents a standard Linear Programming problem that has a computation complexity of $\mathcal{O}(n^3 \log(n))$. When n is large, this can present a huge computational burden. Hence, it may be more beneficial to use the entropic regularization of optimal transport that allows more associations between words by increasing the entropy of the coupling matrix and can be solved faster, in linear time. Our second proposed model thus reads

$$\min_{\gamma \in \Pi\left(d_n^{\ell_1}, d_m^{\ell_2}\right)} \langle A, \gamma \rangle_F - \frac{1}{\lambda} E(\gamma). \tag{2}$$

As in the previous problem, once γ^* is obtained, we estimate the entropic regularized Wasserstein distance (also known as Sinkhorn distance) as

$$W(d_n^{\ell_1}, d_m^{\ell_2}) = \langle A, \gamma^* \rangle_F = \mathrm{tr}(A\gamma^*) - \frac{1}{\lambda} E(\gamma^*).$$

Algorithm 1 summarizes the CLDR process with Wasserstein distance. We also illustrate the effect of regularization, controlled by λ in the OT problem of Eq. (2). Figure 1 presents the obtained coupling matrices and the underlying word matchings between the words of the example we considered above when varying λ. We project the words in 2-D space using t-SNE as our dimensionality

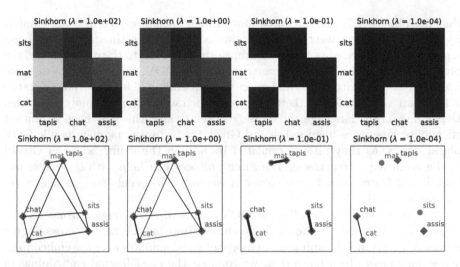

Fig. 1. The impact of entropic regularization on the transport plan. In the top matrices, whiter cells suggest stronger associations in the obtained plan. The association strength is also indicated by wider lines in the bottom figure.

reduction technique.[1] Notice that high λ values lead to the uniform association weights between all the words while the lowest value leads to a complete algorithm failure. For $\lambda = 1$ the corresponding pairs are associated with the bold lines showing that this link is more likely than the other fading lines. Finally, $\lambda = 0.1$ gives the optimal, one-to-one matching. This figure shows that entropic regularization encourages the "soft" associations of words with different degrees of strength. Also, it highlights how OT accounts for the data geometry, as the strongest links occurs between the words that are closest it the space.

Algorithm 1. CLDR with Wasserstein distance.

Data: Query $d_n^{\ell_1}$, Corpus C^{ℓ_2}, Embeddings E^{ℓ_1}, E^{ℓ_2}
if idf **then** Apply idf weights on $d_n^{\ell_1}$, $\forall d_m^{\ell_2} \in C^{\ell_2}$;
L_1-normalize $d_n^{\ell_1}$ and $\forall d_m^{\ell_2} \in C^{\ell_2}$;
for document $d_m^{\ell_2} \in C^{\ell_2}$ **do**
$\quad | \quad dist[m] = Wass(d_n^{\ell_1}, d_m^{\ell_2}, A, \lambda)$; # *Wass* solves Eq. (1) or Eq. (2)*
end
Result: $\arg\min(dist)$: increasing list of distances between $d_n^{\ell_1}$ and $\forall d_m^{\ell_2} \in C^{\ell_2}$

3.3 Out-of-Vocabulary Words

An important limitation when using the dictionaries of embeddings E^{ℓ_1}, E^{ℓ_2} is the out-of-vocabulary (OOV) words. High rates of OOV result in loss of information. To partially overcome this limitation one needs a protocol to handle them. We propose a simple strategy for OOV that is based on the strong assumption

---

[1] We use the Numberbatch embeddings presented in our experiments (Sect. 4).

that morphologically similar words have similar meanings. To measure similarity between words we use the Levenshtein distance, that estimates the minimum number of character edits needed to transform one word to the other. Hence, the protocol we use is as follows: in case a word w^{ℓ} is OOV, we measure its distance from every other word in $\boldsymbol{E}^{\ell}$, and select the embedding of a word whose distance is less than a threshold t. If there are several such words, we randomly select one. Depending on the language and on the dictionary size, this may have significant effects: for languages like English and Greek for instance one may handle OOV plural nouns, as they often formulated by adding the suffix "s" (ς in Greek) in the noun (e.g., tree/trees). The same protocol can help with languages like French and German that have plenty of letters with accents such as acutes (é) and umlauts (ü).

The above strategy handles the OOV words of a language ℓ using its dictionary E^{ℓ}. To fine-tune the available embeddings for the task of cross-lingual retrieval, we extend the argument of morphological similarity for the embeddings across languages. To achieve that, we collapse the cross-lingual embeddings of alike words. Hence, if for two words w^{ℓ_1} and w^{ℓ_2} the Levenshtein distance is zero, we use the embedding of the language with the biggest dictionary size for both. As a result, the English word "transition" and the French word "transition" will use the English embedding. Of course, while several words may be that similar between English and French, there will be fewer, for instance, for English and Finnish or none for English and Greek as they use different alphabets.

We note that the assumption of morphologically similar words having similar meanings and thus embeddings is strong; we do not claim it to be anything more than a heuristic. In fact, one can come up with counter-examples where it fails. We believe, however, that for languages with less resources than English, it is an heuristic that can help overcome the high rates of OOV. Its positive or negative impact for CLDR remains to be empirically validated.

4   Experimental Framework

In our experiments we are interested in CLDR whose aim is to identify corresponding documents written in different languages. Assuming, for instance, that one has access to English and French Wikipedia documents, the goal is to identify the cross-language links between the articles. Traditional retrieval approaches employing bag-of-words representations perform poorly in the task as the vocabularies vary across languages, and words from different languages rarely co-occur.

Datasets. To evaluate the suitability of OT distances for cross-lingual document retrieval we extract four bilingual ($\ell_1 - \ell_2$) Wikipedia datasets: (i) English-French, (ii) English-German, (iii) English-Greek and (iv) English-Finnish. Each dataset defines two retrieval problems: for the first ($\ell_1 \rightarrow \ell_2$) the documents of ℓ_1 are retrieved given queries in ℓ_2; for the second ($\ell_2 \rightarrow \ell_1$) the documents of ℓ_1 are the queries. To construct the Wikipedia excerpts, we use the comparable

Wikipedia corpora of *linguatools*.[2] Following [23], the inter-language links are the golden standard and will be used to calculate the evaluation measures. Compared to "ad hoc" retrieval problems [21] where there are several relevant documents for each query, this is referred to as a "known-search" problem as there is exactly one "right" result for each query [5]. Our datasets comprise 10K pairs of comparable documents; the first 500 are used for evaluating the retrieval approaches. The use of the remaining 9.5K, dubbed "BiLDA Training" is described in the next paragraph. Table 1(a) summarizes these datasets. In the pre-processing steps we lowercase the documents, we remove the stopwords, the punctuation symbols and the numbers. We keep documents with more than five words and, for efficiency, we keep for each document the first 500 words.

Table 1. (a) Statistics for the Wikipedia datasets. W^ℓ is the corpus size in words; V^ℓ is the vocabulary size. (b) The size of the embedding's dictionary.

	Wiki$_{En-Fr}$	Wiki$_{En-Ge}$	Wiki$_{En-Fi}$	Wiki$_{En-Gr}$	Language	$|E|$
$|D|$	10K	10K	10K	10K	En	417,194
V^{ℓ_1}	33,925	33,198	43,230	34,192	Fr	296,986
V^{ℓ_2}	26,602	44,896	31,112	28,192	Ge	129,405
W^{ℓ_1}	36M	36M	58M	39M	Fi	56,899
W^{ℓ_2}	24M	31M	15M	14M	Gr	16,925

Systems. One may distinguish between two families of methods for cross-lingual IR: translation-based and semantic-based. The methods of the first one use a translation mechanism to translate the query from the source language to the target language. Then, any of the known retrieval techniques can be employed. The methods of the second family project both the query and the documents of the target language in a shared semantic space. The calculation of the query-document distances is performed in this shared space.

To demonstrate the advantages of our methods, we present results using systems that rely either on translations or on cross-lingual semantic spaces. Concerning the translation mechanism, we rely on a dictionary-based approach. To generate the dictionaries for each language pair we use Wiktionary and, in particular, the open implementation of [1,2]. For a given word, one may have several translations: we pick the candidate translation according to a unigram language model learned on "BiLDA training" data. In the rest, we compare:[3]

-tf: Euclidean distance between the term-frequency representation of documents. To be applied for CLDR, the query needs to be translated in the language of the target documents.

-idf: Euclidean distance between the idf representation of documents. As with tf, the query needs to be translated.

-nBOW$_x$: nBOW (neural bag-of-words) represents documents by a weighted average of their words' embeddings [4,13]. If x=tf the output is the result of averaging

---

[2] http://linguatools.org/tools/corpora/wikipedia-comparable-corpora/.
[3] We release the code at: https://github.com/balikasg/WassersteinRetrieval.

the embeddings of the occurring words. If $x=$idf, then the embedding of each word is multiplied with the word's inverse document frequency. Having the nBOW representations the distances are calculated using the Euclidean distance. nBOW_x methods can be used both with cross-lingual embeddings, or with mono-lingual embeddings if the query is translated.

–BiLDA: Previous work found the bilingual Latent Dirichlet Allocation (BiLDA) to yield state-of-the-art results, we cite for instance [9,23,26]. BiLDA is trained on comparable corpora and learns aligned per-word topic distributions between two or more languages. During inference it projects unseen documents in the *shared* topic space where cross-lingual distances can be calculated efficiently. We train BiLDA separately for each language pair with 300 topics. We use collapsed Gibbs sampling for inference [23], which we implemented with Numpy [25]. Following previous work, we set the Dirichlet hyper-parameters $\alpha = 50/K$ (K being the number of topics) and $\beta = .01$. We let 200 Gibbs sampling iterations for burn-in and then sample the document distributions every 25 iterations until the 500-th Gibbs iteration. For learning the topics, we used the "BiLDA training" data. Having the per-document representations in the shared space of the topics, we use entropy as distance, following [9].

–Wass_x: is the proposed metric given by Eq. 1. If $x=$tf, it is equivalent to that of [11], as for generating the high-dimensional source and target histograms the terms' frequencies are used. If $x=$idf, the idf weighting scheme is applied.

–Entro_Wass_x is the proposed metric given by Eq. 2. The subscript x reads the same as for the previous approach. We implemented Wass_x and Entro_Wass_x with Scikit-learn [15] using the solvers of POT [8].[4] For the importance of the regularization term in Eq. (2), we performed grid-search $\lambda \in \{10^{-3}, \ldots, 10^2\}$ and found $\lambda = 0.1$ to consistently perform the best.

For the systems that require embeddings (nBOW, Wass, Entro_Wass), we use the Numberbatch pre-trained embeddings of [19].[5] The Numberbatch embeddings are 300-dimensional embeddings for 78 languages, that project words and short expressions of these languages in the same shared space and were shown to achieve state-of-the-art results in cross-lingual tasks [19,20].

Complementary to tf and idf document representations, we also evaluate the heuristic we proposed for OOV words in Sect. 3.3. We select the threshold t for the Levenhstein distances to be 1, and we denote with tf+ and idf+ the settings where the proposed OOV strategy is employed.

Results. As evaluation measure, we report the Mean Reciprocal Rank (MRR) [22] which accounts for the rank of the correct answer in the returned documents. Higher values signify that the golden documents are ranked higher. Table 2 presents the achieved scores for the CLDR problems.

There are several observations from the results of this table. First, notice that the results clearly establish the superiority of the Wasserstein distance for

---

[4] For Entro_Wass we used the sinkhorn2 function with reg=0.1, numItermax=50, method='sinkhorn_stabilized' arguments to prevent numerical errors.

[5] The v17.06 vectors: https://github.com/commonsense/conceptnet-numberbatch.

Table 2. The MRR scores achieved by the systems. With asterisks we to denote our original contributions. $\texttt{Entro\_Wass}_{idf+}$ consistently achieves the best performance by a large margin.

	En → Fr	Fr → En	En → Ge	Ge → En	En → Gr	Gr → En	En → Fi	Fi → En
Systems that rely on topic models								
BiLDA	.559	.468	.560	.536	.559	.508	.622	.436
Systems that rely on translations								
tf	.421	.278	.287	.433	.441	.054	.068	.025
idf	.575	.433	.478	.516	.535	.126	.081	.028
nBOW$_{tf}$	.404	.276	.378	.440	.531	.132	.329	.042
nBOW$_{idf}$	.550	.488	.449	.509	.577	.256	.398	.081
Wass$_{tf}$	.704	.691	.620	.655	.656	.269	.424	.092
Wass$_{idf}$*	.748	.766	.678	.706	.687	.463	.412	.163
Entro_Wass$_{tf}$*	.692	.706	.615	.666	.640	.262	.422	.089
Entro_Wass$_{idf}$*	.745	.794	.675	.720	.683	.467	.425	.171
Systems that rely on cross-lingual embeddings								
nBOW$_{tf}$	.530	.490	.493	.464	.237	.121	.449	.217
nBOW$_{idf}$	.574	.546	.521	.502	.341	.179	.470	.267
Wass$_{tf}$	.744	.748	.660	.681	.404	.407	.582	.434
Wass$_{idf}$*	.778	.784	.703	.718	.507	.465	.620	.479
Entro_Wass$_{tf}$*	.753	.786	.677	.710	.424	.494	.582	.607
Entro_Wass$_{idf}$*	.799	.820	.717	.756	.523	.549	.620	.643
Handling OOV words & cross-lingual embeddings								
nBOW$_{tf+}$	.518	.541	.426	.468	.193	.148	.635	.451
nBOW$_{idf+}$	.653	.659	.597	.592	.407	.349	.693	.544
Wass$_{tf+}$	.815	.836	.788	.801	.675	.435	.845	.731
Wass$_{idf+}$*	.867	.869	.812	.837	.721	.599	.856	.786
Entro_Wass$_{tf+}$*	.830	.856	.796	.812	.718	.555	.851	.816
Entro_Wass$_{idf+}$*	**.875**	**.887**	**.828**	**.855**	**.741**	**.695**	**.864**	**.850**

CLDR. Independently of the representation used ($\texttt{tf}$, $\texttt{idf}$, $\texttt{tf+}$, $\texttt{idf+}$) the performance when the Wasserstein distance is used is substantially better than the other baselines. This is due to the fact that the proposed distances account for the geometry of the data. In this sense, they essentially implement optimal word-alignment algorithms as the calculated transportation cost uses the word representations in order to minimize their transformation from the source to the target document. Although $\texttt{nBOW}$ also uses exactly the same embeddings, it performs a weighted averaging operation that results in information loss.

Comparing the two proposed methods, we notice that the approach with the entropic regularization ($\texttt{Entro\_Wass}$) outperforms in most of the cases its original version $\texttt{Wass}$. This suggests that using regularization in the OT problem improves the performance for our application. As a result, using $\texttt{Entro\_Wass}$ is not only faster and GPU-friendly, but also more accurate. Also, both approaches consistently benefit from the $\texttt{idf}$ weighting scheme. The rest of the baselines, although competitive, perform worse than the proposed approaches.

Another interesting insight stems from the comparison of the translation-based and the semantic-based approaches. The results suggest that the semantic-based approaches that use the Numberbatch embeddings perform better, meaning that the machine translation method we employed introduces more error than the imperfect induction of the embedding spaces. This is also evident by the performance decrease of tf and idf when moving from language pairs with more resources like "En-Fr" to more resource deprived pairs like "En-Fi" or "En-Gr". While one may argue that better results can be achieved with a better-performing translation mechanism, the important outcome of our comparison is that both families of approaches improve when the Wasserstein distance is used. Notice, for instance, the characteristic example of the translation based systems for "Fi→En": tf, idf and their nBOW variants perform poorly (MRR ≤ 0.09), suggesting low-quality translations; still Wass and Entro_Wass achieve remarkable improvements (MRR∼0.17), using the same resources.

Our last comments concern the effect of the OOV protocol. Overall, having such a protocol in place benefits Wass and Entro_Wass as the comparison of the tf and idf with tf+ and idf+ variants suggests. The impact of the heuristic is more evident for the "En-Gr" and "En-Fi" problems that gain several (∼.20) points in terms of MRR. This is also due to the fact that the proposed OOV mechanism reduces the OOV rates as Greek and Finnish have the smallest embeddings dictionary as shown in Table 1b.

5   Conclusions

In this paper, we demonstrated that the Wasserstein distance and its regularized version naturally incorporate term-weighting schemes. We also proposed a novel protocol to handle OOV words based on morphological similarity. Our experiments, carried on eight CLDR datasets, established the superiority of the Wasserstein distance compared to other approaches as well as the interest of integrating entropic regularization to the optimization, and $tf - idf$ coefficients to the word embeddings. Finally, we showed the benefits of our OOV strategy, especially when the size of the embedding's dictionary for a language is small.

Our study opens several avenues for future research. First, we plan to evaluate the generalization of the Wasserstein distances for ad hoc retrieval, using for instance the benchmarks of the CLEF ad hoc news test suites. Further, while we showed that entropic regularization greatly improves the achieved results, it remains to be studied how one can apply other types of regularization to the OT problem. For instance, one could expect that group sparsity inducing regularization applied in the CLDR context can be a promising direction as semantically close words intrinsically form clusters and thus it appears meaningful to encourage the transport within them. Lastly, CLDR with Wasserstein distances is an interesting setting for comparing methods for deriving cross-lingual embeddings as their quality directly impacts the performance on the task.

References

1. Acs, J.: Pivot-based multilingual dictionary building using Wiktionary. In: LREC (2014)
2. Acs, J., Pajkossy, K., Kornai, A.: Building basic vocabulary across 40 languages. In: Sixth Workshop on Building and Using Comparable Corpora@ACL (2013)
3. Benamou, J.D., Carlier, G., Cuturi, M., Nenna, L., Peyré, G.: Iterative Bregman projections for regularized transportation problems. SIAM J. Sci. Comput. **2**(37), A1111–A1138 (2015)
4. Blacoe, W., Lapata, M.: A comparison of vector-based representations for semantic composition. In: EMNLP-CoNLL (2012)
5. Broder, A.: A taxonomy of web search. In: SIGIR. ACM (2002)
6. Courty, N., Flamary, R., Tuia, D.: Domain adaptation with regularized optimal transport. In: Calders, T., Esposito, F., Hüllermeier, E., Meo, R. (eds.) ECML PKDD 2014. LNCS (LNAI), vol. 8724, pp. 274–289. Springer, Heidelberg (2014). https://doi.org/10.1007/978-3-662-44848-9_18
7. Cuturi, M.: Sinkhorn distances: lightspeed computation of optimal transport. In: NIPS, pp. 2292–2300 (2013)
8. Flamary, R., Courty, N.: Pot python optimal transport library (2017)
9. Fukumasu, K., Eguchi, K., Xing, E.P.: Symmetric correspondence topic models for multilingual text analysis. In: NIPS (2012)
10. Kantorovich, L.: On the translocation of masses. C.R. (Doklady) Acad. Sci. URSS(N.S.) **37**(10), 199–201 (1942)
11. Kusner, M.J., Sun, Y., Kolkin, N.I., Weinberger, K.Q., et al.: From word embeddings to document distances. In: ICML (2015)
12. Laclau, C., Redko, I., Matei, B., Bennani, Y., Brault, V.: Co-clustering through optimal transport. In: ICML (2017)
13. Mitchell, J., Lapata, M.: Composition in distributional models of semantics. Cogn. Sci. **34**, 1388–1429 (2010)
14. Monge, G.: Mémoire sur la théorie des déblais et des remblais. Histoire de l'Académie Royale des Sciences, pp. 666–704 (1781)
15. Pedregosa, F., Varoquaux, G., Gramfort, A., et al.: Scikit-learn: machine learning in python. JMLR **12**, 2825–2830 (2011)
16. Richard, S., Paul, K.: Concerning nonnegative matrices and doubly stochastic matrices. Pac. J. Math. **21**, 343–348 (1967)
17. Rubner, Y., Tomasi, C., Guibas, L.J.: A metric for distributions with applications to image databases. In: ICCV (1998)
18. Rubner, Y., Tomasi, C., Guibas, L.J.: The earth mover's distance as a metric for image retrieval. Int. J. Comput. Vis. **40**, 99–121 (2000)
19. Speer, R., Chin, J., Havasi, C.: Conceptnet 5.5: an open multilingual graph of general knowledge. In: AAAI (2017)
20. Speer, R., Lowry-Duda, J.: Conceptnet at semeval-2017 task 2: extending word embeddings with multilingual relational knowledge. arXiv:1704.03560 (2017)
21. Voorhees, E.M.: Overview of TREC 2003. In: TREC (2003)
22. Voorhees, E.M., et al.: The TREC-8 question answering track report. In: TREC (1999)
23. Vulić, I., De Smet, W., Tang, J., Moens, M.F.: Probabilistic topic modeling in multilingual settings: an overview of its methodology and applications. Inf. Process. Manage. **51**, 111–147 (2015)

24. Vulić, I., Moens, M.F.: Bilingual distributed word representations from document-aligned comparable data. J. Artif. Intell. Res. **55**, 953–994 (2016)
25. van der Walt, S., Colbert, S.C., Varoquaux, G.: The NumPy array: a structure for efficient numerical computation. Comput. Sci. Eng. **13**, 22–30 (2011)
26. Wang, Y.C., Wu, C.K., Tsai, R.T.H.: Cross-language article linking with different knowledge bases using bilingual topic model and translation features. Knowl.-Based Syst. **111**, 228–236 (2016)

T-Shaped Mining: A Novel Approach to Talent Finding for Agile Software Teams

Sajad Sotudeh Gharebagh[✉], Peyman Rostami, and Mahmood Neshati

Faculty of Computer Science and Engineering,
Shahid Beheshti University, G.C., Tehran, Iran
{s.sotudeh,p.rostami}@mail.sbu.ac.ir, m_neshati@sbu.ac.ir

Abstract. Human resources management is one of the most overriding parts of organizations. They are always willing to hire individuals who meet their requirements while do not impose high costs on the organization. Hence, most organizations, in particular, those which are engaged in Computer Engineering industry are inclined to find and employ individuals who are characterized by their deep disciplinary knowledge in one single area, and their ability to collaborate across different aspects of projects due to their general knowledge in other areas. Nowadays, Community Question Answering i.e. CQA websites are among the best places to find experts. In this study, we propose two models to find and then rank experts with specialty in a specific skill area, as well as general knowledge in the other skill areas i.e. T-shaped users. We estimate the profile diversity of users in our models to detect those who have the aforementioned feature in CQAs, particularly Stackoverflow. Our experiments on three real test collections generated from Stackoverflow's published data indicate the efficiency of the proposed models in comparison with the state-of-the-art expertise retrieval approach.

1 Introduction

Human resources management (HRM) is one of the most precious assets of any organization. A considerable amount of organizations' budget is dedicated to the employment of human resources. The process of hiring new employees is costly, hence human resources departments are always looking for less costly ways to recruit new members. Recruiting the right person is crucial since if the new employee's skills do not satisfy the minimum requirements of the job position, it might cause a delay or even failure in projects.

There are many available resources for organizations to find the right person for their needs. One of these resources is Community Question Answering (CQA) networks such as Stackoverflow[1] and Stackexchange[2] which are specialized in specific domains [1,2]. CQA networks create a community of users by

---

[1] https://stackoverflow.com.
[2] https://stackexchange.com.

© Springer International Publishing AG, part of Springer Nature 2018
G. Pasi et al. (Eds.): ECIR 2018, LNCS 10772, pp. 411–423, 2018.
https://doi.org/10.1007/978-3-319-76941-7_31

encouraging them to answer and ask questions which are related to different topics. Users can provide feedback to the community by *accepting* answers, giving *up-vote* and *down-vote*, and also *commenting* [3]. Moreover, CQA websites employ gamification approaches to motivate users and accordingly improve the quality of their contribution [4]. As a renowned CQA website, Stackoverflow has become one of the most useful resources for developers and software engineers to find answers to their questions, as well as for recruiters to find talented candidates in specific areas of software engineering [5].

Different elements should be combined together to make a software solution and deliver an end-to-end service. A typical software project is composed of different layers including data access layer, business logic layer, and also the user interface layer. In a software development team, there are also different roles including designers, analyzers, back-end/front-end developers, DBAs (i.e. database administrators), DevOps, and also testers, all who are responsible to develop different facets of a project. Various skills are required to play these roles to deliver a project on time and within budget. As an instance, to develop a flawless Java web-based application, front-end developers should be expert in client-side frameworks/languages such as *javascript, jquery*, and *html* while back-end developers should be professional in dealing with server-side frameworks/languages like *hibernate, spring, j2ee* and so forth.

Nowadays, most IT companies adopt agile methodologies as their software development process [6]. In agile software development, the size of DevOps teams is usually small (typically less than 5) and team members are required to communicate and interact with each other effectively. One approach to form agile teams is to hire people with multiple expertise areas (i.e. full-stack developers). Although this approach can enhance the communication between team members, it imposes high costs on the company since those who are expert in multiple areas are usually experienced professionals and accordingly they demand high salaries. A superior approach to overcome this problem is to appoint people who are expert in a single area and have general knowledge in other areas with which the project is involved [6]. To be more specific, suppose a scenario in which a company wants to get its two positions filled. The first position is involved with back-end technologies whereas the second one is related to front-end. As a matter of fact, this company has the following variety of options to choose:

1. Exploiting a full-stack developer who takes the responsibility of the two positions.
2. Hiring two full-stack developers and assigning them to the positions.
3. Recruiting a "generalizing specialist"[3] for one of the positions, and a full-stack developer for another position.
4. Employing two specialists to fill the two positions.
5. Appointing two generalizing specialists for each of the positions.

---

[3] In our example, a person who is specialized in front-end development, but also has general knowledge in back-end technologies (or vice versa) is called generalizing specialist.

The best choice would be the fifth option. That is, in the first option, the project becomes individual-centric and there is not any backup person for that individual. The second and third options inflict relatively higher costs on the company. That is, a full-stack developer demands much more salary than a generalizing specialist[4]. Moreover, the fourth option is in contrast with the manifesto of modern software engineering as agile teams' members are expected to collaborate across different aspects of projects if needed.

Over the recent years, several research studies have been made on the subject of analyzing user behavior patterns in CQAs [2,7,8]. Following *generalizing specialists* concept introduced in [6,7], in this paper, we propose two models to label Stackoverflow's users as non-experts, T-shaped users, and C-shaped (i.e. Comb-shaped) users. T-shaped users are those with deep knowledge in a particular skill area (the area of specialization, expressed as the vertical bar of the T) and also general knowledge in more than zero area (the areas of generalization, expressed as the horizontal bar of the T) whereas Comp-shaped users, as the name implies, have deep knowledge in several skill areas. As mentioned before, recruiters of agile teams prefer to hire T-shaped candidates rather than C-shaped as T-shaped candidates are more affordable. Additionally, in modern IT industry, T-shaped employees are extremely in demand due to an increasing convergence of technologies and changes in software development. Our experiments indicate that although the state-of-the-art model proposed by Balog *et al.* [9] can retrieve T-shaped users to some extent, it is biased towards C-shaped users. In this study, we propose two models to promote T-shaped users to better ranks, compared to the document-based model proposed in [9].

2   Related Work

Expert finding has been an active area of research in Information Retrieval [10,11]. Different approaches to this problem including candidate-based and document-based models are discussed in [9].

The earliest allusion to T-shaped people was coined by Guest [12]. More recently, with the advent of agile methodologies, a need has emerged for finding T-shaped people who can work interdisciplinarily in software project teams [6], according to the manifesto of the agile software development.

Moreover, Kumar and Pedanekar [7] introduced the idea of mining shapes of user expertise in a Community Question Answering (CQA). They found that expertise in CQA forums often involves making contribution in a variety of areas of expertise rather than a single area. They defined different expertise shapes including I-, T-, and C-shaped, however, in their study, they merely proposed a method to label each candidate, but not to rank them. In our study, we aim to propose a practical retrieval model to find T-shaped users on Stackoverflow and also name expertise of users in our golden set.

---

[4] Assuming that a full-stack developer is expert in multiple areas.

3   Mining the Shape of Expertise

In this section, in order to generate our golden set, we first explain the skill area's concept and the ways of generating them for each data collection. Then, in each skill area, we divide users into three groups according to their knowledge level. Finally, we determine users' shape of expertise based on their knowledge level in the set of their associate skill areas.

3.1   Skill Areas

Each question in Stackoverflow is associated with one or more tags which can be used to explore and search relevant questions. A set of related tags can represent a skill area in terms of programming language. For example, *swing, jtable, user interface, jpanel, and jframe*, which are semantically related tags, form *Java User Interface* skill area. In other words, each skill area contains multiple tags. Recruiters usually prefer to find experts on specific skill areas since skill areas are rather broader topics.[5]

Stackoverflow contains questions and answers which are related to different programming languages. We have divided the entire collection of posts into smaller collections which are related to specific programming languages. In order to identify skill areas, we went through three steps. Firstly, we extracted top-200 frequent tags from each collection. Secondly, we employed agglomerative clustering (average linkage) algorithm to obtain an initial clustering of tags. To set up this algorithm, the similarity between each pair of tags (t_1 and t_2) is calculated using Jaccard Coefficient [13] as follows:

$$Similarity(t_1, t_2) = \frac{\mid Q_{t_1} \bigcap Q_{t_2} \mid}{\mid Q_{t_1} \bigcup Q_{t_2} \mid} \tag{1}$$

where Q_t is the set of questions containing tag t. Finally, in order to have a more precise clustering, we asked a group of recruiters to revise the initial clustering according to the most demanding skill areas they are interested in hiring. Table 1 indicates some of these skill areas for *java*, and *android* collections.[6]

3.2   Knowledge Levels in Stackoverflow

In the realm of Stackoverflow, the accepted answers associated with a skill area can be considered as an evidence of the author's knowledge in the corresponding skill area. For a given skill area *sa*, the knowledge of a user *e* can be rated into three different levels including *beginner, intermediate*, and *advanced*. In order to define these knowledge levels, we define precision and recall of the answers provided by user *e* as follows:

---

[5] For example, recruiters are looking for experts on *User Interface* rather than *jtable* or *jframe* or etc.

[6] The complete list of skill areas with associated tags is made publicly available at http://bit.ly/tshaped-mining.

Table 1. Sample of skill areas for Java and Android collections

Java		Android	
OOP	Spring	User interface	Game
inheritance	spring	textview	unity3d
class	spring-mvc	relativelayout	libgdx
object	spring-security	scrollview	andengine

- *Precision(sa, e)*: The ratio of the accepted answers associated with user e to all answers provided by him on those which are associated with skill area sa.
- *Recall(sa, e)*: The normalized ratio[7] of the accepted answers provided by user e to the total number of accepted answers associated with skill area sa.

To combine these two measures, we have utilized *F-measure* as follows:

$$F = \frac{2 * Recall * Precision}{Recall + Precision} \tag{2}$$

Users of each skill area are sorted in descending order according to their *F-measure* value and then top 5% of the ranking is marked as users who hold advanced knowledge, the next 20% possess intermediate knowledge, and the rest have beginner knowledge on the corresponding skill area. To put it another way, we have considered top **25%** intermediated or advanced users and the rest as beginners.

3.3 Categorization of Users

Stackoverflow's users can be categorized into three groups depending on their count of associated skill areas and depth of knowledge in such areas as follows.[8]

- **Non-expert:** The users who do not have advanced knowledge in any skill areas are known as non-expert users.
- **T-shaped:** Those users who possess advanced knowledge in one single skill area, as well as intermediate knowledge in one or more skill areas, are characterized as T-shaped users.
- **C-shaped:** Those users who hold advanced knowledge in more than one skill areas are known as C-shaped users.

Based on these definitions, we have labeled the users who have posted in three communities of Java, Android, and C#. The categorization of users and the portion of each type in Java collection are shown in Fig. 1. As mentioned in Sect. 1, rather than C-shaped experts, recruiters usually prefer to hire T-shaped experts since they are more affordable. Thus, **T-shaped expert finding** is a practical and industry-motivated problem in expert retrieval.

---

[7] Min-Max Normalization has been applied.

[8] It has to be noted that other shapes of knowledge (e.g. I-shape, Π-shape) can be defined but for the sake of simplicity, we leave them out in this paper.

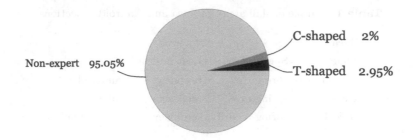

Fig. 1. The share of expertise shapes in Java collection

4   Modeling Expertise Retrieval

Expert finding task is defined as retrieving a list of candidates which are ranked according to a given query (i.e. skill area). In this section, we first introduce the preliminaries and the baseline approach proposed by Balog *et al.* [9] and then describe entropy-based approach alongside the extended entropy-based approach.

4.1   Preliminaries

For a given skill area sa, the problem of detecting T-shaped users can be formulated using generative probabilistic model which expresses the probability of candidate e being T-shaped and relevant to skill area sa i.e. $P(R = 1, T = 1|e, sa)$ where R and T are binary random variables indicating the relevancy and T-shape likeness of candidate e, respectively. By assuming conditional independence between R and T, this likelihood is calculated as shown in Eq. 3

$$P(R = 1, T = 1|e, sa) = P(R = 1|e, sa).P(T = 1|e, sa) \qquad (3)$$

in which $P(R = 1|e, sa)$ denotes the probability that candidate e be relevant to skill area sa and can be obtained using the state-of-the-art document-based model proposed by Balog *et al.* Additionally, $P(T = 1|e, sa)$ indicates the probability of being T-shaped, assuming that candidate e holds an advanced knowledge in skill area sa. A significant amount of this paper is devoted to estimate the likelihood of $P(T = 1|e, sa)$.

4.2   Document-Based Approach (DBA)

As mentioned in previous section, $P(R = 1|e, sa)$ can be determined by the model proposed by Balog *et al.* [9]. This probability is conceptually equivalent to $P(e|sa)$. In this model, for a given skill area sa, the relevance probability of a user e is determined by the following equation:

$$P(R = 1|e, sa) = P(e|sa) = \frac{P(e)P(sa|e)}{P(sa)} \qquad (4)$$

where $P(e)$ denotes the prior relevance probability of user e and is considered to be uniform in document-based model, $P(sa)$ is constant and does not affect ranking and will be ignored in the calculations. Therefore, $P(e|sa)$ can be estimated by $P(sa|e)$ which is calculated using Eq. 5.

$$P(sa|e) = \sum_{d \in D_e} \prod_{t \in sa} \left\{ (1 - \lambda_d).P(t|d) + \lambda_d.P(t) \right\}^{n(t,sa)}.P(d|e) \qquad (5)$$

where sa is the concatenation of the related tags representing skill area sa, $P(sa|e)$ denotes the relevance probability of candidate e to query sa according to the document-based model, D_e is the set of answers associated with candidate e, λ_d is the parameters of the model, and $P(d|e)$ is the binary association strength between document d and candidate e.

4.3   Entropy-Based Approach (EBA)

As mentioned in Sect. 4.1, we aim to estimate $P(T = 1|e, sa)$ which can be obtained using Bayes' Theorem as follows:

$$P(T = 1|e, sa) = \frac{P(T = 1)}{P(e, sa)}.P(e, sa|T = 1) \qquad (6)$$

$$= \frac{P(e)}{P(e, sa)}.P(T = 1|e).P(sa|e, T = 1) \qquad (7)$$

in which $\frac{P(e)}{P(e,sa)}$ is constant for a given candidate e and skill area sa. Therefore, it can be omitted from the calculations, $P(T = 1|e)$ denotes the probability that candidate e is T-shaped. Moreover, $P(sa|e, T = 1)$ is the likelihood that candidate e is T-shaped and have advanced knowledge on skill area sa[9]. In this method, this probability is set to be uniform for all skill areas, for a given candidate e (i.e. $P(sa|e, T = 1) \approx 1$).

In order to estimate $P(T = 1|e)$, we should take two indicators into consideration. First, the expertise of a T-shaped user is deeper than a non-expert user. So, this probability is proportional to the answer counts associated with the user (i.e. $|D_e|$)[10] as each answer is an evidence of his expertise. Second, diversity of a T-shaped user is less than a C-shaped user. Therefore, this probability negatively correlates with user's diversity which can be estimated using his entropy (i.e. $H(e)$).

$$P(T = 1|e) \propto \frac{\log |D_e|}{H(e)} \qquad (8)$$

Entropy is a measure of how "mixed up" a variable is and refers to the impurity of the variables. Indeed, it shows uncertainty or randomness [13]. Answers

---

[9] It is worth noting that this probability semantically exposes that skill area which cause candidate e to be T-shaped.

[10] Logarithm is used to dampen the importance of document count.

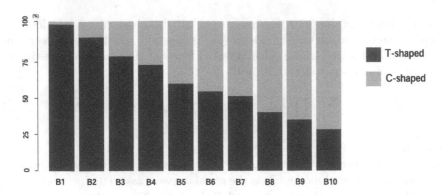

Fig. 2. Share of T-shaped and C-shaped users according to their entropy range fragmented in 10 bins. Non-expert users have been excluded for the sake of simplicity.

associated with a user are usually distributed on several skill areas. Intuitively, the more diverse answers of a user, the higher the uncertainty is and accordingly the associated entropy would be higher. In order to compute Entropy of a user, we follow three steps:

1. Prior to receiving query, we can estimate a distribution of a user's associated skill areas using the following equation.

$$P_{sa,e} = \frac{|D_e \cap D_{sa}|}{|D_e|} \tag{9}$$

   where D_{sa} denotes the all answers associated with skill area sa, and $P_{sa,e}$ is the occurrence probability of answers associated with user e in skill area sa.
2. The uncertainty of user e for each skill area sa is calculated as follows:

$$Uncertainty(sa, e) = -P_{sa,e} \log P_{sa,e} \tag{10}$$

3. Consider n distinct skill areas with probabilities $P_{sa_1,e}$, $P_{sa_2,e}$, $\dots$, $P_{sa_n,e}$. By attributing uncertainty (calculated by Eq. 10) to the i-th skill area, total uncertainty i.e. entropy can be calculated as shown in Eq. 11.

$$H(e) = -\sum_{i=1}^{n} P_{sa_i,e} \log P_{sa_i,e} \tag{11}$$

   As mentioned before, C-shaped users are expected to have a higher diversity than T-shaped users. According to the direct correlation between diversity and entropy, it can be inferred that C-shaped users have much higher entropy than T-shaped. Figure 2 confirms this intuition and illustrates that with increasing amount of entropy, the number of T-shaped users tend to be decreased while the number of C-shaped users tend to be increased, and conversely. In this figure, we clustered expert candidates into 10 equal-length bins according to their entropy value.

4.4   Extended Entropy-Based Approach (XEBA)

As mentioned before, we should estimate the probability $P(T = 1|e, sa)$ to embed into Eq. 3 to obtain an optimal ranking of users. In our first attempt to approximate the mentioned probability, we estimate $P(T = 1|e, sa)$ by $P(T = 1|e)$. In other words, we estimate the probability of being T-shaped for candidate e independent of the given skill area. In this section, we propose a better estimation for the mentioned probability by estimating $P(sa|e, T = 1)$ in Eq. 7 as follows:

$$P(sa|e, T = 1) \approx \frac{P_{sa,e}}{\max(\{P_{sa_i,e} : i = 1, ..., n\})} \tag{12}$$

where $P_{sa,e}$ is computed using Eq. 9, and n denotes the number of skill areas. Mathematically speaking, suppose a T-shaped candidate e who has advanced knowledge in skill area sa'. The probability of $P(sa|e, T = 1)$ for him, according to Eq. 12, is estimated as follows:

$$P(sa|e, T = 1) = \begin{cases} 1 & \text{if} \quad sa = sa' \\ \beta & otherwise \end{cases} \tag{13}$$

in which $\beta \in [0, 1)$ and for a T-shaped user, it approaches 0.

5   Experimental Setup and Evaluation Metrics

In this section, we give details of our datasets and parameter settings.

5.1   Dataset

The dataset we have used is downloaded from Stackoverflow and spans the period from August 2008 up to March 2015 and includes information of about 4 million users and 24,120,523 posts.

Table 2. General statistics of data collections

Dataset	#Q	#A	#U	#SA	#C-shaped	#T-shaped
C#	763,717	1,453,649	84,095	23	1783	2707
Android	638,756	917,924	58,789	18	1074	1830
Java	810,071	1,510,812	83,557	26	1673	2465

To generate three separated data collections, we have exploited questions and their associated answers which are tagged by "C#", "Android", and "Java". It is also worth mentioning that we took the approach discussed in Sect. 3 to generate the golden set. Table 2 represents general statistics of these data collections including number of questions (#Q), answers (#A), users (#U), skill areas (#SA), C-shaped users (#C-shaped), and T-shaped users (#T-shaped).

5.2   Evaluation Metrics and Parameter Setting

For evaluation of the expert finding, we use three metrics to compare the performance of our proposed models with the document-based model. These metrics include Normalized Discounted Cumulative Gain (NDCG), Mean Reciprocal Rank (MRR), and also Expected Reciprocal Rank (ERR). NDCG assigns relevance score (i.e. gain) to each retrieved user based on his shape in golden set. We assign 0 for non-experts' gain, 1 for C-shaped users' gain, and also 2 for T-shaped users' gain. The reciprocal rank of a query is the inverse of the rank of the first relevant result (i.e. first retrieved T-shaped user). The MRR is described as the mean of the reciprocal ranks of results for a set of queries. In addition, we have used the ERR measure to compare the performance of proposed models with baseline. The ERR was introduced in [14] and is inspired by cascade models, the list of returned candidates is read down to the rank in which a T-shaped user is found.

We also tried different settings for λ_d (used in Eq. 5) and finally used $\lambda_d = 0.5$ for all of our experiments.[11]

6   Experimental Results and Analysis

The results of our experiments are summarized in Table 3. According to this table, while the EBA method can slightly improve the NDCG measure on all datasets, in comparison with DBA method, this method can improve MRR and ERR measures significantly. This observation indicates that although EBA method does not affect the whole ranking remarkably, it can retrieve the first T-shaped user in extremely better ranks, in comparison with DBA. In other words, the DBA method is biased towards C-shaped users and retrieves them, rather than T-shaped users, in much better ranks.

Interestingly, the XEBA approach can improve our performance metrics exceedingly in comparison with both the DBA and EBA approaches. This observation indicates that the estimation of $P(sa|e, T = 1)$ is crucial in the ranking of T-shaped candidates in our data collections. Specifically, XEBA method has improved EBA by 7.23%, 16.53%, 11.41% on average of three data collections for NDCG, MRR, and ERR metrics, respectively.

In order to compare proposed models visually, the top 50 retrieved candidates for each method are demonstrated in Fig. 3 for a sample skill area (i.e. *maven*). Non-expert users are shown by light gray, C-shaped by medium gray, and T-shaped by black. With regard to this figure, the XEBA method significantly enhances the ranking of T-shaped users. The EBA method can slightly improve the overall ranking, however, it markedly can retrieve the first T-shaped in a better rank.

---

[11] The implementation of our models is available at http://bit.ly/tshaped-mining.

Table 3. Evaluation and Comparison of the proposed model with Document-based retrieval model. DBA: Document-based approach, EBA: Entropy-based approach, XEBA: Extended entropy-based approach

	Model	NDCG	MRR	ERR
C#	DBA [9]	0.647	0.230	0.334
	EBA	0.658	0.328	0.381
	XEBA	**0.696**	**0.382**	**0.431**
Android	DBA [9]	0.592	0.129	0.191
	EBA	0.616	0.278	0.320
	XEBA	**0.668**	**0.307**	**0.352**
Java	DBA [9]	0.653	0.215	0.402
	EBA	0.655	0.339	0.432
	XEBA	**0.704**	**0.416**	**0.480**
Relative Improvements (%)				
C#	*EBA/DBA*	1.7%	42.61%*	14.07%*
	XEBA/DBA	7.57%*	66.09%*	29.04%*
	XEBA/EBA	5.78%*	16.46%*	13.12%*
Android	*EBA/DBA*	4.05%	115.5%*	67.54%*
	XEBA/DBA	12.83%*	137.98%*	84.29%*
	XEBA/EBA	8.44%*	10.43%*	10.00%*
Java	*EBA/DBA*	0.31%	57.67%*	7.46%*
	XEBA/DBA	7.81%*	93.49%*	19.40%*
	XEBA/EBA	7.48%*	22.71%*	11.11%*

*indicates the improvement is statistically significant.

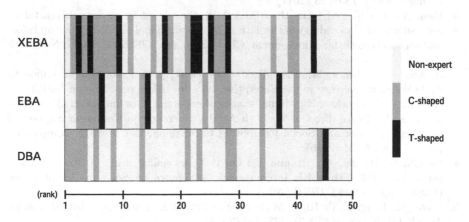

Fig. 3. Comparison of proposed models with document-based approach in terms of retrieved T-shaped users

7   Conclusion

In a typical Community Question Answering (i.e. CQA) platform, each question covers multiple tags which can be used to explore similar questions. The set of similar tags in these networks can be considered skill areas. Users' knowledge levels in the domain of each skill area can be classified into three levels including advanced, intermediate, and beginner. With regard to these levels, users are categorized into three groups which include non-experts, T-shaped users, and C-shaped users. In this paper, we propose a model to retrieve and rank T-shaped users who have advanced knowledge on a single skill area and simultaneously have intermediate (i.e. general) knowledge in other skill areas. Rather than C-shaped users, recruiters are inclined to find and employ T-shaped users as they are more cost-effective and affordable. Furthermore, an increasing convergence of technologies in software development gets companies interested in hiring T-shaped users.

References

1. Dargahi Nobari, A., Sotudeh Gharebagh, S., Neshati, M.: Skill translation models in expert finding. In: Proceedings of the 40th International ACM SIGIR Conference on Research and Development in Information Retrieval, SIGIR 2017, pp. 1057–1060. ACM, New York (2017)
2. Neshati, M., Fallahnejad, Z., Beigy, H.: On dynamicity of expert finding in community question answering. Inf. Process. Manage. 53(5), 1026–1042 (2017)
3. van Dijk, D., Tsagkias, M., de Rijke, M.: Early detection of topical expertise in community question answering. In: SIGIR 2015, pp. 995–998. ACM (2015)
4. Neshati, M.: On early detection of high voted Q&A on stack overflow. Inf. Process. Manage. 53(4), 780–798 (2017)
5. Fang, Y., de Rijke, M., Xie, H.: DDTA 2016: the workshop on data-driven talent acquisition. In: Proceedings of the 25th ACM International on Conference on Information and Knowledge Management, CIKM 2016, pp. 2507–2508. ACM, New York (2016)
6. Ambler, S.W., Lines, M.: Disciplined Agile Delivery: A Practitioner's Guide to Agile Software Delivery in the Enterprise, 1st edn. IBM Press, Boston (2012)
7. Kumar, V., Pedanekar, N.: Mining shapes of expertise in online social Q&A communities. In: Proceedings of the 19th ACM Conference on Computer Supported Cooperative Work and Social Computing Companion, CSCW 2016 Companion, pp. 317–320. ACM (2016)
8. Bazelli, B., Hindle, A., Stroulia, E.: On the personality traits of StackOverflow users. In: 2013 29th IEEE International Conference on Software Maintenance (ICSM), pp. 460–463. IEEE (2013)
9. Balog, K., Fang, Y., de Rijke, M., Serdyukov, P., Si, L.: Expertise retrieval. Found. Trends Inf. Retrieval 6(2–3), 127–256 (2012)
10. Van Gysel, C., de Rijke, M., Worring, M.: Unsupervised, efficient and semantic expertise retrieval. In: Proceedings of the 25th International Conference on World Wide Web, WWW 2016, pp. 1069–1079. International World Wide Web Conferences Steering Committee, Republic and Canton of Geneva (2016)

11. Liang, S., de Rijke, M.: Formal language models for finding groups of experts. Inf. Process. Manage. **52**(4), 529–549 (2016)
12. Guest, D.: The hunt is on for the renaissance man of computing, 17 September 1991
13. Han, J., Kamber, M., Pei, J.: Data Mining: Concepts and Techniques, 3rd edn. Morgan Kaufmann Publishers Inc., San Francisco (2011)
14. Chapelle, O., Metlzer, D., Zhang, Y., Grinspan, P.: Expected reciprocal rank for graded relevance. In: Proceedings of the 18th ACM Conference on Information and Knowledge Management, CIKM 2009, pp. 621–630. ACM, New York (2009)

Modeling Relevance Judgement Inspired by Quantum Weak Measurement

Panpan Wang[1], Tianshu Wang[1], Yuexian Hou[1(✉)], and Dawei Song[2]

[1] School of Computer Science and Technology, Tianjin University, Tianjin, China
{panpan_tju,shugrgr,yxhou}@tju.edu.cn
[2] School of Computing and Communications, The Open University,
Milton Keynes, UK
dawei.song@open.ac.uk

Abstract. Concept in Quantum theory (QT) has been successfully inspired analogous concepts in the field of Information Retrieval (IR). Many IR researchers have employed the QT to investigate cognitive phenomena within user behaviors, and also have verified the existence of quantum-like phenomena in real web search. However, for some complex search task, in which user's information need (IN) is dynamic and hard to be captured, QT currently adopted still can not explain some more complex cognitive phenomena. In this paper, a user experiment is conducted to investigate the variance of relevance judgement, and its results demonstrate that quantum Weak Measurement (WM) is more appropriate than the standard quantum measurement to model relevance judgement. Further, a WM-based session search model (WSM) is presented to model user's dynamic evolving IN. The extensive experiments are tested on the session track of TREC 2013 & 2014 and verify the effectiveness of WSM.

Keywords: Relevance judgement · Quantum weak measurement
Session search

1 Introduction

Relevance judgement in information retrieval (IR) is a task for users to judge whether a document is relevant to a topic. The user's perception of relevance can change over with the evolving of user's information need (IN) in the real search [1]. In order to satisfy higher user's expectation, it is necessary to capture user's IN from a view of dynamic user-based relevance.

Recently, quantum theory (QT) has attracted increasing attention that its concept, for instance, quantum interference, can inspire analogous concepts in the field of IR. van Rijsbergen [2] proposed that QT could be used to axiomatize the geometric, probabilistic and logic-based IR models with a single mathematical formalism in complex Hilbert vector space. Some important notions in IR could also be translated into analogous notions in QT, such as mapping a document into a state vector, regarding each document as a superposition of words or

© Springer International Publishing AG, part of Springer Nature 2018
G. Pasi et al. (Eds.): ECIR 2018, LNCS 10772, pp. 424–436, 2018.
https://doi.org/10.1007/978-3-319-76941-7_32

topics, and replacing the cosine correlation between query and documents with inner product. Following van Rijsbergen's work, a series of QT-based IR models were developed from the perspective of dynamic user-based relevance [3–5].

Wang et al. [6] devoted to exploring the quantum-like phenomena from a real user study perspective. The experiment results showed that: first, there was an apparent judging discrepancy across different users and documents, and empirical data violated the law of total probability; second, many search trials recorded in the user study showed the existence of the order effect in relevance judgement; third, the relevance judgement should be seen as a kind of standard quantum measurement (SQM), which the judgement can trigger the collapse of superposed user's information need (IN) [7]. However, in some more complex search scenarios, users' IN may not have apparent changes after making relevance judgement (measurement): users may have inner conflict (superposition) in mind, or may modify their judgement without external information.

Obviously, SQM can not provide feasible explanations for these phenomena. Quantum **weak measurement (WM)** is the generalization of SQM [8]. A major difference between these two concepts is the variance of the pointer variable in WM which is much larger than the variance in SQM. To explicitly distinguish these two concepts, we call SQM as **strong measurement (SM)** throughout the rest of the paper. We try to explore the relationship between the change of user's IN and relevance judgement from a novel view of WM. For this purpose, we carry out a user study and find the apparent judging discrepancy and the ambivalence of user's cognition. Therefore, we consider that the WM is more suitable for modeling the relevance judgement process and explaining the change of user's IN. Successively, We propose a WM-based session search model (WSM) to capture the dynamic IN for more complex retrieval tasks. In order to obtain the measurement results of WM, i.e. weak value, we construct the representation of past state and future state within a session search. Finally, we use the weak value to calculate the ranking score of evaluated documents. Extensive experiments are conducted on the session track of TREC 2013 & 2014, which shows the effectiveness of our model.

2   Related Works

Session search is a complex and continuous retrieval process, which has attracted many IR researchers. Both quantum and non-quantum based models are developed. Following van Rijsbergen's work, a body of quantum-based frameworks have recently been developed to formalize retrieval models and have achieved significant improvements. Relying upon a generalization of the probability framework of quantum physics, a general framework for interactive IR that is able to capture the full interaction process [9]. Based on the Quantum Language Model (QLM) [10], Session-based Quantum Language Model (SQLM) [11] and Quantum IR Model inspired by Two-state vector formalism (QMT) [12] defined words and dependencies in a quantum probabilistic space and achieved good performance on the session track task.

There are also many non-quantum based models. Luo et al. proposed "win-win search" [13] and Direct Policy Learning (DPL) [14]. Their models are based on Partially Observable Markov Decision Process (POMDP), which is also from the view of user-based relevance.

The complexity for session search tasks comes from the involvement of many more factors than just terms, queries and documents in most existing retrieval algorithms [13]. How to use the history log in session is the key point to construct a effective user model to capture the dynamic IN.

3   Preliminaries of Quantum Theory

In standard QT, the quantum probability space is encapsulated in a Hilbert space $\mathcal{C}^n$. Notions can be represented as unit column vector $|\psi\rangle$ whose conjugate transpose can be expressed as $\langle\psi|$.

Assume that there are two quantum states in a $\mathcal{C}^2$ space system:

$$|-1\rangle \equiv \begin{bmatrix} 1 \\ 0 \end{bmatrix}, |1\rangle \equiv \begin{bmatrix} 0 \\ 1 \end{bmatrix} \tag{1}$$

These states, $|-1\rangle$ and $|1\rangle$, form orthogonal basis. The state of the system can be expressed by:

$$|\psi\rangle = \alpha|-1\rangle + \beta|1\rangle \tag{2}$$

where $|\alpha|^2 + |\beta|^2 = 1$ and now $|\psi\rangle$ is superposed. The concept of superposition resonates with the fuzzy, ambiguous, feelings in many cognitive phenomena. If we use $\{|-1\rangle\langle-1|, |1\rangle\langle1|\}$ to measure $|\psi\rangle$, then with the probability of $|\alpha|^2$ (or $|\beta|^2$) we can get the outcome "-1" (or "1"). After the measurement, $|\psi\rangle$ will collapse to $|-1\rangle$ or $|1\rangle$ respectively.

4   Quantum Weak Measurement

4.1   Strong Measurement and Weak Measurement

In practical physics experiment, measurement is performed by interacting the measured system with the measuring device [15]. The outcome of the measurement is obtained by observing the pointer of the measuring device. The probability distribution of the pointer value is shown in Fig. 1 and X-axis stands for the outcome of the measurement.

Assuming that we have a particle of Eq. 2. The probability distribution of the position of the pointer can be modeled by a Gaussian [15]:

$$P(Q) = (2\pi\Delta^2)^{-1/2}e^{-x^2/2\Delta^2} \tag{3}$$

where x is the pointer variable, Δ^2 is the variance. Supposing the interaction has arisen and the system is interacted with the measuring device. The probability distribution of the pointer variable is:

$$P(Q) = (2\pi\Delta^2)^{-1/2}(\alpha^2 e^{-(x+1)^2/2\Delta^2} + \beta^2 e^{-(x-1)^2/2\Delta^2}) \tag{4}$$

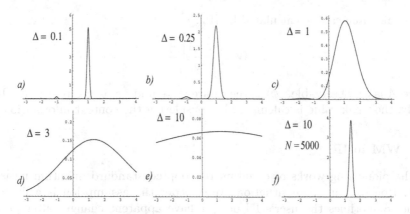

Fig. 1. Shows the probability distribution of the pointer variable in physics experiment [15]. The strength of the measurement is expressed by the uncertainty (variance) of the pointer position, i.e. Δ^2. The smaller Δ^2 is, the "stronger" measurement is. (a)–(e) Measurement on single particle. (f) Weak measurement on the ensemble of 5000 particles. In the strong measurements (a)–(b) the pointer is localized around the precise outcomes, while in the weak measurements (c)–(e) the peak of the distribution becomes smoother (more imprecise) with the increase of uncertainty (variance). In the weak measurements (f) on an ensemble of particles, the peak of the distribution is located in one value, which is the **weak value**.

WM is the generalization of the SM [8]. SM and WM share the same measurement process. However, the biggest difference between the SM and the WM is Δ^2. Generally speaking, Δ^2 of the SM is far smaller than Δ^2 of the WM, which is regarded as the generalization of SM.

In the SM, because the Δ^2 is small, the probability distribution of pointer reading has clear peaks as shown $a - b$ in Fig. 1. After the measurement, the reading of the pointer will yield a value close to one of the two eigenvalues, '−1' and '1', with probability $|\alpha|^2$ and $|\beta|^2$ respectively (like $a - b$ in Fig. 1). If the variable is around '1', $\alpha^2 e^{-(x+1)^2/2\Delta^2}$ will approximately equal to zero while $\beta^2 e^{-(x-1)^2/2\Delta^2}$ will keep effective and vice versa. In this way, the reading of pointer will be centered around one of the eigenvalues with a small variance and the state of the system will collapse.

However, if Δ^2 becomes much larger, the strength of measurement becomes weaker (like $c-e$ in Fig. 1). Since Δ^2 is quite large, the probability distribution of pointer reading will be smoother and may not have clear peaks. And the reading will not be centered around one of the eigenvalues but may point to a range of values. Because of the large Δ^2, both $\alpha^2 e^{-(x+1)^2/2\Delta^2}$ and $\beta^2 e^{-(x-1)^2/2\Delta^2}$ will keep effective after the measurement. Then the system may not collapse into one of the eigenstates but be biased to the corresponding direction of the reading of pointer slightly. The reading of the pointer is vague.

However, by statistical treatment, the distribution of results in statistics will be centered at a value (like f in Fig. 1) again. The value is called weak value W.

The weak value can be calculated by [15]:

$$W = \frac{\langle \phi | A | \psi \rangle}{\langle \phi | \psi \rangle} \tag{5}$$

where A is an observable. $|\psi\rangle$ is the future state and $\langle\phi|$ is the past state. Intuitively, the weak value is calculated by considering the context information.

4.2   WM in IR

All the pioneering works of quantum IR adopted standard quantum measurement. However, in some situation, we consider the assumption is a little too strong. Sometimes the user's IN doesn't have apparent changes after judging the relevance; sometimes after give the answers, the users may still have the inner conflict (superposition) in mind; sometimes the users may also modify their answers without external information. The standard QT can't give feasible explanations for these phenomena but WM can. The WM can explain not only the apparent changes of IN (collapse) but also the slight shift of IN.

To verify the existence of quantum WM in real IR process beyond microscopic physics of particles, we carried out a user experiment. We recruited 20 participants. In the experiment, each user is asked to judge relevance of a query-document pair with a score from '−4' to '4'. The higher the score is, the more relevant the document is and vice versa. In the experiment, for some pairs, we asked the participants to continuously judge them one more time. Then we record whether they change their ideas during the judgement. If user's cognitive state is still superposed after the first judgement, he may change his answer in the continuous judgement. A part of experiment results are shown in Table 1. Each line in Table 1 is a query-document pair.

According to the experiment results, we can find that the real user's judgement have a large variance in several pairs, which show an apparent judging discrepancy. The change rate record the rate of the users who change their answers in the following judgement. From the table, we can figure out that the pairs, which have high change ratio, also have high variance in common.

Though told '4' means 'fully relevant' and '−4' means 'fully irrelevant', users lack of a vivid criterion. The more difficult the judgement is, the more ambiguous the definition of scores are. In this situation, like the measuring process in WM, the user's judgement may have an apparent discrepancy.

"Without external information, why users change their answers?" is a difficult question for standard QT to answer. In standard QT, user's IN will collapse to one of the basic state after the first judgement and if we measure with same measuring bases (same question) the IN should be consistent and stay at the same basic state. However, the reality shows that after the first judgement, sometimes users may still have inner conflicts. In other words, the user's IN is still superposed even after the measurement. Referring to the WM (with large variance), for the specific pairs the users have an apparent discrepancy causing that the IN may not collapse completely but be biased.

Table 1. A part of user experiment result

Query	Session number	Doc number	Change ratio	Mean	Variance	Mode
Heart disease	Session 5, TREC 2013	clueweb12-0000wb-77-27001	0.00	0.95	1.10	1
Planning a road trip	Session 15, TREC 2013	clueweb12-1511wb-28-23959	0.05	0.25	1.15	1
Internet phones VoIP reviews	Session 35, TREC 2013	clueweb12-0705wb-50-30295	0.30	2.30	1.17	2
Gun control laws Obama	Session 41, TREC 2013	clueweb12-1802wb-60-29286	0.35	1.20	2.07	1
Running a half marathon	Session 51, TREC 2013	clueweb12-1516wb-41-17025	0.00	0.05	0.48	0
Tax on junk food	Session 78, TREC 2013	clueweb12-0811wb-09-07205	0.25	−1.25	1.68	−1
Growth of bollywood	Session 1, TREC 2014	clueweb12-0106wb-97-20878	0.05	−1.80	0.80	−2
Fun in Pocono Mountains	Session 2, TREC 2014	clueweb12-0714wb-23-14036	0.05	1.00	0.74	1
Hypnosis for not smoking	Session 13, TREC 2014	clueweb12-0903wb-40-04443	0.10	−1.30	1.17	−1
Students evaluation of teachers	Session 16, TREC 2014	clueweb12-1011wb-50-10156	0.00	0.10	0.52	0
Face transplants patients	Session 38, TREC 2014	clueweb12-0211wb-82-24427	0.00	0.05	0.26	0
Dallas airports parking	Session 58, TREC 2014	clueweb12-0300wb-91-21742	0.10	−1.45	1.32	−2
JP Morgan chase about	Session 70, TREC 2014	clueweb12-0206wb-97-23713	0.00	−0.10	0.63	0
Connecticut Fire Academy	Session 81, TREC 2014	clueweb12-1200tw-13-12106	0.15	1.40	1.73	2
Kenya cooking	Session 83 TREC 2014	clueweb12-0111wb-88-07466	0.00	3.2	0.37	3

If the judgement is easy to make, then the users can give a definitive answer. In this situation, the variance is small and users' IN would collapse. In this situation, user's future is one of the eigenstates and the weak measurement degenerate into the quantum standard measurement. However, if the judgement is hard to make, then the users is hesitant to give definitive answers and they may modify their answers they have just given. In this situation, the variance is large and users' IN may not collapse completely but biased. To summarize, there are a lot of piercing points between the WM and real cognitive process of users. The WM can model more situations in relevance judgement process and we propose WM is more suitable to model the dynamic IN in IR.

5   WM-Based Session Search Model (WSM)

Session search is a challenging IR task that involves multiple queries and interactions to accomplish. We represent the series of queries as $\{q_1, ..., q_i, ..., q_n\}$. The first query in a session q_1 is the start point of a search task, and the last query q_n is the current query we need to handle. Except for q_n, each query has its corresponding interaction. In each session interaction, there are top 10 retrieved documents by the search engine and some of them are clicked and browsed by the end-user. We call the clicked documents, which are browsed over 30 s, as SAT (satisfied) documents [16]. **Notice that:** for simplicity, we omit the normalization factor of a quantum state hereafter. It is sure that every quantum state vector has been normalised in our calculation.

5.1   Representation of Documents and Queries

Doc2vec [17], developed from the Word2vec algorithm [18], is one of the effective unsupervised machine learning algorithm to learn the representation of paragraphs. Doc2vec can model the semantic information of documents. We use the Doc2vec algorithm to learn the distributed representation of documents and queries in a real Hilbert space.

5.2   Representation of Past State

The past state contains current query and Implicit Relevance Feedback (IRF) information [19]. In this paper, IRF is the information in the history log, e.g. clicked documents, dwell time and so on. The changes of user's IN are triggered by WM when browsing the documents and judging the relevance. Since the weak value describes the shift of user's IN in weak measurement, we use the weak value as a weight to accumulate user's IN changes.

$$|\psi\rangle = \theta \cdot |h\rangle + (1 - \theta) \cdot |q_n\rangle \tag{6}$$

where $|h\rangle$ represents IRF information and q_n represents the vector of current query. θ is a hyper-parameter used to balance the weight of current query and historical information. $|h\rangle$ is the accumulation of the IN changes, and can be expressed as

$$|h\rangle = \sum_{j=1}^{m} \frac{\langle q'_j|s_j\rangle\langle s_j|q_j\rangle}{\langle q'_j|q_j\rangle} \cdot |s_j\rangle \tag{7}$$

where m is the number of the SAT documents in the whole session and $|s_j\rangle$ is the embedding of SAT document s_j. In order to simplify the calculation, we use the corresponding adjacent queries, q_j and q'_j, to represent the past state and future state of SAT documents.

The weak value reflect the magnitude of difference between the hereinbefore IN and hereinafter IN. The more a document can bring changes to user's IN, the more important the document is to capture the dynamic IN. Therefore we use the weak value as weights to accumulate the changes. The larger the corresponding weak value is, the more important the document is to describe user's IN.

5.3   Representation of Future State

In real search task, it is unavailable for us to measure the user's future IN. But we can make a reasonable prediction using common used tricks. The future state $\langle\phi|$ in this section contains our prediction for user's possible IN, i.e. PRF information [20]. PRF uses the returned documents in the first round retrieval to improve the result of current query. We pick the top 10 retrieved documents by the first round retrieval and $\langle\phi|$ could be expressed as

$$\langle\phi| = \sum_{i=1}^{10} \frac{1}{1 + \log_2 i} \cdot \langle p_i| \tag{8}$$

where $\langle p_i|$ is the vector of PRF documents p_i.

5.4 Re-ranking

The defination of $|\psi\rangle$ and $\langle\phi|$ remains the superposition in past and future state. After that, the relevance of documents can be measured by the WM. We use the formula of weak value to calculate the relevance of alternative documents.

$$W_i \equiv \frac{\langle\phi|d_i\rangle\langle d_i|\psi\rangle}{\langle\phi|\psi\rangle} \tag{9}$$

where d_i is the vector of the ith documents and W_i is the corresponding measured weak value. Here the larger the weak value is, the more relevant the document considering the context. At last, we return the documents by descending order of weak value.

6 Experiments and Evaluation

6.1 Datasets and Comparisons

Our experiments are conducted on the TREC Session track 2013 and 2014. The document collection is the Clueweb12 (Category B) which contains more than 50 million English webpages. The document collections is indexed by the Indri toolkit[1]. In the indexing process, all words are stemmed with Porters stemmer and stopwords are removed with the normal English stopword list. Documents and queries are trained to vectors with 200 dimensions by Doc2vec algorithms. Other parameters are: $min\_count = 10, window = 10, sample = 1e-4, negative = 5$. The ground truth data is provided by the official of TREC. The parameter θ is set to 0.8 in our model.

To verify the effectiveness of our model, both typical IR models and Quantum-based IR models are used to make comparisons. Some of the models have been mentioned in the previous section and we would not give another introductions here. Language Model (LM) [21] performed by Indri is a widely used open source search engine of the Lemur Project in scientific and research applications. Relevance-based Language Model (RM) is a query expansion approach using PRF [22]. Generalized Language Model (GLM) use the word embeddings [18] to generalize the traditional LM for ad-hoc task [23]. Win-Win is a

---

[1] Website: www.lemurproject.org.

session search framework proposed by Luo et al. in [14]. Note that, nDCG@10 of Win-Win in this paper reports the same result with winwin-short achieved in [14], but MRR was originally not reported. The parameters of all the compared models are set to the best values which is reported in their papers.

6.2 Evaluation Results

We adopt MRR and $nDCG@10$ as the evaluation metrics. The evaluation results show the effectiveness of our model. RM doesn't outperform the baseline model, which prove that PRF is insufficient to improve the performance. Win-Win is non-quantum based model, which capture the dynamic IN by modeling the user's activities in IR process. The results show that our model performs as well as the classical session search model in average evaluation metrics. GLM also uses the word embedding to represent words. However, the model is still static relevance-based, which doesn't make full use of the user's history log, which limits its performance in session search task. The performance of GLM shows that session information can help improve the retrieval performance (Tables 2 and 3).

Table 2. Evaluation results. Significance test has been conducted. Symbol † means $p < 0.05$.

Models	TREC 2013		TREC 2014	
	$nDCG@10$	MRR	$nDCG@10$	MRR
LM	0.1159	0.3943	0.1815	0.4388
RM	0.1162	0.3830	0.1810	0.3967
SQLM	0.1370	0.3906	0.1850	0.4382
QMT	0.1485	0.3942	0.2123	0.4402†
GLM	0.1354	0.3845	0.1901	0.4389
Win-Win	0.1253	–	0.2241†	–
WSM	0.1554†	0.4256†	0.2253	0.4475†

The models based on QT (SQLM and QMT) also improve the retrieval performance. However, QLM-based model are developed from traditional LM. These models combine term dependencies with QT. However, these models still focus on the individual words and neglect semantic information as a fully document. QMT use the Two-State Vector Formalism to model the user's IN. However, they just use the last two query terms which loses precious history information.

Inspired by the WM theory, we can remain the most important information as superposition. The accumulation of user's IN changes provide more complete information for capture the current IN. The formula of weak value takes the context, both past and future information, into account, which provides more complete information for decision making. In other IR models, the KL divergence or the cosine similarity is widely used to score the relevance of the document. In this paper, we use the weak value formula to complete the work of scoring.

Table 3. Performance in difficult session. Significance test has been conducted. Symbol † means $p < 0.05$.

Models	TREC 2013		TREC 2014	
	$nDCG@10$	MRR	$nDCG@10$	MRR
LM	0.0027	0.0410	0.0047	0.0671
RM	0.0034	0.0356	0.0050	0.0581
SQLM	0.0325	0.0852	0.0363	0.1056
QMT	0.0453	0.1223	0.0455	0.1251
GLM	0.0134	0.0556	0.0201	0.0826
WSM	0.0521	0.1546†	0.0462	0.1570†

6.3 Performance in Difficult Session

The difficult session means that the session search is hard for the model to achieve good performance. To verify the effectiveness of our model, we choose some difficult sessions, in which the baseline model (LM) has a low $nDCG@10$ less than 0.05. There are 35 sessions in TREC 2013 and 41 sessions in TREC 2014.

From the table we can figure out that in the difficult session, a series of models perform really badly. Using the history information, the models designed for session search tasks significantly improve the retrieval performance over the baseline model. By the WSM and document vector to integrate the context information, our model gave a good performance. The result gives us confidence that the quantum theory may help us break the bottleneck of IR.

6.4 Performance Grouped by Length

The Table 4 shows the performance of WSM in sessions with different session length. The length of session is the number of interactions in the session. From the table, we can figure out that WSM performs best in medium length session. In short length session, the performance is not so well. Within certain interactions, the longer a session is, the richer implicit information contained in the session log. However, if the length is more than 8, the performance of WSM drops down.

Table 4. Performance grouped by length of a session

Length	TREC 2013		TREC 2014	
	$nDCG@10$	MRR	$nDCG@10$	MRR
1–3	0.2242	0.3803	0.2115	0.3879
4–7	0.2820	0.4481	0.2515	0.4630
8–	0.0841	0.2327	0.0926	0.2876

We think the reason is that: in a long session, the user's may forget partial information of the earlier interactions. As time goes on, the IN changes in the earlier interactions will become inapparent. On the other hand, sometimes long session means that it is hard to return the documents which meet the users' IN. These long sessions may be the difficult sessions which are hard to be retrieved.

7   Conclusion and Future Works

In this paper, we briefly reviewed the pioneering research of quantum IR and proposes a WM-based session search model to capture the dynamic IN in IR. Analogy with the WM, we draw the connection between the WN and relevance judgement in IR, which suggested that WM is more suitable for modeling the changes of IN. The shortcoming in our paper is the use of PRF as the replacement of future information. In empirical experience, PRF sometimes couldn't bring much improvements to the model. There still needs some extra attention to exploring a more suitable and practicable method to construct the future state.

In the future, there are still a lot of unknowns waiting for us to be explored. The model in this paper is a rough and primary proposition. In the future, more researches for the user's cognitive activities in IR will be conducted. We can also improve the performance of WSM by investigating more effective means, e.g. electroencephalography (EEG) and the eye tracker.

Acknowledgements. This work is funded in part by the Key Project of Tianjin Natural Science Foundation (15JCZDJC31100), the Alibaba Innovation Research Foundation 2017, the Major Project of Chinese National Social Science Fund (14ZDB153), MSCA-ITN-ETN - European Training Networks Project (QUARTZ), National Natural Science Foundation of China (Key Program, U1636203) and National Natural Science Foundation of China (U1736103).

References

1. Mizzaro, S.: How many relevances in information retrieval? Interact. Comput. **10**(3), 303–320 (1998)
2. Van Rijsbergen, C.J.: The Geometry of Information Retrieval. Cambridge University Press, Cambridge (2004)
3. Xie, M., Hou, Y., Zhang, P., Li, J., Li, W., Song, D.: Modeling quantum entanglements in quantum language models (2015)
4. Zuccon, G., Azzopardi, L.: Using the quantum probability ranking principle to rank interdependent documents. In: Gurrin, C., He, Y., Kazai, G., Kruschwitz, U., Little, S., Roelleke, T., Rüger, S., van Rijsbergen, K. (eds.) ECIR 2010. LNCS, vol. 5993, pp. 357–369. Springer, Heidelberg (2010). https://doi.org/10.1007/978-3-642-12275-0_32
5. Li, J., Zhang, P., Song, D., Hou, Y.: An adaptive contextual quantum language model. Phys. A **456**, 51–67 (2016)
6. Wang, B., Zhang, P., Li, J., Song, D., Hou, Y., Shang, Z.: Exploration of quantum interference in document relevance judgement discrepancy. Entropy **18**(4), 144 (2016)

7. Ellis, D.: The dilemma of measurement in information retrieval research. J. Am. Soc. Inf. Sci. **47**(1), 23 (1996)

8. Tamir, B., Cohen, E.: Introduction to weak measurements and weak values. Quanta **2**(1), 7–17 (2013)

9. Piwowarski, B., Lalmas, M.: A quantum-based model for interactive information retrieval. In: Azzopardi, L., Kazai, G., Robertson, S., Rüger, S., Shokouhi, M., Song, D., Yilmaz, E. (eds.) ICTIR 2009. LNCS, vol. 5766, pp. 224–231. Springer, Heidelberg (2009). https://doi.org/10.1007/978-3-642-04417-5_20

10. Sordoni, A., Nie, J.Y., Bengio, Y.: Modeling term dependencies with quantum language models for IR. In: Proceedings of the 36th International ACM SIGIR Conference on Research and Development in Information Retrieval, pp. 653–662. ACM (2013)

11. Li, Q., Li, J., Zhang, P., Song, D.: Modeling multi-query retrieval tasks using density matrix transformation. In: Proceedings of the 38th International ACM SIGIR Conference on Research and Development in Information Retrieval, pp. 871–874. ACM (2015)

12. Wang, P., Hou, Y., Li, J., Zhang, Y., Song, D., Li, W.: A quasi-current representation for information needs inspired by two-state vector formalism. Phys. A **482**, 627–637 (2017)

13. Luo, J., Zhang, S., Yang, H.: Win-win search: dual-agent stochastic game in session search. In: Proceedings of the 37th International ACM SIGIR Conference on Research and Development in Information Retrieval, pp. 587–596. ACM (2014)

14. Luo, J., Dong, X., Yang, H.: Session search by direct policy learning. In: Proceedings of the 2015 International Conference on The Theory of Information Retrieval, pp. 261–270. ACM (2015)

15. Aharonov, Y., Vaidman, L.: The two-state vector formalism: an updated review. In: Muga, J., Mayato, R.S., Egusquiza, Í. (eds.) Time in Quantum Mechanics, vol. 734, pp. 399–447. Springer, Heidelberg (2008). https://doi.org/10.1007/978-3-540-73473-4_13

16. Fox, S., Karnawat, K., Mydland, M., Dumais, S., White, T.: Evaluating implicit measures to improve web search. ACM Trans. Inf. Syst. (TOIS) **23**(2), 147–168 (2005)

17. Le, Q.V., Mikolov, T.: Distributed representations of sentences and documents. In: ICML, vol. 14, pp. 1188–1196 (2014)

18. Mikolov, T., Chen, K., Corrado, G., Dean, J.: Efficient estimation of word representations in vector space. arXiv preprint arXiv:1301.3781 (2013)

19. Song, Y., He, L.W.: Optimal rare query suggestion with implicit user feedback. In: Proceedings of the 19th International Conference on World Wide Web, pp. 901–910. ACM (2010)

20. Cao, G., Nie, J.Y., Gao, J., Robertson, S.: Selecting good expansion terms for pseudo-relevance feedback. In: Proceedings of the 31st Annual International ACM SIGIR Conference on Research and Development in Information Retrieval, pp. 243–250. ACM (2008)

21. Zhai, C., Lafferty, J.: Model-based feedback in the language modeling approach to information retrieval. In: Proceedings of the Tenth International Conference on Information and Knowledge Management, pp. 403–410. ACM (2001)

22. Lavrenko, V., Croft, W.B.: Relevance based language models. In: Proceedings of the 24th Annual International ACM SIGIR Conference on Research and Development in Information Retrieval, pp. 120–127. ACM (2001)
23. Ganguly, D., Roy, D., Mitra, M., Jones, G.J.: Word embedding based generalized language model for information retrieval. In: Proceedings of the 38th International ACM SIGIR Conference on Research and Development in Information Retrieval, pp. 795–798. ACM (2015)

Learning/Classification

Active Learning Strategies for Technology Assisted Sensitivity Review

Graham McDonald$^{(\boxtimes)}$ ⓘ, Craig Macdonald ⓘ, and Iadh Ounis ⓘ

University of Glasgow, Glasgow G12 8QQ, UK
g.mcdonald.1@research.gla.ac.uk,
{craig.macdonald,iadh.ounis}@glasgow.ac.uk

Abstract. Government documents must be reviewed to identify and protect any *sensitive* information, such as personal information, before the documents can be released to the public. However, in the era of digital government documents, such as e-mail, traditional sensitivity review procedures are no longer practical, for example due to the volume of documents to be reviewed. Therefore, there is a need for new technology assisted review protocols to integrate automatic sensitivity classification into the sensitivity review process. Moreover, to effectively assist sensitivity review, such assistive technologies must incorporate reviewer feedback to enable sensitivity classifiers to quickly learn and adapt to the sensitivities within a collection, when the types of sensitivity are not known *a priori*. In this work, we present a thorough evaluation of active learning strategies for sensitivity review. Moreover, we present an active learning strategy that integrates reviewer feedback, from sensitive text annotations, to identify features of sensitivity that enable us to learn an effective sensitivity classifier (0.7 Balanced Accuracy) using significantly less reviewer effort, according to the sign test ($p < 0.01$). Moreover, this approach results in a 51% reduction in the number of documents required to be reviewed to achieve the same level of classification accuracy, compared to when the approach is deployed without annotation features.

1 Introduction

At least 95 countries implement Freedom of Information (FOI) laws legislating that governments documents should be *open* to the public[1]. However, many such documents contain *sensitive* information, such as confidential or personal information and, therefore, FOI laws provide *exemptions* to prevent the release of such information. Government documents must, therefore, be sensitivity reviewed to ensure that no exempt information is released.

Historically, sensitivity review has been an exhaustive manual review of all documents being considered for release. However, in the era of born-digital documents such as e-mail, this purely manual review is not feasible [1], for example

---

[1] http://www.right2info.org/access-to-information-laws/access-to-information-laws.

© Springer International Publishing AG, part of Springer Nature 2018
G. Pasi et al. (Eds.): ECIR 2018, LNCS 10772, pp. 439–453, 2018.
https://doi.org/10.1007/978-3-319-76941-7_33

due to the volume of digital documents that are to be reviewed. Recently, automatic sensitivity classification algorithms have been shown to have potential for effectively identifying sensitive information in documents [2–5]. However, the potential consequences from the inadvertent release of sensitive information can be severe, for example if the identity of an informant is made public it can put the informant and their family at risk. Therefore, until automatic sensitivity classification is trusted, all documents that are to be released will continue to be manually reviewed. With this in mind, there is a need for appropriate protocols to integrate sensitivity classifiers into the review process to assist reviewers.

Technology assisted review (TAR), most notably associated with e-discovery [6,7], is a process whereby human reviewers and an Information Retrieval (IR) system actively work together to identify relevant documents. The TAR protocol typically consists of two components, a key-word search system and a learning algorithm. Given a collection of documents and a *request for production*, e.g. "find all documents relating to ..", the TAR system formulates a query[2] to retrieve an initial pool of documents to be manually reviewed and labeled, or *coded*. The labeled pool is then used as a *seed set* to train the learning algorithm. The TAR protocol is then an iterative process where by the learner predicts the k most relevant unlabeled documents which the reviewer labels. The newly labeled documents are added to the training data and the algorithm is re-trained.

The TAR protocol can potentially be adapted to meet the needs of digital sensitivity review. However, in sensitivity review there is no equivalent to the request for production, since the types of sensitivity within the collection are not known *a priori*. Moreover, a judgment of sensitivity is often dependent on the context in which the information is produced and the time at which it is reviewed. Therefore, with this in mind, we propose to derive a representation of the sensitivities within a collection by having a reviewer annotate the specific text in a document that led to the reviewer's decision that the document is sensitive. Moreover, we propose to incorporate this reviewer feedback into the classification model to more quickly learn and adapt to the sensitivities within a collection at the time of review, while using minimal reviewing effort.

One possible strategy for integrating reviewer feedback into classification is active learning [8]. In active learning, the learning algorithm selects the order that documents are presented to a reviewer, with the aim of minimising the reviewer effort that is required to learn an effective classifier. Active learning has previously been shown to be an effective strategy for e-discovery TAR [6] and for *topic-oriented* text classification [9]. However, sensitivity is not topic-oriented [3] and, therefore, it is not obvious which active learning strategy is most appropriate for sensitivity classification.

---

[2] In active learning parlance, "query" usually refers to membership queries i.e. the system poses queries in the form of instances to be reviewed. In this work we use query in the IR sense, i.e. a textual passage used to retrieve relevant documents from an IR system. For membership queries we say that the system suggests documents to be reviewed.

In this work, we simulate the technology assisted sensitivity review process to present a thorough evaluation of active learning strategies for identifying sensitivities within a collection. We test two well-known *uncertainty sampling* active learning strategies from the literature and evaluate, as an active learning strategy, a *semi-automated text classification* [10] approach, that has previously been shown to be effective for increasing the cost-effectiveness of sensitivity reviewers [3]. Moreover, we show that by extending these approaches to incorporate reviewer feedback from sensitive text annotations, we can improve upon the *raw* active learning strategies to develop effective sensitivity classifiers more quickly, i.e. using less reviewer effort.

The contributions of this paper are two fold. Firstly, we provide the first thorough evaluation of active learning strategies for automatic sensitivity classification. Secondly, we present an active learning strategy that integrates reviewer feedback, from sensitive text annotations, to identify features of sensitivity that enable us to learn an effective sensitivity classifier (0.7 Balanced Accuracy) using significantly less reviewing effort, according to the sign test ($p < 0.01$). This approach resulted in a 51% reduction in the number of documents that had to be reviewed to achieve the same level of classification accuracy, compared to when the approach was deployed without annotation features.

The remainder of this paper is structured as follows. Firstly, we present related work in Sect. 2, before presenting the active learning strategies that we evaluate in Sect. 3. We present our experimental setup in Sect. 4 and results in Sect. 5, before, finally, presenting our conclusions in Sect. 6.

2   Related Work

In this section we, firstly, present work relating to automatic sensitivity classification, before discussing technology assisted review and active learning later in the section.

The task of automatically classifying sensitive information that is exempt from release under Freedom of Information (FOI) laws was first introduced by McDonald *et al.* [2]. In that work, the authors presented a proof-of-concept sensitivity classifier for identifying two FOI exemptions. In [2], the authors showed that text classification [11] can provide an effective baseline approach for sensitivity classification, achieving markedly above random effectiveness (0.7372 Balanced Accuracy). In [2], the authors also extended text classification with additional hand-crafted features, such as named entities of interest (e.g. politicians) and a subjective sentences count, which resulted in improved effectiveness for most of the reported metrics (e.g. +5% F_2).

Feature engineering for sensitivity classification was subsequently investigated further by McDonald *et al.* [5]. In that work, the authors constructed document representations using word embeddings to capture semantic relations in the documents, such as *who said what about whom*. In [5], the authors evaluated the effectiveness of these semantic features compared with textual and syntactic features and found that combining semantic and textual features resulted in the largest increases in effectiveness, identifying ~10% more sensitive documents than the baseline approach.

Other works on sensitivity classification have, for example, investigated identifying sequences of sensitive text within documents [4] and selecting an appropriate classifier kernel for sensitivity [12]. However, the approaches mentioned thus far [2,4,5,12] have evaluated sensitivity classification as a 1-shot batch supervised learning process, and therefore relied on there being a pre-judged representative collection with reliably labeled examples of the sensitivities within the collection. This can be problematic for sensitivity classification since, as previously mentioned in Sect. 1, the types of sensitivity in the collection are not known a priori. Therefore, differently from [2,4,5,12], in this work we investigate how to incorporate reviewer feedback into the learning process to quickly learn an effective sensitivity classifier using minimal reviewing effort.

Berardi *et al.* [3] was the first work to investigate optimising the cost-effectiveness of sensitivity reviewers. In that work, the authors evaluated the effectiveness of a *utility-theoretic* [10] semi-automated text classification (SATC) approach, for sensitivity classification. The approach of Berardi *et al.* [10] addresses a scenario in which the underlying state-of-the-art classifier is not effective enough to meet a strict level of accuracy required within an organisation, e.g. reviewing for sensitivity within governments. The approach ranks documents by the expected gain in accuracy that a classification system could expect to achieve by having a reviewer correct mis-classified instances. Berardi *et al.* [3] found that their approach achieved substantial improvements in overall classification ($+3\%$ – $+14\%$ F_2). However, the authors concluded that these improvements were much smaller than their approach had achieved for *topic-oriented* classification tasks, for example in [10]. In this work, we evaluate the utility-theoretic approach of Berardi *et al.* [3,10]. However, differently from those works, which assume that the underlying classifier is state-of-the-art, we evaluate the approach as an active learning strategy to incorporate reviewer feedback into the underlying sensitivity classifier.

Moving on to technology assisted review (TAR), as previously stated in Sect. 1, TAR is an iterative process, whereby a learning algorithm selects batches of documents to be presented to a reviewer to be labeled. The labeled documents are then added to the current training data and the learner is re-trained. This iterative process continues until it is judged that sufficiently many relevant documents have been identified [6]. TAR has been applied to fields such as systematic review for evidence-based medicine [13], test collection construction [14] and, most notably, e-discovery [7,15], where TAR has been shown to be more effective and more efficient than exhaustive manual review [16].

We believe that the TAR protocol can be adapted to meet the needs of sensitivity review. However, there are two noticeable differences in the objectives of TAR, for example in e-discovery, and reviewing for sensitivity. Firstly, the goal of TAR for e-discovery is to identify *close to* all the relevant documents in a collection while minimising the required reviewing effort [6], while in sensitivity review we must identify *all* sensitivities in any documents that are to be released to the public. Secondly, as previously stated in Sect. 1, there is no request for production, or *query*, in sensitivity review. Therefore, in this work we simulate TAR for sensitivity review to incorporate reviewer feedback into the TAR protocol to quickly learn to identity sensitivities from the reviewer feedback. Moreover, we evaluate approaches for selecting documents to be presented to a reviewer so that we can learn the sensitivities using the least reviewing effort possible.

Many TAR approaches deploy an active learning component to select documents to be reviewed. For example, Cormack and Grossman [6] presented an approach called *continuous active learning* and showed that selecting initial training documents through a simple keyword search, and subsequent training documents by continuous active learning, required significantly less (according to a sign test with $p < 0.01$) reviewing effort to achieve any given level of recall, compared to when the learning algorithm did not implement an active learning strategy to select the documents to use for training.

Pool-based active learning is a well known paradigm where by the learner selects documents to be reviewed, and labeled, from a pool of unlabeled documents. The most popular approach to pool-based active learning is uncertainty sampling, which has been extensively studied for developing text classification algorithms [8]. For example, Lewis and Gale [17] evaluated the effectiveness of uncertainty sampling, compared with relevance sampling and random sampling. In that work, the authors found that, for the same amount of labeling effort, uncertainty sampling usually resulted in the most effective classifier compared to the other approaches, when relevant documents are relatively abundant in the collection.

However, the selection of an appropriate active learning strategy is dependent on the nature of both the type of classification task and the task's objective [8]. Moreover, most of the research into active learning for text classification addresses a scenario in which there is a large collection of representative unlabeled examples available. Differently from that scenario, in this work, we investigate how quickly different active learning strategies can effectively learn a classifier for sensitivity classification when the types of sensitivities in a collection are not known a priori.

3   Active Learning Methodologies

In this section, we present the active learning strategies that we evaluate for technology assisted sensitivity review. Firstly, in Sect. 3.1, we provide some preliminary information regarding our methodology for simulating technology assisted sensitivity review and the underlying classifier that we use as a basis for evaluating our active learning approaches. In Sects. 3.2–3.4, we present the active learning strategies that we evaluate.

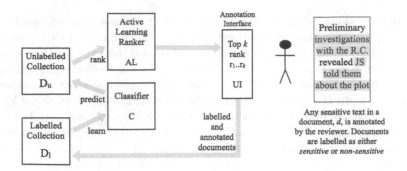

Fig. 1. Technology assisted sensitivity review simulation architecture.

3.1 Preliminaries: Simulating Technology Assisted Sensitivity Review

Figure 1 presents the process that we deploy to simulate technology assisted sensitivity review. The process aims to efficiently solicit sensitivity judgments, for a document collection, D, that can subsequently be used to train a sensitivity classifier. The collection, D, consists of two separate subsets. Firstly, an unlabeled collection, D_u, for which we do not know the collection's sensitivities and, secondly, a labeled collection, D_l, which has been sensitivity reviewed and, therefore, has associated class labels, $l_i, l \in \{sensitive, nonSensitive\}$. Initially, $|D_u| = |D|, |D_l| = 0$ and, moreover, at all times $|D_u| + |D_l| = |D|$. Our review simulation consists of three separate system components. Firstly, an active learning component, AL. At each iteration of the review cycle, AL, is responsible for identifying k documents from D_u that would be likely to provide the most valuable evidence for training a sensitivity classifier, if their associated class labels, $l_1..l_k$, were known. To do this, AL ranks documents, $d_1..d_{|D_u|}, d_i \in D_u$, by means of an active learning strategy, al_j, and selects the top k ranked documents. These top k documents are presented to the reviewer in rank order, $d_1..d_k$, via the second system component, a user interface, UI, that enables the reviewer to label each of the documents with a corresponding class label l_i. For documents that are labeled $l_{sensitive}$, the reviewer also provides text-level annotations, $a_{di}, |a_d| \in \{0..|d_i|\}$, as illustrated in Fig. 1, that indicate which text within the document led to the reviewer's l_i decision. The newly labeled documents, with their associated labels $l_1..l_k$ and annotations a_d are integrated into the labeled document set, D_l. Documents from D_l are then used to train the final system component, a sensitivity classifier, C. For C we select a multinomial naive Bayes (MNB) classifier, since it has been shown to be effective for text classification tasks [18] and, moreover, the model can be easily adapted to integrate different sources of feature evidence by simply weighting the underlying feature's multinomial [9,19]. Once C has been trained, it is deployed and its predicted class labels, $\hat{l}_i$, with a corresponding confidence score, c_i, for the documents in D_u are input to AL to provide evidence of the classifier's current knowledge. The simulation proceeds in this iterative cycle until all documents are labeled, $|D_u| = 0, |D_l| = |D|$.

3.2 Uncertainty Sampling

Uncertainty sampling [17] is a well known set of active learning approaches for evaluating the informativeness of documents in an unlabeled collection. In uncertainty sampling the algorithm tries to identify, and present to a reviewer, the documents in the collection for which the classifier is least certain about their correct class labeling.

In general, uncertainty sampling is a popular set of approaches for active learning since they are relatively easy to implement, are not computationally expensive and have been shown to be effective for many classification tasks [8]. Moreover, when deployed with a classifier that outputs probabilities or confidence scores, the classifier can be viewed as a *black box*. We test two well-known uncertainty sampling approaches, which have previously been shown to be effective for topic-based text classification [8,17]. However, as previously mentioned in Sect. 1, sensitivity is not topic-based and, therefore, we can not presume that they will be effective for sensitivity classification.

The first uncertainty sampling strategy that we evaluate is *entropy based* uncertainty [8]. Entropy uncertainty sampling ranks documents by the sum of their label entropies [20], $H(L) = -\sum_i P(l_i)logP(l_i)$, over all possible labels, l_i. One way to view the intuition of this approach is that it calculates the number of bits it would take to encode the distribution of possible outcomes for L. Therefore, documents with a high $H(L)$ score should provide more information about their assigned label.

The second uncertainty sampling strategy that we evaluate is *margin* sampling [21], $M(d_i, l_1, l_2) = |P(l_1|d_i) - P(l_2|d_i)|$. This approach to uncertainty sampling calculates the margin, or difference, between the classifier predicted probability scores for a document's first and second most likely classification labels. The intuition of margin sampling is that documents with a small margin between the two most likely class prediction probabilities are more ambiguous and, therefore, knowing the class label of these documents would be most beneficial to the classifier.

3.3 Utility

As previously mentioned in Sect. 2, we evaluate the approach of Berardi *et al.* [10] as an active learning strategy for technology assisted sensitivity review. Berardi *et al.*'s approach was designed to rank documents in an order that would achieve the maximum increase in overall classification if a reviewer was to start from the top of the ranking and proceed down the list correcting any mis-classifications until an available reviewing budget had expired. This scenario is different from active learning in that it assumes that the underlying classifier is state-of-the-art and its objective is to produce the most effective ranking for a given reviewing budget. However, we believe that the utility-theoretic approach should perform well as an active leaning strategy for sensitivity classification since it has previously been shown to be able to improve the cost-effectiveness of sensitivity reviewers [3] and, moreover, by feeding the corrected classifications back into the learning process we are, in effect, just closing the loop in the active learning cycle.

The approach's intuition is that in text classification problems where there is an imbalance in the distributions of classification categories, and a metric is chosen to account for this imbalance (e.g. F_2), the improvements in effectiveness, or *gain*, that are derived from correcting a false positive prediction is not the same as that for correcting a false negative prediction. This is important for sensitivity, since the consequences of mis-classifying a sensitive document are much greater than that of a non-sensitive document.

In the case of binary classification, the utility-theoretic measure is defined as $U(d_i) = \sum_e P(e)G(e)$, where $P(e)$ is the probability of an event, e, occurring (i.e. a false negative or a false positive prediction) and $G(e)$ is the gain that can be obtained if that event does occur. To calculate the probability of an event occurring, the approach relies on the underlying classifier's label predictions, $\hat{l}_i$, on documents in D_u to be reliable. The probability of a false negative prediction, given that the classifier has made a negative prediction, is then calculated as $P(FN(d_i)|\hat{l}_i = neg) = 1 - \frac{e^{\sigma c_i}}{e^{\sigma c_i}+1}$, where $\frac{e^{\sigma c_i}}{e^{\sigma c_i}+1}$ is a generalised logistic function that monotonically converts a classifier's confidence score, c, in the range $(-\infty, +\infty)$ to real values in the range $[0.0, 1.0]$. The probability of a false positive occurring is computed analogously.

$G(e)$ is calculated on D_l and $G(FN) \neq G(FP)$. This inequality is reflected in the definitions of the gain functions $G(FN) = \frac{1}{FN}(\frac{2(TP+FN)}{2(TP+FN)+FP} - \frac{2TP}{2TP+FP+FN})$ and $G(FP) = \frac{1}{FP}(\frac{2TP}{2TP+FN} - \frac{2TP}{2TP+FP+FN})$. To compute $G(FN)$ and $G(FP)$ the TP, FP and FN frequency counts are derived by performing a k-fold cross validation on D_l. The corresponding frequencies are then obtained by the maximum-likelihood estimation $\hat{\alpha}^{ML} = \alpha^{D_l} \cdot |D_l|/|D_u|$, $\alpha \in \{TP, FP, FN\}$. Berardi *et al.* provide a thorough examination of the approach in [10], however it is worth noting that when calculating the $\hat{\alpha}^{ML}$ values, to avoid zero counts, Laplace smoothing is applied to each $\hat{\alpha}^{ML}$ in an *on-demand* fashion if any $\hat{\alpha}^{ML} < 1$, resulting in $\hat{\alpha}^{ML} + 1$.

3.4 Sensitivity Annotation Features

The active learning strategies presented in Sects. 3.2 and 3.3 use predictions from the classifier, C, as evidence of the classifier's confidence in correctly classify the unlabeled documents, D_u. However, sensitive information is often only a small passage of text within a document and, therefore, we expect an active learning strategy that integrates term-level features of sensitivity to produce a more confident classifier that, in turn, will enable the active learning strategy to select more informative documents.

With this in mind, in this section, we present three strategies, inspired by Settles [9], that integrate term-level sensitivity features into the active learning process. As shown in Fig. 1, when a document, d_i, is judged to be sensitive, the reviewer annotates the sensitive text within the document, $a_{di}, |a_d| \in \{0..|d_i|\}$. The strategies presented here utilise these document annotations to extend the strategies presented in Sects. 3.2 and 3.3 with informative term-level sensitivity features.

We refer to our first annotation features strategy as *simple* annotation features. The simple strategy assumes that all the terms that a reviewer annotates are equally useful for identifying sensitivity. To integrate term feature importance into the active learning process, we simply increase the prior for the corresponding multinomial in the classifier, C, by a constant value α. This strategy is denoted as +Anno in Sect. 5.

The remaining two strategies make use of the labeled collection of documents, D_l, and the classifier's predictions on the unlabeled documents in D_u to calculate the expected information gain, $IG(f_k) = \sum_{F_k} \sum_i P(F_k, y_i) log \frac{P(F_k,y_i)}{P(F_k)P(y_i)}$, of term features in the unlabeled collection D_u, where $F_k \in \{0,1\}$ indicates the presence or absence of a feature f_k in the class $y_i, y_i = l_i \cup \hat{l}_i$. The first information gain annotation features strategy that we present considers all the term features that are in the intersection of the terms identified by $IG(f_k)$ and the terms annotated by a reviewer, in the current batch of documents being reviewed, as good sensitivity features and increases the prior for the corresponding multinomial in the classifier, C, by α. We refer to this strategy as *information gain* annotation features, denoted as +Anno$_{IG}$ in Sect. 5.

The final annotation features strategy that we evaluate, *annotation pool*, identifies useful sensitivity features through the same process as the previous information gain strategy, except that instead of only considering annotation terms from the current batch of documents being reviewed, a pool of potential sensitivity features is built from all previous annotations and any terms that are in the intersection of the terms identified by $IG(f_k)$ and terms in the annotation pool are considered as being good sensitivity features. This approach is denoted +Anno$_{POOL}$ in Sect. 5.

4   Experimental Setup

In this section, we present our experimental setup for evaluating the effectiveness of active learning strategies for technology assisted sensitivity review. We aim to answer two research questions, namely: **RQ1**; "Which active learning strategy enables the system to learn an effective sensitivity classifier with least reviewer effort?", and **RQ2**; "Which method of integrating a reviewer's annotations feedback is most effective for extending the tested active learning approaches?".

We evaluate our research questions on a test collection, T, of 3801 government documents that have been sensitivity reviewed by government sensitivity reviewers. The collection was assessed for two UK FOI exemptions, namely international relations and personal information. Any documents that contain any exempt information are labeled *sensitive*. The remaining documents are labeled *non-sensitive*, resulting in 502 sensitive documents ($\sim$13%) and 3299 non-sensitive ($\sim$87%).

To ensure the generalisability of our findings, we run our experiments over 25 stratified samples of the collection T. For each sample, we select 2500 documents from T as a training set Tr, which we use for the active learning simulation i.e. $|D_u| + |D_l| = Tr = 2500$. We select 500 documents from T as a held out test set, Te, for evaluating the performance of the classifier, C. We retain the distributions of sensitive and non-sensitive documents from T when generating Tr and Te, resulting in $Tr = \{2150$ non-sensitive, 325 sensitive$\}$ and $Te = \{435$ non-sensitive, 65 sensitive$\}$. We perform a binary classification, *sensitive* vs. *non-sensitive* and report mean scores over 25 samples. To test for statistical significance when evaluating reviewer effort, following [6], we use a sign test with $p < 0.01$.

At each iteration of the active learning cycle, we present the reviewer a new batch of k documents. For our experiments, we set $k = 20$. Previous work has shown that balancing the class distributions when training sensitivity classifiers can lead to a markedly improved model [2,3,5]. Therefore, when integrating newly labeled documents to D_l, we introduce the following constraint: |non-sensitive| $\in D_l \leq (k/2) + $|sensitive| $\in D_l$. We discard documents that violate this constraint[3].

For the utility approach, presented in Sect. 3.3, when estimating $G(FN)$ and $G(FP)$, following Berardi *et al.* [10], we select F_2 as our metric and perform a k-fold cross validation, setting $k = 10$. For the feature labeling approach, presented in Sect. 3.4, when integrating feature importance to the classifier, following [9], we set $\alpha = 50$.

5   Results

In this section, to answer the research questions presented in Sect. 4, we present the results of our active learning classification experiments. Figure 2 presents four plots that show the performance improvements of the learned classifier in terms of Balanced Accuracy (BAC), as evaluated on the held out collection Te. In each of the plots, the x axis shows the required reviewer effort, in number of documents reviewed. In Fig. 2, plot (a) presents the results for the raw Entropy, Margin and Utility approaches, while plot (b) shows each of the approaches extended with the *simple* reviewer annotation features, plot (c) presents the approaches extended with *information gain* annotation features and, finally, plot (d) presents the approaches extended with *annotation pool* features.

Firstly, addressing **RQ1**, we evaluate the effectiveness of each active learning strategy for quickly learning a classifier that can reliably predict sensitivity. From

---

[3] In practice this means that we randomly down-sample the classifier's training data to loosely match the class frequencies. In preliminary experiments this led to uniform improvements across all tested approaches of $\sim$+0.4 Balanced Accuracy, after all documents had been reviewed.

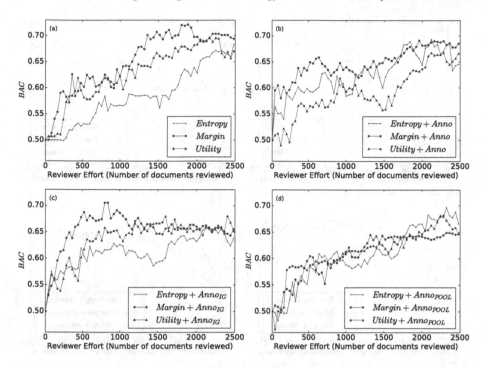

Fig. 2. Reviewer effort vs. classifier effectiveness measured by Balanced Accuracy (BAC). Raw approaches are presented in (a), while (b) presents the approaches extended with *simple* annotation features, (c) presents the approaches extended with *information gain* annotation features, and (d) presents the approaches extended with *annotation pool* features.

Fig. 2(a), we see that the Margin and Utility approaches begin to identify sensitivity noticeably quicker than Entropy, with Margin and Utility resulting BAC scores of 0.59 and 0.57 respectively when only 250 documents have been reviewed, while Entropy results in a random classifier (0.5 BAC). Moreover, the Margin and Utility approaches sustain this additional performance over Entropy for almost the entire review session. As the number of labeled documents increases, particularly when the number of reviewed documents is >1180, we see that Margin shows noticeable improvements compared to the Utility approach. However, the difference between the approaches reduces as the number of reviewed documents approaches 2500. Therefore, in response to **RQ1**, we conclude that the Margin active learning strategy is the best performing strategy when the approaches are not extended with annotation features.

Turning our attention to **RQ2**, Fig. 2(b),(c) and (d), present the active learning approaches with additional annotation features. From Fig. 2(b), we see that the Entropy and Margin approaches with additional *simple* annotation features (+*Anno*) begin to identify sensitivity with markedly less reviewer effort than the approaches on their own (Fig. 2(a)). To achieve 0.6 BAC, Margin + Anno

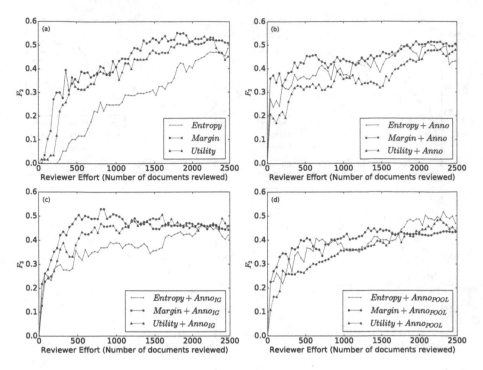

Fig. 3. Reviewer effort vs. Classifier effectiveness measured by F_2. Raw approaches are presented in (a), while (b) presents the approaches extended with *simple* annotation features, (c) presents the approaches extended with *information gain* annotation features, and (d) presents the approaches extended with *annotation pool* features.

required 200 documents to be reviewed while Margin required 400. Moreover, Entropy + Anno achieves 0.6 BAC with significantly less reviewing effort than Entropy, according to a sign test with $p < 0.01$ (400 documents vs. 1800 documents).

In evaluating the overall performance increase that is obtained from additional reviewer annotation features, we note from Fig. 2(c) that *information gain* annotation features enable each of the approaches to develop an effective sensitivity classifier noticeably quicker than the raw approaches in Fig. 2(a). Most notably, Margin sustains its initial gains in classification effectiveness and reaches its peak classification performance ($\sim$0.7 BAC) with significantly less reviewer effort (according to the sign test, $p < 0.01$), requiring only 820 documents to be reviewed as opposed to 1700 when Margin is deployed without annotation features (shown in Fig. 2(a)), therefore, resulting in a 51% reduction in required reviewer effort. However, we note that there is a notable decline in classification performance after this peak.

When classifying sensitive information, there is a much greater penalty from mis-classifying documents that are sensitive than ones that are not. The F_2 metric reflects this asymmetry and, therefore, we present the classification

Table 1. Area under the curve for the BAC and F_2 plots presented in Figs. 2 and 3 respectively.

	BAC	F_2		BAC	F_2		BAC	F_2		BAC	F_2
Entropy	0.5800	0.2675	+Anno	0.6213	0.4070	+Anno$_{IG}$	0.6087	0.3674	+Anno$_{POOL}$	0.6029	0.3924
Margin	0.6480	0.4271	+Anno	0.6432	0.4454	+Anno$_{IG}$	**0.6501**	**0.4503**	+Anno$_{POOL}$	0.6022	0.3963
Utility	0.6236	0.3863	+Anno	0.5871	0.3578	+Anno$_{IG}$	0.6084	0.3551	+Anno$_{POOL}$	0.6317	0.4262

improvements in terms of F_2 in Fig. 3. We can see that the plots in Fig. 3 display similar trends as the BAC plots, with Margin performing best and, moreover, information gain annotation features resulting in an effective classifier with notably less reviewing effort. Therefore, in response to **RQ2**, we conclude that information gain annotation features are most effective for integrating reviewer feedback from sensitivity annotations. We note, however, that the Utility approach is very competitive in terms of F_2 for the raw active learning approaches (Fig. 3(a)) and when extended with information gain annotation features (Fig. 3(c)). This is intuitive since the utility approach is optimised for F_2.

Finally, to provide a measure of overall classification effectiveness, Table 1 presents the Area Under the Curve (AUC) scores for the BAC and F_2 plots presented in Figs. 2 and 3 respectively. As can be seen from Table 1, Margin + Anno$_{IG}$ achieves the best overall classification effectiveness throughout the review simulation. This finding provides extra evidence that the Margin + Anno$_{IG}$ combination can be an effective choice for technology assisted sensitivity review.

6   Conclusions

In this work, we presented a thorough evaluation of active learning strategies for technology assisted sensitivity review. We evaluated two well-known *uncertainty sampling* active learning strategies from the literature and an approach adapted from semi automated text classification, that has previously been shown to be effective for improving the cost-effectiveness of sensitivity reviewers. Moreover, we extended these approaches to integrate term-level reviewer feedback from annotations of sensitive text within documents. We showed that extending Margin uncertainty sampling with high information gain annotation term features enabled us to learn an effective sensitivity classifier (0.7 BAC) using significantly less reviewing effort (according to the sign test with $p < 0.01$), than when the approach was deployed without annotation features, i.e. a 51% reduction in the number of documents that had to be reviewed. Moreover, we found that this approach achieved the best overall classification effectiveness throughout a technology assisted sensitivity review simulation, and conclude that the approach can be an effective choice for quickly learning to classify sensitivity, when the types of sensitivities in a collection are not known *a priori*.

452   G. McDonald et al.

References

1. TNA: The application of technology-assisted review to born-digital records transfer, inquiries and beyond (2016)
2. McDonald, G., Macdonald, C., Ounis, I., Gollins, T.: Towards a classifier for digital sensitivity review. In: de Rijke, M., Kenter, T., de Vries, A.P., Zhai, C.X., de Jong, F., Radinsky, K., Hofmann, K. (eds.) ECIR 2014. LNCS, vol. 8416, pp. 500–506. Springer, Cham (2014). https://doi.org/10.1007/978-3-319-06028-6_48
3. Berardi, G., Esuli, A., Macdonald, C., Ounis, I., Sebastiani, F.: Semi-automated text classification for sensitivity identification. In: Proceedings of the CIKM (2015)
4. McDonald, G., Macdonald, C., Ounis, I.: Using part-of-speech n-grams for sensitive-text classification. In: Proceedings of the ICTIR (2015)
5. McDonald, G., Macdonald, C., Ounis, I.: Enhancing sensitivity classification with semantic features using word embeddings. In: Jose, J.M., Hauff, C., Altıngovde, I.S., Song, D., Albakour, D., Watt, S., Tait, J. (eds.) ECIR 2017. LNCS, vol. 10193, pp. 450–463. Springer, Cham (2017). https://doi.org/10.1007/978-3-319-56608-5_35
6. Cormack, G.V., Grossman, M.R.: Evaluation of machine-learning protocols for technology-assisted review in electronic discovery. In: Proceedings of the SIGIR (2014)
7. Oard, D.W., Baron, J.R., Hedin, B., Lewis, D.D., Tomlinson, S.: Evaluation of information retrieval for e-discovery. Artif. Intell. Law 18(4), 347–386 (2010)
8. Settles, B.: Active learning. Synth. Lect. Artif. Intell. Mach. Learn. 6(1), 1–114 (2012)
9. Settles, B.: Closing the loop: fast, interactive semi-supervised annotation with queries on features and instances. In: Proceedings of the EMNLP (2011)
10. Berardi, G., Esuli, A., Sebastiani, F.: A utility-theoretic ranking method for semi-automated text classification. In: Proceedings of the SIGIR (2012)
11. Sebastiani, F.: Machine learning in automated text categorization. ACM Comput. Surv. 34(1), 1–47 (2002)
12. McDonald, G., García-Pedrajas, N., Macdonald, C., Ounis, I.: A study of svm kernel functions for sensitivity classification ensembles with pos sequences. In: Proceedings of the SIGIR (2017)
13. Lefebvre, C., Manheimer, E., Glanville, J.: Searching for studies. Cochrane handbook for systematic reviews of interventions: Cochrane book series, pp. 95–150 (2008)
14. Sanderson, M., Joho, H.: Forming test collections with no system pooling. In: Proceedings of the SIGIR (2004)
15. Oard, D.W., Hedin, B., Tomlinson, S., Baron, J.R.: Legal track overview. In: Proceedings of the TREC (2008)
16. Grossman, M.R., Cormack, G.V.: Technology-assisted review in E-discovery can be more effective and more efficient than exhaustive manual review. Rich. JL Tech. 17, 1 (2010)
17. Lewis, D.D., Gale, W.A.: A sequential algorithm for training text classifiers. In: Proceedings of the SIGIR (1994)
18. Rennie, J.D., Shih, L., Teevan, J., Karger, D.R.: Tackling the poor assumptions of naive Bayes text classifiers. In: Proceedings of the ICML (2003)
19. McCallum, A., Nigam, K., et al.: Employing EM and pool-based active learning for text classification. In: ICML, vol. 98, pp. 350–358 (1998)

20. Shannon, C.E.: A mathematical theory of communication. BSTJ **27**, 623–656 (1948)

21. Scheffer, T., Decomain, C., Wrobel, S.: Active hidden markov models for information extraction. In: Hoffmann, F., Hand, D.J., Adams, N., Fisher, D., Guimaraes, G. (eds.) IDA 2001. LNCS, vol. 2189, pp. 309–318. Springer, Heidelberg (2001). https://doi.org/10.1007/3-540-44816-0_31

Authorship Verification in the Absence of Explicit Features and Thresholds

Oren Halvani$^{(\boxtimes)}$, Lukas Graner, and Inna Vogel

Fraunhofer Institute for Secure Information Technology,
Rheinstraße 75, 64295 Darmstadt, Germany
Oren.Halvani@SIT.Fraunhofer.de

Abstract. Enhancing information retrieval systems with the ability to take the writing style of people into account opens the door for a number of applications. For example, one can link articles by authorships that can help identifying authors who generate hoaxes and deliberate misinformation in news stories, distributed across different platforms. Authorship verification (AV) is a technique that can be used for this purpose. AV deals with the task to judge, whether two or more documents stem from the same author. The majority of existing AV approaches relies on machine learning concepts based on explicitly defined stylistic features and complex models that involve a fair amount of parameters. Moreover, many existing AV methods are based on explicit thresholds (needed to accept or reject a stated authorship), which are determined on training corpora. We propose a novel parameter-free AV approach, which derives its thresholds for each verification case individually and enables AV in the absence of explicit features and training corpora. In an experimental setup based on eight evaluation corpora (each one from another language) we show that our approach yields competitive results against the current state of the art and other noteworthy AV baselines.

Keywords: One-class · Compression · Intrinsic authorship verification

1 Introduction

Information retrieval (IR) is a well-studied field with many media search applications such as image, speech/music, video or text retrieval. Especially for text, numerous methods and techniques have been proposed to enhance the retrieval process at a fine-grained level. Document classification, for instance, is a well-known technique used by IR systems, where generally the content of the document is the subject of classification. However, in recent years IR systems opened up a new perspective to classify not only the document's content, but also the writing style contained therein [22]. Enhancing IR systems to cope with the writing style opens the door for a wide range of sophisticated applications. People nowadays use social media, which not only provide easy access to news articles, but also allow users to engage with other readers by sharing, commenting and

© Springer International Publishing AG, part of Springer Nature 2018
G. Pasi et al. (Eds.): ECIR 2018, LNCS 10772, pp. 454–465, 2018.
https://doi.org/10.1007/978-3-319-76941-7_34

discussing the content. Unfortunately, such platforms can be misused to create and spread intentionally false information ("fake news"). The wide spread of misinformation can have a negative impact on elections, personalities and the entire society [25]. To counteract this, a smart IR system could filter retrieved documents based on their writing style if they match documents of known authors, who already distributed fake news in the past. One possibility to approach this problem is to make use of authorship verification (AV) methods, which can be seen as similarity methods that operate on the stylistic instead of the semantic layer of documents. However, Potthast et al. have shown that AV alone cannot be used to detect fake news [21]. Still, AV can be used as a filtering mechanism within an IR system.

The goal of AV is to decide, whether a given text $\mathcal{D}_\mathcal{U}$ of an unknown author $\mathcal{U}$ and existing reference texts $\mathbb{D}_\mathcal{A} = \{\mathcal{D}_1, \mathcal{D}_2, \ldots\}$ of a known author $\mathcal{A}$ stem from the same writer such that $\mathcal{U} = \mathcal{A}$ holds. From a machine learning perspective, AV falls into the category of **one-class** classification problems, where the training set comprises only objects of the *target* class $\mathcal{T}$ such that a model must be learned to recognize $\mathcal{T}$ in the presence of *outliers* (all other possible classes). In the context of authorship verification, $\mathcal{T}$ refers to the **true author** $\mathcal{A}$, while the outliers represent **all other authors** $(\neg\mathcal{A})$.

The key challenge that has to be addressed in an AV scenario is how to draw a boundary that captures both $\mathbb{D}_\mathcal{A}$ and unseen texts of $\mathcal{A}$, while simultaneously keeping all texts of $\neg\mathcal{A}$ out of the enclosed space (see Fig. 1). In addition, the question arises how to recognize unseen documents of $\mathcal{A}$, which might differ from $\mathbb{D}_\mathcal{A}$ in terms of topic, genre, chronological stylistic drift or the deliberate attempt to obfuscate his identity. Obviously, determining a threshold that acts as a boundary to cope with this challenge is a nontrivial task.

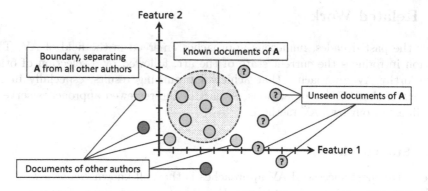

Fig. 1. Authorship verification as an instance of one-class classification problems.

Many existing AV approaches cannot be reimplemented easily due to their complexity or the number of involved parameters, which can be challenging to tune (e. g., [1,7,17]). Other approaches require an appropriate strategy to collect external documents in order to transform the AV task from a one-class to a two-class classification problem (e. g., [14,20]).

We present OCCAV[1], a novel AV method that offers a number of benefits. First, it only requires a compression model i. e., a combination of a compression algorithm and a dissimilarity measure. Since compression libraries are available for the most relevant programming languages, there is no need for reimplementation. Besides this, the dissimilarity measure is very simple and requires only a few lines of code. Second, OCCAV does not involve adjustable parameters and, as a consequence, can be used "out-of-the-box" in an AV scenario without training. Third, the features are generated implicitly by the underlying compression algorithm such that their explicit definition, extraction and optimization is discarded. Fourth, OCCAV determines its threshold (needed to accept or reject the authorship) intrinsically such that learning an appropriate threshold by optimizing it against a training corpus is also avoided. Fifth, OCCAV does not involve any text preprocessing steps, which is the case for some existing AV methods (e. g., [1,2]). Sixth, OCCAV is deterministic in contrast to existing stochastic AV methods (e. g., [14,20]) that involve random subsampling of the feature space and, thus, can lead to unreliable behaviour regarding the AV result. Seventh, in contrast to state of the art methods (e. g., [1,14,20,24]) OCCAV works very fast. On average, it processes a single verification case in less than a second. Hence, the method is applicable for scenarios where runtime matters, as for instance IR systems.

The rest of this paper is structured as follows. In Sect. 2 we review related work and existing approaches that partially serve as baselines in the later experiments. In Sect. 3 we introduce OCCAV and explain in detail the necessary steps to reproduce our results. In Sect. 4 we present our experiments based on OCCAV and the reimplemented baselines and discuss our observations. Finally, we draw our conclusions in Sect. 5 and provide ideas for future work.

2   Related Work

Over the past decades, numerous approaches emerged in the field of AV. This section introduces the current state of the art, followed by a selection of other noteworthy AV approaches that achieved promising results, especially in the context of the PAN[2] competitions. A part of the reviewed approaches serve as baselines for our own AV method.

2.1   State of the Art

One of the most successful AV approaches is the well-known *Impostors Method* (IM), proposed by Koppel and Winter [16]. The term *impostors* refers to external documents, used to represent authors different than the true author $\mathcal{A}$. In this way, the method falls into the category of *extrinsic* AV methods. IM works in a two-stage procedure. First, documents serving as impostors are gathered according to a data collection procedure. This could be accomplished, for example, by

---

[1] One-Class Compression Authorship Verifier.
[2] PAN is a series of scientific events and shared tasks on digital text forensics [28].

querying a search engine for documents, similar in terms of genre or topic. Second, a feature randomization technique is applied iteratively in order to measure the similarity between pairs of documents. If, according to this measure, a suspect is picked out of the set of impostors with sufficient salience, then Koppel and Winter claim the suspect as the author of the disputed document [16]. In this sense, Koppel and Winter transform the AV problem from a one-class to a two-class classification problem ($\mathcal{A}$ vs. $\neg\mathcal{A}$ = impostors). The IM algorithm is presented in detail in [15].

IM gained a lot of attraction in the field of AV and has been reimplemented and extended by researchers in various ways. Seidman introduced *General Impostors* (GI) [24], a slight modification of IM. While the latter performs verification only between two texts, GI considers the case, where multiple documents of $\mathcal{A}$ are available. GI was the overall winning approach of the PAN competition in 2013 [10]. One year later, Khonji and Iraqi proposed a variant of GI named ASGALF [12], where they adopted a modified score measure as well as a larger set of features (letter-/word-/function word-/word shape- and part-of-speech tag-level features). ASGALF was the overall winning approach of the PAN competition in 2014 [27]. Later, Gutierrez et al. proposed an adaptation of GI [6]. Their approach involves an aggregate function, which iterates over document pairs and applies homotopy-based classification[3]. In a nutshell, their modification can be understood as a voting mechanism regarding the randomly selected impostors.

Most recently, Potha and Stamatatos proposed ImpGI, which is also a modification of GI [20]. ImpGI attempts to improve both, (1) *impostor selection* and (2) *ranking information*. Regarding (1), they propose to select external documents with the highest similarity score computed by the min-max (also known as "Ružička") similarity function, instead of selecting impostor documents randomly. The authors argue that this increases the probability of considering challenging impostor documents, which have thematic or stylistic similarity with the known documents. Regarding *ranking information*, they compare the impostor document only with $\mathcal{D}_\mathcal{U}$ (rather than with $\mathcal{D}_\mathcal{A}$ as well). That way, the impostor document is considered as a direct competitor of $\mathcal{D}_\mathcal{A}$. Furthermore, they rank the computed similarities of $\mathcal{D}_\mathcal{A}$ and the impostors and consider the ranking position of $\mathcal{D}_\mathcal{A}$. Potha and Stamatatos evaluated their approach against GI, ASGALF as well as the approach of Gutierrez et al. on the publicly available PAN corpora[4] from 2014 and 2015. Here, it turned out that their method outperformed the other approaches in three out of ten corpora.

2.2 Other Selected Authorship Verification Methods

In contrast to the aforementioned approaches, Jankowska et al. proposed a method which we refer to as CNG [8], that treats AV as a one-class classification problem. Their approach resembles the idea of the *k-center boundary*

---

[3] *Homotopy-based Classification* is used in face recognition, where the goal is to measure the contribution of known faces in the generation of an unknown face [6].

[4] Available under http://pan.webis.de.

method [30], where they use the so-called $\boldsymbol{Common}$ $\boldsymbol{N}$-$\boldsymbol{Grams}$ (CNG) dissimilarity measure, instead of a distance function. For each document the authors extract a profile, comprising the most common character n-grams and their normalized frequencies. Then, for each known document $\mathcal{D} \in \mathbb{D}_{\mathcal{A}}$ the maximum dissimilarity $\text{CNG}_{max}(\mathcal{D})$ to the other known documents is calculated. Furthermore, the authors employ the dissimilarity ratio $r(\mathcal{D}_{\mathcal{U}}, \mathcal{D}) = \frac{\text{CNG}(\mathcal{D}_{\mathcal{U}}, \mathcal{D})}{\text{CNG}_{max}(\mathcal{D})}$ with the following intuition: If $r(\mathcal{D}_{\mathcal{U}}, \mathcal{D})$ is smaller than 1, another known document exists more dissimilar to k than $\mathcal{D}_{\mathcal{U}}$, while $r(\mathcal{D}_{\mathcal{U}}, \mathcal{D})$ greater than 1 meaning, all other known documents are more similar to k than to $\mathcal{D}_{\mathcal{U}}$. Finally, CNG accepts an unknown authorship if $\frac{1}{|\mathbb{D}_{\mathcal{A}}|}\sum_{\mathcal{D} \in \mathbb{D}_{\mathcal{A}}} r(\mathcal{D}_{\mathcal{U}}, \mathcal{D})$ falls below a specific threshold, determined on a training corpus. CNG was the winning approach of the PAN-2013 competition in terms of overall AUC[5] and was selected as a baseline in the subsequent competition [26]. Based on this, we decided to use the CNG method as a baseline in our evaluation.

Similar to the approach of Jankowska et al., Potha and Stamatatos proposed a CNG-based method [19], which we denote by ProfAV. The method differs in a number of ways, as for example, the data representation. Here, the authors concatenate all known documents into a single big document $\mathcal{D}_{\mathcal{A}}$. Then, the α and β most common character n-grams are extracted for $\mathcal{D}_{\mathcal{A}}$ and $\mathcal{D}_{\mathcal{U}}$, respectively. Thus, ProfAV is an asymmetrical AV method in contrast to CNG. Moreover, Potha and Stamatatos proposed three different dissimilarity measures and a parameter to decide which one to use.

3   Proposed Authorship Verification Method: OCCAV

In this section we introduce OCCAV, which aims to decide for a given AV problem $\rho = (\mathcal{D}_{\mathcal{U}}, \mathbb{D}_{\mathcal{A}})$ whether all involved texts were written by the same author ($\mathcal{U} = \mathcal{A}$). First, we explain how the data is represented and what prerequisites we consider. Afterwards, we describe how the similarities between the representations of the texts are measured and, based on these, how the verification decision regarding the authorship is performed.

Data Representation: According to Potha and Stamatatos [19], all authorship attribution (and therefore also authorship verification) methods can be distinguished in *instance-* and *profile-based paradigms*. Instance-based AV methods treat all available texts in $\mathbb{D}_{\mathcal{A}}$ separately such that each $\mathcal{D}_i$ represents an instance of $\mathcal{A}$. On the other hand, profile-based AV methods try to capture the writing style of $\mathcal{A}$ by aggregating all $\mathcal{D}_i \in \mathbb{D}_{\mathcal{A}}$ into a single representation. OCCAV follows the instance-based paradigm, where in case that only one text is available for the true author ($\mathbb{D}_{\mathcal{A}} = \{\mathcal{D}_{\mathcal{A}}\}$), $\mathcal{D}_{\mathcal{A}}$ is split into two halves of (almost) equal length. This procedure was also used by Jankowska et al. in their approach [8]. In the next step, we compress $\mathcal{D}_{\mathcal{U}}$ as well as each text of $\mathbb{D}_{\mathcal{A}}$, which results in the following representations:

---

[5] **Area Under the ROC-Curve.**

$$\mathcal{D}_{\mathcal{U}} \mapsto x$$
$$\mathbb{D}_{\mathcal{A}} = \{\mathcal{D}_1, \mathcal{D}_2, \ldots\} \mapsto Y = \{y_1, y_2, \ldots\}$$

In order to compress the texts we used the PPMd[6] compressor[7]. Our consideration is primarily based on two reasons: First, it has been shown in previous works that PPMd outperforms other existing compressors such as Gzip, BZip2 or LZW [5,18]. Second, our initial experiments have shown that besides these three, using sophisticated compressors such as ZPAQ which, according to a well-known benchmark[8], outperforms PPMd in terms of compression ratio, surprisingly achieve worse results than PPMd.

Computing Similarity: In contrast to existing AV approaches, where distance functions [8,9,19] have been widely used to measure the similarity of texts based on their numerical feature vector representations, OCCAV relies on an alternative measure for compressed files. In our previous and recent work [5] we identified four measures suitable for this purpose: *Normalized Compression Distance, Chen-Li metric, Compression-based Dissimilarity Measure* and *Compression-based Cosine* (CBC) [23]. One finding was that CBC achieved slightly better results than the former three measures, across different corpora [5], which is also the case for the corpora used in this work. Therefore, we decided to rely on CBC which, according to [23], is defined as follows:

$$\mathrm{CBC}(x, y) = 1 - \frac{C(x) + C(y) - C(xy)}{\sqrt{C(x)C(y)}}$$

Here, $C(\cdot)$ denotes the length of a compressed document (or more precisely its size in terms of bytes). Given CBC[9] and all compressed files from the previous step, we calculate a score for each pair (x, y_i) that reflects their dissimilarity. We denote the smallest score $s_{\min}$ and its associated compressed document $y_{\min}$ by:

$$s_{\min} = \min\Big(\{\mathrm{CBC}(x, y_i) \mid y_i \in Y\}\Big) \ \text{ and } \ y_{\min} = \underset{y_i \in Y}{\arg\min}\big(\mathrm{CBC}(x, y_i)\big)$$

Starting from $y_{\min}$, we, again, calculate dissimilarity scores, but this time between $y_{\min}$ and all $y_i \in Y \setminus \{y_{\min}\}$. Then we average these scores and denote the resulting value by s_{avg}. Note that instead of calculating the average we also experimented with the median. However, we could not observe noticeable differences as in five cases both average and median performed equally. However, in two cases *average* performed slightly better, thus, we decided to use it.

Decision Making: Based on the computed scores $s_{\min}$ and s_{avg} we accept (Y) the authorship for a given problem ρ if $s_{\min} < s_{\text{avg}}$ holds. Otherwise, we reject (N) the authorship.

---

[6] **P**rediction by **P**artial **M**atching. Note that **d** refers to the order of the PPM model.

[7] We use Hathcock's library (https://github.com/adamhathcock/sharpcompress).

[8] "Large Text Compression Benchmark" (http://mattmahoney.net/dc/text.html).

[9] Note that this measure is not a metric, as all conditions a metric must satisfy (identity, symmetry and triangle inequality) are not met.

4   Experimental Evaluation

In this section we describe our experimental setup. First, we introduce all involved corpora, followed by the reimplemented AV methods that serve as baselines. Next, we briefly mention a number of important performance measures and which one we choose to evaluate OCCAV against the baselines. Finally, we present the results and summarize our observations.

4.1   Corpora

As an experimental setup we compiled eight corpora[10] (listed in Table 1), which have been crawled from different sources. Each corpus corresponds to one language out of eight (Dutch, English, French, German, Greek, Polish, Spanish and Swedish). The majority of the documents belong to the genre of news articles as these offer a good possibility to link authors to corresponding texts. We partitioned each corpus into a training and test set ($\approx 20/80\%$ split), while each corpus resulted in an even number of Y and N authorship verification problems.

Table 1. All involved corpora used in our experiments. Here, $\mathcal{C}$ denotes a corpus, where tr refers to the training and te to the test set. The number of problems in $\mathcal{C}$ is given by $|\mathcal{C}|$. avg($\mathbb{D}_{\mathcal{A}}$) denotes the average number of known documents in each problem, avg($|\mathcal{D}_{\mathcal{A}}|$) the average character length of each known document $\mathcal{D}_{\mathcal{A}}$ and avg($|\mathcal{D}_{\mathcal{U}}|$) the average character length of each unknown document.

Language	$\mathcal{C}$	$|\mathcal{C}_{tr}|$	$|\mathcal{C}_{te}|$	avg($\mathbb{D}_{\mathcal{A}}$)	avg($|\mathcal{D}_{\mathcal{A}}|$)	avg($|\mathcal{D}_{\mathcal{U}}|$)	Genre
Dutch	Trouw	40	160	3	11013	3637	News articles
English	Enron [4]	16	64	3.5	15541	4393	e-mails
French	L'Express	20	80	2	14644	7341	News articles
German	Gutenberg	20	80	3	5116	5108	Novel excerpts
Greek	Cookpad	24	96	6	8201	1323	Cooking recipes
Polish	Dziennik	20	80	3	17823	6178	News articles
Spanish	El Pais	40	160	3	20202	7024	News articles
Swedish	Svenska	8	32	5	20931	4179	News articles

4.2   Baselines

In order to establish challenging baselines, we decided to reimplement a number of existing AV approaches, partially introduced in Sect. 2. These include CNG [8], ProfAV [19], GLAD [7], GI [24] and ImpGI [20]. As an additional baseline, we also decided to use our previous approach COAV introduced in [5]. Given the training corpora listed in Table 1, we trained each baseline method according to the procedure described in the respective literature.

---

[10] All corpora are available upon request under http://bit.do/ECIR_2018.

4.3   Performance Measures

Compared to authorship attribution, AV is a relatively young and underresearched field[11]. As a consequence, there is no single standardized performance measure, accepted by the community up to the present date. Over the past decade, a variety of performance measures have been used in the AV community, including Accuracy [9], F_1 [29], AUC [19], c@1 [13] or AUC $\cdot$ c@1 [11]. According to the reviewed literature, we observed that numerous existing works consider AUC, most likely, because it is not bounded to a fixed threshold (in contrast to the Accuracy measure). Therefore, we decided also to use it in our evaluation.

4.4   Results

The evaluation results for OCCAV and the six baselines are given, in terms of AUC, in Table 2. As can be seen, OCCAV outperforms all baselines in two out of eight corpora. In five out of eight corpora, it performs very similar to the state of the art (GI and ImpGI). Another observation is that OCCAV's performance behaves stable across all involved languages, genres and topics, in contrast to other baselines such as ProfAV, CNG or GLAD that partially suffer from significant drops. Furthermore, we can infer from Tables 1 and 2, that OCCAV is robust regarding the text lengths of $\mathcal{D}_\mathcal{U}$ and $\mathbb{D}_\mathcal{A}$. In other words, the achieved AUC is high for **near-equal** as well as significantly **different** text lengths.

Table 2. Evaluation results (in terms of AUC) regarding the eight test corpora.

Corpus	OCCAV	COAV	CNG	ProfAV	GLAD	GI	ImpGI	avg(·)
Dutch	0.867	**0.920**	0.745	0.827	0.806	0.831	0.880	0.839
English	0.823	0.878	0.777	0.840	0.849	**0.943**	0.923	0.862
French	0.841	**0.940**	0.698	0.735	0.799	0.929	0.897	0.834
German	0.821	0.903	0.784	0.821	0.863	0.909	**0.932**	0.862
Greek	**0.878**	0.806	0.783	0.761	0.774	0.839	0.748	0.799
Polish	0.937	**0.968**	0.756	0.916	0.854	0.945	0.938	0.902
Spanish	0.908	**0.956**	0.866	0.881	0.907	0.937	0.934	0.913
Swedish	**0.909**	0.782	0.681	0.742	0.194	0.769	0.856	0.705
avg(·)	0.873	**0.894**	0.761	0.815	0.756	**0.888**	0.889	

In analogy to the PAN competition [28], we also computed statistical significance of performance differences between OCCAV and all examined baselines, using *approximate randomization testing*[12]. For this, we compared the accuracies of all involved methods pairwise. The null hypothesis $\mathcal{H}_0$ is that there

---

[11] Evidence for this can be seen by comparing the number of published papers across different bibliographic databases such as *Google Scholar* or *Microsoft Academic*.

[12] We used Van Asch's script (http://www.clips.uantwerpen.be/scripts/art).

is no difference in the output of two methods. Let p denote the p-value computed by the test. This value corresponds to the probability that we measure the observed differences or even stronger differences between the two methods under the assumption that $\mathcal{H}_0$ is true. We indicate $p > 0.05$ (no significant difference) with the symbol "=", $0.01 < p < 0.05$ (significant difference) with "*", $0.001 < p < 0.01$ (very significant difference) with "**", and $p < 0.001$ (highly significant difference) with "***" in Table 3. We can infer from this table that there is no statistically significant difference between OCCAV and the majority of the baselines. While this is not surprising for COAV (since both use PPMd and CBC), it is not clear why it holds for ProfAV, GI and GLAD. Especially, the latter differs from ProfAV and GI, as it relies on many different features such that one would expect at least a significant difference regarding the output. It is also unclear why there is a highly significant difference between GI and ImpGI, since both rely on the same algorithm where the latter has only slight modifications. One possibility is that both methods are stochastic and thus also imply a sort of randomness regarding the output.

Table 3. Pairwise significance tests for the entire evaluation corpus.

	COAV	CNG	ProfAV	GLAD	GI	ImpGI
OCCAV	=	***	=	=	=	***
COAV		***	*	**	=	***
CNG			**	*	***	**
ProfAV				=	=	***
GLAD					*	***
GI						***

4.5 Runtime

Besides the performance of all AV methods we also measured their runtime which, in the context of IR systems, also play an important role. In Table 4 we list the runtime for each method. As can be seen from this table, both methods GI and ImpGI consume longer runtimes, compared to the other methods. The reason for this are the collection of impostors and the similarity calculations for each subset of features. Also interesting to note is that OCCAV performs faster than GI and ImpGI but slower than our previous approach COAV. This is because the latter compress a smaller number of texts (all $\mathcal{D}_i \in \mathbb{D}_A$ are concatenated), while OCCAV compresses each $\mathcal{D}_i$ separately.

Table 4. Average problem runtime (in seconds), averaged over all evaluation corpora.

OCCAV	COAV	CNG	ProfAV	GLAD	GI	ImpGI
0.13	0.031	0.043	0.034	0.462	1.293	3.361

5   Conclusion and Future Work

We presented OCCAV, a novel, simple and parameter-free AV method. The novelty of the proposed approach lies in the fact that it avoids the **explicit** definition and extraction of features as well as the threshold, used to accept or reject a stated authorship in a verification case. OCCAV discards hyperparameter optimization and therefore unfolds its potential in scenarios, where training data is not available. OCCAV is an intrinsic AV method and thus adopts its verification decision based only on the documents of the known author. By this, it omits to collect external documents that serve as representatives of the outlier class.

In contrast to a number of existing AV approaches, OCCAV does not involve any text preprocessing that further reduces its complexity. Besides this, OCCAV is deterministic and thus more reliable in real-world settings, in contrast to existing stochastic AV methods that involve random subsampling of their feature space. Moreover, existing implementations of the underlying compression algorithm are available such that OCCAV can be reimplemented easily. Based on this and the provided eight corpora, our stated results can be reproduced with minimal effort. In an experimental setup based on eight corpora, distributed across eight languages, we have shown that OCCAV yields competitive results compared to current state of the art methods. In two out of eight corpora, OCCAV outperformed all involved baselines.

Beyond the benefits OCCAV offers, it also suffers from the same problem that more sophisticated machine learning concepts (e. g., neural networks) also have: the underlying model cannot be interpreted easily. More precisely, the features are generated implicitly by PPMd and cannot be accessed without drilling out the compressing algorithm. However, this was not the subject in this paper and is left for future work. Another direction for future work is to investigate, whether OCCAV's performance can be increased by using other compressors and similarity measures. In the case that no alternatives will be found, one could also experiment with ensembles, constructed from weak compression models, in the same sense techniques based on *gradient boosting* do.

Acknowledgments. This work was supported by the German Federal Ministry of Education and Research (BMBF) under the project "DORIAN" (Scrutinise and thwart disinformation). We would like to thank Christian Winter and Felix Mayer for their valuable reviews that helped to improve the quality of this paper.

References

1. Bagnall, D.: Author identification using multi-headed recurrent neural networks. In: Working Notes of CLEF 2015 - Conference and Labs of the Evaluation Forum, Toulouse, France, 8–11 September 2015
2. Castillo, E., Cervantes, O., No, D.V., Báez, D.: Author verification using a graph-based representation. Int. J. Comput. Appl. **123**(14), 1–8 (2015)
3. Forner, P., Navigli, R., Tufis, D., Ferro, N. (eds.): Working notes for CLEF 2013 Conference, Valencia, Spain, 23–26 September 2013, CEUR Workshop Proceedings, vol. 1179 (2014). CEUR-WS.org

4. Halvani, O.: Enron Authorship Verification Corpus, Mendeley Data, v1 (2017)
5. Halvani, O., Winter, C., Graner, L.: On the usefulness of compression models for authorship verification. In: Proceedings of the 12th International Conference on Availability, Reliability and Security, ARES 2017, pp. 54:1–54:10 (2017)
6. Hernández, J.G.G., Casillas, J., Ledesma, P., Pineda, G.F., Ruíz, I.V.M.: Homotopy based classification for author verification task: notebook for PAN at CLEF 2015. In: Working Notes of CLEF 2015 - Conference and Labs of the Evaluation Forum, Toulouse, France, 8–11 September 2015
7. Hürlimann, M., Weck, B., von den Berg, E., Šuster, S., Nissim, M.: GLAD: groningen lightweight authorship detection. In: Working Notes of CLEF 2015 - Conference and Labs of the Evaluation Forum, Toulouse, France, 8–11 September 2015
8. Jankowska, M., Keselj, V., Milios, E.E.: Proximity based one-class classification with common N-gram dissimilarity for authorship verification task notebook for PAN at CLEF 2013. In: Forner et al. [3]
9. Noecker Jr., J., Ryan, M.: Distractorless authorship verification. In: Proceedings of the Eight International Conference on Language Resources and Evaluation (LREC 2012), Istanbul, Turkey, May 2012
10. Juola, P., Stamatatos, E.: Overview of the author identification task at PAN 2013. In: Forner et al. [3]
11. Kestemont, M., Stover, J.A., Koppel, M., Karsdorp, F., Daelemans, W.: Authenticating the writings of Julius Caesar. Expert Syst. Appl. 63, 86–96 (2016)
12. Khonji, M., Iraqi, Y.: A slightly-modified GI-based author-verifier with lots of features (ASGALF). In: Working Notes for CLEF 2014 Conference, Sheffield, UK, 15–18 September 2014, pp. 977–983 (2014)
13. Kocher, M., Savoy, J.: A simple and efficient algorithm for authorship verification. J. Assoc. Inf. Sci. Technol. 68(1), 259–269 (2017). https://doi.org/10.1002/asi.23648
14. Koppel, M., Schler, J.: Authorship verification as a one-class classification problem. In: Brodley, C.E. (ed.) Machine Learning, Proceedings of the Twenty-First International Conference (ICML 2004), vol. 69, Banff, Alberta, Canada, 4–8 July 2004. ACM (2004)
15. Koppel, M., Schler, J., Argamon, S.: Authorship attribution in the wild. Lang. Res. Eval. 45(1), 83–94 (2011)
16. Koppel, M., Winter, Y.: Determining if two documents are written by the same author. JASIST 65(1), 178–187 (2014)
17. Moreau, E., Jayapal, A., Lynch, G., Vogel, C.: Author verification: basic stacked generalization applied to predictions from a set of heterogeneous learners-notebook for PAN at CLEF 2015. In: Cappellato, L., Ferro, N., Jones, G., San Juan, E. (eds.) CLEF 2015 Evaluation Labs and Workshop - Working Notes Papers, 8–11 September 2015, Toulouse, France (2015). CEUR-WS.org
18. Nagaprasad, S., Reddy, V., Babu, A.: Authorship attribution based on data compression for telugu text. Int. J. Comput. Appl. 110(1), 1–5 (2015)
19. Potha, N., Stamatatos, E.: A profile-based method for authorship verification. In: Likas, A., Blekas, K., Kalles, D. (eds.) SETN 2014. LNCS (LNAI), vol. 8445, pp. 313–326. Springer, Cham (2014). https://doi.org/10.1007/978-3-319-07064-3_25
20. Potha, N., Stamatatos, E.: An improved *Impostors* method for authorship verification. In: Jones, G.J.F., Lawless, S., Gonzalo, J., Kelly, L., Goeuriot, L., Mandl, T., Cappellato, L., Ferro, N. (eds.) CLEF 2017. LNCS, vol. 10456, pp. 138–144. Springer, Cham (2017). https://doi.org/10.1007/978-3-319-65813-1_14
21. Potthast, M., Kiesel, J., Reinartz, K., Bevendorff, J., Stein, B.: A stylometric inquiry into hyperpartisan and fake news. ArXiv e-prints, February 2017

22. Rexha, A., Kröll, M., Ziak, H., Kern, R.: Extending scientific literature search by including the author's writing style. In: Mayr, P., Frommholz, I., Cabanac, G. (eds.) Proceedings of the Fifth Workshop on Bibliometric-Enhanced Information Retrieval (BIR) Co-located with the 39th European Conference on Information Retrieval (ECIR 2017), Aberdeen, UK, 9th April 2017, CEUR Workshop Proceedings, vol. 1823, pp. 93–100 (2017). CEUR-WS.org

23. Sculley, D., Brodley, C.E.: Compression and machine learning: a new perspective on feature space vectors. In: DCC, pp. 332–332. IEEE Computer Society

24. Seidman, S.: Authorship verification using the impostors method notebook for PAN at CLEF 2013. In: Forner et al. [3]

25. Shu, K., Sliva, A., Wang, S., Tang, J., Liu, H.: Fake news detection on social media: a data mining perspective. ACM SIGKDD Explor. Newsl. 19(1), 22–36 (2017)

26. Stamatatos, E., Daelemans, W., Verhoeven, B., Juola, P., López-López, A., Potthast, M., Stein, B.: Overview of the author identification task at PAN 2015. In: Working Notes of CLEF 2015 - Conference and Labs of the Evaluation forum, Toulouse, France, 8–11 September 2015

27. Stamatatos, E., Daelemans, W., Verhoeven, B., Stein, B., Potthast, M., Juola, P., Sánchez-Pérez, M.A., Barrón-Cedeño, A.: Overview of the author identification task at PAN 2014. In: Working Notes for CLEF 2014 Conference, Sheffield, UK, 15–18 September 2014, pp. 877–897 (2014)

28. Stamatatos, E., Potthast, M., Rangel, F., Rosso, P., Stein, B.: Overview of the PAN/CLEF 2015 evaluation lab. In: Mothe, J., Savoy, J., Kamps, J., Pinel-Sauvagnat, K., Jones, G.J.F., SanJuan, E., Cappellato, L., Ferro, N. (eds.) CLEF 2015. LNCS, vol. 9283, pp. 518–538. Springer, Cham (2015). https://doi.org/10.1007/978-3-319-24027-5_49

29. Stein, B., Lipka, N., Zu Eissen, S.M.: Meta analysis within authorship verification. In: 19th International Workshop on Database and Expert Systems Applications (DEXA 2008), 1–5 September 2008, Turin, Italy, pp. 34–39. IEEE Computer Society (2008)

30. Tax, D.M.J.: One-class classification: concept learning in the absence of counter-examples. Ph.D. thesis (2001)

Efficient Context-Aware K-Nearest Neighbor Search

Mostafa Haghir Chehreghani[1] and Morteza Haghir Chehreghani[2(✉)]

[1] Telecom ParisTech, Paris, France
[2] Chalmers University of Technology, Gothenburg, Sweden
morteza.chehreghani@gmail.com

Abstract. We develop a context-sensitive and linear-time K-nearest neighbor search method, wherein the test object and its neighborhood (in the training dataset) are required to share a similar structure via establishing bilateral relations. Our approach particularly enables to deal with two types of irregularities: (i) when the (test) objects are outliers, i.e. they do not belong to any of the existing structures in the (training) dataset, and (ii) when the structures (e.g. classes) in the dataset have diverse densities. Instead of aiming to capture the correct underlying structure of the whole data, we extract the correct structure in the neighborhood of the test object, which leads to computational efficiency of our search strategy. We investigate the performance of our method on a variety of real-world datasets and demonstrate its superior performance compared to the alternatives.

1 Introduction

K-nearest neighbor (K-NN) search is a very common technique which is widely-used in different contexts such as classification, query processing and retrieval. The goal is to find the K nearest neighbors of a target (test) object from a given (training) dataset. The neighbors might then be processed further to identify the class label of the new object according to the majority in the neighbors. However, the standard K-nearest neighbor search might fail to capture the underlying geometry and structure in data. In particular, the nearest neighbors of an object which is near the boundary between two different classes might include irrelevant training objects that reduce the quality of search [12]. Moreover, different classes might have diverse densities which leads to biased and imprecise results.

Several metric learning approaches have been proposed to address such issues [27,28]. These methods work only with labeled or partially labeled data and might fail to capture the underlying structure as demonstrated in [12]. Manifold learning [1,21,24,29] is another approach for this purpose that computes a possibly low-dimensional embedding that preserves the structure. However, these methods have a restricted scalability, because their computational complexity is

Some part of the work is done at NAVER LABS Europe.

$\mathcal{O}(N^2)$ or even worse, whereas the standard K-nearest neighbor search is linear. In addition, they require tuning the parameters of a kernel very carefully, which is a non-trivial task [13,17]. A good performance is attained only within a very small range of the values of the kernel parameters [26]. *Link-based* measures [3,8] take into account all the *paths* between the objects represented in a graph, in order to compute the effective pairwise distances. These measures are often obtained by inverting the Laplacian of the distance matrix and thus are computationally expensive [7]. A rather similar distance measure, called *Minimax* measure, selects the minimum largest gap among all possible paths between the two objects. This measure has been first investigated for improving the clustering results [6] and then is used for K-nearest neighbor search [11]. The method for K-NN on general graphs [11] requires computing a minimum spanning tree which might take $\mathcal{O}(N^2)$ runtime. A recent work in [4] proposes an efficinet linear time Miniamx K-nearest neighbor search method which additionally provides an outlier detection mechanism.

We particularly consider two main concerns when performing K-nearest neighbor search: (i) when the (test) object is an outlier such that it does not belong to any of the existing classes or structures in the (training) dataset, and (ii) when the structures (e.g. the classes) in the dataset have diverse densities and properties. In our approach, for computing the K nearest neighbors, we do not attempt to capture the correct underlying structure of the whole data. Rather, we aim to extract the correct structure in the neighborhood of the test object. This leads to develop a linear time search algorithm, contrary to the metric learning or manifold learning approaches which are computationally more expensive.

The previous works for tackling these concerns usually consider only one of them. The ENN method computes the coherence of the classes by counting the occurrence of nearest neighbors of the objects inside the same class. Then, the test object is assigned to the class that yields the largest coherence score [23]. However, in addition to high computational cost compared to the standard K-NN, this method only enables to deal with the classes with different densities, i.e. it does not help to identify whether the test object is an outlier. Some other methods, e.g. IKNN [22], MKNN [18] and Adaptive Metric NN (ADAMENN) [5] assign a weight (importance) on the training objects. For instance, IKNN assumes an object is more *informative* if it is close to the test object and far away from the objects with different class labels. In addition to computational challenges, these methods, i) do not provide any outlier detection mechanism, and ii) induce new parameters which may lead to overfitting. For example, ADAMENN requires fixing six parameters in total, whereas the standard K-NN has only one free parameter.

Some other works, compute the K nearest neighbors of the test object solely in order to determine if it is an outlier. The method in [9] computes the N_k score, which counts the frequency that an object occurs inside the K nearest neighbors of other objects. Then, the object is labeled as an outlier if its N_k score is below a predefined threshold. This method is analyzed in [20] and its connection to *anti-hubness* is discussed, as it has been shown in [19] for high-dimensional data too. The method in [20] combines the N_k score of an object with the N_k scores of its K nearest neighbors, in order to improve *discrimination*. However, as the experimental results show, this combination does not yield a significant improvement, thus it is concluded that discrimination is not a main problem in outlier detection. As mentioned, such methods are developed only for the purpose of outlier detection.

We aim to develop a method that addresses both of the concerns, i.e. the existence of diverse training classes and the occurrence of outlier test objects. An outlier is defined as an object that deviates too much from other objects that makes to think it is generated by a different mechanism [10]. In our supervised setup, a test object is an outlier if it does not belong to any of the existing classes in the training dataset. We propose a context-sensitive and linear-time K-nearest neighbor search method based on requiring the test object and its neighborhood (in the training dataset) to share a similar structure and have bilateral relations.[1] We validate our method by performing extensive experiments on several real-world datasets from different domains.

2   Context-Sensitive K-Nearest Neighbor Search

We are given a training dataset which includes the training objects $\mathbf{O}^{train}$, the corresponding measurements $\mathbf{D}^{train}$ and the class labels l_i for each object $i \in \mathbf{O}^{train}$. We assume that the measurements are provided as the pairwise distances between the objects, i.e. $\mathbf{D}^{train}$ constitutes the weights of graph $\mathcal{G}(\mathbf{O}^{train}, \mathbf{D}^{train})$, whose nodes are the objects. This formulation is a generalization of vector representation, where the pairwise distances are for example squared Euclidean distances. We consider K-nearest neighbor search on general graphs, such that the pairwise distances do not necessarily satisfy a metric, i.e. the triangle inequality might be failed. Given a new object $v \in \mathbf{O}^{test}$, the standard K-nearest neighbor search seeks for a subset of K objects in $\mathbf{O}^{train}$ which have a minimal distance to v. We assume that function $d(v, \mathbf{O}^{train})$ computes the pairwise distance between v and the objects in $\mathbf{O}^{train}$. Then, after adding v to $\mathcal{G}$, we obtain graph $\mathcal{G}^+(\mathbf{O}^{train} \cup v, \mathbf{D}^{train} \cup d(v, \mathbf{O}^{train})) \equiv \mathcal{G}^+(\mathbf{O}^+, \mathbf{D}^+)$. Moreover, we assume the graph is symmetric, but it does not need to satisfy the triangle inequality.

Figure 1 demonstrates why a context-sensitive K-nearest neighbor search is necessary. Figure 1(a) illustrates the case where the test object (i.e. the red point) is an outlier, since its neighborhood is significantly sparser than the density of any of the two (training) classes. On the other hand, in Fig. 1(b), the two classes have different densities, such that the test object (i.e. the red point) should belong to the class at the right side whose members are specified by blue dots. However, if we compute its K nearest neighbors, most of the neighbors are selected from the denser class (specified by black dots). These two examples show that computing

---

[1] Note that our approach is orthogonal to the methods such as Minimax distances [4], where they can be conbined and used together.

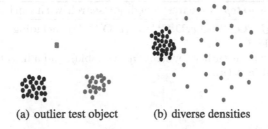

(a) outlier test object   (b) diverse densities

Fig. 1. Demonstration of the need for a context-sensitive K-nearest neighbor search, when the test object might be an outlier (Fig. 1(a)), or the training classes might have diverse densities (Fig. 1(b)). (Color figure online)

the straightforward K neighbors of the test object, as it is done in standard K-nearest neighbor search, is not sufficient. Thereby, we need to obtain more information about the type of the connection between the test object and the objects in the training dataset.

When searching for the K nearest neighbors of a test objects, it is not necessary to take into account the global structure of the whole data, i.e. the objects far from the test object do not play a role. However, on the other hand, looking at the neighbors only from the side of the test object might not be sufficient. It is also necessary to investigate how the test object is seen from the side of its neighbors. Thus, in addition to checking the neighbors of the test object, it is informative to see how the test object is connected to each of its neighbors. This yields to a *bidirectional* analysis of the test object and its neighbors. Since we are seeking for local neighbors, hence, in addition to computing the neighbors of the test object, we also investigate how the test object behaves w.r.t. the neighbors of the training objects. In the following, we propose algorithms that exploit this idea to provide a context-sensitive K-nearest neighbor search.

In Fig. 1(a), the test object should be labeled as an outlier, because its neighbors are very close to each other but far from the test object. One way to deal with such a situation is to check whether the test object itself falls inside the K-nearest neighbors of any (or some) of training objects. If so, then the test object is relevant enough to one of the classes and thus it is not an outlier. Otherwise it is far from the classes and thereby it is labeled as an outlier. This idea is described formally in Algorithm 1, which we will call it the *two-step* method. Function $compute\_K\_dist(i, K, \mathcal{G}^+)$ computes the pairwise distance between object i and its Kth neighbor over graph $\mathcal{G}^+$. The result is stored in $Ndist^+_{i,K}$. In this algorithm, we first obtain $Ndist^+_{i,K}$ for each training object. Then, we investigate if the test object falls inside the neighborhood of a sufficient number of training objects, specified by T. If not, it is labeled as an outlier. Finally, we proceed with the standard K-nearest neighbor search. Thereby, the *two-step* method (being an extension of [9]) deals separately with outlier detection and search.

Algorithm 1. *Two-step* K-nearest neighbor search with outlier detection.

Require: Graph $\mathcal{G}^+(\mathbf{O}^{train} \cup v, \mathbf{D}^{train} \cup d(v, \mathbf{O}^{train}))$ including the test object v.; outlier threshold T.

Ensure: Check if v is an outlier (by *is_outlier* variable), and a list of K nearest neighbors of v stored in $\mathcal{N}(v)$.

1: *is_outlier*= 0
2: $\mathcal{N}(v) = []$
3: $\mathcal{M}(v) = []$
4: **for** i **in** $\mathbf{O}^{train}$ **do**
5:    $Ndist^+_{i,K} = compute\_K\_dist(i, K, \mathcal{G}^+)$
6:    **if** $\mathbf{D}^+_{iv} \leq Ndist^+_{i,K}$ **then**
7:       $\mathcal{M}(v).add(i)$
8:    **end if**
9: **end for**
10: **if** $size(\mathcal{M}(v)) < T$ **then**
11:    *is_outlier*= 1
12: **end if**
13: $Ndist^+_{v,K} = compute\_K\_dist(v, K, \mathcal{G}^+)$
14: **for** i **in** $\mathbf{O}^{train}$ **do**
15:    **if** $\mathbf{D}^+_{iv} \leq Ndist^+_{v,K}$ **then**
16:       $\mathcal{N}(v).add(i)$
17:    **end if**
18: **end for**
19: **return** *is_outlier*, $\mathcal{N}(v)$

Algorithm 2. *Strong* context-sensitive K-nearest neighbor search with outlier detection.

Require: Graph $\mathcal{G}^+(\mathbf{O}^{train} \cup v, \mathbf{D}^{train} \cup d(v, \mathbf{O}^{train}))$ including the test object v.

Ensure: Check if v is an outlier (by *is_outlier*), and a list of K nearest neighbors of v stored in $\mathcal{N}(v)$.

1: *is_outlier*= 0
2: $\mathcal{N}(v) = []$
3: $Ndist^+_{v,K} = compute\_K\_dist(v, K, \mathcal{G}^+)$
4: **for** i **in** $\mathbf{O}^{train}$ **do**
5:    $Ndist^+_{i,K} = compute\_K\_dist(i, K, \mathcal{G}^+)$
6:    **if** $\mathbf{D}^+_{iv} \leq Ndist^+_{v,K}$ **and** $\mathbf{D}^+_{iv} \leq Ndist^+_{i,K}$ **then**
7:       $\mathcal{N}(v).add(i)$
8:    **end if**
9: **end for**
10: **if** $\mathcal{N}(v).isempty() == true$ **then**
11:    *is_outlier*= 1
12: **end if**
13: **return** *is_outlier*, $\mathcal{N}(v)$

However, apart from existence of outlier objects in the test set, the real-world (training) datasets might contain classes with diverse densities, as illustrated in Fig. 1(b). In this example, the test object must be labeled with the class at the right side shown by blue color. However, most of the objects collected in its K nearest neighbors are from the other class. In fact, the left-side class is much denser, which implies that its objects have neighbors from the same class. In other words, the test object does not occur inside the neighborhood of any of the members of this dense class, although it is in average closer to them compared to the members of the sparse class. Thereby, to cope with such cases, we assume there is a *valid* relation between the two (test and training) objects *if and only if each of them occurs inside the K nearest neighborhood of the other object.* This condition ensures both objects to share a similar type of local neighborhood with respect to the overall density and the other objects. Thus, instead of computing the direct neighbors of the test object, we consider only those that mutually have the test object as one of their K nearest neighbors. Algorithm 2 describes this new variant. In this algorithm (called *strong K-nearest neighbor search*), a neighbor of the test object is added if and only if the test object occurs inside the K neighborhood of the neighbor too. A main advantage of Algorithm 2 compared to Algorithm 1 is its adaptation to the local density, whereas both methods provide an outlier detection mechanism.

Efficient implementation. Algorithms 1 and 2 require recomputing the Kth nearest neighbor and the corresponding pairwise distance of each training object on graph $\mathcal{G}^+$ whenever a test object v is being investigated. However, when the test object changes, still, most part of the graph stays unchanged. Thereby, we develop a more efficient method which requires less computation when we perform K-nearest neighbor search on multiple test objects.

Given graph $\mathcal{G}(\mathbf{O}^{train}, \mathbf{D}^{train})$, assume we have already computed the K nearest neighbors of all objects in $\mathbf{O}^{train}$ and the respective distances $Ndist_{i,K}$. Lemma 1 suggests a way to update them whenever a new test object is added.

Lemma 1. *Given graph $\mathcal{G}(\mathbf{O}^{train}, \mathbf{D}^{train})$, the K nearest neighbors of each object $i \in \mathbf{O}^{train}$ and the corresponding distances $Ndist_{i,K}$, by adding a new object v, the $Ndist_{i,K}$'s over the new graph $\mathcal{G}(\mathbf{O}^+, \mathbf{D}^+)$ are obtained by*

$$Ndist_{i,K}^+ = minimum(Ndist_{i,K}, \mathbf{D}_{iv}^+) \quad iff \quad Ndist_{i,K} \neq Ndist_{i,K-1}. \quad (1)$$

Proof. We consider two cases:

1. If $Ndist_{i,K} = Ndist_{i,K-1}$ over graph $\mathcal{G}(\mathbf{O}^{train}, \mathbf{D}^{train})$, then the new $Ndist_{i,K}^+$ on $\mathcal{G}^+$, i.e. after adding v, will not change. Because, even if $\mathbf{D}_{iv}^+ < Ndist_{i,K}$, then $Ndist_{i,K-1}$ will replace $Ndist_{i,K}$ which is equal to that.

2. If $Ndist_{i,K} \neq Ndist_{i,K-1}$, (i.e. when $Ndist_{i,K} < Ndist_{i,K-1}$), then $\mathbf{D}_{iv}^+ < Ndist_{i,K}$ indicates that v is closer to i than the K-th neighbor. Thus, the new $Ndist_{i,K}^+$ is replaced by $\mathbf{D}_{iv}^+$. Hence, in this case we have $Ndist_{i,K}^+ = minimum(Ndist_{i,K}, \mathbf{D}_{iv}^+)$.

Lemma 1 yields a more efficient algorithm whenever the test dataset contain multiple objects. We, first, perform the $K-1$ and K nearest neighbor search and

---

Algorithm 3. Efficient *strong* context-sensitive K-nearest neighbor search with outlier detection.

Require: Graph $\mathcal{G}(\mathbf{O}^{train}, \mathbf{D}^{train})$ and the test set $\{v\}$.
Ensure: Check if the test objects are outlier (by *is_outlier(v)*), and compute a list of K nearest neighbors stored in $\mathcal{N}(v)$ for each test object v.
1: **for** i in $\mathbf{O}^{train}$ **do**
2:    $Ndist_{i,K-1} = compute\_K\_dist(i, K-1, \mathcal{G})$
3:    $Ndist_{i,K} = compute\_K\_dist(i, K, \mathcal{G})$
4: **end for**
5: **for** v in $\mathbf{O}^{test}$ **do**
6:    $is\_outlier(v) = 0$
7:    $\mathcal{N}(v) = []$
8:    Add v to $\mathcal{G}$ to construct $\mathcal{G}^+(\mathbf{O}^+, \mathbf{D}^+)$.
9:    **for** i in $\mathbf{O}^{train}$ **do**
10:       **if** $Ndist_{i,K} \neq Ndist_{i,K-1}$ **then**
11:          $Ndist_{i,K}^+ = minimum(Ndist_{i,K}, \mathbf{D}_{iv}^+)$
12:       **end if**
13:    **end for**
14:    $Ndist_{v,K}^+ = compute\_K\_dist(v, K, \mathcal{G}^+)$
15:    **for** i in $\mathbf{O}^{train}$ **do**
16:       **if** $\mathbf{D}_{iv}^+ \leq Ndist_{v,K}^+$ and $\mathbf{D}_{iv}^+ \leq Ndist_{i,K}^+$ **then**
17:          $\mathcal{N}(v).add(i)$
18:       **end if**
19:    **end for**
20:    **if** $\mathcal{N}(v).isempty() == true$ **then**
21:       $is\_outlier(v) = 1$
22:    **end if**
23: **end for**
24: **return** $is\_outlier, \mathcal{N}(v)$

---

compute the respective distances $Ndist_{i,K-1}$ and $Ndist_{i,K}$ for each object i in $\mathbf{O}^{train}$ (i.e. only over graph $\mathcal{G}(\mathbf{O}^{train}, \mathbf{D}^{train})$).[2] This step is independent from the test objects and can be seen as a training phase. Then, whenever a test object v is added, we use Eq. 1 to update the $Ndist_{i,K}$'s to $Ndist_{i,K}^+$'s. Moreover, we need to compute the K nearest neighbors and the respective distance $Ndist_{v,K}^+$ for the test object. The rest is similar to Algorithm 2. In Algorithm 3, we describe in detail this efficient variant. Compared to Algorithm 2, Algorithm 3 does not require line 5 for computing $Ndist_{i,K}^+$ of each training object i per every new test object v. Instead, we compute them collectively at the beginning. Algorithm 1 can be implemented efficiently in a similar way.

Computational complexity. In all the algorithms, we essentially compute the K nearest neighbors and the respective distance of each training object,

---

[2] Note that if we aim to investigate the algorithm for different K in an incremental manner, then for each K, we only need to compute the Kth nearest neighbor, as this search has been already performed for $K-1$ in the previous step.

in addition to computing them for the test object. Function $compute\_K\_dist(.)$, which computes the pairwise distance between object i and its Kth neighbor, can be efficiently implemented for example via *introselect* algorithm [16]. This algorithm is a combination of Hoare's selection algorithm and median of medians and in average takes $\mathcal{O}(N)$ time, where N is the number of training objects. However, its worst-case performance is $\mathcal{O}(N \log N)$. An alternative approach is to choose the nearest neighbor and its respective distance for K times, such that the runtime will then be $\mathcal{O}(KN)$. In Algorithm 3, for each test object v, we need to i) compute $Ndist^+_{v,K}$ whose runtime is $\mathcal{O}(KN)$, and ii) update the $Ndist_{i,K}$'s, which requires $\mathcal{O}(N)$ time. Thus, the total runtime will be $\mathcal{O}(KN)$ which is equivalent to standard K-nearest neighbor search. Note that computing $Ndist_{i,K}$'s for the training objects (which requires $\mathcal{O}(KN^2)$ time), is a preprocessing phase and can be done off-line. However, in the case that the number of test objects is proportional to the number of training objects (i.e. $|\mathbf{O}^{test}| = \mathcal{O}(N)$), then, this preprocessing will take $\mathcal{O}(KN)$ per test query. Thus, this step does not increase the total time complexity.

Strong K nearest neighbors in clustering. In a rather similar spirit, *strong K nearest neighbors* (called *mutual K nearest neighbors*) have been investigated for clustering [2,14]. However, in this context choosing an appropriate K is very critical and the assumptions are very limiting [14,25]. The main advantage of mutual K- nearest neighbor clustering is supposed to be its ability in separating clusters with different densities. However, consistent to our observations, this ability requires choosing a significantly large K, namely in the order of the number of objects. Such a choice can easily lead to combining even well-separated clusters, particularly when the size of clusters varies. Thereby, *mutual K nearest neighbors* is proposed to identify only the significant cluster seeds [14,25]. Otherwise, if we want to discover large clusters, we end up with combining different clusters. Such an analysis is consistent with our strategy: *strong K-nearest neighbor search* fits well with a *local search*, i.e. to find appropriately the neighborhood of a specific object (or to compute very dense and small regions in the data), rather than a *global search* which aims at identifying comprehensively the global structures in data (i.e. clustering the whole data).

3   Experiments

In this section, we experimentally investigate the performance of different algorithms for computing K-nearest neighbor classification and detecting outliers.

First, we perform our experiments on different subsets of 20 newsgroup collection [15]. (1) DS1: a subset of 20 news group including the following categories: 'comp.graphics', 'comp.os.ms-windows.misc', 'comp.sys.ibm.pc.hardware', 'comp.sys.mac.hardware', 'comp.windows.x', 'misc.forsale'. (2) DS2: a subset of 20 news group including the following categories: 'comp.graphics', 'comp.os.ms-windows.misc', 'comp.sys.ibm.pc.hardware', 'comp.sys.mac.hardware', 'comp.windows.x', 'misc.forsale', 'soc.religion.christian'. (3) DS3: a subset of 20 news group including the following categories: 'rec.autos', 'rec.motorcycles', 'rec.sport.baseball', 'rec.sport.hockey', 'misc.forsale'. (4) DS4: a subset of 20 news group including the following categories: 'rec.autos', 'rec.motorcycles', 'rec.sport.baseball', 'rec.sport.hockey', 'misc.forsale', 'soc.religion.christian'. (5) DS5: a subset of 20 news group including the following categories: 'sci.crypt', 'sci.electronics', 'sci.med', 'misc.forsale'. (6) DS6: a subset of 20 news group including the following categories: 'sci.crypt', 'sci.electronics', 'sci.med', 'misc.forsale', 'soc.religion.christian'. (7) DS7: a subset of 20 news group including the following categories: 'talk.politics.guns', 'talk.politics.mideast', 'talk.politics.misc', 'talk.religion.misc', 'misc.forsale'. (8) DS8: a subset of 20 news group including the following categories: 'talk.politics.guns', 'talk.politics.mideast', 'talk.politics.misc', 'talk.religion.misc', 'misc.forsale', 'soc.religion.christian'.

Additionally, we investigate the algorithms on the following *non-textual* datasets. (9) IRIS: a common dataset with 150 samples from the three species 'setosa', 'virginica' and 'versicolor'. (10) Olivetti: the Olivetti faces dataset from AT&T which contains pictures from 40 individuals and 10 pictures from each. The dimensionality is $4,096$. (11) DIGITS: images of 10 digits each with 64 dimensions ($1,797$ digits in total).

In these datasets, we use either the last category (in DS1, DS3, DS5, DS7, IRIS, Olivetti and DIGITS) or the last two categories (in DS2, DS4, DS6 and DS8) solely for the test set (as outlier objects), to induce some discrepancy between the training and the test datasets. Then, we split the objects (e.g. the documents or the images) of the remaining categories randomly and use 60% for training and keep the rest plus the test-specific categories for test. We observe a consistent behavior with other ratios. For the textual datasets (i.e. the different subsets of 20 news group collection), we compute $1 - cosine(v_i, v_j)$ as the pairwise distance between objects i and j, where v_i indicates the vector of document i obtained according to term-frequency (TF). For non-textual data, we use the pairwise squared Euclidean distances. We measure the accuracy of a classification algorithm by the ratio of the number of correct predictions to the total size of test dataset.

Figures 2 and 3 illustrate the accuracy scores of different algorithms applied to the different datasets, respectively on 20 news group and non-textual datasets. In our experiments, the *two-step2* method refers to first running the method in [20] and then performing K-nearest neighbor classification, similar to the *two-step* method which can be interpreted as an extension of the method in [9] wherein we add classification to the outlier detection ability of the method. We observe, (i) the *strong* K-NN method is almost the best option for all datasets. It performs significantly better, particularly compared to the standard variant. The second choices are the *two-step2* and the *two-step* algorithms, which is due their ability in detecting outliers compared to the other alternatives. (ii) Improvement in the context-sensitive variants (either the *strong*, the *two-step2* or the *two-step* variants) depends on the number of non-matching or outlier objects. A larger discrepancy between the training and test datasets yields a higher improvement.

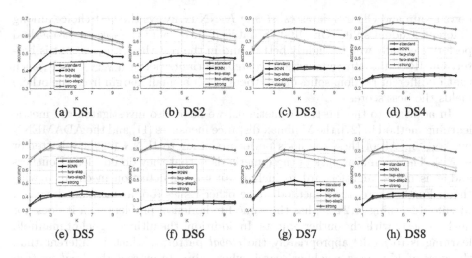

Fig. 2. Accuracy scores of different K-nearest neighbor methods on different 20 newsgroup datasets.

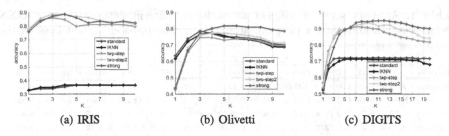

Fig. 3. Accuracy score of different K-NN algorithms on non-textual datasets.

For instance, compare Figs. 2(a), (c), (e) and (g) with Figs. 2(b), (d), (f) and (h). (iii) The *strong* method attains its optimal value at a larger K compared to the other methods. The reason is that it investigates stricter conditions to include a neighbor. Thereby, the true number of neighbors might be less than K, because some of them might not satisfy the bilateral inclusion requirement. (iv) The *two-step2* algorithm performs very similar or only slightly better than the *two-step* algorithm. This observation is consistent with the results in [20] which conclude that improving *discrimination* do not play an important role in outlier detection.[3]

The accuracy scores are obtained by averaging over 50 random splits of the non-outlier set into training and test subsets. We observe a very consistent ranking of different methods among different random splits. Only on very few datasets, and for few splits we see that the ranking changes. More precisely, on 20

---

[3] In our experiments, ENN and IKNN perform very similarly, as both does not provide any outlier detection mechanism. We only report the scores of IKNN.

news group and Olivetti datasets, *strong K*-NN is always the best choice among different random splits. On DIGITS dataset, in 6 splits the *two-step2* method performs equally well or slightly better, and in the rest the *strong* variant is better. On IRIS, the *strong* and the *two-step2* variants perform very closely, such that in different random splits one of them (with a small margin from the other) yields the best score.

In addition to the illustrated methods, we have also investigated the metric learning method in [27], the Minimax distance measures [11] and the ADAMENN method [5]. All of these methods perform similar to IKNN (slightly better or worse). The main drawback of these methods is that they do not detect outliers and it is not straightforward to include an outlier detection mechanism with them. Furthermore, these methods have other issues too, as mentioned earlier. Manifold learning [1, 24, 29] has the same fundamental problem, i.e., it does not deal properly with the outlier objects. In addition, the ultimate goal of manifold learning is to model appropriately the *global* patterns, which is different than the goal of K-nearest neighbor search where aims to extract the *local* pattern near the test object.

Table 1. Accuracy score of different methods on different datasets. *Strong K*-NN performs significantly better than the alternatives. The *two-step* variant is often the second best option.

Dataset	DS1	DS2	DS3	DS4	DS5	DS6	DS7	DS8	DIGITS	Olivetti	IRIS
Standard *K*-NN	0.4553	0.3140	0.4726	0.3250	0.4213	0.2732	0.5830	0.4242	0.7233	0.7857	0.3671
strong K-NN	0.6594	0.7534	0.8110	0.8540	0.7964	0.8486	0.8281	0.8444	0.9504	0.8175	0.8861
two-step K-NN	0.6287	0.7238	0.7530	0.8012	0.7662	0.8109	0.7939	0.7915	0.9143	0.7460	0.8608
two-step2 K-NN	0.6358	0.7270	0.7491	0.8156	0.7709	0.8058	0.8026	0.7855	0.9372	0.7733	0.8895
IKNN	0.5198	0.4800	0.4829	0.3408	0.4438	0.2912	0.6093	0.4436	0.7141	0.7749	0.3671
SVM	0.5406	0.3721	0.5175	0.3595	0.4507	0.2918	0.6143	0.4418	0.6977	0.7930	0.3671
LogReg	0.5372	0.3697	0.5198	0.3595	0.4480	0.2912	0.6143	0.4431	0.6857	0.7891	0.3671

In the following, we investigate two other classification algorithms, Support Vector Machines (SVM) and Logistic Regression (LogReg), and demonstrate how our context sensitive method performs compared to them. For each variant of K-nearest neighbor search method, we report the maximal score with respect to the different values of K. Consistently, in other methods, we report the maximal score with respect to their free parameters. We optimize the results over the regularization factor C and different kernel functions (i.e., radial basis function, linear, polynomial and sigmoid kernels).

Table 1 illustrates the accuracy scores for different methods. We observe that *strong K*-nearest neighbor classification almost always yields the best scores. The *two-step2* method (and similarly the *two-step* variant) are often the second best options and the other methods behave almost equally poor. An exception is the Olivetti dataset where SVM is the second best choice (after the *strong K*-NN) and the other methods perform reasonably well. The reason for the rather poor permanence of SVM and LogReg is that they do not directly provide any outlier detection mechanism. They are appropriate when the training and test datasets have the same distribution, but they are unable to adapt in the case of discrepancy between the training and test distributions. On the other hand, they require a training phase which might be intolerable in some situations, in particular whenever the training set is changing frequently.

4 Conclusion

We developed an efficient method for performing K-nearest neighbor search based on requiring the test object and its neighborhood in the training dataset to share similar patterns and densities. This is attained via including a neighbor if and only if the test object also occurs inside the neighborhood of its neighbors in training dataset. Our approach provides the ability to detect outliers and perform a context-sensitive search, i.e., to deal with the cases where the training classes might have diverse densities. The extensive experiments on several real-world datasets illustrated the superior performance of our method compared to the alternatives.

References

1. Belkin, M., Niyogi, P.: Laplacian eigenmaps for dimensionality reduction and data representation. Neural Comput. **15**(6), 1373–1396 (2003)
2. Brito, M.R., Chávez, E.L., Quiroz, A.J., Yukich, J.E.: Connectivity of the mutual k-nearest-neighbor graph in clustering and outlier detection. Stat. Probab. Lett. **35**(1), 33–42 (1997)
3. Chebotarev, P.: A class of graph-geodetic distances generalizing the shortest-path and the resistance distances. Discrete Appl. Math. **159**(5), 295–302 (2011)
4. Chehreghani, M.H.: K-nearest neighbor search and outlier detection via minimax distances. In: Proceedings of the 2016 SIAM International Conference on Data Mining, pp. 405–413 (2016)
5. Domeniconi, C., Peng, J., Gunopulos, D.: Locally adaptive metric nearest-neighbor classification. IEEE Trans. Pattern Anal. Mach. Intell. **24**(9), 1281–1285 (2002)
6. Fischer, B., Buhmann, J.M.: Path-based clustering for grouping of smooth curves and texture segmentation. IEEE Trans. Pattern Anal. Mach. Intell. **25**(4), 513–518 (2003)
7. Fouss, F., Francoisse, K., Yen, L., Pirotte, A., Saerens, M.: An experimental investigation of kernels on graphs for collaborative recommendation and semisupervised classification. Neural Netw. **31**, 53–72 (2012)
8. Fouss, F., Pirotte, A., Renders, J.-M., Saerens, M.: Random-walk computation of similarities between nodes of a graph with application to collaborative recommendation. IEEE Trans. Knowl. Data Eng. **19**(3), 355–369 (2007)
9. Hautamaki, V., Karkkainen, I., Franti, P.: Outlier detection using k-nearest neighbour graph. In: 17th International Conference on Proceedings of the Pattern Recognition, (ICPR 2004), pp. 430–433 (2004)
10. Hawkins, D.M.: Identification of Outliers. Monographs on Applied Probability and Statistics. Chapman and Hall, London (1980)

11. Kim, K.-H., Choi, S.: Neighbor search with global geometry: a minimax message passing algorithm. In: ICML, pp. 401–408 (2007)
12. Kim, K.-H., Choi, S.: Walking on minimax paths for k-nn search. In: AAAI (2013)
13. Luxburg, U.: A tutorial on spectral clustering. Stat. Comput. **17**(4), 395–416 (2007)
14. Maier, M., Hein, M., von Luxburg, U.: Optimal construction of k-nearest-neighbor graphs for identifying noisy clusters. Theor. Comput. Sci. **410**(19), 1749–1764 (2009)
15. Mitchell, T.M.: Machine Learning, 1st edn. McGraw-Hill Inc., New York (1997)
16. Musser, D.: Introspective sorting and selection algorithms. Softw. Pract. Experience **27**, 983–993 (1997)
17. Nadler, B., Galun, M.: Fundamental limitations of spectral clustering. In: Advanced in Neural Information Processing Systems, vol. 19, pp. 1017–1024 (2007)
18. Parvin, H., Alizadeh, H., Minaei-Bidgoli, B.: MKNN: modified k-nearest neighbor
19. Radovanovic, M., Nanopoulos, A., Ivanovic, M.: Hubs in space: popular nearest neighbors in high-dimensional data. J. Mach. Learn. Res. **11**, 2487–2531 (2010)
20. Radovanovic, M., Nanopoulos, A., Ivanovic, M.: Reverse nearest neighbors in unsupervised distance-based outlier detection. IEEE Trans. Knowl. Data Eng. **27**(5), 1369–1382 (2015)
21. Roweis, S.T., Saul, L.K.: Nonlinear dimensionality reduction by locally linear embedding. Science **290**, 2323–2326 (2000)
22. Song, Y., Huang, J., Zhou, D., Zha, H., Giles, C.L.: IKNN: informative k-nearest neighbor pattern classification. In: 11th European Conference on Principles and Practice of Knowledge Discovery PKDD, pp. 248–264 (2007)
23. Tang, B., He, H.: ENN: extended nearest neighbor method for pattern recognition [research frontier]. IEEE Comp. Int. Mag. **10**(3), 52–60 (2015)
24. Tenenbaum, J.B., de Silva, V., Langford, J.C.: A global geometric framework for nonlinear dimensionality reduction. Science **290**(5500), 2319 (2000)
25. von Luxburg, U.: A tutorial on spectral clustering. Stat. Comput. **17**(4), 395–416 (2007)
26. Wang, F., Zhang, C.: Label propagation through linear neighborhoods. IEEE Trans. Knowl. Data Eng. **20**(1), 55–67 (2008)
27. Weinberger, K.Q., Saul, L.K.: Distance metric learning for large margin nearest neighbor classification. J. Mach. Learn. Res. **10**, 207–244 (2009)
28. Xing, E.P., Jordan, M.I., Russell, S.J., Ng, A.Y.: Distance metric learning with application to clustering with side-information. In: Advances in Neural Information Processing Systems, vol. 15, pp. 521–528. MIT Press (2003)
29. Zhang, Z., Zha, H.: Principal manifolds and nonlinear dimension reduction via local tangent space alignment. SIAM J. Sci. Comput. **26**, 313–338 (2002)

Micro-blogs

An Optimization Approach for Sub-event Detection and Summarization in Twitter

Polykarpos Meladianos[1,2]($\boxtimes$), Christos Xypolopoulos[1], Giannis Nikolentzos[1,2], and Michalis Vazirgiannis[1,2]

[1] Lix, École Polytechnique, Palaiseau, France
christos.xypolopoulos@polytechnique.edu
[2] Athens University of Economics and Business, Athens, Greece
{pmeladianos,nikolentzos,mvazirg}@aueb.gr

Abstract. In this paper, we present a system that generates real-time summaries of events using only posts collected from Twitter. The system both identifies important moments within the event and generates a corresponding textual description. First, the set of tweets posted in a short time interval is represented as a weighted graph-of-words. To identify important moments within an event, the system detects rapid changes in the graphs' edge weights using a convex optimization formulation. The system then extracts a few tweets that best describe the chain of interesting occurrences in the event using a greedy algorithm that maximizes a nondecreasing submodular function. Through extensive experiments on real-world sporting events, we show that the proposed system can effectively capture the sub-events, and that it clearly outperforms the dominant sub-event detection method.

1 Introduction

Twitter is a very popular microblogging service that allows users to post real-time messages known as tweets. Due to its instantaneous nature, Twitter has been established as a major communication medium. Among others, people use the service to report latest news and to comment about real-world events [6]. Users show particular interest in social events such as large parties, political campaigns and sporting events but also for emergency events such as natural disasters and terrorist attacks [3]. Tweets posted by people involved in these events could provide different perspectives regarding the events compared to the ones that appear in traditional media. Besides the above mentioned types of events, a plethora of other events are reported daily on Twitter, the majority of which are not covered systematically by traditional media. An extreme example of such events are those related to the personal wellness of Twitter users [1].

Users are often interested in tracking the evolution of an event that spans a time interval, such as a football match. Most events typically consist of a sequence of important moments or *sub-events* which attract the attention of users. However, due to the massive pace of generated data, accurately detecting

© Springer International Publishing AG, part of Springer Nature 2018
G. Pasi et al. (Eds.): ECIR 2018, LNCS 10772, pp. 481–493, 2018.
https://doi.org/10.1007/978-3-319-76941-7_36

all sub-events of an evolving event is a very challenging task. Besides the identification of the important moments, it is necessary to provide users with a textual description of the events that are being reported on Twitter. The problem of generating a summary for an event using Twitter is related to the problem of multi-document summarization which has been studied extensively in the past [15]. However, instead of a static collection of well-formatted documents, in the Twitter setting, there exists a stream of tweets with noisy content posted by heterogeneous users which makes the summarization task very hard.

From the above, it is clear that event summarization in Twitter can be seen as consisting of two parts: (1) a sub-event detection mechanism capable of identifying the important moments within an event, and (2) a text generating module which creates text descriptions that best summarize a given sub-event. Both these parts have to be tailored to the particularities of microblogging content. In this paper, we propose a novel system that deals with both aforementioned challenges of event summarization in Twitter. Our system decomposes events into time intervals and represents the set of tweets posted during each time interval as a graph. We assume that important moments within an event trigger a rapid change in the vocabulary employed by users and consequently rapid changes in edge weights. To detect important moments within an event, the system uses a convex optimization formulation which accurately determines the amount of change in the edge weights between the current time interval and the previous time intervals. Given the successful detection of an important moment, to generate a textual description, we propose optimizing a submodular function which takes into account the weights of the edges connecting terms of the tweets that have been added into the summary. The function encourages the produced summary to be both representative of the set of tweets and at the same time diverse. The source code of the proposed system is publicly available[1].

The rest of this paper is organized as follows. Section 2 provides an overview of the related work and elaborates our contribution. Section 3 provides a detailed description of our proposed system. Section 4 evaluates the proposed approach. Finally, Sect. 5 summarizes the work.

2 Related Work

Although much effort has been devoted to the problem of event detection in Twitter [2,18,22,23,25], considerably less attention has been paid to the problems of sub-event detection and summarization.

Most existing systems assume that a sharp increase in the volume of status updates corresponds to the occurrence of an important moment within the event. Existing systems employ different approaches to identify such sharp increases in the tweet rate [16,24,26]. To identify important moments, Chierichetti et al. made also use of the retweet rate [5]. Chakrabarti and Punera proposed in [4] a system that does not depend solely on the tweet/retweet rate. Instead, the authors used a modified Hidden Markov Model which detects sub-events based

---

[1] https://bitbucket.org/ksipos/optimization-sub-event-detection.

on both the tweet rate and the word distribution used in tweets. Shen et al. identified the participants of events and used a mixture model to detect sub-events for each participant [20]. Srijith et al. proposed in [21] a sub-event detection approach that is based on hierarchical Dirichlet processes, a probabilistic topic model which can learn latent sub-stories associated with tweets. The work closest to ours is perhaps the one reported in [11]. The authors represent sequences of tweets as graphs and sub-events are identified using the notion of graph degeneracy. In our work, we also build graphs to represent sequences of tweets. However, in contrast to the above work, we propose solving an optimization problem to identify the important moments. Furthermore, in contrast to all existing systems, we design and optimize a novel submodular function in order to generate a summary for each sub-event. The work of Letsios et al. is also related to the proposed approach, but it focuses on the problem of event detection [7]. As regards the summarization task, to extract a representative sentence for each important moment, most systems employed the *tf-idf*-based and graph-based approaches proposed by Sharifi et al. [19]. Mackie et al. compared in [10] several methods for Twitter summarization. The TREC Real-Time Summarization Track has been recently created to foster the development of systems that automatically monitor streams of social media posts to keep users up to date on topics of interest [9].

3   Sub-Event Detection and Summarization in Twitter

In this paper, we developed a system for generating real-time summaries of events by using solely status updates collected from Twitter. The system is composed of several modules, and is illustrated in Fig. 1. In what follows, we present the different modules of the proposed system.

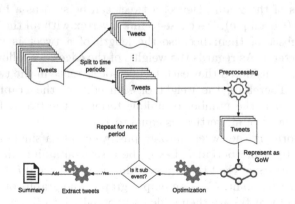

Fig. 1. Overview of our proposed real-time sub-event detection and summarization system.

3.1   Data Preprocessing

In this Section, we give details about the preprocessing steps followed. Data harvested from social media is often noisy and heterogeneous by nature. Taking also into account that social media platforms including Twitter have been infiltrated by various types of unwanted content, such as spam, advertisements and malicious content, it becomes obvious that data preprocessing is a task that should be considered with care.

Given a set of raw tweets $\mathcal{D}_{raw} = \{d_1, d_2, \ldots, d_N\}$, we first remove all retweets and duplicate posts, as they just reproduce the content of other tweets and do not provide additional information. Note that retweets and duplicate tweets constitute a large fraction of the set of tweets $\mathcal{D}_{raw}$. By eliminating them, the proposed system gains both in terms of running time and in terms of performance as these posts increase the noise levels that the system is exposed to. Furthermore, we removed tweets containing "@"-mentions as we assume that in most cases these tweets are not relevant to the event under consideration. All remaining tweets undergo standard text processing tasks including (1) tokenization, (2) stopword removal, (3) punctuation and special character removal, (4) URL removal, and (5) stemming using Porter's algorithm. The preprocessed tweets are then transformed into graphs as described below.

3.2   Graph-of-words Representation

Given the set of preprocessed tweets $\mathcal{D}$, we represent each tweet as a statistical *graph-of-words*, following earlier approaches in keyword extraction [12,13], in summarization [11], and in text categorization [17].

More formally, given the set $\mathcal{D}$ of preprocessed tweets, each tweet corresponds to a sequence of terms. From this sequence we create a graph whose vertices correspond to the unique terms of the tweet. An edge is then drawn between all pairs of vertices of the graph. Hence, a tweet can be seen as a fully-connected graph-of-words (i. e. clique). We chose to link a vertex with all the other vertices instead of a subset of them because the length of a tweet is very short (at most 140 characters). As regards the weights of the edges, we followed an earlier approach and we considered that each term co-occurrence in the tweet is equally important [11]. Therefore, the weight of each edge of the graph is set equal to $1/(n-1)$ where n is the number of unique terms in the tweet. Following this approach, the degree of all vertices is equal to 1.

After transforming the tweets into graphs, we create a single graph G_i that corresponds to the time period i. Let $\mathcal{G}_i$ be a set containing all the graph-of-words representations of the tweets posted during time period i. First, graph G_i is initialized. At this point, G_i is an empty graph. The graphs of $\mathcal{G}_i$ (i. e. graph representations of tweets) are then added sequentially into G_i. Any vertices and edges of these graphs that are not contained in G_i are added to it, while the weights of existing edges are increased by the corresponding weights in these graphs. Hence, pairs of terms that are repeated in many tweets are expected to have a high edge weight between them. Figure 2 illustrates the graph-of-words

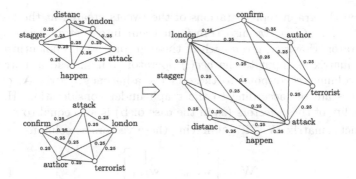

Fig. 2. The graph-of-words representations of two tweets (left), and the single graph that emerges from these two tweets (right).

representations of the following two tweets (left): (1) *"The distance over which the #London attack happened is staggering"*, and (2) *"Confirmed by authorities terrorist attack in London"*. Furthermore, it shows the graph that emerges after merging the two graph-of-words (right).

3.3   Sub-event Detection

This Section describes our proposed system for detecting important moments within an event from a sequence of graphs-of-words. Let $\mathcal{D}_t$ be the set of tweets posted at time period t, and G_t the graph constructed from this set as described above. Let also i be the current time period. We denote the number of unique terms in the set of tweets $\mathcal{D}_i$ by n.

We propose a novel approach which identifies (1) how much the content of the messages posted by users at the current time period i has changed compared to the previous time periods, and (2) if there are any pairs of words whose quantity of appearances was abnormally high. In such a case, the system considers that an important moment has occurred at time period i and the next module is activated to provide the user with a textual description of the corresponding sub-event. The proposed system identifies such interesting occurrences by solving a convex optimization problem. In what follows, we give details about how this optimization problem is formulated.

The first step is to transform graph G_i corresponding to the current time period i into a vector $\mathbf{b} \in \mathbb{R}^{n^2}$. Specifically, vector $\mathbf{b}$ is created by applying the $vec(\cdot)$ operator to the adjacency matrix of G_i. The $vec(\cdot)$ operator creates a column vector from a matrix by stacking the column vectors of the matrix below one another. Therefore, if $\mathbf{A}_i \in \mathbb{R}^{n \times n}$ is the adjacency matrix of G_i, then

$$\mathbf{A}_i = \begin{bmatrix} | & | & & | \\ \mathbf{a}_1 & \mathbf{a}_2 & \ldots & \mathbf{a}_n \\ | & | & & | \end{bmatrix} \quad \text{and} \quad \mathbf{b} = vec(\mathbf{A}_i) = \begin{bmatrix} \mathbf{a}_1 \\ \mathbf{a}_2 \\ \vdots \\ \mathbf{a}_n \end{bmatrix}$$

Then, given the graph representations of the tweets created at the last p time periods $G_{i-p}, \ldots, G_{i-1}$, we also transform them into vectors using again the $vec(\cdot)$ operator. However, given each of these graphs, we do not utilize its own adjacency matrix as input to the $vec(\cdot)$ operator, but instead a matrix whose rows and columns correspond to these of the adjacency matrix $\mathbf{A}_i$ of G_i and whose values are set according to the graph under consideration. Hence, the dimensionality of these vectors (as in the case of $\mathbf{b}$) is also equal to n^2. Finally, we construct a matrix $\mathbf{W} \in \mathbb{R}^{n^2 \times p}$ having these vectors as columns

$$\mathbf{W} = \begin{bmatrix} | & & | \\ \mathbf{w}_{i-p} & \cdots & \mathbf{w}_{i-1} \\ | & & | \end{bmatrix}$$

In other words, for each one of the previous p periods, we extract the weights of the edges connecting all pairs of vertices (for pairs of vertices that are not connected we assume zero weight) of the graph G_i created from tweets posted at the current period and we create a vector $\mathbf{w} \in \mathbb{R}^{n^2}$. The weights are placed in exactly the same order as in vector $\mathbf{b}$. Hence, entries with the same index correspond to the same pair of vertices (i.e. terms). Then, matrix $\mathbf{W}$ is simply a matrix with columns the vectors $\mathbf{w}_{i-p}, \ldots, \mathbf{w}_{i-1}$.

The proposed optimization problem is then formulated as follows

$$\min_{\mathbf{x}} \frac{1}{2} \|\mathbf{W}\mathbf{x} - \mathbf{b}\|_2^2$$
$$\text{s.t.} \quad \mathbf{1}^\top \mathbf{x} = 1 \tag{1}$$
$$x_i \geq 0, \quad \forall i = 1, \ldots, p$$

Let $\mathbf{x}^* \in \mathbb{R}^p$ be the solution of the optimization problem (1). To a certain extent, the values of $\mathbf{x}^*$ indicate how similar is the content of the tweets of the current time period when compared to each one of the previous p periods. For example, if at one of the previous periods users posted tweets describing the same aspect of the event as tweets posted at the current time period, the edge weights corresponding to the same pairs of vertices in the two graphs will have almost equal values. In such a case, the entry of $\mathbf{x}^*$ corresponding to that time period will obtain a value close to 1. In a sense, we can say that our system exploits the fact that the vocabulary of tweets gets more specific when something important happens within an event, and therefore, the weight of the edges between the corresponding terms gets higher.

Let $\mathbf{c} = \mathbf{W}\mathbf{x}^*$. The closer the values of the entries of $\mathbf{c}$ to those of $\mathbf{b}$, the lower the value of the objective function of the optimization problem (1). As mentioned above, we are interested in detecting pairs of terms which co-occur in many posts of the current time period, and which appeared only at a limited number of posts in the previous periods. We assume that such pairs of terms are indicative of major developments of an event. These pairs of terms force the objective function of the optimization problem (1) to take large values. Our detection mechanism is based on that value. However, pairs of terms which co-occurred in many tweets of the previous time periods can also cause the objective function to take large values. To account for this, we set the entries of vector $\mathbf{c}$ that are larger than these of vector $\mathbf{b}$ equal to those of $\mathbf{b}$, that is

$$\mathbf{c}_i = \min(\mathbf{c}_i, \mathbf{b}_i), \quad \forall i \in 1, \ldots, n^2$$

We then proceed to compute the value of the function $\frac{1}{2}||\mathbf{c} - \mathbf{b}||_2^2$. The higher the value of the above formula, the higher the probability that an important moment occurred during the current time period. Hence, to decide if an interesting sub-event has occurred, we compare it against a specified threshold θ (which is learned automatically as described in Sect. 4).

3.4 Summarization

After detecting a sub-event, the summarization module of the proposed system generates a textual description for that sub-event. We propose an extractive summarization algorithm which selects a subset of tweets that contain the most significant concepts of the sub-event. Specifically, given a set of tweets corresponding to the summary of the sub-event, we define a monotone submodular function which rewards both the coverage and the diversity of the set. We then employ a well-known greedy algorithm for optimizing this function [8].

We assume that tweets that contain multiple "important" edges capture most of the details of the sub-event. Given the graph-of-words representation G_i of the tweets posted during a time period i, and a set of tweets $\mathcal{S} \subseteq \mathcal{D}_i$, we define a function f which takes as input a set of tweets (i.e. $\mathcal{S}$) and is equal to the sum of the weights of the edges of G_i that are "covered" by tweets belonging to that set. The function is thus equal to the sum of the weights of all the edges that link all pairs of terms found in the input set of tweets. By maximizing function f given a cardinality constraint, we can generate a summary which is very informative and at the same time diverse. Let $\mathcal{S}$ represent the summary of a sub-event. It is easy to show that when we add a new tweet into our summary, the value of f never decreases (hence, f is monotone nondecreasing), and that it satisfies the property of diminishing returns: given two sets of tweets $\mathcal{A}, \mathcal{B}$ with $\mathcal{A} \subseteq \mathcal{B} \subseteq \mathcal{D} \backslash d$, then it holds that $f(\mathcal{A} \cup d) - f(\mathcal{A}) \geq f(\mathcal{B} \cup d) - f(\mathcal{B})$. Hence, f is a monotone nondecreasing submodular function and we can use a greedy algorithm to compute an approximate solution guaranteed to be within $(1 - 1/e) \approx 0.63$ of the optimal solution [14].

4   Experiments and Evaluation

In this Section, we evaluate the proposed system in the task of sporting event summarization. We have also applied the system to other types of events, such as the terrorist attack that took place near the British Parliament in London on 22 March 2017. Due to space limitations, the empirical results for that event are presented in the supplementary material[2].

---

[2] http://www.db-net.aueb.gr/nikolentzos/files/ecir18suppl.pdf.

Table 1. Summary of the 20 matches from the 2014 and 2010 FIFA World Cups that were used in our experiments.

Match	Actual events	Collected tweets	Preprocessed tweets
ARG - BEL	7	313,803	108,250
ARG - GER	9	824,241	262,112
AUS - NED	12	96,834	25,997
AUS - ESP	9	86,843	13,608
BEL - KOR	7	99,192	32,053
CMR - BRA	11	148,298	35,085
FRA - GER	6	525,725	160,727
FRA - NGA	6	367,899	128,718
GER - ALG	8	712,525	276,227
GER - BRA	12	973,985	295,875
GER - GHA	8	285,804	77,449
GER - USA	7	256,445	86,040
GRE - CIV	10	113,402	51,101
HON - SUI	8	41,539	10,082
MEX - CRO	11	155,549	36,981
NED - CHI	8	95,108	25,819
NED - MEX	10	628,698	217,472
POR - GHA	10	272,389	91,110
GER - SRB	14	45,024	29,062
USA - SVN	12	85,675	53,292
Total	185	6,128,978	2,017,060

Dataset. We evaluated the proposed system on a dataset consisting of football matches from the 2010 and 2014 FIFA World Cups that were collected using the Twitter Streaming API. Specifically, the dataset contains 18 matches from the 2014 FIFA World Cup and 2 matches from the 2010 FIFA World Cup. Table 1 shows statistics of the matches used in our experiments. The Table illustrates both the total number of tweets collected for each match and their resulting number after applying the preprocessing steps described in Sect. 3.1. It also shows the number of sub-events (considered sub-events are described in next Section) that occurred in each match. Overall, the dataset contains 6,128,978 tweets with an average of 306,448 and 100,853 tweets per match before and after preprocessing. Furthermore, the matches contain 185 sub-events in total.

Key Sub-events and Ground-Truth Construction. Following Nichols et al. [16] and Meladianos et al. [11], we considered 8 major sub-event types: (1) goals, (2) own goals, (3) red cards, (4) yellow cards, (5) penalties, (6) match starts, (7) end of matches, and (8) half time breaks. Other types of sub-events such as missed attempts or offsides were discarded as they either depend on the subjective opinion of the person who wrote the summary or their impact in the match outcome is limited and are not broad-casted by Twitter users.

The actual sub-events for each match and their corresponding textual descriptions were collected from the official website of FIFA[3]. To annotate the collected matches, we employed the following approach: we first set the length of the considered time frames equal to 60 seconds. Then, for each time frame, we employed the summarization module of our system to extract a textual description (i. e. a pair of tweets) given the set of tweets posted in this frame. We then had two humans to manually annotate these descriptions. More specifically, tweets that describe the key sub-event types with an intervening period of at most a few minutes between their time frame and the actual sub-event were labeled as positive. Time frames whose extracted textual description contain other sub-events such as missed attempts and substitutions, which are not included in our key sub-event types, were not taken into consideration for the evaluation. In fact, such sub-events provide useful information regarding the match and we would like them to be included into the summary. The remaining time frames were labeled as negative.

Parameter Learning. Since the volume of tweets of the 20 considered football matches varies a lot, we found it necessary to use a different value of threshold θ for each match. Some preliminary experiments provided evidence of a relationship between the optimal value of θ and the number of tweets posted during an event. We followed a supervised learning approach where we split the set of events (i. e. football matches) into a training and a test set. Specifically, we randomly selected 3 matches from the set to serve as training examples, while the remaining 17 matches were placed into the test set. For the 3 matches of the training set, we performed an exhaustive grid search for identifying the optimal value of threshold trying all possible values of $\theta \in [1, 100]$ with step 0.1. The optimal value of threshold was the one that led to the highest f1-score. We observed that there is a strong correlation between the optimum threshold θ and the total number of tweets posted during a match. Therefore, we employed the following linear model comprising of three parameters

$$\hat{\theta} = w_1 x^2 + w_2 x + w_3$$

where x is the number of tweets posted during the match, and w_1, w_2, w_3 are the three parameters. We then computed the values of the parameters that minimize the least-squares-error with regard to the optimal thresholds of the training set. At test time, we used the above model to set the threshold θ of each match based on the total number of tweets related to that match.

Baseline. As mentioned above, most approaches in the literature use tweet rate to detect sub-events. Hence, we implemented a detection system based on post rate to serve as the baseline for our evaluation. Given a stream of tweets,

---

[3] www.fifa.com/worldcup/archive/brazil2014/matches/index.html.

Table 2. Micro- and macro-average precision, recall and f1-scores on the 17 matches of the test set.

Method	Metric					
	Macro-average			Micro-average		
	Precision	Recall	f1-score	Precision	Recall	f1-score
OptSumm	0.76	0.75	0.75	0.73	0.74	0.73
Burst	0.78	0.54	0.64	0.72	0.54	0.62

Table 3. Key sub-event types, their actual numbers and detected numbers over the 17 matches of the test set.

Event type	# actual events	# detected events
Goal	42	42
Own goal	2	2
Penalty	3	3
Red card	3	3
Yellow card	51	15
Match start	17	14
Match end	17	17
Half time	17	17

all tweets posted within a specific time frame are first preprocessed following exactly the same procedure as in the case of the proposed system. Subsequently, the tweeting rate of the current time frame is computed and if it exceeds a specific threshold, the system considers that a sub-event has occurred. The value of the threshold was computed separately for each match using a linear model similar to the one presented above. The parameters of the model were optimized on the same set of matches as in the case of the proposed approach.

Experimental Results. We first evaluate the proposed system (OptSumm) and the baseline approach (Burst) on the task of sub-event detection. For these results, we use the set of manually annotated tweets described above. We report performance using standard measures in information retrieval such as *precision*, *recall* and *f1-score*. Note that for both approaches, the threshold for each match is determined using the parameters learned on the three matches of the training set as described above. Table 2 illustrates the micro- and macro-average precision, recall and f1-scores of the two approaches over the 17 matches of the test set. The proposed system clearly outperforms the baseline. Specifically, OptSumm managed to detect sub-events that could not be detected by Burst, leading to better recall and f1-scores.

We next investigate the ability of the proposed system in detecting sub-events that correspond to different types of plays in the match. Table 3 illustrates

Table 4. Summary of the France vs Nigeria match generated automatically using the proposed system and manually by a journalist on behalf of FIFA.

Our summary	FIFA
Underdog Nigeria vs. European giants France. Going to be a great match!	The match kicks off
Nigeria awarded a free kick in a good position after Matuidi collides with Odemwingie. Nigeria looking decent on the break so far. #fra #nga	Matuidi (France) concedes a free-kick following a challenge on Odemwingie (Nigeria)
France #fra 0-0 Nigeria #nig - Nigeria with Eminike score, but ruled out for offside, good decision 18min	Emenike (Nigeria) is adjudged to be in an offside position
Pogbaaaaaa!!! excellent skill! made that entire move and ended it with a superb volley but keeper made a good save	Pogba (France) sees his effort hit the target
Half time: France 0-0 Nigeria. Goalless in brasília. tight game	The referee brings the first half to an end
54: Blaise Matuidi gets the first yellow card of the game after a nasty challenge	Matuidi (France) is booked by the referee
Nigeria the best team by far. That usually means France will scrape a lucky win	–
Omg! #Benzema so close to scoring, just 2? inches short. Still 0-0 (Nigeria-France) in a suddenly very exciting match!	Benzema (France) sees his effort hit the target
If the French don't score in this game, it would be a miracle for Nigeria. France has been inches away from about 3 goals at this point	–
Goal France! Who else, but the future, Paul Pogba, heading into an open net. Finally les blues score. 1-0, 80th min	Pogba (France) scores!!
Goool! France scores in the 91st min! Partial score, France 2-0 Nigeria. #worldcup goal count - 147	Yobo (Nigeria) scores an own goal!!
Full-time: #fra 2-0 #nga. France book their spot in the quarter-final while Nigeria crash out of the 2014 Fifa world cup	The final whistle sounds

the number of times that each key sub-event was detected compared to the total number of the key sub-events in the 17 matches. The proposed system successfully detected all goals, own goals, penalties and red cards of the 17 matches. This is not surprising since these correspond to primary sub-events which trigger the majority of user tweets. The system also detected all match ends and half times, and almost all match starts. However, the system failed to detect the yellow cards consistently and this may be due to the fact that yellow cards are not of significant impact for the outcome of a match.

As regards the generated summaries, Table 4 compares the summary generated by our system for the match between France and Nigeria with the one created by humans on behalf of FIFA. The proposed summarization module produces very informative and reasonable textual descriptions of the important moments. The quality of the generated summaries remains the same for the other matches of our dataset. We believe that a person can get a great idea of what happened during the match by reading the event summary.

5 Conclusion

In this paper, we presented a system capable of generating real-time summaries of events using only status updates from Twitter. The experiments that we conducted on sporting events showed that our system clearly outperforms the dominant approach on the sub-event detection task, and also generates very informative and readable summaries.

References

1. Akbari, M., Hu, X., Nie, L., Chua, T.S.: From tweets to wellness: wellness event detection from Twitter streams. In: AAAI, pp. 87–93 (2016)
2. Becker, H., Naaman, M., Gravano, L.: Beyond trending topics: real-world event identification on Twitter. ICWSM **11**, 438–441 (2011)
3. Castillo, C.: Big Crisis Data. Cambridge University Press, Cambridge (2016)
4. Chakrabarti, D., Punera, K.: Event summarization using tweets. In: ICWSM, pp. 66–73 (2011)
5. Chierichetti, F., Kleinberg, J., Kumar, R., Mahdian, M., Pandey, S.: Event detection via communication pattern analysis. In: ICWSM, pp. 51–60 (2014)
6. Java, A., Song, X., Finin, T., Tseng, B.: Why we Twitter: understanding microblogging usage and communities. In: SNA-KDD, pp. 56–65 (2007)
7. Letsios, M., Balalau, O.D., Danisch, M., Orsini, E., Sozio, M.: Finding heaviest k-subgraphs and events in social media. In: ICDM Workshops, pp. 113–120 (2016)
8. Lin, H., Bilmes, J.: A Class of submodular functions for document summarization. In: ACL, pp. 510–520 (2011)
9. Lin, J., Roegiest, A., Tan, L., McCreadie, R., Voorhees, E., Diaz, F.: Overview of the TREC 2016 real-time summarization track. In: TREC, vol. 16 (2016)
10. Mackie, S., McCreadie, R., Macdonald, C., Ounis, I.: Comparing algorithms for microblog summarisation. In: Kanoulas, E., Lupu, M., Clough, P., Sanderson, M., Hall, M., Hanbury, A., Toms, E. (eds.) CLEF 2014. LNCS, vol. 8685, pp. 153–159. Springer, Cham (2014). https://doi.org/10.1007/978-3-319-11382-1_15
11. Meladianos, P., Nikolentzos, G., Rousseau, F., Stavrakas, Y., Vazirgiannis, M.: Degeneracy-based real-time sub-event detection in Twitter stream. In: ICWSM, pp. 248–257 (2015)
12. Meladianos, P., Tixier, A.J.P., Nikolentzos, G., Vazirgiannis, M.: Real-time keyword extraction from conversations. In: EACL, pp. 462–467 (2017)

13. Mihalcea, R., Tarau, P.: TextRank: bringing order into texts. In: EMNLP, pp. 404–411 (2004)
14. Nemhauser, G.L., Wolsey, L.A., Fisher, M.L.: An analysis of approximations for maximizing submodular set functions I. Math. Program. **14**(1), 265–294 (1978)
15. Nenkova, A., McKeown, K.: A survey of text summarization techniques. In: Aggarwal, C., Zhai, C. (eds.) Mining Text Data, pp. 43–76. Springer, Boston (2012). https://doi.org/10.1007/978-1-4614-3223-4_3
16. Nichols, J., Mahmud, J., Drews, C.: Summarizing sporting events using Twitter. In: IUI, pp. 189–198 (2012)
17. Nikolentzos, G., Meladianos, P., Rousseau, F., Stavrakas, Y., Vazirgiannis, M.: Shortest-path graph kernels for document similarity. In: EMNLP, pp. 1891–1901 (2017)
18. Petrović, S., Osborne, M., Lavrenko, V.: Streaming first story detection with application to Twitter. In: NAACL-HLT, pp. 181–189 (2010)
19. Sharifi, B., Hutton, M.A., Kalita, J.K.: Experiments in microblog summarization. In: SocialCom, pp. 49–56 (2010)
20. Shen, C., Liu, F., Weng, F., Li, T.: A participant-based approach for event summarization using Twitter streams. In: NAACL-HLT, pp. 1152–1162 (2013)
21. Srijith, P., Hepple, M., Bontcheva, K., Preotiuc-Pietro, D.: Sub-story detection in twitter with hierarchical dirichlet processes. Inf. Process. Manage. **53**, 989–1003 (2016)
22. Walther, M., Kaisser, M.: Geo-spatial event detection in the Twitter stream. In: Serdyukov, P., Braslavski, P., Kuznetsov, S.O., Kamps, J., Rüger, S., Agichtein, E., Segalovich, I., Yilmaz, E. (eds.) ECIR 2013. LNCS, vol. 7814, pp. 356–367. Springer, Heidelberg (2013). https://doi.org/10.1007/978-3-642-36973-5_30
23. Weng, J., Lee, B.S.: Event detection in Twitter. In: ICWSM, pp. 401–408 (2011)
24. Zhao, S., Zhong, L., Wickramasuriya, J., Vasudevan, V.: Human as real-time sensors of social and physical events: a case study of Twitter and sports games. arXiv:1106.4300 (2011)
25. Zhou, X., Chen, L.: Event detection over twitter social media streams. VLDB J. **23**(3), 381–400 (2014)
26. Zubiaga, A., Spina, D., Amigó, E., Gonzalo, J.: Towards real-time summarization of scheduled events from Twitter streams. In: HT, pp. 319–320 (2012)

Spatial Statistics of Term Co-occurrences for Location Prediction of Tweets

Ozer Ozdikis[✉], Heri Ramampiaro, and Kjetil Nørvåg

Norwegian University of Science and Technology, Trondheim, Norway
{ozer.ozdikis,heri,noervaag}@ntnu.no

Abstract. Predicting the locations of non-geotagged tweets is an active research area in geographical information retrieval. In this work, we propose a method to detect term co-occurrences in tweets that exhibit spatial clustering or dispersion tendency with significant deviation from the underlying single-term patterns, and use these co-occurrences to extend the feature space in probabilistic language models. We observe that using term pairs that spatially attract or repel each other yields significant increase in the accuracy of predicted locations. The method we propose relies purely on statistical approaches and spatial point patterns without using external data sources or gazetteers. Evaluations conducted on a large set of multilingual tweets indicate higher accuracy than the existing state-of-the-art methods.

Keywords: Location prediction · Tweet localization
Spatial point patterns · Feature extraction

1 Introduction

Explicit location information in terms of latitude-longitude associated with text messages and photos in social networks provides a valuable resource for a wide range of applications, such as event detection, targeted advertisement, and crisis management. One of the most popular of these social networking platforms is Twitter, which enables users to post 140-character tweets and share them with their followers. Its widespread adoption and the accessibility of tweets through public APIs make it an attractive resource for research. However, despite increasing availability of GPS-enabled mobile devices, geotagged tweets are reported to constitute only 1–3% percent of all tweets [1,2]. As a result, predicting tweet location from its text has recently received considerable attention [1–7].

A widely adopted content-based approach for tweet localization is probabilistic language models. In this approach, the area of interest is partitioned into subregions, and terms in tweets that are posted in these regions are used for the training of text-based classifiers [3]. Specialized feature selection methods that prioritize geo-indicative terms have also been proposed in order to increase the prediction accuracy of these classifiers [5,8,9].

© Springer International Publishing AG, part of Springer Nature 2018
G. Pasi et al. (Eds.): ECIR 2018, LNCS 10772, pp. 494–506, 2018.
https://doi.org/10.1007/978-3-319-76941-7_37

The hypothesis that we investigate in this work is that even if strong location-indicative terms are perfectly identified in tweets, other terms can still be important in the interpretation of spatial information. In other words, if each term in a tweet is considered independent from other terms, probability assignments may give misleading results. The method that we propose in this work explores and evaluates spatial relationships, namely *attraction* and *repulsion*, between co-occurring terms in tweets using spatial point patterns and statistical methods. Selected term pairs (bigrams) with clustering or dispersion tendency with respect to the underlying unigram distributions are included in feature space to improve the accuracy of prediction.

To explain our idea, consider an example where we want to predict the location of a tweet mentioning *heathrow* with high precision, e.g., within a tolerance of 1 km error distance. The term *heathrow* can be considered to provide strong evidence about the location of a tweet, probably supporting the region around the Heathrow Airport in London, which covers a relatively large area. In this example, if that tweet also mentions *terminal*, whose co-occurrence with *heathrow* has stronger clustering tendency than *heathrow* alone, evaluating these two terms together as a new feature can yield predictions closer to the actual location. On the other hand, the phrase *heathrow express* can have an opposite effect (i.e., dispersion) and repel the geographical focus of the tweet to a region away from the airport area. We find such repulsion patterns quite interesting since even if they do not point to a specific place, they can indicate where a region is less likely to be the actual location for a tweet. In this example, the tweet mentioning *heathrow express* is probably posted from somewhere in the city, referring to the train that rides to the airport. Our claim is that such co-occurrences in a tweet can make an attraction or dispersion influence that may affect the geographical interpretation of a single term.

The main contributions of our work can be summarized as follows: (1) we investigate the spatial attraction and repulsion patterns of term co-occurrences, and propose a method to extend the feature space with term pairs having significant clustering or dispersion tendency, (2) we develop statistical techniques that can detect relationships between various types of features including emojis and multilingual texts, (3) we integrate our method with other unigram feature selection techniques to obtain higher prediction accuracies. An important aspect of our approach is that we can achieve the improvement in location prediction using only the tweet text in our analyses, i.e., we do not rely on external data sources, gazetteers, or other tweet metadata.

The remainder of this paper is organized as follows: We present a summary of related work in Sect. 2. We describe our proposed method in Sect. 3, along with a summary of baseline classification and feature selection techniques. Section 4 is devoted to our experiments and evaluation results. Finally, in Sect. 5, we conclude the paper and discuss future research directions.

2   Related Work

Location prediction for tweets can be described as estimating the geographical origin where a tweet is posted from [4,6,10]. Various techniques from the areas of information retrieval, machine learning, and natural language processing have been proposed to make accurate predictions [2–4,7,11–14]. One of the widely adopted techniques to that aim is probabilistic language models. In this technique, probability distributions are assigned for different subregions in an area using the textual content of georeferenced tweets in a training set [5,9,15]. Based on this trained model, per-region probabilities are then determined for non-geotagged tweets to be localized. A significant advantage of content-based approaches is that they can make predictions even in the absence of any other geographical cues [8].

Recent efforts to improve the accuracy of content-based approaches employ feature selection techniques, most of which have previously been used in similar text categorization problems [16]. The objective of these improvements is to determine location indicative terms in tweets by ranking them according to a metric. Top-n ranked features are then used in the training of language models, rather than using the complete vocabulary. Among recent studies in that direction, Cheng et al. [8] determine local words according to an analysis of frequency and dispersion. In [5], the authors experimented with numerous feature selection methods, such as information gain, information gain ratio, χ^2 statistic, geographical spreading, and Ripley's K statistic, and showed that information gain ratio outperforms their benchmark prediction methods in terms of accuracy. In that work, the authors use unigrams and also note that their preliminary results with named entities and higher order n-grams were not satisfactory. In a similar study [9], the authors employed Kernel Density Estimation (KDE) and Ripley's K statistics in order to improve the performance of location estimation for Flickr photos, particularly when only few terms can be selected for prediction. Their experiments revealed that the optimal results using geographical spreading was approximately the same as the optimal results based on KDE and Ripley K. However, geographical spreading showed more sensitivity to the number of features used in prediction.

The main objective in these previous feature selection efforts is to select location-indicative terms and eliminate common words that presumably have no spatial dimension [5]. Our approach essentially differs from these studies by evaluating spatial interactions between term pairs, even if a term appears to have no explicit spatial dimension. The method we adopt in our solution uses Ripley's K function [17], which has widespread usage in characterizing the spatial distribution patterns of objects in two-dimensional space [9,18]. To the best of our knowledge, our work is the first to analyze spatial patterns of term co-occurrences with respect to the underlying term distributions, and use them in the location prediction of tweets.

3   Spatial Co-occurrence Patterns in Location Prediction

In this section, we briefly describe our baseline model, and then explain the details of our location prediction method. Adhering to the probabilistic language model, the region of interest is discretized into mutually exclusive subregions and a Multinomial Naive Bayes (NB) classifier with additive smoothing is trained using terms (unigrams) in tweets in a training set [5,9]. We use Multinomial NB classifier mainly because it incorporates class priors in prediction and is reported to perform well even on scarce training data [5]. We adopt a grid-based approach to define subregions, since we aim to make fine-grained predictions, such as at the level of a place in a city [2,15].

Improvements over this classifier apply feature selection and use only the selected location-indicative terms for training. These methods were categorized as statistical, information-theoretic, and heuristic in [5], and we implemented different methods from each category as our baselines (explained in Sect. 4). Our proposed method can also use the results of these term selection methods, and identify spatially significant bigrams according to the selected unigrams. In the remainder of the section, we explain how we detect spatially significant bigrams and use them in the enhancement of feature space for location prediction.

3.1   Detection of Significant Spatial Co-occurrence Patterns

Ripley's K-function, represented by K_λ, is a statistical method to evaluate spatial patterns of points in a region [9,17,18]. The function calculates a value that is proportional to the number of point pairs that lie within a distance of λ to each other. In practice, it is widely applied to analyze spatial patterns of a set of objects having a certain property in order to determine whether these objects have a clustering or separation tendency.

We use Ripley's K-function to analyze the geographical distribution of specific terms (and term pairs) in tweets based on the latitude-longitude coordinates of these tweets. The K-function is defined as:

$$K_\lambda(X_t) = A \times \frac{|\{(x_i, x_j)|x_i, x_j \in X_t, x_i \neq x_j, d(x_i, x_j) < \lambda\}|}{|X_t|^2} \qquad (1)$$

where $X_t = \{x_1, ..., x_m\}$ with $m = |X_t|$ represents the set of tweets that include the term t, A represents the area of our grid, and $d(x_i, x_j)$ is the distance between two tweets x_i and x_j according to their coordinates. The value of $K_\lambda(X_t)$ is proportional to the number of tweet pairs in X_t that are within a distance of λ to each other. The λ parameter enables the evaluation of spatial relationships at different distance scales.

In an environment where the underlying population distribution is non-homogeneous, the value of the K-function for a specific set of objects may be affected by the population distribution. Therefore, comparison with the underlying point pattern should also be performed in order to evaluate the clustering and dispersion tendency of objects with respect to the population. This is usually achieved by executing a stochastic process, namely the Monte Carlo

---

Algorithm 1. Find co-occurrences with attraction/repulsion w.r.t. feature space

---

1: **Input1:** Set of terms in feature space and set of all terms in the training corpus
2: **Input2:** Distance range λ for Ripley's K-function
3: **Input3:** Number of Monte Carlo simulations to execute, denoted by M
4: **Output:** Set of bigrams B={$\langle t_p, t_c \rangle$| $X_{t_p t_c}$ has either clustering or repulsion tendency with respect to X_{t_p}}
5: **for each** primary term t_p in feature space **do**
6:     Find the set of tweets X_{t_p} that include t_p
7:     **for each** distinct term t_c in the corpus **do**
8:         Find the set of tweets $X_{t_p t_c}$ for which t_p is followed by t_c in the tweet text
9:         Apply K-function in Eq. (1) on $X_{t_p t_c}$ to get $K_\lambda(X_{t_p t_c})$
10:        **for** i=1...M **do**
11:            Randomly sample n tweets from X_{t_p}, where n=$|X_{t_p t_c}|$
12:            Let $X_{t_p}^i$ denote this sample, apply K-function on $X_{t_p}^i$ to find $K_\lambda(X_{t_p}^i)$
13:        **end for**
14:        Calculate upper boundary (u) and lower boundary (l) of envelop using $K_\lambda(X_{t_p}^i)$ values with 0.05 confidence interval
15:        **if** $K_\lambda(X_{t_p t_c}) > u$ **or** $K_\lambda(X_{t_p t_c}) < l$ **then**
16:            Insert tuple $\langle t_p, t_c \rangle$ to the set of selected bigrams B
17:        **end if**
18:     **end for**
19: **end for**

---

simulation [9]. The simulation mainly consists of taking random samples from the population, applying K-function on the samples, and calculating a confidence envelope with upper and lower bounds. A point pattern with K_λ value above the upper bound indicates clustering tendency (attraction), whereas the K_λ values below the lower bound is interpreted as dispersion (repulsion).

Our proposed method employs a similar approach to analyze the spatial patterns of term co-occurrences. However, rather than using the whole tweet set in the corpus as the underlying distribution, we compare the spatial distribution of co-occurring term pairs (bigrams) with the spatial patterns of corresponding single terms (unigrams). In other words, we measure the clustering and dispersion tendency of a bigram with respect to the spatial point pattern of each term in the bigram. This can be considered as a conditional analysis of the bigram's spatial distribution. Our algorithm to find term co-occurrences having significant attraction or repulsion pattern with respect to their unigrams is presented in Algorithm 1. For each unigram in the feature space, which we call *primary term* and denote by t_p, the algorithm finds co-occurring terms t_c in the training corpus that follows a primary term and exerts an attraction or repulsion influence on t_p. Specifically, if the spatial pattern of tweets with bigram $t_p t_c$ has significantly higher K_λ value compared to K_λ values of the tweet samples with t_p alone, $\langle t_p, t_c \rangle$ is regarded to have a clustering tendency in relation to t_p. Similarly, if the $K_\lambda(t_p t_c)$ value is below the lower boundary, $\langle t_p, t_c \rangle$ is selected as a repulsion co-occurrence for t_p. If spatial patterns of tweets that include $t_p t_c$ and t_p have no significant divergence, we do not perform any further analysis on $t_p t_c$. The reason we make a separate definition of primary term is to enable using the aforementioned feature selection methods for unigrams (e.g., χ^2, information gain). That means, a primary term t_p is taken from the feature space, which may be a selected subset of all distinct terms in the corpus, whereas t_c can simply be any term in the corpus.

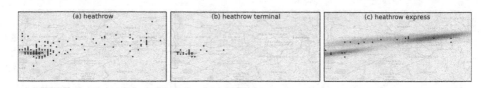

Fig. 1. Tweets mentioning (a) heathrow, (b) heathrow terminal, (c) heathrow express

We present the example in Fig. 1 to explain our findings. The dots in the figure represent locations of tweets in our grid for London area, and shadings in red color are generated by KDE for visualization purposes. Figure 1(a) shows the locations of tweets mentioning *heathrow* in our data set. Although they are slightly scattered, these tweets still exhibit a high concentration around the Heathrow Airport, as expected. Figure 1(b) presents a subset of these tweets, specifically the ones that include the bigram *heathrow terminal*. The density distribution of these tweets noticeably focuses on a more specific region in Heathrow. On the other hand, Fig. 1(c), which is generated for the bigram *heathrow express*, depicts a remarkably different distribution. This is probably due to the fact that people can post tweets about a train line that goes to the airport from places distant to the airport. As a result, although *heathrow* is a strongly descriptive term for location, we observe that its co-occurrences with other terms can change the spatial interpretation remarkably. Our experiments revealed that we can distinguish such attraction and repulsion patterns by applying Algorithm 1.

As noted in [9], when comparing two K-function values, the number of data points that are used in these calculations can affect the results. More specifically, $K_\lambda(X_1)$ and $K_\lambda(X_2)$ would not be comparable if $|X_1|$ and $|X_2|$ were different, since a larger dataset is more likely to yield a higher K_λ value. We do not observe this issue in Algorithm 1, since each simulation takes exactly $n = |X_{t_p t_c}|$ samples from X_{t_p}, as described in line 11. This is enabled by $X_{t_p t_c}$ being a subset of X_{t_p}. This also provides an advantage in terms of computational cost. In fact, except for a few term pairs that co-occur very frequently in our corpus (e.g., *United Kingdom*), we observe that $|X_{t_p t_c}|$ is remarkably lower than $|X_{t_p}|$ in most cases, which resulted in acceptable computation times in our experiments. Moreover, we transform latitude-longitude coordinates of tweets into three-dimensional Euclidean coordinates and index them in a k-d tree [9]. This transformation provided us noticeable performance improvement in the calculation of K_λ values.

The steps in Algorithm 1 describe the analysis of bigrams in the form of $t_p t_c$ (i.e., t_p is followed by t_c). Similar procedures are also executed to identify spatially related bigrams where a primary term t_p is preceded by t_c. Ordering of terms in bigrams should be taken into consideration since we examine the distribution of a bigram conditioned on the distribution of t_p. We explain the effect of ordering on an example. Figure 2(a) and (b) demonstrate the distributions of *stamford* and *bridge*, respectively. Although tweets mentioning *bridge* are

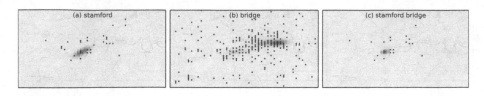

Fig. 2. Distribution of tweets mentioning (a) stamford, (b) bridge, (c) stamford bridge

scattered in a large area, it exhibits a strong clustering tendency when preceded by the term *stamford* in tweets. In other words, when *stamford* and *bridge* are considered as t_c and t_p, respectively, distribution of $t_c t_p$ given in Fig. 2(c) leads to a significant clustering tendency with respect to the distribution of t_p. The density of tweets in (c) actually points to the region around the stadium Stamford Bridge. Similar relationships are also detected between emojis and terms, such as *heathrow* and 🛫, since we primarily use statistical methods in our analyses without applying any restriction on the content of a tweet. The next section explains how we use these detected bigrams in the enhancement of our feature space.

3.2   Enhancement of Feature Space

The enhancement of feature space is performed by adding bigrams from B, which were found by Algorithm 1 above, as additional features to tweets. Specifically, given a tweet x with n terms in its text, denoted by $[t_1^x, t_2^x, ...t_n^x]$, if a bigram $\langle t_i^x, t_{i+1}^x \rangle$ exists in B (i.e., having significant spatial relationship), that bigram is added to the tweet as a new feature. Our rationale in this operation is that, if there is a clustering or dispersion tendency of a bigram $t_i t_j$ with respect to t_i or t_j, we can make more reliable estimations if we also have $t_i t_j$ in the tweet.

We exemplify the expansion operation on a hypothetical tweet with terms $[a,b,c,d]$. Assume that the tuple $\langle a, b \rangle$ exists in B, i.e., a and b were found to have significant spatial relationship in Algorithm 1. In this case, applying expansion on this example tweet results in $[a,b,c,d,\langle a, b \rangle]$. Following this example, if B had also included the pair $\langle b, c \rangle$, that bigram would also be added to produce $[a,b,c,d,\langle a, b \rangle,\langle b, c \rangle]$. This means that we do not impose any restriction about mutual exclusion, and utilize a bigram if it has a spatially significant pattern with respect to its unigrams.

The complete process can be summarized as follows: Using a training tweet set, we obtain B using Algorithm 1, enhance the feature space by the expansion operation, and train the language model for classification. For a new tweet to be localized, we apply the expansion operation for it according to the trained model and estimate its location using a Multinomial NB classifier.

4   Evaluation

In this section, we present the evaluation results of our method and compare with state-of-the-art baselines. We evaluate our methods on regional datasets to predict tweet locations at fine-granular level. Accordingly, we selected the Greater London Area for our experiments and divided the area into equal-size grid cells to form a 100×100 grid. This resulted in cells covering approximately an area of $0.25 \, km^2$ each. We collected public tweets between 3 October 2015 and 20 January 2016 using the Twitter Streaming API[1]. We filtered out tweets without explicit GPS coordinates (i.e., latitude-longitude) and obtained 4,040,775 geotagged tweets posted from our area of interest. Following the common practices in earlier similar studies, we eliminated exact duplicate tweets, Foursquare check-ins, and tweets from possible spammers [5,8,13]. To filter out spammers, we excluded tweets from users with more than 1000 friends or followers or who posted more than 300 tweets in our time window (approximately more than two tweets per day), and tweets with advertisement hashtags (e.g., #job, #realestate), since they have almost the same text and are usually posted from same places [8,13,19]. Finally, we obtained 489,466 unique tweets posted by 100,997 distinct users. We did not apply any restriction on the language of a tweet.

Each tweet in our dataset is assigned to a grid cell according to its GPS coordinates. In our experiments, we used randomly chosen 464,993 tweets for training and the remaining 24,473 for test. Tweet texts are divided into tokens by using Twokenize[2] library. For training data, we discard tokens that appear in less than five tweets, hyperlinks, and single characters to reduce sparsity [3,13]. This yields a total number of 54,752 distinct tokens (unigrams) in our training set. Tokenized tweets in the training set and their assigned grid cells are used in building the probabilistic language models and for analyzing the spatial point patterns of bigrams.

4.1   Evaluation Methodology

We compared our proposed method with different baselines, including the full model (i.e., using all terms in tweets without making prior unigram feature selection) and four feature selection methods. Among a wide range of feature selection techniques, we implemented the following four as our baselines, since they are widely applied in state of the art:

1. IG: Information Gain (*information-theoretic*) [5]
2. IGR: Information Gain Ratio (*information-theoretic*) [5]
3. CHI: χ^2 statistic (*statistical*) [16]
4. GS: Geographical Spreading (*heuristic*) [9]

---

[1] https://dev.twitter.com/overview/api.
[2] https://github.com/brendano/ark-tweet-nlp/.

In our implementations, we followed the descriptions given in the cited papers above. In addition to these five baselines, we also made estimations using class priors [5], which basically finds the grid cell with maximum number of tweets in the training set. This is used to show that assigning all test tweets simply to the most populous place does not yield useful results.

We evaluate the performance of these methods using the following three metrics: (1) *Accuracy:* proportion of tweets in the test set for which the true grid cell is correctly predicted, (2) *Accuracy@n:* proportion of tweets for which the estimated location is at most n kilometers away from the true location, (3) *Median:* median of the distances between the predicted location and the true location for test tweets. The distances in these metrics are calculated based on the centers of predicted and true grid cells for tweets.

For each feature selection method in our baselines, we first apply a ranking of tokens in the training set using the corresponding feature ranking metric. For example, IG ranks all tokens in the training set according to their information gain. Then, the training of a baseline predictor is performed by using top-n features in its ranking, and the location prediction for test tweets are executed with that setting.

We apply our proposed enhancement on each baseline separately. When applied on the full model, Algorithm 1 analyzes all terms in the corpus. When applied on a unigram selection method, the algorithm uses only the top-n unigrams as primary terms (explained in Sect. 3.1). We denote our enhanced methods with suffix *SCoP*, as an abbreviation for *Spatial Co-occurrence Pattern*. For example, $IG_{(n)}$+SCoP represents our enhancement where the top n of unigrams with the highest information gain are used as the feature space in Algorithm 1.

4.2   Evaluation Results and Discussion

Table 1 presents the minimum median error distances obtained by each method, along with the corresponding accuracies. Since each method can achieve its highest accuracy using different top-n features, we also indicate this value with a subscript. For example, $IGR_{(0.6)}$, the most accurate baseline, uses top 60% of unigrams with highest information gain ratio. $IG_{(1.0)}$ means that a feature selection based on information gain does not perform better than the full model for any selection of top-n. We observe that the evaluation results of our baselines are also consistent with the findings in previous studies [5,9].

The table shows that our enhancement (denoted by *SCoP*) on each baseline results in better predictions, even when no prior unigram feature selection is applied (full model). The most accurate predictions are obtained by our enhancement when it is applied on $IGR_{(0.6)}$. In order to analyze the difference in error rates between a baseline and its SCoP enhancement, we employ McNemar's test on their predictions. The results of the test show that the improvement using SCoP is statistically significant for every baseline ($p \ll 0.00001$). In our experiments, the λ distance that we used in the calculation of Ripley-K values is 0.5 (in kilometers). We have also experimented with $\lambda = 2.0$ and obtained similar results. Specifically, when $\lambda = 2.0$ is used, $IGR_{(0.6)}$+SCoP predictions were

Table 1. Comparison of methods with settings minimizing their median error distance

Prediction method	Median (km)	Accuracy	Acc@0.5 km	Acc@1.0 km	Acc@2.0 km
Class prior	3.7743	0.049	0.065	0.119	0.310
Full model	1.4860	0.363	0.391	0.442	0.530
Full Model + SCoP	1.2585	0.408	0.432	0.478	0.558
$IG_{(1.0)}$	1.4860	0.363	0.391	0.442	0.530
$IG_{(1.0)} + SCoP$	1.2585	0.408	0.432	0.478	0.558
$IGR_{(0.6)}$	1.0831	0.407	0.436	0.491	0.586
$\boldsymbol{IGR_{(0.6)} + SCoP}$	**0.7429**	**0.435**	**0.460**	**0.510**	**0.594**
$GS_{(0.3)}$	1.3657	0.389	0.415	0.462	0.535
$GS_{(0.5)} + SCoP$	1.2583	0.414	0.437	0.482	0.558
$CHI_{(0.45)}$	1.2583	0.393	0.422	0.477	0.567
$CHI_{(0.45)} + SCoP$	1.0831	0.424	0.448	0.496	0.575

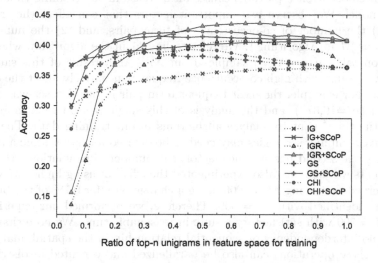

Fig. 3. Accuracies of four baselines and corresponding enhancements using SCoP. Baselines and SCoP enhancements are colored in red and green, respectively. (Color figure online)

made with a median error distance of 0.7430 and an accuracy of 0.432. Since $\lambda = 0.5$ performed slightly better, we demonstrate the results that we found using $\lambda = 0.5$.

Figure 3 presents the detailed accuracies of predictions using different selections of top-n features. The results reveal that our proposed SCoP enhancement improves the accuracies of all baseline (unigram) feature selection methods for every setting of top-n. Among the four baselines, the highest accuracy is obtained by IGR, which is even further improved by applying SCoP on it. The results of GS are also worth discussing. The figure shows that if we had to use only the

5% of features (unigrams) for training, the highest baseline accuracy would be obtained by GS. That means, GS makes the most useful top 5% unigram selection among our baselines, and the figure shows that its predictions are further improved by using our proposed enhancement in GS+SCoP.

These results reveal that we can obtain accuracies with SCoP that could not be obtained by the unigram feature selection methods in our experiments. However, since our method adds new bigrams to the training model, we also analyze the size of increase in feature space. The baseline $IGR_{(0.6)}$, which yields the most accurate predictions among the baselines, uses 32,851 tokens for training (60% of all unigrams). SCoP uses these tokens as primary tokens in Algorithm 1, which detects 10,095 bigrams with significant spatial relationship with respect to the 32,851 unigrams. As a result, $IGR_{(0.6)}$+SCoP uses 42,946 features in total. The number of added bigrams may vary depending on the baseline unigram selection method and the choice of top-n ratio.

Considering the spatial analyses in Algorithm 1, as expected, the detection of spatial co-occurrence patterns causes an increase in the training time of the overall model. We observe two important factors that can affect the training time: (1) the number of co-occurrences of term pairs, and (2) the number of simulations M in Algorithm 1. We refer to line 11 in the algorithm, where the simulation takes $n = |X_{t_p t_c}|$ samples from X_{t_p}. As a result of this sampling strategy, bigrams with high co-occurrence frequency negatively affect the execution time. For example, the most frequent term pair in our dataset was *United Kingdom* (n =16,377), and the analysis of this single bigram took more than 60% of the time spent to analyze all bigrams in our training data. Therefore, alternative sampling strategies may need to be devised for larger-scale analyses. Regarding the second performance factor, the number of simulations M in our experiments was 500. We also experimented the effect of using higher M values, and observed that using $M = 1000$ made a change only for 0.7% of the bigrams that were identified with $M = 500$. Therefore, we performed our experiments using $M = 500$ with satisfactory results in reasonable time. We note that since there is no interdependency or sequential relationship in the spatial analyses of bigrams, these operations can also be parallelized and executed in distributed environments.

Finally, we evaluate the utility of our method by comparing it with a setup in which we do not make any particular selection among bigrams. The total number of distinct bigrams that occur in at least five tweets in our training set is 99,452. We trained the model using all of these bigrams and the unigrams selected by $IGR_{(0.6)}$ (i.e., without making the SCoP analysis in Algorithm 1). Our tests in this experiment resulted in an accuracy of 0.387, which is lower even than the baseline's. Therefore we can conclude that an effective analysis of bigrams, as we proposed in this paper, is critical to obtain accurate predictions.

5   Conclusion

In this paper, we introduced a new approach to detect term pairs that exhibit clustering or dispersion tendency in their geographical distribution in relation to the underlying single-term spatial patterns. We used the detected term pairs to improve probabilistic language models and increase the accuracy of content-based location prediction of tweets. We demonstrated that the effective selection of co-occurring terms yields significant improvement in location prediction accuracy. Using purely statistical methods and spatial point patterns enabled our methods to execute without any dependence on predefined gazetteers or external data sources.

In our future research, we plan to adapt our framework in distributed environments and apply our methods for fine-grained location prediction in global scale. Applying discriminative learning models for location prediction and investigating alternative methods to utilize detected attraction and repulsion patterns are also among our future research directions.

References

1. Li, W., Eickhoff, C., de Vries, A.P.: Geo-spatial domain expertise in microblogs. In: de Rijke, M., Kenter, T., de Vries, A.P., Zhai, C.X., de Jong, F., Radinsky, K., Hofmann, K. (eds.) ECIR 2014. LNCS, vol. 8416, pp. 487–492. Springer, Cham (2014). https://doi.org/10.1007/978-3-319-06028-6_46
2. Paraskevopoulos, P., Palpanas, T.: Where has this tweet come from? Fast and fine-grained geolocalization of non-geotagged tweets. Soc. Netw. Anal. Min. **6**(1), 89 (2016)
3. Melo, F., Martins, B.: Automated geocoding of textual documents: a survey of current approaches. Trans. GIS **21**(1), 3–38 (2017)
4. Zheng, X., Han, J., Sun, A.: A survey of location prediction on Twitter. CoRR abs/1705.03172 (2017)
5. Han, B., Cook, P., Baldwin, T.: Text-based Twitter user geolocation prediction. J. Artif. Int. Res. **49**(1), 451–500 (2014)
6. Priedhorsky, R., Culotta, A., Del Valle, S.Y.: Inferring the origin locations of tweets with quantitative confidence. In: Proceedings of CSCW 2014 (2014)
7. Han, B., Rahimi, A., Derczynski, L., Baldwin, T.: Twitter geolocation prediction shared task of the 2016 workshop on noisy user-generated text. In: Proceedings of W-NUT (2016)
8. Cheng, Z., Caverlee, J., Lee, K.: A content-driven framework for geolocating microblog users. ACM Trans. Intell. Syst. Technol. **4**(1), 2:1–2:27 (2013)
9. Van Laere, O., Quinn, J., Schockaert, S., Dhoedt, B.: Spatially aware term selection for geotagging. IEEE Trans. Knowl. Data Eng. **26**(1), 221–234 (2014)
10. Dredze, M., Osborne, M., Kambadur, P.: Geolocation for Twitter: timing matters. In: Proceedings of HLT-NAACL (2016)
11. Hauff, C., Houben, G.J.: Placing images on the world map: a microblog-based enrichment approach. In: Proceedings of ACM SIGIR 2012, 691–700 (2012)
12. Backstrom, L., Sun, E., Marlow, C.: Find me if you can: improving geographical prediction with social and spatial proximity. In: Proceedings of WWW 2010, pp. 61–70 (2010)
13. Eisenstein, J., O'Connor, B., Smith, N.A., Xing, E.P.: A latent variable model for geographic lexical variation. In: Proceeding of EMNLP 2010, pp. 1277–1287 (2010)
14. Miura, Y., Taniguchi, M., Taniguchi, T., Ohkuma, T.: A simple scalable neural networks based model for geolocation prediction in Twitter. In: Proceedings of W-NUT (2016)

15. O'Hare, N., Murdock, V.: Modeling locations with social media. Inf. Retr. **16**(1), 30–62 (2013)
16. Yang, Y., Pedersen, J.O.: A comparative study on feature selection in text categorization. In: Proceedings of ICML 1997 (1997)
17. Ripley, B.D.: Modelling spatial patterns. J. Roy. Stat. Soc.: Ser. B (Methodol.) **39**(2), 172–212 (1977)
18. Ruocco, M., Ramampiaro, H.: Geo-temporal distribution of tag terms for event-related image retrieval. Inf. Process. Manage. **51**(1), 92–110 (2015)
19. Lee, K., Caverlee, J., Webb, S.: Uncovering social spammers: social honeypots + machine learning. In: Proceedings of ACM SIGIR 2010, pp. 435–442 (2010)

Learning to Leverage Microblog Information for QA Retrieval

Jose Herrera[1]([✉]), Barbara Poblete[1], and Denis Parra[2]

[1] Department of Computer Science, University of Chile, Santiago, Chile
{jherrera,bpoblete}@dcc.uchile.cl
[2] Department of Computer Science, Pontificia Universidad Católica de Chile,
Santiago, Chile
dparra@ing.puc.cl

Abstract. Community Question Answering (cQA) sites have emerged as platforms designed specifically for the exchange of questions and answers among users. Although users tend to find good quality answers in cQA sites, they also engage in a significant volume of QA interactions in other platforms, such as microblog networking sites. This in part is explained because microblog platforms contain up-to-date information on current events, provide rapid information propagation, and have social trust.

Despite the potential of microblog platforms, such as Twitter, for automatic QA retrieval, how to leverage them for this task is not clear. There are unique characteristics that differentiate Twitter from traditional cQA platforms (e.g., short message length, low quality and noisy information), which do not allow to directly apply prior findings in the area. In this work, we address this problem by studying: (1) the feasibility of Twitter as a QA platform and (2) the discriminating features that identify relevant answers to a particular query. In particular, we create a document model at conversation-thread level, which enables us to aggregate microblog information, and set up a learning-to-rank framework, using factoid QA as a proxy task. Our experimental results show microblog data can indeed be used to perform QA retrieval effectively. We identify domain-specific features and combinations of those features that better account for improving QA ranking, achieving a MRR of 0.7795 (improving 62% over our baseline method). In addition, we provide evidence that our method allows to retrieve complex answers to non-factoid questions.

Keywords: Twitter · Question Answering · Learning-to-rank

J. Herrera and B. Poblete have been partially funded by the Millennium Nucleus Center for Semantic Web Research under Grant NC120004. J. Herrera is partially funded by the CONICYT Doctoral Program and D. Parra is funded by FONDECYT under grant 2015/11150783. J. Herrera, B. Poblete and D. Parra have been partially funded by FONDEF under grant ID16I10222.

© Springer International Publishing AG, part of Springer Nature 2018
G. Pasi et al. (Eds.): ECIR 2018, LNCS 10772, pp. 507–520, 2018.
https://doi.org/10.1007/978-3-319-76941-7_38

1   Introduction

Online social networking platforms are designed to facilitate user interaction through the creation of diverse user communities. In particular, community Question Answering (cQA) websites are social platforms that specialize in connecting users that have questions (i.e., information needs expressed as questions in natural language) with other users that can provide answers to them. Examples of these platforms are Y! Answers[1], Stack Exchange[2], among others.

User exchanges in cQA web sites are commonly preserved online indefinitely, becoming in time historic knowledge bases. Once consolidated, these knowledge bases become rich repositories for finding answers to newly formulated questions in the platform, or to queries in search engines, which have been phrased in the form of questions. These resources are important for information retrieval (IR) tasks related to question answering (QA), because questions commonly convey complex information needs, which are difficult to satisfy using traditional IR techniques [16,20].

Despite the significance of cQA platforms, these are not the only way in which users ask and answer questions. Prior work shows that users of the microblog social networking platform Twitter[3], also engage in an important amount of QA, accounting for almost 10% its activity [8,17,26]. For example, Fig. 1 shows two Twitter conversation threads related to the question "Which games are good for ps4?". In this example each conversation provides several relevant answers, which can all be considered as part of the answer to the original question.

Twitter, in particular, has a large user-base[4] and allows users to quickly exchange vast amounts of information through short messages (called *tweets*). Users commonly use Twitter to post real-time status updates, share information on diverse topics and chat with other users. Researchers believe that the immediacy of information on Twitter is one driver for users to engage in QA in this platform, but also because users seek answers from their preferred social network [17,19]. This indicates that the answers sought by Twitter users probably have a temporal, social and/or geographical context, which can be quickly and easily addressed by other users in this network.

The sustained use of Twitter for QA suggests that valuable historic information for automatic QA retrieval could be obtained from this platform. Furthermore, this information could complement that provided by traditional cQA web sites, with more up-to-date and context rich answers.

In this work we addressed the problem of how to effectively leverage Twitter data for QA retrieval. In particular, as an initial approach on this topic, we performed an investigation based on the following three research questions:

---

[1] http://answers.yahoo.com.
[2] http://stackexchange.com.
[3] http://www.twitter.com.
[4] 328 million users in June 2016 (https://about.twitter.com/es/company).

(a) Conversation thread 1.   (b) Conversation thread 2.

Fig. 1. Example of two Twitter conversation threads that answer the question: *"Which games are good for ps4?"*. User suggestions are highlighted in yellow. (Color figure online)

RQ1: Is it possible to retrieve relevant answers to incoming questions using historic Twitter data?

RQ2: Which features are the most important for finding relevant answers to questions using Twitter data?

RQ3: Can Twitter conversation threads be used to answer questions (as opposed to using single tweets) in order to provide answers to complex questions?

We undertook this task by introducing a novel document representation, which considers complete conversation threads as documents. Using a learning-to-rank (LTR) framework we studied which were the most important features for re-ranking relevant conversation threads (referred to as *threads* from now on). We learned and evaluated this approach by using standard factoid QA datasets (i.e., on questions that can be answered by a simple fact). We also performed a manual analysis of the applicability of our approach on non-factoid questions.

We found that, in general, by using a thread-level document representation of microblog data, we are in fact able to re-rank relevant answers in the top positions. Our experiments show that the best single feature set for ranking is parts-of-speech (POS). Furthermore, by combining POS with other features such as: distance-based, social-based, and word embedding features, we are able to achieve a MRR of 0.7795. Improving 18% over the best single performing feature (POS) and 62% over the baseline (BM25). In addition, our manual evaluation of non-factoid questions indicates that our approach is also a very good starting point for answering much more complex and context rich questions.

This document is organized as follows: Sect. 2 discusses relevant prior work in the area, Sect. 3 describes our problem statement and proposed solution.

Section 4 presents our experimental setup, as well as results and discussion, and Sect. 5 conclusions and ideas for future work.

2   Related Work

Previous investigations in QA have identified relevant features for cQA retrieval. However, the particular characteristics of microblog platforms, makes the problem of QA retrieval notably different than that addressed in prior work. For instance, an important difference between microblog and cQA platforms is the length of messages, which requires document models and features that specifically target short text [21]. Another difference, is that users on cQA platforms tend to be more specialized, focusing on one, or very few, topics of interest [9], unlike the Twitter community which tends to be miscellaneous.

QA in non-cQA platforms. There have been studies that analyze how users ask, and what they ask, in non-specific QA social networks. Morris et al. [17] performed a characterization study of questions asked in Facebook and Twitter, and found that the most asked questions are about recommendations, opinions and factual knowledge. Paul et al. [19] found that the most popular questions in Twitter are rhetorical and of factual knowledge. In addition, they observed that roughly 18.7% of the questions received at least one reply; the first within 5–30 min and the remainder within the next 10 h. Zhao et al. [26] extracted features from tweets and built a classifier that distinguishes real questions (i.e., questions that seek answers) from rhetorical questions. In addition, Liu et al. [12] created a taxonomy to describe the types of questions that user ask in Twitter.

LTR and features for QA retrieval. Learning-to-rank (LTR) refers to machine learning techniques for training a model for a ranking task. LTR is widely used in several types of ranking problems in information retrieval (including traditional QA), natural language processing and recommender systems; For example, Duan et al. [3] used LTR to rank tweets according to how informative they were, based on content features such as URLs and length. Molino et al. [15] used LTR to find the *best answers* in Yahoo Answers and Surdenau et al. [22] studied non-factoid questions in the same platform, exploring feature combinations.

We complement prior work by analyzing which features contribute to find relevant answers in Twitter and how they differ from those of cQA platfomrs. In this article we extend our short workshop paper Herrera et al. [7], which introduces the initial problem of using microblogs for QA retrieval. This extended version has been completely rewritten and improved by adding: (1) a formal problem statement, (2) an experimental framework for ranking Twitter conversations for QA retrieval, (3) an experimental evaluation of its effectiveness, and (4) by analyzing the features which contribute most.

3   Microblog Ranking for QA Retrieval

In this section we present our approach for leveraging microblog information for QA retrieval tasks. We model our research problem as that of *re-ranking,* in which the main goal is to rank relevant answers in the top result positions. For our current purpose, we define an answer as *relevant* if it contains a correct answer to the original question. Then, we study which microblog features have the most influence for determining relevant answers. Specifically, we propose an aggregated *thread-level document model*, which considers conversation *threads* as documents for retrieval. This representation allows us to use aggregated information as documents for answering queries, as opposed to using a single tweets. Our hypothesis is that the composition of several tweets into a single thread can provide answers to complex questions.

3.1 Problem Statement for Microblog QA

More formally, let q^* be a question that corresponds to an information need formulated by a user. Let $Q^* = \{q_1, q_2, \ldots, q_n\}$ be the set of possible query formulations of q^*. We define query formulations in Q^* as any variation of the initial input query q^* that allows us to retrieve a set of threads that are candidate answers for q^* [10]. Then, for each $q_i \in Q^*$, we extract all of the threads (documents) that match q_i in a given microblog dataset. In particular, we say that a thread t_i matches query q_i when t_i contains *all of the terms* in q_i. Next, let $\mathcal{T} = \{t_1, t_2, \ldots, t_m\}$ be the set that contains the union of the sets of threads that match the query formulations in Q^*, and therefore by extension, which match q^*.

Hence, the goal of our research is, for a given question q^*, to re-rank the threads in $\mathcal{T}$ according to their relevance for answering q^*. We define this as a re-ranking problem since our starting point are the set of threads in $\mathcal{T}$, which are initially obtained using a very simple retrieval method (detailed in Sect. 4). In other words, our goal is to learn a function $f(q^*, \mathcal{T}) \to \pi$ that produces an optimal permutation $\hat{\pi}$ of the elements in $\mathcal{T}$ for answering the question q^*.

We acknowledge that there are other important tasks besides ranking, in QA retrieval such as, creating the best possible query formulations and selecting the passages within a text that contain the answer to a question [10]. However, at this moment we consider those problems as beyond the current scope of our work.

3.2 Proposed Solution

We propose a solution based on a LTR framework that will allow us to learn the aforementioned function $f(q^*, \mathcal{T})$. In order to identify the features that produce the best ranking, we evaluate several combinations of sets of features using different LTR models. In particular we use the following four LTR models, defined next: MART [6], Ranknet [1], Rankboost [5] and LambdaMart [24].

This solution also requires us to specify: (1) the *query formulations*, (2) the *features* that will be used:

(1) Query formulations. Since our focus is on re-ranking, we select a set of broad query formulations that increase the recall of documents that may contain an answer to the query q^*. In particular, we use the following query formulations:

- q_1: Corresponds to the original question as it was formulated by the user q^*.
- q_2: Corresponds to q^* after lowercase and whitespace normalization, removal of non-alphanumerical characters and terms with only one character.
- q_3: Corresponds to q_2 after the additional removal of stopwords, with the exception of terms in the **6W1H**[5]. For example, the question $q^* =$ "What is the scientific name of tobacco?" becomes $q_3 =$ "what scientific name tobacco".
- q_4: Corresponds to q_3 without the **6WH1**. In the previous example, q^* would be transformed to $q_4 =$ "scientific name tobacco".

(2) Features[6]. We performed a review of the features used in prior work for traditional QA retrieval and adapted those that could be applied to microblog data. In addition, we manually inspected approximately $1,000$ QA threads in Twitter (i.e., threads in which their initial tweet is a question, shown in Fig. 1). This inspection allowed us to identify features that could potentially help determine whether the question in the QA thread was answered inside the thread itself or not. We describe the different types of features next:

- **Distance-based features (D_TFIDF_N, D_WEMB):** These features are based on four well-known distance metrics, between the query q^* and a thread t. *cosine, manhattan, euclidean and jaccard* [2]. We compute these distances using 2 types of vector representations: (i) **D_TFIDF_N**, which are the aforementioned distances between the *tf-idf* vector representations of q^* and t (using n-grams of size $N = \{1, 2, 3\}$), and (ii) **D_WEMB**, which are the same distances but using the word embedding vector representations of q^* and t.
- **Social-based features (SOCIAL):** These features are based on the social interactions observed in a conversation threads (i.e., thread level features). These include: number of replies in a thread, number of different users that participate, fraction of tweets with favorites/retweets/hashtags, number of user mentions, and number of different user mentions.
- **User-based features (USER):** These features are based on properties to the users that participate in a conversation thread. These include: total number of followers and followees of the users that participate in a thread, the fraction of users in the thread that have a verified account, the average *age* of the users in a thread. User age is computed as the difference between date of creation of the Twitter user account and the date when the tweet was posted.
- **Content-based features (CONTENT):** These features are based on the content of a thread. These include: the number of different URLs in the thread, the number of words (removing URLs and punctuation), the length of the thread $\frac{\# words}{\# tweets}$ (considering only words with size ≥ 1), the fraction

---

[5] **6W1H** corresponds to 5WH1 with the addition of the terms "Which" (i.e. Who, What, Where, When, Why, Which and How).

[6] Due to space constrains only a high-level description of the features is provided. However, the detailed list of features is available at https://goo.gl/qqACz5.

of uppercase and lowercase letters, the number of positive/negative/neutral emoticons, and the average number of words in English. Some social-based, user-based and content-based features have been adapted to microblog data from features used in prior works [3, 15].

- **Part-of-speech (POS) features:** These features are based on part-of-speech tagging. In particular we compute the frequency of each high-confidence POS tag in a conversation thread, using the Twitter-specific tagger *TweetNLP* by [18].
- **Representative words feature (REPW):** This feature corresponds to the fraction of *"representative words"* that are contained in a thread. Where a "representative word" is any word which is contained in the top-50% most frequent terms over all threads in the training data (excluding stopwords).
- **Word embedding based features (WEMB_THR, WEMB_Q):** This feature is computed using the explicit vector representation of the query (WEMB_Q) and the thread (WEMB_THR), created using Word2Vec [14] and a pre-trained model on 400 million tweets[7]. Each vector was composed by 300-dimensions and the word embeddings are inferred using a skip-gram architecture.
- **Time-based features (TIME):** These features include time-based characteristics of the thread, such as: time-lapse between the first tweet in the thread and the last, and the average time between tweets in a thread.

4 Experimental Evaluation and Results

Our main goal is to validate the feasibility of retrieving and ranking relevant answers to questions using past information exchanges on Twitter (**RQ1**). In addition, we also want to measure the contribution of the different microblog features for ranking (**RQ2**), and to see if a thread-level document representation can help provide answers to non-factoid questions (**RQ3**). We focus on two types of evaluation, *factoid QA task* and *non-factoid QA task*. The first is a quantitative evaluation based on standard QA ground truth datasets. The second evaluation is more exploratory and is based on a manual evaluation.

4.1 Factoid QA Evaluation Task

We evaluate our approach on factoid questions because there are several benchmark QA datasets available to validate automatically correct answers. Jurafsky and Martin [10] defined as *factoid* question answering those tasks that require one answer, which is a simple fact (e.g., "Who was the first American in space?"). Nevertheless, we use the factoid task as a proxy to our goal of answering more complex questions (i.e., non-factoid).

---

[7] http://www.fredericgodin.com/software/.

Ground-truth dataset. We built a ground truth QA dataset based on datasets provided by the TREC QA challenge[8] (*TREC-8 (1999), TREC-9 (2000), TREC-2004* and *TREC-2005*) and a repository of factoid-curated questions for benchmarking Question Answering systems[9]. All of these datasets, except for TREC-8, provided regular expressions to match correct answers automatically. For the remaining dataset, we manually created regular expressions to match the correct answers. In addition, we manually removed questions that were: time-sensitive (e.g., "What is the population of the Bahamas?"), inaccurate (e.g., "what is the size of Argentina?"), not phrased as a question (e.g., "define thalassemia"), referred to other questions (e.g., "What books did she write?") and questions whose length were over 140 characters. This resulted in a ground-truth dataset of 1,051 factoid questions.

Twitter dataset for LTR factoid questions. In an ideal scenario, our candidate answers would be obtained from a large historical Twitter dataset, or from the complete data-stream. However these types of data repositories were not available to us for this evaluation. We then approximated the ideal retrieval scenario by using the Twitter Search API [10] as an endpoint, which provides access to a sample of the actual data. For each query formulation, described in Sect. 3.2, we retrieved as many tweets as possible up to 1,000 tweets per query. In addition, if the retrieved tweet was part of a conversation thread we also retrieved that thread in full. Overall, we found candidate answers for 491 (47%) of the questions (i.e., the remaining questions had no matching threads, e.g. "What was the ball game of ancient Mayans called?"). To improve the learning process, we removed low relevance threads (i.e., the cosine similarity between thread and query was ≤ 0.3) and threads that had no replies. The resulting dataset contained 33,873 conversation threads with 63,646 tweets. We note that our datasets (ground-truth and Twitter) are publicly available[11].

Baseline Methods. We compare our approach to the following methods, which have been used as baselines in prior cQA studies [15,22,23]:

- **BM25:** The Okapi weighting BM25 is widely used for ranking and searching tasks [13]. We use the BM25 document score in relation to a query. We use $b = 0.75$ and $k_1 = 1.2$ as parameters since they were reported optimal for other IR collections [23].
- **Twitter Search:** This method lists results from Twitter's search interface. Results are obtained by searching for each query in Q^* using the "latest" option, which lists messages from the most recent to the oldest message. The results obtained for each query are then joined in chronological order. However, this method is not reproducible since it works like a "black box" from our perspective.

---

[8] http://trec.nist.gov/data/qamain.html.
[9] https://github.com/brmson/dataset-factoid-curated.
[10] https://dev.twitter.com/rest/public/search.
[11] https://github.com/jotixh/ConversationThreadsTwitter/.

– **REPW:** This method uses the feature REPW, described in Sect. 3, with the best performing LTR model from our experiments. Experimentally, this method behaves as an upper bound of the "Twitter search" method with the advantage that it can be reproduced.

Evaluation methodology. We built several models that rank tweets and conversation threads as potential answers to a set of given factoid questions. We use the LTR software library Ranklib of the LEMUR project[12] for this task. To reduce the probability of obtaining significant differences among the LTR methods only by chance, we relied on bootstrapping [4]. Rather than having a single train/test split of the dataset or cross-validation, which did not work well in practice, we sampled with replacement 30 random collections. Then, each collection was divided into 70% of the questions for training (with their respective tweets/threads) and 30% for testing.

We evaluated different combinations of sets of features in every experiment. For each of these combinations we computed $MRR@10$ and $nDCG@10$. In each case we report the mean value over the 30 bootstrapped collections. We ran the experiments using the default *Ranklib* parameters for the LTR methods used: MART, Ranknet, RankBoost, and LambaMART.

Feature set selection. We propose the following heuristic to find the best combination of features, where f_i is a feature set (e.g. POS features), $F = \{f_1, f_2, \ldots f_n\}$ is the set that contains all of our feature sets, and PBC (initially, $PBC = \emptyset$) is the partial best feature set combination:

1. We run the factoid task evaluation for each feature set in F using each LTR model.
2. We choose the feature set f_i^* which produces the best MRR@10 and add it to the set PBC (i.e., $PBC = PBC \cup f_i^*$) and we remove f_i^* from F (i.e. $F = F - f_i^*$).
3. We run again the same evaluation using the resulting PBC in combination with each remaining feature in F (i.e., $(PBC \cup f_1), (PBC \cup f_2) \ldots (PBC \cup f_n)$).
4. We repeat the process from the step (2) until there is no significant improvement in the MRR@10 value.

Single feature results. Table 1 presents the results of each LTR model trained on different types of features. These results show that features obtained from the text of the messages (POS, WEMB_THR, CONTENT) yield good results compared to, for instance, relying solely on social signals such as replies, likes or retweets (SOCIAL). The single most predictive feature set for ranking answers to factoid questions is *Part-of-Speech* (POS), which significantly outperforms all the other features.

Feature combination results. Table 2 shows the results of several experiments combining feature sets in the LTR framework. The table shows the percent of improvement over the best performing feature set POS (MRR@10 = 0.6587), and

---

[12] https://sourceforge.net/p/lemur/wiki/RankLib/.

Table 1. Factoid task MRR@10 results, mean (μ) and S.D. (σ). POS$^{(2-12)}$ means that POS is significantly better than feature sets 2 (WEMB_THR) to 12 (REPW).

	Feature set	MART		Ranknet		RankBoost		LambdaMart	
		μ	σ	μ	σ	μ	σ	μ	σ
1	POS ($^{2-12}$)	**0.6587**	0.0250	0.5862	0.0266	**0.6730**	0.0377	0.6213	0.0617
2	WEMB_THR ($^{3-12}$)	**0.6202**	0.0296	0.5489	0.0315	0.6013	0.0264	0.5618	0.0388
3	CONTENT ($^{4-12}$)	0.5763	0.0284	0.5694	0.0320	0.5543	0.0230	0.5900	0.0330
4	D_TFIDF_1 ($^{9-12}$)	0.5282	0.0286	0.5299	0.0349	0.5143	0.0311	0.4966	0.0407
5	SOCIAL ($^{8-12}$)	0.5280	0.0284	0.5490	0.0265	0.4766	0.0296	0.5311	0.0424
6	D_WEMB (^{9,11,12})	0.5131	0.0303	0.5278	0.0313	0.5155	0.0337	0.5105	0.0341
7	D_TFIDF_3 (^{9,11,12})	0.5123	0.0331	0.4057	0.0262	0.4716	0.0353	0.4075	0.0280
8	D_TFIDF_2 (^{9,11,12})	0.5083	0.0303	0.4457	0.0277	0.4870	0.0315	0.4338	0.0254
9	USERS	0.4857	0.0223	0.5344	0.0296	0.4883	0.0278	0.5376	0.0503
10	WEMB_Q	0.4942	0.0258	0.4942	0.0258	0.4942	0.0258	0.4942	0.0258
11	TIME	0.4815	0.0428	0.5150	0.0303	0.4942	0.0258	0.5560	0.0315
12	REPW	0.4810	0.0326	0.3651	0.0347	0.4929	0.0328	0.4051	0.0677

Significant differences based on MART pairwise t-tests, $\alpha = .95$, Bonferroni correction.

Table 2. Factoid task, best combinations of features sets, based on MRR@10, and their percent of improvement over the best single feature set (POS).

Combination	MART	Ranknet	RankBoost	LambdaMart
POS	0.6587	0.5862	0.6730	0.6213
POS+D_TFIDF_1	0.6917 ↑ 5%	0.5953	0.6746	0.6200
POS+D_TFIDF_1+SOCIAL	0.7514 ↑ 14%	0.5931	0.6719	0.6361
POS+D_TFIDF_1 +SOCIAL+WEMB_Q	0.7682 ↑ 17%	0.5946	0.6719	0.6464
POS+D_TFIDF_1+SOCIAL+ WEMB_Q+D_TFIDF_3	0.7745 ↑ 18%	0.5904	0.6732	0.6204
POS+D_TFIDF_1+SOCIAL+ WEMB_Q+D_TFIDF_3+REPW	0.7788 ↑ 18%	0.5895	0.6733	0.6415
POS+D_TFIDF_1+SOCIAL+ WEMB_Q+D_TFIDF_3+REPW+TIME	**0.7795** ↑ 18%	0.5867	0.6755	0.6420

we show that a combination with content (D_TFIDF_1, WEMB_Q, D_TFIDF_3, REPW), social and time feature sets can increase the performance up to 18.3% (MRR@10 = 0.7795), showing that these features provide different types of signals for the ranking task.

Methods. Considering both evaluations –on each feature set and over combinations– the best method was MART, specially in the feature set combination results of Table 2. Although LambdaMart is usually presented as the state of the art, there is also recent evidence on non-factoid QA showing MART as the top performing algorithm [25], in line with our results. Notably, all the methods show a strongly correlated behavior in terms of feature set ranking, for the three of them present their best MRR@10 results with the POS feature and their worst

Table 3. Left: Results of our best combination vs. baselines. We improve over Twitter search up to 74.77% (MRR@10) and 29.4% (nDCG@10). **Right:** Datasets description of non-factoid and factoid QA. Size differences justify the need for transfer learning.

Method	MRR@10	nDCG@10
BM25	0.3852	0.4793
Twitter Search	0.4460	0.5625
REPW	0.4810	0.4616
Best comb.	**0.7795**	**0.7279**

	Non-fact	Fact
# of questions	40	491
# of tweets	2,666	63,646
# of threads	386	33,873
% tweets that are part of a thread	87.99%	46,7%
Avg. replies per thread	3.32	0.9

results with the REPW feature (with the exception of RankBoost), as shown in Table 1. This consistent behavior underpins our conclusions in terms of the importance of POS for this task.

Baselines. Results in Table 3 (left). Our best combined LTR model beats all baselines, improving the factoid ranking results by 74.77% in terms of MRR@10 and by 29.4% on nDCG@10 over Twitter search.

4.2 Non-factoid QA Evaluation Task

Non-factoid questions can have more than one answer, and they are usually associated to questions that require opinions, recommendations, or communicate experiences. For example, some non-factoid questions found in Twitter are: "anyone have any GIF maker software or sites they can recommend?" and "Anyone have a remedy for curing a headache?". To perform a preliminary evaluation of our approach for this task, we sampled 40 diverse non-factoid questions from Twitter. We focused on these questions because they represent an important portion (30%) [17] of the questions asked by users and they can be retrieved in a simple way (i.e., we retrieved these questions searching for "recommend ?"). We then obtained candidate answers for each question in the same way as in our factoid task (Sect. 4.1). Next, we used transfer learning (i.e., we use our best factoid QA LTR model) to rank answers for the non-factoid task. The differences in sizes of our datasets, shown in Table 3 (right), justify transferring our existing model, rather than learning a new one from non-factoid data.

Unlike the TREC dataset of factoid questions, we do not have the ground truth of correct answers. Hence, we manually inspected and evaluated the top-15 answers ranked with our approach for each of the 40 non-factoid questions, labeling them as relevant and non-relevant. Table 3 (right) shows the characteristics of both the factoid and non-factoid datasets.

Results. We obtained a $MRR@10 = 0.5802$, which is good compared to results reported recently –MRR = [0.4–0.45] in [25]–, but suboptimal compared to what we obtained in factoid QA task. By further analyzing the data we found that, in average, for every question we retrieved 1.5 threads without any reply, which were also non relevant to the question made. Based on this, we discarded from potential answers those threads without replies (or single tweets). This strategy improved results, $MRR@10 = 0.6675$, with the small trade-off of one question out of 40 for which we could not find answers.

Discussion and Limitations. An interesting result of our evaluation was the high predictive power of part-of-speech features. Previous work on community QA conducted on a large dataset of Yahoo Answers [15] found similar results. We also found a good discriminative effect of coordinating conjunctions (and, but, or, so). Hence, stopwords should not be removed for POS tagging in the factoid QA task using microblog data. However, we also found differences with prior findings on cQA [15], which observed that punctuation was a discriminative feature between relevant and non-relevant answers, unlike our study of QA in microblogs. This is expected since the short nature of microblog text makes people less likely to use punctuation. In addition, Molino et al. [15] used a feature similar to D_WEMB (i.e., the distance between the query and the candidate answers based on word2vec), but this neural-based word embedding feature did not perform well, ranking in 30th place among other features. In our evaluation, we used distances but also the word embeddings directly as features, which yielded excellent results, ranking as the 2nd most important feature set. This indicates that for microblog QA it is better to use the values of the embedded dimensions as features. Our manual inspection of results indicates that transfer learning can be a potential way to perform non-factoid QA, by using a model pre-trained for factoid QA.

A limitation in our work is that, for factoid QA, we could only find answers for about 40% of the questions in our ground-truth dataset. We note our initial factoid dataset, based on TREC challenges, does not have topics related to current events, which are much more likely to be discussed in Twitter [11]. This time gap between our ground-truth questions and our candidate answers, can very likely explain why we were unable to find matching tweets for an important number of questions. Another limitation, which we plan to address in the future, is that we did not study the occurrence of *incorrect answers* within relevant threads.

5 Conclusion and Future Work

In this work we investigated the feasibility of conducting QA using microblog data. We studied several sets of features, ranking methods and we performed a quantitative evaluation on a factoid QA dataset, as well as an informative evaluation with non-factoid questions. Our results validate the potential for using microblog data for factoid and non-factoid QA, identifying the most informative features as well as the best LTR model. In future work we expect to conduct a larger evaluation on non-factoid questions, using new features, and performing a deeper analysis on the effect of certain attributes.

References

1. Burges, C., Shaked, T., Renshaw, E., Lazier, A., Deeds, M., Hamilton, N., Hullender, G.: Learning to rank using gradient descent. In: Proceedings of ICML 2005, pp. 89–96 (2005)
2. Büttcher, S., Clarke, C.L.A., Cormack, G.V.: Information Retrieval -Implementing and Evaluating Search Engines. MIT Press, Cambridge (2010)
3. Duan, Y., Jiang, L., Qin, T., Zhou, M., Shum, H.Y.: An empirical study on learning to rank of Tweets. In: Proceedings of COLING 2010, pp. 295–303 (2010)
4. Efron, B., Tibshirani, R.J.: An Introduction to the Bootstrap. CRC Press, Boca Raton (1994)
5. Freund, Y., Iyer, R., Schapire, R.E., Singer, Y.: An efficient boosting algorithm for combining preferences. J. Mach. Learn. Res. **4**, 933–969 (2003)
6. Friedman, J.H.: Greedy function approximation: a gradient boosting machine. Ann. Stat. **29**(5), 1189–1232 (2001)
7. Herrera, J., Poblete, B., Parra, D.: Retrieving relevant conversations for Q&A on Twitter. In: Proceedings of ACM SIGIR (Workshop of SPS) (2015)
8. Honey, C., Herring, S.C.: Beyond microblogging: conversation and collaboration via Twitter. In: Proceedings of HICSS 2009, pp. 1–10 (2009)
9. Java, A., Song, X., Finin, T., Tseng, B.: Why we Twitter: understanding microblogging usage and communities. In: Proceedings of WebKDD/SNA-KDD 2007, pp. 56–65 (2007)
10. Jurafsky, D., Martin, J.H.: Speech and Language Processing. Prentice Hall, Pearson Education International (2014)
11. Kwak, H., Lee, C., Park, H., Moon, S.: What is Twitter, a social network or a news media? In: Proceedings of WWW 2010, pp. 591–600 (2010)
12. Liu, Z., Jansen, B.J.: A taxonomy for classifying questions asked in social question and answering. In: Proceedings of CHI EA 2015, pp. 1947–1952 (2015)
13. Manning, C.D., Raghavan, P., Schütze, H.: Introduction to Information Retrieval. Cambridge University Press, Cambridge (2008)
14. Mikolov, T., Chen, K., Corrado, G., Dean, J.: Efficient estimation of word representations in vector space. In: Proceedings of ICLR (2013)
15. Molino, P., Aiello, L.M., Lops, P.: Social question answering: textual, user, and network features for best answer prediction. ACM TOIS **35**, 4–40 (2016)
16. Morris, M.R., Teevan, J., Panovich, K.: A comparison of information seeking using search engines and social networks. In: Proceedings of ICWSM 2010, pp. 23–26 (2010)
17. Morris, M.R., Teevan, J., Panovich, K.: What do people ask their social networks, and why?: a survey study of status message Q&A behavior. In: Proceedings of CWSM 2010, pp. 1739–1748 (2010)
18. Owoputi, O., O'Connor, B., Dyer, C., Gimpel, K., Schneider, N., Smith, N.A.: Improved part-of-speech tagging for online conversational text with word clusters. In: Proceedings of ACL (2008)
19. Paul, S.A., Hong, L., Chi, E.H.: Is Twitter a good place for asking questions? a characterization study. In: Proceedings of CWSM 2010, pp. 578–581 (2011)
20. Raban, D.R.: Self-presentation and the value of information in Q&A websites. JASIST **60**(12), 2465–2473 (2009)
21. Sriram, B.: Short text classification in Twitter to improve information filtering. In: Proceedings of ACM SIGIR 2010. ACM (2010)

22. Surdeanu, M., Ciaramita, M., Zaragoza, H.: Learning to rank answers on large online QA collections. In: Proceedings of ACL 2008, pp. 719–727 (2008)
23. Surdeanu, M., Ciaramita, M., Zaragoza, H.: Learning to rank answers to non-factoid questions from web collections. Comput. Linguist. **37**(2), 351–383 (2011)
24. Wu, Q., Burges, C.J.C., Svore, K.M., Gao, J.: Adapting boosting for information retrieval measures. Inf. Retrieval **13**(3), 254–270 (2010)
25. Yang, L., Ai, Q., Spina, D., Chen, R.C., Pang, L., Croft, W.B., Guo, J., Scholer, F.: Beyond factoid QA-effective methods for non-factoid answer sentence retrieval. In: Proceedings of ECIR (2016)
26. Zhao, Z., Mei, Q.: Questions about questions: an empirical analysis of information needs on Twitter. In: Proceedings of WWW 2013, pp. 1545–1556 (2013)

Short Papers

Biomedical Question Answering via Weighted Neural Network Passage Retrieval

Ferenc Galkó and Carsten Eickhoff[✉]

Department of Computer Science, ETH Zurich, Zurich, Switzerland
{ferenc.galko,carsten.eickhoff}@inf.ethz.ch

Abstract. The amount of publicly available biomedical literature has been growing rapidly in recent years, yet question answering systems still struggle to exploit the full potential of this source of data. In a preliminary processing step, many question answering systems rely on retrieval models for identifying relevant documents and passages. This paper proposes a weighted cosine distance retrieval scheme based on neural network word embeddings. Our experiments are based on publicly available data and tasks from the BioASQ biomedical question answering challenge and demonstrate significant performance gains over a wide range of state-of-the-art models.

Keywords: Biomedical question answering · Passage retrieval

1 Introduction

Biomedical question answering is a key task in catering to clinical decision support and personal health information needs. Finding useful information in the extensive collections of scholarly biomedical articles poses a challenge to highly trained practitioners and patients alike [5]. Biomedical questions phrased by experts are usually more specific than common Web search queries, making them difficult to satisfy using general-purpose retrieval models. A typical expert question (for example *"Which enzymes synthesize catecholamines in adrenal glands?"*) shows specific domain knowledge and aims at more than just a general overview about adrenal glands [15]. Instead, the expert is looking for a narrow set of documents or snippets that help answering a specific relation of the queried entities. Since many question answering systems rely heavily on document and passage retrieval, an increased performance in these tasks tends to propagate into significant QA performance gains [13, 16]. While the task of whole-document retrieval for QA is well understood, passage-level retrieval approaches have received comparably less attention [4, 8, 10, 13].

This paper describes a piece of work in progress, focusing on passage-level retrieval for biomedical question answering on the basis of weighted combinations of neural network word embeddings. It makes the following contributions:

© Springer International Publishing AG, part of Springer Nature 2018
G. Pasi et al. (Eds.): ECIR 2018, LNCS 10772, pp. 523–528, 2018.
https://doi.org/10.1007/978-3-319-76941-7_39

(1) We propose a novel approach for weighting query embedding vectors for the cosine distance text matching scheme and show that it outperforms traditional models. (2) We demonstrate significant performance improvements over state-of-the-art neural network retrieval models on a sizable publicly available benchmarking dataset.

The remainder of this paper is structured as follows: Sect. 2 formally introduces the problem domain as well as the proposed method. Section 3 empirically evaluates the merit of our method on the basis of publicly available data and tasks created in the context of the BioAsq biomedical question answering challenge. Finally, Sect. 4 concludes with a summary and an outlook on future directions.

2   Method

Let us assume a textual question Q consisting of individual terms, where each term is represented by a fixed-length vector. To satisfy Q, we rely on a large collection of biomedical documents organized in passages. In order to represent and retain the semantic content of the terms, we project them into semantically meaningful vector spaces, such as induced by Word2Vec [6] or GloVe [11]. In this way both the question $Q = q_1, q_2, \ldots, q_i$ and each passage $P = p_1, p_2, \ldots, p_j$ can be represented as a sequence of fixed-length vectors. In this high-dimensional space, we can now relate individual terms to each other by means of distance functions. A commonly employed measure of relatedness is the cosine of the angle between word vectors as expressed by the cosine distance.

$$cos(q, p) = 1 - \frac{qp}{\|q\|_2 \|p\|_2}$$

Given such a representation as well as measure of similarity, we can now perform a variety of operations ranging from clustering to retrieval [1,3]. In order to represent multi-term units such as entire queries and passages, several authors recommend uniform averaging of word vectors before finally returning a ranked top-k list of closest passages to the query.

$$CD(Q, P) = 1 - \frac{(\frac{1}{|Q|} \sum q_i)(\frac{1}{|P|} \sum p_j)}{\|\frac{1}{|Q|} \sum q_i\|_2 \|\frac{1}{|P|} \sum p_j\|_2}$$

2.1   Weighted Cosine Distance

One serious issue with the well-known cosine distance retrieval approach introduced above is that the uniform average vector is a poor semantic representation for general texts. This is due to the presence and abundance of stop words such as *"the"*, *"a"*, or *"is"* that carry little meaning. By assigning uniform weights to all words, we water down the semantic content of informative words such as *"neurodegenerative"*. To address this problem, non-uniform weighting schemes such as idf can be used. A term's collection-wide inverse document frequency (idf) captures its uniqueness and assigns larger weights to rarer words on a logarithmic scale. By assigning idf instead of uniform weights to the words, a substantial increase in performance can be achieved as we will show in Sect. 3.

$$CD_{idf}(Q, P) = 1 - \frac{(\frac{1}{\sum idf(q_i)} \sum idf(q_i)q_i)(\frac{1}{\sum idf(p_j)} \sum idf(p_j)p_j)}{\|\frac{1}{\sum idf(q_i)} \sum idf(q_i)q_i\|_2 \|\frac{1}{\sum idf(p_j)} \sum idf(p_j)p_j\|_2}$$

While there are alternative approaches for aggregation such as position encoding [14], that can take into consideration the ordering of words, a series of preliminary studies suggest that idf weights perform best on the biomedical QA task. For the sake of brevity, we do not include these experiments in our empirical performance evaluation.

2.2   Adjusted idf Weights

The previously presented idf weights depend solely on the distribution of words in the corpus. Incoming queries, on the other hand, may originate from a different distribution. Using idf scores generated from the document collection to weight query terms might result in a poor representation, since words that appear rarely in documents but frequently in questions can receive an unduly large weight. Question words such as *"what"*, *"when"*, *"where"* are intuitive examples. When idf weights are calculated on a sizable sample of scholarly biomedical articles obtained from PubMed, the weight for *"what"* is 5.05, whereas *"disease"* (2.68), *"protein"* (2.59) and *"artery"* (4.16) end up being much less important despite their greater *de facto* informativeness.

As a consequence, passages containing the word *"what"* will be estimated to be more similar to the question *"What is a degenerate protein?"*, according to our metric, than passages containing the more promising phrase *"degenerate protein"*. In fact, if we were to perform idf-based stopping and only retain the most important components of a question, our example question *"What is a degenerate protein?"* becomes "What degenerate?" with the low-idf component *"is"*, *"a"* and *"protein"* being removed while the desired reduction in this case may have been *"degenerate protein?"* which captures the topical essence of the original question much better.

We address this issue by generating idf weights from an alternative collection, specifically a mixture of large-scale corpora of biomedical [15] and general questions [12]. Generating idf weights from the combined question set results in smaller weights for general terms such as *"what"*, *"which"*, *"when"* and larger weights for rarer, domain-specific terminology such as *"protein"*, *"disease"*, or *"artery"*, making it possible to capture the true intention of bio-medical questions and passages. We refer to this method as CD_q.

3   Experiments

Our empirical performance evaluation is based on documents and questions from the BioASQ 2017 challenge's document and snippet retrieval tasks [7]. The goal in this task is to return the 10 most relevant passages from a collection of 12.8M PubMed abstracts for a specific biomedical question. A training set of 1799 manually curated questions along with relevant passages are provided by the challenge organizers. The test set is comprised of a separate set of 500 questions organized in five equally-sized batches.

3.1   Baselines

To allow for a meaningful system comparison, we include a broad range of traditional as well as state-of-the-art performance baselines, trained and evaluated on the same data and task as our proposed cosine-distance based methods.

RND. This approach returns a random passage from the reference set of highly-ranked documents to create a weak baseline.

MLP. This approach uses position-encoded sentence embeddings that are concatenated and eventually classified in binary fashion in a multi-layer perceptron [4]. This is a re-implementation of a system participating in a previous BioASQ challenge.

MP. The Match Pyramid [9] model generates a similarity matrix from the pairwise word interactions between question and candidate passage. We use a fixed 30 by 30 matrix with zero padding and the identity function as a measure of similarity. In a second step, convolutional filters condense the interaction matrix into a final vector representation for classification.

DRMM. The Deep Relevance Matching Model [2] is based on a similar scheme. It computes pair-wise term interactions between question and candidate passage that are then flattened into fixed-size histograms and discretized and weighted to give the final vector representation.

3.2   Experimental Setup

All (machine learning) methods are trained using five-fold cross validation on the training set. Word embeddings for all methods are computed as length-50 word2vec vectors on the PubMed document corpus [7]. For each question (training and test), a reference set of highly ranked documents is given by the challenge organizers. We split these documents into individual sentences that will serve as our retrieval unit. Negative training examples for machine learning methods are randomly sampled from arbitrary non-relevant documents.

3.3   Results

Table 1 compares the various baselines and cosine distance variants in terms of mean average precision (MAP), Precision, Recall and F_1-scores each at a cut-off rank of 10 retrieved passages. Statistical significance of method differences is determined using a Wilcoxon signed-rank test at $\alpha < 0.05$-level and significant improvements over all baselines are indicated with an asterisk. While questions were originally grouped in batches, here we forego this structure in the interest of

Table 1. Passage retrieval performance for biomedical question answering.

Method	MAP	Precision	Recall	F_1
RND	0.190	0.190	0.289	0.229
MLP [4]	0.226	0.236	0.352	0.282
MP [9]	0.344	0.323	0.470	0.383
DRMM [2]	0.348	0.344	0.510	0.411
CD	0.341	0.339	0.484	0.399
CD_{idf}	0.344	0.348	0.487	0.406
CD_q	0.377*	0.374*	0.519	0.434*

brevity. Batch-level scores showed some variance but displayed the same relative method ranking as the aggregate overview.

Due to the limited length of the reference document list as well as each individual abstract, random sentence selection does surprisingly well and sets a lower limit to method performance. All compared approaches yield meaningful results and significantly outperform this baseline. MLP is the weakest machine learning approach with substantially lower performance scores than those achieved by MP and DRMM. Cosine distance rankings are clearly improved by idf term weighting, lifting their results on a level comparable to that of MP and DRMM. Our adjusted term-weighting scheme following statistics of a separate, more representative, question corpus introduces another improvement in result quality, leading to the best overall results and a significant improvement over all contesting methods.

4   Conclusion

This paper presents a piece of work in progress towards passage retrieval for biomedical question answering via weighted cosine distances. In place of highly parametric end-to-end ranking networks, we devise a number of lean non-parametric weighting schemes that account for the differences in term distribution between document and question corpora. Our experiments on publicly available BioASQ data demonstrate significant improvements over a range of ranking networks. Especially in academic settings where datasets are often not sufficiently large to robustly fit multitudes of neural network parameters, such light-weight architectures are of increased interest.

While we noted the significant difference in term distributions between corpora at a biomedical example, the solution, as such, is not specific to the biomedical domain. In the future, we aim to investigate more formally rigorous ways of accounting for such differences as well as to evaluate them on a wider range of topical domains.

References

1. Brokos, G.-I., Malakasiotis, P., Androutsopoulos, I.: Using centroids of word embeddings and word mover's distance for biomedical document retrieval in question answering. In: Proceedings of the 15th ACL Workshop on Biomedical Natural Language Processing (2016)
2. Guo, J., Fan, Y., Ai, Q., Croft, W.B.: A deep relevance matching model for ad-hoc retrieval. In: CIKM 2016. ACM (2016)
3. Huang, A.: Similarity measures for text document clustering. In: Proceedings of the sixth New Zealand Computer Science Research Student Conference (NZC-SRSC2008), Christchurch, New Zealand, pp. 49–56 (2008)
4. Lee, H.-G., Kim, M., Kim, H., Kim, J., Kwon, S., Seo, J., Choi, J., Kim, Y.-R.: KSAnswer: question-answering system of Kangwon national university and Sogang university in the 2016 BioASQ challenge. In: ACL 2016, p. 45 (2016)
5. Malakasiotis, P., Androutsopoulos, I., Bernadou, A., Chatzidiakou, N., Papaki, E., Constantopoulos, P., Pavlopoulos, I., Krithara, A., Almyrantis, Y., Polychronopoulos, D., Kosmopoulos, A., Balikas, G., Partalas, I., Tsatsaronis, G., Heino, N.: Challenge evaluation report 2 and roadmap. European Commission Report (2014)
6. Mikolov, T., Sutskever, I., Chen, K., Corrado, G.S., Dean, J.: Distributed representations of words and phrases and their compositionality. In: Advances in Neural Information Processing Systems, pp. 3111–3119 (2013)
7. Nentidis, A., Bougiatiotis, K., Krithara, A., Paliouras, G., Kakadiaris, I.: Results of the fifth edition of the BioASQ challenge. In: BioNLP 2017, pp. 48–57. Association for Computational Linguistics, Vancouver, Canada, August 2017
8. Yang, Z., Zhou, Y., Nyberg, E.: Learning to answer biomedical questions: OAQA at BioASQ 4B. In: ACL 2016, p. 23 (2016)
9. Pang, L., Lan, Y., Guo, J., Xu, J., Wan, S., Cheng, X.: Text Matching as Image Recognition. CoRR, abs/1602.06359 (2016)
10. Papagiannopoulou, E., Papanikolaou, Y., Dimitriadis, D., Lagopoulos, S., Tsoumakas, G., Laliotis, M., Markantonatos, N., Vlahavas, I.: Large-scale semantic indexing and question answering in biomedicine. In: Proceedings of the Fourth BioASQ Workshop, pp. 50–54 (2016)
11. Pennington, J., Socher, R., Manning, C.: Glove: global vectors for word representation. In: Proceedings of the 2014 Conference on Empirical Methods in Natural Language Processing (EMNLP), pp. 1532–1543 (2014)
12. Rajpurkar, P., Zhang, J., Lopyrev, K., Liang, P.: SQuAD: 100,000+ questions for machine comprehension of text. In: Proceedings of the Conference on Empirical Methods on Natural Language Processing (EMNLP) (2016)
13. Schulze, F., Schüler, R., Draeger, T., Dummer, D., Ernst, A., Flemming, P., Perscheid, C., Neves, M.: HPI question answering system in BioASQ 2016. In: Proceedings of the Fourth BioASQ workshop at the Conference of the Association for Computational Linguistics, pp. 38–44 (2016)
14. Sukhbaatar, S., Weston, J., Fergus, R., et al.: End-to-end memory networks. In: Advances in Neural Information Processing Systems, pp. 2440–2448 (2015)
15. Tsatsaronis, G., Balikas, G., Malakasiotis, P., Partalas, I., Zschunke, M., Alvers, M.R., Weissenborn, D., Krithara, A., Petridis, S., Polychronopoulos, D., et al.: An overview of the BioASQ large-scale biomedical semantic indexing and question answering competition. BMC Bioinform. 16(1), 138 (2015)
16. Voorhees, E.M.: The TREC question answering track. Nat. Lang. Eng. 7(4), 361–378 (2001)

Topical Stance Detection for Twitter: A Two-Phase LSTM Model Using Attention

Kuntal Dey[1], Ritvik Shrivastava[2(✉)], and Saroj Kaushik[3]

[1] IBM Research, New Delhi, India
kuntadey@in.ibm.com
[2] NSIT, New Delhi, India
ritviks.it@nsit.net.in
[3] IIT Delhi, New Delhi, India
saroj@cse.iitd.ac.in

Abstract. The topical stance detection problem addresses detecting the stance of the text content with respect to a given topic: whether the sentiment of the given text content is in FAVOR of (positive), is AGAINST (negative), or is NONE (neutral) towards the given topic. Using the concept of attention, we develop a two-phase solution. In the first phase, we classify subjectivity - whether a given tweet is neutral or subjective with respect to the given topic. In the second phase, we classify sentiment of the subjective tweets (ignoring the neutral tweets) - whether a given subjective tweet has a FAVOR or AGAINST stance towards the topic. We propose a Long Short-Term memory (LSTM) based deep neural network for each phase, and embed attention at each of the phases. On the SemEval 2016 stance detection Twitter task dataset [7], we obtain a best-case macro F-score of 68.84% and a best-case accuracy of 60.2%, outperforming the existing deep learning based solutions. Our framework, T-PAN, is the first in the topical stance detection literature, that uses deep learning within a two-phase architecture.

1 Introduction

Twitter, a hotbed of user generated content, has recently found traction among the researchers for the problem of topical stance detection. Topical stance detection is the problem of finding whether a given tweet takes a FAVOR (positive), AGAINST (negative) or NONE (neutral) stance towards a given topic. It is at core of the opinion polarity detection and mining problem. The problem is useful to solve in several practical scenarios, such as detecting user stance towards aspects of political, economic and social events, understanding stance-specific information propagation behavior of users *etc.*

1.1 Related Work

Sentiment detection from user-generated content has been a long-standing problem [9]. However, stance detection, where the sentiment (opinion) of the user

© Springer International Publishing AG, part of Springer Nature 2018
G. Pasi et al. (Eds.): ECIR 2018, LNCS 10772, pp. 529–536, 2018.
https://doi.org/10.1007/978-3-319-76941-7_40

is not generic but with respect to a specific topic, has gained research attention only in recent times. A seminal work by Mohammad *et al.* [8], followed by a
SemEval 2016 task [7] conducted by the authors, resulted in starting wide-spread
research in the area.

Different models, including traditional machine learning approaches, genetic
algorithms, and deep learning approaches such as convolutional neural networks
(CNN), recurrent neural networks (RNN) and long short-term memory (LSTM),
were proposed in the SemEval 2016 topical stance detection contest. MITRE
[14] provided the best deep learning solution in the contest, initializing weights
from a 256-dimensional word embeddings learned using the word2vec skip-gram
algorithm [6], followed by a second layer with 128 LSTM units. Among others,
pkudblab [12] and DeepStance [11] use deep CNN models. Augenstein *et al.* [1]
employ a bidirectional attention model.

Some works used a two-phase approach. ECNU [15], in the first phase, determines whether a given tweet is relevant to a given target topic, and in the second
phase, detects orientation (favor/against). The work by ltl.uni-due [13] also uses
a two-level stacked classifier approach using Support Vector Machines (SVM).
Among others, TakeLab [2], mixed machine learning with genetic algorithms.
Other approaches, such as CU-GWU [4] and IUCL-RF [5], employed traditional
machine learning. A shared task has also been proposed recently [10].

The overall average values of F-scores, obtained by the task participants,
ranged from 46.19 at the lower end to all the way up to 67.82 at the higher end.
A recent work was conducted by Du *et al.* [3], the first of its kind that deeply
ingrained the stance words in the architecture and used attention modeling. It
outperformed the deep learning based approaches, attaining F-score of 68.79% as
against the deep-learning state of the art F-score of 67.82%. We further observe
that, the SemEval 2016 tasks were evaluated as a macro average of the F-score
for only the *favor* and *against*, ignoring the *none* (neutral) class. We, however,
perform accuracy measurements against all the three classes as well (in addition
to the F-score that we measure following the traditional literature), and show
that our model outperforms the best-known deep learning system not only for
two-class macro average F-score, but for a full three-class accuracy measure as
well.

1.2   Our Contributions

We propose a two-phase approach, using attention embedding at each phase
and encoding using LSTM. The given SemEval 2016 [7] dataset contains three
classes - FAVOR, AGAINST and NONE. Our work is based on the observation that
messages with neutral stances are usually non-subjective, while the ones with
favor and against stances are usually subjective. Thus, in the first phase of our
two-phase approach, we use a LSTM to detect subjectivity, and classify into
subjective (non-neutral) versus neutral (none). And in the second phase, we
use another LSTM to detect sentiment (favor/against) of the tweets that were
labeled subjective in the earlier phase. Akin to the philosophy of Du *et al.* [3],
we also use an attention model, and deeply embed the topical attention as part

of the input to the classifier. Since a given tweet does not necessarily contain the topic against which the stance is sought for, this step plays an important role in transforming the learning into a topic-specific learning. This is absent in the literature except for Du *et al.* [3]. Our model thus is the first of its kind, that uses a two-phase LSTM-based architecture with attention embedding ingrained.

The contributions of our work are the following.

- We propose T-PAN, a two-phase attention-embedded LSTM-based approach for detecting stance of tweets towards given topics.
- In the first phase, we perform subjectivity analysis of the tweets, using a combination of LSTM and attention embedding.
- In the second phase, we perform sentiment analysis on the subjective tweets, again using a combination of LSTM and attention embedding.
- Empirically, on the SemEval 2016 benchmark dataset, we demonstrate the effectiveness of our system. Our model is novel, and we outperform the deep learning based literature in terms of accuracy (60.2% against 58.7%), as well as F-score (68.84% against 68.79%).

2   Central Idea

2.1   Approach Overview

Table 1 shows a few randomly chosen samples from the training set across topics, to provide the reader with an intuition of the data available. As mentioned earlier, our task comprises of three classes of data: FAVOR, AGAINST and NONE. While *favor* and *against* tweets are often subjective in nature, the *neutral* tweets often are non-subjective. The architecture of our system is presented on Fig. 1.

Our model is a two-phase one. At each phase, there are two components - a bi-directional LSTM and an attention mechanism. The bi-directional LSTM

Table 1. Random examples of tweets of the different stances, for a few of the given target topics

Target	Tweet	Stance
Examples from the **favor** *stance*		
Atheism	Everyone is able to believe in whatever they want. #Freedom	FAVOR
Feminist Movement	@OliviaJeniferx it's not always the guys job. #equality	FAVOR
Examples from the **against** *stance*		
Atheism	Be still. Be patient. Watch and let God work	AGAINST
Feminist Movement	Friendly reminder that the "Gender Pay Gap" is a myth	AGAINST
Examples from the **none** *stance*		
Atheism	Alot of angry people in this world. Peace to all. #love	NONE
Feminist Movement	@sass_unicorn lol! Young male children for	NONE

is used for feature encoding. The attention logic uses augmentation of the word embeddings with target topics, and subsequently passes it through a linear layer for computing attention of each word in the text in the context of the topic under consideration.

2.2 Embedding Augmentation with Target Topics and Determining Attention

To compute attention, we augment the embedding of the constituent words with the average embedding of the target. If the words in a given target topic comprises of word embeddings $\{\tilde{z_1}, \tilde{z_2}, ..., \tilde{z_n}\}$, then we compute the embedding of the target topic $\tilde{z}$ as $\tilde{z} = \frac{\sum_{i=1}^{n} \tilde{z_i}}{|n|}$. The words within the sentence, that have the embeddings $\{z_1, z_2, ..., z_m\}$ of dimension d_z, are thus augmented with dimension $d_{\tilde{z}}$ (the dimension of $\tilde{z}$), and each word gets a new embedding dimension of $d_z + d_{\tilde{z}}$. This is processed as depicted in Fig. 1, by first passing via a linear layer followed by a softmax, and subsequently ingraining the attention derived for each word into the LSTM-encoded features, using a product of the LSTM-embedded features and the output of the linear layer. We note that, while our approach is largely different from Du et al. [3] in terms of the overall system architecture (our approach is two-phase while theirs is one-phase), the philosophy of augmenting each word of the sentence with the average embedding of the target topic words is similar.

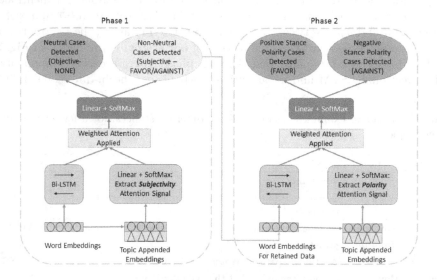

Fig. 1. System architecture diagram

2.3   Training the Models

Using a similar underlying architecture, the first phase is trained for subjectivity, and the second phase for sentiment polarity. Hence, the attention gets trained for subjectivity in the first phase and for polarity in the second phase. The subjective outputs of the first phase are passed through the second phase, while the rest (non-subjective) are assigned a class label NONE and kept aside. We try using both SGD (stochastic gradient descent) as well as Adam optimizers for experiments, and these yield similar effectiveness. We train our model using cross-entropy loss function. The loss of one phase is not propagated to the other.

3   Experiments

3.1   Data Description

We use the benchmark training and test data provided by the SemEval 2016 stance detection task [7]. For self-containment, we reproduce their data, in Table 2. We use the evaluation script they provide, for calculating F-score. Further, since their script only accounts for the FAVOR and AGAINST classes and computes a macro F-score as average of the two (ignoring the NONE class), we develop an additional script to calculate the accuracy using the three classes, as a ratio to the total number of correct predictions to the total test data size. We use PyTorch for programming. We perform data cleaning: net slang removal (for tweet normalization) using an online dictionary[1] and stopword removal using a Stanford NLP resource for stopword removal[2].

Table 2. Data for the SemEval 2016 stance detection task. Target C.C.C. → Climate Change is Concern. Target L.A. → Legalization of Abortion. Table courtesy: [7].

Target	#total	#train	% of instances in Train			#test	% of instances in Test		
			Favor	Against	Neither		Favor	Against	Neither
Atheism	733	513	17.9	59.3	22.8	220	14.5	72.7	12.7
C.C.C.	564	395	53.7	3.8	42.5	169	72.8	6.5	20.7
Feminist movement	949	664	31.6	49.4	19.0	285	20.4	64.2	15.4
Hillary Clinton	984	689	17.1	57.0	25.8	295	15.3	58.3	26.4
L.A.	933	653	18.5	54.4	27.1	280	16.4	67.5	16.1
All	4,163	2,914	25.8	47.9	26.3	1,249	24.3	57.3	18.4

---

[1] http://www.noslang.com/dictionary.

[2] https://nlp.stanford.edu/IR-book/html/htmledition/dropping-common-terms-stop-words-1.html.

3.2   Performance of Our Model T-PAN and Its Constituent Components

Our system delivers commendable performance for detecting the user stances towards the individual topics, as well as, a robust overall performance across the topics. We empirically observe the performance of our T-PAN model. We further examine the performance of different LSTM-based architectures that eventually are composed to develop our end-to-end framework. Table 3 provides the details of the performance attained by the full T-PAN model, as well as the impact of performance of the constituent LSTM blocks and configurations by systematic component ablation.

Table 3. Performance of the different underlying two-phase architectures.

Phase 1	Phase 2	Accuracy
Bi-LSTM	Bi-LSTM	57.08
Bi-LSTM + Tweet Cleaning	Bi-LSTM + Tweet Cleaning	57.61
Bi-LSTM	One-Phase Attention	59.32
Bi-LSTM	One-Phase Attention + Tweet Cleaning	57.53
Bi-LSTM + Tweet Cleaning	One-Phase Attention + Tweet Cleaning	59.85
One-Phase Attention	One-Phase Attention	**60.22**
One-Phase Attention + Tweet Cleaning	One-Phase Attention + Tweet Cleaning	**60.24** (T-PAN)
Our implementation of TAN [3]		58.76

Table 4. Comparing F-scores of different models. A part of the table has been replicated from Du *et al.* [3]. NBOW ← Neural Bag-of-Words. LSTM ← LSTM without target-specific embedding. $LSTM_E$ ← LSTM with target-specific embedding, by [3]. TOP Sem-Eval ← The best-reported systems in SemEval 2016. TAN ← The final output of [3]. T-PAN ← Our framework.

Target	NBOW	LSTM	$LSTM_E$	TOP Sem-Eval	TAN	T-PAN
Atheism	55.12	58.18	59.77	**61.47**	59.33	61.19
C.C.C	39.93	40.05	48.98	41.63	53.59	**66.27**
Feminist Movement	50.21	49.06	52.04	**62.09**	55.77	58.45
Hillary Clinton	55.98	61.84	56.89	57.67	**65.38**	57.48
L.A	55.07	51.03	60.34	57.28	**63.72**	60.21
Overall	60.19	63.21	66.24	67.82	68.79	**68.84**

3.3   Comparing Our System Against the Deep-Learning Literature

As observed in Table 4, our best system (the T-PAN model) outperforms the state of the art that uses deep neural networks for topical stance classification. Out of the five given classes, we perform the best in one class, the *TAN* model [3] outperforms us in two classes and the SemEval tasks perform better than our model (as well as better than the TAN model [3]) for the other two classes.

4   Conclusion

We proposed T-PAN, a two-phase LSTM-based model with attention embedding, for detecting user stance with respect to given topics on Twitter. First, we classified the tweets into two: neutral and non-neutral, where non-neutral comprised of favor and against stances. Second, we classified the tweets labeled as non-neutral in the first phase, into two - favor and against stances. In each phase, we encoded the input sentences in form of a sequence of words using a bi-directional LSTM, and attention embedding. We investigated the impact of embedding topical attention, as well as, the impact of different LSTM architectures, on our approach. We empirically demonstrated the robustness of our framework T-PAN, by delivering the highest-known performance among all the deep learning approaches. Our model is easy to implement, reusable and practicable.

References

1. Augenstein, I., Rocktäschel, T., Vlachos, A., Bontcheva, K.: Stance Detection with Bidirectional Conditional Encoding. arXiv preprint arXiv:1606.05464 (2016)
2. Boltuzic, F., Karan, M., Alagic, D., Šnajder, J.: Takelab at SemEval-2016 task 6: stance classification in tweets using a genetic algorithm based ensemble. In: SemEval, pp. 464–468 (2016)
3. Du, J., Xu, R., He, Y., Gui, L.: Stance classification with target-specific neural attention networks. In: IJCAI, pp. 3988–3994 (2017)
4. Elfardy, H., Diab, M.: CU-GWU perspective at SemEval-2016 task 6: ideological stance detection in informal text. In: SemEval, pp. 434–439 (2016)
5. Liu, C., Li, W., Demarest, B., Chen, Y., Couture, S., Dakota, D., Haduong, N., Kaufman, N., Lamont, A., Pancholi, M., et al.: IUCL at SemEval-2016 task 6: an ensemble model for stance detection in twitter. In: SemEval, pp. 394–400 (2016)
6. Mikolov, T., Chen, K., Corrado, G., Dean, J.: Efficient Estimation of Word Representations in Vector Space. arXiv preprint arXiv:1301.3781 (2013)
7. Mohammad, S.M., Kiritchenko, S., Sobhani, P., Zhu, X., Cherry, C.: SemEval-2016 task 6: detecting stance in tweets. In: Proceedings of SemEval, vol. 16 (2016)
8. Mohammad, S.M., Sobhani, P., Kiritchenko, S.: Stance and Sentiment in Tweets. arXiv preprint arXiv:1605.01655 (2016)
9. Rosenthal, S., Ritter, A., Nakov, P., Stoyanov, V.: SemEval-2014 task 9: sentiment analysis in twitter. In: SemEval 2014, pp. 73–80 (2014)
10. Taulé, M., Martí, M.A., Rangel, F.M., Rosso, P., Bosco, C., Patti, V., et al.: Overview of the task on stance and gender detection in tweets on Catalan independence at IberEval 2017. In: IberEval, CEUR-WS, vol. 1881, pp. 157–177 (2017)
11. Vijayaraghavan, P., Sysoev, I., Vosoughi, S., Roy, D.: Deepstance at SemEval-2016 task 6: detecting stance in tweets using character and word-level CNNs. arXiv preprint arXiv:1606.05694 (2016)
12. Wei, W., Zhang, X., Liu, X., Chen, W., Wang, T.: pkudblab at SemEval-2016 task 6: a specific convolutional neural network system for effective stance detection. In: SemEval, pp. 384–388 (2016)
13. Wojatzki, M., Zesch, T.: ltl.uni-due at SemEval-2016 task 6: stance detection in social media using stacked classifiers. In: SemEval, pp. 428–433 (2016)

14. Zarrella, G., Marsh, A.: Mitre at SemEval-2016 Task 6: Transfer Learning for Stance Detection. arXiv preprint arXiv:1606.03784 (2016)
15. Zhang, Z., Lan, M.: ECNU at SemEval-2016 task 6: relevant or not? supportive or not? a two-step learning system for automatic detecting stance in tweets. In: SemEval, pp. 451–457 (2016)

A Neural Passage Model for Ad-hoc Document Retrieval

Qingyao Ai$^{(\boxtimes)}$, Brendan O'Connor, and W. Bruce Croft

College of Information and Computer Sciences, University of Massachusetts Amherst,
Amherst, MA 01003-9264, USA
{aiqy,brenocon,croft}@cs.umass.edu

Abstract. Traditional statistical retrieval models often treat each document as a whole. In many cases, however, a document is relevant to a query only because a small part of it contain the targeted information. In this work, we propose a neural passage model (NPM) that uses passage-level information to improve the performance of ad-hoc retrieval. Instead of using a single window to extract passages, our model automatically learns to weight passages with different granularities in the training process. We show that the passage-based document ranking paradigm from previous studies can be directly derived from our neural framework. Also, our experiments on a TREC collection showed that the NPM can significantly outperform the existing passage-based retrieval models.

Keywords: Passage-based retrieval model · Neural network

1 Introduction

Ad-hoc retrieval refers to a key problem in Information Retrieval (IR) where documents are ranked according to their assumed relevance to the information need of a specific query formulated by users [1]. In the past decades, statistical models have dominated the research on ad-hoc retrieval. They assume that documents are samples of n-grams, and relevance between a document and a query can be inferred from their statistical relationships. To ensure the reliability of statistical estimation, most models treat each document as single piece of text.

There are, however, multiple reasons that motivate us not to treat a document as a whole. First, there are many cases where the document is relevant to a query only because a small part of it contains the information pertaining to the user's need. The ranking score between the query and these documents would be relatively low if we construct the model with statistics based on the whole documents. Second, because reading takes time, sometimes it is more desirable to retrieve a small paragraph that answers the query rather than a relevant document with thousands of words. For instance, we do not need a linux textbook to answer a query "linux copy command".

Given these observations, IR researchers tried to extract and incorporate relevance information from different granularities for ad-hoc document retrieval.

© Springer International Publishing AG, part of Springer Nature 2018
G. Pasi et al. (Eds.): ECIR 2018, LNCS 10772, pp. 537–543, 2018.
https://doi.org/10.1007/978-3-319-76941-7_41

One simple but effective method is to cut long documents into pieces and construct retrieval models based on small passages. Those passage-based retrieval models are able to identify the subtopics of a document and therefore capture the relevance information with finer granularities. Also, they can extract relevant passages from documents and provide important support for query-based summarization and answer sentence generation.

Nonetheless, the development of passage-based retrieval models is limited because of two reasons. First, as far as we know, there is no universal definition of passages in IR. Most previous studies extracted passages from documents with a fixed-length window. This method, however, is not optimal as the best passage size varies according to both corpus properties and query characteristics. Second, aggregating information from passages with different granularities is difficult. The importance of passages depends on multiple factors including the structure of documents and the clarity of queries. For example, Bendersky and Kurland [2] noticed that passage-level information is not as useful on highly homogeneous documents as it is on heterogeneous documents. A simple weighting strategy without considering these factors is likely to fail in practice.

In this paper, we focus on addressing these challenges with a unified neural network framework. Specifically, we develop a convolution neural network that extracts and aggregates relevance information from passages with different sizes. In contrast to previous passage-based retrieval models, our neural passage model takes passages with multiple sizes simultaneously and learns to weight them with a fusion network based on both document and query features. Also, our neural passage model is highly expressive as the state-of-the-art passage-based retrieval models can be incorporated into our model as special cases. We conducted empirical experiments on TREC collections to show the effectiveness of the neural passage model and visualized the network weights to analyze the effect of passages with different granularities.

2 Related Work

Passage Extraction. Previous studies have explored three types of passage definitions: structure-based, topic-based and window-based passages. Structure-based passage extraction identifies passage boundaries with author-provided marking such as empty line, indentation etc. [7]. Topic-based passage extraction, such as TextTiling [3], divides documents into coherent passages with each passage corresponding to a specific topic. Despite its intuitive motivation, this approach is not widely used because identifying topic drift in documents is hard and computationally expensive. Instead, the most widely used methods extract passages with overlapped or non-overlapped windows [10].

Passage-based Retrieval Model. Most passage-based retrieval models in previous studies are unigram models constructed on window-based passages with fixed length. Liu and Croft [5] applied the language modeling approach [6] on overlapped-window passages and ranked documents with their maximum passage language score. Bendersky and Kurland [2] combined passage-level language models with document-level language models and weighted them according to the homogeneity of each document. To the best of our knowledge, our work is the first study that incorporates a neural network framework for passage-based retrieval models.

3   Neural Passage Model

In this section, we describe how to formulate the passage-based document ranking with a neural framework and aggregate information from passages with different granularities in our neural passage model (NPM). The overall model structure is shown in Fig. 1.

Passage-based Document Ranking. Passage-based retrieval models use passages as representatives for each document and rank documents according to their passage scores. Specifically, given a query q and a passage g extracted with a fixed-length window, the score of g is the maximum log likelihood of observing q given g's unigram language model as

$$\log P(q|g) = \sum_{t \in q} \log P(t|g) = \sum_{t \in q} \log((1 - \lambda_c)\frac{tf_{t,g}}{n} + \lambda_c\frac{cf_t}{|C|}) \tag{1}$$

where $tf_{t,g}$ is the count of t in g, cf_t is the corpus frequency of t, $|C|$ is the length of the corpus and λ_c is a smoothing parameter.

Assuming that passages can serve as proxies of documents [2], the ranking score of a document d under the passage-based document ranking framework should be $Score(q, d) = \log \sum_{g \in d} P(g|d) \cdot P(q|g)$. Intuitively, $P(g|d)$ could be a uniform distribution since all passages are extracted following the same methodology. However, averaging passage scores produces poor retrieval performance

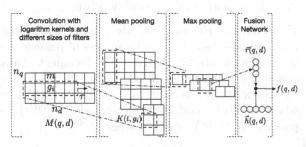

Fig. 1. The structure of the NPM.

in practice and the state-of-the-art models adopt a winner-take-all strategy that only uses the best passage to compute document scores [2,5]:

$$Score(q, d) = \max_{g \in d} \log P(q|g) \tag{2}$$

Passage Extraction with a Convolution Layer. Given a fixed length window, window-based passages are extracted by sliding the window along the document with fixed step size. Formally, given a document d with length n_d, the set of extracted passages $G(d)$ with window size m and step size τ is

$G(d) = \{g_i | i \in [0, \lfloor \frac{n_d}{\tau} \rfloor]\}$ where g_i represents the ith passage starting from the $i \cdot \tau$th term in d with size m. Let n_q be the length of query q, then the matching of terms in q and d is a matrix $M(q, d) \in \mathbb{R}^{n_q \times n_d}$ in which $M(q, d)_{i,j}$ represents the matching between the ith term in q and the jth term in d. In this work, we define the matching of two terms as a binary variable (1 if the two terms are same and 0 if they are not). Let $K(t, g_i)$ be the score of g_i given term t in q, then the extraction of window-based passages can be achieved with a convolution filter with size m, stride τ and kernel K over $M(q, d)$. Passages with different granularities can be directly extracted with different sizes of filters.

Language Modeling as a Logarithm Kernel. Let $M(t, g_i)$ be the binary matching matrix for term t and g_i with shape $(1, m)$ and $W \in \mathbb{R}^m$, $b_t \in \mathbb{R}$ be the parameters for a logarithm convolution kernel K, then we define the passage model score for q and g_i as

$$Score(q, g_i) = \sum_{t \in q} K(t, g_i) = \sum_{t \in q} \log(W \cdot M(t, g_i) + b_t) \tag{3}$$

Let W be a vector of 1 and b_t be $\frac{\lambda_c \cdot m \cdot cf_t}{(1 - \lambda_c)|C|}$, then $K(t, g_i)$ is equal to $\log P(t|g_i)$ in Eq. 1 plus a bias ($\log \frac{m}{1 - \lambda_c}$). Thus, the term-based language modeling approach can be viewed as a logarithm activation function over the linear projection of $M(t, g_i)$. Further, if we implement the sum of query term scores with a mean pooling and the winner-take-all strategy in Eq. 2 with a max pooling, then the passage-based document ranking framework can be completely expressed with a three-layer convolution neural network.

Aggregating Passage Models with a Fusion Network. Bendersky and Kurland observed that the usefulness of passage level information varies on different documents [2]. To consider document characteristics, they proposed to combine the passage models with document models using document homogeneity scores $h^{[M]}(d)$ as $Score(q, d) = h^{[M]}(d)P(q|d) + (1 - h^{[M]}(d)) \max_{g \in d} P(q|g)$ where $h^{[M]}(d)$ could be length-based ($h^{[length]}$), entropy-based ($h^{[ent]}$), inter-passage ($h^{[intPsg]}$) or doc-passage ($h^{[docPsg]}$):

$$h^{[length]}(d) = 1 - \frac{\log n_d - \min_{d_i \in C} \log n_{d_i}}{\max_{d_i \in C} \log n_{d_i} - \min_{d_i \in C} \log n_{d_i}}$$

$$h^{[ent]}(d) = 1 + \frac{\sum_{t' \in d} P(t'|d) \log(P(t'|d))}{\log n_d} \tag{4}$$

$$h^{[intPsg]} = \frac{2}{\lceil \frac{n_d}{\tau} \rceil (\lceil \frac{n_d}{\tau} \rceil - 1)} \sum_{i < j; g_i, g_j \in d} \cos(g_i, g_j), \quad h^{[docPsg]} = \frac{1}{\lceil \frac{n_d}{\tau} \rceil} \sum_{g_i \in d} \cos(d, g_i)$$

where $\cos(d, g_i)$ is the cosine similarity between the tf.idf vector of d and g_i.

Inspired by the design of homogeneity scores and studies on query performance prediction [9], we propose a fusion network that aggregates scores from passages according to both document properties and query characteristics. We extract features for queries and concatenate them with the homogeneity features to form a fusion feature vector $\mathbf{h}(q, d)$. For each query term, we extract their inverse document/corpus frequency and a clarity score [9] defined as $SCQ_t = (1+\log(cf_t))\log(1+idf_t)$ where idf_t is the inverse document frequency of t. For each feature, we compute the sum, standard deviation, maximum, minimum, arithmetic/geometric/harmonic mean and coefficient of variation for $t \in q$. We also include a list feature as the average scores of top 2,000 documents retrieved by the language modeling approach. Suppose that $h(q,d) \in \mathbb{R}^\beta$ and let $r(q,d) \in \mathbb{R}^\alpha$ be a vector where each dimension denotes a score from one convolution filter, then the final ranking score $f(q,d)$ is computed as

$$Score(q,d) = f(q,d) = \tanh\left(r(q,d)^T \cdot \phi(h(q,d)) + b_R\right) \qquad (5)$$

where $\phi(h(q,d)) = \frac{\exp(W_R^i \cdot h(q,d))}{\sum_{j=1}^\alpha \exp(W_R^j \cdot h(q,d))}$ and $W_R \in \mathbb{R}^{\alpha \times \beta}$, $b_R \in \mathbb{R}$ are parameters learned in the training process.

Table 1. The performance of passage-based retrieval models. $*$, $+$ means significant differences over MSP[base] and MSP[docPsg] with passage size $(150, \infty)$ respectively.

	MAP	NDCG@20	Precison@20	MAP	NDCG@20	Precison@20
	Passage Size $(50, \infty)$			Passage Size $(150, \infty)$		
MSP[base]	0.193	0.317	0.288	0.207	0.335	0.302
MSP[length]	0.210	0.333	0.298	0.223*	0.355*	0.315*
MSP[ent]	0.209	0.338	0.304	0.216*	0.349*	0.314*
MSP[interPsg]	0.204	0.329	0.296	0.215*	0.346*	0.310*
MSP[docPsg]	0.206	0.331	0.296	0.226*	0.362*	0.312*
	Passage Size $(50, 150, \infty)$					
NPM[doc]	0.255*+	0.412*+	0.366*+	-	-	-
NPM[query]	**0.256*+**	**0.416*+**	**0.369*+**	-	-	-
NPM[doc+query]	0.255*+	0.413*+	0.367*+	-	-	-

4   Experiment and Results

In this section, we describe our experiments on Robust04 with 5-fold cross validation [4]. For efficient evaluation, we conducted an initial retrieval with the query likelihood model [6] and performed re-ranking on the top 2,000 documents. We reported MAP, NDCG@20, Precision@20 and used Fisher randomization test [8] ($p < 0.05$) to measure the statistical significance. Our baselines include the max-scoring language passage model [5](MSP[base]) and the state-of-the-art passage-based retrieval model with passage weighting [2] – the MSP with length scores (MSP[length]), the MSP with entropy scores (MSP[ent]), the MSP with inter-passage scores (MSP[interPsg]) and the MSP with doc-passage scores (MSP[docPsg]). We follow the same parameter settings used by Bendersky and Kurland [2] and tested all models with passage size 50 and 150 separately. We used filters with length 50, 150 and ∞ for NPMs and set τ as the half of the filter lengths. The filter with length 50 extracts the same passages used in MSP models with passage size 50, and the filter with length ∞ treats the whole document as a single passage. Notice that the MSP with passage weighting [2] also uses sizes 50 (or 150) and ∞ to combines the scores of passages and the whole document. We tested the NPMs with document homogeneity features (NPM[doc]), query features (NPM[query]) and both (NPM[doc+query]). Due to the limit of Robust04, we only have 249 labeled queries, which are far from enough to train a robust convolution kernel with hundreds of parameters. Therefore, we fixed the convolution kernels as discussed in Sect. 3.

Overall Performance. Table 1 shows the results of our baselines and the NPM models with passage size 50 and 150. As we can see, the variations of MSP significantly outperformed MSP[base] with the same passage size, and the MSP models with passage size 150 performed better than MSP with passage size 50. Compared to MSP models, the NPM models showed superior performance on all reported metrics. As discussed in Sect. 3, MSP models can be viewed as special cases of the NPM with predefined parameters. With passage size 50, the MSP[base] model is actually a NPM model with filter length 50 and no fusion layer; and the MSP with homogeneity weighting is a NPM model with filter lengths 50, ∞ and a linear fusion with document homogeneity scores. From this perspective, the NPM model is more powerful than MSP models as it automatically learns to weight passages according to document/query features.

Weights of Passages. Figure 2 shows the means of passage weights $\phi(\boldsymbol{h}(q, d))$ on all query-doc pairs for NPM[query], NPM[doc] and NPM[doc+query]. In our experiments, the passages with size ∞ are the most important passages in the NPMs, but the scores from smaller passages also impact the final ranking. Although the MAP of the NPM[query] and NPM[doc] are close, their passage weights are different and they performed differently on 211 of 249 queries on Robust04. This indicates that, when evaluating a document with respect to multiple queries, all passages could be useful to determine the document's relevance; when evaluating multiple documents with respect to one query, models with ∞ passage size are more reliable in discriminating relevant documents from irrelevant ones.

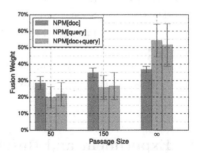

Fig. 2. The fusion weights for passages in the NPM averaged over query-doc pairs.

5 Conclusion and Future Work

In this paper, we proposed a neural network model for passage-based ad-hoc retrieval. We view the extraction of passages as a convolution process and develop a fusion network to aggregate information from passages with different granularities. Our model is highly expressive as previous passage-based retrieval models can be incorporated into it as special cases. Due to the limit of our data, we used binary matching matrix and deprived the freedom of the NPM to learn convolution kernels automatically. We will explore its potential to discover new matching patterns from more complex signals and heterogeneous datasets.

Acknowledgments. This work was supported in part by the Center for Intelligent Information Retrieval and in part by NSF IIS-1160894. Any opinions, findings and conclusions or recommendations expressed in this material are those of the authors and do not necessarily reflect those of the sponsor.

References

1. Baeza-Yates, R., Ribeiro-Neto, B., et al.: Modern Information Retrieval, vol. 463. ACM Press, New York (1999)
2. Bendersky, M., Kurland, O.: Utilizing passage-based language models for document retrieval. In: Macdonald, C., Ounis, I., Plachouras, V., Ruthven, I., White, R.W. (eds.) ECIR 2008. LNCS, vol. 4956, pp. 162–174. Springer, Heidelberg (2008). https://doi.org/10.1007/978-3-540-78646-7_17
3. Hearst, M.A.: Texttiling: a quantitative approach to discourse segmentation. Technical report, Citeseer (1993)
4. Huston, S., Croft, W.B.: A comparison of retrieval models using term dependencies. In: CIKM 2014, pp. 111–120. ACM (2014)
5. Liu, X., Croft, W.B.: Passage retrieval based on language models. In: CIKM 2002, pp. 375–382. ACM (2002)
6. Ponte, J.M., Croft, W.B.: A language modeling approach to information retrieval. In: SIGIR 1998, pp. 275–281. ACM (1998)
7. Salton, G., Allan, J., Buckley, C.: Approaches to passage retrieval in full text information systems. In: SIGIR 1993, pp. 49–58. ACM (1993)
8. Smucker, M.D., Allan, J., Carterette, B.: A comparison of statistical significance tests for information retrieval evaluation. In: CIKM 2007, pp. 623–632. ACM (2007)
9. Zhao, Y., Scholer, F., Tsegay, Y.: Effective pre-retrieval query performance prediction using similarity and variability evidence. In: Macdonald, C., Ounis, I., Plachouras, V., Ruthven, I., White, R.W. (eds.) ECIR 2008. LNCS, vol. 4956, pp. 52–64. Springer, Heidelberg (2008). https://doi.org/10.1007/978-3-540-78646-7_8
10. Zobel, J., Moffat, A., Wilkinson, R., Sacks-Davis, R.: Efficient retrieval of partial documents. Inf. Process. Manag. **31**(3), 361–377 (1995)

On the Cost of Negation for Dynamic Pruning

Joel Mackenzie[1]([✉]), Craig Macdonald[2][iD], Falk Scholer[1][iD],
and J. Shane Culpepper[1][iD]

[1] RMIT University, Melbourne, Australia
joel.mackenzie@rmit.edu.au
[2] University of Glasgow, Glasgow, Scotland, UK

Abstract. Negated query terms allow documents containing such terms to be filtered out of a search results list, supporting disambiguation. In this work, the effect of negation on the efficiency of disjunctive, top-k retrieval is examined. First, we show how negation can be integrated efficiently into two popular dynamic pruning algorithms. Then, we explore the efficiency of our approach, and show that while often efficient, negation can negatively impact the dynamic pruning effectiveness for certain queries.

Keywords: Dynamic pruning · Query semantics · Negation
Efficiency

1 Introduction

Modern Information Retrieval (IR) systems are extremely complex, often using a multi-stage approach to support both efficient and effective retrieval. In order to improve effectiveness, user queries can be rewritten into improved representations through query expansion, query reduction, named entity recognition, and other advanced rewriting strategies [1,2]. These strategies can make use of the *term negation* operator [3], which allows the results to be filtered so that documents containing negated terms are not returned by the IR system. In particular, this operator is useful for disambiguating query terms, allowing irrelevant results to be filtered from the result list. Furthermore, while negation is not commonly used directly by users of web search systems [4], many large scale search systems still explicitly support the negation operator as an advanced search feature, including Google,[1] Bing,[2] and Twitter [5], where users can manually negate query terms if desired.

2 Dynamic Pruning Strategies

Dynamic pruning strategies are often used in large scale search systems to allow efficient *candidate generation*, where typically hundreds or thousands of

---

[1] https://support.google.com/websearch?p=adv_operators (Accessed Oct. 2017).
[2] https://msdn.microsoft.com/library/ff795633.aspx (Accessed Oct. 2017).

© Springer International Publishing AG, part of Springer Nature 2018
G. Pasi et al. (Eds.): ECIR 2018, LNCS 10772, pp. 544–549, 2018.
https://doi.org/10.1007/978-3-319-76941-7_42

potentially relevant documents are quickly identified for further consideration [6]. In this work, we focus on two popular *Document-at-a-Time* (DAAT) dynamic pruning strategies, namely WAND [7] and BMW [8].

The WAND algorithm provides efficient traversal of document-ordered postings lists by storing the *upper-bound* score that the ranking function can contribute for each given term (U_t). WAND uses the score of the lowest scoring document in the current result set as a *threshold* (θ). Then, the values of U_t are used to estimate an upper-bound score that a document may achieve, and only documents with an upper-bound score greater than θ are evaluated, allowing redundant documents to be skipped.

Improving the WAND algorithm, Ding and Suel [8] proposed the *Block-Max* WAND (BMW) algorithm. Instead of just storing the U_t score, BMW also stores the maximum score for each *block* in each postings list $(U_{b,t})$. Query processing is the same as WAND, but after a potential (*pivot*) document has been found, the block scores are used to refine the upper-bound score, to ensure the potential score of the pivot is still greater than θ. If not, the block max scores can be used to induce additional skipping, resulting in faster retrieval. We refer the reader to the work of Broder et al. [7], Ding and Suel [8], and Petri et al. [9] for more information on the workings of these algorithms.

Integrating Negation. We now explain how negation can be efficiently supported in modern dynamic pruning algorithms. Firstly, the query parser is modified to recognise the term negation operator (-). Next, the query processing framework is modified to maintain two sets of postings lists: those to be scored, and those which are negated. Next, an efficient function called isNegated is implemented, which, when given the set of negated postings and a document identifier (DocID), returns true if the document contains a negated term, and false otherwise. This function efficiently skips to a compressed block that may contain the given document identifier, decompresses the block, and probes for the candidate DocID. If found, the function returns true. Otherwise, the next negated list is considered. This process is repeated for all negated lists which have cursors before the pivot document. Note that, similar to WAND, postings are sorted from current smallest to largest document identifier, as this allows early exiting from the isNegated function.

For the WAND algorithm, processing proceeds as usual, except that once a pivot document has been found, the pivot is checked for negated terms using the isNegated function. If it does contain negated terms, we select a new pivot. Otherwise, we proceed to score the document. We denote this algorithm as N-WAND in our experiments and discussion. Figure 1 shows a pictorial example of N-WAND pivoting and the isNegated function.

For the BMW algorithm, we propose two versions to address negated query terms. The first version will select a pivot document, and perform the refined block-max check. If the block-max check passes, the document is then examined for negated terms. We denote this version N-BMW$_{v1}$. In the second version, we switch the order of the negation test and the block-max check. That is, we select a pivot, and then ensure that it does not contain any negated terms. If the pivot contains no negated terms, we then continue to the block-max check. Otherwise, we select a new pivot. This algorithm is denoted as N-BMW$_{v2}$.

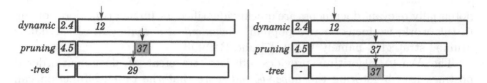

Fig. 1. An example of N-Wand processing for the query *"-tree dynamic pruning"* with a heap threshold of $\theta = 5.1$. Left shows the pivot selection: document 37 is the next document that can potentially make the heap since $2.4 + 4.5 > \theta$. Right shows the check negation function: the negated list is searched for the next DocID ≥ 37. Since 37 was found in the negated list, a new pivot will be selected.

3   Experiments

Experimental Setup. Experiments are conducted on the standard TREC ClueWeb09B collection, which contains 50 million web documents. We use a custom implementation of the Wand and Bmw algorithms [10], modified as discussed above, and the code is made available for reproducibility.[3] All timings are performed in-memory on an otherwise idle Red Hat Enterprise Linux Server (v7.2) with two Intel Xeon E5-2690 v3 CPUs, and 256GB of RAM. All reported timings use a single core only, and are the average of 3 runs. The query log consists of 317 queries extracted from the Excite query logs from 1997, 1999 and 2001 [4]. We extracted all queries containing negated terms, removed illegal characters, and then s-stemmed and stopped the queries. We also created a copy of this query set that has the negated terms removed, which allows us to compare the negated algorithms with the plain Wand and Bmw algorithms. For example, consider the query *"fish net -stocking"*; the corresponding plain query would be *"fish net"*. For each query, the top-k documents are ranked and retrieved using a BM25 ranking model.

Comparing Negation and Plain Disjunction. For each algorithm, we retrieve the top $k = \{10, 100, 1{,}000, 10{,}000\}$ documents. The resulting response times are shown in Fig. 2. Firstly we note that, similar to the disjunctive algorithms, the efficiency of processing queries with negated terms decreases as the value of k increases, and the N-Bmw variants outperform N-Wand for all values of k. In addition, adding negated terms makes processing slower than plain disjunctive processing, although the gap is quite small. This is not surprising, as negation involves an additional check for every pivot document that is considered by Wand or Bmw. To further explore the overhead caused by negated terms, we profile each of the algorithms (Table 1). Interestingly, we find that on average, the negated algorithms score fewer postings than when processing the corresponding plain query. Therefore, the cost of negation is not in the scoring of postings, but rather, in the negation check and pivot selection aspects of the algorithms.

---

[3] http://github.com/JMMackenzie/DaaT-Negation.

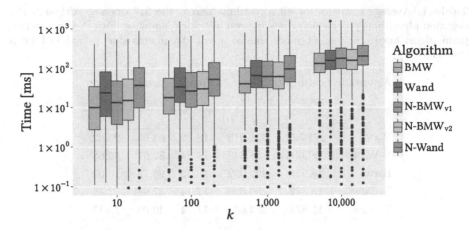

Fig. 2. Comparing all BMW and WAND algorithms across all queries for varying values of k. The plain BMW and WAND algorithms run each query by simply ignoring the negated terms.

Further Exploration of Dynamic Pruning Power. Examining Table 1 further, it is clear that the algorithms that process negated terms have *lower* thresholds than the corresponding plain algorithms. Since some documents will not make the heap due to containing negated terms, the heap threshold does not rise as quickly or as highly it would in a plain disjunctive setting. A lower threshold means less pruning power, which results in more pivot documents being considered, more negation checks, and less skipping across the DocID space.

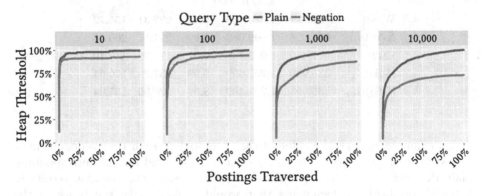

Fig. 3. Comparing the heap threshold of the negated query to the heap threshold of the plain query for the query pair *"silver city -new mexico"* and *"silver city mexico"*. Clearly, by negating documents containing the term *new*, fewer high scoring documents make it into the heap, leading to a reduced threshold, and less ability for dynamic pruning to occur. As k increases, so too does the gap between the plain and negated threshold.

Table 1. Average statistics for all algorithms across the 317 queries. Although the negation algorithms process fewer postings on average (compared with their respective plain algorithms), they select more unique pivot documents due to reduced heap thresholds.

Algorithm	Postings processed	Unique pivots	Final θ	Mean time	Median time
		$k = 10$			
WAND	271,533	611,013	18.32	63.63	23.62
N-WAND	226,605	628,349	17.74	82.91	36.68
BMW	18,958	222,032	18.32	28.73	10.03
N-BMW$_{v1}$	16,872	250,090	17.74	37.98	13.46
N-BMW$_{v2}$	16,873	254,634	17.74	40.03	15.33
		$k = 100$			
WAND	412,657	795,176	16.25	86.50	33.65
N-WAND	357,645	836,650	15.64	112.93	51.72
BMW	51,279	351,743	16.25	42.59	17.93
N-BMW$_{v1}$	46,046	389,226	15.64	60.65	26.34
N-BMW$_{v2}$	46,050	396,606	15.64	62.91	29.80
		$k = 1,000$			
WAND	607,186	1,051,990	13.02	124.24	64.39
N-WAND	535,137	1,125,580	12.27	167.54	94.58
BMW	159,800	571,003	13.02	77.51	39.60
N-BMW$_{v1}$	147,730	642,714	12.27	110.29	61.31
N-BMW$_{v2}$	147,742	650,942	12.27	106.74	60.70
		$k = 10,000$			
WAND	1,085,950	1,574,320	9.47	222.31	155.72
N-WAND	985,028	1,696,080	8.99	281.17	200.51
BMW	577,460	1,073,130	9.47	172.92	130.10
N-BMW$_{v1}$	535,980	1,201,890	8.99	237.29	176.04
N-BMW$_{v2}$	535,995	1,209,400	8.99	220.96	157.04

As an example, we plot the value of the threshold, θ, each time a document is scored for a single query, "*silver city -new mexico*", and for the corresponding plain disjunction, "*silver city mexico*" (Fig. 3). Clearly, the negation results in a lower threshold, as documents that would usually make the top-k in the disjunctive processing may contain negated terms. This explains why N-BMW$_{v2}$ outperforms N-BMW$_{v1}$ for larger values of k; N-BMW$_{v2}$ performs less redundant block-max checks than N-BMW$_{v1}$, especially as the density of negated terms in high scoring documents increases. Similar observations were made by Petri et al. [9], where the authors explored the impact of various ranking functions on the dynamic pruning power of both WAND and BMW.

4   Conclusion and Future Work

In this work, we presented an initial investigation on the impact that negation has on modern dynamic pruning algorithms. We first show how to integrate efficient negation into modern dynamic pruning strategies, namely WAND [7] and BMW [8], and then compare and contrast the efficiency of such extensions across a log of around 300 real queries. Our results demonstrate that although negation can be supported efficiently most of the time, occasionally queries can negatively impact the effectiveness of dynamic pruning. Future work will compare the N-WAND and N-BMW approaches with similar extensions to the MAXSCORE [11, 12] algorithm and the recently proposed Variable-BMW algorithms [13]. A more comprehensive exploration of the role of negation in query substitutions [1] for large scale search is a captivating problem that we leave for future work.

Acknowledgements. This work was supported by the Australian Research Council's *Discovery Projects* Scheme (DP170102231), an Australian Government Research Training Program Scholarship, and a grant from the Mozilla Foundation.

References

1. Jones, R., Rey, B., Madani, O., Greiner, W.: Generating query substitutions. In: Proceedings WWW, pp. 387–396 (2006)
2. Macdonald, C., Tonellotto, N., Ounis, I.: Efficient & effective selective query rewriting with efficiency predictions. In: Proceedings SIGIR, pp. 495–504 (2017)
3. Kim, Y., Seo, J., Croft, W.B.: Automatic Boolean query suggestion for professional search. In: Proceedings SIGIR, pp. 825–834 (2011)
4. Spink, A., Wolfram, D., Jansen, M.B.J., Saracevic, T.: Searching the web: the public and their queries. JASIST **52**(3), 226–234 (2001)
5. Busch, M., Gade, K., Larson, B., Lok, P., Luckenbill, S., Lin, J.: Earlybird: real-time search at Twitter. In: Proceedings ICDE, pp. 1360–1369 (2012)
6. Macdonald, C., Santos, R.L.T., Ounis, I.: The whens and hows of learning to rank for web search. Inf. Retriev. **16**(5), 584–628 (2013)
7. Broder, A.Z., Carmel, D., Herscovici, M., Soffer, A., Zien, J.Y.: Efficient query evaluation using a two-level retrieval process. In: Proceedings CIKM, pp. 426–434 (2003)
8. Ding, S., Suel, T.: Faster top-k document retrieval using block-max indexes. In: Proceedings SIGIR, pp. 993–1002 (2016)
9. Petri, M., Culpepper, J.S., Moffat, A.: Exploring the magic of WAND. In: Proceedings ADCS, pp. 58–65 (2013)
10. Crane, M., Culpepper, J.S., Lin, J., Mackenzie, J., Trotman, A.: A comparison of Document-at-a-Time and Score-at-a-Time query evaluation. In: Proceedings WSDM, pp. 201–210 (2017)
11. Turtle, H., Flood, J.: Query evaluation: strategies and optimizations. Inf. Proc. Man. **31**(6), 831–850 (1995)
12. Strohman, T., Turtle, H., Croft, W.B.: Optimization strategies for complex queries. In: Proceedings SIGIR, pp. 219–225 (2005)
13. Mallia, A., Ottaviano, G., Porciani, E., Tonellotto, N., Venturini, R.: Faster blockmax wand with variable-sized blocks. In: Proceedings SIGIR, pp. 625–634 (2017)

A Meta-Evaluation of Evaluation Methods for Diversified Search

Suneel Kumar Kingrani, Mark Levene, and Dell Zhang[✉]

Birkbeck, University of London, Malet Street, London WC1E 7HX, UK
{suneel,mark}@dcs.bbk.ac.uk, dell.z@ieee.org

Abstract. For the evaluation of diversified search results, a number of different methods have been proposed in the literature. Prior to making use of such evaluation methods, it is important to have a good understanding of how diversity and relevance contribute to the performance metric of each method. In this paper, we use the statistical technique ANOVA to analyse and compare three representative evaluation methods for diversified search, namely α-nDCG, MAP-IA, and ERR-IA, on the TREC-2009 Web track dataset. It is shown that the performance scores provided by those evaluation methods can indeed reflect two crucial aspects of diversity — richness and evenness — as well as relevance, though to different degrees.

1 Introduction

The same query could be submitted to a search engine by users from different backgrounds and with different information needs. When this occurs, the search engine should present users with relevant and diversified results that can cover multiple aspects or subtopics of the query. For more than a decade, there has been a surge of research in the diversification of search results [3,13,17,19]. The main objective of such research is to deal with the ambiguity of query or the multiplicity of user intent.

To evaluate the performance of diversified search, a variety of metrics have been proposed in recent years, such as α-nDCG [9], MAP-IA [1], and ERR-IA [5] which generalise the corresponding traditional IR metrics [15] to capture both the *diversity* and the *relevance* of search results. In this paper, we aim to investigate exactly how the above mentioned three representative performance metrics for diversified search are determined by diversity and relevance, using the Analysis of Variance (ANOVA) [10].

2 Related Work

The widely used IR performance metric nDCG [12] measures the accumulated usefulness ("gain") of the ranked result list with the gain of each relevant document discounted at lower positions. Clarke et al. proposed its extended version α-nDCG [9] to evaluate diversified search results. It takes into account not only

© Springer International Publishing AG, part of Springer Nature 2018
G. Pasi et al. (Eds.): ECIR 2018, LNCS 10772, pp. 550–555, 2018.
https://doi.org/10.1007/978-3-319-76941-7_43

the position at which a relevant document is ranked but also the subtopics contained in that document, and uses a parameter $\alpha \in [0, 1)$ to control the severity of redundancy penalisation. Specifically, α-nDCG for the top-k search results is the discounted cumulative gain α-DCG$[k]$ normalised by its "ideal" value, and DCG$[k]$ can be calculated as:

$$\alpha\text{-DCG}[k] = \sum_{i=1}^{k} \frac{\sum_{s=1}^{N} g_{i,s}(1 - \alpha)^{\sum_{j=1}^{i-1} g_{j,s}}}{\log_2(i + 1)}, \tag{1}$$

where N is the total number of distinct subtopics, and $g_{i,s}$ is the human judgement for whether subtopic s is present or not in document i.

Agrawal et al. [1] proposed an approach to generalising traditional IR performance metrics for the search results of a query with multiple subtopics (user intents). The idea is to calculate the given performance metric for each subtopic separately, and then aggregate those scores based on the probability distribution of subtopics for the query. Extending the traditional IR performance metrics MAP [15] and ERR [6] in this way, we get their diversified versions:

$$\text{MAP-IA} = \sum_{s=1}^{N} P(s) \cdot \text{MAP}_s \quad \text{and} \quad \text{ERR-IA} = \sum_{s=1}^{N} P(s) \cdot \text{ERR}_s, \tag{2}$$

where N is the total number of distinct subtopics, $P(s)$ is the probability or weight of subtopic s, while MAP_s and ERR_s are the MAP and ERR scores for subtopic s respectively.

The previous studies most similar to our work are those from Clarke et al. [8] and Chandar et al. [4] which attempt to compare evaluation methods in the context of diversified search. The former assumes simple cascade models of user behaviour, while the latter measures diversity just by the subtopic recall — s-Recall [18] — which may not reveal the full picture of diversity.

3 Meta-Evaluation

3.1 Factors

To examine the diversity of search results for a query, it is important to consider not only the number of distinct subtopics but also the relative abundance of the subtopics present in the search result set. Drawing an analogy between subtopics and species, we would like to borrow two measures from ecology [2,14,16] — richness and evenness — to describe the above two complementary dimensions of diversity respectively. The measure of richness on its own cannot provide a complete picture of diversity, as it does not account for the varying proportions of different species in a population. For example, intuitively, one wild-flower field with 500 daisies and 500 dandelions should be more diverse than another wild-flower field with 999 daisies and 1 dandelions — although they both have the same richness (two species), evidently the first field has much higher evenness than the second field.

Formally, we define the two measures, richness and evenness, in the context of diversified search, as follows. The richness of the search result set for a query (topic) could be just defined as the amount of distinct subtopics appeared in the set. In order to make the value of richness comparable across queries, we choose to use not the absolute number of distinct subtopics but the relative proportion of distinct subtopics:

$$richness = R/N, \qquad (3)$$

where R is the number of distinct subtopics covered by the given search result set for a query, while N is the total number of distinct subtopics relevant to that query. This proportionate version of richness is actually equivalent to the *s-Recall* proposed by Zhai et al. [18]. The value of (proportionate) richness is obviously between 0 and 1. The evenness of the search result set for a query (topic) refers to how close in numbers each subtopic in the set is, i.e., it quantifies how evenly the search results are spread over the subtopics. For example, a search result set having 5 results from subtopic u and 5 results from subtopic v should have greater evenness than a search result set having 2 results from subtopic u and 8 results from subtopic v. Mathematically, the value of evenness is calculated as the normalised diversity:

$$evenness = D/D_{\max}, \qquad (4)$$

where D is a diversity index, and $D_{\max}$ is the maximum possible value of D. Here, we use the well-known *inverse Simpson's diversity index* [11]:

$$D = \left(\sum_{s=1}^{R} p_s^2 \right)^{-1}. \qquad (5)$$

where R is the number of distinct subtopics covered by the search result set, and p_s is the proportion of subtopic s within the search result set. In this case, it can be proved that $D_{\max}$ is equal to R, which happens when all the subtopics appear in the search result set with equal frequencies $\frac{1}{R}$. The value of evenness is greater than 0, and less than or equal to 1.

For the purpose of assessing the *relevance* of search results, we can simply use the Precision@k measure [15], as in [4].

3.2 Data

The dataset used for our experiments comes from TREC-2009 Web track diversity task [7] which have also been used in previous studies [4,8]. This dataset includes 50 topics, each of which consists of a set of subtopics representing different user needs.

3.3 Experiments

The evaluation methods for diversified search, including α-nDCG, MAP-IA, and ERR-IA, must be able to capture not only the relevance of search results but also the diversity of search results in terms of both richness and evenness. The statistical technique, Analysis of Variance (ANOVA) [10], provides the perfect tool to gain insight into how each of these three factors (richness, evenness, and relevance) contributes to the overall performance measured by an evaluation method.

In our experiments, the dependent variable for the ANOVA would be the performance score given by α-nDCG[1], MAP-IA, or ERR-IA. Regarding the independent variables (richness, evenness, and relevance), since the real IR system outputs submitted to the TREC-2009 Web track could not account for all the possible scenarios that we would like to investigate, we generated a number of synthetic search result sets via a simulation process similar to the "*Rel+Div*" setting in [4]. Given a query (topic) in our dataset, we randomly sampled 10 documents form the full *qrels* file [7] to create such artificial document rankings that satisfy one of the $3^3 = 27$ different experimental conditions for top-10 search results: low/medium/high *richness*, low/medium/high *evenness*, and low/medium/high *relevance*, where the category labels low, medium, and high correspond to the value ranges 0.0–0.3, 0.3–0.6, and 0.6–1.0 respectively. The simulation process would continue until for each of the 50 queries (topics) we had generated 10 search result sets (rankings) per experimental condition. Therefore, the ANOVA for each evaluation method would have $50 \times 10 \times 27 = 13500$ data points to analyse.

3.4 Results

The statistical significance results of the ANOVA are shown in Table 1. It can be seen that all those performance metrics, α-nDCG, MAP-IA, and ERR-IA, would be influenced heavily by the individual factors — richness, evenness, and relevance — with almost zero p-values, but not so much by their interactions. This confirms that the chosen three factors are relatively independent (untangled) aspects of a system's performance for diversified search.

Table 1. The statistical significance results of the ANOVA.

Component	α-nDCG		MAP-IA		ERR-IA	
	F	p-value	F	p-value	F	p-value
Richness	362.4	0.00	590.9	0.00	253.7	0.00
Evenness	480.0	0.00	521.7	0.00	282.7	0.00
Relevance	465.7	0.00	285.0	0.00	397.0	0.00
*Richness * evenness*	10.8	0.00	2.9	0.03	0.9	0.46
*Richness * relevance*	3.5	0.01	5.3	0.00	5.8	0.00
*Evenness * relevance*	4.3	0.00	0.3	0.91	3.2	0.01
*Richness * evenness * relevance*	0.8	0.53	1.4	0.24	2.4	0.05

---

[1] The parameter α for α-nDCG was set to 0.5, the default value used in the TREC-2009 Web track diversity task.

Table 2. The variance decomposition results of the ANOVA.

Component	α-nDCG		MAP-IA		ERR-IA	
	SSE	(%)	SSE	(%)	SSE	(%)
Richness	13.1	(8%)	8.2	(13%)	2.0	(6%)
Evenness	17.4	(11%)	7.2	(11%)	2.3	(7%)
Relevance	16.9	(10%)	3.9	(6%)	3.2	(9%)
*Richness * evenness*	0.6	(0%)	0.1	(0%)	0.0	(0%)
*Richness * relevance*	0.3	(0%)	0.1	(0%)	0.1	(0%)
*Evenness * relevance*	0.3	(0%)	0.0	(0%)	0.1	(0%)
*Richness * evenness * relevance*	0.1	(0%)	0.0	(0%)	0.0	(0%)
Residual	117.0	(71%)	44.6	(70%)	25.9	(77%)

Furthermore, Table 2 shows the variance decomposition results of the ANOVA, where SSE stands for the sum of squared errors. It seems that MAP-IA reflects more richness than the other two performance metrics, as the change of richness accounts for 13% of the total variability in MAP-IA which is substantially higher than 8% in α-nDCG and 6% in ERR-IA. On the other hand, evenness is probably reflected better by α-nDCG or MAP-IA than ERR-IA, as the change of evenness accounts for 11% of the total variability in α-nDCG and MAP-IA but only 7% in ERR-IA. In terms of relevance, α-nDCG looks the most accurate indicator, because 10% of its total variability is attributed to the change of relevance, which is followed by 9% in ERR-IA and 6% in MAP-IA. The "residual" component which comprises everything about the performance metric unexplained by the proposed independent variables (factors) occupies a high proportion of the total variability, which suggests that the difficulty of the query (topic) and also the specific ranking algorithm still play the major roles in determining performance scores.

4   Conclusion

In this paper, we have shown using the ANOVA that the three representative evaluation methods for diversified search, α-nDCG, MAP-IA and ERR-IA, do reflect two crucial aspects of diversity — richness and evenness — as well as relevance, though to different degrees.

References

1. Agrawal, R., Gollapudi, S., Halverson, A., Ieong, S.: Diversifying search results. In: Proceedings of the 2nd International Conference on Web Search and Web Data Mining (WSDM), Barcelona, Spain, pp. 5–14 (2009)
2. Begon, M., Harper, J.L., Townsend, C.R.: Ecology: Individuals, Populations, and Communities, 3rd edn. John Wiley & Sons, Hoboken (1996)

3. Carbonell, J.G., Goldstein, J.: The use of MMR, diversity-based reranking for reordering documents and producing summaries. In: Proceedings of the 21st Annual International ACM SIGIR Conference on Research and Development in Information Retrieval (SIGIR), Melbourne, Australia, pp. 335–336 (1998)
4. Chandar, P., Carterette, B.: Analysis of various evaluation measures for diversity. In: Proceedings of the DDR Workshop, Dublin, Ireland, pp. 21–28 (2011)
5. Chapelle, O., Ji, S., Liao, C., Velipasaoglu, E., Lai, L., Wu, S.L.: Intent-based diversification of web search results: metrics and algorithms. Inf. Retr. **14**(6), 572–592 (2011)
6. Chapelle, O., Metlzer, D., Zhang, Y., Grinspan, P.: Expected reciprocal rank for graded relevance. In: Proceedings of the 18th ACM Conference on Information and Knowledge Management (CIKM), Hong Kong, China, pp. 621–630 (2009)
7. Clarke, C.L.A., Craswell, N., Soboroff, I.: Overview of the TREC 2009 web track. In: Proceedings of The 18th Text REtrieval Conference (TREC), Gaithersburg, MD, USA (2009)
8. Clarke, C.L.A., Craswell, N., Soboroff, I., Ashkan, A.: A comparative analysis of cascade measures for novelty and diversity. In: Proceedings of the 4th International Conference on Web Search and Web Data Mining (WSDM), Hong Kong, China, pp. 75–84 (2011)
9. Clarke, C.L.A., Kolla, M., Cormack, G.V., Vechtomova, O., Ashkan, A., Buttcher, S., MacKinnon, I.: Novelty and diversity in information retrieval evaluation. In: Proceedings of the 31st Annual International ACM SIGIR Conference on Research and Development in Information Retrieval (SIGIR), Singapore, pp. 659–666 (2008)
10. Gamst, G., Meyers, L.S., Guarino, A.: Analysis of Variance Designs: A Conceptual and Computational Approach with SPSS and SAS. Cambridge University Press, New York (2008)
11. Hill, M.O.: Diversity and evenness: a unifying notation and its consequences. Ecology **54**(2), 427–432 (1973)
12. Järvelin, K., Kekäläinen, J.: Cumulated gain-based evaluation of IR techniques. ACM Trans. Inf. Syst. (TOIS) **20**(4), 422–446 (2002)
13. Kingrani, S.K., Levene, M., Zhang, D.: Diversity analysis of web search results. In: Proceedings of the ACM Web Science Conference (WebSci), Oxford, UK, pp. 43:1–43:2 (2015)
14. Magurran, A.E.: Ecological Diversity and Its Measurement. Princeton University Press, Princeton (1988)
15. Manning, C.D., Raghavan, P., Schütze, H.: Introduction to Information Retrieval. Cambridge University Press, Cambridge (2008)
16. Pielou, E.C.: An Introduction to Mathematical Ecology. Wiley-Interscience, New York (1969)
17. Santos, R.L., Macdonald, C., Ounis, I.: Search result diversification. Found. Trends Inf. Retr. **9**(1), 1–90 (2015)
18. Zhai, C., Cohen, W.W., Lafferty, J.D.: Beyond independent relevance: methods and evaluation metrics for subtopic retrieval. In: Proceedings of the 26th Annual International ACM SIGIR Conference on Research and Development in Information Retrieval (SIGIR), Toronto, Canada, pp. 10–17 (2003)
19. Zuccon, G., Azzopardi, L., Zhang, D., Wang, J.: Top-k retrieval using facility location analysis. In: Baeza-Yates, R., de Vries, A.P., Zaragoza, H., Cambazoglu, B.B., Murdock, V., Lempel, R., Silvestri, F. (eds.) ECIR 2012. LNCS, vol. 7224, pp. 305–316. Springer, Heidelberg (2012). https://doi.org/10.1007/978-3-642-28997-2_26

Co-training for Extraction of Adverse Drug Reaction Mentions from Tweets

Shashank Gupta[1], Manish Gupta[1(✉)], Vasudeva Varma[1], Sachin Pawar[2],
Nitin Ramrakhiyani[2], and Girish Keshav Palshikar[2]

[1] International Institute of Information Technology-Hyderabad, Hyderabad, India
shashank.gupta@research.iiit.ac.in, {manish.gupta,vv}@iiit.ac.in
[2] TCS Research, Pune, India
{sachin7.p,nitin.ramrakhiyani,gk.palshikar}@tcs.com

Abstract. Adverse drug reactions (ADRs) are one of the leading causes
of mortality in health care. Current ADR surveillance systems are often
associated with a substantial time lag before such events are officially
published. On the other hand, online social media such as Twitter contain information about ADR events in real-time, much before any official reporting. Current state-of-the-art methods in ADR mention extraction use Recurrent Neural Networks (RNN), which typically need large
labeled corpora. Towards this end, we propose a semi-supervised method
based on co-training which can exploit a large pool of unlabeled tweets
to augment the limited supervised training data, and as a result enhance
the performance. Experiments with ∼0.1M tweets show that the proposed approach outperforms the state-of-the-art methods for the ADR
mention extraction task by ∼5% in terms of F1 score.

Keywords: Semi-supervised learning · Pharma-covigilance
Co-training

1 Introduction

Estimates show that Adverse Drug Reactions (ADRs) are the fourth leading
cause of deaths in the United States ahead of cardiac diseases, diabetes, AIDS
and other fatal diseases[1]. Hence, it necessitates the monitoring and detection
of such adverse events to minimize the potential health risks. Typically, post-
marketing drug safety surveillance (also called as pharmacovigilance) is conducted to identify ADRs after a drug's release. Such surveys rely on formal
reporting systems such as Federal Drug Administration's Adverse Event Reporting System (FAERS)[2]. However, often a large fraction (∼94%) of the actual ADR

M. Gupta is also a Principal Applied Scientist at Microsoft.
[1] https://ethics.harvard.edu/blog/new-prescription-drugs-major-health-risk-few-
offsetting-advantages.
[2] http://bit.ly/2xnu7pE.

© Springer International Publishing AG, part of Springer Nature 2018
G. Pasi et al. (Eds.): ECIR 2018, LNCS 10772, pp. 556–562, 2018.
https://doi.org/10.1007/978-3-319-76941-7_44

instances are under-reported in such systems [9]. Social media presents a plausible alternative to such systems, given its wide userbase. A recent study [6] shows that Twitter has three times more ADRs reported as compared to FAERS.

Earlier work in this direction focused on feature based pipeline followed by a sequence classifier [12]. More recent works are based on Deep Neural Networks [4,13]. Deep learning based methods [5,11] typically rely on the presence of a large annotated corpora, due to their large number of free parameters. Due to the high cost associated with tagging ADR mentions in a social media post and limited availability of labeled datasets, it is hard to train a deep neural network effectively for such a task. In this work, we attempt to address this problem and propose a novel semi-supervised method based on co-training [2] which can harness a large pool of unlabeled related tweets, which are more economical to collect than ADR annotated tweets.

2   Approach

In this section, we define the ADR extraction problem, discuss the supervised ADR extraction method, and then propose our semi-supervised co-training method.

2.1   Problem Definition

The problem of ADR extraction can be defined as follows. Given a social media post in the form of a word sequence $x = x_1...., x_n$, where n is the maximum sequence length, predict an output sequence $y_1, ..., y_n$, where each y_i is encoded using standard sequence labeling encoding scheme such as the IO encoding similar to that used in [4].

2.2   Supervised ADR Extraction

We choose the model described in [4] for modeling the ADR extraction task. Given an input word sequence x, a bi-directional LSTM transducer (bi-LSTM) [8] is employed to capture complex sequential dependencies. Formally, at each time-step t, the bi-LSTM transducer attempts to model the task as follows.

$$h_t = \text{bi-LSTM}(e_t, h_{t-1}) \tag{1}$$

where $h_t \in \mathcal{R}^{(2 \times d_h)}$, is the hidden unit representation of the bi-LSTM with d_h being the hidden unit size. Since it is a concatenation of hidden units of a forward sequence LSTM and a backward sequence LSTM, its overall dimension is $2 \times d_h$. e_t is the embedding vector corresponding to the input word x_t extracted from a pre-trained word embedding lookup table.

$$y_t = \text{softmax}(W \times h_t + b) \tag{2}$$

where $y_t \in \mathcal{R}^{d_l}$, is the output vector at each time-step which encodes the probability distribution over the number of possible output labels (d_l) at each time-step of the sequence. $W \in \mathcal{R}^{d_l * d_h}$ and $b \in \mathcal{R}^{d_l}$ are weight vectors for the affine transformation. Finally, the cross entropy loss function for the task is defined as follows.

$$L_{ADR} = -\sum_{t=1}^{n} \sum_{i=1}^{d_l} \hat{y}_{t_i} \log y_{t_i} \tag{3}$$

where $\hat{y}_t$ is the one-hot representation of the actual label at time-step t.

2.3   Co-training Method for ADR Extraction

Algorithm 1 outlines the method for semi-supervised co-training. Co-training [2] requires two feature views of the dataset, which in our case are: (1) word2vec embeddings trained on a generic tweet corpus [7] followed by a bi-LSTM feature extractor and (2) word2vec embeddings trained on domain specific (drug-related) tweet corpus (described in the Sect. 3) followed by a bidirectional Gated Recurrent Unit (bi-GRU) transducer [3]. At each step of co-training, the transducers M_1 and M_2 are trained on their respective views minimizing the ADR training loss (Lines 4 to 5 of Algorithm 1). Each sample from the unlabeled example pool is scored using a scoring function computed as follows. First, the current transducer is used to decode/infer output label distribution for each word in the unlabeled sample. For each word in the output sequence, we simply choose the output label which has the maximum probability. We filter out the data sample if the transducer does not output even a single ADR label for any word in the sample. If there is at least one word labeled as ADR, we compute the score for the sample as the multiplication of the ADR probabilities for the ADR-labeled words in the sample normalized by the number of ADR words. If this confidence score of the sample is greater than some pre-defined threshold τ, the sample is added to the training set of the other transducer along with its output labels as generated by the transducer (Lines 7 to 10). Due to this cross-exchange of training data, both transducers work in synergy and learn from mistakes of each other.

Algorithm 1. Co-training Method for ADR Extraction

   Input U: Large collection of unlabeled tweets, τ: Threshold for co-training
               D^1_{ADR}, D^2_{ADR}: Two views of the ADR annotated data
   Output   Model parameters $\theta^{LSTM}, \theta^{GRU}$
1: $T^1, T^2 \leftarrow D^1_{ADR}, D^2_{ADR}$
2: Initialize model parameters, $\theta^{LSTM}, \theta^{GRU}$ randomly.
3: **while** *(stopping criteria is not met)* **do**
4:      $M^1 \leftarrow$ train bi-LSTM on T^1 (minimize L^1_{ADR})
5:      $M^2 \leftarrow$ train bi-GRU on T^2 (minimize L^2_{ADR})
6:      **for** $i \leftarrow 1, |U|$ **do**
7:         **if** $M^1.score(U_i) \geq \tau$ **then**
8:            $T^2 \leftarrow T^2 \cup \{U_i\}, U \leftarrow U - U_i$
9:         **if** $M^2.score(U_i) \geq \tau$ **then**
10:        $T^1 \leftarrow T^1 \cup \{U_i\}, U \leftarrow U - U_i$

---

3   Experiments

In this section, we discuss details of the datasets, implementation details and experimentation results.

3.1   Datasets

We use two datasets for evaluation detailed as follows.

(1) **Twitter ADR:** The *Twitter ADR* dataset, described in [4], contains 957 tweets posted between 2007 and 2010, with mention annotations of ADR and some other medical entities. Due to Twitter's license agreement, authors released only tweet ids with their corresponding mention span annotations. At the time of collection of the original tweets using Twitter API, we were able to collect only 639 tweets with 1526 ADR mentions.

(2) **TwiMed:** The *TwiMed* dataset, described in [1], contains 1000 tweets with mention annotation of Symptoms from drug (ADR) and other mention annotations posted in 2015. Due to Twitter's license agreement, we were able to extract 663 tweets only with 1091 ADR mentions.

Unlabeled Tweets: For semi-supervised learning, we collected ∼0.1M tweets using the keywords as drug-names and ADR lexicon publicly available[3]. This filtering step ensures that all collected tweets have at least one drug-name occurrence and one ADR phrase. The tweets were posted in 2015 and have no ADR mentions labeled.

3.2   Implementation Details

For implementation of the model, we use the popular Python deep learning toolkits: Keras[4] and TensorFlow[5]. Training data for each fold is divided according to 90:10% train-validation split.

Pre-processing: As part of text pre-processing, all HTML links and user mentions are normalized to a single token respectively. Special characters and emoticons are removed, and each tweet is padded with the maximum tweet length in the corpus.

Hyper-parameter settings for the two views: The hyper-parameter settings for the two views as required by the co-training method are as follows.

View 1: For the first view we use bi-LSTM transducer, with the hyper-parameter setting similar to the one reported in [4]. Word embedding dimension is set to 400.

---

[3] http://diego.asu.edu/downloads.
[4] https://keras.io/.
[5] https://www.tensorflow.org/.

View 2: For the second view, we use bi-GRU transducer with input as word2vec word embeddings trained on the unlabeled drug-related tweets described in the previous section. The word embedding dimension is set to 300.

For both transducers, the hidden unit dimension (d_h) is set to 500. The number of output units (d_l) is 4. We use Adam [10] as optimizer with a learning rate of 0.001 and a batch size of 32.

Co-training Parameters: For the co-training methods, confidence threshold value is empirically set to 0.5. The stopping criteria for the co-training kicks in when the number of iterations reaches 5 or if the unlabeled tweets pool is exhausted, whichever occurs first. The number of epochs are set to a maximum with 25, with early-stopping employed if validation loss drops for more than 3 epochs.

3.3 Results

The results of various methods are presented in Table 1 for the Twitter ADR and the TwiMed datasets respectively. For the ADR task, to encode the output labels we use the IO encoding scheme where each word is labeled with one of the following labels: (1) I-ADR (ADR mention), (2) I-Other (mention category other than ADR), (3) O (others), (4) PAD (padding token). Since our entity of interest is ADR, we report the results on ADR only. An example tweet annotated with IO-encoding is as follows. "$@BLENDOS_O$ $Lamictal_O$ and_O $trileptal_O$ and_O $seroquel_O$ of_O $course_O$ the_O $seroquel_O$ I_O $take_O$ in_O $severe_O$ $situations_O$ $because_O$ $weight_{I-ADR}$ $gain_{I-ADR}$ is_O not_O $cool_O$". For performance evaluation, we use approximate-matching [14], which is used popularly in biomedical entity extraction tasks [4,12]. We report the F1-score, Precision and Recall computed using approximate matching as follows.

$$\text{Precision} = \frac{\#ADR \text{ approx. matched}}{\#ADR \text{ spans predicted}}, \text{ Recall} = \frac{\#ADR \text{ approx. matched}}{\#ADR \text{ spans in total}} \quad (4)$$

The F1-score is the harmonic-mean of the Precision and Recall values. All results are reported using 10-fold cross-validation along with the standard deviation across the folds. Our baseline methods are bi-LSTM transducer [4] with traditional word embeddings and the current state-of-the-art bi-LSTM transducer which uses traditional word embeddings augmented with knowledge-graph based embeddings [13]. For both the datasets, it should be noted that Cocos et al. [4] used RMSProp as an optimizer, and since we are using Adam for all our methods, so for a fair comparison we also report the baseline results with Adam. The corresponding results are reported in the first two rows of Table 1. It is clear that re-implementation with Adam optimizer enhances the performance, which is consistent with the general consensus around Adam optimizer. The KB-embedding baseline [13] replaces word embeddings of the medical entities in the sentence with the corresponding embeddings learned from a knowledge-base. The corresponding results can be seen in row 3. It is clear that adding

Table 1. Accuracy Comparison for Various Methods (along with Std. Deviation)

Method	Twitter ADR dataset			TwiMed dataset		
	Precision	Recall	F1-score	Precision	Recall	F1-score
Baseline [4]	0.7067 ± 0.057	0.7207 ± 0.074	0.7102 ± 0.049	0.6120 ± 0.116	0.5149 ± 0.099	0.5601 ± 0.100
Baseline with Adam	0.7065 ± 0.058	0.7576 ± 0.083	0.7272 ± 0.051	$\mathbf{0.6281 \pm 0.094}$	0.5614 ± 0.110	0.5859 ± 0.079
KB-Embedding baseline [13]	0.7171 ± 0.058	0.7713 ± 0.091	0.7397 ± 0.055	0.5960 ± 0.081	0.6144 ± 0.068	0.6042 ± 0.060
Co-training (5k)	0.7247 ± 0.056	0.7770 ± 0.082	0.7488 ± 0.063	0.5806 ± 0.093	0.6746 ± 0.078	$\mathbf{0.6192 \pm 0.066}$
Co-training (10k)	0.7288 ± 0.041	$\mathbf{0.8238 \pm 0.064}$	0.7719 ± 0.040	0.5484 ± 0.092	0.6355 ± 0.113	0.5851 ± 0.090
Co-training (25k)	0.7181 ± 0.035	0.8005 ± 0.048	0.7561 ± 0.031	0.5774 ± 0.082	0.6425 ± 0.076	0.6051 ± 0.066
Co-training (50k)	0.7207 ± 0.034	0.7870 ± 0.042	0.7516 ± 0.029	0.5420 ± 0.054	0.6342 ± 0.061	0.5836 ± 0.053
Co-training (75k)	0.7478 ± 0.062	0.8033 ± 0.053	0.7730 ± 0.047	0.5525 ± 0.059	$\mathbf{0.6875 \pm 0.069}$	0.6110 ± 0.056
Co-training (100k)	$\mathbf{0.7514 \pm 0.053}$	0.8045 ± 0.056	$\mathbf{0.7754 \pm 0.042}$	0.5548 ± 0.064	0.6786 ± 0.058	0.6081 ± 0.048

KB-based embeddings enhances the performance over the baseline, due to the external knowledge added in the form of KB embeddings.

The results for our methods are presented from row 4 onwards. It is clear that the co-training method outperforms the baseline by a significant margin. It clearly indicates the efficacy of semi-supervised learning when the labeled data is scarce.

Effect of Unlabeled Data Size: We also analyze the effect of the size of the unlabeled tweet dataset on the method's performance. The results are presented from row 4 onwards. The results are fairly constant as unlabeled data size is varied, indicating the robustness of the method.

4   Conclusions

In this paper, we proposed a semi-supervised co-training based learning based methods to tackle the problem of labeled data scarcity for adverse drug reaction mention extraction task. Our method uses large unlabeled drug related tweets to augment the limited existing ADR extraction datasets providing superior results in comparison to pure supervised learning based methods. We analyzed the method on two popular ADR extraction datasets, and it demonstrates superior results as compared to the state-of-the-art methods in ADR extraction.

References

1. Alvaro, N., Miyao, Y., Collier, N.: TwiMed: Twitter and PubMed comparable corpus of drugs, diseases, symptoms, and their relations. JMIR Publ. Health Surveill. **3**(2), 24 (2017)
2. Blum, A., Mitchell, T.: Combining labeled and unlabeled data with co-training. In: Proceedings of the Eleventh Annual Conference on Computational Learning Theory, pp. 92–100. ACM (1998)

3. Cho, K., Van Merriënboer, B., Bahdanau, D., Bengio, Y.: On the Properties of Neural Machine Translation: Encoder-Decoder Approaches. arXiv preprint arXiv:1409.1259 (2014)

4. Cocos, A., Fiks, A.G., Masino, A.J.: Deep learning for pharmacovigilance: recurrent neural network architectures for labeling adverse drug reactions in Twitter posts. JAMIA **24**(4), 813–82 (2017)

5. Collobert, R., Weston, J., Bottou, L., Karlen, M., Kavukcuoglu, K., Kuksa, P.: Natural language processing (almost) from scratch. JMLR **12**(Aug), 2493–2537 (2011)

6. Freifeld, C.C., Brownstein, J.S., Menone, C.M., Bao, W., Filice, R., Kass-Hout, T., Dasgupta, N.: Digital drug safety surveillance: monitoring pharmaceutical products in Twitter. Drug Saf. **37**(5), 343–350 (2014)

7. Godin, F., Vandersmissen, B., De Neve, W., Van de Walle, R.: Multimedia Lab@ ACL W-Nut NER shared task: named entity recognition for Twitter microposts using distributed word representations. In: 2015 ACL-ICJNLP, pp. 146–153 (2015)

8. Graves, A.: Sequence Transduction with Recurrent Neural Networks. CoRR abs/1211.3711 (2012)

9. Hazell, L., Shakir, S.A.: Under-reporting of adverse drug reactions: a systematic review. Pharmacoepidemiol. Drug Saf. **14**, S184–S185 (2005)

10. Kingma, D., Ba, J.: Adam: A Method for Stochastic Optimization. arXiv:1412.6980 (2014)

11. LeCun, Y., Bengio, Y., Hinton, G.: Deep learning. Nature **521**(7553), 436–444 (2015)

12. Nikfarjam, A., Sarker, A., OConnor, K., Ginn, R., Gonzalez, G.: Pharmacovigilance from social media: mining adverse drug reaction mentions using sequence labeling with word embedding cluster features. JAMIA **22**(3), 671–681 (2015)

13. Stanovsky, G., Gruhl, D., Mendes, P.N.: Recognizing mentions of adverse drug reaction in social media using knowledge-infused recurrent models. In: EACL, pp. 142–151 (2017)

14. Tsai, R.T.H., Wu, S.H., Chou, W.C., Lin, Y.C., He, D., Hsiang, J., Sung, T.Y., Hsu, W.L.: Various criteria in the evaluation of biomedical named entity recognition. BMC Bioinf. **7**(1), 92 (2006)

Concept Embedding for Information Retrieval

Karam Abdulahhad(✉)

GESIS - Leibniz Institute for the Social Sciences, Cologne, Germany
karam.abdulahhad@gesis.org

Abstract. Concepts are used to solve the term-mismatch problem. However, we need an effective similarity measure between concepts. Word embedding presents a promising solution. We present in this study three approaches to build concepts vectors based on words vectors. We use a vector-based measure to estimate inter-concepts similarity. Our experiments show promising results. Furthermore, words and concepts become comparable. This could be used to improve conceptual indexing process.

1 Introduction

Conceptual indexing includes the process of annotating raw text by concepts[1] of a particular knowledge source [1]. It is used to represent the content of documents and queries by more *informative* terms, namely concepts rather than words. Annotating text by concepts is used to solve the term-mismatch problem by considering the semantic of text rather than its form [1]. For example, the two terms "cancer" and "malignant neoplastic disease" correspond to the same concept (*synset*) in WordNet[2]. However, using concepts instead of words has some side-effects. First, the process of annotating text by concepts is a potential source of noise, e.g. "x-ray" corresponds to more than 6 different concepts in UMLS[3], so which one best fits the original textual content. Second, to better solve the term-mismatch problem, we need to exploit the relations between concepts. Hence, we need a way to quantify these relations. However, inter-concepts similarity is still problematic and non easy to measure [11], because the similarity between two concepts depends on the relation between them. Since relations are different in semantic, e.g. *is-a*, *part-of*, etc., and have different properties, e.g. symmetric or not, there is no standard way on how to quantify relations, where it is task-dependent. For example, for one task the *is-a* relation is much useful than *part-of*, but for another task the *part-of* is much useful, and so on.

Word embedding [10,12] has recently proved its effectiveness for several NLP tasks. It is also studied in Information Retrieval (IR), where word embedding is

---

[1] Concepts have many definitions [1]. A concept here refers to a category ID that encompasses synonymous words and phrases, e.g. UMLS concepts, WordNet synsets.
[2] wordnet.princeton.edu.
[3] www.nlm.nih.gov/research/umls/.

© Springer International Publishing AG, part of Springer Nature 2018
G. Pasi et al. (Eds.): ECIR 2018, LNCS 10772, pp. 563–569, 2018.
https://doi.org/10.1007/978-3-319-76941-7_45

used for ad-hoc retrieval [7], query expansion [6,14], or text similarity [8]. Some features make word embedding potentially useful for IR, where a word is a low-dimensional numerical vector rather than a sequence of characters and algebraic operations between vectors reflect semantic relatedness between words [10].

Concept embedding takes word embedding to a higher level. It is the process of representing concepts by low-dimensional vectors of real numbers. Through concept embedding, one can keep the advantages of both conceptual indexing and word embedding, and at the same time avoid some of conceptual indexing disadvantages. More precisely, by using concepts vectors, on the one hand, we exploit concepts to reduce the term-mismatch effect, and on the other hand, we avoid complexities related to relation-based inter-concept similarity, and measure the similarity between two concepts by comparing their corresponding vectors.

In this study, we propose a way to generate concept embedding based on word embedding. Then, we use concepts vectors in classical IR models. It is worth mentioning that we do not aim in this study to compare different approaches of tackling term-mismatch. Hence, we do not report in results any comparison between our approach and approaches like: pseudo-relevance feedback or word based expansion. The main goal of this paper is to check the *profitability of using concept embedding, and the adaptability of vector based concept similarity to IR.*

2   Related Works

De Vine et al. [5] build medical concept embedding through replacing the textual content of documents by their corresponding medical concepts, and then training *word2vec* [10] on the new corpus, which is now a sequence of concepts. At the end of the process, they obtain a vector representation for each concept appeared in the corpus. Choi et al. [2] use a similar approach to obtain concepts vectors, except that they use temporal information from medical claims to adapt the definition of context window of *word2vec* to medical data. Both approaches build vectors for the concepts that only appear in the corpus and not for all concepts of the corresponding knowledge resources. Furthermore, if we build word embedding vectors of the same corpus, then concepts vectors and words vectors will not be comparable, because they are represented in different vector spaces.

Several studies proposed to use word embedding to represent more informative elements rather than a single word. Clinchant et al. [3] use Fisher kernel to aggregate words vectors of a document to build a document vector. Le et al. [9] extend *word2vec* to be able to compute paragraph-level embedding. Zamani et al. [13] optimize a query language model to estimate the embedding vector of a query, where averaging query's words vectors is a special case of their approach.

Concerning the inter-concepts similarity, many approaches have been used in literature [11]. They can be categorized [11]: 1- *path-based* measures, which depend on the length and the nature of the path that links two concepts within a knowledge resource; 2- *information content* measures, which use some corpus-based statistics to estimate the information content of a concept, and then measuring similarity; and 3- *vectors-based* measures, which depend on the ability to represent concepts by vectors, where cos is the main measure in this category.

3   Concept Embedding

We present in this study three methods for concept embedding based on word embedding. The difference between these methods is the additional information that is used, beside word embedding vectors, to build concepts vectors.

Flat embedding (FEmb): In this method, we do not use any additional information rather than word embedding vectors. The main hypothesis here is that any concept can be mapped to a set of words. Hence, the embedding vector of a concept c is a function F of the vectors of its words. For example, in Word-Net the two words "snake" and "serpent" belong to the "S01729333" synset, so $\overrightarrow{S01729333} = F(\overrightarrow{snake}, \overrightarrow{serpent})$, where F is any function able to merge several vectors in only one vector, e.g. vectors addition, vectors average, etc.

Hierarchical embedding (HEmb): Beside word embedding vectors, we use in this method the internal structural information of each concept. This method is initially proposed to deal with UMLS medical concepts, but it is applicable to any resource exhibiting similar concept structure. In UMLS, each *concept c* consists of several *terms*, which represent the different forms of text that could be used to express the underlying meaning of c, and each term could appear in different lexical variations or *strings*, where a string can be mapped to either a word or a set of words. Therefore, we have a hierarchy related to each concept. Assume that a concept c consisting of two terms t_1, t_2, and each term t_i consists of two strings s_1^i, s_2^i. In this case, $\overrightarrow{c} = F\left(F\left(s_1^1, s_2^1\right), F\left(s_1^2, s_2^2\right)\right)$.

Weighted embedding (WEmb): This method is an extension of *FEmb*, where we incorporate external statistical information. More precisely, instead of equally treating the words of each concept, we attach a weight indicating their relative importance, i.e. $\overrightarrow{c} = F(\alpha_1 \overrightarrow{w_1}, \ldots, \alpha_n \overrightarrow{w_n})$, where α_i is the weight of w_i.

Evaluation strategy: Since our goal is to study the profitability of concept embedding for IR, we evaluate the retrieval performance improvement of an IR model that is able to incorporate inter-terms similarity. We use the model of [4]:

$$RSV(d, q) = \sum_{c \in q} weight_q(c) \times sim(c, c^*) \times weight_d(c^*) \tag{1}$$

where $sim(c, c^*)$ is the similarity between two concepts, and c^* is the closest document concept to the query concept c according to the similarity measure sim. If the query concept c also belongs to d, then $c^* = c$ and $sim(c, c^*) = 1$. We use several definitions for sim, some of them are vector based and some are not. By this way we can see if concept embedding vectors are useful for IR.

4   Experimental Setup

Generating word embedding: We generate concept embedding vectors based on word embedding vectors. To obtain words vectors, we train *word2vec* on open access *PubMed Central* collection[4], with the following configurations: vector size 500, continuous bag of words, window size 8, and negative sampling is set to 25.

Generating concept embedding: We apply our approach to *UMLS2017AA* medical concepts, and we only consider the concepts that have English content. Assume the following example for clarification. The concept *C0004238* (denoted c) has two textual forms or terms: *L0004238* (denoted t_1) and *L0004327* (denoted t_2). Term t_1 appears in two lexical variations: singular *S0016668*="atrial fibrilliation" (denoted s_1^1), and plural *S0016669*="atrial fibrilliations" (denoted s_2^1). The same for term t_2 which corresponds to two strings s_1^2 and s_2^2. By tokenizing, we transform each string s_j^i to a set of words $W_{s_j^i}$ (*we remove duplication*).

In **FEmb**, the concept vector is: $\vec{c} = avg(\vec{w_1}, \ldots, \vec{w_l})$, where $\vec{w_i}$ is the word embedding vector of word w_i, $w_i \in \bigcup_{i,j} W_{s_j^i}$, and avg returns the average of a set of vectors. For **HEmb**, the concept vector is: $\vec{c} = avg(avg(t_1), \ldots, avg(t_m))$, where $avg(t_i) = avg(avg(s_1^i), \ldots, avg(s_k^i))$, $avg(s_j^i) = avg(\vec{w_1}, \ldots, \vec{w_l})$, and $w \in W_{s_j^i}$. Concerning **WEmb**, we follow the same approach as *FEmb*, except that we compute the weighted average $wavg$ instead of average avg. More precisely, $\vec{c} = wavg(\vec{w_1}, \ldots, \vec{w_l}) = \frac{1}{l} \sum_w \alpha_w \vec{w}$, where l is the number of words. The weight α_w of a word w is its *idf* score in *PubMed*, namely $\alpha_w = \ln(\frac{N+1}{n})$, where N is the number of documents in *PubMed* and n is document frequency of w.

We generate fixed random vectors for missing words, which means, if a missing word w appears in several concepts, we use the same randomly generated vector. For the *idf*-weight of missing words, we tested several options: assuming that the word is too popular ($n = N$), too rare ($n = 1$), or in between ($n = \frac{N}{2}$). The three approaches give similar performance; therefore, we only report the first option where $n = N$, which means, a poor *idf* score.

Test collections: To evaluate our proposal, we use ad-hoc image-based corpus of ImageCLEF (www.imageclef.org) of years 2011 (*clef11*) and 2012 (*clef12*), where documents are captions of medical images with short queries. *clef11* has 230K documents and 30 queries. *clef12* contains 300K documents and 21 queries (we removed query 14 because it is not mapped to any concept). Documents and queries are mapped to UMLS concepts using MetaMap (metamap.nlm.nih.gov).

IR model and concept similarity: There are three components to be described in the IR model of (1). The weight of concepts in documents and queries, and the similarity between concepts. To compute the weight of a concept in a document or a query, we apply two classical IR weighting schema: Pivoted Normalization or BM25 [1]. For both models, we use standard parameters values reported in [1]. To compute the similarity between concepts $sim(c, c')$, we use two measures. The first one is compatible with the vector representation of concepts:

$$sim(c, c') = \begin{cases} 0 & \cos(\theta) \leq 0 \\ \beta \times \cos^2(\theta) & \text{otherwise} \end{cases} \quad (2)$$

---

[4] www.ncbi.nlm.nih.gov/pmc/, *PubMed* collection contains: 1177879 vocabularies.

where θ is the angle between the two vectors $\overrightarrow{c}$ and $\overrightarrow{c'}$, and β is a tuning parameter. We optimized the value of β on *clef11* but it is applied to all collections. In our results, we only report the retrieval performance of $\beta = 0.5$. In addition, we only consider the similarity when $\cos(\theta) > 0$, i.e. we ignore the concepts that could have an opposite meaning. We use $\cos^2(\theta)$ instead of $\cos(\theta)$, because it is more discriminant, especially for small angles $\theta \in [-\frac{\pi}{4}, \frac{\pi}{4}]$. For comparison, we use Leacock measure [11], which depends on the length of the path of *is-a* relations between two concepts in UMLS.

5   Evaluation

Table 1 shows results for *clef11* and *clef12*. *FEmb*, *HEmb*, and *WEmb* refer to our approaches to build concepts vectors, where the similarity measure between concepts is (2). *NoEmb* refer to deal with concepts rather than concepts vectors, where *Leacock* refer to the similarity between concepts, whereas, we do not incorporate similarity in *NoSim*. ∗ and † refer to a statistically significant difference with *NoEmb_NoSim* and *NoEmb_Leacock*, respectively, according to Fisher Randomization test with ($\alpha < 0.05$).

Table 1. Experimental results for *clef11* and *clef12* collections

	clef11				clef12			
	piv		bm25		piv		bm25	
	MAP	P@10	MAP	P@10	MAP	P@10	MAP	P@10
NoEmb-NoSim	0.1096	0.2300	0.1552	0.3100	0.0978	0.1381	0.1083	0.1571
NoEmb-Leacock	0.1085	0.2267	0.1505	0.2933	0.0927	0.1429	0.1064	**0.1667**
FEmb_Eq2	0.1089	**0.2333**	0.1608	**0.3167**	0.0934	0.1429	0.1119	0.1524
HEmb_Eq2	0.1111	0.2100	0.1640∗†	0.3133	0.0987	**0.1524**	0.1140∗	0.1619
WEmb_Eq2	**0.1137**∗	0.2267	**0.1654**∗†	0.3133	**0.1012**	0.1476	**0.1154**∗	0.1571

Table 1 shows that exploiting relations between concepts and using a relation-based similarity measure introduce noise, where the MAP of *NoEmb-Leacock* is lower than the MAP of *NoEmb-NoSim* for both IR models and in both collections. P@10 is also lower in *clef11* and slightly better in *clef12*.

WEmb gives the best MAP among our approaches, where we use external statistical knowledge beside word embedding vectors. The comparison of *WEmb* to *NoEmb-NoSim* shows that representing concepts by vectors and using vector based inter-concept similarity improve the results. In 3 out of 4 cases the improvement is statistically significant. Moreover, there is no degradation in P@10. If we compare *WEmb* with *NoEmb-Leacock*, we see that there is a small gain of MAP (for *clef11* and *bm25* the gain is statistically significant), and without corrupting P@10. Our approaches to represent concepts by vectors, and use vector-based similarity, improve MAP without corrupting P@10, i.e. the approaches are able to improve results without introducing noise. The only exception is *HEmb*, where building concepts vectors considers the same word several times if it appears in several strings of the same concept, which represents a possible source of noise.

6 Conclusion

We presented three approaches to build concept embedding vectors based on pre-trained word embedding vectors. We used concepts vectors along with a vector-based similarity to improve IR performance. The results are promising, where the overall performance is improved without losing the absolute precision.

This study can be extended by achieving more in depth free parameters tuning, especially for vector size. Furthermore, we mainly compare the performance of a path-based measure, i.e. Leacock, to a vector-based measure (2) [11]. However, we can also compare the results to content-based measures [11].

Both words and concepts are represented in the same vector space, so they are comparable. It is thus possible to compare concepts to the original textual content of documents. This is helpful to either achieve conceptual indexing or to improve the quality of some conceptual indexing methods like MetaMap by filtering out non-related or noisy concepts.

References

1. Abdulahhad, K.: Information Retrieval (IR) modeling by logic and lattice. Application to Conceptual IR. Theses. Université de Grenoble, May 2014
2. Choi, Y., Chiu, C.Y.I., Sontag, D.: Learning low-dimensional representations of medical concepts. In: AMIA Summits on Translational Science Proceedings 2016, p. 41 (2016)
3. Clinchant, S., Perronnin, F.: Aggregating continuous word embeddings for information retrieval. In: Proceedings of the Workshop on Continuous Vector Space Models and their Compositionality, pp. 100–109 (2013)
4. Crestani, F.: Exploiting the similarity of non-matching terms at retrievaltime. Inf. Retr. **2**, 27–47 (2000)
5. De Vine, L., Zuccon, G., Koopman, B., Sitbon, L., Bruza, P.: Medical semantic similarity with a neural language model. In: Proceedings of the 23rd ACM International Conference on Conference on Information and Knowledge Management, CIKM 2014, pp. 1819–1822 (2014)
6. Diaz, F., Mitra, B., Craswell, N.: Query expansion with locally-trained word embeddings. CoRR abs/1605.07891 (2016). http://arxiv.org/abs/1605.07891
7. Guo, J., Fan, Y., Ai, Q., Croft, W.B.: Semantic matching by non-linear word transportation for information retrieval. In: the 25th ACM International on Conference on Information and Knowledge Management, CIKM 2016, pp. 701–710 (2016)
8. Kenter, T., de Rijke, M.: Short text similarity with word embeddings. In: Proceedings of the 24th ACM International on Conference on Information and Knowledge Management, CIKM 2015, pp. 1411–1420 (2015)
9. Le, Q., Mikolov, T.: Distributed representations of sentences and documents. In: Proceedings of the 31st International Conference on International Conference on Machine Learning, ICML 2014, vol. 32, pp. II-1188–II-1196 (2014). JMLR.org

10. Mikolov, T., Sutskever, I., Chen, K., Corrado, G., Dean, J.: Distributed representations of words and phrases and their compositionality. In: 26th International Conference on Neural Information Processing Systems, NIPS 2013, pp. 3111–3119 (2013)

11. Pedersen, T., Pakhomov, S.V.S., Patwardhan, S., Chute, C.G.: Measures of semantic similarity and relatedness in the biomedical domain. J. Biomed. Inform. **40**(3), 288–299 (2007)

12. Pennington, J., Socher, R., Manning, C.D.: Glove: global vectors for word representation. EMNLP **14**, 1532–1543 (2014)

13. Zamani, H., Croft, W.B.: Estimating embedding vectors for queries. In: Proceedings of the 2016 ACM International Conference on the Theory of Information Retrieval, ICTIR 2016, pp. 123–132 (2016)

14. Zuccon, G., Koopman, B., Bruza, P., Azzopardi, L.: Integrating and evaluating neural word embeddings in information retrieval. In: Proceedings of the 20th Australasian Document Computing Symposium, ADCS 2015, pp. 12:1–12:8 (2015)

Patient-Age Extraction for Clinical Reports Retrieval

Rúben Ramalho, André Mourão, and João Magalhães[✉]

NOVA LINCS, Faculdade de Ciências e Tecnologia,
Universidade NOVA Lisboa, Caparica, Portugal
ras.ramalho@gmail.com, a.mourao@campus.fct.unl.pt,
jm.magalhaes@fct.unl.pt

Abstract. Patient demographics are of great importance in clinical decision processes for both diagnosis, tests and treatments. Natural language is the standard in clinical case reports, however, numerical concepts, such as age, do not show their full potential when treated as text tokens. In this paper, we consider the patient age as a numerical dimension and investigate several Kernel methods to smooth a temporal retrieval model. We extract patient age from the clinical case narrative and extend a Dirichlet language to include the temporal dimension. Experimental results on a clinical decision support task, showed that our proposal achieves a relative improvement of 5.7% at the top 10 retrieved documents over a time agnostic baseline.

1 Introduction

In the past decade more attention as been drawn towards modelling explicit concepts such as patient demographics in clinical information systems. The incorporation of NLP techniques in clinical decision systems has shown several successes [2], leading to an increase in the number of NLP systems for automatic information extraction. These contributions range from temporal relations extraction [8], to the extraction of named entities, such as diseases [7].

In the Clinical Decision Support track of the TREC conference series, all clinical reports contain rich temporal information. However, the best performing team in 2015 [1], did not explicitly model patient age but instead adopted a Markov random field model and jointly optimised the weights of statistical and semantic concepts from different sources. These strategies capture patient data like disease and symptoms, but do not take full advantage of temporal data. If we naively consider representing numerical features as n-grams (e.g. *70 years old*, *71 yo*) it becomes clear that this is an unfruitful pursuit because n-grams representation does not allow a retrieval model to integrate temporal algebra. The system proposed by the *DUTH* team [3] tried to leverage demographics, including age, and explicitly incorporated this information into the retrieval system. Their age extraction approach is a rule based method that classify documents/queries according to age groups, e.g. child ($age < 16$), and use these categories as expansion concepts.

© Springer International Publishing AG, part of Springer Nature 2018
G. Pasi et al. (Eds.): ECIR 2018, LNCS 10772, pp. 570–576, 2018.
https://doi.org/10.1007/978-3-319-76941-7_46

We argue that using patient age, require further investigation about how to embedded temporal data. This is our key motivation - the idea put forward by this article is that we can extract the patient age from the narrative and use this added information to improve the model performance. To successfully deliver this idea, we include the patient age as a new structured field in the retrieval model and assess different Kernel methods for smoothing temporal computations.

2   Patient-Age Extraction from Clinical Documents

Age expressions are rather explicit in the narrative and do not span multiple sentences. In our approach, we first pre-process the narrative by normalising numerical expressions (e.g. replacing *fifteen* with the numerical value 15) and segmenting the document D into individual sentences $S_1, S_2, \cdots, S_m$, that are posteriorly tokenized into individual tokens $X_1, X_2, \cdots, X_k$ for k length sentences. For this tasks we relied on the CoreNLP toolset [4]. Tokens of each sentence S_i are then arranged into n-grams ϕ_j, which we formally represent as $\phi_j = \{X_j, X_{j+1}, \cdots, X_{j+n}\}$, that is, ϕ_j is a sequence of tokens starting at position j and finishing at position $j + n$.

After the narrative is pre-processed, individual n-grams ϕ_i with a constant size of $n = 10$ are considered and a detector of the form $p(Y_i|\phi_i, \theta)$, estimates the probability of observing age information on that n-gram. Following this process, we analyse the whole narrative and identify the n-gram that is most likely to contain information about the age of the patient. Formally, we compute

$$\hat{\phi} = \arg \max_{\phi_j} p(age_j|\theta, \phi_j \in D)$$

which determines n-gram that we regard as the most likely to contain the patient age. In Algorithm 1, we present the algorithm to extract the patient-age, which includes the temporal expression detector and the patient age tagger.

2.1   Detecting Temporal Expressions

Relying on the Bayesian inference framework and insight that the presence of *age* is a binary variable with outcomes (*true*, *false*), a natural choice is to model *age* as a set of Bernoulli trials. This yields the convenient likelihood function

$$\mathcal{L}(\theta|\phi) = \prod_{i=1}^{n} \mathrm{Bern}\left(age_i|\delta(\phi_i, \theta)\right),$$

where δ is a $[0, 1]$ bounded functional of the n-gram that expresses our model assumptions and parameter θ. Maximisation of the above function would be the path in the presence of annotated data. However, confronted with the lack of a labelled sample we relaxed the model by assuming each token is conditionally independent given its outcome (bag of words model), thus arriving at the estimator

$$p(age|\phi, \theta) = \frac{|\phi \cap \theta_r|}{n},$$

Algorithm 1. Patient-age extraction

Data: Document Narrative
Result: Patient Age
$w := \{\}$;
for *Sentence* $\in$ *Document* **do**
    for *n-gram* $\in$ *Sentence* **do**
        n-gram.score $:= p(age|\phi, \theta)$;
        w.add(*n*-gram);

    for *n-gram* $\in$ *Sorted(w.score, "DESC")* **do**
        for *token* $\in$ *n-gram* **do**
            if *not previous(token)* $\in$ *prefix_x and not next(token)* $\in$ *suffix_x* **then**
                return token;
            else
                continue;

return ∞;

---

where θ_r is a set of tokens that we deemed relevant; in other words, it is a reference language model that contains words commonly found nearby ages (e.g. 'years', 'old'). Finally, estimation is now reduced to the count of tokens that appear in both language models normalised by n.

2.2 Temporal (Patient-Age) Tagger

Having exposed the inner workings of the n-gram selector, we turn to the problem of identifying the patient age within a set of tokens; this is a rule based process and takes into account both token suffixes and prefixes. Isolated analysis of numerical tokens does not yield much information about their meaning.

For that purpose we established a sufficient condition for a token to be considered the target token: the token must be numeric, within the domain $[0, 120]$ (be a plausible age) and its suffix or prefix are not in the set of exclusion terms; exclusion terms are used to filter out numerical tokens that do not convey age information, such as, *5* MM, *10 persons* or *60* s.

3   Ranking with Smoothed Patient-Age

In the query likelihood model [5] a Dirichlet language model is constructed for each document and we make the assumption that a given query q has been generated by a latent language model d. Given this assumption, documents are ranked by their likelihood of having generated q. After applying Bayes theorem and dropping the normalising constant we have:

$$p(d \,|\, q) \propto p(q \,|\, d)p(d),$$

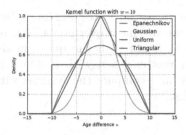

Fig. 1. Different Kernel functions.

Table 1. Kernel parametrisation.

Kernel	Expression
Uniform	$\frac{1}{2} \mathbf{1}_{\{|u| \leq w\}}$
Normal	$\exp\left(\frac{-u^2}{2w^2}\right)$
Triangular	$\left(1 - |u|\right) \mathbf{1}_{\{|u| \leq w\}}$
Epanechnikov	$\frac{(w^2 - u^2)}{w + w^2 \cdot \frac{4}{3}} \mathbf{1}_{\{|u| \leq w\}}$

where $p(q|d)$ is usually represented as a multinomial unigram language model. Delving deeper into $p(q|d)$ is out of the scope of this article, for an solid discussion of existing approaches please consult [9].

In our method we consider that a query q with patient age q_a has also been generated by an unknown language model d, thus resulting in the query likelihood model $p(d|q, q_a)$. Assuming independence between q and q_a, we can rewrite the model as

$$p(d \mid q, q_a) \propto p(q, q_a \mid d)p(d) \propto p(q \mid d)p(d)p(q_a \mid d) \propto p(d \mid q)p(q_a \mid d),$$

which yields a much more convenient formulation. This ranking modification may be rewritten as the product of any other query likelihood model (Fig. 1).

Different alternatives are available to convey structure to $p(q_a|d)$ – the probability that a query with an given patient-age was generated from document d. Assuming a Gaussian process, we could say that d generates q_a according to a normal distribution with mean d_a (age in document): $Q_a \sim \mathbb{N}(d_a, w)$. In this form $p(q_a|d)$ becomes the likelihood of q_a being observed given d_a:

$$P(q_a \mid d) = \mathcal{L}(d_a \mid q_a) = \frac{1}{\sqrt{2\pi w^2}} \exp\left[-\frac{(q_a - d_a)^2}{2w^2}\right] \propto \exp\left[-\frac{(q_a - d_a)^2}{2w^2}\right],$$

where w is a model parameter that expresses the expected variability in the generation process. We also considered the Kernel smoothers of Table 1[1], besides the Gaussian distribution. All the models were defined in terms of u: the difference between estimated query age and estimated document age ($u = \hat{q}_a - \hat{d}_a$).

4 Evaluation

4.1 Experimental Setup

Corpus. The corpus consists of a sample of 733,138 articles from the Open Access Subset of PubMed Central (PMC), an online digital database of freely available full-text biomedical literature. This collection was used in the TREC Clinical Decision Support track [6]. In this paper we used 30 medical clinical case descriptions as queries (they describe the patients history and symptoms) and the full set of relevant articles.

---
[1] Where $\mathbf{1}$ represents the indicator function.

Protocol. The test hypothesis is that age is an important factor and its explicit inclusion in a clinical IR setup will lead to an improvement over non age-aware systems. Two systems were defined, a control system with the query model $p(d \,|\, q)$ and a system whose ranking function was enriched with patient age information $p(d \,|\, q, q_a)$. We proceeded to assess and compare the systems with Accuracy, Precision and Mean Average Precision.

4.2   Results and Discussion

Patient-Age Extraction Baseline. Given the specialised scope of the problem of age extraction for medical corpora we evaluate our age extractor against a simple statistical baseline that tries to randomly guess if an article contains age information and its location in the document. A simple baseline for the task of binary classification is the conditional expectation: $P(Detection|Doc) = \frac{1}{2}$. That is, assuming equally distributed classes a classifier that always says "age is present" is correct half the time. For the task of tagging we assume that the average document length is 1000 words, we can now estimate the accuracy of a random guesser to be: $P(Location|Doc) = \frac{1}{1000}$. Since the task of tagging and detection are independent tasks, the joint probability of correct detection and correct tagging, $P(Detection, Location|Doc)$, becomes:

$$P(Det.|Doc) \cdot P(Loc.|Doc) = 0.0005$$

Experiment: Patient-Age extraction. For the purpose of assessing the degree to which the extractor is reliable on the detection and tagging of patient-age in documents, a sample of 50 random documents (holdout) were manually labelled and the temporal detector and localiser evaluated. The results of our evaluation, Table 2, indicates that the proposed method is significantly superior to a random baseline. The temporal tagging task is not as a good as the temporal expression detection because it requires a more exhaustive language model, containing more expressions or richer rules to locate the presence of patient-age mentions.

Table 2. Extraction accuracy results.

	Baseline	Age
Temp. exp. detection	0.500	0.880
Temp. tagger	0.001	0.778
Temp. inf. extraction	0.001	0.685

Experiment: Patient-Age-Aware Retrieval. Now that we assessed the age extraction accuracy, we turn our attention to the influence of age information in a clinical-case retrieval task. In this experimental setting, the extracted age (with

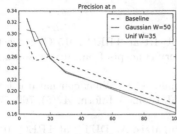

Fig. 2. Precision at n.

Table 3. P@10 and MAP results.

Kernel	w	P@10	MAP
None	-	0.287	**0.097**
Gaussian	65	0.297	0.085
Epanechnikov	65	0.2833	0.076
Uniform	35	**0.303**	0.072
Triangular	65	0.260	0.066

an accuracy of 68.45%) was incorporated on an information retrieval system both on the document and query sides. The precision of the two systems in terms of Mean Average Precision (MAP) and Precision at 10 (P@10) are summarised on Table 3. Results suggest that age-aware systems are an important dimension in clinical retrieval systems. The top ranked documents are consistently more relevant than the baseline, independently of the used Kernel. While both Uniform and Gaussian Kernel exhibited a moderate improvement over the baseline, the Uniform Kernel was more precise at 10 and the Gaussian Kernel did better in terms of MAP. Top-rank high-precision methods are a corner stone for many other methods such as pseudo-relevance feedback. Thus, retrieval models that rely on temporal data with Uniform or Gaussian smoothing will improve the overall performance of clinical decision support systems (Fig. 2).

5   Conclusions

We started by exposing the importance of patient demographics in a clinical setting, leading to the need of new query likelihood models that consider age-as-a-field. The extraction of patient age as a structured field from the clinical narrative and the proposed retrieval model achieved a significant improvement: the full age-aware retrieval system achieved a 5.7% in improvement P@10, over a time agnostic baseline. This improvement relied on the performance of the age extraction method (68.45%), which can be improved with a better parsing of the age expressions, thus, bringing further gains in terms of top-rank high-precision models.

Acknowledgements. This work has been partially funded by the NOVA LINCS project ref. UID/CEC/04516/2013.

References

1. Balaneshinkordan, S., Kotov, A., Xisto, R.: WSU-IR at TREC 2015 CDS track: joint weighting of explicit and latent medical query concepts from diverse sources. In: TREC (2015)
2. Demner-Fushman, D., Chapman, W.W., McDonald, C.J.: What can natural language processing do for clinical decision support? J. Biomed. Inform. **42**(5), 760–772 (2009)
3. Drosatos, G., Roumeliotis, S., Kaldoudi, E., Arampatzis, A.: DUTH at TREC 2015 clinical decision support track. In: TREC (2015)
4. Manning, C.D., Surdeanu, M., Bauer, J., Finkel, J., Bethard, S.J., McClosky, D.: The Stanford CoreNLP natural language processing toolkit. In: ACL System Demonstrations (2014)
5. Ponte, J.M., Croft, W.B.: A language modeling approach to information retrieval. In: ACM SIGIR (1998)
6. Roberts, K., Simpson, M.S., Voorhees, E.M., Hersh, W.R.: Overview of the TREC 2015 clinical decision support track. In: TREC (2015)
7. Savova, G.K., Masanz, J.J., Ogren, P.V., et al.: Mayo clinical text analysis and knowledge extraction system (cTAKES): architecture, component evaluation and applications. J. Am. Med. Inform. Assoc. **17**(5), 507–513 (2010)
8. Sun, W., Rumshisky, A., Uzuner, O.: Evaluating temporal relations in clinical text: 2012 i2b2 challenge. J. Am. Med. Inform. Assoc. **20**(5), 806–813 (2013)
9. Zhai, C., Lafferty, J.: A study of smoothing methods for language models applied to ad hoc information retrieval. In: ACM SIGIR (2001)

Inverted List Caching for Topical Index Shards

Zhuyun Dai$^{(\boxtimes)}$ and Jamie Callan

Language Technologies Institute, Carnegie Mellon University, Pittsburgh, USA
{zhuyund,callan}@cs.cmu.edu

Abstract. Selective search is a distributed retrieval architecture that intentionally creates skewed postings and access patterns. This work shows that the well-known QtfDf inverted list caching algorithm is as effective with topically-partitioned indexes as it is with randomly-partitioned indexes. It also shows that a mixed global-local strategy reduces total I/O without harming query hit rates.

1 Introduction

Search engines use caching to reduce computation and disk access. One form of caching, *list caching*, keeps the inverted lists of frequent query terms in memory. Usually it is used by distributed search architectures that *randomly* partition the corpus into index shards, maintain one list cache per shard, and assign one shard per processor. Thus postings are distributed uniformly across index shards, list caches, and processors.

Selective search is a distributed search architecture that partitions a corpus into many *topic-based* index shards and routes each query to the few shards that are most likely to have relevant documents [1,4,5]. Topic-based indexes skew postings distributions and access patterns. Few index shards are searched per query, so it is not necessary to have a one-to-one mapping between shards and processors. Instead, multiple topic-based index shards are assigned to each processor. These differences create a substantially different caching environment - one in which data and access patterns are more skewed, and multiple index shard caches compete for a processor's RAM.

It is an open question whether inverted list caching is effective with topic-based index shards. This paper investigates the behavior of the well-known QtfDf static caching algorithm [2] when used with selective search. First it investigates whether topically indexed shards reduce the impact of QtfDf caching. Second, it investigates whether using shard-wise query log information can improve list caching for selective search.

2 Related Work

Selective search partitions a collection into topic-based index shards that have skewed vocabularies and access patterns. Resource selection algorithms

© Springer International Publishing AG, part of Springer Nature 2018
G. Pasi et al. (Eds.): ECIR 2018, LNCS 10772, pp. 577–583, 2018.
https://doi.org/10.1007/978-3-319-76941-7_47

(e.g., Rank-S [5], Taily [1]) select shards to search for each query. Shards are searched in parallel; results are merged to form a final ranking. Computational costs are lower than with traditional distributed retrieval because fewer and smaller shards are searched for any query [4,5].

Inverted list caching has been studied extensively [2,6–8]. Most studies considered a centralized system or a traditional distributed system in which queries are run on all shards. QtfDf [2] is a well-known static caching policy for posting lists. It caches the posting lists of the terms with the highest values of the ratio $\frac{qtf}{df}$, where qtf denotes the frequency of the term in a query log, and df the document frequency of the term. Experiments showed that although QtfDf is a static policy, it can outperform dynamic policies because the query term vocabulary is stable over time [2].

3   Caching Strategies

A distributed system has a total of C CPU cores. Each core has a cache of size m. With random document partitioning, there are C shards, and each core serves one shard. With topical partitioning, there are P shards, typically with $P \gg C$, and each core serves multiple shards. The cache is a key-value data structure. The key is (shard s, term t), and the corresponding value is the posting list of term t in shard s.

QtfDf [2] uses term frequency in a query log (qtf) and document frequency in the corpus (df) to select terms for the cache. It is a *global* strategy because it does not use shard-specific statistics. We refer to it as QtfDf-G to distinguish it from local methods.

$$\text{QtfDF-G}(t) = \frac{qtf(t)}{df(t)} \tag{1}$$

Each CPU core caches the posting lists of the highest scoring terms from its shard(s). When a CPU core hosts multiple shards, they share its cache. Each *topical* shard uses a different amount of cache because a term's posting list lengths vary across shards. We found it to be more effective than forcing topical shards to use equal amounts of cache. Thus, topical and random partitions do not cache identically under QtfDf-G.

When shards are partitioned randomly, global qtf and df are representative for individual shards. However, topical shards have skewed contents and access patterns. A term may not be equally valuable to each shard. The second strategy estimates the value of term t to shard s.

$$\text{QtfDf-L}(t, s) = \frac{qtf_s(t)}{df_s(t)} \tag{2}$$

$qtf_s(t)$ is the term frequency of term t in queries submitted to shard s. $df_s(t)$ is the number of documents in shard s that contain term t. Each core ranks its (shard s, term t) pairs, and loads the highest scoring posting lists into the cache until it is full.

4   Experimental Methodology

Document collection: We used ClueWeb09-B which has 50M documents, 96M unique terms, and 109GB of uncompressed postings. Each posting is a pair of (docid, tf).

Selective Search settings followed prior research. The collection was distributed on two 8-core machines (16 cores in total). For exhaustive search, each CPU core served one index partition; documents were distributed *randomly* across 16 partitions. For selective search, the document collection was partitioned into 123 *topical* index shards using QKLD-QInit[1] [3]. Shards were assigned to the 16 cores by random and log-based policies [4]. Taily [1] was used for resource selection with the default parameters. It selected 3.9 shards per query on average. For higher recall rate, we also tested Taily with a lower threshold, which selected an average of 8.5 shards per query.

Query log: We used the AOL query log. Queries from the first 3 weeks were used to populate caches; queries from the next 3 weeks were used for testing. Queries that appeared less than 10 times were filtered out. The training queries had 0.3 million unique terms covering 86 GB of posting lists.

Evaluation: Two metrics were used. **Term hit rate** is the percentage of times a requested term was in the cache for a selected shard. **Query hit rate** is the percentage of queries that had *all* terms in cache for *all* selected shards. Term hit rate measures the effect on posting list I/O. Query hit rate measures the effect on response times. It is harder to achieve high query hit rate than high term hit rate.

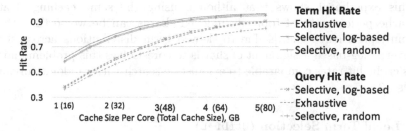

Fig. 1. Term hit rate (solid lines) and query hit rate (dashed lines) of QtfDF-G under different shard assignment policies. Average number of shards searched per query: $n = 3.9$

5   Experimental Results

5.1   Global Term Selection (QtfDf-G)

First we tested the global method QtfDf-G on random (*exhaustive*) and topical (*selective*) shards. There is a many-to-one mapping of topical shards to cores,

---

[1] Obtained   from   http://boston.lti.cs.cmu.edu/appendices/CIKM16-Dai/CW09B-cent1/.

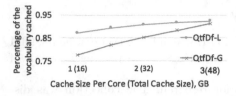

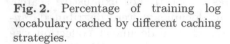

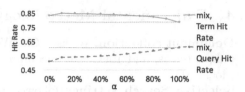

Fig. 2. Percentage of training log vocabulary cached by different caching strategies.

Fig. 3. Hit rate of mixed caching. α: percent of cache used by QtfDf-G. Cache size: 2 GB.

thus selective search requires a shard assignment policy. Most prior research randomly assigned shards to cores [5], but log-based policy improves throughput by assigning shards by their popularity [4]; we tested both.

QtfDf-G is equally effective for exhaustive search and selective search with *log-based* assignment (Fig. 1). QtfDf-G is less effective for selective search with *random* shard assignment, especially for query hit rate. Similar trends were observed in the high-recall setting that searched an average of $n = 8.5$ shards per query (not shown).

Random assignment had a lower hit rate because topical shards have skewed posting distributions. Shards with long inverted lists could be assigned to the same core, forcing them to compete for cache space. Log-based assignment reduced this competition by spreading popular shards across different cores. We found that log-based shard-to-core assignment enables caching of 10% more vocabulary than with random assignment.

This experiment shows that although using the same caching strategy, the cache performance of topically indexed shards can be worse than that of randomly-partitioned shards due to skewed term distributions across shards. However, a log-based assignment of shards to cores solves this problem and produces cache hit rates comparable to exhaustive search with randomly-organized shards.

5.2   Local Term Selection (QtfDf-L)

The second experiment explored the use of local term and query distributions to improve caching for *selective* search. It compares the cache hit rates of global (QtfDf-G) and local (QtfDf-L) strategies. Table 1 shows the hit rates of different strategies.

QtfDf-L produced a substantially higher **term hit rate** than QtfDf-G for small caches. The local-ranking term hit rate was 18% higher than the global-ranking hit rate for the cache size m = 1 GB; this gap gradually decreased when cache size increased.

If QtFDf-G decides to cache a term such as 'apple', it caches the postings from all shards. QtfDf-L may cache the postings from only a few shards (e.g., 'cooking' and 'technology' topics), saving space to cache other terms from other shards. Figure 2 shows that QtfDf-L (partially) caches almost the whole vocabulary, while `QtfDf-G` (fully) caches a smaller vocabulary. This difference is stronger when the cache size is small.

When few shards are searched ($n = 3.9$), the local method `QtfDf-L` has equal or higher **query hit rate** than the global method `QtfDf-G`. However, when the query runs on more shards ($n = 8.5$), `QtfDf-L`'s query hit rate drops substantially. The next two experiments examined this behavior.

5.3 Rank-Biased Local Term Selection (rQtfDf-L)

One cause of cache misses in `QtfDF-L` is the mismatch between the caching strategy and the resource selection algorithm. Resource selection favors shards that have many documents containing query term t (high $df_s(t)$). However, `QtfDf-L` penalizes high $df_s(t)$, thus t is not always cached for the most-relevant shards.

To align caching strategy with resource selection, we propose a rank-biased strategy, $\texttt{rQtfDf-L}(t,s) = \frac{rqtf_s(t)}{df_s(t)}$, where $rqtf_s(t)$ is a discounted query term frequency based on how resource selection ranks shards. It is defined as $rqtf_s(t) = \sum_{q \in Q_s \& t \in q} 1/\log(r_s^q + 1)$, where r_s^q is the rank of shard s for query q given by resource selection. Each occurrence of term t in the shard's local query log Q_s is discounted by $1/\log(r_s^q + 1)$. Term t is more likely to be cached if the shard ranks highly for queries that contains t.

Table 1 reports the hit rates of `rQtfDf-L`. `rQtfDf-L` significantly improved the query hit rate of `QtfDf-L` for small cache sizes. In the high precision setting,

Table 1. Hit rate of `QtfDf-G`, `QtfDf-L` and `rQtfDf-L`. n: average number of shards searched. † and ⋆ indicate statistically significant improvement over `QtfDf-G` and `QtfDF-L`, respectively. Methods are compared by a pairwise permutation test with $p = 0.05$.

	$n = 3.9$ (High precision)						$n = 8.5$ (High recall)					
	Term hit rate			Query hit rate			Term hit Rate			Query hit rate		
Cache	1 GB	2 GB	3 GB	1 GB	2 GB	3 GB	1 GB	2 GB	3 GB	1 GB	2 GB	3 GB
QtfDf-G	0.59	0.78	0.89	0.38	0.61	0.76	0.57	0.79	0.90	**0.38**⋆	**0.60**⋆	**0.76**⋆
QtfDf-L	0.68†	0.82†	0.89	0.40	0.61	0.75	**0.67**†	**0.84**†	0.90	0.27	0.49	0.67
rQtfDf-L	**0.70**†	**0.85**†	**0.91**	**0.43**†⋆	**0.63**	**0.77**	0.66†	**0.84**†	**0.91**	0.31⋆	0.53⋆	0.68

Table 2. Hit rate of `QtfDf-G`, `rQtfDf-L`, and `mix` caching strategy. †: statistically significant improvement over `QtfDf-G` using pairwise permutation test with $p = 0.05$. ᵃ: equivalence to `QtfDf-G` using a noninferiority test with 5% margin and 95% confidence interval.

	$n = 3.9$ (High precision)						$n = 8.5$ (High recall)					
	Term hit rate			Query hit rate			Term hit rate			Query hit rate		
Cache	1 GB	2 GB	3 GB	1 GB	2 GB	3 GB	1 GB	2 GB	3 GB	1 GB	2 GB	3 GB
QtfDf-G	0.59	0.78	0.89	0.38	0.61	0.76	0.57	0.79	0.90	**0.38**	**0.60**	**0.76**
rQtfDf-L	**0.70**†	0.85†	0.91ᵃ	**0.43**†	0.63ᵃ	0.77ᵃ	**0.66**†	**0.84**†	0.91ᵃ	0.31	0.53	0.68
mix	**0.70**†	**0.86**†	**0.92**ᵃ	0.42†	**0.64**ᵃ	**0.78**ᵃ	0.64†	**0.84**†	**0.92**ᵃ	0.36ᵃ	0.58ᵃ	0.74ᵃ

rQtfDF-L has the best cache performance. In the high recall setting, it improved QtfDf-L's query hit rate by up to 15%, but is still lower than QtfDf-G. For term hit rate, rQtfDf-L maintains the performance of QtfDf-L; both local methods are significantly better than the global method for small cache sizes.

5.4   Mixed Caching

The second reason that local term selection methods have lower query hit rates is that they do not cache *general* terms. Local methods favor terms that characterize the shard. It does not cache terms such as 'main', 'cheap', and 'American' that are not topical, but instead commonly modify the scope of other query terms. These general terms occur in a large portion of the query traffic, causing partial miss for many queries.

The last experiment combined global and local methods by allocating part of the cache to each method. QtfDf-G selected terms cached by all shards. rQtfDf-L selected terms cached by individual shards. The percentage of cache used by QtfDf-G was α.

Figure 3 shows the hit rates as a function of α. First there is an increase in term hit rate, meaning that the terms with the highest global QtfDf-G scores are of high importance and should always be cached. Term hit rate then remains higher than rQtfDf-L until 70% of the cache is used by globally-selected terms. Term hit rate is stable because local caching can maintain a useful set of cached terms even with fairly small cache sizes (Fig. 2). On the other hand, query hit rate increased almost linearly as the cache size used by the global method (α) grew. This result suggests that a small local cache is sufficient to maintain term hit rate, whereas a larger global cache is required to provide a good query hit rate. In other words, there is a small vocabulary of important shard-specific terms, and a larger vocabulary of important topic-independent terms.

Table 2 shows hit rates for a mixed strategy that devotes $\alpha = 70\%$ of the cache to global caching. In the high-precision setting, mix slightly improved rQtfDf-L; they both outperformed the QtfDf-G baseline. In the high-recall setting, the query hit rate for mix was almost equal to the QtfDf-G baseline, however the term hit rate was improved by over 10% for small cache sizes. Thus, response time remains the same for most users (as determined by query hit rate), but there is less disk activity and I/O (as determined by term hit rate). A parallel system is likely to have moderately improved throughput.

6   Conclusion

Selective search uses topical index shards and selects just a few shards per query, which skews the distributions of postings and queries across index shards. It has been an open question how typical inverted list caching algorithms perform in this architecture.

Our experiments show that the skewed distribution of postings can cause lower query hit rate in topically indexed shards. However, this effect can be eliminated by using a log-based shard assignment policy to spread popular shards across different cores, which prior research showed also improves load balancing and query throughput.

Global and local term selection methods provide choices to system designers using selective search. Global selection is best for user response time (as determined by query hit rate). Local selection using shard-specific information significantly reduces the posting list I/O (as determined by term hit rate) for small cache sizes, which increases throughput in parallel systems; however the query hit rate is lower compared with the global method when larger number of shards are searched. A rank-biased local variant that incorporates the preferences of resource selection significantly improves the query hit rate for small cache sizes.

A mixed strategy allows trade-offs between user response time and system throughput. Devoting a small portion of the cache to the local method maintains the response time of the global method for most users, but reduces system-wide posting list I/O.

Acknowledgments. This research was supported by National Science Foundation grant IIS-1302206. Any opinions, findings, and conclusions in this paper are the authors' and do not necessarily reflect those of the sponsors.

References

1. Aly, R., Hiemstra, D., Demeester, T.: Taily: Shard selection using the tail of score distributions. In: Proceedings of SIGIR (2013)
2. Baeza-Yates, R., Gionis, A., Junqueira, F., Murdock, V., Plachouras, V., Silvestri, F.: The impact of caching on search engines. In: Proceedings of SIGIR (2007)
3. Dai, Z., Xiong, C., Callan, J.: Query-biased partitioning for selective search. In: Proceedings of CIKM (2016)
4. Kim, Y., Callan, J., Culpepper, J., Alistair, M.: Load-balancing in distributed selective search. In: Proceedings of SIGIR (2016)
5. Kulkarni, A.: Efficient and effective large-scale search. Ph.D. thesis, Carnegie Mellon University (2013)
6. Marin, M., Gil-Costa, V., Gomez-Pantoja, C.: New caching techniques for web search engines. In: Proceedings of HPDC (2010)
7. Saraiva, P.C., Silva de Moura, E., Ziviani, N., Meira, W., Fonseca, R., Riberio-Neto, B.: Rank-preserving two-level caching for scalable search engines. In: Proceedings of SIGIR (2001)
8. Zhang, J., Long, X., Suel, T.: Performance of compressed inverted list caching in search engines. In: Proceedings of WWW (2008)

ALF-200k: Towards Extensive Multimodal Analyses of Music Tracks and Playlists

Eva Zangerle$^{(\boxtimes)}$, Michael Tschuggnall, Stefan Wurzinger, and Günther Specht

Department of Computer Science, Universität Innsbruck, Innsbruck, Austria
{eva.zangerle,michael.tschuggnall,stefan.wurzinger,
guenther.specht}@uibk.ac.at

Abstract. In recent years, approaches in music information retrieval have been based on multimodal analyses of music incorporating audio as well as lyrics features. Because most of those approaches are lacking reusable, high-quality datasets, in this work we propose ALF-200k, a publicly available, novel dataset including 176 audio and lyrics features of more than 200,000 tracks and their attribution to more than 11,000 user-created playlists. While the dataset is of general purpose and thus, may be used in experiments for diverse music information retrieval problems, we present a first multimodal study on playlist features and particularly analyze, which type of features are shared within specific playlists and thus, characterize it. We show that while acoustic features act as the major glue between tracks contained in a playlists, also lyrics features are a powerful means to attribute tracks to playlists.

Keywords: Music information retrieval · Multimodal dataset
Lyrics features · Audio features · Playlist analyses · Classification

1 Introduction

With the advent of music streaming platforms, the way users consume music has changed fundamentally. Users stream music from large online music collections and listen to it using a variety of devices [1]. Platforms like Spotify[1] naturally also provide means for creating personal playlists. The analysis of such playlists has mostly been performed from either an (automatic) playlist generation perspective (e.g., [2]) or an organizational perspective (e.g., [3]). Also, features extracted from tracks (either from the track's audio signal or from metadata) have been utilized for music classification tasks (e.g., [4,5]). Similarly, multimodal approaches that combine audio with lyrics features have been proposed for tasks like genre classification (e.g., [5]) or emotion recognition (e.g., [6]). Nevertheless, especially when incorporating song lyrics, most of the datasets used either lack quantity or quality, due to the variety and quality of online lyrics sources. In this work,

---

[1] https://www.spotify.com, accessed October 2017.

© Springer International Publishing AG, part of Springer Nature 2018
G. Pasi et al. (Eds.): ECIR 2018, LNCS 10772, pp. 584–590, 2018.
https://doi.org/10.1007/978-3-319-76941-7_48

we at first present the ALF-200k dataset (ALF stands for Acoustic and Lyrics Features), a novel dataset tackling this problem by providing an extensive set of more than 200,000 music tracks together with their occurrences in user's playlists and 176 pre-computed audio and lyrics features. As a first case study incorporating this dataset, we set out to analyze user-generated playlists regarding features that are shared among the tracks within a given playlist, i.e., features that characterize the playlist, are utilized implicitly and in an presumably unconscious manner. Particularly, we perform a multimodal classification task on the characteristics of playlists gathered from Spotify and analyze these in regards to their predictive power. By modeling the analyses as a classification task on a per-playlist basis, we show that acoustic features act as the major glue between tracks contained in the same playlist. We foresee that the dataset and the proposed collection approach may contribute to improving the collection of correct and comprehensive lyrics and audio features in future research.

2 Related Work

Multimodal approaches incorporating both the audio signal and lyrics have been shown to perform well [5–7] for genre classification. Mayer et al. [5,7] use rhyme, part-of-speech, bag-of-words and text statistics for genre classification. They showed that lyrics features can be used orthogonally to audio features and that they can be beneficial in determining different genres. Other approaches solely rely on features extracted from lyrics: Fell and Sporleder [8] propose an n-gram model incorporating vocabulary, style, semantics and song structure for genre classification. Hu et al. [6] propose to use basic text features, lyrics content (bag-of-words), linguistic features, psychological categories, contained sentiment and text-stylistic features for music emotion recognition. However, for none of these tasks datasets of sufficient size are publicly available. The dataset most similar to ALF-200k, the Million Song Dataset (MSD) [9], features one million songs, according artists, last.fm tags and similarities. The musixmatch extension to the MSD dataset provides a mapping between the MSD dataset and lyrics on the musixmatch platform. However, while the MSD contains audio features, no lyrics features are provided. On the other side, solely lyrics features have been utilized for mood detection, e.g., by the MoodyLyrics dataset [10]. Nevertheless, to our knowledge, the proposed large-scale ALF-200k dataset is novel in that it combines rich lyrics and audio features at scale.

3 Dataset

In the following, we present the methods utilized for creating the ALF-200k dataset. To foster reproducibility and repeatability, we make our code and data publicly available on GitHub[2].

---

[2] https://github.com/dbis-uibk/ALF200k.

Generally, we aim to curate a dataset containing tracks, respective lyrics and audio features and playlists of users containing these tracks. We therefore rely on the dataset collected by Pichl et al. [11], which contains 18,000 playlists created by 1,016 users, resulting in a total of 670,000 distinct tracks.

As for the corresponding lyrics features, we propose the following crawling method to ensure reliable, correct and complete lyrics for the analyses: At first, we utilize the provided Spofiy IDs of Pichl's dataset to gather artist names and titles of the according tracks. Along the lines of previous research [4,5], we subsequently search for corresponding lyrics on the following user-contributed lyrics databases. Concretely, we utilize ChartLyrics, LYRICSnMUSIC, LyricWikia, eLyrics.net, LYRICSMODE, METROLYRICS, Mp3lyrics, SING365, SONG-LYRICS and Songtexte.com. While the former three platforms provide an API that allows for gathering lyrics based on artist name and track title, the latter seven do not provide any interface and hence, have to be scraped by gathering the HTML code of the underlying websites. After having gathered the lyrics from the proposed platforms, all tracks with non-English lyrics are removed as a number of features are not available for other languages (e.g., uncommon or slang words). In a next step, we clean the obtained lyrics by removing non-UTF8 characters, superfluous white-spaces and also by removing typical characteristics of online lyrics like track structure annotations (e.g., verse/chorus/interlude/...), references and abbreviations of repetitions (e.g., "–Chorus (x2)–"), annotations of background voices (e.g., "yeah yeah yeah") or track remarks (e.g.,"written by" or "Duration: 3:01"). Subsequently, we incorporate only lyrics into the ALF-200k dataset that are confirmed by at least three of the crawled lyrics platforms. Therefore, we compute the similarity of all found lyrics versions for a given track and rely on word bigrams as a representation of each crawled lyrics. Next, we apply the Jaccard similarity coefficient on the set of word bigrams representing the lyrics for all pairs of lyrics. Finally, we choose the version for which at least three sources share a high similarity according to an empirically estimated threshold. If less than three sources confirm a specific lyrics variant, the respective track is removed from the dataset as it would not be possible to reliably extract lyrics features from it. This presents us with a total of 226,747 lyrics.

As we also aim to include extracted features in the dataset, we rely on *audio and lyrics features* to represent tracks, as these have been shown to be orthogonal and beneficial in multimodal approaches [5,6,12]. As for *audio content descriptors* of tracks, we rely on standard acoustic features retrieved via the Spotify Track API[3]. These content features are extracted and aggregated from the audio signal of a track and comprise: *danceability* (how suitable a track is for dancing), *energy* (perceived intensity and activity), *speechiness* (presence of spoken words in a track), *acousticness* (confidence whether track is acoustic), *instrumentalness* (prediction whether track contains no vocals), *tempo* (in beats per minute), *valence* (musical positiveness conveyed), *liveness* (prediction whether track was recorded live or in studio), *duration* (total time of track) and *loudness* (sound intensity in decibels). Besides acoustic features, we also incorporate more

---

[3] https://developer.spotify.com/web-api/, accessed October 2017.

than hundred different *lyrics features* which have been shown to be beneficial for track classifications [4,8]. We thereby included four different types of lyrics features: lexical [6], linguistic [5,8,13], syntactic [8] and semantic [6,8,14,15] features. Due to space constraints, we provide a detailed overview of all features in Table 1 and refer the interested reader to the according papers.

Table 1. Extracted Lyrics Features (# refers to the number of features contained; bag-of-word features (marked with *) are counted as one feature each, despite that they amount to hundreds of features depending on the lyrics).

Type	#	Features
Acoustic (AU)	10	danceability, energy, speechiness, liveness, acousticness, valence, tempo, duration, loudness, instrumentalness
Lexical (LX)	34	bag-of-words* (4), token count, unique token ratios (3), avg. token length, repeated token ratio, hapax dis-/tris-/legomenon, unique tokens/line, avg. tokens/line, line counts (5), words/lines/chars per min., punctuation and digit ratios (9), stop words ratio, stop words/line
Linguistic (LI)	39	uncommon words ratios (2), slang words ratio, lemma ratio, Rhyme Analyzer features (24), echoisms (3), repetitive structures (8)
Semantic (SE)	55	Regressive imagery (RI) conceptual thought features (7), RI emotion features (7), RI primordial thought features (29), SentiStrength scores (3), AFINN scores (4), Opinion Lexicon scores, VADER scores (4)
Syntactic (SY)	38	POS bag-of-words*, pronouns frequencies (7), POS frequencies (6), text chunks (23), past tense ratio

4 Case Study: User Playlist Characteristics

As a first case study based on the ALF-200k dataset, we are interested in finding features that are shared among tracks within playlists. Therefore, we apply the following method: for each playlist of size s (i.e., playlists containing s tracks), we add s random tracks that are not contained in the original playlist. This allows us to evaluate the binary classification performance by measuring the accuracy at which any given track in the test set was predicted to be part of the playlist or not. By utilizing 5-fold cross-evaluation, the performance of classifiers is measured by computing the average classification accuracy, averaged across all folds. We rely on a set of standard classification approaches provided by the Weka framework [16]: BayesNet, Naïve Bayes, KNN, SVM with different kernels (linear, C-SVM, nu-SVM), J48 decision trees and PART, utilizing the respective standard parameter configurations.

In a preliminary experiment, we determined the minimum required length of a playlist to contain enough reasonable data (tracks) and removed all playlists that do not fulfill the minimum required playlist size (8 tracks), which results in a dataset comprising 7,903 playlists. For each playlist in the dataset, we apply a 5-fold cross-validation and measure the prediction accuracy.

Table 2. Average accuracies for each classifier, sorted by maximum accuracy (Max.)

Feature set	BayesNet	J48	kNN	LibLinear	LibSVM (C)	LibSVM (nu)	Naïve Bayes	PART	Max.
AU	0.65	0.66	0.66	0.70	0.62	0.70	0.65	0.66	**0.70**
all features	0.69	0.62	0.58	0.68	0.55	0.68	0.69	0.62	**0.69**
AU+LX+LI+SE	0.69	0.62	0.57	0.67	0.55	0.67	0.69	0.62	**0.69**
AU+LX+LI	0.69	0.63	0.56	0.67	0.55	0.67	0.69	0.63	**0.69**
AU+LX	0.69	0.63	0.55	0.66	0.54	0.67	0.69	0.63	0.69
LX+LI+SE+SY	0.66	0.60	0.57	0.67	0.55	0.67	0.66	0.60	**0.67**
LX+LI+SE	0.66	0.60	0.56	0.66	0.55	0.66	0.66	0.60	**0.66**
LX+LI	0.65	0.60	0.55	0.65	0.54	0.66	0.65	0.60	**0.66**
LX	0.65	0.60	0.54	0.65	0.54	0.65	0.64	0.60	**0.65**
LI	0.57	0.58	0.57	0.61	0.53	0.61	0.57	0.58	**0.61**
SY	0.59	0.57	0.57	0.60	0.52	0.60	0.59	0.57	**0.60**
SE	0.55	0.56	0.55	0.57	0.51	0.58	0.55	0.56	**0.58**
Baseline	*0.50*	*0.50*	*0.50*	*0.50*	*0.50*	*0.50*	*0.50*	*0.50*	*0.50*

Table 2 lists the average accuracies of all individual feature sets and combinations thereof. Being in line with previous findings (e.g., [5]), the best result is achieved by the SVMs with linear and nu-kernel and reaches 70% by utilizing only acoustic features (AU), slightly outperforming the set of all available features. At a first glance, this indicates that acoustic features represent the main characteristic that holds playlists together. Except for Naïve Bayes and BayesNet, AU reached the best accuracy values for all classifiers. As can be seen in Table 2, the feature sets achieving the worst accuracy results are SE, SY and LI. These findings suggest that—when inspected individually—semantic (e.g., contained sentiment or psychological categories), syntactic or linguistic features (uncommon or slang words) are not able to fully capture what actually makes the tracks of a playlist cohesive. Nevertheless, the best combination without relying on acoustic metrics, i.e., combining only textual features extracted from song lyrics, gains an accuracy of 67%, which is only slightly inferior to the best result. Thus, our preliminary experiments demonstrate that lyrics within playlists are homogeneous to a substantial extent and that they can be used to attribute tracks to playlists. Finally, we also note that the analysis conducted is not able to capture user-specific contextual motivations to put certain tracks in a playlist (besides the mere characteristics of tracks). Users may also create playlists to remember certain events and the music they associate with this occasion as, e.g., their wedding or holidays, where the cohesive features of the playlist do not necessarily lie in the track's characteristics, but rather in the perceived emotion and evoked memories. Nevertheless, we believe that this study can provide interesting and relevant insights into the composition of playlists on streaming platforms from a multimodal perspective.

5 Conclusion and Future Work

In this paper, we presented ALF-200k, a novel, publicly available dataset for multimodal music classification problems, containing over 200,000 tracks including precomputed audio features as well as hundreds of metrics extracted from high-quality lyrics. In an exemplary case study we analyzed multimodal features, particularly focusing on detecting features that are shared within playlists and hence, characterize those playlists. As for future work, we are highly interested in utilizing and learning from the dataset to be able to automatically group playlists per genre, user and also per context [17]. Furthermore, we aim to evaluate the characteristics of Spotify playlists to the quality criterion applied for playlist recommendation tasks [2].

References

1. Zhang, B., Kreitz, G., Isaksson, M., Ubillos, J., Urdaneta, G., Pouwelse, J.A., Epema, D.: Understanding user behavior in spotify. In: 2013 Proceedings of IEEE INFOCOM, pp. 220–224, April 2013
2. Bonnin, G., Jannach, D.: Automated generation of music playlists: survey and experiments. ACM Comput. Surv. (CSUR) 47(2), 26 (2015)
3. Kamalzadeh, M., Baur, D., Möller, T.: A survey on music listening and management behaviours. In: Proceedings of ISMIR 2012 (2012)
4. Hu, X., Downie, J.S., Ehmann, A.F.: Lyric text mining in music mood classification. Am. Music 183(5,049), 2–209 (2009)
5. Mayer, R., Neumayer, R., Rauber, A.: Combination of audio and lyrics features for genre classification in digital audio collections. In: Proceedings of ACM MM, pp. 159–168 (2008)
6. Hu, X., Downie, J.S.: Improving mood classification in music digital libraries by combining lyrics and audio. In: Proceedings of JCDL 2010, pp. 159–168. ACM (2010)
7. Mayer, R., Neumayer, R., Rauber, A.: Rhyme and style features for musical genre classification by song lyrics. In: Proceedings of ISMIR 2008, pp. 337–342 (2008)
8. Fell, M., Sporleder, C.: Lyrics-based analysis and classification of music. In: COLING 2014, pp. 620–631 (2014)
9. Bertin-Mahieux, T., Ellis, D.P., Whitman, B., Lamere, P.: The million song dataset. In: Ismir, vol. 2, p. 10 (2011)
10. Çano, E., Morisio, M.: Moodylyrics: A sentiment annotated lyrics dataset. In: Proceedings of the 2017 International Conference on Intelligent Systems, Metaheuristics & Swarm Intelligence, ISMSI 2017, pp. 118–124. ACM, New York (2017)
11. Pichl, M., Zangerle, E., Specht, G.: Understanding playlist creation on music streaming platforms. In: Proceedings of IEEE Symposium on Multimedia. IEEE (2016)

12. Laurier, C., Grivolla, J., Herrera, P.: Multimodal music mood classification using audio and lyrics. In: Proceedings of ICMLA 2008, pp. 688–693. IEEE (2008)
13. Hirjee, H., Brown, D.G.: Rhyme analyzer: An analysis tool for rap lyrics. In: Proceedings of ISMIR 2010 (2010)
14. Martindale, C.: Romantic progression: the psychology of literary history (1976)
15. Ribeiro, F.N., Araújo, M., Gonçalves, P., André Gonçalves, M., Benevenuto, F.: Sentibench - a benchmark comparison of state-of-the-practice sentiment analysis methods. EPJ Data Sci. **5**(1), 23 (2016)
16. Hall, M., Frank, E., Holmes, G., Pfahringer, B., Reutemann, P., Witten, I.H.: The weka data mining software: an update. ACM SIGKDD Explor. Newsl. **11**(1), 10–18 (2009)
17. Pichl, M., Zangerle, E., Specht, G.: Towards a context-aware music recommendation approach: what is hidden in the playlist name? In: Proceedings of ICDM Workshops, pp. 1360–1365 (2015)

Proposing Contextually Relevant Quotes
for Images

Shivali Goel(✉), Rishi Madhok(✉), and Shweta Garg(✉)

Delhi Technological University, Delhi, India
{shivali_bt2k14,rishi_bt2k14,shweta_bt2k14}@dtu.ac.in

Abstract. Due to the rise in deep learning techniques used for the task of automatic image captioning, it is now possible to generate natural language descriptions of images and their regions. However, these captions are often too plain and simple. Most users on social media and other micro blogging websites use flowery language and quote like captions to describe the pictures they post online. We propose an algorithm that uses a combination of deep learning and natural language processing techniques to provide contextually relevant quotes for any given input image. We also present a new dataset, QUOTES500K, with the goal of advancing research requiring large dataset of quotes. Our dataset contains five hundred thousand (500K) quotes along with the author name and their category tags.

Keywords: Computer vision · Natural language processing
Automatic image captioning · Deep learning

1 Introduction

Nowadays, as more and more pictures are uploaded to social media websites like Facebook or Instagram, the users often caption their pictures with deep meaningful quotes. Even after spending a lot of time and effort online, an appropriate quote may not be found. Therefore, we present a model which provides contextually relevant quotes for a given input image, thereby saving user's time and suggesting better meaningful quotes that the user may skip while searching.

Previous work related to generation of simple natural language description of images includes that of [6] which was one of the first approaches to use Neural Networks for this task. Other approaches [1,7] involved the use of an encoder decoder based architecture, similar to one used in machine translation to generate captions. The Neural network based models were further enhanced by adding an attention module [3,12]. Prior to the use of neural networks, template-based description generation techniques were used [4,5,10]. Most existing work computing the similarity of two strings considers only syntactic similarities measures

---

S. Goel, R. Madhok and S. Garg—Contributed equally to this work.

G. Pasi et al. (Eds.): ECIR 2018, LNCS 10772, pp. 591–597, 2018.
https://doi.org/10.1007/978-3-319-76941-7_49

like the number of common words or n-grams [9], Jaccard Similarity [2], and Hidden Markov Model-based measure [11] etc.

To the best of our knowledge, this is the first work that delves into the combination of a quote search algorithm together with a neural image captioning system in order to suggest fancy captions for pictures.

2 Research Methodology

The proposed algorithm is divided into 3 parts. The first part takes as input, an image given by the user, and outputs a description of the image in natural language. The second part of the algorithm takes the description generated in the first part and then forms a list of candidate quotes, which are relevant for the image, using our QUOTES500K dataset. The third step measures the similarity between each candidate quote and the description generated in the first step. Finally, the three most relevant quotes are returned as the output for the input image. The architecture of the proposed algorithm is shown in Fig. 1. Each of the steps are explained in detail in the following subsections.

2.1 Generating Natural Language Description of Images

In the first step of the algorithm, we built an end to end neural network consisting of object detection and caption generation modules to generate captions. Output from the penultimate layer of the pre-trained model of VGG16 from Keras was used to extract features of the images in the dataset. This was fed into our image model consisting of two hidden layers. Output from this module was further passed on sequentially to the captioning module consisting of a LSTM layer which used words and image I as inputs and predicted one word of the caption at a time, considering the context of image observed and of the preceding words $p(S_t|I, S_0, ..., S_{t-1})$. We trained the model using Stochastic Gradient Descent on the Flickr8K Dataset [13] for a total of 50 epochs reaching a loss value of 1.693. Batch size of 256 and categorical cross entropy loss function was chosen for our experiment. The caption so generated was given as input to the second step of the algorithm.

2.2 Proposing a Set of Candidate Quotes

The second step of the algorithm filtered out and generated a list of candidate quotes i.e. quotes relevant to the image from our QUOTES500K dataset. To prepare this list, comparison was made recursively between two strings — the caption (fixed in all iterations) and a quote from the dataset until all quotes in it were compared.

Formally, let $C = \{c_1, c_2, ... c_n\}$ and $Q = \{q_1, q_2, ... q_m\}$ denote the caption and the quote string respectively. The initial steps included tokenizing the string, removal of stop words and stemming of the words, after which we were left with a bag of words, $bow = \{w_1, w_2, ... w_n\}$, for each string. We then applied the

Part-of-Speech (POS) Tagger to extract the nouns from these bag of words. A word w_i was selected if $\forall i$, $POS(w_i) == NN \mid NNP \mid NNS \mid NNPS$.

Let $N = \{n_1, n_2, \ldots n_m\}$ denote the *noun_list*, which is the set of nouns for each string that were extracted from the previous step. Hence, N_C denotes the *noun_list* for the caption and N_Q denotes the *noun_list* for the quote. For each noun n_i in *noun_list*, we extracted the synonyms using the `PyDictionary` package and made a list of lists called the *syn_list* denoted by W_C for the caption and list of synonyms W_Q for the quote. Denoted by $W_C = \{\{c_{11}, c_{12}, \ldots c_{1p}\}, \{c_{21}, c_{22}, \ldots c_{2q}\}, \ldots \{c_{n1}, c_{n2}, \ldots c_{nr}\}\}$ and $W_Q = \{q_{11}, q_{12}, \ldots q_{1p}, q_{21}, q_{22}, \ldots q_{2q}, \ldots q_{n1}, q_{n2}, \ldots q_{nr}\}$.

The score s, was then calculated by searching if any word c_{ij} in W_C was present in W_Q, where c_{ij} is the jth word for the ith synonym list in the *syn_list*. If c_{ij} was present, the score was updated by one. An important point to note here is that if any one of the synonym word c_{ij} of the ith list in W_C found a match in W_Q, the iterator was moved forward to the next list, $(i+1)^{th}$ list of W_C, i.e. we did not look for a match for the other synonyms in the ith synonym list. Mathematically, for each quote Q_k, the score s_k (initially $s_k = 0$) was given by,

$$\forall i, j \quad If\, c_{ij}\, in\, W_Q :$$
$$s_k = s_k + 1 \tag{1}$$
$$i = i + 1$$

Then the quotes with the maximum score i.e. the quotes with score equal to the length of N_C, the *noun_list* for the caption, were listed out as relevant quotes in a list, denoted by *quote_list*, which was further passed as input to the third step of the algorithm.

$$If\, s_k == len(N_C) :$$
$$quote\_list.append(s_k) \tag{2}$$

2.3 Semantic Comparative Analysis of Vectorized Candidate Quotes

In the final step of the algorithm, each quote in the candidate *quote_list*, obtained from the previous step was encoded to a 4800 dimensional vector space. For this, we followed the approach of unsupervised learning of the sentence encoder described in [8]. The generic and distributed nature of the GRU (Gated Recurrent Unit) based encoder provided a perfect match for our application as we are not limiting our focus to a particular genre or domain while suggesting quotes.

Let candidate $quote\_list = \{Q_1, Q_2, \ldots Q_n\}$, in which Q_i denotes a quote in this list and n is the total number of candidate quotes. Then the Vectorization Process is given by, $Q_i \rightarrow V(Q_i)$, where $V(Q_i)$ denotes the Vectorized form of Q_i and $V(Q_i) = [f_1, f_2, f_3, \ldots f_{4800}]$, where f_i is a floating point number. The Image Caption obtained from first step was also encoded to a vector using the same vectorization process, $C \rightarrow V(C)$ where C is the caption obtained from step 1.

Semantic relatedness between C and Q_i was then found by using Cosine Similarity as the similarity measure, $S(x, y)$

$$S(x, y) = cos(V(Q_i), V(C_i)) = \frac{V(Q_i) \cdot V(C_i)}{||V(Q_i)|| \cdot ||V(C_i)||} \tag{3}$$

Other similarity measures were also implemented such as Jaccard Similarity, Euclidean Distance, Manhattan Distance, Minkowski Distance and Tanimoto Similarity. However, the best results were obtained by using the Cosine Similarity Measure. This step of our algorithm also explains the relevance of choosing to build a natural language caption generator in step 1 instead of a simple object detection module. This step further prunes the list of candidate quotes which contain key objects from the images (refer step 2) according to the semantic relatedness with caption generated in the first step.

Finally, the top 3 quotes from the *quote_list* having the maximum similarity measure were reported.

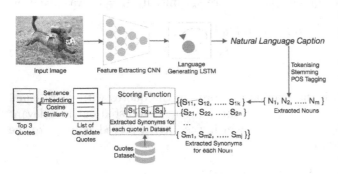

Fig. 1. Architecture of the proposed algorithm

Table 1. Meta data for dataset

Attribute	Value
Quotes	499,709
Language used	English
Total #Tags	7117
Avg #Tags/Quote	5
Min #Tags/Quote	1
Max #Tags/Quote	303
#Unique Authors	95,815

Table 2. An example of a row entry in the dataset

Quote	Author	Tags
A friend is someone who knows all about you and still loves you	Elbert Hubbard	friend, friendship, knowledge, love

3   QUOTES500K Dataset

Since, there was no publicly available large dataset on Quotes, we prepared a dataset of our own. For this, we used the Python package — BeautifulSoup, to crawl quotes from various popular websites — Goodreads, Brainyquotes, Famousquotesandauthors and Curatedquotes. All quotes, except the ones in English, were removed using the Python package — langdetect. The final dataset prepared was in the csv file format which contains three columns — the quote, the author of the quote and the category tags for that quote. Examples of tags include — love, life, philosophy, motivation, family etc. These tags help in describing the various categories that a particular quote belongs to. The total number of quotes in our final dataset after crawling and further cleaning of the dataset was approximately equal to five hundred thousand (**500K**) quotes. We have made this dataset publicly available, so that it can be used by fellow researchers for educational and research purposes. The link to the dataset is given in the footnote[1]. The meta data for this dataset is shown in Table 1 and an example of a row entry in the dataset is shown in Table 2.

4   Results and Discussion

A total of 500 images from the Flickr8K Dataset were tested and provided with the top 3 quotes along with their similarity measures. Performance of the proposed algorithm was analyzed on the QUOTES500K dataset, using both qualitative as well as quantitative measures.

4.1   Qualitative Analysis

The results of our proposed algorithm on two of the images, out of a total of 500 test images, are shown in Table 3. As seen from the table, the top 3 quotes provided for each of the two images display contextual coherency as well as semantic relatedness and hence can be classified as context aware quotes for these images. For instance, the top quote provided for the first image, having similarity measure 77.3%, beautifully expresses the meaning of the image. Even though, the second quote for the same image is not as good as the first one, but still the nouns, "mountain" and "man" are retained by our model and the quote displays contextual coherency.

4.2   Quantitative Analysis (User Study)

To test the usefulness of our results, a user study of 100 users was conducted. To avoid the bias in our evaluation, the users were deliberately selected across various age groups ranging from 10 years to 60 years. Each user was presented with a web portal wherein the user was shown an image and the top 3 most relevant quotes for that image. The user was then presented with a choice to rate the quotes as either "Relevant Quotes" or "Irrelevant Quotes". This experiment was performed for 500 images and for each image, 100 users evaluated the quotes generated by the model. Table 3 shows the results obtained from our model for 2 images. The overall accuracy obtained for our model was 83.2 % which was calculated as the average positive response for the 500 test images.

---

[1] Link to QUOTES500K Dataset.

Table 3. Examples of results obtained (where +ve: Positive Evaluation and -ve: Negative Evaluation)

Image with Caption Generated	Top 3 Best Quotes	Similarity (in %)	Evaluation +ve	Evaluation -ve
a man is standing on a hill	Change is the watchword of progression. When we tire of well-worn ways, we seek for new. This restless craving in the souls of men spurs them to climb, and to seek the mountain view	77.3		
	Men trip not on mountains, they stumble on stones	74.1	87	13
	Most of life is routine-dull and grubby, but routine is the mountain that keeps a man going. If you wait for inspiration you'll be standing on the corner after the parade is a mile down the street	73.9		
a bird is flying through the trees	But Hopes are Shy Birds flying at a great distance seldom reached by the best of Guns	75.2		
	Perfect as the wing of a bird may be, it will never enable the bird to fly if unsupported by the air. Facts are the air of science. Without them a man of science can never rise	73.9	81	19
	It may be hard for an egg to turn into a bird: it would be a jolly sight harder for it to learn to fly while remaining an egg. We are like eggs at present. And you cannot go on indefinitely being just an ordinary, decent egg. We must be hatched or go bad	72.8		

5   Conclusion

In this paper, we presented a dataset of approximately 500 K quotes which would significantly advance the research in providing Context Aware Quotes for Images. Furthermore, an algorithm was proposed which provides contextually relevant quotes for an image. The results obtained from our algorithm were evaluated by using both qualitative as well as quantitative measures.

References

1. Bahdanau, D., Cho, K., Bengio, Y.: Neural machine translation by jointly learning to align and translate. CoRR abs/1409.0473 (2014)
2. Chaudhuri, S., Ganti, V., Kaushik, R.: A primitive operator for similarity joins in data cleaning. In: 22nd International Conference on Data Engineering (ICDE 2006), p. 5, April 2006
3. Denil, M., Bazzani, L., Larochelle, H., de Freitas, N.: Learning where to attend with deep architectures for image tracking. Neural Comput. **24**(8), 2151–2184 (2012)

4. Elliott, D., Keller, F.: Image description using visual dependency representations. In: EMNLP, pp. 1292–1302. ACL (2013)
5. Farhadi, A., Hejrati, M., Sadeghi, M.A., Young, P., Rashtchian, C., Hockenmaier, J., Forsyth, D.: Every picture tells a story: generating sentences from images. In: Daniilidis, K., Maragos, P., Paragios, N. (eds.) ECCV 2010. LNCS, vol. 6314, pp. 15–29. Springer, Heidelberg (2010). https://doi.org/10.1007/978-3-642-15561-1_2
6. Kiros, R., Salakhutdinov, R., Zemel, R.: Multimodal neural language models. In: Xing, E.P., Jebara, T. (eds.) Proceedings of the 31st International Conference on Machine Learning. Proceedings of Machine Learning Research, vol. 32, pp. 595–603. PMLR, Bejing, China, 22–24 June 2014
7. Kiros, R., Salakhutdinov, R., Zemel, R.S.: Unifying visual-semantic embeddings with multimodal neural language models. CoRR abs/1411.2539 (2014)
8. Kiros, R., Zhu, Y., Salakhutdinov, R., Zemel, R.S., Torralba, A., Urtasun, R., Fidler, S.: Skip-thought vectors. CoRR abs/1506.06726 (2015)
9. Kondrak, G.: N-Gram similarity and distance. In: Consens, M., Navarro, G. (eds.) SPIRE 2005. LNCS, vol. 3772, pp. 115–126. Springer, Heidelberg (2005). https://doi.org/10.1007/11575832_13
10. Li, S., Kulkarni, G., Berg, T., Berg, A., Choi, Y.: Composing simple image descriptions using web-scale N-grams, pp. 220–228 (2011)
11. Miller, D.R.H., Leek, T., Schwartz, R.M.: A hidden Markov model information retrieval system. In: Proceedings of the 22nd Annual International ACM SIGIR Conference on Research and Development in Information Retrieval, SIGIR 1999, pp. 214–221. ACM (1999)
12. Mnih, V., Heess, N., Graves, A., Kavukcuoglu, K.: Recurrent models of visual attention. In: Advances in Neural Information Processing Systems, vol. 27, pp. 2204–2212 (2014)
13. Rashtchian, C., Young, P., Hodosh, M., Hockenmaier, J.: Collecting image annotations using Amazon's Mechanical Turk. In: Proceedings of the NAACL HLT 2010 Workshop on Creating Speech and Language Data with Amazon's Mechanical Turk, CSLDAMT 2010, pp. 139–147. Association for Computational Linguistics (2010)

To Cite, or Not to Cite? Detecting Citation Contexts in Text

Michael Färber[1](✉)(iD), Alexander Thiemann[1], and Adam Jatowt[2]

[1] University of Freiburg, Freiburg im Breisgau, Germany
michael.faerber@cs.uni-freiburg.de, mail@athiemann.net
[2] Kyoto University, Kyoto, Japan
adam@dl.kuis.kyoto-u.ac.jp

Abstract. Recommending citations for scientific texts and other texts such as news articles has recently attracted considerable amount of attention. However, typically, the existing approaches for citation recommendation do not explicitly incorporate the question of whether a given context (e.g., a sentence), for which citations are to be recommended, actually "deserves" citations. Determining the "cite-worthiness" for each potential citation context as a step before the actual citation recommendation is beneficial, as (1) it can reduce the number of costly recommendation computations to a minimum, and (2) it can more closely approximate human-citing behavior, since neither too many nor too few recommendations are provided to the user. In this paper, we present a method based on a convolutional recurrent neural network for classifying potential citation contexts. Our experiments show that we can significantly outperform the baseline solution [1] and reduce the number of citation recommendations to about 1/10.

Keywords: Citation context · Citation recommendation
Recommender systems · Deep learning

1 Motivation

Due to a variety of reasons, such as supporting claims and arguments or giving attribution to authors, scientific works must contain appropriate citations to other works [2]. Citing properly is a challenging task: Not only all works leading to new results should be cited, but also adding citations to further explain concepts and ideas often helps the reader to correctly understand the goals and ideas of a paper. Finding a good balance between not too many and not too few citations is rather time consuming and requires years of practice in scientific writing. This issue also appears in the context of recommending citations automatically. Recommending citations for scientific texts and other texts such as news articles has recently attracted a considerable amount of attention, due to the dramatic increase in the number of published papers. Existing citation recommendation approaches, however, do not explicitly incorporate the question

© Springer International Publishing AG, part of Springer Nature 2018
G. Pasi et al. (Eds.): ECIR 2018, LNCS 10772, pp. 598–603, 2018.
https://doi.org/10.1007/978-3-319-76941-7_50

of whether a given context (e.g., a sentence), for which citations should be recommended, actually "deserves" citations. If this is the case, we call it an actual *citation context*. In this paper, we approach this as a classification task and call it *citation context detection*. It can be regarded as a step before the actual recommendation of relevant citations. Note that the task of citation recommendation and, hence, citation context detection, is not limited to scientific texts such as publications, but can be applied to any text for which citations are needed, such as news texts or encyclopedic articles like those in Wikipedia.

Sugiyama et al. [1] provide the first approach to the presented research problem. In this paper, we show that the problem can be solved more effectively by a convolutional recurrent neural network. Our experiments reveal the superiority of our approach and offer insights into human-citing behavior.

2 Related Work

Citation context characterization and classification. Explicitly classifying potential citation contexts with respect to cite-worthiness has been carried out by Sugiyama et al. [1] by means of an SVM approach. However, they only report the accuracy of results and do not address the high imbalance of negative to positive instances. In [3], the authors focus on the distributions of citation locations in publications, although they only provide visual analyses and no prediction model. Angrosh et al. [4] only consider sentences in related work sections and classify them into 13 classes. We cannot apply their classification scheme or approach, as we do not only consider related work sections, so our sentences are of a different nature. Citation contexts have also been studied in further respects. Most prominently, the citation function has been analyzed and predicted [5], and the citation importance [6] and further linguistic characteristics such as the discourse structure of citation contexts [7] were also analyzed.

Citation recommendation. For citation recommendation, a variety of approaches have been proposed (see [8], but we note that most listed methods approach the paper recommendation task, which differs from the citation recommendation by using information from the entire paper instead of short citation contexts only). Most recent approaches typically utilize neural networks and learning-to-rank-frameworks [9–11]. Typically, the citation contexts are already predetermined. Using our approach, such prerequisites are no longer necessary. We believe that having a flexible approach that determines the placement of citations themselves is more user-friendly and can be applied in a variety of scenarios. Furthermore, in this way we can reduce the number of citation recommendations, as we only focus on the cite-worthy contexts for the recommendation.

3 Citation Context Classification

We now describe our approach to identify citation contexts. As potential citation contexts, we use single sentences since sentences are a natural unit for expressing

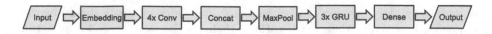

Fig. 1. Our network used for citation context detection.

statements and since prior studies have revealed that there is no single optimal choice for a citation context unit (such as sentences or a fixed window) [12]. Our method attempts to classify sentences into "needing citations" and "not needing citations." As an underlying method, we use a convolutional recurrent neural network (CRNN) [13], as it was shown to be a good fit for text classification. A pure recurrent neural network (RNN) classifier is biased: Later words dominate compared to earlier words. This is unnatural for documents, as important information could be spread in the text. This problem can be addressed with convoluational neural networks (CNNs), but picking a good window size is challenging, as it could lead to the loss of important information. Thus, we combine both methods to obtain state-of-the-art performance.

The full architecture of the CRNN[1] – visible in Fig. 1 – consists of four convolutional layers with 128 hidden states with a filter size of 1, 2, 3, and 5. Next is a concatenation step followed by max pooling. After the convolutional part, the recurrent part consists of three gated recurrent unit (GRU) [14] layers with a (recurrent) dropout of 0.2. Finally, a densely connected layer with a softmax activation function and two outputs provides the final classification.

4   Evaluation

4.1   Evaluation Data Sets

For evaluating our approach, the contents of publications are needed. Note that many available scholarly data sets either only cover the citation contexts and not all sentences of the publications (e.g., see CiteSeerX) or only cover meta-information about the publications, such as the citation network. After reviewing scholarly data sets, we decided on using the following:[2]

arXiv CS [15] is a data set of over 9M sentences extracted from all computer science publications hosted at arXiv.org. This data set was constructed from the $T_{E}X$ files provided by the authors. Since each citation is explicitly given via a cite command in $T_{E}X$, for this data set we can ensure that we do not miss any citations and that we always link to the correct reference. This makes this data set to be of relatively high quality.

Scholarly Dataset 2[3] contains about 100k publications in PDF format from the ACM Digital Library. By using this data set, we can evaluate the impact of having PDFs as input, which, in reality, is often the case.

---

[1] The source code is available online at https://github.com/agrafix/grabcite-net.
[2] All data sets are available online at http://citation-recommendation.org/publications.
[3] http://www.comp.nus.edu.sg/~sugiyama/SchPaperRecData.html.

ACL-ARC[4] is a widely used corpus of scholarly publications about computational linguistics. In order to compare against Sugiyama et al.'s approach [1], we use the 2008 version, which contains 10,921 papers.

For transforming the PDF files of Scholarly and ACL-ARC into plaintext files, we use IceCite [16], which is a state-of-the-art information extraction tool for scientific publications.

4.2   Evaluation of Citation Context Detection

Training data. We build training data by iterating over all sentences in our input data (i.e., plaintext of publications), detecting if a citation marker is present, and labeling the sentences accordingly before removing the citation markers from the sentences. This gives us heavily imbalanced training data: 10% of all sentences contain at least one reference and 90% do not. We solve this imbalance by oversampling for all NNs and by undersampling for all SVM approaches. We use the pretrained GloVe word embeddings ("GloVe 6B") for our NNs. For arXiv CS (since it is the cleanest data set), we also try our own word embeddings trained via auto encoding.

Methods. We use the following approaches for a comparison:

- **SVM.** Following Sugiyama et al. [1], we use a Support Vector Machine (SVM) and try out different feature settings listed in Table 3.
- **CNN.** Secondly, we try a convolutional neural network (CNN).
- **RNN.** Then, a recurrent neural network (RNN) is also used.
- **CRNN.** This is our approach proposed in Sect. 3 using five epochs with a batch size of 64.

Evaluation results. The results are given in Table 1. When considering accuracy as metric, already simple approaches like $SVM_{doc2vec}$ perform very well and outperform NNs. However, accuracy is not a very suitable metric in the case of unbalanced data (as given here); for instance, using a SVM with doc2vec as the only feature achieved very good accuracy scores, too, but low F1 scores.

Sugiyama's approach [1], which is based on a SVM with noun phrases or the indication about citations in the neighboring sentences as a feature, can be outperformed by many of our approaches. The better accuracy value of our SVM_{NPs} for ACL-ARC compared to SVM [1] may be a result from improved PDF-to-text conversion.

Considering the F1 scores, our NNs outperform the SVM approaches. Interestingly, both the CNN and the RNN perform roughly the same as the CRNN. This indicates that for citation context detection, the ordering of the words is not that important and that the word embeddings themselves already have a strong signal for the classification. This is plausible if we consider that named entities and abstract concepts rather than complete statements are cited.

---

[4] http://acl-arc.comp.nus.edu.sg/.

Table 1. Results of classifying sentences regarding their cite-worthiness (P: Precision, R: Recall, F1: F1 score, A: Accuracy).

	ACL-ARC				Scholarly				arXiv CS			
	P	R	F1	A	P	R	F1	A	P	R	F1	A
SVM of [1]	-	-	-	0.882	-	-	-	-	-	-	-	-
SVM$_{TF-IDF}$	0.049	0.052	0.051	0.938	0.043	0.009	0.015	0.969	0.100	**0.955**	0.180	0.131
SVM$_{POS}$	0.061	**0.658**	0.112	0.670	0.050	0.680	0.093	0.662	0.191	0.631	0.293	0.695
SVM$_{NPs}$	0.034	0.050	0.041	0.926	0.019	0.001	0.002	**0.973**	0.094	0.048	0.063	0.858
SVM$_{doc2vec}$	0.086	0.004	0.008	**0.967**	0.026	**1.000**	0.050	0.028	0.140	0.016	0.029	**0.892**
SVM$_{PER}$	0.049	0.099	0.066	0.912	0.116	0.137	0.126	0.951	0.338	0.186	0.240	0.882
SVM$_{Cits}$	0.083	0.578	0.145	0.786	0.068	0.509	0.120	0.809	0.199	0.724	0.313	0.681
CNN$_{GloVe}$	**0.196**	0.269	**0.227**	0.941	**0.227**	0.792	**0.329**	0.812	**0.433**	0.709	**0.538**	0.870
RNN$_{GloVe}$	0.171	0.317	0.222	0.928	0.181	0.823	0.322	0.811	0.400	0.785	0.530	0.851
CRNN$_{GloVe}$	0.182	0.260	0.214	0.930	0.207	0.763	0.326	0.807	0.376	0.750	0.501	0.841

Table 2. Results for arXiv data set using custom word embeddings instead of pre-trained GloVe word embeddings.

	P	R	F1	A
CNN$_{custom}$	0.418	0.724	0.530	0.863
RNN$_{custom}$	0.393	**0.790**	0.525	0.849
CRNN$_{custom}$	**0.430**	0.715	**0.537**	**0.869**

Table 3. Features used for our SVM approach.

Name	Description
SVM$_{TF-IDF}$	TF-IDF
SVM$_{POS}$	# POS tags
SVM$_{NPs}$	BOW of extracted noun phrases
SVM$_{doc2vec}$	doc2vec
SVM$_{PER}$	Contains person accord. to Stanford NER
SVM$_{Cits}$	# citations in neighb. sentences (up to 5)

We can observe significant differences in the evaluation scores for all approaches between the different data sets. The reason is likely to be the varying quality of the data sets. The contents of ACL-ARC and Scholarly are extracted from PDFs and are thus quite noisy (especially ACL-ARC), while arXiv CS remains comparatively clean. Thus, we can assume that arXiv CS best reflects the actual citing behavior. For this data set (and for other data sets), the results of the NN approaches only vary very little. If instead of pretrained GloVe word embeddings, our own embeddings are trained – as done exemplarily for the arXiv data set – we achieve considerably better precision and F1 scores for the CRNN (see Table 2), while the results for the CNN slightly decrease and remain stable for the RNN. Hence, under the assumption of having a larger training corpus in the future, the CRNN$_{custom}$ seems to be one of the most promising approaches. In total, we achieve F1 scores of over 0.5 for all NNs on the arXiv CS data set, making citation context detection attractive to be applied in actual systems.

5   Conclusion and Outlook

Existing citation recommendation approaches do not incorporate the question of whether a given citation context actually deserves citations. In this paper, we address this question and build a classifier that can determine the "cite-worthiness" to a considerable degree. As a result, we can reduce the number of costly citation recommendation computations to a minimum (about 1/10), since recommendations need to be computed only for cite-worthy contexts and since only about 10% of the sentences in our data sets contain citations. Our experiments on three data sets show that we can significantly outperform the existing solution of Sugiyama et al. [1]. For future work, we plan to consider additional features using the papers' meta-data.

Acknowledgements. Michael Färber is an International Research Fellow of the Japan Society for the Promotion of Science (JSPS). The work was partially supported by MIC SCOPE (171507010). The Titan Xp used for this research was donated by the NVIDIA Corporation.

References

1. Sugiyama, K., Kumar, T., Kan, M.Y., Tripathi, R.C.: Identifying citing sentences in research papers using supervised learning. In: CAMP 2010, pp. 67–72. IEEE (2010)
2. Teufel, S., Siddharthan, A., Tidhar, D.: An annotation scheme for citation function. In: SIGdial 2009, pp. 80–87 (2009)
3. Hu, Z., Chen, C., Liu, Z.: Where are citations located in the body of scientific articles? A study of the distributions of citation locations. J. Inf. **7**(4), 887–896 (2013)
4. Angrosh, M.A., Cranefield, S., Stanger, N.: Context identification of sentences in related work sections using a conditional random field: towards intelligent digital libraries. In: JCDL 2010, pp. 293–302 (2010)
5. Teufel, S., Siddharthan, A., Tidhar, D.: Automatic classification of citation function. In: EMNLP 2007, pp. 103–110 (2006)
6. Valenzuela, M., Ha, V., Etzioni, O.: Identifying meaningful citations. In: SBD 2015 (2015)
7. Fisas, B., Saggion, H., Ronzano, F.: On the discoursive structure of computer graphics research papers. In: LAW@NAACL-HLT 2015, pp. 42–51 (2015)
8. Beel, J., Gipp, B., Langer, S., Breitinger, C.: Research-paper recommender systems: a literature survey. Int. J. Digit. Libr. **17**(4), 305–338 (2016)
9. Ebesu, T., Fang, Y.: Neural citation network for context-aware citation recommendation. In: SIGIR 2017, pp. 1093–1096 (2017)
10. Jiang, Z., Liu, X., Gao, L.: Chronological citation recommendation with information-need shifting. In: CIKM 2015, pp. 1291–1300 (2015)
11. Huang, W., Wu, Z., Chen, L., Mitra, P., Giles, C.L.: A neural probabilistic model for context based citation recommendation. In: AAAI 2015, pp. 2404–2410 (2015)
12. Alvarez, M.H., Gómez, J.M.: Survey about citation context analysis: tasks, techniques, and resources. Nat. Lang. Eng. **22**(3), 327–349 (2016)
13. Lai, S., Xu, L., Liu, K., Zhao, J.: Recurrent convolutional neural networks for text classification. In: AAAI 2015, pp. 2267–2273 (2015)
14. Zhou, G., Wu, J., Zhang, C., Zhou, Z.: Minimal gated unit for recurrent neural networks. CoRR abs/1603.09420 (2016)
15. Färber, M., Thiemann, A., Jatowt, A.: A high-quality gold standard for citation-based tasks. In: LREC 2018 (2018)
16. Bast, H., Korzen, C.: A benchmark and evaluation for text extraction from PDF. In: JCDL 2017, pp. 99–108 (2017)

Automated Assistance in E-commerce:
An Approach Based on
Category-Sensitive Retrieval

Anirban Majumder[1], Abhay Pande[1], Kondalarao Vonteru[2],
Abhishek Gangwar[2], Subhadeep Maji[1], Pankaj Bhatia[1], and Pawan Goyal[2(✉)]

[1] Flipkart Internet Pvt. Ltd., Bengaluru, India
{majumder.a,abhay.pande,subhadeep.m,pankaj.bhatia}@flipkart.com
[2] IIT Kharagpur, Kharagpur, India
sunnysai12345@iitkgp.ac.in, abhishek.g307@gmail.com,
pawang@cse.iitkgp.ac.in

Abstract. This paper aims towards building an automated conversational assistant to help customers in an e-commerce scenario. Our dataset consists of live chat messages between human agents and buyers. These chats belong to many different issue types and we build a multi-instance SVM classifier to automatically classify these chats into the corresponding issue types. We further use this insight to append the category information obtained from the classifier to an LSTM based architecture to be able to provide appropriate responses given an utterance by a human agent. We find that using class information along with the base dual encoder model helps in improving the quality of the retrieved responses in terms of BLEU scores. Human judgement experiments validate that using class information is able to bring out relevant messages in top-3 and top-5 responses much more number of times than the base model that does not use the class information.

1 Introduction

To build a conversational agent and/or chatbot with sufficient artificial intelligence has always been long cherished goal for researchers and practitioners. It is very challenging for computers to do a coherent, continuous and meaningful conversation with humans. These automatic conversation models are of great importance to a wide variety of applications, starting from open-domain entertaining chatbots which can naturally and meaningfully converse with humans on open-domain topics, to goal-oriented technical support systems which can assist users towards completion of a task. In an open domain setting, user can take conversation in any direction, there is not a well defined intention or goal. In a closed domain setting, the domain of inputs and outputs is somewhat limited as the user is trying to achieve very specific goal. These systems need to fulfill their specific task as efficiently as possible.

Customer support in the e-commerce scenario sees customers reaching out with a wide variety of inquiries like the status of the order, the return process,

© Springer International Publishing AG, part of Springer Nature 2018
G. Pasi et al. (Eds.): ECIR 2018, LNCS 10772, pp. 604–610, 2018.
https://doi.org/10.1007/978-3-319-76941-7_51

delays in processing of refund and offer inquiries. As the business scales, the number of contacts from the customers about such inquiries also increase at the same rate. An automated conversational agent that can help address customer queries, goes a long way in providing cost-effectiveness as well as scalability.

While the traditional systems required a lot of domain expertise to be crafted manually [1], in the recent years, an increasing amount of research has happened to build purely data-driven conversational systems. These systems mainly use two types of approaches. *Retrieval-based models* use a repository of either predefined responses or context-response pairs along with a retrieval/ranking mechanism to pick an appropriate response with the help of the input and context. For instance, Lowe et al. [2] introduced the Ubuntu Dialogue corpus, and also presented an LSTM based framework to provide a score to any candidate response given an input message. Yan et al. [3] proposed a retrieval based approach that can also leverage on unstructured documents in addition to the context-response pairs. Yan et al. [4] introduced a chat companion system, which given the human utterances as queries, responds with corresponding replies retrieved and highly ranked from a massive conversational data repository. They perform ranking with and without using the context for multi-turn and single-turn conversations, respectively. *Generative models*, on the other hand, do not rely on predefined responses. Instead, they generate new responses. There have been a number of related attempts to address the problem using generative models with the help of neural networks. Sequence to Sequence Learning [5] uses a multi layered Long Short-Term Memory (LSTM) [6] to encode the input sequence to a vector of a fixed dimensionality, which is used to generate (decode) the response. Sordoni et al. [7] proposed to encode the message along with context in a recurrent language model based architecture. Yao et al. [8] modeled both attention and intention processes for generating natural responses.

E-commerce is a closed domain, where typical conversation is a mix of standard responses and context based dynamic responses to customer queries. Our work, as one of the very first experiments with conversational agents in e-commerce domain, demonstrates experiments with retrieval approaches, where we evaluate the hypothesis that in e-commerce domain, leveraging the issue type classes with existing approaches improves the quality of responses.

2   Dataset Description

Our dataset consists of real time chats that took place between human agents and the buyers during issue resolution by customer care for Flipkart[1] between the months of July to December, 2016. One such example chat is shown below:

---

[1] https://www.flipkart.com/.

> Customer:Hi
> Auto:Hi , I'm ⟨ consultant name⟩. We had spoken earlier and I'll be happy to help again.
> Customer: I will go for replacement.
> Customer: For my product.
> CX: Sure, I will escalate to cancel the refund request so within next 48 h it will be canceled and
> we will intimate you about this over email on your registered email id.

Messages starting with "Customer" indicate what the customer said, while those starting with "CX" indicate what the customer support consultant said. Messages with "Auto" are automated messages sent by the system. Every chat interaction between the customer and the human agent is categorized into a group of issue types and sub-issue types by human agents. These issue types and sub-issue types identify the nature of the problem being faced by the buyer. We sampled the data for the month of July and found that while there were 24 issue types, only 13 issue types had significant number of chat sessions, including an issue type to classify spams, appreciations or incomplete requests. The remaining 12 issue types covered 93% of the non spam/incomplete chats in the month of July. A distribution among these 13 issue types is shown in Fig. 1.

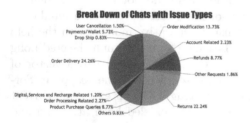

Fig. 1. The distribution of issue types

Table 1. Properties of the e-commerce chat corpus used in this study

# Chat sessions	39,000
# Context-response pairs	448,335
# Issue types	13
Average # words per context	13.65
Average # words per response	20.62

The dataset we used for our experiments consists of **448,335** context-response pairs, taken from chats during the months of July to December. Table 1 shows the details of our dataset. We convert all consecutive occurrences of "Customer" into context and those of "CX" into utterance to get a context-response pair. We have contextual information such as dates, product names, consumer names and customer agent names. In pre-processing steps, we replace the contextual information above by placeholders. The data also contains generic responses and greetings which are noise to our training purposes and we had them removed in our pre-processing steps.

Issue-type Classifier: In the e-commerce domain, a conversation can be classified to various issue types. This class information can help the agents/bot to narrow down to the problem faster and get to the appropriate response. For our experiments, we use a classifier trained on a small dataset annotated with issue type labels. However, these annotations are at a chat level rather than individual context-response pair. Further, different sequence pairs in a chat can belong to different issue types. The classifier that we used is based on the multi-instance

SVM proposed by Andrews et al. [9]. The multi-instance SVM is a variation of SVM which specializes on training labels for a bag of instances with one or more labels and extrapolates that to a bag of single instance.

The SVM classifier was trained on a dataset of size 80 K, completely disjoint from the dataset for the LSTM model. The training of the model was done on the annotated dataset. We used word embeddings weighed with the tf-idf scores of the word as features for classification. We use cross validation to evaluate the results of these classifiers which gave an accuracy of 80% for the above-mentioned 13 classes. We also performed a manual evaluation on a small held-out dataset of 1 K instances and the accuracy was found to be close to 70%.

3   Automating Retrieval of Conversational Responses

We used the retrieval model architecture used in [2], an LSTM based recurrent neural network, which tries to learn representations of the context and response and scores how appropriate a response is for a given context. During the training phase, the model encodes the context and response using an LSTM. The authors of [2] first obtain **context embeddings** from the given input context (LSTM hidden state corresponding to the last input), and then learn a layered perceptron network to obtain the **model response embeddings** from this context embedding. Now, to rank the candidate responses, the model [2] first embeds them using LSTM using the same approach as above, and then gives a probability score for each context-response pair based on how well these response embeddings match the model response embeddings. The loss is computed using cross entropy of the targets and scores computed.

Using Chat Category Information: Our hypothesis is that using the class information together with the context can help in providing a better response. We aim to use our issue-type classifier to classify these conversations into the particular issue type. Below are some example contexts.

Context 1: i have an order but i cancelled because of changing my address but now i didnt get my refund

Context 2: i have an order of ¡prod name¿ but i have to change my address can u help

For context 1 above, the customer talks about cancelling an order due to the change in delivery address, and for the next context, he just wants to change the address. From the text, it seems that both of these contexts come into the **Order Modification** class, but the first context actually belongs to class **Refunds**. Here, the class information can help in proper response retrieval.

To incorporate this class information, we use exactly the same retrieval-based architecture as mentioned above, except that the class information, as obtained from the automated classifier, is provided as input. One standard method is to provide it as input to the initial states of the recurrent neural network. Introducing the classifier as the initial states, we ensure that the class information is used along with the context to encode the response as well as the context embeddings. Formally,

$$c_0 = g(class_i) \mid class_i \in \mathbb{I} \tag{1}$$

$$h_0^{'} = \langle h_0, c_0 \rangle \tag{2}$$

where the $class_i$ denotes the classifier output as the most appropriate issue type, which belongs to the set of the issue types $\mathbb{I}$ and $g(.)$ provides the vector representation of the class information using one-hot encoding. The initial states of the LSTM contain the initial hidden and cell state. We replace the initial cell state with the class information. We will explain the results of this variation of the model and compare with the base model in Sect. 4.

4   Experiments

We used 222,209 context-response pairs as obtained after some preprocessing steps for training. The basic preprocessing steps involved greetings or chit-chats removals, removal of very short context-response pairs[2] as well as replacing various entities such as product names, dates, customer names with generic entity tags like 'PRODUCT', 'DATE' etc. We trained all our models for 22 K epochs in a standard GPU machine which took less than 1 h.

We also created a test dataset of 10 K context-response pairs from the actual chats, completely disjoint from the training set. This set was created after performing the same pre-processing steps. Since the retrieval model only ranks the candidate responses, Okapi BM25 was used to retrieve the top 10 responses from the training set of context-response pairs corresponding to the given context. The LSTM used for the model was build on the Tensorflow library.

Evaluation Metrics: As per the earlier works, we also evaluate the performance of our models using BLEU scores [10] for top-k responses ranked by our model with reference to the ground truth. We use T1-BLEU, T3-BLEU and T5-BLEU to denote the BLEU scores based on top 1, 3 and 5 responses.

Experimental Results: Table 2 shows the results of our evaluation. We see that adding class information to the base retrieval model helps in improving the performance for all the three cases (top 1, 3 and 5). The improvements were highest for the top retrieved sentence.

Manual Evaluation: We also perform a human judgment experiment to verify the importance of the class information on LSTM ranking with 100 randomly chosen context-response pairs. 2 participants, different from the authors, were given 50 contexts each, where every context had 5 responses as ranked by the two retrieval models: LSTM and LSTM+class info. The participants were asked to answer whether a relevant response appears within top 3 (and top 5) of these ranked lists, without disclosing the identities of the two models. Table 3 shows the results of this experiment. This clearly tells that the class information helps to bring relevant response in top-3 and 5 retrieved results.

---

[2] We fixed the minimum length of these context-response pair to be 4. These context-response pairs correspond to almost 39 K chats.

Table 2. Comparison between different competing models based on Bleu scores

Models	T1 BLEU	T3 BLEU	T5 BLEU
BM-25+LSTM ranking [2]	22.62	28.48	31.72
[2]+Class info	**24.21**	**29.59**	**32.22**

Table 3. Comparison between different retrieval models as per human judgement experiments

Models	Rel. top 3	Rel. top 5
Base retrieval model [2]	26	32
With class info	**34**	**43**

5 Discussions

From Tables 2 and 3, we can see that the model with class information performs better than the base model. The following example can validate our claim further. Without the class information, the bot is not sure about the context and replied asking about the product name. With class information ("**Refunds**"), the bot replies properly with the details of refund status.

> **Customer :** i have an order but i cancelled because of changing my address but now i didnt get my refund
>
> **Without class information: Bot :** you are welcome is there anything else that i can help you with today ?
>
> **With class information: Bot :** according to bank procedure this takes ⟨no of days⟩ to refund the amount as i see that the order get cancelled on the ⟨ date ⟩ so i kindly request you to allow another ⟨no of days⟩ for refund to be done

While we get some good results from the introduction of the classes, on error analysis, we observe that error in classification phase can lead to an incorrect response. In the follwoing example, the context was wrongly classified to the class **"Returns"** instead of **"Refunds"** and the model produces an incorrect response:

> **Customer :** Today please tell filpkart, please today refund
>
> **Bot with class info :** i would like to inform you that as the product has been delivered to you i request you to please contact the brand for this issue

6 Conclusions and Future Work

This paper presented a framework for automating conversational responses to assist the customers in an e-commerce scenario. In addition to being one of the first studies exploring conversational agents in e-commerce domain, novelty of our approach lies in using the issue type information, as obtained by a multi-instance classifier, to enhance the existing LSTM-based retrieval model. Experimental results suggest that it indeed helps in improving the quality of the responses. Future work would involve experiments with other architectures, as well as adding more information from the chat context, as well as product/sale context.

References

1. Young, S.J.: Talking to machines (statistically speaking). In: INTERSPEECH (2002)
2. Lowe, R., Pow, N., Serban, I., Pineau, J.: The ubuntu dialogue corpus: a large dataset for research in unstructured multi-turn dialogue systems. arXiv preprint arXiv:1506.08909 (2015)
3. Yan, Z., Duan, N., Bao, J., Chen, P., Zhou, M., Li, Z., Zhou, J.: Docchat: an information retrieval approach for chatbot engines using unstructured documents. In: ACL (2016)
4. Yan, R., Song, Y., Zhou, X., Wu, H.: "Shall i be your chat companion?": Towards an online human-computer conversation system. In: CIKM 2016, New York, NY, USA, pp. 649–658. ACM (2016)
5. Sutskever, I., Vinyals, O., Le, Q.V.: Sequence to sequence learning with neural networks. In: NIPS, pp. 3104–3112 (2014)
6. Hochreiter, S., Schmidhuber, J.: Long short-term memory. Neural Comput. 9(8), 1735–1780 (1997)
7. Sordoni, A., Galley, M., Auli, M., Brockett, C., Ji, Y., Mitchell, M., Nie, J.Y., Gao, J., Dolan, B.: A neural network approach to context-sensitive generation of conversational responses. arXiv preprint arXiv:1506.06714 (2015)
8. Yao, K., Zweig, G., Peng, B.: Attention with intention for a neural network conversation model. arXiv preprint arXiv:1510.08565 (2015)
9. Andrews, S., Tsochantaridis, I., Hofmann, T.: Support vector machines for multiple-instance learning. In: NIPS, pp. 561–568. MIT Press (2003)
10. Papineni, K., Roukos, S., Ward, T., Zhu, W.J.: Bleu: a method for automatic evaluation of machine translation. In: ACL, pp. 311–318 (2002)

Neural Multi-step Reasoning
for Question Answering
on Semi-structured Tables

Till Haug, Octavian-Eugen Ganea[✉], and Paulina Grnarova

Department of Computer Science, ETH Zurich, Zürich, Switzerland
till@veezoo.com, {octavian.ganea,paulina.grnarova}@inf.ethz.ch

Abstract. We explore neural network models for answering multi-step reasoning questions that operate on semi-structured tables. Challenges arise from deep logical compositionality and domain openness. Our approach is weakly supervised, trained on question-answer-table triples. It generates human readable *logical forms* from natural language questions, which are then ranked based on word and character convolutional neural networks. A model ensemble achieved at the moment of publication state-of-the-art score on the WikiTableQuestions dataset.

1 Introduction

Teaching computers to answer complex natural language questions requires sophisticated reasoning and human language understanding. We investigate generic natural language interfaces for simple arithmetic questions on semi-structured tables. Typical questions for this task are topic independent and may require performing multiple discrete operations such as aggregation, comparison, superlatives or arithmetics.

We propose a weakly supervised neural model that eliminates the need for expensive feature engineering in the candidate ranking stage. Each natural language question is translated using the method of [8] into a set of machine understandable candidate representations, called *logical forms* or *programs*. Then, the most likely such program is retrieved in two steps: (i) using a simple algorithm, logical forms are transformed back into *paraphrases* (textual representations) understandable by non-expert users, (ii) next, these strings are further embedded together with their respective questions in a jointly learned vector space using convolutional neural networks over character and word embeddings. Multi-layer neural networks and bilinear mappings are further employed as effective similarity measures and combined to score the candidate interpretations. Finally, the highest ranked logical form is executed against the input data to retrieve the answer. Our method uses only weak-supervision from question-answer-table input triples, without requiring expensive annotations of gold logical forms.

We empirically test our approach on a series of experiments on WikiTableQuestions, to our knowledge the only dataset designed for this task. An ensemble of our best models reached state-of-the-art accuracy of 38.7% at the moment of publication.

© Springer International Publishing AG, part of Springer Nature 2018
G. Pasi et al. (Eds.): ECIR 2018, LNCS 10772, pp. 611–617, 2018.
https://doi.org/10.1007/978-3-319-76941-7_52

2 Related Work

We briefly mention here two main types of QA systems related to our task[1]: semantic parsing-based and embedding-based. *Semantic parsing-based* methods perform a functional parse of the question that is further converted to a machine understandable program and executed on a knowledgebase or database. For QA on semi-structured tables with multi-compositional queries, [8] generate and rank candidate logical forms with a log-linear model, resorting to hand-crafted features for scoring. As opposed, we learn neural features for each question and the paraphrase of each candidate logical form. Paraphrases and hand-crafted features have successfully facilitated semantic parsers targeting simple factoid [1] and compositional questions [10]. Compositional questions are also the focus of [7] that construct logical forms from the question embedding through operations parametrized by RNNs, thus losing interpretability. A similar fully neural, end-to-end differentiable network was proposed by [11].

Embedding-based methods determine compatibility between a question-answer pair using embeddings in a shared vector space [2]. Embedding learning using deep learning architectures has been widely explored in other domains, e.g. in the context of sentiment classification [3].

3 Model

We describe our QA system. For every question q: (i) a set of candidate logical forms $\{z_i\}_{i=1,...,n_q}$ is generated using the method of [8]; (ii) each such candidate program z_i is transformed in an interpretable textual representation t_i; (iii) all t_i's are jointly embedded with q in the same vector space and scored using a neural similarity function; (iv) the logical form z_i^* corresponding to the highest ranked t_i^* is selected as the machine-understandable translation of question q and executed on the input table to retrieve the final answer. Our contributions are the novel models that perform steps (ii) and (iii), while for step (i) we rely on the work of [8] (henceforth: PL2015).

3.1 Candidate Logical Form Generation

We generate a set of candidate logical forms from a question using the method of [8]. Only briefly, we review this method. Specifically, a question is parsed into a set of candidate logical forms using a semantic parser that recursively applies deduction rules. Logical forms are represented in Lambda DCS form [6] and can be executed on a table to yield an answer. An example of a question and its correct logical form are below:

```
How many people attended the last Rolling Stones concert?
R[λx[Attendance.Number.x]].argmax(Act.RollingStones,Index).
```

---

[1] An extensive list of open-domain QA publications can be found here: https://aclweb.org/aclwiki/Question_Answering_(State_of_the_art).

3.2 Converting Logical Forms to Text

In Algorithm 1 we describe how logical forms are transformed into interpretable textual representations called "paraphrases". We choose to embed paraphrases in low dimensional vectors and compare these against the question embedding. Working directly with paraphrases instead of logical forms is a design choice, justified by their interpretability, comprehensibility (understandability by non-technical users) and empirical accuracy gains. Our method recursively traverses the tree representation of the logical form starting at the root. For example, the correct candidate logical form for the question mentioned in Sect. 3.1, namely How many people attended the last Rolling Stones concert?, is mapped to the paraphrase Attendance as number of last table row where act is Rolling Stones.

3.3 Joint Embedding Model

We embed the question together with the paraphrases of candidate logical forms in a jointly learned vector space. We use two convolutional neural networks (CNNs) for question and paraphrase embeddings, on top of which a max-pooling operation is applied. The CNNs receive as input token embeddings obtained as described below.

Algorithm 1. Recursive paraphrasing of a Lambda DCS logical form. The + operation means string concatenation with spaces. Lambda DCS language is detailed in [6].

```
 1: procedure PARAPHRASE(z)          ▷ z is the root of a Lambda DCS logical form
 2:     switch z do
 3:         case Aggregation                                    ▷ e.g. count, max, min...
 4:             t ← AGGREGATION(z) + PARAPHRASE(z.child)
 5:         case Join                 ▷ join on relations, e.g. λx.Country(x, Australia)
 6:             t ← PARAPHRASE(z.relation) + PARAPHRASE(z.child)
 7:         case Reverse                               ▷ reverses a binary relation
 8:             t ← PARAPHRASE(z.child)
 9:         case LambdaFormula                         ▷ lambda expression λx.[...]
10:             t ← PARAPHRASE(z.body)
11:         case Arithmetic or Merge               ▷ e.g. plus, minus, union...
12:             t ← PARAPHRASE(z.left) + OPERATION(z) + PARAPHRASE(z.right)
13:         case Superlative                             ▷ e.g. argmax(x, value)
14:             t ← OPERATION(z) + PARAPHRASE(z.value) + PARAPHRASE(z.relation)
15:         case Value                                           ▷ i.e. constants
16:             t ← z.value
17:     return t       ▷ t is the textual paraphrase of the Lambda DCS logical form
18: end procedure
```

Token Embedding. The embedding of an input word sequence (e.g. question, paraphrase) is depicted in Fig. 1 and is similar to [4]. Every token is parametrized by learnable word and character embeddings. The latter help dealing with unknown tokens (e.g. rare words, misspellings, numbers or dates). Token vectors are then obtained using a CNN (with multiple filter widths) over the constituent characters, followed by a max-over-time pooling layer and concatenation with the word vector.

Sentence Embedding. We map both the question q and the paraphrase t into a joint vector space using sentence embeddings obtained from two jointly trained CNNs. CNNs' filters span a different number of tokens from a width set L. For each filter width $l \in L$, we learn n different filters, each of dimension $\mathbb{R}^{l \times d}$, where d is the word embedding size. After the convolution layer, we apply a max-over-time pooling on the resulting feature matrices which yields, per filter-width, a vector of dimension n. Next, we concatenate the resulting max-over-time pooling vectors of the different filter-widths in L to form our sentence embedding. The final sentence embedding size is $n|L|$.

Neural Similarity Measures. Let $u, v \in \mathbb{R}^d$ be the sentence embeddings of question q and of paraphrase t. We experiment with the following similarity scores: (i) DOTPRODUCT: $u^T v$; (ii) BILIN: $u^T S v$, with $S \in \mathbb{R}^{d \times d}$ being a trainable matrix; (iii) FC: u and v concatenated, followed by two sequential fully connected layers with ELU non-linearities; (iv) FC-BILIN: weighted average of BILIN and FC. These models define parametrized similarity scoring functions $\Phi : Q \times T \to \mathbb{R}$, where Q is the set of natural language questions and T is the set of paraphrases of logical forms.

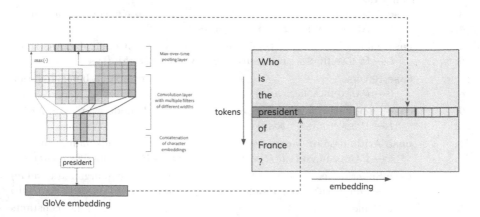

Fig. 1. Conversion of a sentence into a token embedding matrix using word embeddings and char CNNs. The resulting matrix is again fed to a CNN/RNN (not pictured) to produce a sentence embedding. Figure inspired from [4].

3.4 Training Algorithm

For training, we build two sets $\mathcal{P}$ (positive) and $\mathcal{N}$ (negative) consisting of all pairs $(q, t) \in Q \times T$ of questions and paraphrases of candidate logical forms generated as described in Sect. 3.1. A pair is positive or negative if its logical form gives the correct or respectively incorrect gold answer when executed on the corresponding table. During training, we use the ranking hinge loss function (with margin θ):

$$L(\mathcal{P}, \mathcal{N}) = \sum_{p \in \mathcal{P}} \sum_{n \in \mathcal{N}} \max(0, \theta - \Phi(p) + \Phi(n))$$

4 Experiments

Dataset: For training and testing we use the train-validation-test split of WikiTableQuestions [8], a dataset containing 22,033 pairs of questions and answers based on 2,108 Wikipedia tables. This dataset is also used by our baselines, [7,8]. Tables are not shared across these splits, which requires models to generalize to unseen data. We obtain about 3.8 million training triples (q, t, l), where l is a binary indicator of whether the logical form gives the correct gold answer when executed on the corresponding table. 76.7% of the questions have at least one correct candidate logical form when generated with the model of [8].

Table 1. Precision@1 of various baselines and our models on the WikiTableQuestions dataset.

Baseline Systems	P@1
Neural Programmer [7] (single model)	34.2%
Neural Programmer [7] (15 ensemble models)	37.7%
PL2015 [8]	37.1%

Our Models	P@1
CNN-DOTPRODUCT	31.4%
CNN-BILIN	20.4%
CNN-FC	30.4%
RNN-FC-BILIN	29.6%
CNN-FC-BILIN (best single model)	34.8%
CNN-FC-BILIN (15 ensemble models)	**38.7%**

Training Details: Our models are implemented using TensorFlow and trained on a single Tesla P100 GPU. Training takes approximately 6 h. We initialize word vectors with 200 dimensional GloVe ([9]) pre-trained vectors. For the character CNN we use widths spanning 1, 2 and 3 characters. The sentence embedding CNNs use widths of $L = \{2, 4, 6, 8\}$. The fully connected layers in the FC models have 500 hidden neurons, which we regularize using 0.8-dropout. The loss margin θ is set to 0.2. Optimization is done using Adam [5] with a learning rate of 7e-4. Hyperparameters are tunned on the development data split of the WikiTableQuestions table. We choose the best performing model on the validation set using early stopping.

Table 2. Example of common errors of our model.

Question	Paraphrase (gold vs predicted)
Which association entered last?	association of last row
	association of row with highest number of joining year
What is the total of all the medals?	count all rows
	number of total of nation is total
How many episodes were originally aired before December 1965?	count original air date as date <= 12 1965
	count original air date as date < 12 1965

Results: Experimental results are shown in Table 1. Our best performing single model is FC-BILIN with CNNs, Intuitively, BILIN and FC are able to extract different interaction features between the two input vectors, while their linear combination retains the best of both models. An ensemble of 15 single CNN-FC-BILIN models was setting (at the moment of publication) a new state-of-the-art precision@1 for this dataset: 38.7%. This shows that the same model initialized differently can learn different features. We also experimented with recurrent neural networks (RNNs) for the sentence embedding since these are known to capture word order better than CNNs. However, RNN-FC-BILIN performs worse than its CNN variant.

There are a few reasons that contributed to the low accuracy obtained on this task by various methods (including ours) compared to other NLP problems: weak supervision, small training size and a high percentage of unanswerable questions.

Error Analysis: The questions our models do not answer correctly can be split into two categories: either a correct logical form is not generated, or our scoring models do not rank the correct one at the top. We perform a qualitative analysis presented in Table 2 to reveal common question types our models often rank incorrectly. The first two examples show questions whose correct logical form depends on the structure of the table. In these cases a bias towards the more general logical form is often exhibited. The third example shows that our model has difficulty distinguishing operands with slight modification (e.g. smaller and smaller equals), which may be due to weak-supervision.

Ablation Studies: For a better understanding of our model, we investigate the usefulness of various components with an ablation study shown in Table 3. In particular, we emphasize that replacing the paraphrasing stage with the raw strings of the Lambda DCS expressions resulted in lower precision@1, which confirms the utility of this stage.

Analysis of Correct Answers: We analyze how well our best single model performs on various question types. For this, we manually annotate 80 randomly chosen questions that are correctly answered by our model and report statistics in Table 3.

Table 3. Ablation studies. Left: Component contributions to our model. Right: Types of questions answered correctly by our system.

System	P@1	System	Amount
CNN-FC-BILIN	34.1%	Lookup	10.8%
w/o Dropout	33.3%	Aggregation & Next/Previous	39.8%
w/o Char Embeddings	33.8%		
w/o GloVe (random init)	32.4%	Superlatives	30.1%
w/o Paraphrasing	33.1%	Arithmetic & Comparisons	19.3%

5   Conclusion

In this paper we propose a neural network QA system for semi-structured tables that eliminates the need for manually designed features. Experiments show that an ensemble of our models reaches competitive accuracy on the WikiTableQuestions dataset, thus indicating its capability to answer complex, multi-compositional questions. Our code is available at https://github.com/dalab/neural_qa.

References

1. Berant, J., Liang, P.: Semantic parsing via paraphrasing. In: ACL, vol. 1, pp. 1415–1425 (2014)
2. Bordes, A., Chopra, S., Weston, J.: Question answering with subgraph embeddings. arXiv preprint arXiv:1406.3676 (2014)
3. Kim, Y.: Convolutional neural networks for sentence classification. In: EMNLP. Citeseer (2014)
4. Kim, Y., Jernite, Y., Sontag, D., Rush, A.M.: Character-aware neural language models. In: Proceedings of the Thirtieth AAAI Conference on Artificial Intelligence, pp. 2741–2749. AAAI Press (2016)
5. Kingma, D., Ba, J.: Adam: A method for stochastic optimization. arXiv preprint arXiv:1412.6980 (2014)
6. Liang, P.: Lambda dependency-based compositional semantics. arXiv:1309.4408 (2013)
7. Neelakantan, P., Le, Q.V., Abadi, M., McCallum, A., Amodei, D.: Learning a natural language interface with neural programmer. arXiv:1611.08945 (2016)
8. Pasupat, P., Liang, P.: Compositional semantic parsing on semi-structured tables. In: Proceedings of the Annual Meeting of the Association for Computational Linguistics. Citeseer (2015)
9. Pennington, J., Socher, R., Manning, C.D.: Global vectors for word representation. In: EMNLP, vol. 14, pp. 1532–1543 (2014)
10. Wang, Y., Berant, J., Liang, P., et al.: Building a semantic parser overnight. In: ACL, vol. 1, pp. 1332–1342 (2015)
11. Yin, P., Lu, Z., Li, H., Kao, B.: Neural enquirer: learning to query tables in natural language. In: Proceedings of the Twenty-Fifth International Joint Conference on Artificial Intelligence, pp. 2308–2314. AAAI Press (2016)

Document Ranking Applied to Second Language Learning

Rodrigo Wilkens[✉], Leonardo Zilio, and Cédrick Fairon

Université catholique de Louvain, Louvain-la-Neuve, Belgium
{rodrigo.wilkens,leonardo.zilio,cedrick.fairon}@uclouvain.be

Abstract. This paper addresses the needs of language learners and teachers by combining keyword-based search and language level information on an algorithm that can rank documents by pertinence to the required topic (keywords) and adequacy to the user's language level. We conducted several experiments using the EF-CAMDAT corpus (annotated for topic and level) and we observed that the best ranking results were an average of BM25 and linguistic information. We also saw that the grammar of level C1 is the best indicator for level. Finally, we proposed a customization for prioritizing beginner or intermediate levels at the top of the rank.

1 Introduction

The retrieval of interesting documents based on a given set of keywords is a trivial event in the everyday life of almost everyone and it also applies to the activity of learning a new language, especially when this kind of behavior is encouraged or requested by the language teacher [1]. As indicated by [2], the pedagogical approach of flooding learners with language information is extensively used by language teachers. However, the search for relevant reading material takes time and effort, and, as a result, teachers often resort to schoolbooks, which are readily available, but have a limited text choice and are possibly not up-to-date and hardly in line with students interests.

When the task concerns only the search for keywords, most search engines can provide the users with a good set of texts that will suffice for their needs, but some of the texts will not be accessible to those with a lower language proficiency [3]. As such, when dealing with the need for documents that, at the same time, are relevant according to the given keywords and match certain constraints of text complexity, then further re-ranking of the documents may be needed to address these new criteria. This could be the case, for instance, when the user has a low-level reading capacity, has a disease that affects reading, or is a language learner. The latter case is the main focus of this paper.

By retrieving documents that are relevant in terms of both topic and language complexity it would be possible to address the needs of a broader range of users. From the teachers' or learners' perspective, it would be possible to retrieve texts that are better tailored to their needs, whether being more engaging for the

© Springer International Publishing AG, part of Springer Nature 2018
G. Pasi et al. (Eds.): ECIR 2018, LNCS 10772, pp. 618–624, 2018.
https://doi.org/10.1007/978-3-319-76941-7_53

learners or presenting relevant information for an extra class task. As such, the use of level and topic-oriented document retrieval could better inform the acquisition of second language in both fronts.

Taking all of this into consideration, this paper aims at presenting a system that retrieves documents based on a keyword-oriented query and ranks them according to topic and language level. We further hypothesize that both text content and level are complementary for the ranking process.

This paper is divided as follows: Sect. 2 presents search engines that address some level of language complexity together with topic; Sect. 3 describes the corpus that was used and the methodology employed in the experiments; Sect. 4 shows the results of our experiments; and, finally, Sect. 5 contains our final remarks and perspectives for future work.

2 Related Work

There are several systems that attempt to retrieve texts based on topic of interest while also supporting language learning activities. Here we present a selected few, which address topic and, in varying degrees, language level: SourceFinder [4], REAP (Reader-Specific Lexical Practice for Improved Reading Comprehension) [5], FLAIR (Form-Focused Linguistically Aware Information Retrieval) [2], and READ-X [6]. These systems are similar in purpose, but vary in their methods, which can be better compared by observing the following dimensions: text source, text representation, language model, and target user.

Regarding the text source, they can be divided in systems based on the Web (REAP, FLAIR, and READ-X) or on closed corpus (SourceFinder). The Web-based systems delegate the document retrieval to a Web-based search engine, and then re-rank the results. In contrast, closed corpus systems run their own search criteria over the whole corpus. The closed corpus approach can have a lot of information annotated off-line, but disposes of fewer documents in comparison to the Web-based approach.

Moving on to the dimension of text representation, some systems (REAP, FLAIR, and SourceFinder) index and rank the documents according to a series of linguistic information, while systems like READ-X use only one type of measure. Linguistic information can involve, among others, part-of-speech tags, length-based counts, and readability measures. Topic annotation is usually automatic in these systems, but they differ in the source and number of tags.

An important dimension for all of the systems is how they model the learners language skills. Most of them model the learners capacity in terms of a readability measure, but others further explore this dimension by using a language learning framework (e.g. FLAIR is strongly based on the CEFR [7]). The readability measures are used in different ways: for instance, READ-X annotates the texts with classic measures and presents them to the user as parameters for queries, while REAP applies readability measures that use dictionary information.

Finally, there is the intended target user. READ-X does not specify its target users, but they need to understand about readability measures for interpreting the output and selecting the adequate thresholds. SourceFinder and FLAIR focus on a teachers audience, providing support in their activities, while REAP focuses on both sides of the spectrum, and is designed for both teachers and learners. All of these systems require that the users have a deep understanding of their language model. While this assumption is acceptable for language teachers, it is not for the learners.

3   Methodology

In this paper, we want to use document ranking to address the needs of learners and teachers for texts that correspond to a topic, but that are also adequate for a specific language level. In this section, we present the procedures that we carried out and the resources we used for achieving our main goal.

The European Union developed the Common European Framework of Reference for Language Learning (CEFR) [7], which provides guidance for the teaching of foreign languages. The CEFR presents a description of communication goals that a learner should achieve in each of six main levels: A1, A2, B1, B2, C1, and C2, which are further subdivided into three levels.

For evaluating the system's ranking algorithm, we selected the EF-CAMDAT [8] corpus, which presents a collection of texts produced by learners of English from different levels of proficiency, according to the CEFR. The corpus is divided according to the CEFR and contains a total of 532 thousand documents (33 million tokens) by 83,385 learners from 137 countries. Each document is linked to a specific topic (e.g., writing an invitation, or describing an experience), and similar topics are generally addressed on a single CEFR level.

For testing our hypothesis that text content and level are complementary for ranking documents, we selected only documents that belong to the same topic, but spreading across different, non-adjacent CEFR levels. This selection was to ensure that the documents were not too close to each other in terms of level, so as to avoid errors derived from bias in human judgment. By applying these strong constraints for the selection of documents, we obtained three valid topics. From each of them, we extracted the first 10 noun keywords based on the higher log-likelihood [9].

For ranking the documents by keywords, we opted for BM25 [10], which is a bag-of-words retrieval function that ranks documents using a query of terms. We used the Lucene implementation of BM25 [11], but, for ranking the documents by language level, we substituted the score function of BM25 for linguistic information. The full corpus was annotated with linguistic information, such as grammar level, vocabulary level [1], and three readability measures (Dale-Chall Score [12], Flesch-Kincaid measure [13], and Gunning Fog Index [14]). After that, we proceeded with three experiments for ranking the documents:

---

[1] For the grammar and vocabulary of level C1, we used the inversed score, because we expect that an A-level document would present them in a lesser degree.

Ranking score: We evaluated the results of different ranking algorithms (BM25 alone, linguistic information alone, and our model, which is the average of the previous scores) to observe which one renders the best results.

Linguistic information: We evaluated which linguistic information could best predict the language level of a document, a task that is hard to approximate automatically.

Search customization by language level: We proposed a method for ranking the documents in a way that a selected language level is given preference.

All of these experiments were evaluated by Spearman's rank correlation according to: (1) the rank of BM25 for topic, and (2) the rank of documents by level assessment scores. We also evaluate the results using precision at ten (p@10): for topic, we identified as correct when a document had the expected topic, and, for level, we used the lowest language level retrieved. As terms of comparison, we considered both extremes: the standard BM25 score, which prioritizes the topic, and the scores from linguistic information, which prioritize the level.

4   Evaluation

The results for the three experiments can be seen in Table 1. The table presents correlation and precision at ten (p@10) of a given feature in relation to the topic of the retrieved documents and their language levels.

At the top of the table we have information about the ranking score experiment. Comparing to BM25, the average model was a bit worse in terms of topic, but improved the level ranking in correlation and had 100% of p@10 for both level and topic. On the theoretical level, this confirms our hypothesis that averaging BM25 and linguistic information improves the document ranking for the purposes of addressing language level. In a more practical approach, the next experiment addresses the extent to which linguistic measures can approximate language level.

Regarding the linguistic information experiment, in which we used linguistic features to model the level, we can see that, in terms of p@10, many features present 100% for both level and topic. As such, for a learner that is only interested in the top documents, the average model can achieve excellent results using several features individually. However, when we consider the correlation, which takes into account all of the retrieved documents, we see that the best feature, specially for level, was the grammar of C1. Comparing to the pure linguistic model, the average model also consistently improves the scores both in correlation and in p@10. These results further support the hypothesis for the average model, even in a practical setting, when level assessment is not provided and linguistic features are the only clues for approximating it.

For the last experiment, the ranking using a specific level, we focused on beginner and intermediate levels[2]. We modeled the beginner level in the same

---

[2] In this work we did not address advanced-level learners, because they are supposed to use traditional search engine systems.

way as in the previous experiments, and, for the intermediate level, we averaged the score from the beginner level with its inverse, which is assumed to be the score of advanced level, for targeting the middle of the CEFR spectrum. The results showed that we had mainly intermediate-level documents as best representatives of the topic queries, and, as such, the BM25 score achieves better results both in correlation and in p@10. Even so, if we want to ensure that intermediate-level documents are prioritized, the average model presented an improvement over isolated linguistic features. In a more practical scenario, the best feature to approximate intermediate level would be Dale-Chall score [12]. The results (on the columns Beginner Level and Intermediate Level of Table 1) show that the ranking equation did not have a negative impact in terms of topic, and achieved a high correlation in terms of level.

Table 1. Evaluation of document retrieval by topic and level

Model	Beginner level								Intermediate level							
	Avg model				Linguistic model				Avg model				Linguistic model			
	Correlation		p@10 (%)		Correlation		p@10 (%)		Correlation		p@10 (%)		Correlation		p@10 (%)	
	Topic	Level	Topic	Level	Topic	Level	Topic	Level	Topic	Level	Topic	Level	Topic	Level	Topic	Level
$BM25$	100	56	100	56	100	33	100	33	100	100	100	67	100	100	100	33
$Level$	96	69	100	100	86	75	87	100	96	69	100	50	86	74	87	7
$A1_{gram}$	95	61	100	100	82	59	100	100	81	26	100	67	45	-2	63	23
$A1_{vocab}$	94	57	100	100	77	49	100	100	81	26	100	67	42	-4	30	3
$A2_{gram}$	75	26	90	100	14	-10	43	100	79	50	100	60	27	31	20	7
$A2_{vocab}$	93	56	100	100	76	48	93	100	83	31	97	67	46	1	3	3
$B1_{gram}$	74	25	70	100	11	-11	37	100	80	44	100	43	30	21	17	3
$B1_{vocab}$	87	42	100	100	54	22	67	100	79	23	97	33	33	-10	33	3
$B2_{gram}$	68	30	100	100	-5	-4	90	100	80	60	83	57	28	48	10	20
$B2_{vocab}$	80	35	100	100	29	6	70	100	83	51	83	50	38	34	7	3
$C1_{gram}$	95	69	87	100	79	76	13	100	79	21	97	67	36	-11	17	3
$C1_{vocab}$	90	69	90	100	64	72	10	100	81	39	90	53	36	13	10	3
[12]	87	41	100	100	55	19	93	100	92	60	100	67	70	55	87	13
[13]	94	56	100	100	79	49	100	100	93	48	100	60	73	35	30	20
[14]	92	50	100	100	71	37	100	100	94	54	100	67	79	46	100	33

5   Conclusion

In this paper, we combined language learning with traditional document retrieval. We used a language learners corpus, which contains documents annotated with language level and topic. The experiments with ranking documents by topic and level showed us that averaging a traditional information retrieval algorithm and linguistic information renders better results from a language learning perspective. Among all linguistic features, we observed that the grammar of level C1 had better results for both topic and level in terms of correlation. Finally, when trying to customize the document ranking for prioritizing an intermediate language level, we concluded that the use of Dale-Chall score can approximate the best result, while also largely ensuring a correspondence with the language level order.

In general, the results of our experiments show that by averaging topic and language level, we have a complementary document ranking process. A more language level-oriented ranking of the documents tends to not affect much the ranking by topic, while allowing for a result that is useful for a learner.

In future studies, we intend to investigate more informative linguistic measures by applying a level assessment classification of documents (e.g., [15]). We also want to research the filtering of documents by language level assessment.

Acknowledgements. We would like to thank the Walloon Region (Projects BEWARE n. 1510637 and 1610378) for support, and Altissia International for research collaboration.

References

1. Purcell, K., Rainie, L., Heaps, A., Buchanan, J., Friedrich, L., Jacklin, A., Chen, C., Zickuhr, K.: How teens do research in the digital world. Pew Internet & American Life Project (2012)
2. Chinkina, M., Kannan, M., Meurers, D.: Online information retrieval for language learning. In: 2016 ACL, p. 7 (2016)
3. Vajjala, S., Meurers, D.: On the applicability of readability models to web texts. In: 2013 ACL, p. 59 (2013)
4. Passonneau, R., Hemat, L., Plante, J., Sheehan, K.M.: Electronic sources as input to gre® reading comprehension item development: sourcefinder prototype evaluation. ETS Res. Rep. Ser. **2002**(1) (2002). 66 pages
5. Collins-Thompson, K., Callan, J.: Information retrieval for language tutoring: an overview of the reap project. In: 27th Annual International ACM SIGIR Conference on Research and Development in Information Retrieval, pp. 544–545. ACM (2004)
6. Miltsakaki, E., Troutt, A.: Read-x: Automatic evaluation of reading difficulty of web text. In: E-Learn: World Conference on E-Learning in Corporate, Government, Healthcare, and Higher Education, Association for the Advancement of Computing in Education (AACE), pp. 7280–7286 (2007)
7. Verhelst, N., Van Avermaet, P., Takala, S., Figueras, N., North, B.: Common European Framework of Reference for Languages: Learning, Teaching, Assessment. Cambridge University Press, Cambridge (2009)
8. Geertzen, J., Alexopoulou, T., Korhonen, A.: Automatic linguistic annotation of large scale l2 databases: the ef-cambridge open language database (EFCAMDAT). In: Cascadilla Proceedings Project on 31st Second Language Research Forum. Somerville, MA (2013)
9. Rayson, P., Garside, R.: Comparing corpora using frequency profiling. In: Workshop on Comparing Corpora, pp. 1–6. ACL (2000)
10. Robertson, S.E., Walker, S.: Okapi/keenbow at TREC-8. In: TREC, pp. 151–162 (1999)
11. Pérez-Iglesias, J., Pérez-Agüera, J.R., Fresno, V., Feinstein, Y.Z.: Integrating the probabilistic models bm25/bm25f into lucene. preprint arXiv:0911.5046 (2009)
12. Dale, E., Chall, J.S.: A formula for predicting readability: instructions. Educ. Res. Bull. **27**, 37–54 (1948)
13. Kincaid, J.P., Fishburne Jr., R.P., Rogers, R.L., Chissom, B.S.: Derivation of new readability formulas (automated readability index, fog count and flesch reading ease formula) for navy enlisted personnel. Technical report, Naval Technical Training Command Millington TN Research Branch (1975)

14. Gunning, R.: The Technique of Clear Writing. Mcgraw-Hill, NY (1968)
15. Heilman, M., Collins-Thompson, K., Callan, J., Eskenazi, M.: Combining lexical and grammatical features to improve readability measures for first and second language texts. In: HLT 2007: The Conference of the NAACL, pp. 460–467 (2007)

Generating High-Quality Query Suggestion Candidates for Task-Based Search

Heng Ding[1,2(⊠)], Shuo Zhang[2], Darío Garigliotti[2], and Krisztian Balog[2]

[1] Wuhan University, Wuhan, China
hengding@whu.edu.cn
[2] University of Stavanger, Stavanger, Norway
{shuo.zhang,dario.garigliotti,krisztian.balog}@uis.no

Abstract. We address the task of generating query suggestions for task-based search. The current state of the art relies heavily on suggestions provided by a major search engine. In this paper, we solve the task without reliance on search engines. Specifically, we focus on the first step of a two-stage pipeline approach, which is dedicated to the generation of query suggestion candidates. We present three methods for generating candidate suggestions and apply them on multiple information sources. Using a purpose-built test collection, we find that these methods are able to generate high-quality suggestion candidates.

1 Introduction

Query suggestions, recommending a list of relevant queries to an initial user input, are an integral part of modern search engines [8]. Accordingly, this task has received considerable attention over the last decade [1,2,7]. Traditional approaches, however, do not consider the larger underlying task the user is trying to accomplish. In this paper, we focus on generating query suggestions for supporting task-based search. Specifically, we follow the problem definition of the *task understanding* task from the TREC Tasks track: given an initial query, the system should return a ranked list of suggestions "that represent the set of all tasks a user who submitted the query may be looking for" [14]. Thus, the overall goal is to provide a complete coverage of aspects (subtasks) for an initial query, while avoiding redundancy.

We envisage a user interface where task-based query suggestions are presented once the user has issued an initial query; see Fig. 1. These query suggestions come in two flavors: *query completions* and *query refinements*. The difference is that the former are prefixed by the initial query, while the latter are not. It is an open question whether a unified method can produce suggestions in both flavors, or rather specialized models are required. The best published work on task-based query suggestions, that we know of, is by Garigliotti and Balog [4], who use a probabilistic generative model to combine keyphrase-based suggestions extracted from multiple information sources. Nevertheless, they rely

© Springer International Publishing AG, part of Springer Nature 2018
G. Pasi et al. (Eds.): ECIR 2018, LNCS 10772, pp. 625–631, 2018.
https://doi.org/10.1007/978-3-319-76941-7_54

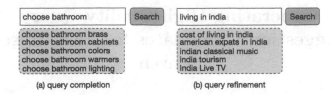

(a) query completion   (b) query refinement

Fig. 1. Examples of query suggestions to support task-based search.

heavily on Google's query suggestion service. Thus, another main challenge in our work is to solve this task without relying on suggestions provided by a major web search engine (and possibly even without using a query log).

Following the pipeline architecture widely adopted in general query suggestion systems [7,12], we propose a two-step approach consisting of *suggestion generation* and *suggestion ranking* steps. In this paper, we focus exclusively on the first component. Our aim is to generate sufficiently many high-quality query suggestion candidates. The subsequent ranking step will then produce the final ordering of suggestions by reranking these candidates (and ensuring their diversity with respect to the possible subtasks). The first research question we address is: *Can existing query suggestion methods generate high-quality query suggestions for task-based search?* Specifically, we employ the popular suffix [7], neural language model [10], and sequence-to-sequence [12] approaches to generate candidate suggestions. The second research question we ask is: *What are useful information sources for each method?* We are particularly interested in finding out how a task-oriented knowledge base (Know-How [9]) and a community question answering site (WikiAnswers [3]) fare against using a query log (AOL [11]). We find that the sequence-to-sequence approach performs best among the tested three methods. As for data sources, we observe that, as expected, the query log is the highest performing one among all. Nevertheless, the other two also provide valuable suggestion candidates that cannot be generated from the query log. Overall, we find that all method-source configurations contribute unique suggestions, and thus it is beneficial to combine them.

2   Query Suggestion Generation

Given a task-related initial query, q_0, we aim to generate a list of query suggestions that cover all the possible subtasks related to the task the user is trying to achieve. For a given suggestion q, let $P(q|q_0)$ denote the probability of that suggestion. Below, we present three methods from the literature. The first two methods are specialized only in producing query completions, while the third one is able to handle both query completions and refinements. Due to space constraints, only brief descriptions are given; we refer to the respective publications for further details.

2.1   Popular Suffix Model

The *popular suffix model* [7] generates suggestions using frequently observed suffixes mined from a query log. The method generates a suggestion q by extending the input query q_0 with a popular suffix s, i.e., $q = q_0 \oplus s$, where $\oplus$ denotes the concatenation operator. The query likelihood is based on the popularity of suffix s: $P(q|q_0) = pop(s)$, where $pop(s)$ denotes the relative frequency of s occurring in the query log.

2.2   Neural Language Model

Neural language models (NLMs) can be used for generating query suggestions [10]. Suggestion q is created by extending the input query q_0 character by character: $q = q_0 \oplus \mathbf{s} = (c_1, \ldots, c_n, c_{n+1}, \ldots, c_m)$, where $c_1, \ldots, c_n$ are the characters of q_0, and $\mathbf{s} = (c_{n+1}, \ldots, c_m)$ are characters generated by the neural model. Given a sequence of previous characters $\mathbf{c} = (c_1, \ldots, c_i)$, the model generates the next character $(i \geq n)$ according to: $P(c_{i+1}|\mathbf{c}) = softmax(h_i)$, where the hidden state vector at time i is computed using $h_i = f(x_i, h_{i-1})$. Here, f denotes a special unit, e.g., a long short-term memory (LSTM) [5]; x_i is the vector representation of the ith character of suggestion q and is taken to be $x_i = \sigma(c_i)$, where σ denotes a mapping function from a character to its vector representation. Finally, the query likelihood is estimated according to:

$$P(q|q_0) = \prod_{j=n}^{m-1} P(c_{j+1}|c_1, \ldots, c_j).$$

Our implementation uses a network of 512 hidden units. We initialize the word-embedded vector with the pre-trained vector from Bing queries.[1] Beam search width is set to 30.

2.3   Sequence to Sequence Model

The sequence-to-sequence model (Seq2Seq) [6] aims to directly model the conditional probability $P(w'_1, \ldots, w'_m | w_1, \ldots, w_n)$ of translating a source sequence $(w_1, \ldots, w_n)$ to a target sequence $(w'_1, \ldots, w'_m)$. Thus, it lends itself naturally to implement our query suggestion task using Seq2Seq, by letting the initial query be the source sequence $q_0 = (w_1, \ldots, w_n)$ and the suggestion be the target sequence $q = (w'_1, \ldots, w'_m)$. Typically, a Seq2Seq model consists of two main components: an encoder and a decoder.

The *encoder* is a recurrent neural network (RNN) to compute a context vector representation c for the original query q_0. The hidden state vector of the encoder RNN at time $i \in [1..n]$ is given by: $h_i = f(w_i, h_{i-1})$, where w_i is the ith word of the input query q_0, and f is a special unit (LSTM). The context vector

---

[1] https://www.microsoft.com/en-us/download/details.aspx?id=52597.

representation is updated by $c = \phi(h_1, h_2, \ldots, h_n)$, where ϕ is an operation choosing the last state h_n.

The *decoder* is another RNN to decompress the context vector c and output the suggestion, $q = (w_1', \ldots, w_m')$, through a conditional language model. The hidden state vector of the decoder RNN at time $i \in [1..m]$ is given by $h_i' = f'(h_{i-1}', w_{i-1}', c)$, where w_i' is the ith word of the suggestion q, and f' is a special unit (LSTM). The language model is given by: $P(w_i'|w_1', \ldots, w_{i-1}', q_0) = g(h_i', w_{i-1}', c)$, where g is a softmax classifier. Finally, the Seq2Seq model estimates the suggestion likelihood according to:

$$P(q|q_0) = \prod_{j=1}^{m-1} P(w_{j+1}'|w_1', \ldots, w_j', q_0) .$$

We use a bidirectional GRU unit with size 100 for encoder RNNs, and a GRU unit with size 200 for decoder RNNs. We employ an Adam optimizer with an initial learning rate of 10^{-4} and a dropout rate of 0.5. Beam search width is set to 100.

3   Data Sources

We consider three independent information sources. For the PopSuffix and NLM methods, we need a collection of short texts, $\mathcal{C}$. For Seq2Seq, we need pairs of question-suggestion pairs, $\langle \mathcal{Q}, \mathcal{S} \rangle$.

- *AOL query log* [11]: a large query log that includes queries along with anonymous user identity and timestamp. We extract all queries from the log as $\mathcal{C}$. We detect sessions (each session including multiple queries) using the same criterion as in [12]. Then, we pair queries in the same session to obtain $\langle \mathcal{Q}, \mathcal{S} \rangle$, where $\mathcal{Q}$ denotes a set of queries and $\mathcal{S}$ are suggestions paired against $\mathcal{Q}$. In order to obtain more pairs, we extract all proper prefixes from the query, and pair them together. For example, given a query "make a pancake", we can construct two pairs ⟨ "make", "make a pancake" ⟩ and ⟨ "make a", "make a pancake" ⟩. This way, we end up with a total of 112 K prefix-query pairs.
- *KnowHow* [9]: a knowledge base that consists of two and half million entries. Each triple $\langle s, p, o \rangle$ represents a fact about a human task, where the subject s denotes a task (e.g., "make a pancake") and the object o is a subtask (e.g., "prepare the mix"). We collect all subjects and objects as $\mathcal{C}$, and take all (142K) subject-object pairs to form $\langle \mathcal{Q}, \mathcal{S} \rangle$. Additionally, prefixes from tasks (i.e., subjects and objects) are extracted to get more pairs, the same way as it is done for the AOL query log.
- *WikiAnswers* [3]: a collection of questions scraped from WikiAnswers.com.[2] We detect task-related questions using a simple heuristic, namely, that a task-related question often starts with question constructions "how do you"

---

[2]   http://www.answers.com/Q/.

or "how to." These question constructions are removed from the questions to obtain $\mathcal{C}$ (e.g., "how to change gmail password" → "change gmail password"). This source can only be used for the PopSuffix and NLM methods, as it does not provide pairs for Seq2Seq.

4   Experiment

We design and conduct an experiment to answers our research questions. First, we collect a pool of candidate suggestions, by applying the methods on different sources. Second, we collect annotations for each of these suggestions via crowdsourcing. Finally, we report and analyze the results.

4.1   Pool Construction

We consider all queries (100 in total) from the TREC 2015 and 2016 Tasks tracks [13,14]. We combine the proposed methods (Sect. 2) with various information sources (Sect. 3) for suggesting candidates. We shall write s-m to denote a particular configuration that uses method m with source s. In addition, we also include the suggestions generated by (i) the keyphrase-based query suggestion system [4], and (ii) the Google Query Suggestion Service (referred to as *Google API* for short. Two pools are constructed, one for query completions (QC) and one for query refinements (QR). The pool depth is 20, that is, we consider (up to) the top-20 suggestions for each method.

4.2   Crowdsourcing

Due to the fact that many of our candidate suggestions lack assessments in the TREC ground truth (and thus are considered irrelevant), we obtain relevance assessments for all suggestions via crowdsourcing. Specifically, we use the Crowdflower platform, where a dynamic number of annotators (3–5) are asked to label each suggestion as relevant or non-relevant. A suggestion is relevant if it targets for some more specific aspect of the original query, i.e., the suggestion must be related to the original query intent. It does not have to be perfectly correct grammatically, as long as the intent behind is clearly understandable. The final label is taken to be the majority vote among the assessors. A total of 12,790 QC and 9,608 QR suggestions are annotated, at the total expense of 692$. Further details are available in the online appendix.[3]

4.3   Results and Analysis

Table 1 presents a comparison of methods in terms of precision at cutoff points 10 and 20 (P@10 and P@20). Overall, we find that our methods can generate high-quality query suggestions for task-based search. Our best numbers are

---

[3] http://bit.ly/2BnSjhR.

Table 1. Precision for candidate suggestions generated by different configurations. For QC methods, we also report on recall (R) and cumulative recall (CR).

Method	QC				QR	
	P@10	P@20	R	CR	P@10	P@20
AOL-PopSuffix	0.257	0.245	0.168	0.168	-	-
KnowHow-PopSuffix	0.195	0.170	0.102	0.256	-	-
WikiAnswers-PopSuffix	0.181	0.167	0.101	0.333	-	-
AOL-NLM	0.256	0.241	0.170	0.474	-	-
KnowHow-NLM	0.166	0.147	0.108	0.575	-	-
WikiAnswers-NLM	0.163	0.121	0.088	0.650	-	-
AOL-Seq2Seq	0.283	0.181	0.156	0.765	0.043	0.031
KnowHow-Seq2Seq	0.158	0.111	0.079	0.813	0.206	0.148
Keyphrase-based [4]	0.321	0.239	0.130	-	0.575	0.504
Google API	0.267	0.134	0.078	-	0.289	0.145

comparable to that of the Google API. It should, however, be noted that the Google API can only generate a limited number of suggestions for each query. The keyphrase-based method [4] is the highest performing of all; it is expected, as it combines multiple information sources, including suggestions from Google.

We find that the AOL query log is the best source for generating QC suggestions, while KnowHow works well for QR. AOL-Seq2Seq performs poorly on QR; this is because only 2% of all training instances in $\langle \mathcal{Q}, \mathcal{S} \rangle$ are QR pairs. For QC suggestions, the performance of AOL-PopSuffix and AOL-NLM are close to that of the Google API, while AOL-Seq2Seq even outperforms it. For QR suggestions, the performance of AOL-Seq2Seq is close to that of the Google API in terms of P@20, but is lower on P@10. In general, our system is able to produce more suggestions than what the (public) Google API provides.

Additionally, we also evaluate the recall of each QC method, using all relevant suggestions found as our recall base. We further report on cumulative recall, i.e., for line i of Table 1 it is the recall of methods from lines $1..i$ combined together. We observe that each configuration brings a considerable improvement to cumulative recall. This shows that they generate unique query suggestions. For example, given the query "choose bathroom", our system generates unique query suggestions from different methods and sources, e.g., "choose bathroom marks" (WikiAnswers-NLM), "choose bathroom supply" (AOL-NLM), "choose bathroom for your children" (KnowHow-NLM), "choose bathroom grout" and "choose bathroom appliances" (KnowHow-Seq2Seq), which are beyond what the Google API and the keyphrase-based system [4] provide. Therefore it is beneficial to combine suggestions from multiple configurations for further reranking.

5   Conclusions

We have addressed the task of generating query suggestions that can assist users in completing their tasks. We have focused on the first component of a two-step pipeline, dedicated to creating suggestion candidates. For this, we have considered several methods and multiple information sources. We have based our evaluation on the TREC Tasks track, and collected a large number of annotations via crowdsourcing. Our results have shown that we are able to generate high-quality suggestion candidates. We have further observed that the different methods and information sources lead to distinct candidates.

As our next step, we will focus on the second component of the pipeline, namely, suggestions ranking. As part of this component, we also plan to address the specific issue of subtasks coverage, i.e., improving the diversity of query suggestions.

References

1. Bhatia, S., Majumdar, D., Mitra, P.: Query suggestions in the absence of query logs. In: Proceedings of SIGIR 2011, pp. 795–804 (2011)
2. Cai, F., de Rijke, M.: A Survey of Query Auto Completion in Information Retrieval. Now Publishers Inc., Hanover (2016)
3. Fader, A., Zettlemoyer, L., Etzioni, O.: Open question answering over curated and extracted knowledge bases. In: Proceedings of KDD 2014 (2014)
4. Garigliotti, D., Balog, K.: Generating query suggestions to support task-based search. In: Proceedings of SIGIR 2017, pp. 1153–1156 (2017)
5. Hochreiter, S., Schmidhuber, J.: Long short-term memory. Neural Comput. **9**, 1735–1780 (1997)
6. Luong, M., Pham, H., Manning, C.D.: Effective Approaches to Attention-based Neural Machine Translation (2015)
7. Mitra, B., Craswell, N.: Query auto-completion for rare prefixes. In: Proceedings of CIKM 2015, pp. 1755–1758 (2015)
8. Ozertem, U., Chapelle, O., Donmez, P., Velipasaoglu, E.: Learning to suggest: a machine learning framework for ranking query suggestions. In: Proceedings of SIGIR, vol. 12, 25–34 (2012)
9. Pareti, P., Testu, B., Ichise, R., Klein, E., Barker, A.: Integrating Know-How into the Linked Data Cloud (2016)
10. Park, D.H., Chiba, R.: A neural language model for query auto-completion. In: Proceedings of SIGIR 2017, pp. 1189–1192 (2017)
11. Pass, G., Chowdhury, A., Torgeson, C.: A picture of search. In: Proceedings of InfoScale 2006 (2006)
12. Sordoni, A., Bengio, Y., Vahabi, H., Lioma, C., Grue Simonsen, J., Nie, J.-Y.: A hierarchical recurrent encoder-decoder for generative context-aware query suggestion. In: Proceedings of CIKM 2015, pp. 553–562 (2015)
13. Verma, M., Kanoulas, E., Yilmaz, E., Mehrotra, R., Carterette, B., Craswell, N., Bailey, P.: Overview of the TREC tasks track 2016. In: Proceedings of TREC 2016 (2016)
14. Yilmaz, E., Verma, M., Mehrotra, R., Kanoulas, E., Carterette, B., Craswell, N.: Overview of the TREC: tasks track. In: Proceedings of TREC 2015 (2015)

An Incremental Approach
for Collaborative Filtering
in Streaming Scenarios

Rama Syamala Sreepada$^{(\boxtimes)}$ and Bidyut Kr. Patra

Department of Computer Science and Engineering,
National Institute of Technology Rourkela, Rourkela 769008, Odisha, India
{515cs1002,patrabk}@nitrkl.ac.in

Abstract. The crux of a recommendation engine is to process users ratings and provide personalized suggestions to the user. However, processing the ratings and providing recommendations in real time still remains challenging, when there is a perpetual influx of new ratings. Traditional approaches fail to accommodate the new streamlined ratings and update the users' preferences on the fly. In this paper, we address this challenge of streaming data without compromising accuracy and efficiency of recommender system. We identify the affected users and incrementally update their vital statistics after each new rating. We propose an incremental similarity measure for finding neighbors who play an important role in personalizing recommendations for active user. Experimental results on real-world datasets show that the proposed approach outperforms the state-of-the-art techniques in terms of accuracy and execution time.

Keywords: Collaborative filtering · Personalized recommendation
Streamlined ratings · Tendency based approach · Incremental updates

1 Introduction

Collaborative filtering (CF) techniques have been very successful and popular in providing recommendations to the users. CF techniques are broadly categorized into two types: Neighborhood based approach (User based CF and Item based CF) and Model based approach. The user-based (UB) and item-based (IB) CF approaches are one of the earliest methods in recommender systems [1]. On the other hand, model based CF techniques are proven to be more accurate in learning the user's and item's features through building a model using machine learning and other techniques [4]. In [3], Cacheda *et al.* proposed a tendency-based technique to further improve the recommendation accuracy and computational cost.

The traditional techniques are not equipped to process the incremental ratings in real time. As the new ratings arrive, the users' taste/interests can change *w.r.t.* time *i.e.* the users' preferences calculated earlier could become obsolete,

© Springer International Publishing AG, part of Springer Nature 2018
G. Pasi et al. (Eds.): ECIR 2018, LNCS 10772, pp. 632–637, 2018.
https://doi.org/10.1007/978-3-319-76941-7_55

leading to a concept drift in the recommender system. A feasible solution is to incrementally update the preferences with the arrival of each new rating. In this direction, Huang *et al.* proposed an item-neighborhood based approach which prunes the probable dissimilar items to reduce the number of similarity computations [2]. Subbian *et al.* proposed a probabilistic neighborhood based approach to compute the similarity between the items [5]. This approach approximates similarity using Locality Sensitive Hashing where hash functions are applied on the user indices. The approaches proposed in [2,5] compute only an approximation but not the actual similarity which can affect the accuracy of recommender system.

In this paper, we introduce an incremental approach termed as *IncRec* to update the tendency of user and item in real-time. Subsequently, we propose an incremental user similarity measure to personalize the item tendency *w.r.t* the active user. We simulated a streaming environment on two real-world datasets (Yahoo! Music and Movielens 10M) and the experimental results show that the proposed technique outperforms tendency based approach and the state-of-the-art techniques in terms of accuracy and execution time.

2   Background and Proposed Approach

As discussed in the previous section, tendency based approach is found to be more accurate and efficient compared to the traditional CF approaches [3]. Tendency based approach computes two important statistics namely, *user tendency* and *item tendency*. The user tendency (τ_u) is computed in terms of the aggregate rating deviation from the user's rating to each rated item's average rating. Likewise, item tendency (τ_i) is computed (Eq. 1).

$$\tau_u = \frac{\sum_{i \in I_u}(r_{ui} - \bar{r}_i)}{|I_u|}, \qquad \tau_i = \frac{\sum_{u \in U_i}(r_{ui} - \bar{r}_u)}{|U_i|} \qquad (1)$$

where r_{ui} is the rating of user u on item i, $\bar{r}_u$, $\bar{r}_i$ are the average ratings of the user and item respectively, I_u is the set of items rated by the user, and U_i is the set of users who rated the item i. The final rating of an unrated item is calculated based on the tendency statistics and average rating of the active user. Although this approach needs lesser computational time, it does not update the tendencies in streaming environment. It can be observed from Eq. 1 that the item tendency (τ_i) remains unchanged across the users. This leads to "non-personalized" recommendations. Also, it cannot address the challenges posed by streaming environment. In this paper, we address these drawbacks by proposing an incremental approach which is effective in streaming scenarios.

2.1   IncRec: Proposed Approach

As discussed in the previous section, tendency based approach has two major shortcomings: (a) inability to accommodate the streamlined ratings (b) generalized item tendency computation leading to non-personalized recommendations.

We overcome the first shortcoming by incrementally updating user and item tendencies. Secondly, we introduce an incremental similarity update approach to dynamically personalize the item tendency $w.r.t.$ the active user.

Incremental Tendency Approach: To accommodate the streamlined ratings and to update the preferences of the users (and items), we propose to incrementally update the tendency of users and items. Equation 1 of user u's tendency (τ_u) is split as shown in Eq. 2.

$$\tau_u = \frac{\sum_{i \in I_u}(r_{ui} - \bar{r}_i)}{|I_u|} = \frac{\sum_{i \in I_u}(r_{ui}) - \sum_{i \in I_u}(\bar{r}_i)}{C_u(u)} = \frac{A_u - B_u}{C_u(u)} \qquad (2)$$

where, $C_u(u)$ is the number of items rated by user u. To accommodate incremental updates, we store the item i's mean rating $(\bar{r}_i)$, the number of users who rated item i $(C_i(i))$, the aggregate rating value of user u $(A_u = \sum_{i \in I_u}(r_{ui}))$ and aggregate mean rating of the items rated by user $(B_u = \sum_{i \in I_u}(\bar{r}_i))$. In a streaming setting, active user u might face two scenarios: (1) user u newly rates item x with rating r_{ux}, (2) user v rates item x which has been previously rated by user u. In order to update user u's tendency (τ_u), we need to update the terms A, B and $C_u(u)$. In the first scenario, the aggregate rating value of user u is updated as $A'_u = A_u + r_{ux}$. Cardinality of item x $(C_i(x))$ and user u $(C_u(u))$ are incremented by 1. Subsequently, item x's mean rating $(\bar{r}_x)$, aggregate mean of the items rated by user u (B_u) and finally the user tendency τ'_u are incrementally updated as shown below.

$$\bar{r}'_x = \frac{C_i(x) * \bar{r}_x + r_{ux}}{C_i(x) + 1}, \qquad B'_u = B_u + \bar{r}'_x, \qquad \tau'_u = \frac{A'_u - B'_u}{C'_u(u)}$$

In the second scenario, user v's tendency (τ'_v) is updated by modifying item x mean rating $(\bar{r}'_x)$, A'_v, B'_v, $C'_u(v)$, $C'_i(x)$ (as discussed above). Apart from user v's tendency, user u's tendency (τ'_u) needs to be updated. The term B_u is modified as $B'_u = B_u + \bar{r}'_x - \bar{r}_x$, where $\bar{r}_x$ is the mean of item x before arrival of rating r_{vx} and $\bar{r}'_x$ is the new mean of item x. Subsequently, user u tendency is updated as: $\tau'_u = (A_u - B'_u)/C_u(u)$. It can be noted that A_u, $C_u(u)$ remain unchanged as this item has already been rated by user u. Likewise, the tendency of all the users who previously rated item x is incrementally updated. In a similar fashion, computation of item tendency (τ_i) is split and updated incrementally[1]. In the following section, we address the problem of non-personalization in tendency based approach.

Personalized Item Tendency with Incremental Similarity Update: Let U_i be the set of users who rated target item i. We find the set of neighbors of active user u_a who rated target item. Let U_k be the neighbor set $(U_k \subset U_i)$. We utilize the neighbors' ratings to personalize the item tendency $w.r.t.$ the active user. In real–time scenarios, Pearson Correlation, Jaccard similarity measure cannot be used as these are computationally cumbersome. Therefore, we propose

---

[1] Due to space constraint, we omit the incremental item tendency computation in this paper.

an incremental approach (inspired from Jaccard similarity) which captures the interest/disinterest of the users. The proposed similarity measure between two users, u and v is computed as shown in Eq. 3.

$$sim(u, v) = \frac{|L_{sim}(u, v)| + |D_{sim}(u, v)|}{|I_u \cup I_v|} = \frac{P(u, v)}{C(u) + C(v) - I_{corated}(u, v)} \qquad (3)$$

where $L_{sim}(u, v)$ is the set of items liked by users u and v, $D_{sim}(u, v)$ is the set of items disliked by users u and v. I_u, I_v are the set of items rated by users u and v, respectively and $I_{corated}(u, v)$ is the number of co-rated items. In the case of streamlined ratings, recomputing the set of liked/disliked items for each user pair is expensive. Therefore, we propose an incremental approach in updating the similarity of only those users who are affected with the new ratings. Let r_{ui} be newly liked item i $(r_{ui} > \theta)$ and a user set V who liked this item in the past. The similarity value between user u and user v $(\forall v \in V)$ is computed by incrementally updating the terms: $P'(u, v) = P(u, v)+1$, $C'_u(u) = C_u(u)+1$ and $I'_{corated} = I_{corated} + 1$, as shown in Eq. 4. In the similar fashion, the similarity can be updated if $r_{ui} < \theta$. Therefore, for each new rating in the streamline, the similarity values of the affected users are incrementally updated (detailed algorithm is shown in https://goo.gl/hSNgz7). To support the incremental similarity updates, we only store the most relevant information such as list of users who liked (and disliked) each item. In order to reduce the number of database access and processing time, we store cardinality of each liked (and disliked) item, number of items liked (and disliked) by each user separately.

$$sim'(u, v) = \frac{P(u, v) + 1}{(C_u(u) + 1) + C_u(v) - (I_{corated}(u, v) + 1)} \qquad (4)$$

After obtaining the neighbors (U_k) of the active user, we utilize the ratings of these neighbors to compute the personalized item tendency (τ_i^p) w.r.t. active user as shown in Eq. 5. Likewise, the mean of the target item $(\bar{r}_i^p)$ is personalized w.r.t. the active user (Eq. 5).

$$\tau_i^p = \frac{\sum_{\acute{u} \in U_k} (r_{\acute{u}i} - \bar{r}_{\acute{u}})}{|U_k|}, \qquad \bar{r}_i^p = \frac{\sum_{\acute{u} \in U_k} r_{\acute{u}i}}{|U_k|} \qquad (5)$$

where $r_{\acute{u}i}$ is the rating provided by a neighbor $\acute{u}$ on the target item i and $\bar{r}_{\acute{u}}$ is the mean rating of the neighbor $\acute{u}$. Finally, the overall rating of the active user on the target item is predicted using the active user's tendency and target item's personalized tendency as discussed in [3].

3   Experimental Results

In this paper, real-world datasets, MovieLens 10M (ML 10M) and Yahoo! Music (YM) are used to evaluate our approach. To replicate a streamlined environment, we shuffled the rating dataset and divided it into five equal parts. In this first phase, the first part of the dataset is used on training and the second part is used for testing. In the second phase, the ratings in the second part are incrementally added to the training set and the third part of the dataset is used for testing purpose. This process is repeated in remaining phases. We used 5% of the samples in each phase for parameter setting. Mean Absolute Error (MAE) and recently proposed Good Items MAE (GIMAE) [3] are used to evaluate the approaches. We compare our approach (IncRec) with Tendency based CF [3], Regularized SVD [4], traditional neighborhood based approach (UBCF with Jaccard), CF approaches for streaming data (StreamRec [5] and TencentRec [2]). It can be noted that all the experiments are conducted on a single CPU system.

Execution Time in Streaming Scenarios: We report the execution time of all the approaches including IncRec in Fig. 1(a) and (b) on ML 10M and YM datasets, respectively. Results of UBCF with Jaccard are not reported as this technique takes significantly longer execution time (three order of magnitude) than the proposed approach. It can be noted (from Fig. 1) that recently proposed CF approaches

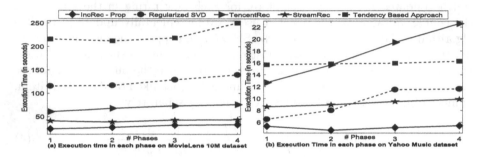

Fig. 1. Execution time in each phase on YM and ML 10M datasets

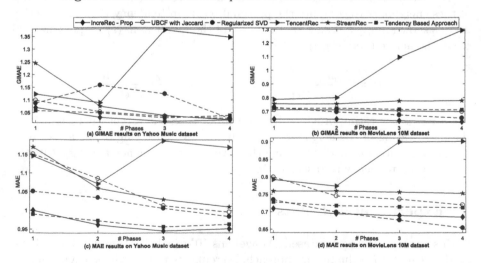

Fig. 2. Accuracy results on MovieLens YM and ML 10M datasets

(TencentRec and StreamRec) have lesser execution time than tendency based CF. However, the proposed approach outperforms all the approaches in each phase.

GIMAE and MAE Results: GIMAE focuses on the error incurred on the predictions of good items (relevant) ratings [3]. The results of YM and ML10M are plotted in Fig. 2(a) and (b), respectively. On YM dataset, StreamRec performs better than the existing techniques in the first two phases. Regularized SVD performs better than the existing approaches in third and forth phases. However, the proposed approach outperforms all the existing techniques in all the phases. Similar trends are found on ML 10M dataset. The MAE results of all the approaches are shown in Fig. 2(c) and (d). On YM dataset, tendency based approach incurred least MAE in the first phase. As the number of ratings increases, the proposed approach performs significantly better than tendency based CF and other existing techniques. On ML 10M dataset, the proposed approach outperforms the existing techniques in the first and second phase. Regularized SVD is found to be superior than all the approaches in the third and fourth phases. However, Regularized SVD cannot be used in a streamlined setting as the execution time is two order of magnitude slower than the proposed approach. To summarize, the proposed approach performs significantly better than the existing techniques in terms of execution time, GIMAE and MAE.

4 Conclusion

In this paper, we proposed an incremental CF approach to tackle real time streaming scenarios. Existing streaming algorithms in CF framework usually sacrifice accuracy for speed-up. However, experimental results validate that our approach is faster and more accurate compared to many existing techniques on real–world datasets.

Acknowledgment. This work is partially funded by Visvesvaraya Ph.D Scheme, Govt. of India.

References

1. Ricci, F., Rokach, L., Shapira, B., Kantor, P.B.: Recommender Systems Handbook, 2nd edn. Springer, New York (2015). https://doi.org/10.1007/978-0-387-85820-3
2. Huang, Y., Cui, B., Zhang, W., Jiang, J., Xu, Y.: Tencentrec: real-time stream recommendation in practice. In: Proceedings of the SIGMOD International Conference on Management of Data, pp. 227–238. ACM (2015)
3. Cacheda, F., Carneiro, V., Fernndez, D., Formoso, V.: Comparison of collaborative filtering algorithms: limitations of current techniques and proposals for scalable, high-performance recommender systems. ACM Trans. Web 5(1), 2:1–2:33 (2011)
4. Paterek, A.: Improving regularized singular value decomposition for collaborative filtering. In: Proceedings of KDD Cup and Workshop, pp. 5–8. ACM (2007)
5. Subbian, K., Aggarwal, C., Hegde, K.: Recommendations for streaming data. In: Proceedings of the International on Conference on Information and Knowledge Management, pp. 2185–2190. ACM (2016)

Collection-Document Summaries

Nils Witt[1]([⊠]), Michael Granitzer[2], and Christin Seifert[2]

[1] ZBW-Leibniz Information Centre for Economics,
Düsternbrooker Weg 120, 24105 Kiel, Germany
n.witt@zbw.eu
[2] University of Passau, Innstraße 32, 94032 Passau, Germany
{Michael.Granitzer,Christin.Seifert}@uni-passau.de
http://www.zbw.eu

Abstract. Learning something new from a text requires the reader to build on existing knowledge and add new material at the same time. Therefore, we propose collection-document (CDS) summaries that highlight commonalities and differences between a collection (or a single document) and a single document. We devise evaluation metrics that do not require human judgement, and three algorithms for extracting CDS that are based on single-document keyword-extraction methods. Our evaluation shows that different algorithms have different strengths, e.g. TF-IDF based approach best describes document overlap while the adaption of Rake provides keywords with a broad topical coverage. The proposed criteria and procedure can be used to evaluate document-collection summaries without annotated corpora or provide additional insight in an evaluation with human-generated ground truth.

Keywords: Collection-document summaries · Text summarization

1 Introduction

Learning from educational or scientific texts requires readers to integrate new concepts into their existing background knowledge [1]. In the case of digital libraries this means that every search result has to be judged on existing, new and additional information compared to already acquired knowledge of the user. In digital libraries, this judgment is usually based on explicit summary information about the search result in questions, such as title and abstract and does not include explicit information on what is new and what has already been covered by previous searches or the user's private library. Similarities between a document collection and a document can be measured with a qualitative values (e.g. [4]) and quantitatively judged using single-instance summaries (e.g. [6]). Both, however, cannot provide comprehensive, explicit summaries about what content is covered in both, the collection and the document (commonalities) and what content is new in the document compared to the collection (novelties). In this paper we propose *collection-document summaries*, i.e., textual summaries that stress differences and commonalities between a collection of documents and candidate documents. Concretely, the contributions of this paper are the following:

© Springer International Publishing AG, part of Springer Nature 2018
G. Pasi et al. (Eds.): ECIR 2018, LNCS 10772, pp. 638–643, 2018.
https://doi.org/10.1007/978-3-319-76941-7_56

- We identify requirements for keyword-based collection-document summaries.
- Based on the requirements, we propose evaluation metrics for collection-document summaries that do not require human-centric ground-truth.
- Provide baseline algorithms for collection-document summaries by adapting single-document summarizations methods.

The collection-document summaries are intended to be directly consumed by users, for instance, to help them judge the suitability of a search result. Due to the lack of available training data and the required effort to collect it, we aim for a automatic evaluation that does not require human-centered ground-truth. The focus for collection-document summaries is on transparency for users, but they could also be used as features in recommendation and retrieval algorithms.

2   Related Work

Automatic *text summarization* aims to generate short-length text covering the most important concepts and topics of the text [2]. Text summaries can either be sentences, phrases or keyphrases, and the content of the summary can either be chosen from the document itself (extractive summaries) or generated anew based on the document (abstractive summaries) [5]. Most methods for text summarization either focus on single-documents or adapt single-document methods to multiple documents. *Multi-document* summarization aims to summarize a collection of textual documents [9]. Methods for multi-document summarizaiton include using single-document methods on super-documents (concatenation of all documents from a collection) or averaging the results for single-document methods over the collection [9]. This work relates to multi-document summarization as follows: we also extract summaries for collections of documents, but output the differences and commonalities of a candidate document (not in the collection) to the collection in terms of keyphrases. *Keyphrase extraction* attempts to extract phrases that concessively and most appropriately cover the concepts of the text [3]. In this work, we extend keyphrase extractions to collection-document summaries, by postprocessing the results of two well known-keyphrase extraction methods, namely TextRank [6] and Rake [8] and comparing the results with a simple baseline considering TF-IDF term weights in the vectorspace-model.

3   Collection-Document Summaries

We define Collection-Document Summaries (CDS) as summarization of a collection of documents and a document, representing how the document's content differs and which content it has in common with the collection. Similarly, we can also compare two documents (i.e., as a collection containing a single document). Consider the scenario of a person accessing a new field by reading literature. The reader has already read n papers ($D = \{d_0, ..., d_n\}$) and wants to decide whether to read the paper d_c next. In that scenario the reader is interested to find documents that have some known content to start with and also have some content

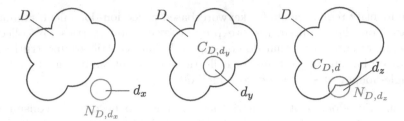

Fig. 1. Types relationships between collections D and documents. Left: d_x differs from D ($C_{D,d_x} = \emptyset$), center: d_y is similar to D ($N_{D,d_y} = \emptyset$), right: collection and document share some concepts, i.e. $N_{D,d_z} \neq \emptyset$ and $C_{D,d_z} \neq \emptyset$

that is new to the reader. In other words, the reader is looking for documents with both, commonalities and novelties:

- **Commonalities:** d_c contains concepts that are also contained in D. These are concepts the reader is already familiar with.
- **Novelties:** d_c contained concepts that are not contained in D. These are the concepts the reader is going to encounter when reading d_c.

Few commonalities and many novelties indicate a big conceptual gap between D and d_c. The reader may have difficulties to grasp the content of d_c. Few novelties and many commonalities on the other hand indicate d_c lacks worthwhile content. We assume the reader is interested in documents with a balanced amount of novelties and commonalities, which may not always be true (e.g. when known concepts are to be revived). While generally, commonalities and novelties as conceptual views on the documents can be represented in multiple ways (e.g. subparts of an ontology), in the remainder of this paper we assume that commonalities and novelties are represented as words. Therefore, we define CDS as follows: *The collection-document summary of a collection D and a document d (i.e. $\Delta(D, d)$) is the pair $(C_{D,d}, N_{D,d})$ where $C_{D,d}$ represents the common keywords and $N_{D,d}$ the novel keywords of documentd with respect to D.*

Figure 1 shows three types of relations between collections and documents. We will motivate and discuss desired properties of CDS and then propose according evaluation measures for these properties in the next section.

- **Comparability:** a document d_c similar to the collection D should introduce no (or only few) new keywords, i.e., if the content of d is already covered by the collection this should be reflected in the keywords.
- **Differentiability:** a document d_c that is not similar to the collection D introduces new keywords, i.e., the difference should be visible by viewing the keywords.
- **Diversity:** the keywords of either commonalities or novelties of a document should cover all concepts that the document deals with.
- **Specificity:** the keywords of either commonalities or novelties should be specific rather than abstract, e.g. *university education* is preferred over *education*.

– **Utility:** The above criteria are necessary but not sufficient, as they do not assess whether the results are meaningful for users. Generally, it requires humans to assess whether CDS are meaningful for a given task, standard metrics to measure utility are precision, recall and F1 w.r.t. to the human-annotated ground-truth.

4   Experiments

In the experiments we evaluated three different algorithms for DCS with the criteria presented in Sect. 3. Source code and data sets are publicly available[1].

Data Sets. The data set consists of 140,341 scientific papers from the economic domain available in the digital collection EconStor[2]. The data set contains information about author, paper abstract, paper type, publication year, venue and a set of JEL-classification codes[3] in meta-data fields. For our experiments we selected those papers that have an abstract and at least one JEL code assigned resulting in 67,813 documents. We annotated phrase candidates of at most 3 terms using the phrase collocation detection described by Mikolov et al. [7]. We constructed artificial user collections D containing k documents with the following property: All documents in the collection must have at least n JEL codes in common, where $n \in \{1, \ldots, 8\}$ is an agreement parameter. Additionally, we randomly generate documents d_x and d_y with the following properties: d_x must have all JEL codes present in the collection and d_y must not have any of the collection's JEL codes (cf. Fig. 1 for a visualization). We chose the agreement on JEL codes for constructing the collection and determining the similarities because JEL codes provide an abstract, topical view on the documents, comprise multiple topics and are high-quality human-annotated meta-data fields. The parameters were set to $k = 10$ and $n = 5$ in our experiments.

Algorithms. For the simple baseline, ΔTF, we rank the words of a documents by their TF-IDF score and select the upper 20% of that list. For ΔTR, we applied TextRank [6] on the documents, keeping the top 20% of the words. We used the TextRank implementation of the Python summa library. For $\Delta Rake$, we used Rake [8] from the Python library rake_nltk. All the algorithms create a set of keywords for a single document. The keywords for a collection were derived using the set union operator for all documents in a collection. The *commonalities* $C(D, d)$ were calculated as the set intersection between the keywords of the collection D and the document d. *Novelties* $N(D, d)$ were calculated by subtracting the set of keywords of the collection D from the set of keywords from the document d (Table 1).

Evaluation Measures. We measure **Comparability** and **Differentiability** as the size of the keyword overlap: $\frac{kw_m(d_c) \cap kw_m(D)}{kw_m(d_c)}$, where, in the case of comparability $d_c = d_x$ and in the case of differentiability $d_c = d_y$ (cf. Fig. 1). $kw_m(d)$

[1] https://doi.org/10.5281/zenodo.1133311.
[2] https://www.econstor.eu, last accessed 2017-10-27.
[3] https://www.aeaweb.org/jel/guide/jel.php, last accessed 2017-10-27.

Table 1. Example keywords.

$\Delta Rake$	ΔTF	ΔTR
unanticipated reform, major change, cultural conditions, mothers income, order births, favorable institutional, strong labor market attachment	compensate, hampered, births, essentially, unanticipated, unfavorable, mothers	fully compensated, essential incentives, mothers, largely driven, earlier

Table 2. Overview of results. Showing mean and variance aggregated for all measures

Method	Keywords per doc	Comparability	Differentiability	Specificity	Diversity
$\Delta Rake$	13.5 ± 6.6	0.37 ± 0.04	**0.10 ± 0.03**	$2.9\% \pm 0.4\%$	**0.60 ± 0.10**
ΔTF	6.2 ± 2.9	**0.50 ± 0.06**	0.13 ± 0.04	**1.2% ± 0.3%**	0.15 ± 0.03
ΔTR	3.6 ± 2.2	0.45 ± 0.03	0.17 ± 0.09	$2.2\% \pm 0.8\%$	0.18 ± 0.06
Samples		100	100	500	10,000

is the number of keywords extracted by method m on the document d. For **Diversity** we construct a binary JEL code-keyword matrix (M) for each keyword extraction algorithm on the entire data set. Each entry m_{ij} in M indicates whether a specific JEL code i occurs in at least t documents for which keyword i has also been extracted. The parameter t is set to 10 for Rake, 20 for TF-IDF and 5 for TextRank in the experiments. These values were obtained by manual optimization. Thus, the columns of M contain representations of keywords in terms of JEL codes. In a second step, given a candidate document, keywords are extracted and their respective columns of M are combined by logical OR yielding a vector v. The ground-truth JEL codes for the document are compared to the candidate vector v using Jaccard similarity. To measure the **Specificity** we generate two disjoint collections, i.e. two collections that do not share any JEL code. Afterwards the keywords of both collections are extracted and the set intersection and set symmetric difference are computed. The intuition behind this is, that, since the two collections share no JEL codes, they are topically different. Hence, for keyword extractors that generate specific keywords the intersection should be empty. Keywords in the intersection are expected to be unspecific. This measure is normalize by the amount of generated keywords. We divide the intersection size by the size of the symmetric difference.

5   Results

The results of our experiments are summarized in Table 2. We see that the average amount of keywords produced varies considerably, with Rake producing too many keywords given the assumption that the results should be consumed by people. ΔTF scores best at *Comparability* and achieves proper results in *Differentiability*, leading to the larges gap between these two related measures.

That means that ΔTF is the preferred method to model the assumption depicted in Fig. 1. Presumably, $\Delta Rake$'s bad *Comparability* performance can be partially explained by its much larger number of unique keywords, which makes matching keywords less probable whereby the comparability score drops. $\Delta Rake$'s bad *Specificity* performance is surprising, as it has the largest repertoire of keywords available, which should allow it extract specific keywords. ΔTF on the other hand performs much better albeit its much smaller keyword repertoire. High *Diversity* scores indicate that the keywords a method extracts are good classification features to predict the JEL codes of documents. This is the measure where $\Delta Rake$ excels, due to its multi-token keywords (cf. Table 2) and probably also because of the higher keyword per document count.

6   Summary

We have introduced the notion of collection-document summaries and identified criteria by which the quality of those summaries can be measured. Furthermore, we have conducted experiments with three keyword extraction methods. The applied keyword extraction methods are state-of-the-art methods for single document summarization, and therefore should be considered a lower bound baseline for collection-document summarization. Future work includes the devision of new algorithms, for instance by combining $\Delta Rake$ (best diversity) and ΔTF (best comparability) and an evaluation of the methods on a human-generated ground-truth to answer the question about the utility of the extracted keywords.

References

1. Eddy, M.D.: Fallible or inerrant? a belated review of the constructivist's bible. Br. J. Hist. Sci. **37**(1), 93–98 (2004). Jan Golinski, making natural knowledge: Constructivism and the history of science. Cambridge history of science
2. Gambhir, M., Gupta, V.: Recent automatic text summarization techniques: a survey. Artif. Intell. Rev. **47**(1), 1–66 (2017)
3. Hasan, K.S., Ng, V.: Automatic keyphrase extraction: a survey of the state of the art. In: ACL, vol. 1, pp. 1262–1273 (2014)
4. Huang, A.: Similarity measures for text document clustering. In: Proceedings of the New Zealand Computer Science Research Student Conference, pp. 49–56 (2008)
5. Mani, I.: Advances in Automatic Text Summarization. MIT Press, Cambridge (1999)
6. Mihalcea, R., Tarau, P.: TextRank: bringing order into texts. In: Proceedings of Conference on Empirical Methods in Natural Language Processing, Barcelona, Spain (2004)
7. Mikolov, T., Sutskever, I., Chen, K., Corrado, G., Dean, J.: Distributed representations of words and phrases and their compositionality. In: Proceedings of the International Conference on Neural Information Processing Systems, NIPS 2013, vol. 2, pp. 3111–3119 (2013)
8. Rose, S., Engel, D., Cramer, N., Cowley, W.: Automatic Keyword Extraction from Individual Documents. Wiley, New York (2010). pp. 1–20
9. Verma, R.M., Lee, D.: Extractive summarization: limits, compression, generalized model and heuristics. CoRR abs/1704.05550 (2017)

Towards an Understanding of Entity-Oriented Search Intents

Darío Garigliotti and Krisztian Balog[(✉)]

University of Stavanger, Stavanger, Norway
{dario.garigliotti,krisztian.balog}@uis.no

Abstract. Entity-oriented search deals with a wide variety of information needs, from displaying direct answers to interacting with services. In this work, we aim to understand what are prominent entity-oriented search intents and how they can be fulfilled. We develop a scheme of entity intent categories, and use them to annotate a sample of queries. Specifically, we annotate unique query refiners on the level of entity types. We observe that, on average, over half of those refiners seek to interact with a service, while over a quarter of the refiners search for information that may be looked up in a knowledge base.

1 Introduction

A large portion of information needs in web search look for specific entities [11]. Entities are natural units for organizing information, and can provide not only more focused responses, but often immediate answers [9]. Another type of entity-bearing queries is more transaction-oriented. Either trying to book a flight or looking for tickets for an concert, just to mention two popular examples, users are often engaged to fulfill information needs by interacting with a third-party service or application. There has been an increasing focus on supporting task-based search [7], and on modeling actionable knowledge; see, e.g., the dedicated vocabulary for actions in the schema.org ontology, and the NTCIR AKG task.[1] These developments display the interest and efforts towards transforming search engines into actions-guided task completion assistants [1]. In this work, we are interested in studying one particular type of information needs, namely, entity-oriented searches. Specifically, we want to answer a question arising from this web landscape: *what do entity-oriented queries ask for?* Furthermore, which of those searches can be fulfilled by looking up direct answers from a knowledge base, and which would require to interact with external services?

Most entity-oriented queries consist of an entity name, complemented with context terms, i.e., *refiners*, to express the underlying intent of the user [11]. Examples of these queries are *"the rock movies"* and *"london book a hotel"*. Our main objective is to understand entity-related search intents by studying those refiners. Specifically, we represent refiners on the level of entity types.

---

[1] http://ntcirakg.github.io/tasks.html.

© Springer International Publishing AG, part of Springer Nature 2018
G. Pasi et al. (Eds.): ECIR 2018, LNCS 10772, pp. 644–650, 2018.
https://doi.org/10.1007/978-3-319-76941-7_57

Just like entity types boost the disambiguation of known entities and the group-ing of emerging ones [10], these type-level characterizations of entity refiners would favor knowledge abstraction and generalization. As an example, by rep-resenting with *[city]* any entity of the type city, we want to categorize a refiner, e.g., "rentals", in the type-level query *"[city] rentals"*. Then, we categorize these type-level refiners using an intent classification scheme. Our classification scheme comprises four main categories: property, website, service, and other.

We perform this study without having direct access to past usage data or query logs. To overcome the absence of such data, we exploit query suggestions from a major search engine API. This strategy has been employed successfully in previous work for various applications [3,5]. After acquiring query suggestions for entities of a given type, they are aggregated to extract type-level refiners. Then, for a representative sample of 50 Freebase types, we collect human annotations for those refiners with respect to the classification scheme we developed.

Our main findings show that, on average, more than a half of all unique type-level refiners correspond to interacting with external services, while over a quarter of them look for information that may be looked up in a knowledge base. Another contribution of this work is a large collection of type-level refiners, annotated with intent categories. The resources developed within this paper are made available at http://bit.ly/ecir2018-intents.

2   Related Work

Broder's categorization of information needs is broadly accepted and is the most commonly used one for web search [4], with further refinements, e.g., in [6,13]. We strive for a similar high-level categorization of intents, but specifically for entity-oriented search queries. Previous work has identified high-level patterns from web search queries. For example, according to Lin et al. [8], a query can be classified as an entity, an entity plus a refiner (e.g., *"emma stone 2017"*), a category, a category plus a refiner (e.g., *"doctors in barcelona"*), a website, or other sort of query. Such classification relies merely on lexico-syntactic forms and lacks a more semantically-grounded distinction.

Search intents have been studied in previous work. Reinanda et al. [12] explore entity aspects in user interaction log data. Beyond finding aspects by comparing clustering methods over refiners, they address the tasks of ranking the intents for a given entity independently from a query and recommending aspects. Unlike them, we (i) operate with individual query refiners (i.e., with-out clustering them together), (ii) model entity intents at the level of types, (iii) always consider entities in queries, and (iv) perform our study in the absence of search logs.

3   Approach

This section describes the process we followed for understanding entity-oriented search intents. An *entity-oriented* or *entity-bearing query* is a query that consists of an entity name possibly complemented with a refiner, usually as a suffix. Here, by *entity* we mean an individual with its own independent existence, uniquely identified in a knowledge base [2]. More than just a syntactic complement, a *refiner* is a complementary surface form expressing an underlying user *intent* in relation with the entity. As an example, consider the entity *keens steakhouse* (a restaurant) in the search query *"keens steakhouse menu."* The refiner "menu" expresses the intent of reading the restaurant's menu. To understand what these entity-bearing queries ask for, we characterize the refiners on the level of *entity types*, where an entity type is a semantic class that groups entities together with common characteristics. For example, one of the types of *Albert Einstein* in Freebase is `award_winner`.

Our approach, to be detailed in the next subsections, can be summarized as follows. We collect refiners for a set of prominent entities, and aggregate them across entity types to obtain type-level refiners. Next, we develop a classification scheme of *intent categories*, with a focus on how to fulfill the intent expressed by a type-level refiner. Finally, we annotate a representative sample of entity types with intent categories, and obtain a corpus of prominent type-level refiners assigned to those categories.

3.1   Collecting Refiners

We use the type system of Freebase. It is a two-layer categorization system, where types on the leaf level are grouped under high-level domains. Specifically, we use the latest public Freebase dump (2015-03-31), discarding domains meant for administering the Freebase service itself (e.g., `base`, `common`).

We focus on prominent entities, since in this way we benefit from observing a larger and more representative selection of information needs. As the criterion of an entity prominence, we rely on Wikistats page views.[2] This dataset registers the number of times its English Wikipedia article has been requested. We set empirically a prominence threshold of 3,000 page views per article over a span of one year (from June 2015 to May 2016). Given a Freebase type, we select it if it covers at least 100 entities with a prominence above the threshold. Applying these criteria, the selected set contains 634 types.

In a second step, we collect query suggestions from the Google Suggestions API for at most top 1,000 entities per type according to the above prominence criteria. Then, we replace the name of the entity by its type in each query suggestion. This can be viewed as getting queries where a refiner complements the type. For example, the type-level query *"[travel destination] map"* is obtained from all queries for popular travel destinations, e.g., *"sydney map"* and *"paris map"*. Finally, we retain only those refiners that occur in at least 5 suggestions for the given type. This leads to a total of 2,688 distinct type-level refiners for 631 types.

---

[2] https://dumps.wikimedia.org/other/pagecounts-ez/.

3.2   Classification Scheme

To address our main goal of understanding entity-related search intents, we need a suitable scheme to classify the entity intents. After a close inspection of the type-level refiners, we define the following scheme of *intent categories*. These categories are focused on how (and from which type of source) the information need can be fulfilled.

- **Property**: The refiner looks for a specific entity property or attribute that can be looked up in a knowledge base. For example, "children" in the query *"angelina jolie children"* or "opening times" in *"at&t stadium opening times"*.
- **Website**: The refiner is about reaching a specific website or application. For example, "twitter" in the query *"karpathy twitter"*. This category is a rough equivalent of navigational queries in [4].
- **Service**: The refiner expresses the need to interact with a service, possibly by redirecting to an external site or app. For example, "menu" in the query *"keens steakhouse menu"* would indicate the need for accessing to an external site for reading the restaurant's menu. As another example, "new album" in *"eric clapton new album"* looks for a service to read about, or listen to, or buy the new album.
- **Other**: None of the previous ones is applicable. For example, "batman" in the query *"christian bale batman"* serves to disambiguate the person's role of interest.

3.3   Annotation

We need to sample a set of representative types, since it is unfeasible to annotate all types in the knowledge base. From the set of 631 types, we perform stratified sampling as follows. We sort the types by the total aggregated frequencies of refiners. We delimit 5 roughly equally-sized intervals by the splitting values of 1,500, 3,000, 6,000, and 8,500 refiners per type; we randomly pick 10 types from each interval. We annotate data for this final set of 50 representative Freebase types.

We used crowdsourcing to annotate type-level refiners with intent categories. Specifically, using the Crowdflower platform, for each annotation instance we presented workers with the query, indicating its entity type and refiner, and asked them to select one of the four intent categories. A total of 5,301 unique instances (type-level refiners) were annotated, each by at least 3 judges (5 at most, if necessary to reach a majority agreement, using dynamic judgments). We paid ¢5 per batch, comprising 11 annotation instances. We ensured quality by requiring a minimum accuracy of 80%, a minimum time of 20 s per batch, and a minimum confidence threshold of 0.7. For each type, we only retain an annotated refiner if at least three annotators agreed on the majority category. This leads to a total of 2,313 unique refiners.

4   Results and Analysis

Figure 1 presents the number of refiners classified per each category, for the 50 sampled types, grouped in one plot for each of the 5 intervals of the stratified sampling. Since the final set of types was sampled from types with prominent entities, this ordering, given by the number of refiners, in a way also reflects the prominence of types.

We obtain a distribution of entity intent categories per type after normalizing the frequency of each category by the total of refiners for that type. From the average proportions in these distributions, we can answer our initial questions. A 54.06% of unique entity-oriented queries are to be fulfilled by interacting with some external service or app, meanwhile, 28.6% look for direct answers from a knowledge base. Further, 5.34% of the type-level refiners represent an attempt to reach a website, while 12.08% of them do not fit into any of the previous three categories.

The types with the largest proportion of *service* intents are **netflix_genre** (with refiners, e.g., "videos", "live"), **election** ("map", "polls"), **football_match** ("video", "highlights"), and **music_album**. The *property* intent category covers refiners that are of a more static nature, e.g., **chemical compound** (with refiners like "structural formula", "molecular weight"), **political_party** ("slogan", "president"), **star** ("type of star", "temperature"), or **tower** ("hours", "height"); only the first one is a very prominent type. Most of the entity types exhibit a non-empty proportion of *website* intents. Among all the types, this category exceeds the average proportion, e.g., for **organization**, **business_operation**, **hotel** and **blogger**. The most frequent website refiners in the whole corpus are "wikipedia", "twitter", "facebook", and "youtube". For a few types like **muscle**, **election**,

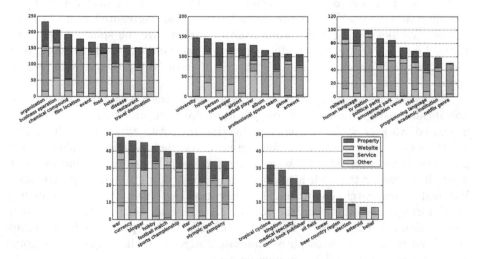

Fig. 1. Distributions of intent categories for the sampled types. Note that the y-axis scales differ.

Table 1. Examples of refiners for each intent category, for each (stratified) type group.

Entity type	Intent category			
	Property	Website	Service	Other
comic_book_publisher	logo, address	wiki, website, twitter	submissions, publishing, comics	movies
tower	height, address, opening hours	wiki	tickets, restaurant	collapse
war	deaths, results, cause	youtube, wikipedia, reddit, quizlet	video, uniforms, pictures, documentary	ap euro, in hindi
academic_institution	logo, email, notable alumni	wiki, login, twitter, portal	scholarships, ranking, map, library, jobs	baseball
automotive_company	stock, logo, ceo, address	wikipedia, website, linkedin, facebook	parts, careers, investor relations	india, inc
programming_language	syntax, ide	wikipedia, wiki, github	jobs, examples, interview questions	3, 2017
restaurant	phone number, owner, location	yelp, twitter, app, tripadvisor, groupon	wine list, vouchers, recipes, menu prices	sf, nj, nyc
music_album	value, cast, release date	youtube, wikipedia, amazon, imdb	zip download, video, ukulele chords, tracklist	2015, lp
person	son, salary, real name	youtube, instagram, snapchat	tour, quotes, photos, new album	sr, now, ww2
travel_destination	zip code, train station	craigslist	weather radar, vacation, tours, things to do	today, nj

belief, or medical_speciality, all in the lowest populated groups, no website refiner is present. A marginal proportion of refiners are classified as having the *other* intent. A few exceptional cases with large proportions of other intents are, e.g., business_operation and house (where the refiner is usually a location), or basketball_player (for which many refiners refer mostly to an NBA franchise, e.g., "lakers"). Table 1 provides additional examples for a selection of types.

5   Conclusions and Future Work

The study performed in this work has lead to a better understanding of what entity-oriented queries ask for. We have developed a classification scheme to categorize entity-oriented search intents and annotated a representative sample of type-level refiners using this scheme. We have found that, on average, more than a half of those are to be fulfilled by interaction with services; another large proportion of information needs look for direct answers from a knowledge base. Several lines of future work arise from our study. One of them is to develop a method for automatic intent categorization. Another direction is the clustering of refiners which express the same underlying intent. Finally, we seek to extend our approach to be able to capture tail entities and intents.

References

1. Balog, K.: Task-completion engines: a vision with a plan. In: Proceedings of the 1st International Workshop on Supporting Complex Search Tasks (2015)
2. Balog, K.: Entity retrieval. Encyclopedia of Database Systems, pp. 1–6. Springer, New York (2017). https://doi.org/10.1007/978-1-4899-7993-3
3. Benetka, J.R., Balog, K., Nørvåg, K.: Anticipating information needs based on check-in activity. In: Proceedings of WSDM, pp. 41–50 (2017)
4. Broder, A.: A taxonomy of web search. SIGIR Forum **36**(2), 3–10 (2002)
5. Fourney, A., Mann, R., Terry, M.: Characterizing the usability of interactive applications through query log analysis. In: Proceedings of CHI, pp. 1817–1826 (2011)
6. Jansen, B.J., Booth, D.L., Spink, A.: Determining the informational, navigational, and transactional intent of web queries. Inf. Process. Manage. **44**(3), 1251–1266 (2008)
7. Kelly, D., Arguello, J., Capra, R.: NSF workshop on task-based information search systems. SIGIR Forum **47**(2), 116–127 (2013)
8. Lin, T., Pantel, P., Gamon, M., Kannan, A., Fuxman, A.: Active objects: actions for entity-centric search. In: Proceedings of WWW, pp. 589–598 (2012)
9. Mika, A.: Entity search on the web. In: Proceedings of WWW, pp. 1231–1232 (2013)
10. Nakashole, N., Tylenda, T., Weikum, G.: Fine-grained semantic typing of emerging entities. In: Proceedings of ACL, pp. 1488–1497 (2013)
11. Pound, J., Mika, P., Zaragoza, H.: Ad-hoc object retrieval in the web of data. In: Proceedings of WWW, pp. 771–780 (2010)
12. Reinanda, R., Meij, E., de Rijke, M.: Mining, ranking and recommending entity aspects. In: Proceedings of SIGIR, pp. 263–272 (2015)
13. Rose, D.E., Levinson, D.: Understanding user goals in web search. In: Proceedings of WWW, pp. 13–19 (2004)

Towards Measuring Content Coordination
in Microblogs

Dmitri Roussinov[✉]

University of Strathclyde, 16 Richmond Street, Glasgow G11XQ, UK
dmitri.roussinov@strath.ac.uk

Abstract. The value of microblogging services (such as Twitter) and social
networks (such as Facebook) in disseminating and discussing important events
is currently under serious threat from automated or human contributors
employed to distort information. While detecting coordinated attacks by their
behaviour (e.g. different accounts posting the same images or links, fake profiles, etc.) has been already explored, here we look at detecting coordination in
the content (words, phrases, sentences). We are proposing a metric capable of
capturing the differences between organic and coordinated posts, which is based
on the estimated probability of coincidentally repeating a word sequence. Our
simulation results support our conjecture that only when the metric takes the
context and the properties of the repeated sequence into consideration, it is
capable of separating organic and coordinated content. We also demonstrate
how those context-specific adjustments can be obtained using existing resources.

Keywords: Language models · Simulating text · Online bots and trolls

1 Introduction

Recent media reports discovered massive efforts by various political groups worldwide
to over-represent their support by employing automated or paid-human contributors [3].
For example, the 2016 US presidential election witnessed use of automated bots on both
sides, with 5:1 ratio for the winner [7]. The Islamic State of Iraq and the Levant (ISIL or
ISIS) has been noted to use coordinated bots [11]. Twenty (*20*) percent of all the internet
comments in China are believed to be made by paid pro-government trolls [12]. Russian
government spends millions of dollars every year on similar activities [13].

As we further elaborate in our "Related Work" section, several methods to detect
coordination in microblogging activities have been proposed. However, they are so far
based only on troll's behaviour and profile characteristics. Meanwhile, several studies
noted the occurrence of identical word sequences that can potentially serve as tell-tales
of ongoing coordination in the content, for example the use of the same *6* word
sequence ("Ukrainians killed him out of jealousy") in Twitter right after a Russian
opposition leader's assassination [5] or *14*-word sequence ("How Chris Coons budget
works- uses tax $ 2 attend dinners and fashion shows") to smear a US democratic
senator Chris Coons [10]. While repeating those sequences indeed looks suspicious, we
still don't know *what are the properties (e.g. the minimum length, rarity of the words
used, number of repetitions, etc.) of the repeated sequence to be suspicious* since

© Springer International Publishing AG, part of Springer Nature 2018
G. Pasi et al. (Eds.): ECIR 2018, LNCS 10772, pp. 651–656, 2018.
https://doi.org/10.1007/978-3-319-76941-7_58

repetitions happen in organic (not-coordinated) communication as well. For example, several tweets wrote "Earthquake hits central Alaska", when such event indeed occurred on May 7th, 2017. Is this suspicious? While we do not claim to provide complete answers to those questions here, we still make several important steps towards it by providing a framework for future work. Our contributions are the following: (1) we propose to model the classes of repetitions rather than individual suspicious sequences. (2) By using simulation and counter-examples, we demonstrate that without taking the context of the post (topic) into consideration, repetitions from organic communication may look unjustifiably suspicious (false positives). (3) We propose the necessary context adjustments that allow separating organic coincidences from coordinated ones.

The next section presents the related work, followed by the description of our framework. The "Conclusions…" section summarizes our findings and possible future directions.

2   Related Work

Distinguishing automated from human accounts has been successfully tackled by several research projects, e.g. [1, 2, 4, 10], which offered various successful machine learning detection methods that are based on the *behaviour* of the coordinated bots (trolls, users, actors, etc.), such as posting the same links or same digital object or using fake profiles or being somehow associated with other, already detected trolling accounts. However, the behaviour-only methods would not solve the problem for the following reason: as those methods become known through their publications, the bots' coordinators will simply modify their behaviour to avoid being caught. As an alternative, here we focus on detecting coordination in the content, since it is intrinsically inseparable from the bots' purpose: to amplify a certain message by artificially inflating the presence of certain content. A closely related problem of *plagiarism detection* has been receiving significant scrutiny resulting in a number of useful tools [9], however they are not known to involve quantitative estimates, but rather treat any repetitions as suspicious. Also, potentially relevant to the task here are the algorithms on review-spam detection, e.g. [6], but it is still a different task since the review spamming accounts are typically short lived, the reviews themselves are much longer than the microblog posts, and the contributors are not connected into a network.

3   Simulating Repetitions in Text

The task we are trying to solve here is formally the following: given a "suspicious" repeating sequence of words (n-gram), estimate the probability of occurrence of this sequence more than once in organic (not coordinated) set of short documents (tweets, posts, etc.). If this probability is very low (e.g. <0.0001), then we may claim with a high certainty that coordination is taking place. The suspect string is typically identified by a manual investigation [5] or by automatically applying a clustering algorithm [1, 2]. Once the suspect sequence is identified, we can try to estimate the probability of generating it by applying a language model [8], e.g. using Microsoft's n-grams service (https://azure.microsoft.com/en-us/services/cognitive-services/), trained on the portion of WWW indexed by their search engine (Bing). For the suspicious sentence from [5] it gives us 4.8×10^{-16}, thus accidentally repeating it in entire Twitter even once is highly unlikely. But how can we generalize from this to the other suspect sequences? For better generalization, we suggest to model the classes of repetitions rather than specific sequences, so we can distinguish between types of repetitions that are suspects and those that do happen in organic posts. Thus, we define a *repetition class C(n, p)* as a repeated sub-sequence of n content-bearing words (n-gram), with p equal to their maximum probability of occurrence. We would intuitively expect the classes with large n and small p to be suspect, while the repetitions in the classes with small n and large p to be quite common. Ignoring the stopwords is justifiable since their use is determined by grammatical relationships between the content-bearing words around them. Based on the same Bing's n-grams statistics, the repetition class for the example sequence above will be $C(3, 3.4 \times 10^{-5})$, where 3.4×10^{-5} is the probability of occurrence of the word *killed* as the most frequent out of the three context-bearing words (*ukranians, killed, jealousy*). Table 1 lists some of example sequences from real trolling attacks reported in the prior works and from organic communication, along with the parameters defining their repetition classes.

Table 1. Examples of repeated sequences from trolling attacks and organic communication along with their repetition class parameters.

	n	p
Coordinated		
Ukrainians killed him out of jealousy	3	3.4×10^{-5}
Ukrainians killed him out of jealousy. He stole a girlfriend from one of them	5	3.6×10^{-5}
How Chris Coons budget works- uses tax $ 2 attend dinners and fashion shows	10	1.7×10^{-4}
Organic		
Earthquake hits central Alaska	4	1.6×10^{-4}
16 foreigners among 39 killed in Istanbul nightclub	4	3.5×10^{-5}
Spanish prosecutors have charged Catalan cabinet	5	7.1×10^{-5}

To estimate the probabilities of occurrences within those repetition classes, we run a simulation by sampling n-grams from a uniform distribution matching the class probabilities (p) and the lengths (n). We generated *1000* "tweets" of a typical size (*10* words), and looked for repetitions using a hash table. We also obtained similar results by using the Zipf distribution with several sets of typical parameters, but omitting them here due to space limitations. We <u>did not observe occurrences in any of the classes defined by those examples</u> in any of *10,000* simulation runs. This suggests that our metric based on a notion of a repetition class correctly identifies the examples from coordinated attacks as suspicious (probability of happening in organic posts < *1/10000*). But the metric also erroneously identifies all the sequences from organic communication as suspect, thus underestimating the probability of repetition. In reality, the tweets are not random utterances as they are typically posted about certain events. Thus, their word distributions are strongly affected by the topic. For example, according to a search run over an indexed copy of Wikipedia, the probability of the word *jealousy* increases almost *200* fold when the document already has the word *killed*.

This observation can be quantified by introducing the probability adjustment factors $a(w|T)$ for each word w estimated as the ratio of the probability of occurrence within a particular topic T to the probability of occurrence in the corpus (regardless of a topic):

$$a(w|T) = \frac{p(w|T)}{p(w)}, \tag{1}$$

where T is a topic defined by a boolean query (e.g. "*assassination AND russia*" here). The probability of a word occurrence conditional on the topic T is estimated as $p(w|T) = \frac{\#(w \, AND \, T)}{\#(T)}$, where $\#(q)$ is the number of documents in the corpus matching the query q. The probability of a document having the word w is estimated as $p(w) = \frac{\#(w)}{W}$, where W is the total number of documents in the corpus, or can be obtained from Bing's n-grams. Alternatively, for the words closely related to the topic, the adjustments can be estimated empirically by running the search queries defining the topic T (or the related hashtags) in Twitter and counting the occurrences of those words in the returned tweets. For the words defining the topic itself (e.g. *alaska, earthquake*), those probabilities typically range between *0.05* and *0.1*. Table 2 shows the same examples of repetitions with their classes adjusted for the context. The last column (S) shows the number of runs in which any repetitions within that class occurred out of all *10,000* simulation runs. The following can be observed: (1) The repetitions classes corresponding to the examples from the organic posts do indeed happen, and, as a result, those repetitions will not be flagged as suspicious. (2) The classes of repetitions corresponding to the first sentence from the coordinated attack ("Ukrainians...")

Table 2. Examples of repeated sequences from trolling attacks and organic communication along with their repetition class parameters adjusted for the context. The last column is the number of simulation runs S, out of *10000*, in which any repetitions in that class occurred.

	n	p	S
Coordinated			
Ukrainians killed him out of jealousy	3	7.2×10^{-3}	10000
Ukrainians killed him out of jealousy. He stole a girlfriend from one of them	5	7.2×10^{-3}	0
How Chris Coons budget works- uses tax $ 2 attend dinners and fashion shows	10	1.23×10^{-3}	0
Organic			
Earthquake hits central Alaska	4	1.2×10^{-2}	3911
16 foreigners among 39 killed in Istanbul nightclub	4	2.3×10^{-2}	5677
Spanish prosecutors have charged Catalan cabinet	5	1.0×10^{-2}	38

also happen, and, thus this sentence alone <u>may not</u> serve as sufficient evidence of coordination contrary to the investigators' in [5] claim. (3) Only when combined with the sequence immediately following it in the posts under investigation (next line in the table), the entire sequence belongs to the classes of repetitions that do not happen in organic posts. (4) The sentence about Chris Coons falls into a class of repetitions which signals a coordinated attack.

Table 3 presents additional simulation runs for various repetition classes. It is possible to observe the following: (1) Repetitions with $n = 3$ (repeating a sequence with 3 content bearing words) are normally not suspicious (happen in organic communication) <u>unless</u> all those words are rare ($p < 5 \times 10^{-5}$, which generally means <u>not</u> among 10000 most frequent words) (2) For $n = 4$, any repetition is suspect (does not happen in organic communication) unless it includes the words highly associated with the topic ($p > .01$, which is often the case with the words defining the topic itself, e.g. *earthquake* and *alaska* here. (3) Repeating a sequence of 5 or more content bearing words is always suspicious, <u>regardless of the magnitudes of the context adjustments</u>. While the occurrence estimates we obtained using Zipf distribution are somewhat smaller, they support the same observations.

Table 3. Simulation results for various repetition classes: Number of trials out of 10000 in which repetitions happen.

n-gram length	n = 3	n = 4	n = 5	n = 6	n = 7
p					
.00005	0	0	0	0	0
.0001	3	0	0	0	0
.0003	31	0	0	0	0
.0005	54	0	0	0	0
.001	126	0	0	0	0
.003	7861	27	0	0	0
.005	10000	47	4	0	0
.01	10000	3911	38	0	0

4 Conclusions and Future Work

Our estimates and numeric simulations here demonstrate that it is possible to quantify coordination in the content, which potentially leads to exposing unwelcome activities in the microblogging posts (e.g. Twitter), and, thus, reducing the damage that it inflicts. We have proposed a metric that is based on modeling repetitions within a class rather than trying to model repeating individual sequences. This study also suggests that context-specific adjustments are necessary and demonstrates how they can be obtained based on a training corpus. We have illustrated this on several examples from past works and selected real microblog posts, leaving room for future more powerful approaches, such as those based on machine learning models and more formal evaluation.

Acknowledgements. I would like to thank N. Puchnina for providing a number of valuable suggestions.

References

1. Beutel, A., Xu, W., Guruswami, V., Palow, C., Faloutsos, C.: Copy-catch: stopping group attacks by spotting lockstep behavior in social networks. In.: 22nd International Conference on World Wide Web, pp. 119–130 (2013)
2. Cao, Q., Yang, X., Yu, J., Palow, C.: Uncovering large groups of active malicious accounts in online social networks. In.: ACM SIGSAC Conference on Computer and Communications Security, pp. 477–488. ACM (2014)
3. Ferrara, E., Varol, O., Davis, C., Menczer, F., Flammini, A.: The rise of social bots. Commun. ACM **59**(7), 96–104 (2016)
4. Ferrara, E., Varol, O., Menczer, F., Flammini, A.: Detection of promoted social media campaigns. In.: Tenth International AAAI Conference on Web and Social Media (2016)
5. Global Voices: Social Network Analysis Reveals Full Scale of Kremlin's Twitter Bot Campaign. Global Voices, 02 April 2015 (2015)
6. Jindal, N., Liu, B.: Review spam detection. In: WWW Conference 2007, pp. 1189–1190 (2007)
7. Howard, P.N., Kollanyi, B., Woolley, S.: Bots and Automation over Twitter during the US Election. Zugriff am, 14 (2016)
8. Ponte, J.M., Croft, W.B.: A language modelling approach to information retrieval. In.: Research and Development in Information Retrieval, pp. 275–281 (1998)
9. Potthast, M., Eiselt, A., Barrón-Cedeño, A., Stein, B., Rosso, P.: Overview of the 3rd International Competition on Plagiarism Detection, Notebook Papers of CLEF LABs and Workshops (2011)
10. Ratkiewicz, J., Conover, M.D., Meiss, M., Gonçalves, B., Flammini, A., Menczer, F.: Detecting and tracking political abuse in social media. In: ICWSM (2011)
11. Shane, S., Hubbard, B.: ISIS Displaying a Deft Command of Varied Media. New York Times (2014)
12. Simon, J.: The New Censorship: Inside the Global Battle for Media Freedom. Columbia University Press, New York (2015)
13. Sindelar, D.: The Kremlin's Troll Army. The Atlantic, 12 August 2014

Long-Span Language Models
for Query-Focused Unsupervised
Extractive Text Summarization

Mittul Singh[1] , Arunav Mishra[2]([✉]), Youssef Oualil[1], Klaus Berberich[2],
and Dietrich Klakow[1]

[1] Spoken Language Systems (LSV), Saarland Informatics Campus,
Saarbrücken, Germany
msingh@lsv.uni-saarland.de
[2] Max Planck Institute for Informatics, Saarland Informatics Campus,
Saarbrücken, Germany
amishra@mpi-inf.mpg.de

Abstract. Effective unsupervised query-focused extractive summarization systems use query-specific features along with short-range language
models (LMs) in sentence ranking and selection summarization subtasks. We hypothesize that applying long-span n-gram-based and neural
LMs that better capture larger context can help improve these subtasks.
Hence, we outline the first attempt to apply long-span models to a query-
focused summarization task in an unsupervised setting. We also propose
the *A*cross *S*entence *B*oundary LSTM-based LMs, *ASB*LSTM and
*biASB*LSTM, that is geared towards the query-focused summarization
subtasks. Intrinsic and extrinsic experiments on a real word corpus with
100 Wikipedia event descriptions as queries show that using the long-
span models applied in an integer linear programming (ILP) formulation
of MMR criterion are the most effective against several state-of-the-art
baseline methods from the literature.

1 Introduction and Background

Extractive text summarization system has been traditionally considered as an
effective tool to address *information overloading* by facilitating efficient consumption of information that spans across multiple documents. Broadly, an
extractive multi-document summarization task can be setup as *supervised*: having example summaries to design and train systems, or *unsupervised*: having
access only to a large text corpus.

In the supervised setting, recent approaches [1,2] that leverage deep neural
network-based models (e.g., attention-based encoder-decoder LSTM [1]) require
large number of training examples. Moreover, often the interpretability is poor
thereby limiting their usages as black boxes. This issue makes it hard to gather
insights into the methods which limit the scope of improvement. However, with
enough examples, neural network-based models with end-to-end training have
recently shown impressive results.

© Springer International Publishing AG, part of Springer Nature 2018
G. Pasi et al. (Eds.): ECIR 2018, LNCS 10772, pp. 657–664, 2018.
https://doi.org/10.1007/978-3-319-76941-7_59

On the other hand, recent unsupervised approaches have utilized query-specific information with unigram Language Models (LMs) [3], and architectures such as Restricted Boltzmann Machines [4,5]) for effective performance. Though state-of-the-art in extractive summarization, similar techniques that rely on document and corpus statistics have earlier been outperformed by unsupervised long-span neural LMs on similar sentence ranking tasks [6,7] such as in question answering [8].

A typical unsupervised system that operates on a large corpus implements two stages [3–5]: (1) *sentence ranking* to generate a candidate set of sentences from the entire corpus; and (2) *sentence selection* from the candidate set to compose a (length budgeted) summary. In the sentence selection stage, traditional unsupervised extractive multi-document summarization systems address a *global inference problem* [9,10] that aims to generate a length-budgeted summary with the candidate sentences that maximize the overall *relevance* while avoiding informational *redundancy*. In the past, several Integer Linear Programming (ILP)-based approaches [9,11] based on the popular Maximal Marginal Relevance (MMR) [12] criterion have been shown to achieve high-quality results. Such an objective comes with explicit relevance and redundancy functions that can leverage LMs [3]. However, these ILP-based approaches work for summarizing a small number of pre-selected documents and do not scale to larger number input documents (e.g., entire corpus). Thus, a preliminary sentence ranking step is required to generate a smaller set of (top-k ranked) candidate sentences.

In this paper, we aim to study the effectiveness of *long-span-based neural sentence LMs* [13] for an unsupervised query-focused extractive multi-document summarization task. In a long-span LM, word probabilities are estimated by considering long-range dependencies within a large local context (e.g. few surrounding sentences, passages, or entire source document) in contrast to short-context models that use word independence (e.g., count-based LMs) and Markovian restrictions that use the previous word (e.g., n-gram LMs). Specifically, we address the problem recently proposed by [3] inputs short (single sentence) Wikipedia event descriptions as queries, and outputs a focused extractive summary from a longitudinal collection. In this task, we focus on developing long-span LMs that are robust to variable sentence lengths; and effective for computing relevance to query and inter-sentence redundancies for global inference.

We make the following key contributions in this paper: (1) To the best of our knowledge, we are the first to incorporate across sentence-based [8] and LSTM-based long-span LMs for an unsupervised query-focused extractive summarization task. For sentence selection, we extend the ILP-based approach to incorporate the proposed LM. For application of long-span LM to this ILP, sentence *relevance* is computed by scoring the candidate sentences conditioned on the query whereas, for *inter-sentence redundancy*, we propose comparing query words given the candidate sentences (Sect. 2) to allow comparison of arbitrary length sentences. (2) We present two *Across Sentence Boundary* LSTM-based LMs: *ASBLSTM* and *biAS-BLSTM* (Sect. 2.1), that build over an LSTM-based LM for our task. (3) Intrinsic and extrinsic experiments are performed (Sect. 3) to evaluate the effectiveness of the LMs with a test query set containing 100 Wikipedia event descriptions released by [3] on the English Gigaword corpus [14].

2   Approach

We next describe the two summarization stages and the proposed language model.

Sentence Ranking: As the first stage, the primary goal is to generate a candidate set CS of sentences to reduce the search space during the summarization process. For this purpose, we employ two steps: **(1)** For a given event description as a query q, we retrieve a set of top-k documents using query likelihood retrieval framework as pseudo-relevance feedback [15]. **(2)** Then, sentences within the retrieved documents are ranked according to the generative probability $P(s|q)$ where s is a candidate sentence. Finally, we consider the top-100 sentences as the CS.

Sentence Selection: In this second stage, we solve an ILP to select sentences from the generated CS with the criterion [11]: **Max** $\sum_i \lambda rel_i \xi_i - (1-\lambda) \sum_{j \neq i} red_{ij} M_{ij}$, where rel_i is the relevance score of a sentence; and red_{ij} represents the redundancy between two candidate sentences. ξ_i and M_{ij} are indicator variables and λ controls the importance of relevance and redundancy. We refer to [11] for full details on the ILP.

In our setup, rel_i computes the likelihood $P(s|q)$ of generating a sentence s from a given query q. The redundancy red_{ij} between two sentences is computed as the Jensen-Shanon Divergence (JSD) between sentence LMs. In the ILP objective, while computing red_{ij} between two sentences, we find that the long-span LMs, if estimated naïvely, suffer from lack of query-relevant context (e.g., terms that are semantically related to the query terms) which leads to poor estimate of redundancy w.r.t the query. Thus, we propose to compute JSD between query LMs conditioned on the candidate sentences to estimate their redundancy. For the given query q; two candidates s_i and s_j; and the respective LMs $P(q|s_i)$ and $P(q|s_j)$, $red_{ij} = -JSD(P(q|s_i)||P(q|s_j))$. Intuitively, calculating redundancy in such a manner compares the predictive nature of different query terms given the sentences of arbitrary lengths.

2.1   ASBLSTM Language Model

LSTM-based methods have shown impressive improvements in language modelling tasks (e.g., Machine Translation and Speech Recognition) in comparison to standard count-based methods [15]. The main advantage of LSTMs is a single state that encodes the global linguistic context and controls its longevity using a forget gate. However, the individual hidden state in LSTM tends to lose the long-term context [13] which also becomes essential for query-based sentence ranking tasks.

To reduce this loss, [8] have incorporated across sentence information explicitly. These LMs learn to trigger words across sentences instead of just the within-sentence triggers. Intuitively, in such a triggering scheme a sentence is less divergent (or more relevant) to an adjacent sentence (query) if the words in the sentence predict words in the adjacent sentence with a higher probability. However, in a standard LSTM architecture, the recurrent state focusses more on the within-sentence words as triggers while losing the information around the sentence. Thus, implying that sentences with more within-sentence triggers are heavily boosted while not considering the impact of across sentence triggers, which is more relevant for a query-specific setup. We address this issue by introducing an extra memory state (as shown in Fig. 1) into the architecture that stores the LSTM state of the previous sentence s_{-1} and uses this state while scoring the current sentence s, hence, calculating $P(s|s_{-1})$. We refer to this LM as the Across Sentence Boundary-based LSTM or *ASBLSTM*.

Here, we assume that each sentence is represented by the hidden layer state achieved at the end of the sentence. Hence, the previous sentence information is contained in the hidden state $(H_{s_{-1}})$ observed at the end of the previous sentence. Using this hidden state, the output layer (O_t) is defined as the combination of within-sentence context captured by the present hidden state H_t and the across-sentence boundary context captured by the hidden state $H_{s_{-1}}$ as, $O_t = g_{softmax}(WH_t + W_sH_{s_{-1}})$. At inference, we compute $P(s|q)$ and $P(q|s)$ for sentence ranking and selection stages. For the query-specific long-span LMs, the query is used as the previous sentence of a candidate s for computing $P(s|q)$, whereas their order is reversed while computing $P(s|q)$.

3   Experimental Evaluation

Data: Test query set contains 100 random event descriptions that happened between 1987 and 2007 with Wikipedia articles central to the events as the gold standard summaries. This test set was publicly released by [3]. Our target is English Gigaword corpus [14] with about 9 million news articles published between 1994 to 2008 taken from four different news sources. We evaluate the 250 worded system-generated summaries against the gold standard Wikipedia articles using standard Rouge-2 and -SU4 measures. We make our data publicly available[1].

The disparate quantity of text between the gold standard Wikipedia articles and the system generated summaries result in low Rouge scores. Thus, we perform one-tailed student's t-test over the Rouge-SU4 scores. Significant improvements at levels 0.05 and 0.01 are indicated by △ and ▲ while decrements by ▽ and ▼.

Methods: Table 1 describes the different baseline methods under comparison.

Implementation: Using long-span LMs for this summarization system, we restrict the LM vocabulary to 80000 most frequent words. This vocabulary is

---

[1] http://resources.mpi-inf.mpg.de/d5/asblstmSumm.

Table 1. Baseline schemes and language models for sentence ranking and selection.

Language Models	*Uni*	Dirichlet smoothed unigram LM proposed by [3]
	ASB	An across boundary n-gram sentence LM proposed by [8]. For training, we look at a window of one previous sentence in the source document
	LSTM	A stateful *LSTM*-based language model estimated with projection and hidden layers containing 200 nodes, and a vocabulary-sized output layer for sentence selection. The stateful LSTM variant initializes the first hidden state with the last hidden state of the previous sentence
	biASB, *biLSTM,* *biASBLSTM*	The **bi-** suffix is added to denote that the above described LMs and *ABSLSTM* LM introduced in Sect. 2.1 are trained using both left and right context. Training using bi-directional context alleviates the inherent boosting of sentences that *follow* query words within a document
Schemes	*Random*	This approach selects sentences form *CS* at uniform random
	MEAD	Uses centroid based *MEAD D*, lead-based *MEAD LB* [16] that come with the open source framework
	Rdh	Uses Dirichlet smoothed unigram LMs for sentence selection with corpus background model [3]
	Gil	Maximizes query-salient *words* with an ILP [10] using an advanced unigram query LM [3]

then appended with words included in the test queries. Rest of the words are replaced by an out-of-vocabulary symbol. This modification allows for constraining the parametric size of LMs leading to a faster processing of the data. Our LMs are trained on the pseudo-relevant documents for all the queries, allowing the LMs to learn the triggering information but staying agnostic to explicit query information. Only at the time of inference, query text is used in LMs to help score candidate sentences.

Sentence Ranking: Since the relevance judgments for sentence in a candidate set *CS* generated with an LM is not available; we design an extrinsic experiment in contrast to an information retrieval style evaluation. We leverage the notion that a *CS* containing more query-informative sentences will lead to generating better summaries. First, we create a *CS* by reranking using a LM and selecting top-100 sentences. Then this is input to an *Oracle* genetic algorithm proposed by Riedhammer et al. [17] that is aware of gold standard Wikipedia articles to generate the best possible summary.

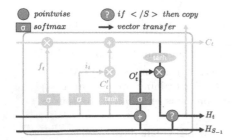

- ● pointwise
- σ softmax
- ? if < /S > then copy
- → vector transfer

Fig. 1. ASBLSTM block with additional $H_{S_{-1}}$ state. Last hidden state of the previous sentence, $H_{S_{-1}}$, is updated with separator symbol $(< /\text{S} >)$

Table 2. Oracle summary results with significance test against the *ASB*.

LM used for CS	Rouge-2 F1	Rouge-SU4 F1
Uni	0.0406	0.0542 (▽)
ASB	**0.0415**	**0.0549** (-)
LSTM	0.0377	0.0516 (▲)
biLSTM	0.0383	0.0509 (▲)
ASBLSTM	0.0395	0.0527 (▲)
biASBLSTM	0.0390	0.0516 (▲)

Table 3. Sentence selection results with significance tests against $(Rdh/\ Gil)$.

LM used for CS	LM used for selection	Rouge-2 F1	Rouge-SU4 F1
	Random	0.0293	0.0388 (▾/▾)
	MEAD LB	0.0287	0.0400 (▾/▾)
	MEAD D	0.0305	0.0422 (▾/▾)
	Rdh	0.0364	0.0510 (-/△)
	Gil	0.0381	0.0531 (△/-)
Uni	*ASB*	0.0382	0.0548 (▲/△)
	biASB	0.0383	0.0549 (▲/△)
	LSTM	0.0388	**0.0553** (▲/△)
	biLSTM	0.0387	0.0550 (▲/△)
	ASBLSTM	0.0387	0.0547 (▲/△)
	biASBLSTM	**0.0389**	0.0552 (▲/▲)
ASB	*ASB*	0.0371	0.0534 (-/△)
	biASB	0.0376	0.0537 (△/△)
	LSTM	0.0393	0.0550 (▲/▲)
	biLSTM	0.0392	**0.0552** (▲/▲)
	ASBLSTM	0.0396	0.0547 (▲/▲)
	biASBLSTM	**0.0399**	**0.0552** (▲/▲)
ASBLSTM	*ASB*	0.0378	0.0536 (△/△)
	biASB	0.0377	0.0539 (△/△)
	LSTM	0.0371	0.0527 (-/-)
	biLSTM	0.0369	0.0524 (-/-)
	ASBLSTM	0.0368	0.0523 (-/-)
	biASBLSTM	**0.0383**	**0.0538** (▲/△)

Table 2, reports the Rouge scores of the oracle summarizer with different candidate sets CS as input. The CS generated using simpler unigram and n-gram based ASB outperform those generated with LSTM-based models. A recent work [18] argues that the neural LSTM models are more suited for the semantically oriented task rather than retrieval-based ranking tasks, where n-gram-based LMs work better. Our finding conforms with this argument during sentence ranking stage.

Sentence Selection: Table 3 reports the performance of the different schemes using the CS generated with estimated LMs. We find that using *biASBLSTM* LM for sentence selection across all the CS proves to be most effective. As expected, the *Random* to be the worst. *MEAD_LB* considers only the documents' lead paragraphs and is also not able to achieve good scores. *MEAD_D* represents centroid based method and performs significantly worse than the ILP-based *Rdh* and *Gil* methods. The best combination is to use an n-gram-based *ASB*-based sentence LM for the ranking, and *biASBLSTM* LM for selection stage. This shows significant improvements over the state-of-the-art coverage-based *Gil* and MMR-style *Rdh* by approximately 5% and 10% in Rouge-2 F1.

In summary, LSTMs have been shown to capture semantic relations from within the text. In summarization, such semantic relations can better model the notion of inter-sentence redundancy [18]. This observation is also reflected in our experiments where we find that adding more context, in fact, improves the quality of short text summaries.

4 Conclusion

In this paper, we proposed applying long-span models to query-focused unsupervised text summarization. We presented the *ASBLSTM* LM for the sentence ranking and selection stages of summarization. In summary, the *ASBLSTM* LM outperform other models in for the sentence selection stage. A scheme that uses an n-gram-based across sentence boundary (ASB) LM for sentence ranking and *biASBLSTM* LM for selection stage of summarization, demonstrated to be the most effective.

References

1. Cao, Z., et al.: AttSum: joint learning of focusing and summarization with neural attention. arXiv preprint arXiv:1604.00125 (2016)
2. Nallapati, R., et al.: Classify or select: neural architectures for extractive document summarization. arXiv preprint arXiv:1611.04244 (2016)
3. Mishra, A., et al.: Event digest: a holistic view on past events. In: SIGIR (2016)
4. Zhong, S., et al.: Query-oriented unsupervised multi-document summarization via deep learning model. Expert Syst. Appl. **42**(21) (2015)
5. Yousefi-Azar, M., et al.: Text summarization using unsupervised deep learning. Expert Syst. Appl. **68**, 93–105 (2017)
6. Cao, Z., et al.: Ranking with recursive neural networks and its application to multi-document summarization. In: AAAI (2015)
7. Palangi, H., et al.: Deep sentence embedding using long short-term memory networks: analysis and application to information retrieval. In: TASLP (2016)
8. Momtazi, S., et al.: Trained trigger language model for sentence retrieval in QA: bridging the vocabulary gap. In: CIKM (2011)
9. McDonald, R.: A study of global inference algorithms in multi-document summarization. In: Amati, G., Carpineto, C., Romano, G. (eds.) ECIR 2007. LNCS, vol. 4425, pp. 557–564. Springer, Heidelberg (2007). https://doi.org/10.1007/978-3-540-71496-5_51
10. Gillick, D., et al.: A scalable global model for summarization. In: ILP-NAACL-HLT (2009)
11. Riedhammer, K., et al.: Long story short - global unsupervised models for keyphrase based meeting summarization. Speech Commun. **52**(10), 801–815 (2010)
12. Carbonell, J., et al.: The use of MMR, diversity-based reranking for reordering documents and producing summaries. In: SIGIR (1998)
13. Oualil, Y., et al.: Long-short range context neural networks for language modeling. arXiv preprint arXiv:1708.06555 (2017)
14. English Gigaword Corpus. https://catalog.ldc.upenn.edu/ldc2003t05
15. Zhai, C.X., et al.: Statistical language models for information retrieval. Synth. Lect. Hum. Lang. Technol. **1**(1), 1–141 (2008)

16. Radev, D.R., et al.: MEAD-a platform for multidocument multilingual text summarization. In: LREC (2004)
17. Riedhammer, K., et al.: Packing the meeting summarization knapsack. In: INTERSPEECH (2008)
18. Guo, J., et al.: A deep relevance matching model for ad-hoc retrieval. In: CIKM (2016)

Topic-Association Mining for User Interest Detection

Anil Kumar Trikha, Fattane Zarrinkalam[✉], and Ebrahim Bagheri

Laboratory for Systems, Software and Semantics (LS³),
Ryerson University, Toronto, Canada
{atrikha,fzarrinkalam,bagheri}@ryerson.ca

Abstract. The accurate identification of user interests on Twitter can lead to more efficient procurement of targeted content for the users. While the analysis of user content has engaged with on Twitter is a rich source for detecting the user's interests, prior research have shown that it may not be sufficient. There have been work that attempt to identify a user's *implicit interests*, i.e., those topics that could interest the user but the user has not engaged with them in the past. Prior work has shown that *topic semantic relatedness* is an important feature for determining users' implicit interests. In this paper, we explore the possibility of identifying users' implicit interests solely based on topic association through frequent pattern mining without regard for the semantics of the topics. We show in our experiments that topic association is a strong feature for determining users' implicit interests.

1 Introduction

User interest detection techniques that automatically identify users' interests towards active topics on Twitter have become an emerging research area in the recent years, primarily due to its potential to improve the quality of higher-level applications such as news recommendation [1] and retweet prediction [2], among others. Most of the existing work in the field of user interest detection are focused on extracting explicit interests via analyzing textual contents shared by the users [4,12]. Based on the fact that the majority of users in social networks are not very active (free-riders), their available content is sparse and does not reveal sufficient clues about their interests. To address this challenge, there have been work dedicated to inferring implicit interests of users [7,9]. Implicit interests are those potential interests that might be relevant and interesting for the user but the user has not engaged in them explicitly in the past [11].

Several authors have indicated that interaction patterns between users and topics are among the important clues for determining implicit interests [8,10]. To systematically investigate the suitability of users' interaction patterns and topic relatedness on the quality of implicit interest detection, a graph-based link prediction scheme is proposed in [11], which combines these two factors into a unified representation model. Based on the experiments, the authors found that topic

© Springer International Publishing AG, part of Springer Nature 2018
G. Pasi et al. (Eds.): ECIR 2018, LNCS 10772, pp. 665–671, 2018.
https://doi.org/10.1007/978-3-319-76941-7_60

relatedness is a contributing factor that can accurately uncover implicit interests of users. In other words, users on Twitter are more likely to be interested in topics that are conceptually similar to the topics that they have explicitly engaged with in the past. On the basis of this finding, in this paper, we are interested in topic association as a means to infer implicit interests of users by turning the implicit interest detection problem into a frequent pattern mining problem. Frequent pattern mining (FPM) is a widely adopted data mining technique that has mostly contributed to the discovery of co-occurrences and associations between items of a dataset. FPM methods have already been used in the field of social network analysis for finding hidden patterns in social data [5,6]. In line with these works, we apply FPM in the context of implicit interest detection to extract the association between topics on Twitter and subsequently infer users' implicit interests. We then build the interest profile of a user considering both her explicit and implicit interests.

2   Proposed Approach

We study the problem of inferring user interest profiles towards active topics on Twitter, within a given time interval, which can formally be defined as follows:

Definition 1 (Interest Profile). *Given a set of K topics $\mathbb{Z}$, an interest profile of a user $u \in \mathbb{U}$ in time interval T, called $P^T(u)$, is represented by a vector of weights over K topics, i.e., $(f_u(z_1), ..., f_u(z_K))$, where $f_u(z_k)$ denotes the degree of u's interest in topic $z_k \in \mathbb{Z}$ at time interval T. A user topic profile is normalized using $L1 - norm$.*

Our proposed approach performs the following three steps to infer the interest profile of users: *(1)* Inferring users' explicit interests by extracting information from the content that the users have shared on Twitter; *(2)* Generating frequent patterns based on the collective set of users' explicit interests in order to understand the relation between topics in a given time interval T; and, *(3)* Augmentation that incorporates additional implicit interests into a user's interest profile based on the frequent patterns learnt in Step 2. These three steps are described in the following.

2.1   Inferring User Explicit Interests

The interests that are observable in a user's tweets are known as *explicit* interests. User explicit interest detection methods from Twitter have been studied in the literature and therefore are not the focus of our work and we are able to work with any topic and interest detection method to extract topics $\mathbb{Z}$ and the explicit interest profile of each user u toward these topics in time interval T, denoted as $P_E^T(u) = (f_u^E(z_1), ..., f_u^E(z_K))$. Considering $\mathbb{M}$, the set of available microposts, it is possible to extract topics $\mathbb{Z}$ using Latent Dirichlet Allocation (LDA), the *de facto* standard in topic modeling. As suggested in [11,12], to obtain better topics from Twitter without modifying the standard topic detection methods, we annotate the text of each tweet with Wikipedia concepts using an existing semantic annotator. Next, given the published or retweeted microposts of a user u, $\mathbb{M}_u$, we initially divide $\mathbb{M}_u$ into N segments based on a uniform time interval T, $\mathbb{M}_u = \{\mathbb{M}_u^1, \mathbb{M}_u^2, ..., \mathbb{M}_u^N\}$. Then, we aggregate all concepts extracted from each tweet segment of a user into a single document and apply LDA on the collection of such documents to discover K topics $\mathbb{Z}$, and explicit interest profile of each user u in each time interval T, i.e., $P_E^T(u)$.

2.2 Discovering Frequent Topic Patterns

Given the collective set of users' explicit interests, i.e. $\{P_E^T(u)|1 \leq T \leq N, u \in \mathbb{U}\}$, in this section, we aim at utilizing FPM methods to find closely related topics that frequently co-occur within the explicit interests of our user set. To do so, we treat topics $\mathbb{Z}$ as items and use the explicit interest profile of each user in a given time interval T, $P_E^T(u) = (f_u^E(z_1), ..., f_u^E(z_K))$, to form a transaction τ. Thus each transaction τ consists of the set of topics that a user is explicitly interested in at time T, i.e., $\tau = \{z|f_u^E(z) > 0\}$. Then, we apply an FPM method to mine the transactional database built based on Definition 2 to calculate the frequent topic patterns in time interval T, denoted as $FP_{\mathbb{Z}}^T$.

Definition 2 (Transactional Database). *The transactional database for time interval T, denoted as TDB^T, includes the collective set of all users' explicit interests in time interval T and L past time intervals, i.e. $TDB^T = \{P_E^t(u)|T - L \leq t \leq T, u \in \mathbb{U}\}$.*

In Definition 2, in order to be able to study the impact of considering historical user interests on the performance of extracted frequent topic patterns, we also add the historical explicit interest profile of all users in L past time intervals to the transactional database. Appropriate algorithms like *Apriori, Eclat* and *FP-Growth* have been developed to efficiently discover frequent patterns. In this work, we utilize the FP-Growth algorithm as an efficient method which mines frequent patterns without costly candidate generation. It has been experimentally shown in [3] that FP-Growth algorithm has the best performance among the others and is thus the most scalable.

Now, given the explicit interest detection method described in Sect. 2.1 is based on dividing the user's data into N discrete time intervals, we perform the above process for each of the intervals T. This will produce $\{FP_{\mathbb{Z}}^T|1 \leq T \leq N\}$, which is the input of our augmentation method to build interest profile of each user in each time interval T. For example, $s = \{z_{35}, z_{88}\}$ is the most frequent topic-set extracted for December 10, 2010. Topic $z_{35} = \{$Mixtape, Hip_hop_music, Rapping, Kanye_West, Jay-Z, Remix$\}$ refers to the hip-hop music collaboratively produced by American rappers *Jay-Z* and *Kanye West* and topic $z_{88} = \{$Lady_Gaga, Song, Album, Concert, Canadian_Hot_100$\}$ refers to the concert of *Lady Gaga* in Canada. It is clear that these two topics are related to music and the users who are explicitly interested in z_{35} could potentially also be interested in z_{88}.

2.3   Interest Profile Augmentation

In this section, to build the interest profile of a user u in time interval T, $P^T(u)$, as defined in Definition 1, we augment the explicit interest profile of the user $P_E^T(u)$, using FP_Z^T, the frequent topic patterns in time interval T, based on Algorithm 1. As shown in the Algorithm, given $P_E^T(u) = (f_u^E(z_1), ..., f_u^E(z_K))$, we take each topic z which is of explicit interest to user u, i.e., $f_u^E(z) > 0$, and search FP_Z^T to find any topic-set s which includes topic z (Lines 3 to 5). If such a topic-set s exists, we take the other topics in s and add them to the interest profile of user u, $P^T(u)$ (Line 7). At the end of this process, the explicit interest profile of each user in each time interval is augmented with additional interests from the frequent topic patterns.

3   Experiments

We use the publicly available Twitter dataset [1] that includes 3M tweets posted by approximately 135K users, starting from Nov. 1st and lasting for two months until Dec. 31st 2010. As mentioned in Sect. 2.1, we annotated the text of each tweet with Wikipedia concepts using the TagMe RESTful API, which resulted in 350,731 unique concepts. Then, we applied the Gensim implementation of LDA to extract topics and explicit interests of users over these topics in each time interval T. The number of topics is set to 100 and the length of time interval T is set to 1 day.

Evaluation Methodology and Metric. Adopted from [9], we deploy a retweet prediction application for evaluation. Since the main goal is not to propose a retweet prediction system, the authors have adopted a simple algorithm which is only based on user interest profiles. To do so, given the tweets of two consecutive time intervals, i.e., T_1 and T_2, for a user u, her interest profile $P^{T_1}(u)$ is built based on the tweets that she has published or retweeted in time interval T_1. Further, the tweets that she has retweeted in time interval T_2 are considered to be the ground truth for that user in order to evaluate the results of the retweet prediction application. For user u, to predict a retweet, the tweets of her followed users from whom she has retweeted at least one tweet in time interval T_2 are considered as candidates, and the topic similarity between a candidate tweet and the user interest profile of user u is computed as described in [9].

Then, we rank the tweets based on the similarity scores in descending order. By comparing the ranked list of candidate tweets with the ones that are in the ground truth, we evaluate the quality of retweet prediction, and therefore determine how successfully the interests of a user have been identified. We adopt Mean Average Precision (MAP) as our evaluation metric.

Comparison Methods. we consider the following user interest detection methods for comparison: (1) **EUI**: In this method, the **Explicit User Interest** detection method described in Sect. 2.1, is used to build user interest profiles. (2) **Zarrin's Model**: This method builds user interest profiles based on combining

---

Algorithm 1. Augmentation process

Input: $P_E^T(u), FP_Z^T; 1 \leq T \leq N$
Output: $P^T(u); 1 \leq T \leq N$

1: **for** $P_E^T(u) = (f_u^E(z_1), ..., f_u^E(z_K)) : 1 \leq$
   $T \leq N$ **do**
2:    $P^T(u) \leftarrow P_E^T(u)$
3:    **for** $z \in \{z | f_u^E(z) \geq 0\}$ **do**
4:        **for** $s \in FP_Z^T$ **do**
5:            **if** $z \in s$ **then**
6:                **for** $x \in s$ **do**
7:                    $f_u(x) \leftarrow 1$
8: **return** $P^T(u)$

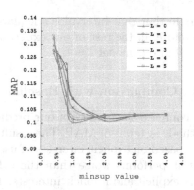

Fig. 1. Effect of the value of minSup and L on the performance of the proposed model.

Explicit and Implicit Interest profiles. In this method, we build $P^T(u)$, by augmenting $P_E^T(u)$, explicit interests of user u at T, with the implicit interest of user u to each topic z that she is not explicitly interested in, i.e., the value of $f_u^E(z_k)$ is equal to 0. To infer the implicit interests, we follow the proposed link prediction method described in [11]. Based on the results in [11], we selected the best configuration of this paper i.e., S that considers the semantic relatedness between topics and Adamic/Adar as link prediction method, for comparison here. (3) **Wang's Model**: This method which is proposed by Wang et al. [9] learns interest profile of user u, i.e., $P^T(u)$, based on a link structure regularization framework that consider both user explicit interest and the relationship between users to detect implicit interests.

3.1   Effect of Parameters

By setting the value of the minimum support threshold *minsup* in the frequent pattern mining process, it is possible to generate variable number of patterns as needed. Further, as described in Sect. 2.2, L denotes the number of historical time intervals included in the transactional database to extract frequent topic patterns in each time interval. Here, we investigate the impact of these parameters on the quality of our proposed method by changing the value of *minsup* from 0.4% to 4% and the size of L from 0 to 5. The results are reported in Fig. 1. Based on results, the quality of prediction results in terms of MAP has significantly decreased by increasing the value of minsup value from 0.4% to 4%. When minsup is set low, the number of frequent topic-sets increases dramatically.

Thus, it can be concluded that increasing the number of frequent topic patterns leads to user interest profiles that are richer for predicting relevant tweets to a given user. As another observation, it can be seen that considering the historical data of users does not have a significant impact on the increase or decrease of the quality of prediction results. This means that to infer the interest profile of users in each time interval, considering the information provided by users in that time interval is adequate to extract the relatedness between topics. Therefore, in the rest of our experiments, we set the minsup value to 0.4% and L to 0 in our model.

3.2 Comparison with Baseline Methods

We compare the quality of our predicted results with the results of comparison methods in terms of MAP. The results are reported in Table 1. The EUI model is only based on explicit interests of users. Based on the results, it can be observed that it performs worse than the other methods which are all based on both users explicit and implicit interests. This means that incorporating user implicit interests in addition to their explicit interests leads to user profiles that are more accurate for predicting relevant tweets to a given user. In other words, the content generated by users does not reveal sufficient clues to extract all of the users' interests. Therefore, the incorporation of the indirect association between topics or relationships between users can lead to a more accurate representation of users' interests and consequently improve the quality of recommendations.

Based on the results, Both Zarrin's and our model which utilize some form of association between topics to extract implicit interests of users outperforms Wang's model that utilizes the relationship between the users. In line with results reported in [11], this can indicate that finding topic association has a higher influence on identifying users' implicit interests as opposed to considering users' social connections. Based on the Zarrin's model, a user is interested in topics that are conceptually similar to the topics that they have explicitly engaged with. Therefore, it calculates the semantic relatedness between topics based on their constituent concepts and then applies link prediction to infer implicit interests of each user. However, in our proposed model, given the explicit interests of all the users, the implicit interest detection problem is converted into a frequent pattern mining problem to extract relationships between topics. As shown in Table 1, our model builds more accurate user profiles which contribute to improved quality of retweet prediction. This shows that frequent pattern mining methods that do not consider the semantics of topics and only focus on topic co-occurrence can also capture topic association to an accurate degree.

Table 1. Performance comparisons. * shows significant difference over baselines at p-value < 0.01.

Method	EUI	Zarrin's model	Wang's model	Our model
MAP	0.078	0.096	0.080	0.134*

4   Conclusion and Future Work

In this paper, we proposed an approach for identifying user interests over a set of topics on Twitter, considering both their explicit and implicit interests.

We model the problem of inferring implicit interests as a frequent pattern mining problem to extract the association between topics and subsequently augmenting explicit interests of users. As future work, based on the fact that users are interested in topics that are conceptually similar, we intend to include semantic similarity between topics in our framework, and infer interest profile of users considering both association and semantic similarity of topics.

References

1. Abel, F., Gao, Q., Houben, G., Tao, K.: Analyzing user modeling on twitter for personalized news recommendations. In: UMAP, pp. 1–12, (2011)
2. Feng, W., Wang, J.: Retweet or not?: personalized tweet re-ranking. In: WSDM, pp. 577–586 (2013)
3. Garg, K., Kumar, D.: Comparing the performance of frequent pattern mining algorithms. Int. J. Comput. Appl. **69**(25), 21–28 (2013)
4. Kapanipathi, P., Jain, P., Venkataramani, C., Sheth, A.: User interests identification on twitter using a hierarchical knowledge base. In: Presutti, V., d'Amato, C., Gandon, F., d'Aquin, M., Staab, S., Tordai, A. (eds.) ESWC 2014. LNCS, vol. 8465, pp. 99–113. Springer, Cham (2014). https://doi.org/10.1007/978-3-319-07443-6_8
5. Moosavi, S.A., Jalali, M., Misaghian, N., Shamshirband, S., Anisi, M.H.: Community detection in social networks using user frequent pattern mining. Knowl. Inf. Syst. **51**(1), 159–186 (2017)
6. Petkos, G., Papadopoulos, S., Aiello, L.M., Skraba, R., Kompatsiaris, Y.: A soft frequent pattern mining approach for textual topic detection. In: WIMS, pp. 25:1–25:10 (2014)
7. Piao, G., Breslin, J.G.: Inferring user interests for passive users on twitter by leveraging followee biographies. In: Jose, J.M., Hauff, C., Altıngovde, I.S., Song, D., Albakour, D., Watt, S., Tait, J. (eds.) ECIR 2017. LNCS, vol. 10193, pp. 122–133. Springer, Cham (2017). https://doi.org/10.1007/978-3-319-56608-5_10
8. Shen, W., Wang, J., Luo, P., Wang, M.: Linking named entities in tweets with knowledge base via user interest modeling. In: KDD, pp. 68–76 (2013)
9. Wang, J., Zhao, W.X., He, Y., Li, X.: Infer user interests via link structure regularization. ACM TIST **5**(2), 23:1–23:22 (2014)
10. Wen, Z., Lin, C.: Improving user interest inference from social neighbors. In: CIKM, pp. 1001–1006 (2011)
11. Zarrinkalam, F., Fani, H., Bagheri, E., Kahani, M.: Inferring implicit topical interests on Twitter. In: Ferro, N., Crestani, F., Moens, M.-F., Mothe, J., Silvestri, F., Di Nunzio, G.M., Hauff, C., Silvello, G. (eds.) ECIR 2016. LNCS, vol. 9626, pp. 479–491. Springer, Cham (2016). https://doi.org/10.1007/978-3-319-30671-1_35
12. Zarrinkalam, F., Fani, H., Bagheri, E., Kahani, M., Du, W.: Semantics-enabled user interest detection from Twitter. In: WI-IAT, pp. 469–476 (2015)

A Study of an Automatic Stopping Strategy for Technologically Assisted Medical Reviews

Giorgio Maria Di Nunzio$^{(\boxtimes)}$

Department of Information Engineering, University of Padua, Padua, Italy
giorgiomaria.dinunzio@unipd.it

Abstract. Systematic medical reviews are a method to collect the findings from multiple studies in a reliable way. Given budget and time constraints, limiting the recall of a search may undermine the quality of a review to such a degree that the validity of its findings is questionable. In this paper, we investigate a variable threshold approach to tackle the problem of a total recall task in medical reviews proposed by a Cross-Language Evaluation Forum (CLEF) eHealth lab in 2017. Compared to the official results submitted to the CLEF eHealth task, our approach performed consistently better over all the range of thresholds considered achieving a recall greater than 0.95 with 25,000 documents less than the best performing systems. The runs and the source code to generate the analyses of this paper are available at the following GitHub repository (https://github.com/gmdn/ECIR2018).

1 Introduction

The large and growing number of published studies makes the task of identifying relevant documents for the realization of systematic reviews complex and time consuming [7]. In particular, for healthcare providers, researchers, and policy makers, "it is unlikely that they will have the time, skills and resources to find, appraise and interpret all this evidence and to incorporate it into healthcare decisions."[1] International evaluation campaigns have recently organized labs in order to study this problem in terms of the evaluation, through controlled simulation, of methods designed to achieve very high recall [4,9] and, in particular, for technology assisted reviews in empirical medicine [3,5].

In this paper, we investigate a variable threshold approach to tackle the problem of a total recall task based on the active learning framework proposed by [6]. We propose a stopping strategy based on the geometry of the two-dimensional space of documents [1] that uses the relevance feedback information given by the expert to automatically estimate the number of documents that need to be read in order to declare the review complete. The paper is organized as follows: in Sect. 2, we define the framework and the stopping strategy; in Sect. 3, we present the experiments and the results. In Sect. 4, we give our final remarks.

---

[1] Cochrane Handbook for Systematic Reviews of Interventions http://handbook-5-1.cochrane.org.

© Springer International Publishing AG, part of Springer Nature 2018
G. Pasi et al. (Eds.): ECIR 2018, LNCS 10772, pp. 672–677, 2018.
https://doi.org/10.1007/978-3-319-76941-7_61

2   Two Dimensional BM25

The two-dimensional BM25 merges the two dimensional representation of probabilistic models [1] with the BM25 model [8] in the following way: the 'classical' relevance weight of each term $w_i^{BM25}(tf)$ is decomposed into two parts, $w_i^{BM25,\mathcal{R}}(tf)$ and $w_i^{BM25,\mathcal{NR}}(tf)$, capturing the explicit relevance feedback given by the experts for each document that has been judged as either relevant, $\mathcal{R}$, or non-relevant, $\mathcal{NR}$, during the systematic review process [6]. Then, each document d is represented by two coordinates computed as:

$$P(d|\mathcal{R}) = \sum_{w_i \in d} w_i^{BM25,\mathcal{R}}(tf), \quad P(d|\mathcal{NR}) = \sum_{w_i \in d} w_i^{BM25,\mathcal{NR}}(tf) \qquad (1)$$

The two-dimensional representation gives you the advantage of transforming the problem of classifying (or ranking) documents into an intuitive geometric problem consisting in finding the decision line

$$\underbrace{P(d|\mathcal{NR})}_{y} < m\underbrace{P(d|\mathcal{R})}_{x} + q \qquad (2)$$

that separates the relevant and non-relevant documents in the best possible way. The parameters m and q can be optimized according to a particular goal [2] (for example, the values $m = 1$ and $q = 0$ correspond to a zero-one loss function of a cost-sensitive framework).

In this paper, we iteratively adapt the values of m and q according to the relevance feedback given by an expert for each document reviewed in order to automatically filter out all the non-relevant documents and decide when to stop the reviewing process. In particular, we extend the work of [6] by adding a stopping strategy based on the following steps:

1. Given a set of documents, choose a subset (percent) of documents that will be read and a maximum number of documents (threshold) that an expert is willing to judge during this step;
2. Rank the documents according to the BM25 scores and obtain from the expert a relevance judgement of the top ranked documents (no more than the threshold set in step 1);
3. Find the interpolating line (parameters m_{rel} and q_{rel}) of the relevant documents, and find the line with the same slope m_{rel} passing through the least relevant document (parameter q_{least}), as shown in Fig. 1a;
4. If there is any un-judged document below the line $y < m_{rel}x + q_{least}$ (see Eq. 2), as shown in Fig. 1b, repeat from step 1;
5. If there are no un-judged documents below the line, ask the expert an 'extra' effort to judge some more documents until the proportion of non-relevant documents over the total number of judged documents exceeds a fixed value.

In step 5, we use a geometric progression to compute the number of extra documents to judge. For the n-th extra round the expert will judge $m = threshold/2^n$ documents. The reviewing process stops when (i) $m = 1$ or (ii) the proportion of non-relevant documents over the total number of judged documents exceeds 90% (a precision of 0.1), see for example Fig. 1c.

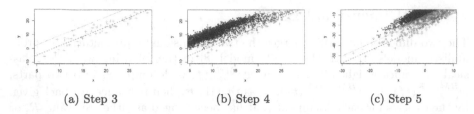

(a) Step 3   (b) Step 4   (c) Step 5

Fig. 1. Visualization of three steps of the stopping strategy. Green crosses are relevant documents, red circles non-relevant documents, black circles un-judged documents. The reviewing process should stop not before experts have judged all the documents in the area between the two lines, see Step 5. (Color figure online)

3   Experiments and Results

The dataset provided by the Technological Assisted Reviews in Empirical Medicine Task at CLEF 2017 is based on 50 systematic reviews conducted by Cochrane experts on Diagnostic Test Accuracy (DTA). The dataset consists of: a set of 50 topics (20 training and 30 test) and, for each topic, the set of PubMed Document Identifiers (PIDs) returned by running the query in MEDLINE, as well as the relevance judgements for both abstracts and documents [3,5]. We also used the publicly available results of all the participants to the task[2] as a baseline for our analyses. In our experiments, we extended the original source code provided by the authors of [6] in order to study how different values of the percentage of documents per topic and the threshold of the maximum number of documents impact on the performance of the stopping strategy (for a detailed analysis of the data pre-processing phase refer to [6]). We varied the percentage from 10% to 50% (with a step of 10%) in combination with a threshold from 100 to 1000 documents (with a step of 100). We did not use any topic of the training set and used only abstract relevance judgements. We report the recall obtained for each combination in Table 1. In the figures described in the following sections, we use a continuous blue line to highlight the performance of the five runs with threshold 500 documents (runid 5, 15, 25, 35 and 45 in Table 1) since this combination showed a good balance between recall (often between 0.92 and 0.94, and consistently above 0.70 per topic) and number of documents judged (between 33,000 to 41,000 documents).

Recall at Number of Documents Shown. In Fig. 2, we show the performance of the 19 official runs of the CLEF eHealth task with colored points and a dashed line that highlights the Pareto frontier[3] of the best runs. The black dots represents all the fifty combinations of runs of our stopping strategy. Our approach dominates the Pareto frontier across all the range of values of documents shown. For high recall values (greater than 0.95), our solution uses 25,000

---

[2] https://github.com/leifos/tar.
[3] https://en.wikipedia.org/wiki/Pareto_efficiency.

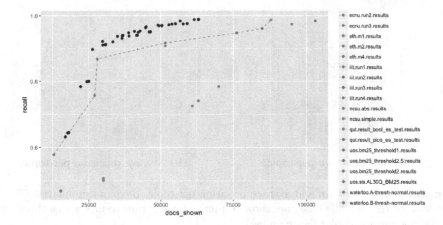

Fig. 2. Pareto frontier of CLEF 2017 of recall at documents shown. Colored dots are the official runs. Black dots are the two-dimensional approach using the stopping strategy which dominate the best performances of the official runs.

documents less than the best systems. In Fig. 3, we analyze a breakdown of the results per topic using boxplots to summarize the values of the recall at number of documents shown per topic (in Fig. 3a the runs of CLEF 2017, in Fig. 3b the runs of the stopping strategy). The median recall per topic of the stopping strategy is consistently greater than the median of the official results, and the small interquartile range shows that we can achieve a 100% recall with any combination of percentage and threshold.

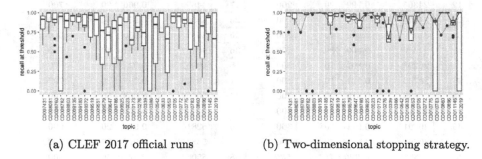

(a) CLEF 2017 official runs   (b) Two-dimensional stopping strategy.

Fig. 3. Boxplots of recall at threshold (number of documents shown) for the official CLEF runs (a) and for the 50 runs that use stopping strategy (b). Blue lines indicate runs with a threshold at 500 documents. (Color figure online)

Documents Shown and Average Precision per Topic. In Fig. 4, we show a comparison of the performances of the CLEF 2017 runs (boxplots) with the stopping strategy with a threshold of 500 documents (blue line). In general, the

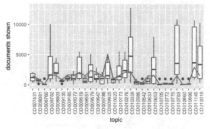

(a) Documents shown per topic.

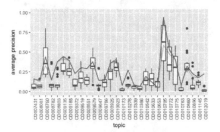

(b) Average precision per topic.

Fig. 4. Documents shown and average precision per topic. Boxplots summarize CLEF official runs results while blue lines show stopping strategy runs results with a threshold of 500 documents. (Color figure online)

Table 1. Recall at number of documents shown by percent of documents per topic and maximum number of documents per topic. The 50 results are split in two parts for space limits and ordered by number of documents shown. The performances of runs 5, 15, 25, 35, 45, formatted in italics, are highlighted in blue in Figs. 3b, 4a, and 4b.

runid	docs shown	recall	percent	threshold	runid	docs shown	recall	percent	threshold
1	16936	0.632	10	100	*25*	*41486*	*0.940*	*30*	*500*
11	17687	0.644	20	100	*35*	*41792*	*0.940*	*40*	*500*
21	18064	0.645	30	100	*45*	*41792*	*0.940*	*50*	*500*
31	18100	0.645	40	100	16	42109	0.949	20	600
41	18100	0.645	50	100	9	42771	0.953	10	900
2	22177	0.784	10	200	10	44711	0.963	10	1000
12	24305	0.799	20	200	17	44899	0.966	20	700
22	24876	0.800	30	200	26	45969	0.950	30	600
32	24989	0.800	40	200	46	46427	0.950	50	600
42	24989	0.800	50	200	36	46427	0.951	40	600
3	26298	0.897	10	300	18	49062	0.971	20	800
4	29746	0.921	10	400	27	49414	0.969	30	700
13	29836	0.910	20	300	47	50454	0.969	50	700
23	30619	0.911	30	300	37	50454	0.970	40	700
43	30791	0.911	50	300	19	52816	0.972	20	900
33	30791	0.912	40	300	28	53877	0.974	30	800
5	*33293*	*0.920*	*10*	*500*	38	55064	0.975	40	800
14	34869	0.937	20	400	48	55064	0.975	50	800
6	36431	0.931	10	600	20	56687	0.984	20	1000
24	36517	0.939	30	400	29	57741	0.975	30	900
44	36739	0.939	50	400	39	59120	0.976	40	900
34	36739	0.940	40	400	49	59120	0.976	50	900
7	38759	0.948	10	700	30	61612	0.988	30	1000
15	*38885*	*0.938*	*20*	*500*	50	63073	0.988	50	1000
8	40770	0.952	10	800	40	63073	0.989	40	1000

number of documents shown for the runs using this threshold is in line with the median of the official runs, while average precision is greater than the values of the third quartile of the official runs.

4 Conclusions and Future Work

In this paper, we studied an automatic stopping strategy for the problem of technologically assisted reviews based on the two-dimensional BM25 approach. We evaluated the proposed solution on a standard CLEF collection based of PUBMED abstracts. Compared to the official runs submitted to the CLEF eHealth task, our approach performed consistently better over all the range of thresholds considered and achieved a recall greater than 0.95 with 25,000 documents less than the best performing systems.

Despite the significant improvements compared to the best runs of the task, there is still room for improvements especially if we consider that none of the twenty topics of the training set was taken into account. In addition, we observed an underestimation of the difficulty of some topics which lead to an insufficient number of documents to assess and consequently to a relatively low recall.

References

1. Di Nunzio, G.M.: A new decision to take for cost-sensitive Naïve Bayes classifiers. Inf. Process. Manage. **50**(5), 653–674 (2014)
2. Di Nunzio, G.M.: Interactive text categorisation: the geometry of likelihood spaces. In: Lai, C., Giuliani, A., Semeraro, G. (eds.) Information Filtering and Retrieval. SCI, vol. 668, pp. 13–34. Springer, Cham (2017). https://doi.org/10.1007/978-3-319-46135-9_2
3. Goeuriot, L., et al.: CLEF 2017 eHealth evaluation lab overview. In: Jones, G.J.F., et al. (eds.) CLEF 2017. LNCS, vol. 10456, pp. 291–303. Springer, Cham (2017). https://doi.org/10.1007/978-3-319-65813-1_26
4. Grossman, M.R., Cormack, G.V., Roegiest, A.: TREC 2016 total recall track overview. In: Proceedings of the Twenty-Fifth TREC 2016, Gaithersburg, Maryland, USA, 15–18 November (2016)
5. Kanoulas, E., Li, D., Azzopardi, L., Spijker, R.: CLEF 2017 technologically assisted reviews in empirical medicine overview. In: Working Notes of CLEF 2017 - Conference and Labs of the Evaluation Forum, CEUR Workshop Proceedings, Dublin, Ireland, 11–14 September 2017. CEUR-WS.org (2017)
6. Di Nunzio, G.M., Beghini, F., Vezzani, F., Henrot, G.: An interactive two-dimensional approach to query aspects rewriting in systematic reviews. IMS Unipd At CLEF eHealth Task 2. In: Working Notes of CLEF 2017 - Conference and Labs of the Evaluation Forum, Dublin, Ireland, 11–14 September (2017)
7. O'Mara-Eves, A., Thomas, J., McNaught, J., Miwa, M., Ananiadou, S.: Using text mining for study identification in systematic reviews: a systematic review of current approaches. Syst. Rev. **4**(1), 5 (2015)
8. Robertson, S.E., Zaragoza, H.: The probabilistic relevance framework: BM25 and beyond. Found. Trends Inf. Retr. **3**(4), 333–389 (2009)
9. Roegiest, A., Cormack, G.V., Grossman, M.R., Clarke, C.L.A.: TREC 2015 total recall track overview. In: Proceedings of the Twenty-fourth TREC 2015, Gaithersburg, Maryland, USA, 17–20 November (2015)

Improving Deep Learning for Multiple Choice Question Answering with Candidate Contexts

Bogdan Nicula[✉], Stefan Ruseti, and Traian Rebedea

University Politehnica of Bucharest, 313 Splaiul Independentei,
060042 Bucharest, Romania
bogdan.nicula22@gmail.com,{stefan.ruseti,traian.rebedea}@cs.pub.ro

Abstract. Deep learning solutions have been widely used lately for improving question answering systems, especially as the amount of training data has increased. However, these solutions have been developed for specific tasks, when both the question and the candidate answers are long enough for the deep learning models to provide a better text representation and a more complex similarity function. For multiple choice questions that have short answers, information retrieval solutions are still largely used. In this paper we propose a novel deep learning model that determines the correct answer by combining the representation of each question-candidate answer pair with candidate contexts extracted from Wikipedia using a search engine.

1 Introduction

Until recently, open-domain question answering (QA) systems have mainly used retrieval-based strategies for determining the correct answer in large collections of documents [9], knowledge bases [3], or both [4]. However, these systems provide good accuracy only for simple factoid questions as they assume that shallow processing of the texts suffices and few or no inferences are necessary to determine the correct answer.

The advancements in deep learning and the advent of larger training corpora for open domain QA have given rise to a wide range of deep neural network models that provide better results than previous information retrieval (IR) approaches [10,15]. However, these solutions usually tackle specific QA tasks, like answer selection [14] or community QA [11], where both the question and the candidate answers are longer pieces of text or when the answer should be selected from a given document. In both cases, deep learning models have two main objectives: to find better representations for the question and the possible answer and to assess the similarity between the two computed representations in a more complex way.

However, neither deep learning or retrieval based models alone have been successful for answering non-factoid multiple choice questions that have short and somewhat similar candidate answers. One such dataset has been provided

G. Pasi et al. (Eds.): ECIR 2018, LNCS 10772, pp. 678–683, 2018.
https://doi.org/10.1007/978-3-319-76941-7_62

by Allen AI for the Aristo challenge and it contains more than 2, 000 8th grade science questions with four candidate answers [1]. In this paper, we propose a novel deep learning architecture which first employs a retrieval based model to generate candidate contexts for each question-candidate answer pair and then uses the contexts to determine the correct response.

The paper continues with a short overview of existing solutions for this task, which aim to improve the simple retrieval-based baseline. Then we present our proposed method, introducing several deep network architectures that make use of question-answer pairs and corresponding candidate contexts. The paper ends with a comparison between the proposed method and existing IR and deep learning solutions, highlighting the role of candidate contexts and fostering the concluding remarks.

2   Related Work

A simple IR baseline proposed for The Allen AI Science Challenge is to use a search engine that looks for each question-candidate answer pair in a large collection of texts and picks the top scoring answer. Several improvements for this baseline have been proposed. For example, Aristo uses a combination of text similarity, statistical information - Pointwise Mutual Information (PMI) and a simple Support Vector Machine (SVM) ranker based on word embeddings, and structured knowledge [2].

Sachan et al. [13] propose a knowledge rich method, employing a wide range of lexical, semantic and discourse features extracted from science textbooks, study guides, science dictionaries, and knowledge tables together with a latent structured SVM (LSSVM) to rank the candidate answers. They show that their method surpasses several baselines, including the basic IR baseline previously mentioned, as well as some deep neural networks such as QANTA [7] and Long Short-Term Memory (LSTM) [6].

The most similar approach to the one presented in this paper makes use of a complex method for finding justifications in supporting scientific texts and dictionaries for each question-candidate answer pair [8]. The justifications are meant not only to improve the QA system's accuracy, but also to provide a human readable explanation for the selected answer. Two classifiers are compared for selecting the correct answer using the justifications.

Our work is original in two ways. First, we propose the first deep learning architecture which outperforms the IR baseline for this task, which is not the case for other neural approaches [8]. We also manage to obtain this result without using a voting approach based on several classifiers [2]. Second, although our model requires additional data for retrieving candidates, it does not use any knowledge rich data sources [13] or hand-crafted features [8].

3   Method

We propose several neural network architectures designed for answering non-factoid multiple choice questions, where additional data is required to

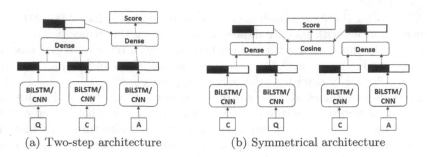

(a) Two-step architecture   (b) Symmetrical architecture

Fig. 1. Deep learning architectures for multiple choice question answering with context candidates (Q - question, A - answer, C - context)

determine the correct answer. They receive a question-candidate answer pair as input and predict whether the answer is correct. The question and answers are pre-processed by removing stop-words and lemmatizing the words using NLTK[1]. Then, each word in the question and in the answer is converted into a 300-dimensional word embedding using GloVe [12] vectors pretrained on Wikipedia.

Two simple neural network architectures were implemented as deep learning baselines. First, they generate a separate representation for the question and for each candidate answer and then compute the likelihood of the answer being correct by either using either cosine similarity between the two vector representations or by employing a final dense layer applied on the concatenated vectors. The first architecture uses convolutional neural networks (CNNs), similar to QANTA [7], to create the question/answer embeddings, while the latter uses bidirectional LSTMs (BiLSTMs) [5].

The main issue with the previous approach is twofold. First, as many candidate answers are rather short, the computed representation for the answer may be insufficient. Second, there are many situations where the question alone does not provide enough context to select the correct answer. As an improvement, in addition to the question-candidate answer pairs, we added contextual information. This allows a more informed decision for questions that require additional information to determine the correct answer. The candidate contexts are obtained by querying a paragraph-level Lucene[2] index of Wikipedia with a query that contains the question concatenated to the answer. The neural networks designed for this case learn to predict how likely is an association between a question, an answer and a context.

Similar to the two baselines, we implemented the candidate contexts architecture both with CNNs and BiLSTMs. Furthermore, we tested two different approaches regarding the underlying neural architecture as presented in Fig. 1. Both architectures generate separate representations for the question (Q), the answer (A), and the context (C) starting from the individual word embeddings.

---

[1] http://www.nltk.org.

[2] https://lucene.apache.org/core/.

The difference consists in the way these representations are combined to determine the score for the current question-candidate answer pair and the context. One version uses a two-step approach in which the context and the question embeddings are combined with a dense layer and the result is then combined with the answer in order to make the prediction (see Fig. 1a). The other one uses a symmetrical architecture, where the context is combined with both the question and the answer separately, and then the two results are used to make a prediction, as shown in Fig. 1b.

4   Results

The dataset used for training the deep learning models and for assessing the performance of all solutions was proposed for The Allen AI Science Challenge[3]. The corpus was available only for the users enrolled in the competition and it is not publicly available. In addition, we also used another dataset from Allen AI containing science questions from elementary and middle school levels. Together, the 2 datasets contain approximately 5,000 questions, each having 4 choices and one correct answer. Tests were made on both the original dataset with 2,500 questions proposed for the challenge and on the extended 5,000 questions corpus.

Because only one of the 4 answers is correct, the training set had 3 negative samples for each positive. In order to have a balanced training set, we have randomly oversampled from the positive class. Thus, for 2,500 questions and 10,000 question-candidate answer pairs, we obtained 15,000 entries in the dataset, half positive and half negative. However, the entire dataset was split into separate questions used for training, validation and testing with a ratio of 85:10:5 (resulting in 2,125 questions for train, 250 for validation, and 125 for test).

The only metric we used for evaluation is accuracy (or P@1, the percentage of questions answered correctly). Besides accuracy, we also analyzed some performance aspects: the number of epochs necessary for the network to learn and the time required for a training epoch for each architecture. As a first results, all models converged after being trained for 50 epochs, using minibatches of 256.

The CNN models used one layer of convolutional filters of different sizes (3/5 for the answer, 5/7 for the question, and 9 for the context) followed by a max-pooling layer. The BiLSTM models used a 60-dimensional hidden state for question, answer, and context. The text representation is obtained by concatenating the last output from the forward and backward LSTMs. In all cases, the candidate context was generated by concatenating the first 5 hits returned by the Lucene index. All these hyperparameters were chosen to maximize the accuracy on the validation set. Based on this metric, we also decided to use a dense layer instead of cosine for the deep learning baselines.

All solutions, including the IR baseline, compute a score for each question-candidate answer pair and the selected answer is the one with the highest score.

---

[3] More info online at The Allen AI Science Challenge, https://www.kaggle.com/c/the-allen-ai-science-challenge.

Table 1. Evaluation of the proposed methods on the test dataset. Training time is measured per epoch for deep learning approaches. The training was done on a Nvidia 960M GPU.

Model	Parameters	Training time	Accuracy
Models without context			
IR baseline	N/A	N/A	27%
BiLSTM	347000	14 s	39%
CNN	394000	0.7 s	40%
Models with context			
Symmetrical BiLSTM	568000	78 s	41%
Symmetrical CNN	2420000	2 s	42%
Two-step BiLSTM	544000	76 s	42%
Two-step CNN	1436000	2 s	**44%**

As the entire dataset is rather small, training deep learning solutions can be a problem even when oversampling the minority (positive) class. Thus, the test set contains only 125 questions (the first ones in the dataset).

As can be easily seen in Table 1, all neural architectures (CNNs, biLSTMs) with candidate context performed better than their corresponding baselines. Moreover, for this dataset CNNs outperform LSTMs without attention which is in line with previous research using deep learning for answer selection [15]. Related to the two proposed architectures using candidate context, the best result is provided by the two-step model probably due to the smaller number of parameters. Thus, the two-step CNN model achieves 44% accuracy on the test set thus significantly improving both the IR and deep learning baselines.

We also trained our models on the extended dataset with 5, 000 questions. However, we noticed a decrease in performance in this case. This probably happens because we tested only on the 125 questions from the original dataset. Unfortunately, comparisons to prior work are not very relevant. The top reported results are 44% for a private 1, 000 questions test set [8] and 47.84% for using the last 500 questions in the current dataset for test [13]. However, both works use a multitude of additional knowledge sources and hand-crafted features.

5   Conclusions

In this paper we introduced a novel deep learning architecture aimed at answering non-factoid multiple choice questions where the answers are rather short and the correct answer cannot be determined from the question-answer pairs alone. The main contribution of the neural models is that they incorporate a method of combining contextual information, retrieved from a Lucene index of Wikipedia for each question-answer pair, with existing deep learning models used in QA.

The results show that the best neural architectures with candidate context provide a relative increase over all baseline of over 10%, proving that the context information allows the systems to have a better representation of each question-answer pair. More, CNNs proved to be better than non-attentional LSTMs, with or without candidate context.

The proposed architecture can be improved either by employing more complex models for computing the similarity between the question and candidate answer, transformed through the context, or by determining a better way to choose and to represent the candidate context. However, training more complex models will not be possible without larger datasets.

References

1. Clark, P.: Elementary school science and math tests as a driver for AI: take the Aristo challenge! In: AAAI, pp. 4019–4021 (2015)
2. Clark, P., Etzioni, O., Khot, T., Sabharwal, A., Tafjord, O., Turney, P.D., Khashabi, D.: Combining retrieval, statistics, and inference to answer elementary science questions. In: AAAI, pp. 2580–2586 (2016)
3. Fader, A., Zettlemoyer, L., Etzioni, O.: Open question answering over curated and extracted knowledge bases. In: Proceedings of the 20th ACM SIGKDD Conference on Knowledge Discovery and Data Mining, pp. 1156–1165. ACM (2014)
4. Ferrucci, D., Brown, E., Chu-Carroll, J., Fan, J., Gondek, D., Kalyanpur, A.A., Lally, A., Murdock, J.W., Nyberg, E., Prager, J., et al.: Building watson: an overview of the deepQA project. AI Mag. **31**(3), 59–79 (2010)
5. Graves, A., Schmidhuber, J.: Framewise phoneme classification with bidirectional LSTM and other neural network architectures. Neural Netw. **18**(5), 602–610 (2005)
6. Hochreiter, S., Schmidhuber, J.: Long short-term memory. Neural Comput. **9**(8), 1735–1780 (1997)
7. Iyyer, M., Boyd-Graber, J.L., Claudino, L.M.B., Socher, R., Daumé III, H.: A neural network for factoid question answering over paragraphs. In: EMNLP, pp. 633–644 (2014)
8. Jansen, P., Sharp, R., Surdeanu, M., Clark, P.: Framing QA as building and ranking intersentence answer justifications. Comput. Linguist. **43**, 407–449 (2017)
9. Ko, J., Nyberg, E., Si, L.: A probabilistic graphical model for joint answer ranking in question answering. In: Proceedings of SIGIR 2007, pp. 343–350. ACM (2007)
10. Lin, J., Zhou, X., Hu, B., Xiang, Y., Wang, X.: ICRC-HIT: a deep learning based comment sequence labeling system for answer selection challenge. In: SemEval-2015, p. 210 (2015)
11. Oh, J.H., Torisawa, K., Kruengkrai, C., Iida, R., Kloetzer, J.: Multi-column convolutional neural networks with causality-attention for why-question answering. In: Proceedings of the Tenth ACM International Conference on Web Search and Data Mining, pp. 415–424. ACM (2017)
12. Pennington, J., Socher, R., Manning, C.D.: GloVe: global vectors for word representation. In: Proceedings of the 2014 Conference on Empirical Methods in Natural Language Processing, pp. 1532–1543 (2014)
13. Sachan, M., Dubey, A., Xing, E.P.: Science question answering using instructional materials. arXiv preprint arXiv:1602.04375 (2016)
14. Severyn, A., Moschitti, A.: Learning to rank short text pairs with convolutional deep neural networks. In: Proceedings of the SIGIR 2015, pp. 373–382. ACM (2015)
15. Tan, M., Santos, C.D., Xiang, B., Zhou, B.: LSTM-based deep learning models for non-factoid answer selection. arXiv preprint arXiv:1511.04108 (2015)

A Text Feature Based Automatic Keyword Extraction Method for Single Documents

Ricardo Campos[1,2]($\boxtimes$) (iD), Vítor Mangaravite[2] (iD), Arian Pasquali[2] (iD),
Alípio Mário Jorge[2,3] (iD), Célia Nunes[4] (iD), and Adam Jatowt[5] (iD)

[1] Polytechnic Institute of Tomar, Tomar, Portugal
ricardo.campos@ipt.pt
[2] LIAAD – INESC TEC, Porto, Portugal
{vima,arrp}@inesctec.pt
[3] DCC – FCUP, University of Porto, Porto, Portugal
amjorge@fc.up.pt
[4] University of Beira Interior, Covilhã, Portugal
celian@ubi.pt
[5] Kyoto University, Kyoto, Japan
adam@dl.kuis.kyoto-u.ac.jp

Abstract. In this work, we propose a lightweight approach for keyword extraction and ranking based on an unsupervised methodology to select the most important keywords of a single document. To understand the merits of our proposal, we compare it against RAKE, TextRank and SingleRank methods (three well-known unsupervised approaches) and the baseline TF.IDF, over four different collections to illustrate the generality of our approach. The experimental results suggest that extracting keywords from documents using our method results in a superior effectiveness when compared to similar approaches.

Keywords: Keyword extraction · Information extraction · Feature extraction

1 Introduction and Related Work

With the massive explosion of data, manually processing documents turned out to be an impossible task. As a direct consequence, several automatic solutions have emerged over the last few years, some following a supervised approach, of which a well-known example is KEA [11], others following an unsupervised methodology [5–7, 9], with TextRank [7], Rake [8], and SingleRank [10] being probably the most well-known solutions. Although the above-mentioned works offer first insights into how this problem can be answered, the task of extracting keywords is yet to be solved. In this work, we present an alternative approach that attempts to overcome the results of the above-mentioned works, while not being dependent on an external source or on linguistic tools. We follow an unsupervised methodology supported by a heuristic approach, which can easily scale to different collections, domains, and languages in a short time span. Our contributions are as follows: (1) we propose an unsupervised keyword extraction method named Yake! which builds upon text statistical features, to extract keywords (both single-word and multi-word terms) from single documents,

© Springer International Publishing AG, part of Springer Nature 2018
G. Pasi et al. (Eds.): ECIR 2018, LNCS 10772, pp. 684–691, 2018.
https://doi.org/10.1007/978-3-319-76941-7_63

thus without the need to rely on a document collection; (2) YAKE! may work for domains and languages for which there are no ready methods as it neither requires a training corpora, nor it depends on any external sources (such as WordNet) or linguistic tools (e.g., NER or PoS taggers).

2 YAKE! Architecture

The proposed method has four main components: (1) Text pre-processing; (2) Feature extraction; (3) Individual term weighting; (4) Candidate keywords generation.

2.1 Text Pre-processing

In the text pre-processing phase, we apply a tokenization process which splits the text into individual terms whenever an empty space or a special character (e.g., brackets, comma, period, etc.) delimiter is found.

2.2 Feature Extraction

Second, we devise a set of five features to capture the characteristics of each individual term. Although these features may be applied to any language, they are particularly suited to Western ones for which some characteristics we devise are particularly tuned. This is the case of *word casing*, which, in Western languages, reflects an important signal of a word. The features we considered are: (1) *Casing*; (2) *Word Position*; (3) *Word Frequency*; (4) *Word Relatedness to Context*; and (5) *Word DifSentence*. A more detailed study of the individual contribution of each of these features, will be conducted in the future. In the following we shortly describe each one of them.

2.2.1 Casing (W_{Case})
In this work, we give particular attention to any word starting with a capital letter (excluding ones at the beginning of sentences) or to any acronym (that is, where all letters of the word are capital) under the assumption that these words tend to be more relevant. Instead of counting them twice we only consider the maximum occurrence within the two of them. Equation 1 reflects this casing aspect:

$$W_{Case} = \frac{\max(TF(U(w)), TF(A(w)))}{\log_2(TF(w))} \tag{1}$$

where $TF(U(w))$ is the number of times the candidate word w starts with an uppercase letter, $TF(A(w))$ is the number of times the candidate word w is marked as an acronym and $TF(w)$ is the frequency of *w*.

2.2.2 Word Position ($W_{Position}$)
Considering the positions of the sentence where the word occurs may be an important feature for the keyword extraction process as the early parts of documents (especially, scientific and news publications) tend to contain a high rate of relevant keywords. We calculate this weight using the following Equation:

$$W_{\text{Position}} = \log_2(\log_2(2 + Median(Sen_w))) \tag{2}$$

where Sen_w indicates the positions of the set of sentences where the word w occurs, and Median is the median function. The result is an increasing function, where values tend to increase smoothly as words are positioned at the end of the document, meaning that the more often a word occurs at the beginning of a document the less its W_{Position} value. Conversely, words positioned more often at the end of the document (likely less relevant) will be given a higher $W_{\text{Positional}}$ value. Note that a value of 2, is considered in the equation to guarantee that $W_{\text{Positional}} > 0$.

2.2.3 Word Frequency (W_{Freq})

This feature indicates the frequency of the word w within the document. It reflects the belief that the higher the frequency, the more important the word is. To prevent a bias towards high-frequency in long documents the TF value of a word w is divided by the mean of the frequencies (MeanTF) plus one time their standard deviation (σ) as in Eq. 3. Our purpose is to score all those words that are above the mean of the terms (balanced by the degree of dispersion given by the standard deviation).

$$W_{\text{Freq}} = \frac{\text{TF}(w)}{\text{MeanTF} + 1 * \sigma} \tag{3}$$

2.2.4 Word Relatedness to Context (W_{Rel})

W_{Rel} quantifies the extent to which a word resembles the characteristics of a stopword. To compute this measure, we resort to the number of different terms that occur in a window of size n to the left (and right) side of the candidate word. The more the number of different terms that co-occur with the candidate word (on both sides), the more meaningless the candidate word is likely to be. W_{Rel} is defined in Eq. 4:

$$W_{\text{Rel}} = \left(0.5 + \left(\left(\text{WL} * \frac{\text{TF}(w)}{\text{MaxTF}}\right) + \text{PL}\right)\right) + \left(0.5 + \left(\left(\text{WR} * \frac{\text{TF}(w)}{\text{MaxTF}}\right) + \text{PR}\right)\right) \tag{4}$$

More precisely, WL [WR] measures the ratio between the number of different words that co-occur with the candidate word (on the left [right] hand side) and the number of words that it co-occurs with. TF(w) is the frequency of the word with regards to the maximum term frequency within all words (MaxTF), and PL [PR] measures the ratio between the number of different words that co-occur with the candidate word (on the left [right] hand side) and the MaxTF. In practical terms, the more insignificant the candidate word is, the higher the score of this feature will be. Thus, stopwords-like terms will easily obtain higher scores.

2.2.5 Word DifSentence ($W_{\text{DifSentence}}$)

This feature quantifies how often a candidate word appears within different sentences. It is computed using the following equation:

$$W_{DifSentence} = \frac{SF(w)}{\#Sentences} \tag{5}$$

where $SF(w)$ is the sentence frequency of the word (w), i.e., the number of sentences where (w) appears, and $\#Sentences$ is the total number of sentences in the text.

2.3 Individual Term Weighting

In the third step, we heuristically combine all these features into a single measure (see Eq. 6) such that each term is assigned a weight $S(w)$. The smaller the value $S(w)$, the more important the word (w) would be. This weight will feed the process of generating keywords to be explained in the next section.

$$S(w) = \frac{W_{Rel} * W_{Position}}{W_{Case} + \frac{W_{Freq}}{W_{Rel}} + \frac{W_{DifSentence}}{W_{Rel}}} \tag{6}$$

By looking at the equation, we can observe that both W_{Freq} and $W_{DifSentence}$ are offset by W_{Rel}. The motivation behind this offset is to assign a high weight to words that appear frequently and appear in many sentences (likely indicative of their importance) as long as the word is relevant (i.e., for which W_{Rel} is low). Indeed, some words may occur plenty of times and in many sentences and yet be useless (e.g., stopwords or similar). These terms should be penalized. Likewise, the position of a word in sentences occurring at the top of a document is an important feature that is taken into account, when multiplying $W_{Rel} * W_{Position}$.

2.4 Candidate Keyword List Generation

The fact that a keyword may consist of more than one word, forces us to consider a further step where the final score of a keyword (be it one, two or n-terms) is determined. To collect the candidate keywords, we consider a sliding window of 3-g, generating a contiguous sequence of 1, 2 and 3-g candidate keywords. In addition, keywords beginning or ending with a stopword will not be considered. It is also important to mention that no conditions are set in respect to the minimum frequency or sentence frequency that a candidate keyword must have. This means that we can have a keyword considered as significant/insignificant with either one occurrence or with multiple occurrences. Each candidate keyword will then be assigned a final $S(kw)$, such that the smaller the score the more meaningful the keyword will be. Equation 7 formalizes this:

$$S(kw) = \frac{\prod_{w \in kw} S(w)}{TF(kw) * \left(1 + \sum_{w \in kw} S(w)\right)} \tag{7}$$

where $S(kw)$ is the score of a candidate keyword with a maximum size of 3 terms, determined by multiplying (in the numerator) the score of the first term of the candidate keyword by the subsequent scores of the remaining terms, such that the smaller this multiplication the more meaningful the keyword will be. This is divided by the sum of the S(w) scores to average out with respect to the length of the keyword, such that longer n-grams do not benefit just because they have a higher n. The result is further divided by TF(kw)- term frequency of the keyword - to penalize less frequent candidates. The final step in determining suitable candidate keywords is to eliminate similar candidates. For this, we use the *Levenshtein distance* [4] which measures the similarity between two strings. Among the strings considered similar (in our case, two strings are considered similar if their Levenshtein distance is above a given threshold) we keep the one that has the lowest S(kw) score. Finally, the method will output a list of potential relevant keywords, formed by 1, 2 or 3-g, such that the lower the S(kw) score the more important the keyword will be.

3   Evaluation

To evaluate the effectiveness of our method and its generality, we tested it on 4 different datasets characterized by different sizes of documents, data and languages: SemEval2010 [3], 500N-KPCrowd-v1.1 [5], WICC [1] and Schutz2008 [9]. SemEval2010 [3], which is probably one of the most well-known collections in this kind of evaluation, consists of 244 full scientific computer science papers ranging from 6 to 8 pages collected from ACM (8,020 tokens per document on average, the longest documents used in our experiments). Schutz2008 [9] in turn, consists of 1,231 papers, but this time belonging to the medical domain (selected from PubMed Central). A different collection is 500N-KPCrowd-v1.1 [5], which despite containing short documents (393 tokens per document on average) represents a different type of data: 500 English broadcast news stories from 10 different categories. Finally, WICC [2], is a Spanish dataset composed of 1,640 computer scientific articles published between 1999 and 2012 (which makes this not only the largest collection among all the datasets considered, but also a different one due to its language). In our experiments, we retrieve keywords with a maximum keyword size of 3-g and make use of a stopword corpus list. In addition, we consider a Levenshtein threshold of 0.8. To have a fair evaluation, we compare our method against TextRank[1] [7], RAKE[2] [8] and SingleRank (See footnote 2) [10], which are the state-of-the-art of unsupervised approaches. In addition, we also compare against TF.IDF (See footnote 2) which, despite being unsupervised, demands the existence of more than one document. Note that, unlike our method, TextRank, SingleRank and the implementation we used for TF.IDF make use of a PoS tagger. A python implementation of YAKE! is also available at PyPi[3]. This will enable researchers not only to test our method but also to compare their approach against ours, thus guaranteeing the reproducibility of the research. An online version and API of our method is also available here: http://bit.ly/YakeDemoECIR2018 [2].

---

[1] Implementation available at http://www.hlt.utdallas.edu/~saidul/code.html.

[2] Implementation available at https://github.com/zelandiya/RAKE-tutorial.

[3] Implementation available at https://pypi.python.org/pypi/yake.

3.1 Results

For the task of evaluating our proposal, we follow the traditional match evaluation scheme. That is, for each single document, we exactly match the keywords in the ground truth with those retrieved by tested methods, and calculate precision, recall, and F1-score. In the experiments, we assess the effectiveness over top 10 keywords retrieved by each method under f different collections to study the effect of document length, different types of data and languages. Table 1 presents the results for all datasets. We apply a paired sample t-test considering a significance level of 0.05. ▼ indicates a statistically significant improvement of the results of YAKE! method over corresponding baselines.

Table 1. SemEval2010, Schutz2008, 500N-KPCrowd and WICC results

Method	SemEval2010			Schutz2008			500N-KPCrowd			WICC		
	P	R	F1	P	R	F1	P	R	F1	P	R	F1
YAKE!	0.153	0.103	0.123	0.217	0.058	0.091	0.251	0.063	0.101	0.050	0.141	0.073
TextRank	0.101▼	0.067▼	0.081▼	0.198▼	0.052▼	0.082▼	0.265	0.063	0.103	0.018▼	0.058▼	0.027▼
TF.IDF	0.036▼	0.023▼	0.028▼	0.100▼	0.028▼	0.043▼	0.223▼	0.060▼	0.095▼	0.026▼	0.067▼	0.037▼
SingleRank	0.035▼	0.022▼	0.027▼	0.082▼	0.024▼	0.037▼	0.190▼	0.054▼	0.084▼	0.017▼	0.045▼	0.024▼
RAKE	0.007▼	0.004▼	0.005▼	0.013▼	0.004▼	0.006▼	0.120▼	0.038▼	0.058▼	0.004▼	0.012▼	0.006▼

The results illustrate the effectiveness of the proposed method, with YAKE! achieving both higher precision, recall and F1-M in comparison to the baselines. Overall, one can note that except for the 500N dataset and the TextRank method, for which the t-test does not show any significant improvement of one method over another, all the remaining results reflect the fact that YAKE! method achieves better and statistically significant results. Among all the datasets, the best results are achieved in the 500N dataset by TextRank with 0.265 for P@10, followed very closely by YAKE! with 0.251. However, as previously referred there is no statistically significant difference between any of the two methods, meaning that they are rather equivalent, beyond being evidently superior when compared to the remaining methods.

The results further confirm that regardless the type of data, YAKE! method tends to have relatively stable effectiveness. To study this effect, one can look at the results of Schutz2008, which in contrast to 500N (a collection of broadcast news) is composed of full text research articles. Despite a slight drop, results are still relatively good with YAKE! achieving 0.217 of P@10 still significantly better than the 0.198 achieved by the TextRank method.

The effect of the document length is then studied by running our experiments under the SemEval2010 collection (the largest documents here studied −8.020 per document on average, twice the double of the Schtuz2008 collection and 25 times more than the 500N). Although the results have dropped considerably, still, they are significantly better than the ones of the second-best approach (TextRank), which only achieves 0.101 of P@10 (significantly lower than 0.154 obtained by our method). This proves that, although our method performs better than any of the baselines, still the effect of document length significantly impacts the results obtained. This should be studied in the future. It is also important to stress out that, while TextRank depends and benefits from NLP techniques, such as PoS taggers, YAKE! simply takes as input a set of plain features extracted from the text. These may be understood as an advantage over baselines anchored on PoS taggers, which may not be available or may perform poorly for some minor languages for which there is either a lack of interest or lack of open source tools.

Finally, we wanted to evaluate the effectiveness of our method under a different language. To this regard we consider the WIC dataset. Once again YAKE! returns the best results, although in this case a score of only 0.05 of P@10 has been obtained, which has much to do with the fact that only 3.57 of gold keywords per document have been defined in the collection (the smallest number of gold keywords among all the datasets). While, one may be tempted to claim that the results are quite low when compared to other IR tasks, it should be taken into account that unlike other IR core research areas, the realm of keyword extraction is a different one, with the tendency to have lower scores. One of the reasons for this is that an exact match between the ground-truth and the methods keyword is usually used as a rule-of-thumb thus impeding partial matches. An additional reason is that some of the keywords of the ground-truth cannot simply be found in the text, thus making it impossible to have an exact match. Thus, any increase in the effectiveness of current solutions, would always represent a significant contribution over state-of-the-art solutions.

4 Conclusions

In this paper, we propose a novel approach to extract keywords from single documents. Based on the experiments, we could confirm that YAKE! achieves better results in comparison to four state-of-the-art unsupervised keyword extraction algorithms, over a large number of text documents in four different datasets. Unlike supervised approaches, which require a training corpus, YAKE! is fully unsupervised. Moreover, the fact that it only leverages features drawn from the text itself together with its independence with regards to natural language processing techniques makes it suitable for other text collections, including different domains and languages. As future work, we plan to investigate how our method performs in comparison with the most popular supervised approaches like KEA [11].

Acknowledgements. This work is partially funded by the ERDF through the COMPETE 2020 Programme within project POCI-01-0145-FEDER-006961, and by National Funds through the FCT as part of project UID/EEA/50014/2013 and of project UID/MAT/00212/2013. It was also financed by MIC SCOPE (171507010) and by Project "TEC4Growth - Pervasive Intelligence, Enhancers and Proofs of Concept with Industrial Impact/NORTE-01-0145-FEDER-000020" which is financed by the NORTE 2020, under the PORTUGAL 2020, and through the ERDF.

References

1. Aquino, G., Lanzarini, L.: Keyword identification in Spanish documents using neural networks. J. Comput. Sci. Technol. **15**(2), 55–60 (2015)
2. Campos, R., Mangaravite, V., Pasquali, A., Jorge, A., Nunes, C., Jatowt, A.: YAKE! collection-independent automatic keyword extractor. In: Pasi, G., Piwowarski, B., Azzopardi, L., Hanbury, A. (eds.) ECIR 2018, LNCS, vol. 10772, pp. 806–810. Springer, Cham (2018)
3. Kim, S., Medelyan, O., Kan, M.-Y., Baldwin, T.: SemEval-2010 task 5: automatic keyphrase extraction from scientific articles. In: SemEval 2010, Sweden, pp. 21–26 (2010)
4. Levenshtein, V.: Binary codes capable of correcting deletions, insertions, and reversals. Sov. Phys. Dokl. **10**(8), 707–710 (1966)
5. Marujo, L., Viveiros, M., Neto, J.: Keyphrase cloud generation of broadcast news. In: arXiv (2013)
6. Matsuo, Y., Ishizuka, M.: Keyword extraction from a single document using word co-occurrence statistical information. J. Artif. Intell. Tools **13**(1), 157–169 (2004)
7. Mihalcea, R., Tarau, P.: TextRank: bringing order into texts. In: EMNLP 2004, pp. 404–411 (2004)
8. Rose, S., Engel, D., Cramer, N., Cowley, W.: Automatic Keyword Extraction from Individual Documents. Text Mining: Theory and Applications. Wiley, Chichester (2010)
9. Schutz, A.T.: Keyphrase extraction from single documents in the open domain exploiting linguistic and statistical methods. Master thesis, National University of Ireland (2008)
10. Wan, X., Xiao, J.: Single document keyphrase extraction using neighborhood knowledge. In: AAAI 2008, 13–17 July, pp. 855–860 (2008)
11. Witten, I., Paynter, G., Frank, E., Gutwin, C., Nevill-Manning, C.: KEA: practical automatic keyphrase extraction. In: Proceedings of the JCDL 2004, 7–11 June, pp. 254–255 (1999)

Simplified Hybrid Approach for Detection of Semantic Orientations in Economic Texts

Jan Štihec[1], Martin Žnidaršič[2], and Senja Pollak[2(✉)]

[1] Faculty of Economics, University of Ljubljana, Ljubljana, Slovenia
stihec.jan@gmail.com
[2] Jožef Stefan Institute, Ljubljana, Slovenia
{martin.znidarsic,senja.pollak}@ijs.si

Abstract. The aim of this work is to reproduce the approach to detecting semantic orientations in economic texts that was presented in the paper Good Debt or Bad Debt: Detecting Semantic Orientations in Economic Texts by Malo et al. The approach employs the Linearized Phrase Structure model for sentence level classification of short economic texts into a positive, negative or neutral category from investor's perspective and yields state-of-the-art results. The proposed method employs both rule based linguistic models and machine learning. Where possible we follow the same approach as described in the original paper, with some documented modifications. Our solution is simplified in at least two aspects, but its performance is comparable to the original and overall remains better than the reported results of other benchmark algorithms mentioned in the original paper. The differences between the two models and results are described in detail and lead to conclusion that the original approach is to a large extent repeatable and that our simplified version does not overly sacrifice performance for generalizability.

1 Introduction

The analysis of market sentiments and of approaches to measuring them is an interesting topic and a subject of many recent studies, e.g. [1–4]. The idea of using media texts as a source of investors' sentiments spurred attempts of predictive approaches, in which financial news are used to roughly predict the evolution of market sentiments and the behavior of market participants [5]. Most automatic sentiment analysis approaches rely on lexicons or word lists as the primary source of sentiment cues. A high quality of domain specific lexicons—in our case from financial domain—is therefore essential for successful sentiment analysis [6,7]. Methods for sentiment analysis that employ machine learning [1,3], on the other hand, do not require lexicons, but rely on availability of annotated datasets. Again, domain specific datasets are crucial for the best possible results.

The paper *Good Debt or Bad Debt: Detecting Semantic Orientations in Economic Texts* by Malo et al. [7] presents an approach to detecting semantic orientation in texts which uses both linguistic models and machine learning

G. Pasi et al. (Eds.): ECIR 2018, LNCS 10772, pp. 692–698, 2018.
https://doi.org/10.1007/978-3-319-76941-7_64

techniques to categorize texts and improves classification performance compared to a selection of benchmark approaches. The algorithm focuses on categorizing short economic texts and benefits from the domain-specific knowledge through usage of lexicons that are adapted to financial domain. Inclusion of finance-specific lexicons and consideration of interactions between financial entities and direction giving expressions are the two main additions of the original work compared to less specific approaches (e.g. [8]).

There are three main contributions of the original paper as described by the authors: (I) a collection of approximately 5,000 human-annotated sentences[1] (financial and economic news texts) which can serve as high quality training data for statistical techniques in financial sentiment analysis, (II) enriched financial domain specific lexicon which consists of general entities that are based on general [9] and financial [6] polarity lexicons, finance specific concepts, directionalities and polarity influencers which modify the polarity of financial concepts. (III) The finance-adapted Linearized Phrase Structure (LPS) model approach.

Our aim was to re-implement the methodology as described in the original paper. Where possible we use the same approach as the authors, but in some aspects our solution differs from the original work. We used the same phrase bank of annotated sentences, but in our experiments we use only the subsets with 100% and >75% annotator agreement. We attempted to achieve similar sentence categorization accuracy with a simpler and more easily repeatable approach. The main difference is that we opted for simple POS tagging and window-based approach instead of one based on syntactic parse trees. The main reason for this decision is related to the generalisability of the reimplemented approach. On the one hand, we plan to adapt the approach also for one of the less-resourced languages, where syntactic sentence parsing is not available, and secondly, we want to cover several text types, including tweets, which, due to informal language, represent a challenge for sentence parsing. Other differences are due to the unavailability of some of the resources that were used by the authors.

This paper will follow our re-implementation steps, present the approach and describe what are the key differences from the original one. Our results using the same data set are compared to the original algorithm and other benchmark algorithms that are listed in the original paper.

2   Following the LPS Approach

The original approach could shortly be described as follows. Firstly the phrase structure is recovered from an incoming sentence. Entity recognizers are then used to transform the phrase structure into a sequence of lexicon entities taking into account the preset entity pruning rules. Each sequence is interpreted as a representative of an equivalence class of phrases with similar features. Lastly a linear multi-label classifier is used to associate the sequences with corresponding

---

[1] The financial phrase bank was annotated by 16 annotators and is split into 4 subsets depending on the level of annotator agreement (100%, >75%, >66%, >50%).

semantic orientations which were indicated by the annotators. In this paper, we present our implementation in the same order of steps.

2.1 Finance Domain Lexicons

Inclusion of finance specific lexicons is a key element of the approach and greatly contributes to its performance. The lexicon collection consists of four entity classes: general entities, financial entities, direction and polarity influencers. The statistics of the number of elements in each category are provided in Table 1.

General entities or general expressions with polarity are obtained from the Multi-perspective Question answering (MPQA) corpus of opinion annotations [9–11]. The MPQA lexicon includes single words with information on degree of subjectivity, prior polarity, part-of-speech and lemma. The MPQA dictionary is merged and augmented with a finance specific wordlist [6,12]. The financial domain sentiment overrides the prior polarity defined in MPQA dictionary when overlaps are encountered. The general entity list is obtained as described in the original paper with the exception of using the updated financial dictionary[2] from 2015 [12], which results in our list being slightly longer (see Table 1).

The second lexicon class is comprised of *financial entities* categorized as positive-if-up or negative-if-up concepts. For instance *operating profit* is defined as a positive-if-up whereas *operating loss* is a negative-if-up concept. Semantic orientation of an entity can depend on its association with other entities. To take this into account, a special financial dictionary is used, which is used to assess how a financial concept's semantic orientation is modified by direction of events. Financial entities list is the point, where our approach differs from the original which uses 347 commonly encountered financial concepts extracted from Investopedia's[3] list of financial terms. Since the list used in the original paper was not available, we manually surveyed the phrase bank sentences for most commonly used financial terms and completed our list[4] this way (193 categorized concepts).

The third and fourth lexicon entity classes are *direction* and *polarity influencers*. As detection of prior polarities is not sufficient for solving phrase level sentiment analysis [11], directional and polarity influencing entities are used to better determine contextual semantic orientation. For example *increase* is a direction entity categorized as -up- and *decrease* is categorized as -down-. Our lexicon includes only directional entities (directionalities), which determine whether the belonging financial entity is increasing or decreasing - going up or down, hence the usage of positive-if-up or negative-if-up concepts. The list of directionalities is acquired from four Harvard IV word lists from the General Inquirer categories; decrease and fall for down terms and increase and rise for up terms. This is the same as described in the original paper, with a slight statistical difference since the authors enriched their list by adding some recently developed expressions which we did not include as the source was not provided.

---

[2] Available at https://www3.nd.edu/~mcdonald/Word_Lists.html.
[3] https://www.investopedia.com/.
[4] Our list is online: http://kt.ijs.si/data/finentities/financial_entities.zip.

The original implementation also includes 4 groups of polarity influencers; reversal, modal, litigious and uncertain. These are not included in our lexicon, since the original paper does not provide their sources neither their exact role in the analysis of the phrase structure.

Table 1. Financial lexicon statistics (LPS refers to the original approach by Malo et al. and reimp-LPS to our approach).

Entity class	reimp-LPS	LPS
General entity	10903	9469
Financial entity	193	347
Direction	272	314
Polarity influencer		435

2.2 Entity Detection

After the financial lexicon construction, the next step in the process is the phrase structure analysis. First, the phrase is traversed by entity-recognizers which aim to produce a sequence of matched lexicon entities. For every candidate expression, which is either a unigram or a n-gram in case of some financial entities and directionalitites, the algorithm searches for a corresponding financial lexicon entry. If an expression is not found in any of the lexicon categories, it is marked as a neutral entity. This process results in every word or a sequence of words in a sentence being represented as one of the following entity types: positive, negative, neutral, positive if up, negative if up, up, down. For example, for sentence (i), the corresponding representation is provided in (ii).

(i) *The number of collection errors fell considerably, and operations speeded up.*
(ii) $['nevt','nevt','nevt','negifup','down','nevt','nevt','posifup','up']$

As seen in the above example, *collection errors* and *operations* are recognized as financial entities while *fell* and *speeded up* are directionalities. Other words are neutral entities.

2.3 Entity Pruning

The second step is the application of the entity pruning rules to the extracted entity sequence. The first rule is to merge successive neutral entities into one neutral entity. The second rule is merging an entity whose polarity is modified by the direction giving entities with directionalities into an entity with modified polarity. The original paper uses a phrase tree parser to apply the correct influencer to the polarity modified entity. We used a simpler approach: a set word length window in which, if both are present, a polarity influencer and an entity whose polarity is modified by the influencer, are merged into an entity with modified polarity. The window size parameter was manually experimentally

Table 2. Label distribution and accuracy comparison.

	label	label dst.	reimp-LPS	LPS	SVM-MPQA	W-Loughran
100% agreement	positive	0.252	0.742	0.869	0.746	0.755
	negative	0.134	0.71	0.951	0.87	0.849
	neutral	0.614	0.869	0.828	0.652	0.625
	overall	**1.000**	**0.816**	**0.855**	**0.704**	**0.688**
>75% agreement	positive	0.257	0.717	0.836	0.744	0.758
	negative	0.122	0.684	0.945	0.886	0.863
	neutral	0.621	0.842	0.792	0.657	0.636
	overall	**1.000**	**0.787**	**0.821**	**0.707**	**0.695**

assessed on a limited subset of sentences and set to seven. The nearest polarity influencer is applied, if there are multiple occurrences in the window.

After the entity pruning step the algorithm produces a sequence of the following possible entity types; positive, negative, neutral, negative up, negative down, positive up, positive down. The last transformation for the above example (i) and its entity representation (ii), is provided in example (iii) below. As it can be seen, successive neutral entities are merged into one and both financial entities are merged with matching directionalities into polarity modified entities.

(iii) $['nevt','negdown','nevt','posup']$

2.4   Learning Mechanism

After all the incoming sentences are represented as sequences of entities, we have to correctly associate the representations with their semantic orientation as provided in the annotations. To solve this problem we use the multiclass SVM approach with one-versus-one strategy, as proposed in the original paper. We code each entity type as a bit sequence of length m, where m is the number of alternative entity types. For the purpose of learning, each incoming sentence is represented as a bit sequence of length equal to the maximum number of entities in a phrase times the number of alternative entity types. To achieve the same number of entities, we add neutral entities at the end of each phrase as necessary.

2.5   Results

We tested our algorithm using the same data set that was used in the original paper, specifically on the datasets of sentences with 100% and 75% annotator agreement (2,259 and 3,448 items, respectively). The label distribution is shown in Table 2. All results are computed using 10-fold cross-validation as in the original paper. We could not, however, exactly reproduce data separation into folds (which data item is in exactly which fold). Table 2 shows our results compared to the results reported for the original implementation and the two benchmarks.

The results show that our reimplementation is performing slightly worse in terms of overall accuracy, which can be contributed to not using the polarity influencers category, employing the word length window approach and considering less financial entities[5]. Our algorithm appears to be more accurate at categorizing neutral sentences and less accurate at negative and positive sentence detection.

3   Conclusion

The availability of an annotated dataset that allows comparison of results and good performance of the approach that was presented in the original paper were two main reasons that inspired our attempt to re-implement the algorithm. For the most part the original approach is well described and easy to repeat. The authors have provided the financial phrase bank dataset and detailed information about most of the financial lexicon, which can be constructed using the same lists and dictionaries as in the original paper. In parts, where a complete reproducibility was not possible, we used our own best solutions. The first key difference is in the financial lexicon, where the polarity influencing entities were left out, since their list was not made available by the authors and the financial entities, which we manually extracted from the phrase bank in contrast to the authors which extracted their list from Investopedia. The second notable difference is the choice of not using a parser, when extracting entities from incoming sentences. For the reasons of language and text type generalizability we opted for the window based approach on a POS tagged corpus. Differences in the implementation do contribute to different accuracy scores when comparing the two algorithms, but also make our approach easier to repeat and reproduce.

Acknowledgments. We acknowledge financial support from the Slovenian Research Agency for research core funding No. P2-0103 and the research project *Influence of formal and informal corporate communications on capital markets*, No. J5-7387.

References

1. Oh, C., Sheng, O.: Investigating predictive power of stock micro blog sentiment in forecasting future stock price directional movement. In: ICIS (2011)
2. Bollen, J., Mao, H., Pepe, A.: Modeling public mood and emotion: Twitter sentiment and socio-economic phenomena. In: ICWSM 2011, pp. 450–453 (2011)
3. Smailović, J., Grčar, M., Lavrač, N., Žnidaršič, M.: Stream-based active learning for sentiment analysis in the financial domain. Inf. Sci. **285**, 181–203 (2014)
4. Cortis, K., Freitas, A., Daudert, T., Huerlimann, M., Zarrouk, M., Handschuh, S., Davis, B.: SemEval-2017 Task 5: Fine-grained sentiment analysis on financial microblogs and news. In: Proceedings of the 11th International Workshop on Semantic Evaluation (SemEval-2017), pp. 519–535 (2017)

---

[5] We tested the effect of a smaller set of entities by experimenting with randomly halved set, which on average (10 runs) caused the accuracy to drop for 6.2%.

5. Mitra, G., Mitra, L.: The Handbook of News Analytics in Finance, vol. 596. Wiley, Hoboken (2011)
6. Loughran, T., McDonald, B.: When is a liability not a liability? Textual analysis, dictionaries, and 10-Ks. J. Finan. **66**(1), 35–65 (2011)
7. Malo, P., Sinha, A., Korhonen, P., Wallenius, J., Takala, P.: Good debt or bad debt: Detecting semantic orientations in economic texts. J. Assoc. Inf. Sci. Technol. **65**(4), 782–796 (2014)
8. Moilanen, K., Pulman, S.: Sentiment composition. In: Proceedings of RANLP, vol. 7, pp. 378–382 (2007)
9. Wiebe, J., Wilson, T., Cardie, C.: Annotating expressions of opinions and emotions in language. Lang. Resour. Eval. **39**(2), 165–210 (2005)
10. Wilson, T.A.: Fine-grained subjectivity and sentiment analysis: Recognizing the intensity, polarity, and attitudes of private states. University of Pittsburgh (2008)
11. Wilson, T., Wiebe, J., Hoffmann, P.: Recognizing contextual polarity: an exploration of features for phrase-level sentiment analysis. Comput. Linguist. **35**(3), 399–433 (2009)
12. Bodnaruk, A., Loughran, T., McDonald, B.: Using 10-K text to gauge financial constraints. J. Financ. Quant. Anal. **50**(4), 623–646 (2015)

Towards Maximising Openness in Digital Sensitivity Review Using Reviewing Time Predictions

Graham McDonald$^{(\boxtimes)}$ (ID), Craig Macdonald (ID), and Iadh Ounis (ID)

University of Glasgow, Glasgow G12 8QQ, UK
g.mcdonald.1@research.gla.ac.uk,
{Craig.Macdonald,Iadh.Ounis}@glasgow.ac.uk

Abstract. The adoption of born-digital documents, such as email, by governments, such as in the UK and USA, has resulted in a large backlog of born-digital documents that must be *sensitivity reviewed* before they can be *opened* to the public, to ensure that no sensitive information is released, e.g. personal or confidential information. However, it is not practical to review all of the backlog with the available reviewing resources and, therefore, there is a need for automatic techniques to increase the number of documents that can be opened within a fixed reviewing time budget. In this paper, we conduct a user study and use the log data to build models to predict reviewing times for an average sensitivity reviewer. Moreover, we show that using our reviewing time predictions to select the order that documents are reviewed can markedly increase the ratio of reviewed documents that are released to the public, e.g. +30% for collections with high levels of sensitivity, compared to reviewing by shortest document first. This, in turn, increases the total number of documents that are opened to the public within a fixed reviewing time budget, e.g. an extra 200 documents in 100 hours reviewing.

1 Introduction

Sensitivity review is the manual process of reviewing government documents that are to be transferred, or *opened*, to the public domain, to ensure that no *sensitive* information is released, e.g. personal or confidential information. However, existing sensitivity review processes are not practical for the review of born-digital documents, such as email, due to the volume of documents that are created. For example, in the UK, some government departments have reported having a backlog of 190 TB of emails [1][1]. A significant portion of this backlog will be selected for transfer to the public archive and, hence, will need to be sensitivity reviewed.

Technology assisted review (TAR), most notably associated with e-discovery [2], has the potential to alleviate some of the barriers to digital sensitivity

---

[1] In the UK, fifty government departments are expected to transfer born-digital documents to the public archive by 2021 [1].

© Springer International Publishing AG, part of Springer Nature 2018
G. Pasi et al. (Eds.): ECIR 2018, LNCS 10772, pp. 699–706, 2018.
https://doi.org/10.1007/978-3-319-76941-7_65

review [3]. However, it is generally accepted that all government documents that are to be opened will continue to be manually reviewed until reviewers develop trust in TAR technologies [3]. Moreover, even with the adoption of TAR, the volume of documents to be reviewed is expected to be much greater than the available reviewing time [3] and, therefore, there is a need for strategies to prioritise the review of the documents that are most likely to be released, and to increase the overall number of documents that are opened to the public within the available reviewing time budget.

In this work, we conduct a user study and use the log data to study how government archivists sensitivity review born-digital documents. Moreover, we use the reviewers' interactions to predict the time an average reviewer would require to review a specific document. Furthermore, using simulated collections containing varying distributions of sensitive information, we compare the effectiveness of four ranking strategies for maximising openness within an available reviewing time budget. We show that by ranking documents by their predicted reviewing times, we can markedly increase the mean hourly ratio of reviewed documents that are released to the public (+30% for collections with high levels of sensitivity). This, in turn, will enable government departments to release more of the backlog of documents. For example, on a collection in which 70% of documents contain some portion of sensitive information, for 100 hours of reviewing we expect an extra 200 documents to be released. This will substantially increase the total number of documents that can be opened by each government department.

2   Related Work

Assisting the sensitivity review of digital government documents has received some attention in the literature in recent years [4–9]. Most of that work has focused on developing classification algorithms for identifying sensitivity, either at the document level [5,7] or sensitive text within documents [9]. Berardi et al. [8] investigated improving the cost-effectiveness of sensitivity reviewers by deploying a utility-theoretic [10] *semi-automatic* text classification approach to identify a ranking strategy that can maximise the overall classification effectiveness when a reviewer corrects a portion of mis-classified documents, i.e. to minimise the number of mis-classified documents released to the public (when a portion of the released documents are not manually reviewed) by having reviewers review the documents that are most likely to be mis-classified.

Differently from the work of Berardi et al., in this work, we model the time a reviewer is likely to take to review a document, to increase the number of documents that can be released to the public within a fixed reviewing time budget, when all documents that are released must first be manually reviewed.

Predicting reviewing times is a complex task, as there are many variables that can lead to large variations in reviewing times, such as document length, the complexity of documents or a reviewer's reading speed. Jethani and Smucker [11] modeled the average time to review as a function of document length. In that

Table 1. The generated test collection. Document length is measured by number of words. Reviewing time and Normalised Dwell Time (NDT) are measured in seconds.

	docs	% sensitive	Avg. length	Avg. review time	Avg. NDT
Training data	184	9.63	824.6	321.05	297.88
Test data	181	17.4	710.3	385.77	333.38

work, the authors learned a linear model to predict reviewing times and found that the model accounted for 26% of the variance in reviewing times, when a reviewer had to review an entire document to make a decision (as is the case for sensitivity review). This is a relatively good result since, in [11], there is a large variance in the times taken to judge documents of similar lengths. In this work, we also use a linear model to predict document reviewing times. However, differently from Jethani and Smucker, we use the reviewing time predictions to select effective ranking strategies for technology assisted digital sensitivity review.

Damessie *et al.* [12] used a reviewer's dwell time, i.e. the time from a reviewer first viewing a document until the reviewer records a relevance judgment, to study the relationship between the time taken to assess relevance and (1) topic difficulty, (2) the degree of relevance and (3) the presentation order. To normalise for the differences in the reading speeds of reviewers, they proposed *normalised dwell time* (NDT) to measure the reviewing time of an average reviewer. Differently from Damessie *et al.* [12], in this work, we use NDT to predict the number of documents an average reviewer can review within a fixed reviewing time budget and, moreover, to maximise the number of documents that are opened to the public within the available budget.

3 Digital Sensitivity Reviewer Study

Study Design and Participants: 16 volunteers from the official UK government archive were asked to sensitivity review a collection of digital government documents. The volunteers were familiar with sensitivity review, however, they were provided detailed guidance regarding (1) the scope of the task that they were being asked to perform and (2) the software deployed in the task to collect sensitivity reviews.

The collection used in the study contains real sensitivities, as defined by the UK Freedom of Information Act. Reviewers were asked to identify any documents containing personal information or international relations sensitivities[2]. In addition to recording judgements at the document level, reviewers were asked

---

[2] Sections 40 and 27 are representative of the most frequent types of sensitivities in UK government documents. 92% of paper *records* (i.e. documents, photographs, etc.) that were closed between 10/02/05 and 30/04/14 were closed due to Personal or National Interest sensitivities [1].

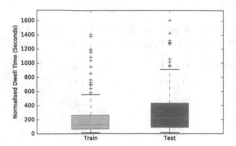

Fig. 1. Normalised Dwell Time (NDT) distributions in seconds for the training and test data.

Table 2. Reviewing time predictions. R^2, adjusted R^2 (R^2_{Adj}) and root mean squared error (RMSE) for the test data predictions.

Feature set	R^2	R^2_{Adj}	RMSE
Decision	0.0537	0.0483	297.72
Surface	0.1095	0.0942	288.81
Complexity	−0.0639	−0.0822	315.68
Decision+Surface	0.2599	**0.2385**	263.29
Decision+Complexity	0.0898	0.0635	291.97
Surface+Complexity	0.1087	0.0722	288.94
All Features	**0.2714**	0.2326	**261.23**

to annotate any sensitive text in a document. Non-sensitive documents could simply be identified as such.

Reviewers were provided a web-based interface to navigate the collection and record sensitivities. To ascertain the duration taken to review, we logged the time when a document was loaded to view, t_0, and when a judgement was saved, t_1. The reviewing time, rt, for a document, d, is then calculated as $rt(d) = t_1 - t_0$. Judgements could also be revisited. For revisited documents, we calculate reviewing time as $rt(d) = \sum_{i=1}^{n} t_{1i} - t_{0i}$, where n is the number of times the document was viewed and judged.

461 documents were reviewed in total. 62 documents were judged as being sensitive and 399 as not-sensitive. The mean number of documents reviewed by a reviewer was 28.8, with a range of 5 to 199 and standard deviation of $\sigma = 45.4$.

Generated Test Collection: We use the collected reviews to generate a test collection for developing our models. To ensure that reviewers had committed to the task, we only included reviews from reviewers who (1) made at least 10 judgements, and (2) recorded sensitivity annotations. This resulted in 11 reviewers contributing to the test collection. Additionally, since we could not control for reviewers taking breaks, we removed documents that took longer than 2 hours to review.

Each reviewer's reviews were ordered by the order that they were judged and we then split the reviews so that the first 50% of a reviewer's reviews contribute to the training data and the later 50% contribute to the test data. Table 1 provides an overview of the training and test data for the generated test collection.

4   Predicting Reviewing Time

Developing the reviewing times prediction model: As a measure of the time that an average reviewer would be expected to take to review a particular document, we deploy an approach proposed by Damessie *et al.* [12] that accounts for variations in reviewers reading speeds, namely *normalised dwell time* (NDT). The NDT for a document, d, is defined as $NDT = exp^{(\log(time)+\mu-\mu_\alpha)}$, where $\log(time)$ is the log of the time taken to review d, μ is the global mean reviewing time calculated over all documents for all reviewers, and μ_α is the mean reviewing time for the reviewer who reviewed d. However, since calculating NDT relies on the means μ and μ_α, we learn a linear regression model to predict a document's NDT using three sets of features, as follows:

The first set of features represent aspects of a reviewer's *decision* process when making a sensitivity judgement: (1) the number of documents that a reviewer has reviewed prior to the current document; and (2) whether the document is sensitive or not[3]. The second set of features are document *surface* features: (1) the number of sentences in a document; (2) total prepositions, such as *at, with* or *from*; (3) total number of syllables; and (4) the ratio of unique words/total words.

The last set of features that we test are standard readability metrics that represent the *complexity*, or reading difficulty, of a document: (1) Simple Measure of Gobbledygook (SMOG) [13] is a simple readability metric based on the number of polysyllabic words per sentence within a 30-sentence sample from a document; (2) the Automated Readability Index (ARI) [14] is a weighted sum of the mean words per sentence and the mean number of characters per word; (3) the Coleman-Liau Index [15] is a weighted sum of the avg. number of characters per 100 words and the average number of sentences per 100 words; (4) the Gunning Fog Index [16] is a weighted sum of the avg. sentence length and the percentage of *complex* words. In total, ten features were used to build our reviewing time prediction model.

Model Effectiveness: Table 2 presents the results of our reviewing time predictions. We select root mean squared error (RMSE) as our main metric as it provides an absolute measure of variance, in seconds, for our predictions. Additionally, we report R^2, defined as $R^2 = 1 - \frac{\sum_i (y_i - \hat{y}_i)^2}{\sum_i (y_i - \bar{y})^2}$, where y is a document's NDT, $\bar{y}$ is the mean NDT of all documents and $\hat{y}$ is a document's predicted NDT. R^2 measures the amount of variation in the data that is explained by the learned model. It has an upper bound of 1, obtained by a perfect model, and can be negative since the model can be arbitrarily worse. We also report adjusted R^2, $R^2_{Adj} = 1 - \frac{(1 - R^2)(n - 1)}{n - k - 1}$, where n is the number of documents and k is the number of features. R^2_{Adj} enables a fair comparison between models with different numbers of features, i.e. when a new feature is added to a model R^2_{Adj} increases only if the model improves more than would be expected by chance.

As can be seen from Table 2, deploying all three feature sets results in a RMSE of 261.23 ($\sim$4 min). 261.23 RMSE provides relatively good predictions since, as can be seen from Fig. 1 which presents the distribution of NDT in the training and test data, although the median NDT in the test data is $\sim$200 seconds, there are many outlier documents with NDT in the range of 600 to 1600 seconds and, therefore, the model performs well at predicting the reviewing time for documents that take longer to review. Table 2 also shows that R^2_{Adj} for our model deploying all three feature sets is 0.23, i.e. 23% of the variance in NDT in the test data is explained by the model. This is in line with the 0.26 R^2_{Adj} observed

---

[3] In a production environment, when predicting a document's reviewing time, this feature must be supplied by a sensitivity classifier, e.g. [7].

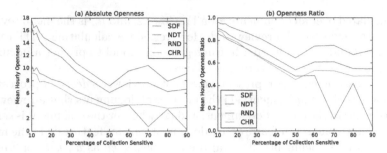

Fig. 2. (a) Number of documents opened per hour. (b) Ratio of reviewed documents opened.

by Jethani and Smucker [11] when reviewers were required to read an entire document to make a relevance judgment. This gives us additional confidence that our model provides relatively good predictions and we, therefore, select this configuration for evaluating ranking strategies in the remainder of this paper.

5   Strategies for Maximising Openness

In this section, we present how our reviewing time prediction model can be used to increase the number of documents that a reviewer can review and release in a given time period, we denote this as the achieved *openness*. Moreover, we evaluate how effective our model is depending on the amount of sensitivity that is in a collection. To do this, we simulate collections with varying distributions of sensitivity by sampling with replacement from the test data to fit the desired sensitivity distribution. We generate nine separate collections, ranging from 10%–90% sensitive data, where for each collection, C, $\sum_{i=0} NDT(d_i) = 1hour$, $d_i \in C$. We select one hour as our reviewing time budget, since it is straightforward to reason about larger time periods from this basis. Moreover, to ensure the generalisability of our findings, we generate 100 example collections for each distribution. Therefore, in this section we report mean values over 100 * 1 hour samples of the test data presented in Sect. 3 and Fig. 1.

We evaluate our shortest predicted reviewing time approach (NDT) against three baseline approaches, namely: random (RND); shortest document first (SDF), this strategy naively assumes that shorter documents take less time to review; and chronological (CHR), a strategy currently deployed by sensitivity reviewers.

Figure 2 presents the effectiveness of each of the four ranking strategies on collections of varying sensitivity distributions. Firstly, from Fig. 2(a), we note that ordering documents by their expected time to review (NDT), consistently results in more documents being released to the public than the next best approach, i.e. shortest document first (SDF). This shows that the complexities of reviewing a document for sensitivity are not strongly correlated with document length. Secondly, we note that the improvements in openness are fairly consistent when < 50% of the collection is sensitive. However, when the collection has high levels of sensitivity, NDT can result in higher relative gains in openness. Figure 2(b) presents the ratio of reviewed documents that were released. As can be seen from Fig. 2(b), for a collection that is 60%–70% sensitive, NDT results in a 30% increase in the ratio of reviewed documents that are actually opened, e.g. for a collection in which 70% of documents contain some portion of sensitive information our NDT ranking strategy would result in an extra 200 documents being released for 100 hours of reviewing time. This, in turn, will enable government departments to substantially reduce the backlog awaiting review by increasing the total number of documents that can be opened to the public within the available reviewing time budget.

6   Conclusions

In this work, we presented an approach for predicting the time taken to sensitivity review digital government documents. Moreover, we showed that by using these reviewing time predictions to select the order that documents are presented to reviewers, we can notably increase the rate at which documents are released to the public. Presenting documents based on their predicted reviewing times resulted in a 30% increase in the proportion of reviewed documents that were released when the collection contained 60%–70% sensitive documents. As future work, we will expand this approach to meet other reviewing objectives, such as quickly identifying specific types of sensitivity.

References

1. TNA: The digital landscape in government 2014–2015: business intelligence review (2016)
2. Oard, D.W., Baron, J.R., Hedin, B., Lewis, D.D., Tomlinson, S.: Evaluation of information retrieval for e-discovery. Artif. Intell. Law **18**(4), 347–386 (2010)
3. TNA: The application of technology-assisted review to born-digital records transfer (2016)
4. Gollins, T., McDonald, G., Macdonald, C., Ounis, I.: On using information retrieval for the selection and sensitivity review of digital public records. In: Proceedings of PIR@SIGIR (2014)
5. McDonald, G., Macdonald, C., Ounis, I., Gollins, T.: Towards a classifier for digital sensitivity review. In: Proceedings of ECIR (2014)
6. Elragal, A., Päivärinta, T.: Opening digital archives and collections with emerging data analytics technology: a research agenda. Tidsskriftet Arkiv **8**(1), 1–15 (2017)
7. McDonald, G., Macdonald, C., Ounis, I.: Enhancing sensitivity classification with semantic features using word embeddings. In: Proceedings of ECIR (2017)
8. Berardi, G., Esuli, A., Macdonald, C., Ounis, I., Sebastiani, F.: Semi-automated text classification for sensitivity identification. In: Proceedings of CIKM (2015)
9. McDonald, G., Macdonald, C., Ounis, I.: Using part-of-speech n-grams for sensitive-text classification. In: Proceedings of ICTIR (2015)
10. Berardi, G., Esuli, A., Sebastiani, F.: A utility-theoretic ranking method for semi-automated text classification. In: Proceedings of SIGIR (2012)

11. Jethani, C.P., Smucker, M.D.: Modeling the time to judge document relevance. In: Proceedings of SIGIR (2010)
12. Damessie, T.T., Scholer, F., Culpepper, J.S.: The influence of topic difficulty, relevance level, and document ordering on relevance judging. In: Proceedings of ADCS (2016)
13. Mc Laughlin, G.H.: SMOG grading - a new readability formula. J. Reading **12**(8), 639–646 (1969)
14. Senter, R., Smith, E.A.: Automated readability index. Technical report, DTIC (1967)
15. Coleman, M., Liau, T.L.: A computer readability formula designed for machine scoring. J. Appl. Psychol. **60**(2), 283 (1975)
16. Gunning, R.: The Technique of Clear Writing. McGraw-Hill, New York (1952)

Towards a Unified Supervised Approach for Ranking Triples of Type-Like Relations

Mahsa S. Shahshahani[1]([✉]), Faegheh Hasibi[2], Hamed Zamani[3], and Azadeh Shakery[1]

[1] School of ECE, College of Engineering, University of Tehran, Tehran, Iran
{ms.shahshahani,shakery}@ut.ac.ir
[2] Norwegian University of Science and Technology, Trondheim, Norway
faegheh.hasibi@ntnu.no
[3] Center for Intelligent Information Retrieval, University of Massachusetts Amherst, Amherst, USA
zamani@cs.umass.edu

Abstract. Knowledge bases play a crucial role in modern search engines and provide users with information about entities. A knowledge base may contain many facts (i.e., RDF triples) about an entity, but only a handful of them are of significance for a searcher. Identifying and ranking these RDF triples is essential for various applications of search engines, such as entity ranking and summarization. In this paper, we present the first effort towards a *unified supervised approach* to rank triples from various type-like relations in knowledge bases. We evaluate our approach using the recently released test collections from the WSDM Cup 2017 and demonstrate the effectiveness of the proposed approach despite the fact that no relation-specific feature is used.

Keywords: Knowledge bases · Triple scoring · Entity facts

1 Introduction

Knowledge bases (KBs) are now a commodity in modern search engines and various semantic search systems. They are structured repositories of entities (such as people, locations, and organizations), where the knowledge about entities is stored in the form of ⟨*subject, predicate, object*⟩ triples, referred to as RDF triples. While knowledge bases contain a large amount of RDF triples about entities, only a handful of them might be of significance for a searcher. Ranking and scoring of these triples is a common step in various semantic search applications, such as entity summarization [5] and generating content for entity cards [3]. Consider for example RDF triples related to the *profession* of the entity *Oscar Wilde*, where the top-ranked profession can be displayed next to his name in the entity card. Another example application is incorporating triple scores for answering queries

© Springer International Publishing AG, part of Springer Nature 2018
G. Pasi et al. (Eds.): ECIR 2018, LNCS 10772, pp. 707–714, 2018.
https://doi.org/10.1007/978-3-319-76941-7_66

such as "german politicians", where the ideal answer should contain entities with politician and German as their primary *profession* and *nationality*, respectively.

The triple scoring task has been defined as "computing a score that measures the relevance of the statement expressed by the triple compared to other triples from the same relation" [1]. The task has been introduced by Bast et al. [1] and further received attention at the WSDM Cup 2017 [7]. It is specifically focused on the ranking of triples related to the type-like relations, i.e., the triples belonging to an abstract group or type. For example, considering *profession* as a type-like relation, the following scores can be obtained:

⟨Oscar Wilde,	profession,	Playwright⟩	1.0
⟨Oscar Wilde,	profession,	Poet⟩	0.4
⟨John F. Kennedy,	profession,	Politician⟩	1.0
⟨John F. Kennedy,	profession,	Author⟩	0.2

A number of supervised approaches have been proposed for the triple scoring task at the WSDM Cup 2017. In this paper, we argue that these approaches all suffer from a fundamental drawback: different feature sets are extracted for different relations. This is not a desired solution for many real-world scenarios. The reason is that knowledge bases contain a large number of type-like relations that makes extracting a separate set of features for each relation infeasible. This calls for a *unified* approach that can be used for every type-like relation. Designing a unified supervised approach for this task is the main motivation of this paper. To this aim, we propose a set of features that can be extracted irrespective of a specific type-like relation. We further train a learning to rank model based on the defined features. Our experiments on the WSDM Cup 2017 dataset suggest that although the proposed approach does not use relation-specific features, our approach performs on a par with the winners of the WSDM Cup. We further demonstrate that in addition to relation-independent features, a single model trained on triples of different relations can bring solid performance. In essence, a single model can suffice to achieve good performance, sometimes even better than the specific purpose-learned models, and this is due to the availability of more training data. We also study the importance of each defined feature in our experiments. This paper presents the first efforts on unified supervised approaches for the triple scoring task and we believe our findings can smooth the path towards effective methods for real-world scenarios.

2   A Unified Supervised Approach for Triple Ranking

Problem Statement. Let $T(e, r) = \{t_1, t_2, \cdots, t_n\}$ be the set of all triples from the type-like relation r for entity e, where each triple $t_i = \langle e, r, . \rangle$ denotes "a relation between an entity and an abstract group or type" [1]. For instance, ⟨*Abraham Lincoln, profession, Statesman*⟩ is a triple from the type-like relation "profession". These relations can be extracted from knowledge bases, such as DBpedia and Freebase. The aim of the task is to *rank* or *score* the triples in a given set $T(e, r)$.

Table 1. The relation-independent features extracted for a triple $\langle e_1, r, e_2 \rangle$ where e_1 and e_2 are two entities and r denotes a relation between e_1 and e_2.

ID	Feature	Description
1	inSent	Presence of e_2's name in the first sentence of e_1's Wikipedia document
2	inPar	Presence of e_2's name in the first paragraph of e_1's Wikipedia document
3	TF-Par	The TF of e_2's name in the first paragraph of e_1's Wikipedia document
4	EFirstPar	Is e_2 the first entity mentioned in the first paragraph of e_1's Wikipedia document or not.
5	#Rel	Total number of relations for e_1
6	#CmnWords	Number of common words between the first paragraph of the Wikipedia documents of e_1 and e_2
7	#UniqWords	Number of unique common words between the first paragraph of the Wikipedia documents of e_1 and e_2
8	LenPar$_2$	Length of the first paragraph of e_2's Wikipedia document
9	LenPar$_1$	Length of the first paragraph of e_1's Wikipedia document
10	PosSent	Position of e_2 in the first sentence of e_1's Wikipedia document
11	NormPosSent	Feature #10/total number of relations for e_1
12	PosPar	Position of e_2 in the first paragraph of e_1's Wikipedia document
13	NormPosPar	Feature #12/total number of relations for e_1
14	CosSim	Cosine similarity between the embedding vectors of the first paragraph of the Wikipedia documents of e_1 and e_2

Approach. A number of supervised ranking approaches have been proposed for this task in the WSDM Cup 2017 [2]. The main shortcoming of these approaches is that they are relation-dependent. Therefore, these approaches can only be applied to a single relation. While there is a general consensus that supervised methods outperform unsupervised ones, no general supervised method has been proposed to date for the triple scoring task; only a set of unsupervised approaches has been proposed in [1]. This motivates us to propose a *unified* supervised approach, which is necessary for real-world applications, such as search engines. Our approach is to use a learning to rank framework with relation-independent features, which can be used for all type-like relations in a knowledge base.

Features. In total, we defined a set of 14 features that can be extracted from each triple of type-like relations. The features are listed in Table 1. As shown in the table, the features extracted for each triple $\langle e_1, r, e_2 \rangle$ are all relation-independent, i.e., independent of r, which is necessary for a unified approach. A number of these features have been used in prior work [6]. The only assumption here is that there exists a short textual description, i.e., a paragraph, for each entity. In our

experiments, we use the first paragraph of the Wikipedia documents corresponding to the entities. The feature set consists of term counting (e.g., feature #1),
semantic matching (e.g., feature #14), graph information from the knowledge
base (e.g., feature #5), and hybrid features (e.g., feature #11). The features are
either binary or numerical. Note that in the first three features, each entity might
have multiple names, e.g., US, USA, and United States, and appearance of at
least one of them is sufficient. For the last feature, the average word embedding
vector for all terms in the paragraph is calculated as the paragraph's embedding
vector.

3   Experimental Setup

Data. We evaluate our approach using two type-like relations: profession and
nationality. The dataset contains 1,387 triples (1,028 for profession and 359 for
nationality). The entities and relations were extracted from Freebase. A relevance
score from $\{0, 1, \cdots, 7\}$ has been assigned to each triple via crowdsourcing as
described in [1]. This dataset has been used in the WSDM Cup 2017 [7]. Similar
to the WSDM Cup setup, we use about half of these triples (i.e., 677 triples) as
training set and the remaining part as the test set. Following [1], we show the
generecity of our approach using the *profession* and *nationality* relations; experimenting with other type-like relations requires building new test collections,
which is not the focus of this paper but is an obvious future direction.

To extract the features, all documents were stemmed using the Porter stemmer and were stopped using the standard INQUERY stopword list. We indexed
the documents using the Lemur toolkit.[1] For feature #14, we used the pre-
trained embedding vectors with 300 dimensions learned by GloVe on Wikipedia
dump 2014 plus Gigawords 5.[2] All features were normalized using $l2$ normalization. The parameters of the learning algorithms were set using 5-fold cross-
validation over training set.

Evaluation Metrics. To evaluate our models, we consider two score-based metrics: average score difference (ASD) and accuracy. The former is calculated based
on the average of the absolute difference between the predicted and the ground
truth scores, while the latter is the ratio of predicted scores with the difference of
at most 2 from the ground truth score. To be consistent with the runs submitted
to the WSDM Cup 2017, the predicted scores should be integers, and thus the
scores were rounded.

Since predicting accurate rankings, rather than scoring, might be the main
objective in many applications (cf. Sect. 1), we also consider ranking-based metrics in our experiments. We use Kendall's τ distance and normalized discounted
cumulative gain (NDCG) with three different cutoffs, i.e., 1, 3, and 5. NDCG@1
lets us know how accurate is the highest ranked triple, while the other ones
demonstrate the quality of the generated ranked list. The Kendall's τ distance

---

[1] See https://www.lemurproject.org/lemur.php.
[2] See https://nlp.stanford.edu/projects/glove/.

as well as the score-based metrics (ASD and accuracy) have been used in the WSDM Cup 2017 [7] as evaluation metrics. For accuracy and NDCG, higher values are better, while lower ASD and τ distance show better performance. Statistical significant differences between the results are determined using the paired t-test with 95% confidence interval.

4   Results and Discussion

We explored different learning to rank models including point-wise (linear regression and gradient boosting regression trees (GBRT)), pair-wise (AdaRank and RankBoost), and list-wise (ListNet and LambdaMART) ones. Although pair-wise and list-wise approaches often outperform point-wise models in many ranking scenarios, we observed that GBRT achieves the highest performance in our experiments. We attribute this to the limited amount of training data available for the task. The learning curve (see Fig. 1) also validates that the amount of training data provided by the WSDM Cup 2017 is not enough for training. In the following experiments, we focus on GBRT as a well-performing ranking model for our task. For the sake of space, we do not report the results achieved by different ranking models.

Comparison with baselines. In the next set of experiments, we compare our approach with the winners of the WSDM Cup 2017. The results are reported in Table 2. Bokchoy [4], Cress [6], and Goosefoot [8] respectively achieved the best accuracy, ASD, and the Kendall's τ distance at the WSDM Cup 2017. According to Table 2, our model outperforms all the models in terms of ASD and performs comparably in terms of the Kendall's τ distance. Our model also performs better than Cress and Goosefoot, in terms of accuracy. It is notable that all the baselines use relation-specific methods. These results suggest that although our approach only uses relation-independent features, the performance is still good compared to the winners of the WSDM Cup 2017, and even better in terms of ASD.

Single model vs. separate models for the relations. We show that we can achieve a solid performance by only using relation-independent features, but the models were trained separately for each relation. An important research question here is how does our model perform if we do not train different models for different relations? To address this question, we combine the training data of the two relations and learn a single GBRT model, which is further used for both relations. Table 3 presents the results and signifies that our single generic model performs on a par with the ones trained for a single relation, in terms of accuracy, ASD, and τ. No significant difference is observed. Interestingly, although different relations may have different feature distributions, putting all of them together and training a single model lead to a better performance in terms of NDCG@k for different k. The reason is due to the limited amount of training data.

Table 2. Comparison with the top-performing systems at the WSDM Cup 2017.

Method	Acc	ASD	Kendall
Bokchoy [4]	**0.87**	1.63	0.33
Cress [6]	0.78	1.61	0.32
Goosefoot [8]	0.75	1.78	**0.31**
Our [GBRT]	0.80	**1.57**	0.32

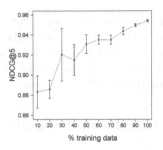

Fig. 1. Learning curve trained for all relations

Table 3. Performance of single model vs. separate models for different relations.

Model	Test relation	Acc	ASD	Kendall	NDCG@1	NDCG@3	NDCG@5
Separate models	Profession	0.79	1.55	0.27	0.8827	0.9288	0.9396
	Nationality	0.85	1.64	0.39	0.8113	0.9453	0.9453
	Total	0.80	1.57	0.32	0.8649	0.9394	0.9472
Single model	Profession	0.79	1.55	0.28	0.8947	0.9341	0.9465
	Nationality	0.81	1.52	0.39	0.8858	0.9670	0.9670
	Total	0.80	1.54	0.33	0.8905	0.9479	0.9547

Learning curve. To analyze performance variations with different amounts of data, we create subsets of the whole training set, for 10 different sizes ranging from 10% to 100% of the instances. We repeat this random sampling process 10 times for each size. The learning curve in terms of NDCG@5[3] is plotted in Fig. 1. According to the figure, by increasing the amount of training data, the average performance increases and the standard deviation decreases. The learning curve is generally increasing and it does not get stable; which demonstrate that providing more training data would lead to even better ranking performance.

Feature analysis. To analyze the importance of each feature, we performed forward selection based on the Gini index and report the ranking metrics after selection of each feature. The results for both profession and nationality relations are plotted in Fig. 2. According to the plots, the paragraph lengths, semantic similarity based on word embeddings, and some term matching features are considered as the best features. The term matching features and semantic similarity are also among the best features for the nationality relation.

---

[3] For the sake of space, we only consider NDCG@5 as the evaluation metric in this experiment. The learning curves with respect to the other metrics, e.g., ASD, are also similar.

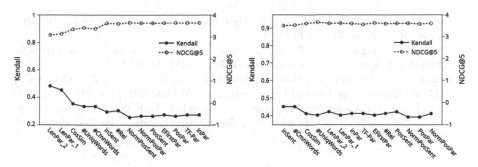

Fig. 2. Feature importance for profession (left) and nationality (right) relations.

5   Conclusions and Future Work

We proposed a *unified* supervised approach for ranking triples of type-like relations. Our approach is based on a set of relation-independent features and a learning to rank model. Our experiments on the WSDM Cup 2017 dataset suggested that good performance can be achieved using only relation-independent features. An interesting finding of the paper is that there is no need to train the model on different training sets for different relations, at least for the considered relations. Due to the lack of available data, we were only able to study two relations, which may not be sufficient to make general claims. This preliminary research magnifies the importance of developing relation-independent approaches and smooths the path towards studying unified approaches for the triple scoring task. Studying this task for several other relations will be the first step in our future work. Another avenue is to work on a unified semi-supervised approach for this task, as it is unlikely to have access to labeled training data for all relations.

Acknowledgements. This work was partially supported in part by the Center for Intelligent Information Retrieval. Any opinions, findings and conclusions or recommendations expressed in this material are those of the authors and do not necessarily reflect those of the sponsors.

References

1. Bast, H., Buchhold, B., Haussmann, E.: Relevance scores for triples from type-like relations. In: SIGIR 2015, pp. 243–252 (2015)
2. Bast, H., Buchhold, B., Haussmann, E.: Overview of the triple scoring task at the WSDM Cup 2017. In: WSDM Cup (2017)
3. Bota, H., Zhou, K., Jose, J.M.: Playing your cards right: the effect of entity cards on search behaviour and workload. In: CHIIR 2016, pp. 131–140 (2016)
4. Ding, B., Wang, Q., Wang, B.: Leveraging text and knowledge bases for triple scoring: an ensemble approach - the Bokchoy triple scorer at WSDM Cup 2017. In: WSDM Cup (2017)

5. Hasibi, F., Balog, K., Bratsberg, S.E.: Dynamic factual summaries for entity cards. In: SIGIR 2017, pp. 773–782 (2017)
6. Hasibi, F., Garigliotti, D., Zhang, S., Balog, K.: Supervised ranking of triples for type-like relations - the cress triple scorer. In: WSDM Cup (2017)
7. Heindorf, S., Potthast, M., Bast, H., Buchhold, B., Haussmann, E.: WSDM Cup 2017: vandalism detection and triple scoring. In: WSDM 2017, pp. 827–828 (2017)
8. Zmiycharov, V., Alexandrov, D., Nakov, P., Koychev, I., Kiprov, Y.: Finding people's professions and nationalities using distant supervision - the Goosefoot triple scorer. In: WSDM Cup (2017)

ParsTime: Rule-Based Extraction and Normalization of Persian Temporal Expressions

Behrooz Mansouri[1,2(✉)] ⓘ, Mohammad Sadegh Zahedi[1,2],
Ricardo Campos[3,4] ⓘ, Mojgan Farhoodi[1], and Maseud Rahgozar[2]

[1] Information Technology Faculty,
Iran Telecommunication Research Center, Tehran, Iran
{b.mansouri, s.zahedi, farhoodi}@itrc.ac.ir
[2] Database Research Group, Control and Intelligent Processing Center
of Excellence, School of Electrical and Computer Engineering,
University of Tehran, Tehran, Iran
rahgozar@ut.ac.ir
[3] Polytechnic Institute of Tomar, Tomar, Portugal
ricardo.campos@ipt.pt
[4] LIAAD – INESC TEC, Porto, Portugal
rncampos@inescporto.pt

Abstract. Extraction and normalization of temporal expressions are essential for many NLP tasks. While a considerable effort has been put on this task over the last few years, most of the research has been conducted on the English domain, and only a few works have been developed on other languages. In this paper, we present ParsTime, a tagger for temporal expressions in Persian (Farsi) documents. ParsTime is a rule-based system that extracts and normalizes Persian temporal expressions according to the TIMEX3 annotation standard. Our experimental results show that ParsTime can identify temporal expressions in Persian texts with an F1-score 0.89. As an additional contribution we make available our code to the research community.

Keywords: Temporal tagger · Time normalization · Pattern matching

1 Introduction

Extracting temporal information from text plays an important role in natural language processing and information retrieval tasks such as text summarization and temporal query classification [1]. It can also be used by search engines for tasks like query auto completion, result ranking or query classification [2], to name but a few.

The first step to extract temporal information from text, is to recognize temporal expressions and to convert them into a standard annotation. To conduct this, temporal taggers are usually used. Over the last few years, a considerable number of different time taggers were proposed for the English language. GUTime [3] was developed by the Georgetown University as part of TARSQI toolkit, with the purpose to improve question answering systems towards temporally-based questions. HeidelTime [4],

© Springer International Publishing AG, part of Springer Nature 2018
G. Pasi et al. (Eds.): ECIR 2018, LNCS 10772, pp. 715–721, 2018.
https://doi.org/10.1007/978-3-319-76941-7_67

introduced by Strötgen and Gertz, is probably one of the most well-known approaches and the best performing system in task A for English of the TempEval-2 challenge (http://semeval2.fbk.eu). Another well-known system is that of Chang and Manning, who presented SU-Time [5] as part of Stanford CoreNLP. Besides English, temporal taggers have been addressed for only a few other languages. Li et al. [6] for example, proposed, a Chinese temporal parser for extracting and normalizing temporal information using HeidelTime architecture. Strötgen et al. [7] in turn, introduced the temporal tagger for Arabic, Italian, Spanish and Vietnamese using the same architecture. Many researches in NLP and information retrieval have developed research considering the Persian language [8–10]. However, no one so far, has developed a temporal tagger devoted to this important Indo-European language, which is one the most dominant in the Middle East, spoken in several countries like Iran, Tajikistan and Afghanistan. In this paper, we propose a first attempt on this matter, by introducing a temporal tagger for detecting Persian temporal expressions within a text. ParsTime is a rule-based temporal tagger that can identify and normalize different Persian temporal expressions within a text with high precision. These expressions may refer to different types such as date, time, duration or set. Besides the expression type, the calendar type (Georgian, Hijri or Jalali) is also identified by ParsTime, an additional challenge when compared to Western Languages which only refer to a single calendar type. ParsTime was evaluated under two different datasets achieving an F1-score of 0.89, which is in line with the results obtained by other temporal taggers in a diversity of languages. As a further outcome of our research, we also make available an implementation of our method in Java which we made publicly available on GitHub[1].

2 Method Description

2.1 Types of Temporal Expressions

TIMEX3 [13] is a well-known annotation scheme for temporal expressions which is usually used in this kind of task. In TIMEX3, each temporal expression has two attributes, Type and Value. A Type can be one of the four types: Date, Time, Duration and Set. A Value, instead, corresponds to a normalized temporal value, which is a normalized way of referring to a temporal expression. For example, for the Time "ربع به هفت یک" (a quarter to seven) a normalized value would be 2017-10-25T06:45. (With reference date as 2017-10-25). Like in many other languages, Persian also contains a huge variety of temporal expressions. In this work, we consider the following four types defined in TIMEX3 [13]:

- Date: This expression points to a calendar time. It may point to a day, week, month, season or year. ParsTime is able to extract both explicit temporal expressions (e.g., "اکتبر 2017" (October 2017)), as well as relative temporal expressions (e.g., "روز گذشته" (yesterday)). While extracting and normalizing the first type is straightforward, for the latter we need the reference time. As a rule-of-thumb we

---

[1] https://github.com/BehroozMansouri/ParsTime.

will consider document publish time, or the last date extracted from the text, whenever the first one cannot be found.

- Time: This kind of expression refers to a time of the day. It might refer to an exact time, such as "اکتبر 1 صبح 9" (9:00 am October 1), or an approximate time, such as, "صبح شنبه" (Saturday morning).
- Duration: This type of expression describes a duration (interval). The start and end points of duration might be exactly mentioned, as in expression "از اکتبر تا نوامبر" (From October to November). The duration itself, may be exact or an approximation. For example, "طی این هفته" (during this week) has implicit boundaries, while the expression "تا ساعاتی دیگر" (until next few hours) does not have a certain end point.
- Set: This expression represents periodic temporal sets. In other words, it refers to a temporal signal that occurs on a regular basis, such as "هر دوشنبه" (every Monday). Table 1 provides an example of some temporal expressions for each of these four types with normalized TIMEX3 value.

Table 1. Persian temporal expression example with translation and a ref. time of 2017-10-25

Type	Example	Translation	Normalization Value
Date	هفده سپتامبر 2017	September seventeenth 2017	2017-09-17
Time	یک ربع به هفت	a quarter to seven	2017-10-25T06:45
Interval	چهار روز آینده	the next four days	P4D
Set	عصر هر سه‌شنبه	every Tuesday afternoon	XXXX-WXX-2TAF

2.2 ParsTime Architecture

In this section, we introduce the architecture underlying the ParsTime method. In summary, it takes a text as an input and split its sentences. Each sentence is then temporally annotated. It is then matched against a list of predefined temporal patterns and normalized if it matches a pattern, before outputting it with TIMEX3 annotation. Figure 1 indicates the workflow of ParsTime. In the following we describe each of these steps in more detail.

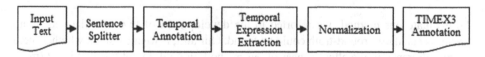

Fig. 1. ParsTime workflow.

- Pre-Processing: First, a preprocessing step is needed to prepare the input text for the coming stages. Like in many other NLP systems, the input text is first tokenized. In our method we resort to the ParsiPardaz toolkit [11].
- Temporal Annotation: Before matching the input text against predefined patterns, tokenized text coming from the previous section, is first temporally annotated with 12 predefined tags (including day, month, season, numbers, etc.) in a similar fashion as a standard PoS tagger. For instance, "2018 ژانویه" (January 2018) is annotated as

"Month Number"; ژانویه annotated as a "Month", and 2018 is annotated as "Number". This is an important stage, as temporal patterns will be defined (in the next stage) based on these temporal tags notations. To conduct this process, we resort to define a set of expression resources for each temporal tag. For example, for the temporal tag "Month", a list of months from "January" to "December" (in addition to a list of months in Hijri and Jalali calendar) is defined. Each term in the input text is matched against the related list for each temporal tag.

- Temporal expression extraction: After the input text is temporally annotated, annotated tokens are matched against a list of predefined temporal expression patterns to identify temporal expressions. These patterns are defined based on the temporal tags introduced in the previous step. For instance, the defined pattern "Month + Number", can detect temporal expressions such as "ژانویه 2018" (January 2018). As mentioned in Sect. 2.1, four types of temporal expressions can be recognized by ParsTime. For each of these categories, a considerable number of patterns is defined: 149 for Date, 74 for Time, 97 for Duration and 26 for Set, totalizing 346 temporal expression patterns. This number is considerably higher than other languages due to the complexity of the Persian language. For instance, HeidelTime [4] has defined a reduced number of 248 patterns for the English language. It should be noticed that some of the patterns are subset of others, such that, when ParsTime recognizes the shorter pattern, it continues to check if the text contains a longer pattern or not. If the longer pattern is not recognized, then the shorter one is extracted. For example, for the expression "فردا صبح" (Tomorrow morning), the pattern "RD" (Relative day) is recognized, but as we also have the pattern "RD + PD" (Relative Day + Part of Day), ParsTime continues and identifies the longer pattern.

- Normalizing: The final step of this workflow is to normalize the extracted temporal expression. Every extracted temporal expression E_i has two attributes and can be viewed as a two-tuple $E_i = \langle V_i, T_i \rangle$, where V_i is the normalized value that indicates the temporal semantic of an expression as defined by TIMEX3 standard, and T_i the type of temporal expression (Date, Time, Duration or Set).

3 Evaluation

To evaluate our method, we resorted to the development of two new datasets as no previous standard ground-truth could be found. Our aim, is to understand whether there is any difference in the effectiveness of our method, upon different types of input (more formal or more informal temporal expressions). For the first one, we relied on a news dataset, a kind of dataset that is usually used for this kind of tasks. For this purpose, we selected Hamshahri [12], a news dataset which covers a wide range of news in Persian language, including politics, entertainment and sports from a ten-year period, spanning from 1996 to 2006. We then randomly selected 2000 news articles domain from this dataset and asked 4 students to tag the temporal expressions using TIMEX3 annotations. An inter-rater reliability analysis using the Fleiss Kappa statistics was performed to determine consistency among the editors. Overall, the annotators obtained about 0.82 of agreement level, which represents a high agreement between editors. For the second dataset, we relied on search engine query logs, to capture a more informal nature of temporal expressions. Query logs are a useful resource of this kind of data, as search engine users usually tend to be more relaxed when specifying their temporal intents. To this purpose, we asked the very same 4 students to select 250 unique queries (totalizing 1000 queries) containing temporal expressions from the query log records of a Persian search engine, Parsijoo. For each student, a unique period of query log records was provided and they were asked to select and annotate 250 queries (using TIMEX3 annotations) that contains at least one temporal expression. Both of these datasets are publicly available[2]. For the task of evaluating the results, we computed the precision, recall and F1-scores for the extraction of temporal expressions. For the type and value attributes, only the correctly identified temporal expressions are considered (the ratio of correct guesses for both type and value). Table 2 reports the results for each dataset.

Table 2. ParsTime performance on Hamshahri corpus and Parsijoo query log records.

Dataset	Extraction			Attribute	
	Precision	Recall	F1-score	Value	Type
Hamshahri	0.92	0.86	0.89	0.84	0.93
Parsijoo	0.91	0.85	0.88	0.82	0.92

The results achieved by ParsTime are quite satisfying and in line with the results reported by other temporal taggers for different languages (see Table 3). For both formal and informal temporal expressions, the results of ParsTime performance was nearly the same. To better understand the results we looked at the errors and noticed that, wrong input format was the main reason that affects the recall. Wrong spelling, wrong punctuation (and spacing) and the rare words used in queries were the main causes that ParsTime was unable to detect some of temporal expressions. For example, in the expression "غروبگاه شنبه" (Saturday evening), only day patterns is recognized, as the term "غروبگاه" (uncommon word meaning evening) is a rare term and was not defined in patterns. Also, our error analysis reveals that, one of the main reasons that may affect the precision of ParsTime, may be related to ambiguous words which are

Table 3. Effectiveness of ParsTime and other temporal taggers.

Temporal tagger	Extraction			Attribute	
	Precision	Recall	F1-score	Value	Type
ParsTime	0.92	0.86	0.89	0.84	0.93
SuTime (English)	0.88	0.96	0.92	0.82	0.92
HeidelTime (English)	0.90	0.82	0.86	0.85	0.96
HeidelTime (Spanish)	0.96	0.84	0.90	0.85	0.87
HeidelTime (Arabic)	0.95	0.83.8	0.89	–	–

---

[2] http://dbrg.ut.ac.ir/ParsTime.

often common in Persian. For instance, the word "بهمن" is a name of a Persian Month but also the name of Football team. Considering this ambiguousness, in the expression "تیم بهمن سال 93" (Team Bahman 93), a temporal pattern "Month + Number" is wrongly recognized.

4 Conclusion

In this paper, we presented ParsTime, the first temporal tagger for extracting and normalizing Persian temporal expression from texts. ParsTime is a rule-based system that can extract different types of temporal expressions including date, time, duration and set. Our experimental results, over two newly created TIMEX3 annotated datasets, show that ParsTime achieved high F1-score. As an additional contribution to the research community we also make available a Java version of our method. This will enable researchers to use our system despite guaranteeing the reproducibility of our research. In future work, we plan to provide resources for detecting implicit temporal expressions, such as "Rio Olympics", which implicitly refer to 5-21 August 2016.

Acknowledgement. This research was supported by Persian native search engine program from Iran Telecommunication Research Center (ITRC). It was also partially funded by the ERDF through the COMPETE 2020 Programme within project POCI-01-0145-FEDER-006961, and by National Funds through the FCT as part of project UID/EEA/50014/2013.

References

1. Campos, R., Dias, G., Jorge, A.M., Jatowt, A.: Survey of temporal information retrieval and related applications. ACM Comput. Surv. (CSUR) **47**(2), 15 (2015)
2. Mansouri, B., Zahedi, M., Rahgozar, M., Oroumchian, F., Campos, R.: Learning temporal ambiguity in web search queries. In: ACM Conference (CIKM), pp. 6–10 (2017)
3. Verhagen, M., Mani, I., Sauri, R., Knippen, R., Jang, S.B., Littman, J., Pustejovsky, J.: Automating temporal annotation with TARSQI. In: Proceedings of the ACL 2005 on Interactive Poster and Demonstration Sessions, pp. 81–84. Association for Computational Linguistics (2005)
4. Strötgen, J., Gertz, M.: Heideltime: high quality rule-based extraction and normalization of temporal expressions. In: Proceedings of the 5th International Workshop on Semantic Evaluation, pp. 321–324. Association for Computational Linguistics (2010)
5. Chang, A.X., Manning, C.D.: Sutime: a library for recognizing and normalizing time expressions. In: LREC, vol. 2012, pp. 3735–3740 (2012)
6. Li, H., Strötgen, J., Zell, J., Gertz, M.: Chinese temporal tagging with heidel time. In: EACL, vol. 2014, pp. 133–137 (2014)
7. Strötgen, J., Armiti, A., Van Canh, T., Zell, J., Gertz, M.: Time for more languages: temporal tagging of Arabic, Italian, Spanish, and Vietnamese. ACM Trans. Asian Lang. Inf. Process. (TALIP) **13**(1), 1 (2014)
8. Zahedi, M., Mansouri, B., Moradkhani, S., Farhoodi, M., Oroumchian, F.: How questions are posed to a search engine? an empiricial analysis of question queries in a large scale Persian search engine log. In: 2017 3rd International Conference on Web Research (ICWR), pp. 84–89. IEEE (2017)

9. Zahedi, M., Aleahmad, A., Rahgozar, M., Oroumchian, F., Bozorgi, A.: Time sensitive blog retrieval using temporal properties of queries. J. Inf. Sci. **43**(1), 103–112 (2017)
10. Mansouri, B., Zahedi, M., Rahgozar, M., Campos, R.: Detecting seasonal queries using time series and content features. In: Proceedings of the 2017 ACM on International Conference on the Theory of Information Retrieval, pp. 279–300. ACM (2017)
11. Sarabi, Z., Mahyar, H., Farhoodi, M.: ParsiPardaz: Persian language processing toolkit. In: 3rd International eConference on Computer and Knowledge Engineering (ICCKE), pp. 73–79. IEEE (2013)
12. AleAhmad, A., Amiri, H., Darrudi, E., Rahgozar, M., Oroumchian, F.: Hamshahri: a standard Persian text collection. Knowl. Based Syst. **22**(5), 382–387 (2009)
13. Pustejovsky, J., Castano, J.M., Ingria, R., Sauri, R., Gaizauskas, R.J., Setzer, A., Radev, D.R.: TimeML: robust specification of event and temporal expressions in text. In: New Directions in Question Answering, vol. 3, pp. 28–34 (2003)

Active Search for High Recall: A Non-stationary Extension of Thompson Sampling

Jean-Michel Renders[(✉)]

Naver Labs Europe, Meylan, France
jean-michel.renders@naverlabs.com

Abstract. We consider the problem of Active Search, where a maximum of relevant objects - ideally all relevant objects - should be retrieved with the minimum effort or minimum time. Typically, there are two main challenges to face when tackling this problem: first, the class of relevant objects has often low prevalence and, secondly, this class can be multi-faceted or multi-modal: objects could be relevant for completely different reasons. To solve this problem and its associated issues, we propose an approach based on a non-stationary (aka restless) extension of Thompson Sampling, a well-known strategy for Multi-Armed Bandits problems. The collection is first soft-clustered into a finite set of components and a posterior distribution of getting a relevant object inside each cluster is updated after receiving the user feedback about the proposed instances. The "next instance" selection strategy is a mixed, two-level decision process, where both the soft clusters and their instances are considered. This method can be considered as an insurance, where the cost of the insurance is an extra exploration effort in the short run, for achieving a nearly "total" recall with less efforts in the long run.

1 Introduction

Contrary to Active Learning, Active Search does not aim at building the best possible classifier with the minimum number of labelled instances, but simply aims at discovering virtually all the instances of the positive class (assuming a binary classification problem) with the minimum "reviewing" effort or cost. The collection – or pool – of objects to search in is assumed to be known in advance and the setting is, in some way, similar to the Transductive Learning setting, but with an on-line, incremental, "recall-oriented" perspective. Consequently, Active Search algorithms can be very different from traditional Active Learning algorithms. Active Search applications could be found in numerous domains: fraud detection, compliance monitoring, e-discovery, systematic medical reviews, prior art search when filing a patent, *etc.*

Recently, some pieces of work [5,9] have focused on developing Active Search strategies that significantly depart from Active Learning, by emphasizing the "Total Recall" aspect and the "Continuous Active Learning" setting. The main

© Springer International Publishing AG, part of Springer Nature 2018
G. Pasi et al. (Eds.): ECIR 2018, LNCS 10772, pp. 722–728, 2018.
https://doi.org/10.1007/978-3-319-76941-7_68

idea of this family of works is to greedily select the next instances as the ones with the largest estimated probabilities of belonging to the relevant class (these probabilities are given by a classifier incrementally trained from all labelled instances up to the current time).

In addition to belonging to a low prevalence class, relevant objects can take multiple forms or facets: the landscape of the positive class is often "multimodal". When considering these challenges – unbalanced class distribution and multi-modality of the relevant class –, there is a clear need to control the exploration-exploitation trade-off if we want to improve the "baseline" greedy approach and make it more robust: typically, the search starts with a small number of "seed" instances and these seeds rarely cover all modes and facets of the positive class. So, a greedy selection approach runs the risk of missing large areas of relevant instances, when the corresponding facets are not covered (or hardly reachable from) the instances reviewed and labelled as positive up to the current time. At the early stage of the search, it can be useful to spend some effort in exploring diverse regions of the instance space, provided that these regions could offer potentially relevant elements in the long run.

The use of Multi-Armed Bandits (MAB) appears as a natural choice to solve this exploration-exploitation trade-off. Instead of considering the problem as an instance recommendation problem with a binary response as in [3] and solving it by using contextual bandit strategies, we follow an alternative strategy that turns out to be more efficient in our use cases. This alternative consists in discretizing the structure of the feature space into a finite set of clusters and in relying on the cluster structure to manage the exploitation-exploration trade-off. More precisely, the idea is to consider each cluster as an arm of a MAB and to focus on the most promising ones, while ensuring that the selection strategy covers all facets or clusters of the instance space. However, we face the problem of dealing with non-stationary (or restless) mortal bandits, as the reward distribution of a cluster gradually declines each time we "exploit" it.

2 Related Work

Even if there is a large literature on Active Learning, the case of Active Search has not received the same attention. One of the first works on this problem was presented by [6] who proposed a Bayesian approach that requires computations exponential in the number of look-ahead steps, which makes this method impractical for large collections. Several pieces of work have approached Active Search as exploration on graphs [7] or also graph learning [8]. However, the mostly used state-of-the-art method remains the greedy (one single step ahead) approach: one of the most striking examples of this is the "Continuous Active Learning" concept of [5], implemented in the form of the AutoTAR system in the field of e-Discovery, and extended further in [9]. This vein of work relies on variants around a main baseline, which greedily selects the next instances as the ones with the largest probability scores with respect to the relevant class. The probability scores are estimated by a classifier trained incrementally.

The use of MAB in Active Learning is not new and relatively well studied. A relatively common way to solve Active Learning with MAB is to cluster the instances on the pool and consider that each cluster is an arm [2,4]. In this case, the payoff distribution for each arm is non-stationary since the probability of finding relevant instances in a cluster decreases as the cluster is exploited. This could be solved by assuming a known fixed trend as in [1].

Finally, our work is not the only one to use MAB for Active Search. Even if initially formulated as an item recommendation problem, the MAB-based approach of [3] could be used as such for Active Search; in this piece of work, the "next item selection" problem is expressed as a contextual bandit, where each instance is an arm; the expected reward (or relevance label) of an instance is expressed as a logistic regression model, whose parameters are sampled following a Thompson sampling strategy from the posterior distribution updated each time a new label is collected. We have implemented this method for our collections and the results were extremely weak due to the fact that the method is not adapted to sparse high-dimensional data and requires a lot of exploration, often exceeding the budget and reaching the collection size.

3   Proposed Method

Traditional Active Search methods are typically greedy: they are looking for the most promising instances, i.e. the ones with the highest conditional probabilities as estimated by a classifier trained on the instances reviewed up to the current time. However, greedy strategies are likely to fail when the relevant class is "multi-modal" or "multi-faceted", in other words when they are multiple, well distinct and possible unbalanced ways of being relevant. As the greedy strategies introduce a high selection bias when incrementally building the training set, it could be that the selection algorithm will miss important sectors of the relevant category, because they are relative far from the positively-labelled training instances. In practice, this risk strongly depends on the quality of the seed set: the seed set should be diverse enough and have a good coverage of the different facets of relevance, but unfortunately this guarantee is hard to obtain.

We translate the Active Search problem into a MAB setting, by considering clusters of pool instances as arms. But this choice has the particularity that the algorithm can "exhaust" a cluster and that, consequently, the related arm will "die" once all instances in the cluster have been reviewed and labelled. Intuitively, we face a diminishing return issue: the retrieval rate of relevant objects decreases as we are exploiting the cluster. Note that, in our approach, we rely on soft clustering so that an instance can belong to multiple clusters with different degrees of membership. This renders the approach more robust with respect to a particular clustering method but, on the other side, we have to adapt the MAB algorithms for a non-standard reward scheme: the reward obtained for a particular instance should be re-assigned to multiple clusters (or arms) with an appropriate weighting. Moreover, as it is rather usual in Active Search to proceed with batches and not with single instance proposals, we have adapted the MAB algorithms so that they can provide us with a "batch" of arm trials and update the arms' sufficient statistics (i.e. the posterior distribution of the reward) accordingly.

Due to lack of space, we only give an intuitive description of the algorithm, redirecting the reader the full version of this paper[1] for a formal description of the algorithm.

The first step of the algorithm is the creation of an initial training set from the seed set: a few positive instances discovered by any means and a random sample from the pool set temporarily labelled as "negative" (which introduces very low label noise as the relevance class has low prevalence). It then iteratively creates a batch of instances from the pool using a non-stationary, batch extension of the "Thompson sampling" MAB algorithm, asks the reviewer to label them, removes them from the pool, updates the reward distribution estimates; and it finally retrains the classifier based on the labelled instances and a random sample re-drawn from the pool, temporarily labelled as negative. Actually, the MAB algorithm is a two-level process, where the algorithm first samples B times (the batch size) a "conversion rate" for each cluster/arm from a Beta posterior distribution (the conversion rate is the probability of a high-score member of the cluster to be annotated as a relevant instance) and, secondly, selects an instance that maximizes the probability of being relevant, given the sampled conversion rates of the clusters it belongs to.

Let's first focus on the update of the posterior distributions. The posterior distribution of the "conversion rate" of each cluster is a Beta distribution with parameters (S_k, F_k), initialised with $S_k = F_k = 0.5$ for all arms k (Jeffreys' prior). When receiving the label (or, equivalently, the reward) of the instances selected at the previous round, the binary reward is re-distributed over the clusters with a weight equal to the membership value of the instance with respect to the cluster. Updating the posterior is done using a forgetting factor γ, discounting the previous (weighted) success/failure counts by a factor γ. Assuming that, at round t, the conversion rate of cluster k follows a Beta distribution with parameters $(S_k^{(t)}, F_k^{(t)})$, then the posterior distribution of the conversion rate at round $(t+1)$ is $Beta(S_k^{(t+1)}, F_k^{(t+1)})$ with: $S_k^{(t+1)} = \gamma S_k^{(t)} + \sum_{i \in B^{(t)}} r_i \mu_{i,k}$ and $F_k^{(t+1)} = \gamma F_k^{(t)} + \sum_{i \in B^{(t)}} (1 - r_i)\mu_{i,k}$, where $B^{(t)}$ is the batch of instances selected at round t, $\mu_{i,k}$ is the membership value of i in cluster k and r_i is the binary reward (i.e. $0/1$ label) of instance i.

Let's now focus on the double-stage selection process. For a batch of size B, we repeat B times the following steps: for each arm/cluster, draw a value θ_k from the Beta distribution associated to the cluster: $\theta_k \propto Beta(S_k^{(t)}, F_k^{(t)})$; this value should be interpreted as the parameter (the mean) of a Bernoulli distribution modelling the arm reward distribution; in this work, we use an "Optimistic Bayesian sampling" variant, where one does not allow θ_k to be smaller than the empirical discounted mean of the arm reward (based on observations up to the current round): $\theta_k^* = \max(\frac{S_k^{(t)}}{S_k^{(t)} + F_k^{(t)}}, \theta_k)$. The algorithm then chooses the

---

[1] https://arxiv.org/abs/1712.09550.

instance i^* such that: $i^* = \mathrm{argmax}_{i \in U} \, \pi_i^{(t)} \sum_k \mu_{i,k} \theta_k^*$, with U the set of unlabelled instances, $\pi_i^{(t)}$, the probability that instance i is relevant, as estimated by the current classifier using labelled instances up to round t. Intuitively, this criterion selects the instance that has the best "optimistic" chance of being converted towards a real relevant instance. The exploration effect relies on the sampling from $Beta(S_k^t, F_k^t)$, which can potentially favour less explored clusters as their posterior distribution is less peaked. Of course, when generating the batch of instances, once an instance has been selected, it is removed from the pool and will not be selected again.

4   Experiments and Results

Due to lack of space, we present the experimental results for a single use case, but the extended version of this paper also describes two other use cases (one in e-discovery and one in multi-media diversity-focused retrieval). The collection used here is the Reuters RCV1 Corpus, considering only classes (or topics) at the first level of the hierarchy with a prevalence less than 10% and having at least three children. These classes are C17, C18, E14, E31, E51 and G15. The reason for selecting these classes is that, by construction, they consist of multiple diverse sub-classes and, consequently, are "multi-faceted". The classifier used in all experiments is a L2-regularised Logistic Regression, based on the tokenised, TF-IDF-weighted, L2-normalised bag-of-word representation of the documents. As soft clustering method, we used LDA (Latent Dirichlet Analysis). We fixed the discounting factor γ to 0.99 and the number of arms/clusters K (the number of latent components in LDA) to 200. These hyper-parameters were tuned on an "unused" topic (M14).

The performance measure is simply the proportion of the collection to be reviewed to reach certain levels of recall, focusing on the high recall values. For each topic, we have performed 10 different runs with different seed sets; each seed set consisted of three random relevant instances of the topic. We have limited the reviewing budget to 40% of the collection. The values given in the Table 1 are the average over these 10 runs.

Table 1. Reuters RCV1 collection. Percentage of the collection to be reviewed to reach a Recall level

Topic	Baseline			Proposed method		
	Recall = 0.9	Recall = 0.95	Recall = 0.99	Recall = 0.9	Recall = 0.95	Recall = 0.99
C17	7.92%	12.27%	25.97%	8.15%	11.9%	19.48%
C18	6.81%	8.96%	15.87%	6.91%	8.75%	11.21%
E14	1.19%	2.99%	12.63%	1.48%	2.81%	10.71%
E31	0.93%	1.67%	17.06%	1.13%	1.61%	12.34%
E51	6.67%	10.42%	19.85%	6.81%	10.01%	13.45%
G15	2.39%	2.98%	5.36%	2.5%	2.92%	4.12%

Note that we also tried to use a variant of the greedy strategy, based on the Maximal Marginal Relevance (MMR) method; but the latter performed worse than our method, because it is still favouring exploration and diversity at the late stage of the search (cf. the extended version).

There are several important observations that we can make from these experimental results. First, if the requested level of recall is relatively low, the baseline is still the best choice. But, for a sufficiently high recall, the exploration effort spent during the early phases of the search starts to be beneficial and our method outperforms the baseline. The "break-even" point between the two strategies depends on the collection and on the topic. In some way, our method can be considered as an "insurance" to be able to reach efficiently a high recall without forgetting significant segments of relevant instances; the cost of this insurance is the extra effort spent in exploring diverse clusters during the search. A second observation, which is more visible on other collections and topics not presented here, is that the beneficial effect seems to decrease, and even to disappear, for extreme values of recall. The most likely reason of this sudden decline is the label noise: some irrelevant instances, incorrectly labelled as relevant, are virtually unreachable from any classifier built from (correctly labelled) positive instances.

5 Conclusions and Future Works

This paper considers the Active Search problem as a resource allocation task in an uncertain environment and handles it in a way similar to what is done for petroleum drilling and ore mining projects. By soft-clustering the landscape of the instance feature space and using sampling strategies based on MAB, the method proposed here should be considered as an insurance to decrease the risk of missing a significant amount of relevant objects when the task is to achieve high recall of a low-prevalence, multi-faceted relevant class. Future works will focus on analysing the cost/benefit ratio depending on the task and the collection to be processed.

References

1. Bouneffouf, D., Fraud, R.: Multi-armed bandit problem with known trend. Neurocomputing **205**, 16–21 (2016)
2. Bouneffouf, D., Laroche, R., Urvoy, T., Feraud, R., Allesiardo, R.: Contextual bandit for active learning: active Thompson sampling. In: Loo, C.K., Yap, K.S., Wong, K.W., Teoh, A., Huang, K. (eds.) ICONIP 2014. LNCS, vol. 8834, pp. 405–412. Springer, Cham (2014). https://doi.org/10.1007/978-3-319-12637-1_51
3. Chapelle, O., Manavoglu, E., Rosales, R.: Simple and scalable response prediction for display advertising. ACM TIST **5**(4), 1–34 (2014)
4. Collet, T., Pietquin, O.: Optimism in active learning. Comput. Intell. Neurosci. **2015**, 94 (2015)

5. Cormack, G.V., Grossman, M.R.: Scalability of continuous active learning for reliable high-recall text classification. In: Proceedings of CIKM 2016, pp. 1039–1048 (2016)
6. Garnett, R., Krishnamurthy, Y., Xiong, X., Schneider, J.G., Mann, R.P.: Bayesian optimal active search and surveying. In: Proceedings of ICML 2012 (2012)
7. Ma, Y., Huang, T.-K., Schneider, J.G.: Active search and bandits on graphs using sigma-optimality. In: Proceedings of the 21st Conference on Uncertainty in Artificial Intelligence, UAI 2015, pp. 542–551 (2015)
8. Pfeiffer III, J.J., Neville, J., Bennett, P.N.: Active exploration in networks: using probabilistic relationships for learning and inference. In: Proceedings of CIKM 2014, pp. 639–648 (2014)
9. Zhang, H., Lin, W., Wang, Y., Clarke, C.L.A., Smucker, M.D.: Waterlooclarke: TREC: total recall track. In: Proceedings of the Twenty-Fourth Text REtrieval Conference, TREC 2015 (2015)

Classifying Short Descriptions
of Past Events

Yasunobu Sumikawa[1]([⊠]) and Adam Jatowt[2]

[1] Department of Information Sciences, Tokyo University of Science, Tokyo, Japan
ysumikawa@acm.org
[2] Department of Social Informatics, Kyoto University, Kyoto, Japan
adam@dl.kuis.kyoto-u.ac.jp

Abstract. Mentions and brief descriptions of events often appear in a variety of document genres such as news articles containing references to related events, historical accounts or biographies. While event categorization has been previously studied, it was usually done on entire news articles or longer event descriptions. In this work we focus on short descriptions of historical events which are typically in the form of one or a few sentences. We categorize them into 9 general event categories using a range of diverse features and report F-measure close to 80%.

Keywords: Event classification · Short event descriptions
Digital history

1 Introduction

Many past events are referred to in texts in the form of brief references, typically, a sentence or few sentences long. For example, a news article on a recent earthquake can briefly refer to a past earthquake to provide necessary background information. A document about the history of a city would typically mention several past events that affected or occurred in that city. Note that brief descriptions of retrospective events do not need to occur within longer texts, but they may be standalone such as in the timelines or event lists. For example, the Wikipedia's Current Portal[1] contains lists of significant events in each month where every event is usually described by a single brief list item. Table 1 shows examples of event descriptions in the Wikipedia's Current Portal.

We focus in this work on the problem of categorizing short descriptions of important, retrospective events. Correctly understanding event mentions could have many applications. For example, by being able to tell the category of mentioned events one could better understand as well as represent the intricate network of related events thanks to studying which past event types are mentioned in news articles. Furthermore, the lists of historical events or timelines could be structured by organizing the events based on their semantic categories.

---

[1] https://en.wikipedia.org/wiki/Portal:Current_events.

© Springer International Publishing AG, part of Springer Nature 2018
G. Pasi et al. (Eds.): ECIR 2018, LNCS 10772, pp. 729–736, 2018.
https://doi.org/10.1007/978-3-319-76941-7_69

Table 1. Average lengths, sizes and examples of event descriptions for all the categories. The abbreviated names of classes are used: Armed Conflicts and Attacks (AA), Arts and Culture (AC), Business and Economy (BE), Disasters and Accidents (DA), Health and Environment (HE), Law and Crime (LC), Politics and Elections (PE), Science and Technology (ST) and Sport (S)

Class	Ave. len.	Num. of events	Example
AA	23.6	8,886	Bombs across Iraq detonate, killing 18 people
AC	22.9	1,800	The Beatles release their back catalogue on iTunes
BE	23.6	2,517	Brazil's economy falls into recession
DA	23.1	4,961	A bus crashes into a ravine in Tibet, killing at least 44 people
HE	28.7	487	The number of Zika virus infected in Singapore rises above 40
LC	27.5	4,984	The Constitutional Council of France upholds a ban on fracking
PE	25.2	5,517	Voters in Costa Rica go to the polls for a general election
ST	24.6	1,066	Iran successfully puts the Fajr satellite in orbit using a Safir-B1 rocket
S	23.3	2,400	The Winter Olympics in Sochi, Russia officially concludes

Equipped with knowledge on the categories of past event mentions one could also foster collective memory studies [1] as well as support search methods for finding historical events. Finally, the classification technique could be used for constructing thematic timelines or event lists (e.g., list of disasters/accidents in Asia, timeline of armed conflicts in USA).

Note that the task is not trivial. Prior literature on event classification typically focused on entire news articles which usually contain sufficient amount of text for effective category assignment [4]. In our case, events can be just passing mentions or can be only briefly described with little content available for their automatic classification. The main challenge lies in the scarcity of data, the ambiguity of expressions and variety of diverse means in which events can be referred to. Furthermore, oftentimes, in realistic scenarios, events are not called by their explicit names, or, they may have no known names[2]. Consequently, their automatic detection using NER tools is problematic. To provide sufficient data we use a range of features including ones computed from external knowledge bases like Wikipedia and VerbNet, ones based on lexical analysis as well as ones based on distributional word representation using neural networks. We make an assumption that the context of such descriptions (e.g., surrounding sentences in

---

[2] Usually, only very popular or important events have own names.

original text) is not available to cover also the case of standalone descriptions. Hence we rely only on the event description itself.

In prior literature, two kinds of approaches for short text classifiers were assumed. On one hand, context information was added to feature vectors built from text content. For example, Sriram et al.'s [10] approach classifies tweets by using author information, url and hashtags of tweets. Nie et al. [8] use Naive Bayes classifier equipped with texts, image and video contents for Q&A classification. Lee et al. [6] classifies queries using user-click behavior to identify user goals in web search. On the other hand, several studies have used external knowledge bases such as Wikipedia. Zelikovitz and Marquez [12] trained a classifier with LSA based on Wikipedia data, and Phan et al. [9] proposed a generalized framework of classifiers with topic model. This framework first trains the topic model on texts of an external resource. Explicit Semantic Analysis (ESA) was applied in [11] to map short texts to Wikipedia articles.

In our study, we classify brief mentions of past events using Wikipedia and other external resources, and we investigate which features (e.g., entities, actions, etc.) are best suited for this task. We test the proposed categorization method on the set of 32,362 short descriptions of events from the last 6 years, where descriptions contain on average 25 words. Our approach achieves on average 79.7% F-measure value which should be sufficient for some applications.

2   Data Collection

Event Classes. We use 9 general event classes[3]: `Armed Conflicts` and `Attacks` (AA), `Arts` and `Culture` (AC), `Business` and `Economy` (BE), `Disasters` and `Accidents` (DA), `Health` and `Environment` (HE), `Law` and `Crime` (LC), `Politics` and `Elections` (PE), `Science` and `Technology` (ST) and `Sport` (S). They were described in [4] as a proposal of a comprehensive event class list based on definitions and guidelines used by Wikipedia editors. Although the authors investigated also automatic classification, they did it on entire news articles and using only simple features (TF-IDF).

Dataset. We collected 32,618 typed events from the *Current Events* portal of Wikipedia. Their timespan ranges from 2010/1/1 to 2016/12/17. The average lengths of event descriptions per individual category are shown in Table 1. On average, for all the classes, the descriptions contain 25 words, though the length can be as short as 10 words. Note that the *Current Events* portal of Wikipedia contains also quite large number (precisely, 69,554) of unlabeled events which we did not collect. They occurred between AD1 and Dec. 2016. Their automatic labeling (or at least automatic support of manual annotation) would be an opportunity for the application of the developed classifiers.

---

[3] See Table 1 for examples of events in each class.

3   Methodology

In this section we list the features used for constructing the classifiers and give the intuition behind their choice.

Term based features. We first create TF-IDF vectors (F_1) from all the event descriptions to consider representative terms of events.

Latent semantic based features. To capture latent semantic structures present in text we use both Doc2Vec [5] (F_2) and LSA [2] (F_3).

Verbs based features. Verbs are the essence of events indicating what actions were carried. We then map verbs to VerbNet[4] to obtain their semantic classes (e.g., `destroy` class contains `demolish`, `ruin` verbs among others) and count the number of collected classes to use as features (F_4). There are 429 semantic groups of verbs allowing to organize events by their common actions. We use Stanford POS tagger to collect verbs.

Entity type based features. Most event descriptions contain entities (e.g., persons, organizations, or places), which are actors, locations where the events occurred, important stakeholders and so on. Certain entity types can be strongly associated with particular event classes (e.g., the occurrence of company or organization may suggest `Business & Economy` type of an event). Similarly, the lack of a particular entity type can be suggestive for a certain event type (e.g., the lack of any location mention correlates with low probability of `Disasters & Accidents` type of an event). Furthermore, different combinations of entity types can be indicative of different event classes. We then detect and generalize entities by their types. In order to select and type the entities mentioned in descriptions we apply Yodie [3] - a named entity recognition and disambiguation (NERD) tool. Finally, we count how many entity types an event description contains (F_5).

Head entity/verb based features. Often the head entity is the main actor in the event, hence, we use its type as an additional feature (F_6). We also extract the head verb which is likely to denote the main action of the event and we map it to VerbNet for obtaining its class (F_7) as the semantic representation of the action of the head entity.

Concept based features. We use Wikipedia as a knowledge base to find similar events to the target event as well as to capture semantic concepts underlying an event description. For mapping descriptions into Wikipedia we employ ESA [7] which outputs Wikipedia articles ranked based on their correspondence to the input text. Many of the returned articles are actually about past events similar to the target event.

Next, we collect categories of the obtained Wikipedia articles. In particular, we fetch all the categories of the top-10 articles given by ESA. Then we use terms from the category names (F_8) and ones from the titles of the top articles (F_9) as additional representation of the target event based on TF-IDF weighting.

---

[4] https://verbs.colorado.edu/~mpalmer/projects/verbnet.html.

Finally, we combine all the features and perform feature selection to avoid sparsity. In particular, we select k-important ($k = 2,000^5$) features by using the forests of trees[6].

4 Experimental Evaluation

4.1 Setup

For constructing the feature set for our dataset we have collected 17,503 entities (on average, 2.5 per event), 72,540 Wikipedia articles/concepts (on average, 10 per event) and 116,809 Wikipedia categories (22.5 per event, on average). We trained and tested three kinds of classifiers, SVM with RBF kernel, Naive Bayes Classifier (NB) and Random Forests (RFs), on the nine feature groups in a One-vs-All classifier mode using 10-fold cross-validation. The dimension sizes of LSA and Doc2Vec were set to 300 after experimenting with different numbers on a small held-out dataset.

We compare our approach with classifiers proposed in [4,9]. The former one achieves event classification by training SVM on TF-IDF weighted BOW vectors. The latter is one of the most widely used algorithms for short text classification. That method adds hidden topics into feature vectors in addition to term based features, and trains a MaxEnt classifier.

In addition, for a deeper investigation of our approach we train SVMs separately on each feature group that we collected to analyze how much the particular features and the feature selection can improve the results.

4.2 Discussions of Results

Table 2 compares F-measures of our approaches with that of baselines. SVM equipped with all the features achieves the best results for almost all the classes as well as on the whole dataset. Looking at the micro-average ROC curves (see Fig. 1) for the three classifiers trained on all the feature groups and for the two baselines we see that SVM indeed performs the best among all the compared classifiers. We then focus on SVM in the rest of our analysis.

Looking at Table 2 again we can obtain the detailed analysis of F-measure on the feature group level. The conclusion is that combining all the features improves F-measure for almost all the classes. Especially, the F-measures for AC, BE, and ST are improved over 10% compared with the best results of individual feature groups.

Next, Fig. 2 shows the precision, recall and F-measure per each class. The results are reasonably high (a bit less than 80%) for total Precision, Recall, and F-measures. Weaker results for HE and AC classes are likely due to relatively small size of training data for these classes as indicated in Table 1.

---

[5] This value was empirically chosen based on analyzing the results on the small held-out development dataset.

[6] http://scikit-learn.org/stable/index.html.

Table 2. F-measures for SVM obtained when using individual feature groups vs. all features used together for SVM, NB and RFs settings for each class.

Class	SVM with individual feature groups										Proposed methods		
	F_1 ([4])	F_2	F_3	F_4	F_5	F_6	F_7	F_8	F_9	[9]	All+NB	All+RFs	All+SVM
AA	68.2%	10.1%	83.8%	52.5%	28.5%	23.7%	0.0%	69.6%	28.6%	52.3%	50.0%	46.4%	**85.3%**
AC	21.6%	9.1%	41.4%	13.0%	6.2%	0.0%	0.0%	44.8%	7.3%	**79.9%**	70.0%	68.4%	59.7%
BE	36.5%	3.7%	66.2%	21.8%	7.6%	0.0%	0.0%	59.8%	1.4%	73.3%	61.9%	58.2%	**75.5%**
DA	65.5%	22.0%	83.8%	37.5%	1.8%	30.3%	0.0%	68.3%	9.1%	84.4%	64.7%	60.1%	**88.4%**
HE	42.9%	4.4%	54.5%	8.2%	2.2%	3.8%	0.0%	25.9%	3.3%	**89.3%**	72.2%	66.5%	54.0%
LC	42.0%	11.4%	68.3%	33.2%	14.5%	0.0%	0.0%	46.8%	10.2%	65.2%	49.0%	44.4%	**72.0%**
PE	52.7%	15.0%	72.6%	39.0%	20.3%	0.0%	0.0%	61.9%	9.7%	65.5%	58.6%	50.0%	**77.6%**
ST	31.3%	6.8%	58.6%	8.9%	5.0%	0.0%	0.0%	63.6%	0.0%	8.3%	43.0%	44.9%	**71.8%**
S	66.7%	8.2%	85.0%	36.5%	14.4%	0.0%	13.7%	81.8%	10.5%	57.3%	50.3%	52.0%	**89.3%**
Total	*54.5%*	*27.2%*	*74.7%*	*40.9%*	*28.5%*	*11.6%*	*1.0%*	*62.6%*	*46.0%*	*64.0%*	*58.3%*	*54.6%*	***79.7%***

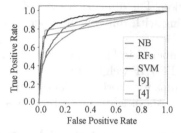

Fig. 1. ROC curves.

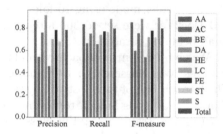

Fig. 2. Results of the proposed approach with SVM.

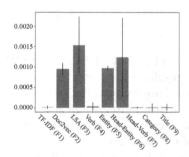

Fig. 3. Feature importance. (Color figure online)

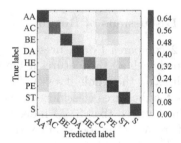

Fig. 4. Confusion matrix on 3,260 randomly sampled events.

In Fig. 3 we show average importance values (blue bars) and standard deviations (black lines) of our features. We can see that LSA and entities (especially the head ones) play importance roles in short event description classification. TF-IDF, verbs, texts of Wikipedia articles' titles and categories were not very important for this task.

Finally, in Fig. 4, we analyze how the classifier mistakes the classes. AC tends to be often confused with PE, while LC and PE tend to be often mistaken with

AA. The reason could be that actors in these events are often nations, countries, regions or persons referred to by their nationalities (e.g., "A Japanese scientist"). HE events are sometimes mistakenly classified as ST, likely due to the discoveries in medicine and biology or similar areas. In addition, HE and DA events sometimes occur due to the same trigger. For example, the outbreak of Zika virus in 2016 caused death of many people (HE event) but also the decrease in the population of bees (DA event).

5 Conclusions

It is quite common to briefly refer to past events. Understanding categories of referred events can have many applications including support for building historical analogy models, across-time connection of events/entities or structuring longer text collections such as Wikipedia (e.g., year related articles). In this paper we introduce classification technique for short, retrospective descriptions of events and report satisfactory results over the dataset of 32k event descriptions.

Acknowledgments. This work was supported in part by MEXT Grant-in-Aids (#17H 01828 and #17K12792) and MIC SCOPE (#171507010).

References

1. Au Yeung, C.M., Jatowt, A.: Studying how the past is remembered: towards computational history through large scale text mining. In: CIKM 2011, pp. 1231–1240 (2011)
2. Deerwester, S., Dumais, S.T., Furnas, G.W., Thomas, K.L., Harshman, R.: Indexing by latent semantic analysis. J. Am. Soc. Inform. Sci. **41**(6), 391–407 (1990)
3. Gorrell, G., Petrak, J., Bontcheva, K.: Using @Twitter conventions to improve #LOD-based named entity disambiguation. In: Gandon, F., Sabou, M., Sack, H., d'Amato, C., Cudré-Mauroux, P., Zimmermann, A. (eds.) ESWC 2015. LNCS, vol. 9088, pp. 171–186. Springer, Cham (2015). https://doi.org/10.1007/978-3-319-18818-8_11
4. Košmerlj, A., Belyaeva, E., Leban, G., Grobelnik, M., Fortuna, B.: Towards a complete event type taxonomy. In: WWW 2015 Companion, pp. 899–902. ACM, New York (2015)
5. Le, Q., Mikolov, T.: Distributed representations of sentences and documents. In: ICML 2014, vol. 32, pp. 1188–1196, Bejing, China, 22–24 June 2014
6. Lee, U., Liu, Z., Cho, J.: Automatic identification of user goals in web search. In: WWW 2005, pp. 391–400. ACM, New York (2005)
7. Chang, M.W., Ratinov, L.A., Roth, D., Srikumar, V.: Importance of semantic representation: dataless classification. In: AAAI, p. 7 (2008)
8. Nie, L., Wang, M., Zha, Z., Li, G., Chua, T.S.: Multimedia answering: enriching text qa with media information. In: SIGIR 2011, pp. 695–704. ACM, New York (2011)
9. Phan, X.H., Nguyen, L.M., Horiguchi, S.: Learning to classify short and sparse text & web with hidden topics from large-scale data collections. In: WWW 2008, pp. 91–100. ACM, New York (2008)

10. Sriram, B., Fuhry, D., Demir, E., Ferhatosmanoglu, H., Demirbas, M.: Short text classification in twitter to improve information filtering. In: SIGIR 2010, pp. 841–842. ACM, New York (2010)
11. Sun, X., Wang, H., Yu, Y.: Towards effective short text deep classification. In: SIGIR 2011, pp. 1143–1144. ACM, New York (2011)
12. Zelikovitz, S., Marquez, F.: Transductive learning for short-text classification problems using latent semantic indexing. Int. J. Pattern Recognit Artif Intell. **19**(2), 146–163 (2005)

Stopword Detection for Streaming Content

Hossein Fani[1,2]([✉]) [ID], Masoud Bashari[2], Fattane Zarrinkalam[2],
Ebrahim Bagheri[2], and Feras Al-Obeidat[3]

[1] Faculty of Computer Science, University of New Brunswick,
Fredericton, Saint John, Canada
hosseinfani@gmail.com
[2] Laboratory for Systems, Software and Semantics (LS[3]),
Ryerson University, Toronto, Canada
[3] College of Technological Innovation,
Zayed University, Dubai, United Arab Emirates

Abstract. The removal of stopwords is an important preprocessing step
in many natural language processing tasks, which can lead to enhanced
performance and execution time. Many existing methods either rely on
a predefined list of stopwords or compute word significance based on
metrics such as tf-idf. The objective of our work in this paper is to identify
stopwords, in an unsupervised way, for streaming textual corpora such
as Twitter, which have a temporal nature. We propose to consider and
model the dynamics of a word within the streaming corpus to identify
the ones that are less likely to be informative or discriminative. Our
work is based on the discrete wavelet transform (DWT) of word signals
in order to extract two features, namely *scale* and *energy*. We show that
our proposed approach is effective in identifying stopwords and improves
the quality of topics in the task of topic detection.

1 Introduction

Stopwords are non informative or noise tokens which can and should be
removed during any preprocessing step. Removing stopwords not only decreases
computational complexity [15] but can also improve the quality of the final
output in document processing tasks such as clustering [11], indexing [14],
and event detection [8]. There is existing work that focus on removing task-
specific non-informative words from a collection of documents [10,13]. In these
approaches, words are assumed to be features and off-the-shelf feature selection
methods such as mutual information are used to identify the most important
features (words) with respect to the target task. As a result, in the same text
corpus, the list of unimportant words depends on the target task and is not
reusable in other tasks. On the contrary, our goal is to devise a systematic procedure to identify an effective list of stopwords which would be agnostic to the
underlying task. In this vein, Salton et al. [14] proposed the tf-idf word weighting scheme in information retrieval and document indexing. According to the

© Springer International Publishing AG, part of Springer Nature 2018
G. Pasi et al. (Eds.): ECIR 2018, LNCS 10772, pp. 737–743, 2018.
https://doi.org/10.1007/978-3-319-76941-7_70

tf-idf scheme, a stopword's *average* tf-idf value over all documents is either very high or very low. The latter means that the word is common to all documents and, consequently cannot discriminate documents from each other while the former indicates that the word is rare and does not contribute significantly to the corpus. Tf-idf has seen many variations since its inception, e.g., term frequency-proportional document frequency (tf-pdf) [4] and has been applied in different domains and languages such as [3,5,13], but none has incorporated the temporal stream-based (real-time) nature of textual collections in online social networks to identify stopwords.

Online social content are inherently temporal, short and noisy about transient topics that are generated in bursts by users and fade away shortly afterwards. It is in contrast to long formal documents that are often well structured and the whole corpus is available in advance. Researchers have looked at ways through which word significance can be computed so as to dynamically identify stopwords in different periods of time. For instance, Weng et al. [15] and Fani et al. [7] employ signal cross-correlation to filter out trivial words. He et al. [8] use spectral analysis and perform word feature clustering for the same purpose. Both methods distinguish themselves from the traditional way of stopword detection by taking the temporal dimension into account. Inspired from these works, in this paper, we propose a discrete wavelet transform (DWT) of word signals for identifying stopwords and explore its effectiveness in the context of topic detection on Twitter.

To the best of our knowledge, the two work presented in [7,8] are the state of the art in identifying stopwords based on the temporal dynamics of words in a temporal corpus and as such they comprise our baselines with which we evaluate our proposed approach. We also include tf-idf to our baselines as the most common domain independent best practice for finding stopwords.

2   Proposed Approach

The main objective of our work is to identify stopwords in a temporal stream of textual content. To this end, we first build word signals based on their occurrences sampled across time intervals as follows:

Definition 1. *(Word Signal) The signal of a word w is a sequence $X^w = [x^w_{1:T}]$ where each element x^w_t shows the number of times w occurred in time interval t.*

We transform the word signals from the time domain to the time-scale (frequency) domains by discrete wavelet transform (DWT) in order to extract two features, namely *scale* and *energy*. This is similar to the spectral analysis done by He et al. [8] where Fourier transform is used to extract power and periodicity features from word signals. However, while Fourier transform only informs us about the frequency spectrum of the signal, we opt for wavelet transform for its ability to disclose temporal extent of the signal (time) in addition to the frequency spectrum (scale is inversely proportional to frequency, i.e., scale $\propto \frac{1}{freq.}$). Fourier transform does not reveal bursty changes efficiently since it represents a signal as the sum of sine waves, which are not localized in time. To accurately analyze signals with rich dynamics in time and frequency, wavelets are used. A wavelet is a rapidly decaying, wave-like oscillation that has zero mean. Unlike sinusoids, which extend to infinity, a wavelet exists for a finite duration. The foundational representation of the discrete wavelet transform is [9]:

$$W_\Psi\{X^w\}(a,b) = \frac{1}{\sqrt{2^a}} \sum_{t=0}^{T} x_t^w \Psi(t - 2^a b) \tag{1}$$

where x_t^w is the value of word signal at time t, Ψ is the base (mother) wavelet, and 2^a and b define the scale and shifting (translation) in dyadic sampling, respectively with $a \in \mathbb{Z}^+, b \in \mathbb{Z}$. This kind of sampling eliminates redundancy in coefficients and yields the same number of wavelet coefficients as the length of the input word signal, i.e., T, for *all* scales. Otherwise, in normal sampling the number of wavelet coefficients would be T for *each* scale. Also Ψ is chosen to be Mexican-hat for its high temporal resolution.

To figure out which scales play a more important role to reconstruct the entire original word signal, we calculate the total energy at each scale, which is a sum of wavelet coefficients over the whole signal through shifting variable b, as follows [9]:

$$E_\Psi\{X^w\}(a) = \sum_b |W_\Psi\{X^w\}(a,b)|^2 \tag{2}$$

Next, given energy of word signals in different scales, we classify the words into four categories: (1) *Significant* words are such words that have signals of high energy in low scales (high frequency); (2) *Common* words' are such words that have signals of high level of energy in high scales (low frequency). This means they have approximately constant behaviour with high amplitude over different time intervals; (3) *Noise* words have a low occurrence and hence low energy in low scales (high frequency); (4) *Rare* words denote the words that have low energy in high scales (low frequency).

3   Experiments

We evaluate our wavelet-based approach (Wvl) against three baselines; Fourier-based (Ftp) [8], cross-correlation (Xcr) [7], average tf-idf, and random selection in finding high quality emerging topics using latent Dirichlet allocation (LDA) topic modelling. We hypothesize that stopword removal produce topics with higher quality in the task of topic detection. Conversely, we conclude that higher quality topics imply better stopword removal assuming the dataset and the topic detection method are both kept constant in our experiments.

To evaluate the performance of the proposed wavelet-based (Wvl) approach and the baselines, we first identify the stopwords by each method for Abel et al.'s Twitter dataset [1]. This dataset consists of 3M tweets posted by 135,731 unique users between Nov. 1 and Dec. 31, 2010. In each approach, we create a list of all

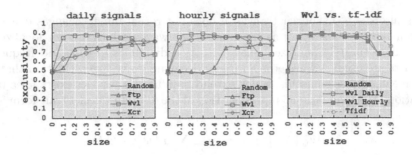

Fig. 1. Average exclusivity of topics.

words ordered by their significance score and incrementally build the stopword list of from the least to the most significant words. This has been initiated from the empty list of size zero, i.e., no stopwords, increasing by 10% at each iteration to 90% of all words. Also all temporal approaches, i.e., Wvl, Ftp, and Xcr, are examined using both hourly and daily sampling rates.

We then apply LDA, the *de facto* standard in topic modeling, to detect emerging topics using mallet[1] after removing the stopwords. The number of topics has been already investigated and set to 50 for the this dataset [6]. Finally, we use *exclusivity, specificity,* and *coherence* as quality metrics for the output topics. It should be noted that perplexity and the likes which measure the quality of the topic modeling approaches, not the topics, cannot be applied here. In the following, we explain each metric followed by a discussion on the respective results.

Exclusivity [2] measures to what extent the top words for a topic do not appear as top words in other topics, as follows:

$$e(z) = \frac{1}{|z|} \sum_{i=1}^{|z|} \frac{Pr(w_i|z)}{\sum_{z' \neq z} Pr(w_i|z')} \tag{3}$$

where z is a topic consisting of a list of $|z| = 50$ words ordered by probability of occurring in the topic and $P(w|z)$ is the probability of word w occurring in topic z.

Figure 1 shows the exclusivity values under varying size of stopword lists. The figure reveals that our proposed wavelet-based (Wvl) approach reaches the highest exclusivity while removing less words (10%) in both daily and hourly signals indicating best stopword removal among all the temporal competitors. Further, in contrast to Wvl, Fourier-based (Ftp) and cross-correlation (Xcr), deliver high sensitivity to the signal sampling rate. As shown in Fig. 1, Ftp in daily signals outperforms its hourly version as it reaches its peak sooner where 20% of the words are removed as stopwords. However, Xcr excels in hourly signals compared to its daily version. This points to the robustness of the proposed

---

[1] mallet.cs.umass.edu/topics.php.

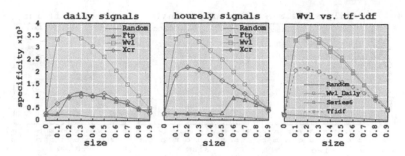

Fig. 2. Average specificity of topics.

Wvl approach with regards to sampling rate. By comparing average tf-idf (non-temporal) and Wvl (best temporal), exclusivity scores are very competitive.

Specificity[2] measures the effective number of words in each topic. The idea is to count topic's words while applying the probability that a word is contributing to a topic. Higher specificity implies that words are distributed in different topics exclusively and are not conforming to the uniform distribution. Specificity is calculated as:

$$s(z) = \sum_{w \in z} \frac{1}{Pr^2(w|z)} \tag{4}$$

In Fig. 2, we show the average specificity values over all topics for the proposed Wvl approach and the baselines. As evident, the Wvl approach significantly excels the specificity score both in daily and hourly signals when removing 20% of words as stopwords in comparison to not only the temporal baselines but also the average tf-idf.

Coherence [12] measures whether the words in a topic tend to co-occur together. The score is the sum of log probability that a document contains at least one instance of a the higher-ranked and a lower-ranked word pair. Formally,

$$c(z) = \sum_{i=1}^{|z|} \sum_{j<i} log \frac{D(w_i, w_j) + \epsilon}{D(w_i) + \epsilon} \tag{5}$$

where ϵ is a smoothing parameter set to 0.01, $D(w)$ is the number of documents that contain at least one token of type w, $D(w_i, w_j)$ is the number of documents that contain at least one w_i and one w_j. Since these scores are log probabilities, they are negative. High values (closer to zero) indicate that words of the topic tend to co-occur more often; hence, the topic is coherent. Figure 3 shows the average coherence values for the identified LDA topics. A first look reveals that our proposed Wvl is the worst and random selection of words as stopwords outperforms all the other approaches. As such, we manually inspect this seemingly unintuitive phenomenon and found out that stopwords which are common to all documents such as 'rt', 'http', and 'us' contribute to almost all

---

[2] mallet.cs.umass.edu/diagnostics.php.

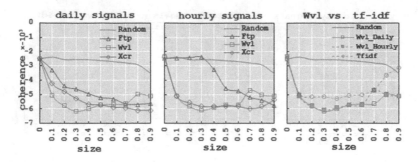

Fig. 3. Average coherence of topics.

topics with a high probability (top words of almost all topics). Removing such words greatly reduces coherence as shown in Fig. 3. In other words, in the task of stopword removal, the lower coherence to an extremum the better. With this respect, our proposed Wvl is the best and the hourly cross-correlation is the runner up.

4   Concluding Remarks

In this paper, we propose a wavelet-based approach for identifying stopwords from streaming textual corpora with temporal nature in an unsupervised way. We employ discrete wavelet transform to model the dynamics of a word within the streaming corpus. We identify informative words based on their energy over different scales of word signals. We showed that our proposed approach is able to improve *exclusivity* and *specificity* of topics learnt based on LDA when the identified stopwords were removed from the corpus. Further, we observed that the current definition of the *coherence* metric is ineffective for the task of stopword removal. Common words happen to be in the top words of almost all topics, the removal of which drops *coherence* significantly. For future work, we will investigate the potential performance improvements of our proposed method in the context of higher-level applications such as user interest modeling, news recommendation and community detection.

References

1. Abel, F., Gao, Q., Houben, G.-J., Tao, K.: Analyzing user modeling on twitter for personalized news recommendations. In: Konstan, J.A., Conejo, R., Marzo, J.L., Oliver, N. (eds.) UMAP 2011. LNCS, vol. 6787, pp. 1–12. Springer, Heidelberg (2011). https://doi.org/10.1007/978-3-642-22362-4_1
2. Bischof, J.M., Airoldi, E.M.: Summarizing topical content with word frequency and exclusivity. In: ICML 2012, pp. 9–16, USA. Omnipress (2012)
3. Blanchard, A.: Understanding and customizing stopword lists for enhanced patent mapping. World Patent Inf. **29**(4), 308 (2007)

4. Bun, K.K., Ishizuka, M.: Emerging topic tracking system in WWW. Knowl. Based Syst. **19**(3), 164–171 (2006)
5. Darwish, K., Magdy, W., Mourad, A.: Language processing for Arabic microblog retrieval. In: CIKM 2012, pp. 2427–2430 (2012)
6. Fani, H., Bagheri, E., Zarrinkalam, F., Zhao, X., Du, W.: Finding diachronic like-minded users. Comput. Intell. (2017). https://doi.org/10.1111/coin.12117
7. Fani, H., Zarrinkalam, F., Zhao, X., Feng, Y., Bagheri, E., Du, W.: Temporal identification of latent communities on Twitter. CoRR, abs/1509.04227 (2015)
8. He, Q., Chang, K., Lim, E.: Analyzing feature trajectories for event detection. In: SIGIR 2007, pp. 207–214 (2007)
9. Kaiser, G.: A Friendly Guide to Wavelets. Birkhauser Boston Inc., Cambridge (1994)
10. Klatt, B., Krogmann, K., Kuttruff, V.: Developing stop word lists for natural language program analysis. Softwaretechnik-Trends **34**(2) (2014)
11. Li, X., Chen, J., Zaïane, O.R.: Text document topical recursive clustering and automatic labeling of a hierarchy of document clusters. In: PAKDD, pp. 197–208 (2013)
12. Mimno, D.M., Wallach, H.M., Talley, E.M., Leenders, M., McCallum, A.: Optimizing semantic coherence in topic models. In: EMNLP, pp. 262–272 (2011)
13. Popova, S., Krivosheeva, T., Korenevsky, M.: Automatic stop list generation for clustering recognition results of call center recordings. In: Ronzhin, A., Potapova, R., Delic, V. (eds.) SPECOM 2014. LNCS (LNAI), vol. 8773, pp. 137–144. Springer, Cham (2014). https://doi.org/10.1007/978-3-319-11581-8_17
14. Salton, G., Buckley, C.: Term-weighting approaches in automatic text retrieval. Inf. Process. Manage. **24**(5), 513–523 (1988)
15. Weng, J., Lee, B.: Event detection in Twitter. In: ICWSM 2011 (2011)

Generating Adversarial Text Samples

Suranjana Samanta and Sameep Mehta[(⊠)]

IBM Research, Delhi, India
{suransam,sameepmehta}@in.ibm.com

Abstract. Adversarial samples are strategically modified samples, which are crafted with the purpose of fooling a trained classifier. In this paper, we propose a new method of crafting adversarial text samples by modification of the original samples. Modifications of the original text samples are done by deleting or replacing the important or salient words in the text or by introducing new words in the text sample. While crafting adversarial samples, one of the key constraint is to generate meaningful sentences which can at pass off as legitimate from the language (English) viewpoint. Experimental results on IMDB movie review dataset for sentiment analysis and Twitter dataset for gender detection show the efficacy of our proposed method.

Keywords: Adversarial text mining · Machine learning
NLP · Resilient analytics

1 Introduction

An adversarial sample [4,6] can be defined as one which appears to be drawn from a particular class by humans but is assigned a different class label by the classifier. In this paper, we focus on generation of adversarial samples in text domain. Majority of the prior works in the field of synthesizing adversarial samples consider images data to work with [2,6,9,10]. A minor change in the pixel values of an image, does generate a meaningful image and also the change is negligible to human eyes. These properties make the synthesize of a new image or modification of an image relatively easier, when compared to a more structured data like text. The discreet nature of the Word2Vec set makes it difficult to map any arbitrary vector to a valid word in the vocabulary. Another important aspect to generate adversarial text sample is to maintain the syntactic meeting of the sample along with the grammar, so that the resultant sample is not gibberish. All these challenges, makes the problem of adversarial text crafting very challenging and there has been little work [7] for adversarial text crafting in the text domain.

The proposed algorithm works well in presence of sub-categorization like genre in IMDB data. Words like *'good'*, *'excellent'*, *'like'* etc. indicates a positive review irrespective of the genre of the movie. However, there exists some words which are specific to the genre of the movie. For example: consider the sentence *"The movie was hilarious"*. This indicates a positive sentiment for a comedy

© Springer International Publishing AG, part of Springer Nature 2018
G. Pasi et al. (Eds.): ECIR 2018, LNCS 10772, pp. 744–749, 2018.
https://doi.org/10.1007/978-3-319-76941-7_71

movie. But the same sentence denotes a negative sentiment for a horror movie. Therefore, the word *'hilarious'* contributes to the sentiment of the review based on the genre of the movie. We leverage this property to generate samples where words are still drawn from same distribution as the input training data but across genres. Such samples will be difficult to detect using simple word distribution based checks.

Often presence of adverbs can alter the sentiment of a review text. For example: consider the sentence: "The movie was fair". This sentiment may belong to either of the class. However, when we add the adverb 'extremely' and change the sentence to "The movie was extremely fair", it indicates that the movie was poor and should be treated as a negative sentiment.

2 Related Work

Fast Gradient Sign Method (FGSM) is one of the most popular method of creating adversarial sample images from existing ones. The method was introduced by Goodfellow et al. [2]. The noise is generated with the help of the gradient of the cost function of the classifier, with respect to the input image.

Papernot et al. [10], an attacker can successfully generate adversarial samples by training a own model using similar input data and create adversarial samples using FGSM method. These adversarial samples can confuse the deployed classifier with high probability. Therefore, the attacker can be successful with almost no information of the deployed model. Papernot et al. introduced the term 'gradient masking' and show its application on real world images like traffic signs, where the algorithm performs well.

Liang et al. [7] proposed an algorithm to produce adversarial text samples using character level CNN [11]. It shows the problem of directly using algorithms like FGSM for crafting adversarial text samples. The results were gibberish texts, which can be easily identified by humans as noisy samples. The codebase TextFool [5] has been shared to craft adversarial text sample, along these ideas, which mostly modify existing words in the text sample by its synonyms and typos. Apart from the above mentioned work, there has been couple of notable works in the area of adversarial sample crafting in the text domain. Hossein et al. [3] shows that strategical insertion of punctuations with some selected words can fool a classifier and toxic comments can be bypassed when a model is used as a filer. However, the modified texts are easily detectable by humans and may not qualify as a good example of adversarial sample. Crafting of adversarial samples in text domain is still at nascent stage compared to its image counterpart.

3 Proposed Method

We propose three modifications to alter a regular input into an adversarial sample: (i) replacement, (ii) insertion and (iii) removal of words into the text. We aim to change the class-label of the sample by minimum number of alteration.

A. Calculate contribution of each word towards determining class-label: A word in the sentence is highly contributing if its removal will change the class probability value to a large extent. Computing exact contribution of each word using traditional posterior probability ideas will be time consuming process. Therefore, we use the concept of FGSM [2] to approximate the contribution of a word w_k, which can be calculated as:

$$\mathcal{C}_F(w_k, y) = -\nabla_{w_k} J(F, s, y) \tag{1}$$

where, y is the true class of the sentence s containing the word w_k and J is the cost function of the classifier F in hand: $F(s) = y$. Since, $-\nabla_s J(F, s, y)$ denotes the adjustment to be made in the input s, to obtain the minimum cost function during training, it is a good way to determine the importance of each of the features for the particular classifier F.

B. Build candidate pool $\mathcal{P}$ for each word in sample text

Synonyms and typos: For each word, we build a candidate pool, which consists of the words with which the current word can be replaced with. For example: the word 'good' in the sentence 'The movie was good.' can be replaced with 'nice', 'decent' etc. Here we are considering the synonyms of the word 'good'. Another possible way to extend the candidate pool is to consider the possible typos [5] that may happen while typing the review. However, if we introduce too many typos the input text s can turn meaningless.

Sub Category specific keywords: Apart from using synonyms and typos, we also consider sub-category specific keywords. Let, δ_i denote the set of distinctive keywords for the ith class. Since we are dealing with two class problem, let the set of distinctive words for class 1 and 2 be denoted as δ_1 and δ_2 respectively. Now, to consider the sub-category or the genre information, we examine the set of distinctive words for the two classes from the texts belonging to each of the genres separately. Let these sets be denoted as $\delta_{1,k}$ and $\delta_{2,k}$ for the two classes of samples for the particular genre k. Now, using these sets of keywords, we add terms in the candidate pool as:

$$\mathcal{P} = \mathcal{P} \cup \{\delta_j \cap \delta_{i,k}\} \tag{2}$$

where, $i, j \in \{1, 2\}$, $i \neq j$ and k is the genre class of the word for which we are building the candidate pool $\mathcal{P}$ of words.

C. Crafting the adversarial sample: We consider three different approaches to change the given text sample s at each iteration, so that the modified sample s' flips its class label. Let at any iteration, the word chosen for modification is denoted as w_i. The heuristics for modifying the text s are:

Removal of words: We check if the word under consideration i.e. w_i is an adverb or not. If w_i is an adverb and its contribution score $\mathcal{C}_F(w_i, y)$ is considerably high, then we remove the word w_i from s to get the modified sample s'. The motivation behind this heuristics is that adverbs put an emphasis on the meaning of the sentences, and often do not alter the grammar of the sentence.

Addition of words: If the first step of modification is not satisfied, we select a word from $\mathcal{P}$, the candidate pool of w_i, using FGSM. If the selected word is from $\mathcal{P}$ is p_j, then the following condition is being satisfied:

$$j = \underset{k}{\text{argmin}}\, \mathcal{C}_F(p_k, y) \;\; \forall k \qquad (3)$$

Now, if w_i is an adjective and p_j is an adverb, we modify s by inserting p_j just before w_i to get s'.

Replacement of word: In case the conditions of the first two steps are not satisfied, we replace w_i with p_j in s to obtain s'. Now, if p_j is obtained from the genre specific keywords, then we consider it for replacement only if the parts of speech of both w_i and p_j are the same. Otherwise, we select the next best word from $\mathcal{P}$ and do the replacement. The matching of the parts of speech is necessary to avoid detection of the modified sample s' by the humans as an adversarial sample. Also, this condition ensures that the grammar of the sentence does not gets corrupted too much.

   We keep on changing each word at a time in s to obtain s', unless the class-label of s and s' becomes different. Since we consider words in the order of their contribution score $\mathcal{C}_F(w_i, y)$, we are crafting adversarial samples in the least possible changes using the idea of greedy method. Once we obtain the adversarial samples, we re-train our existing text classifier to make it more robust to adversarial attacks, as done in [2].

4   Experimental Results

We show our experimental results on two datasets (i) IMDB movie review dataset [8] for sentiment analysis and (ii) Twitter dataset for gender classification [1]. We compare our method with the existing method, TextFool, as described in [7] using Twitter dataset. We use CNNs as base classifier which has been shown to produce good classification accuracy for text.

4.1   IMDB Movie Review Sentiment Analysis

The IMDB movie review dataset consists of the reviews for different movies along with genre and class-label (positive or negative sentiment) and the url from which the review has been taken. The dataset is divided into training and testing sets, each containing 22500 positive samples and 22500 negative samples. Table 1 presents the quantitative performance metrics and comparison to TextFool. The accuracy in 3rd row in Table 1 is the baseline, with which we compare other variations considered in the experimentation. To show the effectiveness of using genre specific keywords in the candidate pool during the process of crafting, we show all the above mentioned values in Table 1 when the genre specific keywords were not considered in the candidate pool (4th column).

   Apart from the model performance, we also observe the number of test samples converted to their corresponding adversarial samples successfully (5th row

Table 1. Performance results on IMDB movie review dataset.

	TextFool [7]	Proposed method using genre specific keywords	Proposed method w/o using genre specific keywords
CNN trained with original training set			
Accu. using original test set	74.53	74.53	74.53
Accu. using adversarial test set	74.13	32.55	57.31
%-age of perturbed samples	0.64	90.64	42.76
CNN re-trained with perturbed training set			
Accu. using original test set	68.14	78.00	78.81
Accu. using adversarial test set	68.08	78.46	78.21

in Table 1). The proposed algorithm was able to manipulate a large fraction of input data as opposed to existing methods. This metric is useful for text because unlike images, all text cannot be converted to adversarial counterparts while respecting grammar, spellings and preserving word level features. The seventh and eighth row in Table 1 show the accuracy of the original test set and the perturbed test set after re-training the CNN with adversarial samples. The difference in the accuracies in these two rows are very less, which shows that the model has generalized well after retraining.

4.2   Twitter Data for Gender Prediction

This dataset [1] contains Twitter comments on various topics from male and female users, and the task is to predict the user's gender. The dataset contains 20050 instances and 26 features. Though the Twitter dataset is much smaller in size and the tweets are also shorter in length than that in the IMDB dataset, but it is very challenging as the tweets contain lots of urls and hash tag descriptions. We consider two features of the dataset, 'texts' and 'description', for determining the gender. Out of 20050 instances, we only consider the instances where the feature 'gender confidence' score is 1. This leaves us with 10020 samples, out

Table 2. Performance results on Twitter gender classification dataset.

	TextFool	Proposed method
CNN trained with original training set		
Accu. using original test set	63.43	63.43
Accu. using adversarial test set	63.23	50.10
Ratio of perturbed samples	0.03	58.53
CNN re-trained with perturbed training set		
Accu. using original test set	61.82	61.68
Accu. using adversarial test set	61.34	60.07

of which 8016 samples are used for training and 2004 number of samples are used for testing purpose. We present the performance metrics for Twitter data in Table 2. It is evident that our method not only has been able to produce a larger number of adversarial samples, but they are also close in terms of semantic meaning of the text from their original form.

5   Conclusion

In this paper we describe a method to modify an input text to create an adversarial sample. To our best of knowledge, this is one of the first papers to synthesize adversarial samples from complicated text samples. Unlike images, in texts, a number of conditions must be satisfied while doing the modification steps to ensure the preservation of semantic meaning and the grammar of the text sample. In this paper, we are considering each word at a time and selecting the most appropriate modification in a greedy way. However, a much better approach is to consider each of the sentences in the text sample and modifying it which eventually confuses the classifier. The steps adopted for modifications are heuristic in nature, which can be improved and automated further to obtain better results.

References

1. CloudFlower: Twitter gender classification dataset. https://www.kaggle.com/crowdflower/twitter-user-gender-classification (2013)
2. Goodfellow, I.J., Shlens, J., Szegedy, C.: Explaining and harnessing adversarial examples. CoRR (2014). http://arxiv.org/abs/1412.6572
3. Hosseini, H., Kannan, S., Zhang, B., Poovendran, R.: Deceiving Google's perspective API built for detecting toxic comments. ArXiv e-prints (2017)
4. Kos, J., Fischer, I., Song, D.: Adversarial examples for generative models. ArXiv e-prints (2017)
5. Kulynych, B.: Project title (2017). https://github.com/bogdan-kulynych/textfool
6. Kurakin, A., Goodfellow, I.J., Bengio, S.: Adversarial examples in the physical world. CoRR (2016). http://arxiv.org/abs/1607.02533
7. Liang, B., Li, H., Su, M., Bian, P., Li, X., Shi, W.: Deep text classification can be fooled. ArXiv e-prints (2017)
8. Maas, A.L., Daly, R.E., Pham, P.T., Huang, D., Ng, A.Y., Potts, C.: Learning word vectors for sentiment analysis. In: Association for Computational Linguistics: Human Language Technologies (2011)
9. Papernot, N., McDaniel, P.D., Goodfellow, I.J.: Transferability in machine learning: from phenomena to black-box attacks using adversarial samples. CoRR (2016). http://arxiv.org/abs/1605.07277
10. Papernot, N., McDaniel, P.D., Goodfellow, I.J., Jha, S., Celik, Z.B., Swami, A.: Practical black-box attacks against machine learning. In: Asia Conference on Computer and Communications Security (2017)
11. Zhang, X., Zhao, J.J., LeCun, Y.: Character-level convolutional networks for text classification. In: Neural Information Processing Systems (2015)

Controlling Information Aggregation
for Complex Question Answering

Heeyoung Kwon[1]([⊠]), Harsh Trivedi[1], Peter Jansen[2], Mihai Surdeanu[2],
and Niranjan Balasubramanian[1]

[1] Stony Brook University, Stony Brook, NY 11790, USA
{heekwon,hjtrivedi,niranjan}@cs.stonybrook.edu
[2] University of Arizona, Tucson, AZ 85721, USA
pajansen@email.arizona.edu, msurdeanu@email.arizona.edu

Abstract. Complex question answering, the task of answering complex natural language questions that rely on inference, requires the aggregation of information from multiple sources. Automatic aggregation often fails because it combines semantically unrelated facts leading to bad inferences. This paper proposes methods to address this inference drift problem. In particular, the paper develops unsupervised and supervised mechanisms to control random walks on Open Information Extraction (OIE) knowledge graphs. Empirical evaluation on an elementary science exam benchmark shows that the proposed methods enables effective aggregation even over larger graphs and demonstrates the complementary value of information aggregation for answering complex questions.

1 Introduction

Question answering (QA), i.e., finding short answers to natural language questions, is one of the most important but challenging tasks on the road towards natural language understanding (Etzioni 2011). QA methods are moving beyond tackling simple factoid questions to more complex questions, which require aggregating and reasoning over multiple pieces of information. The elementary science exam benchmark, for example, includes questions that test the student's ability to reason over connected facts (Clark et al. 2013; Clark and Etzioni 2016; Jansen et al. 2016). Such aggregation however is prone to inference drift, where semantically unrelated pieces of information are combined resulting in spurious inferences. This is especially true when reasoning over knowledge graphs built using shallow semantic representations as Open Information Extraction (OIE) triples (Mausam et al. 2012). To mitigate this issue, recent QA methods that operate over OIE graphs limit themselves to reasoning with artificially short paths (Khot et al. 2017; Khashabi et al. 2017).

Rather than limit inference to smaller paths or sub-graphs, this paper proposes methods that allow for reasoning over larger graphs while still avoiding inference drifts. To this end, this paper formulates complex QA as a random walk method over knowledge graphs formed with OIE triples. Each random

© Springer International Publishing AG, part of Springer Nature 2018
G. Pasi et al. (Eds.): ECIR 2018, LNCS 10772, pp. 750–757, 2018.
https://doi.org/10.1007/978-3-319-76941-7_72

walk begins at some question nodes, and aggregates information by following relation edges to reach other nodes. For example, for a question asking if an iron nail is a conductor or an insulator, a random walk explaining the positive answer to this question begins with iron nail, and links facts such as (iron nail, is made of, iron), (iron, is a, metal), and (metals, are, electric conductors), all obtained from different sentences. Answer nodes are then ranked by the random walk scores computed over these paths. It is easy to imagine how this type of linking can lead to inference drift, where semantically irrelevant pieces of information for QA are combined so that causing spurious inferences, especially on large graphs.

Our main idea to minimize inference drift is to actively guide the random walks to stay on paths that are relevant to the question context. We explore two instantiations of this idea with PageRank (Haveliwala 2002): (i) We develop unsupervised estimations of node importance, edge and teleportation probabilities to guide the random walks. These are estimated based on some measure of similarity of the nodes and edges with respect to the overall question context. (ii) We develop a novel supervised PageRank formulation that directly estimates the node importance, edge and teleportation probabilities. We use a set of retrieval features to model node and edge importance and use supervision to combine these in order to directly maximize the end QA performance.

Our empirical evaluations on a standard science exam benchmark shows that: (i) controlled aggregation with drift-sensitive PageRank yields up to +3.2% gain in accuracy (precision@1) over standard topic-sensitive PageRank (TPR) (ii) the new supervised formulation yields even better results with a +3.7% gain over TPR, (iii) aggregation over sentences with drift-sensitive methods improves over a sentence-only model in an ensemble with a +2.0% gain. Overall, the results show that aggregation can be useful for complex questions when it is controlled carefully to stay in the question context.

2   Graph-Based Reasoning for QA

Complex QA can be formulated as a reasoning problem over knowledge graphs that aggregate related facts. Given an appropriate graph, finding an answer translates into a traversal of relevant facts that lead to the answer. In the case of multiple-choice questions, this can be cast as a ranking problem among nodes corresponding to the candidate answer choices. The idea is to induce a knowledge sub-graph that connects the question related nodes to candidate answer choices and then assess the strength of these connections. The implicit assumption is that a correct answer has stronger connections to the question nodes than incorrect answers. Figure 1 illustrates with an example question from a science exam. Nodes in this graph are question terms (blue), answer choices (green for correct and pink for incorrect), and other concepts that link the question and answer choices through OIE relations. As it shows, with automatically extracted OIE relations, the graph can include nodes and relations that cause inference drift (shown via the red node and dashed arrows). For example, even when the

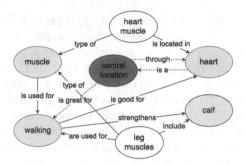

Fig. 1. OIE based knowledge graph for the question: `Which muscle is used for walking?` (A) `heart` (B) `calf` (Color figure online)

random walk begins from a question node, say "walking", inference can erroneously link (`central location, is great for, walking`), (`heart, is a, central location`), and (`heart, is a type of, muscle`) to conclude that (`heart, is used for, walking`).

We adapt Topic-sensitive PageRank (TPR) to assess the importance of the answer nodes. Consider a walker who starts from one of the question nodes and follows the graph structure. At any node i, the walker can either choose one of the outgoing edges with probability $(1 - d_i)$ or ignore the outgoing edges and instead teleport to one of the question nodes (i.e., the seed nodes) with probability d_i. The seed probability v_j specifies the likelihood that the random walker would choose to teleport to the node j (we discuss later multiple strategies for initializing seed probabilities). Intuitively, we use the seed probabilities to bias the random walks towards question nodes. This ensures that nodes well connected to question nodes will get higher PageRank (PR) scores.

Given the transition matrix A, the seed probabilities vector v, and the teleportation probabilities vector d, the PR vector π at time step $t + 1$ is: $\pi^{(t+1)} = (1 - d)A\pi^t + dv$. The answer choice that receives the highest PR score is returned as the answer.

2.1 Drift-Sensitive Page Rank

Even with TPR walks that are already biased towards question nodes, inference drift remains an issue. Effective solutions should ensure that reasoning stays close to paths that are relevant to specific question context. We propose two such solutions here.

Unsupervised Estimation. Our first method introduces node-specific teleportation probabilities that enhance the likelihood that teleportation actions land on nodes related to the question context. The method operates as follows. (i) We estimate *transition probabilities* by scoring the edges based on the relevance score given to the source sentence by the information retrieval model.

These scores are then normalized by their sum to turn them into probabilities. (ii) We estimate *node-specific teleportation* based on their importance within the question context. That is, when a walk reaches a high-relevance node, we want it to continue. When it reaches a low relevance node, however, we want it to teleport and restart from one of the question nodes. Thus we want the teleportation probability to be low when the relevance is high and vice versa. We experiment with the degree of the node and the similarity of the node to the question context as surrogates to estimate relevance (see Sect. 3). (iii) We use *seed probabilities* to capture relative node importance. Specifically, we use normalized scores from the abstractness-concreteness norms, a measure that was previously shown to be useful for measuring whether a word in the question is likely to be a key focus word (Jansen et al. 2017).

Supervised Page Rank. Our second method learns a parameterization of TPR that improves the ranking of correct answers. PageRank's effectiveness depends on the quality of the graph structure and the probability estimates used in the random walk: the transition (A), seed (v), and teleportation (d) probabilities. Rather than using independent estimation methods, we develop a supervised PR method that learns a non-linear function, via a two layer feed forward network, which combines well-known information retrieval features to compute the estimates. This supervised version can be written as a parametrization[1] of TPR as shown as follows: $\pi^{(t+1)} = (1 - d)A_\phi \pi^t + dv_\theta$, where θ is the parameter vector for seed probability features and ϕ the parameter vector for edge probability features. For seed probability estimation, we use three features: (1) focus word weights (Jansen et al. 2017), (2) Rocchio query expansion weights (Rocchio 1971), a traditional measure of relative term importance, and (3) an entropy based discriminativeness measure, that ensures a question word that occurs in all answer choices receives a low score, and one that occurs in only one answer choice gets a high score. For edge features, we use (1) IR score, which is retrieval score of triple from IR system when queried with question text, (2, 3) Context Score of source and target (word-wise word2vec similarity between question and each node with entailment score for out-of-vocabulary), (3, 4) Entropy based discriminativeness score of source and target, (5) Triple confidence score from OIE extraction, (6–9) boolean features for edge type (QI, IA, AI, II)[2].

Let x_s denote the seed probability features for a node s, and let z_{st} denote the edge probability features for an edge (s, t). Then, we have: $v_\theta(u) = \frac{f_\theta(x_s)}{\sum_{u \in G} f_\theta(x_u)}$ and $A_\phi(s, t) = \frac{g_\phi(z_{st})}{\sum_{e_{sq} \in G} g_\phi(z_{sq})}$. The functions $f_\theta(.)$ and $g_\phi(.)$ are feed-forward networks with 1 hidden layer of 3 nodes. Output layer has 1 node and Rectified Linear Unit (ReLU) is used for the activation units. Weights θ and ϕ control node specific reset probabilities and edge specific transition probabilities in neural

---

[1] This is inspired by the work of Gao et al. (2011), who use a linear parametrization but for a single graph problem using a different modeling approach.

[2] QI denotes edge between Question and Intermediate Node.

networks f and g respectively. Training finds parameters that maximize the following function:

$$\arg\max_{\theta,\phi} \sum_q \left((log(\pi(a_c)) - log\left(\sum_{a_i \in ans(q)} (\pi(a_i)) \right) \right)$$

This objective cannot be optimized in closed form but since it is differentiable we use Adam, a stochastic optimization method (Kingma and Ba 2014).

In contrast to the unsupervised setting, the bidirectional edges were unhelpful in the supervised setting due to stability issues in training. We report results on the graph where we remove backward edges (IQ, AQ and AI) for improved convergence.

3   Evaluation

We evaluate our approach for aggregation on a standard science exam benchmark[3], which consists of multiple choice questions for 6th-9th grade science. The dataset set includes 2,068 questions for training, of which we reserve 485 as a development set (dev) and a blind test set of 1,639 questions. We build knowledge graphs using OIE triples from five relevant corpora including study guides, an openly available textbook, and study flashcards totaling 588,472 triples.

3.1   Drift-Sensitive PageRank

Table 1 shows the accuracy of the drift-sensitive PageRank methods. All models were run on graphs built over top K sentences returned by a sentence retrieval model (Jansen et al. 2017). We set K = 40 based on dev set performance (see Table 2 for other K values). In order not to end up with several fragmented graphs, we introduced entailment edges. We only kept the high scored edges and the threshold is tuned also based on dev set performance. All drift-sensitive

Table 1. Comparison of different drift-sensitive methods: focus - Using abstract/concreteness norms based probabilities. quest. sim. - Using question similarity for teleportation probabilities

Method	Seeds	Teleportation	Test	Reference
Page rank	None	Uniform	35.51	
TPR	Uniform	Uniform	38.26	
drift-sensitive	Focus	Uniform	40.33	(A)
	Focus	Quest. sim.	41.49	(B)
	Sup.	Sup.	42.34	Sup.

[3] https://www.kaggle.com/c/the-allen-aiscience-challenge.

methods perform better than regular PR (top row) and TPR (second row). Using question words as seeds with uniform probabilities provides a +3.17 gain in accuracy over TPR. Using focus word weights as seed probabilities yields an additional +1.56 points. Using teleportation estimates based on question context (lower chance of teleportation when node is more similar to question) gives a +2.8 points gain. Supervised PageRank outperforms all variants with a +3.66 gain over basic TPR and +0.85 gain over the best unsupervised method.

3.2 Graph Sizes

We evaluated performance with knowledge graphs of different sizes by varying the number of source sentences (Table 2). Drift-sensitive methods perform better than the TPR across all data sizes, with more pronounced differences when the graph sizes are larger. Inference drift is more likely in large graphs and controlling random walks with drift-sensitive methods helps in these cases.

Table 2. Performance on different graph sizes using top X sentences to construct the graph

Method	Top 10	Top 20	Top 30	Top 40	Top 50
TPR	39.54	40.63	41.31	38.26	38.68
Unsupervised	41.00	41.55	42.46	40.33	39.84
Supervised	41.30	42.22	41.80	42.34	42.40

3.3 Utility of Aggregation

Aggregation is not only useful on its own, but it provides complementary benefits in an ensemble with non-aggregation based methods. We built an ensemble model with a strong non-aggregation model, a supervised sentence-retrieval model trained in (Jansen et al. 2017). The ensemble simply chooses between the two models based on their scores and the number of nodes and edges in the concept graphs. Most drift-sensitive models add value in this simple ensemble, with the supervised model providing a +2 points gain over the sentence model thus showing the complementary value of aggregation for complex questions (Table 3).

Table 3. Results of aggregation with a sentence retrieval model (sent)

Method	Sent	Sent + (A)	Sent + (B)	Sent + Sup.
Accuracy	43.44	44.30	45.58	45.45

4   Related Work

For the elementary science benchmarks with complex questions, a range of infer-ence methods have been explored, including probabilistic reasoning with first-order logic (Khot et al. 2015), constraints-based inference with semantic and shallow semantic structures (Clark et al. 2016; Khot et al. 2017), and graph-based methods with syntactic alignment and lexical semantics (Sharp et al. 2015). Sharp et al. (2015) show a method for aggregating information from multiple sentences using syntactic structure based alignments. Khot et al. (2017) show how OIE can benefit a constraint-based inference mechanism that tightly con-trols how multiple short facts can be combined. Fried et al. (2015) show that graphs built using words or syntactic dependencies can aggregate knowledge to improve performance on a community question answering task, but that long graph traversals lead to "semantic drift" and decreased performance. Our work builds on these ideas but explores the utility of the different graph methods explicitly through the use of page rank based methods.

5   Conclusions

Aggregating information from multiple texts is critical for answering complex questions. However, aggregation introduces spurious inferences, especially when using shallow semantic representations such as Open Information Extraction graphs. This forces models to limit reasoning to smaller paths or sub-graphs. Instead this paper introduces drift-sensitive variants of PageRank that allow for effective reasoning over large graphs. By controlling the random walks to stay on question contexts, the drift-sensitive methods achieve substantial gains over standard topic-sensitive PageRank and provides gains in an ensemble with a sentence-level model.

References

Etzioni, O.: Search needs a shake-up. Nature **476**(7358), 25–26 (2011)

Clark, P., Harrison, P., Balasubramanian, N.: A study of the knowledge base require-ments for passing an elementary science test. In: AKBC@CIKM (2013)

Clark, P., Etzioni, O.: My computer is an honor student - but how intelligent is it? Standardized tests as a measure of AI. AI Mag. **37**, 5–12 (2016)

Jansen, P., Balasubramanian, N., Surdeanu, M., Clark, P.: What's in an explanation? Characterizing knowledge and inference requirements for elementary science exams. In: COLING (2016)

Mausam, Schmitz, M., Soderland, S., Bart, R., Etzioni, O.: Open language learning for information extraction. In: EMNLP-CoNLL (2012)

Khot, T., Sabharwal, A., Clark, P.: Answering complex questions using open informa-tion extraction. CoRR, abs/1704.05572 (2017)

Khashabi, D., Sabharwal, T.K.A., Roth, D.: Learning what is essential in questions. In: COLING (2017)

Haveliwala, T.H.: Topic-sensitive pagerank. In: WWW (2002)

Jansen, P., Sharp, R., Surdeanu, M., Clark, P.: Framing QA as building and ranking intersentence answer justifications. In: Computational Linguistics (2017)

Gao, B., Liu, T.-Y., Wei, W., Wang, T., Li, H.: Semi-supervised ranking on very large graphs with rich metadata. In: ACM SIGKDD, pp. 96–104 (2011)

Rocchio, J.J.: Relevance feedback in information retrieval. In: Salton: The SMART Retrieval System: Experiments in Automatic Document Processing (1971)

Kingma, D., Ba, J.: Adam: a method for stochastic optimization. arXiv preprint arXiv:1412.6980 (2014)

Khot, T., Balasubramanian, N., Gribkoff, E., Sabharwal, A., Clark, P., Etzioni, O.: Exploring Markov logic networks for question answering. In: EMNLP (2015)

Clark, P., Etzioni, O., Khot, T., Sabharwal, A., Tafjord, O., Turney, P.D., Khashabi, D.: Combining retrieval, statistics, and inference to answer elementary science questions. In: AAAI (2016)

Sharp, R., Jansen, P., Surdeanu, M., Clark, P.: Spinning straw into gold: using free text to train monolingual alignment models for non-factoid question answering. In: HLT-NAACL (2015)

Fried, D., Jansen, P., Hahn-Powell, G., Surdeanu, M., Clark, P.: Higher-order lexical semantic models for non-factoid answer reranking. Trans. Assoc. Comput. Linguist. **3**, 197–210 (2015)

Local Is Good: A Fast Citation Recommendation Approach

Haofeng Jia and Erik Saule[✉]

Department of Computer Science, UNC Charlotte, Charlotte, USA
{hjia1,esaule}@uncc.edu

Abstract. Finding relevant research works from the large number of published articles has become a nontrivial problem. In this paper, we consider the problem of citation recommendation where the query is a set of seed papers. Collaborative filtering and PaperRank are classical approaches for this task. Previous work has shown PaperRank achieves better recommendation in experiments. However, the running time of PaperRank typically depends on the size of input graph and thus tends to be expensive. Here we explore LocRank, a local ranking method on the subgraph induced by the ego network of the vertices in the query. We experimentally demonstrate that LocRank is as effective as PaperRank while being 15x faster than PaperRank and 6x faster than collaborative filtering.

1 Introduction

Scientists around the world have published tens of millions of research papers, and the number of new papers has been increasing with time. At the same time, literature search became an essential task performed daily by thousands of researcher around the world. Finding relevant research works from the large number of published articles has become a nontrivial problem.

There are several developed paper recommendation systems based on keywords, such as Google Scholar or Mendeley. However, the drawbacks of keyword based systems are obvious: first of all, the vocabulary gap between the query and the relevant documents might result in poor performance; moreover, a simple string of keywords might not be enough to convey the information needs of researchers. There are many cases where a keyword query is either over broad, returning many articles that are loosely relevant to what the researcher actual need, or too narrow, filtering many potentially relevant articles out or returning nothing at all.

Researchers have devoted efforts on citation recommendation based on a set of seed papers [1–10], which can avoid above mentioned problems. Intuitively, most approaches rely on the citation graph to recommend relevant papers. In general, there are two classic frameworks: collaborative filtering (CF) [1,2] and random walk with restart [3,7,9], of which PaperRank [3] is the most representative work. The different approaches to recommending academic papers have been extensively surveyed [11].

© Springer International Publishing AG, part of Springer Nature 2018
G. Pasi et al. (Eds.): ECIR 2018, LNCS 10772, pp. 758–764, 2018.
https://doi.org/10.1007/978-3-319-76941-7_73

Previous work has shown PaperRank achieves better performance on recall than CF, while CF is more efficient. The running time of PaperRank typically depends on the size of input graph and thus tends to be more expensive. Collaborative filtering essentially computes the weighted co-citation relationships and thus does not need to take the global graph into account. However, for a practical citation recommendation system, both quality and speed of recommendation are important.

To this end, we explore LocRank, a local method of which the running time depends on the size of the local subgraph of the query. Experiments demonstrate the LocRank provide recommendation of comparable quality to PaperRank; however, LocRank is 15x faster than PaperRank and 6x faster than CF.

2 A Local Ranking Method

2.1 Citation Recommendation

Let $G = (V, E)$ be the citation graph, with n papers $V = \{v_1, \ldots, v_n\}$. In G, each edge $e \in E$ represents a citation relationship between two papers. We use $Ref(v)$ and $Cit(v)$ to denote the reference set of and citation set to v, respectively. And $Adj(v)$ is used to denote the union of $Ref(v)$ and $Cit(v)$. We will use an superscript denoting the graph used when it is not obvious.

In this paper, we focus on the citation recommendation problem assuming that researchers already know a set of relevant papers; this assumption is not uncommon and is made by systems such as the Advisor [9]. Therefore, the recommendation task can be formalized as:

Citation Recommendation. Given a set of seed papers S as a query, return a list of papers ranked by relevance to the ones in S.

Collaborative Filtering [1] has been proven to be an effective idea for most recommendation problems. For citation recommendation, a ratings matrix is built using the adjacent matrix of citation graph, where citing papers correspond to users and citations correspond to items. A pseudo target paper that cites all seed papers is added to the matrix. Then, CF computes the cosine similarity of all papers with the target paper and identify x *peer* papers, having the highest similarity to the target paper. Then each paper is scored by summing the similarity of the peer paper that cites it.

PaperRank [3] is a biased random walk proposed to recommend papers based on citation graph. In particular, the restarts probability from any paper will be distributed to only the seed papers.

2.2 LocRank

In general, PaperRank tends to find those papers which are "close" to the seeds and have high eigenvector centralities. Collaborative filtering essentially focuses on a local co-citation relationships of seeds papers. Now we present LocRank, a method that can provide high quality recommendations at low computational cost.

We define a local induced subgraph of a query q: $G_q = (V_q, E_q)$, where V_q contains all nodes in S and any node which is a neighbor of at least one seed paper:

$$V_q = S \cup S_n$$

where S_n denotes

$$S_n = \bigcup_{s \in S} \{v : v \in Adj(s)\}$$

E_q remains all citation relationships between nodes in V_q. In other words, G_q is the subgraph induced by the distance 1 neighborhood of the seed papers.

Given the induced subgraph G_q, LocRank computes a random walk on G_q. It assumes a random walker in paper v continues to a neighbor with a damping factor d, and with probability $(1 - d)$ it restarts at one of the seed papers in S. The edges are followed proportionally to the weight of that edge w_{ji} which is often set to 1, but can be set to the number of time paper i is referenced by paper j.

$$R(v_i) = (1 - d)\frac{1}{|S|} + d \times \sum_{v_j \in Adj^{G_q}(v_i)} \frac{w_{ji}}{\sum_{v_k \in Adj^{G_q}(v_j)} w_{jk}} R(v_j)$$

After convergence, LocRank returns papers based on $R(v), v \in S_n$.

3   Expriments

3.1   Data Preparation

There are many publicly available academic data sets. However, none of them is perfect for the citation recommendation task. For instance, Microsoft Academic Graph[1] [12] contains abundant information from various disciplines but it is fairly noisy: some important attributes are missing or wrong. In contrast, the records in DBLP are much more reliable although it does not contain citation information.

To obtain a clean and comprehensive academic dataset, we match Microsoft Academic Graph, CiteSeerX[2] and DBLP[3] [13] datasets through DOI and titles for their complementary advantages. In this way, we derive a corpus of Computer Science papers consisting of 2,035,246 Papers and 12,439,090 Citations.

3.2   Experiment Setup

In order to simulate the typical usecase where a researcher is writing a paper and tries to find some more references, we design the random-hide experiment. First of all, a query paper q with 20 to 200 references and published between

---

[1] https://www.microsoft.com/en-us/research/project/microsoft-academic-graph/.
[2] http://citeseerx.ist.psu.edu/.
[3] http://dblp.uni-trier.de/xml/.

2005 to 2010 is randomly (uniformly) selected from the dataset. We then remove the query paper q and all papers published after q from the citation graph to simulate the time when the query paper was being written. Instead of using hide-one strategy [1,2], we randomly hide 10% of the references and place them in the *hidden set*. This set of hidden papers is used as ground truth to recommend. The remaining papers are used as the set of seed papers.

For evaluating effectiveness of recommendation algorithm, we use *recall@k*, the ratio of hidden papers appearing in top k of the recommended list. Following results are based on the performance for 2,500 independent randomly selected queries.

The codes are written in C++. The graphs are represented in Compressed Row Storage format for compact storage. The codes are compiled with g++ 4.8.2 with option -O3. The codes are run on 1 core of an Intel(R) Xeon CPU E-5-2623 @ 3.00 GHz processor.

3.3   Results on Efficiency

Table 1 shows the average runtime for a single query and the 95% confidence intervals. In general, LocRank is around 15x faster than PaperRank and even 6x faster than CF. We compare LocRank with PaperRank and CF respectively as follows.

Table 1. Performance on Effciency (in second)

Method	Mean runtime	95% Confidence interval	
CF	1.0070	1.0067	1.0073
PaperRank	2.4796	2.4621	2.4972
LocRank	0.1674	0.1671	0.1677

LocRank vs. PaperRank. Basically, LocRank is much faster than PaperRank because of the following two reasons: First of all, LocRank is a ranking method of which the runtime only depends on the size of local induced graph, while PaperRank is a global ranking method; Secondly, a local induced graph tends to have a smaller diameter, which means LocRank can reach the convergence within less iterations.

LocRank vs. CF. Although LocRank and CF are both local approaches for citation recommendation, CF needs to sort the neighbors based on their similarities with the pseudo target paper for each query.

From the confidence interval in Table 1, it is easy to see that the standard deviation of runtime of PaperRank is larger. This is because we remove the query paper q and all papers published after q from the citation graph to simulate the time when the query paper was being written, so the size of the global graph

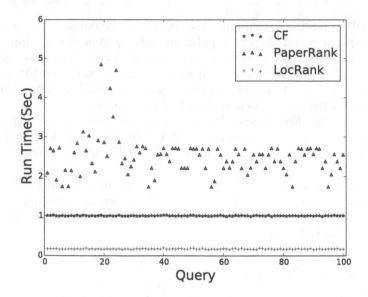

Fig. 1. Runtime for 100 instance queries

is different for queries with different publication date. In Fig. 1, we show the runtime for 100 randomly sampled independent queries. The fluctuations in the runtime of PaperRank corresponds its larger standard deviation.

3.4 Results on Effectiveness

Although LocRank is a ranking method on local induced graph, the quality of recommendation of LocRank is competitive with classical methods. Table 2 shows the results on mean recall of 2,500 independent queries. 95% confidence interval at top 50 demonstrates the difference between PaperRank and LocRank is not statistically significant, while LocRank and PaperRank are statistically significantly better than CF.

Table 2. Performance on Recall

Method	@10	@20	@30	@40	@50	95% Confidence Interval@50	
CF	0.1894	0.2669	0.3158	0.3597	0.3917	0.3775	0.4059
PaperRank	0.2344	0.3260	0.3895	0.4360	0.4715	0.4573	0.4857
LocRank	0.2395	0.3281	0.3900	0.4338	0.4679	0.4537	0.4821

Essentially, the LocRank is a tradeoff between the upper bound of recall and the efficiency. It turns out that LocRank has an equivalent ability to find hidden papers as PaperRank, which demonstrates that many findable hidden papers are actually neighbors of seed papers. Even though PaperRank could find 100% of the hidden paper, it is reasonable that the hidden papers that are not directly connected to a seed paper are hard to find for current citation recommendation systems. Some recent work has shown that popular methods are poor to find loosely connected hidden papers [10]. There are many reasons that can result in this phenomenon. For example, a machine learning paper will typically cite previous works on the same problem, at the same time, it is likely to cite a bioinformatics paper to demonstrate the wide applications of the proposed method. In this case, it is hard to infer this bioinformatics paper based on the rest of references.

4   Conclusion and Future Work

This paper explores LocRank, a local method for recommending paper in a citation recommendation problem. The runtime of LocRank depends on the size of the local subgraph of the query. Experiments show that LocRank is 15x faster than PaperRank and 6x faster than CF. At the same time, the performance on recall demonstrates that LocRank is still as effective as PaperRank, and better than CF. In the future, it would be interesting to investigate other local methods that can achieve cheerful performance on both efficiency and effectiveness. In particular closeness centrality and k-core decomposition of the local graph seems relevant and easy to compute.

Acknowledgments. This material is based upon work supported by the National Science Foundation under Grant No. 1652442.

References

1. McNee, S.M., Albert, I., Cosley, D., Gopalkrishnan, P., Lam, S.K., Rashid, A.M., Konstan, J.A., Riedl, J.: On the recommending of citations for research papers. In: Proceedings of CSCW, pp. 116–125 (2002)
2. Torres, R., McNee, S.M., Abel, M., Konstan, J.A., Riedl, J.: Enhancing digital libraries with techlens+. In: Proceedings of JCDL, pp. 228–236 (2004)
3. Gori, M., Pucci, A.: Research paper recommender systems: a random-walk based approach. In: Proceedings of Web Intelligence, pp. 778–781 (2006)
4. Ekstrand, M.D., Kannan, P., Stemper, J.A., Butler, J.T., Konstan, J.A., Riedl, J.T.: Automatically building research reading lists. In: Proceedings of RecSys, pp. 159–166 (2010)
5. El-Arini, K., Guestrin, C.: Beyond keyword search: discovering relevant scientific literature. In: Proceedings of KDD, pp. 439–447 (2011)
6. Golshan, B., Lappas, T., Terzi, E.: Sofia search: a tool for automating related-work search. In: Proceedings of SIGMOD, pp. 621–624 (2012)
7. Caragea, C., Silvescu, A., Mitra, P., Giles, C.L.: Can't see the forest for the trees?: a citation recommendation system. In: Proceedings of JCDL, pp. 111–114 (2013)
8. Küçüktunç, O., Kaya, K., Saule, E., Çatalyürek, Ü.V.: Fast recommendation on bibliographic networks. In: Proceedings of ASONAM (2012)

9. Küçüktunç, O., Saule, E., Kaya, K., Çatalyürek, Ü.V.: Towards a personalized, scalable, and exploratory academic recommendation service. In: Proceedings of ASONAM (2013)
10. Jia, H., Saule, E.: An analysis of citation recommender systems: beyond the obvious. In: Proceedings of ASONAM (2017)
11. Beel, J., Gipp, B., Langer, S., Breitinger, C.: Research-paper recommender systems: a literature survey. Int. J. Digit. Libr. **17**(4), 305–338 (2016)
12. Sinha, A., Shen, Z., Song, Y., Ma, H., Eide, D., Hsu, B.J.P., Wang, K.: An overview of Microsoft Academic Service (MAS) and applications. In: Proceedings of WWW, pp. 243–246 (2015)
13. Ley, M.: DBLP - some lessons learned. PVLDB **2**(2), 1493–1500 (2009)

On Refining Twitter Lists as Ground Truth Data for Multi-community User Classification

Ting Su[⊠] , Anjie Fang, Richard McCreadie, Craig Macdonald,
and Iadh Ounis

University of Glasgow, Glasgow, UK
{t.su.2,a.fang.1}@research.gla.ac.uk,
{richard.mccreadie,craig.macdonald,iadh.ounis}@glasgow.ac.uk

Abstract. To help scholars and businesses understand and analyse Twitter users, it is useful to have classifiers that can identify the communities that a given user belongs to, e.g. business or politics. Obtaining high quality training data is an important step towards producing an effective multi-community classifier. An efficient approach for creating such ground truth data is to extract users from existing public Twitter lists, where those lists represent different communities, e.g. a list of journalists. However, ground truth datasets obtained using such lists can be noisy, since not all users that belong to a community are good training examples for that community. In this paper, we conduct a thorough failure analysis of a ground truth dataset generated using Twitter lists. We discuss how some categories of users collected from these Twitter public lists could negatively affect the classification performance and therefore should not be used for training. Through experiments with 3 classifiers and 5 communities, we show that removing ambiguous users based on their tweets and profile can indeed result in a 10% increase in F1 performance.

1 Introduction

Due to the popularity of social media platforms, such as Twitter, people with different backgrounds can express their views towards topics during events (e.g. elections). Indeed, such platforms have become a major channel to share ideas and opinions [1]. In previous studies, researchers have used text corpora (e.g. articles, books or speeches) to analyse how social groups influence one another (e.g. how journalists influence the public [2]). However, the emergence of social media as a popular communication medium and the relative ease of collecting large volumes of user and post data, provides new opportunities for researchers to better analyse how social groups/communities interact. On the other hand, users do not explicitly specify their social group/community affiliations. Hence, researchers need to resort to automatic approaches to infer this at scale. One popular means to classify Twitter users into different groups is to train learned classification models [3–5]. These approaches require a high quality training dataset

© Springer International Publishing AG, part of Springer Nature 2018
G. Pasi et al. (Eds.): ECIR 2018, LNCS 10772, pp. 765–772, 2018.
https://doi.org/10.1007/978-3-319-76941-7_74

Table 1. Lists used to extract users for each community.

Community	Lists used	# users
ACA	Higher Ed Thought Leaders(@MSCollegeOpp), Edu-Scholars(@sesp_nu) Favourite academics(@AcademiaObscura), Northwestern(@sesp_nu), SESP Alumni(@sesp_nu), STEM Academic Tweeters(@LSEImpactBlog), The Academy(@AcademicsSay), Harvard(@hkslibrary)	3592
BSN	Social CEOs on Twitter(@debweinstein), Tech, Startups & Biz(@crblev) Tech Startup Founders(@realtimetouch), Top CEO's(@chrisgeorge187), Awesome Entrepreneurs(@vincentdignan)	3013
MDA	Mirror Political Journos(@MirrorPolitics), BBC News Official(@BBCNews) sunday-mirror(@DailyMirror), Financial Tweets(@TIME), TIME Staff(@TIME), Sun accounts(@TheSun), Sun people(@TheSun), Mirror reporters/columist(@DailyMirror), BBC Asian Network(@BBC), BBC News(@BBC), Business staff(@guardian), Observer staff(@guardian), Money staff(@guardian),Technology staff (@guardian), Politics staff(@guardian), News staff (@guardian)	1242
PLT	UK MPs(@TwitterGov), US Governors(@TwitterGov), US Senate(@TwitterGov), US House(@TwitterGov), Senators(@CSPAN), New Members of Congress(@CSPAN)	1899
CTZ (celebrities)	celebrity(@mashable),the-celebrity-list(@buzzedition) celebrities(@GALUXSEE)	774
CTZ (normal users)	N/A	800

to produce effective models, especially when it comes to difficult multi-class classification tasks, such as community classification. However, topical overlap between communities and ambiguous user affiliations makes training accurate and generalisable models challenging. Unsupervised clustering can also be used for community identification task [6], however the clusters obtained may not reflect predefined notions of communities, and hence supervised methods are of interest.

In this paper, we aim to produce a reliable dataset that can be used to effectively perform community classification of Twitter users into five classes:

- **Academic (ACA)**: users involved in research and/or teaching.
- **Business (BSN)**: company executives, managers and other white-collar workers.
- **Media (MDA)**: journalists or reporters working for news-media or as freelancers.
- **Politics (PLT)**: politically-active users, e.g. members of parliament or activists.
- **Citizen (CTZ)**: users who do not belongs to any of the 4 other classes.

Our goal is to cover these five particular roles in political events, based on users' jobs and/or social roles. However, in reality, users may have multiple roles, temporally/permanently switch roles, or act as if they have different roles. This can make it challenging to conduct accurate training and generalise models to categorise users based on their profile and past tweets. This also makes existing methods, which are often based on crowdsourced data (e.g. [7]) or automatic user behaviour analysis based on predefined rules (e.g. hashtag usage [8], words in profiles [9], location and name [10]) unsuitable, because human-labelled data is expensive, and user classification based on their behaviours can be vague when classes overlap with each other.

Hence, as a first step towards producing high quality training data for multi-community classification, we examine the effectiveness of a list-based approach [11], as well as investigate where it tends to fail. We first collect Twitter lists representing our target communities, and then crawl the posts and profiles for each user in those lists. We analyse these lists, with the aim of removing users that might make poor quality training examples, producing several (more refined) datasets. We then train several supervised community classification models based on the original and 5 refined datasets, and compare their performance when tested on a separate gold-standard human-labelled dataset. In this way, we establish the raw performance of models produced by the list-based approach, as well as show how removing potentially problematic users leads to better classification models that can increase performance by up to 10%. Finally, we discuss the main issues observed when relying on Twitter lists for use as training data.

Thus, the contributions of this paper are as follows: (1) We conduct a failure analysis of a training dataset for multi-community classification that is automatically generated from Twitter public lists. (2) We discuss four categories of problematic users collected from these Twitter public lists, and empirically show that they can negatively affect classification performance when used for training.

2   Analysis of Community Lists and Users

To evaluate the effectiveness of list-based approaches when used for training community classification models, we construct an evaluation dataset and analyse categories of users that may cause issues when training.

Dataset Collection: To create an initial dataset, we extract users from existing public lists on Twitter. For each community, the lists we use, and the number of users obtained from lists per class are shown in Table 1. These lists were selected based on their descriptions. For example, `Edu-Scholars` is described as "A selection of the nation's most influential academics in education" by its creator. Hence, we consider its users within our ACA category. The outlier is the 'citizen' (CTZ) class, for which we first extract celebrity users from lists as examples of citizens. Normal citizens are not likely to be collected in any public lists, hence we also randomly sample users with <200 followers along with the celebrities.

User Analysis: Having collected users belonging to each community, we randomly select 1000 users per community to form an initial and balanced dataset for training classifiers, and manually analyse those users based on their Twitter profiles and previous posts. Based on this analysis, we identify four categories of users that might be problematic if used as examples to train a community classifier. The prevalence of each category within the communities is shown in Table 2.

1. **Users with ambiguous descriptions. (Category 1).** We observe that there are a subset of users that would clearly label themselves as a member of a community, but in practice mainly tweeted/shared content on other unrelated topics. These off-community tweets can confuse text-based classifiers as they include words that may be associated to other communities.
2. **Users retweeting/sharing links without adding comments. (Category 2).** We observe that there is an active group of users who only retweet the community-related popular topics (e.g. others tweets, links, links with title, etc.), but without their own opinions. Since links and words used in article titles tend to exist among all communities, including users who only tweet about such topics when training may add noise to the resultant classification model.
3. **Users (re)tweet useless content. (Category 3).** We observe that some users only make tweets containing 'useless' content, such as motivational pictures or quotes, tweets generated by other platforms, and advertisements links. Such tweets can contain highly duplicated content and off-community words, which can potentially reinforce classifiers with false features, and may lead to weak classifier models.
4. **Non-active users. (Category 4).** Public lists can be quite old and unmaintained, and hence can include users that have been inactive for years. Using users that have been inactive for an extended period of time may be problematic for training purposes, as the types of discussion topics that help distinguish a community change over time. Hence, training on old users/tweets may hinder the development of accurate classification methods.

3   Investigating the Impact of User Filtering

Having produced a tweet dataset for community classification and identified some potential issues that might arise when using it for training, we now examine if the issues we have identified do indeed impact upon classification performance.

3.1   Experimental Setup

Methodology. To evaluate community classification, we train classification models based on the dataset discussed above. As discussed above, we randomly

Table 2. Number of users in each categories.

	ACA	BSN	CTZ	MDA	PLT
AllUsers	1000	1000	1000	1000	1000
Category 1	323	411	27	298	2
Category 2	47	111	54	194	1
Category 3	27	80	49	30	0
Category 4	5	19	108	10	4
Non-English	38	33	60	5	0

Table 3. Number of users in training & test datasets.

	ACA	BSN	CTZ	MDA	PLT
AllUsers	1000	1000	1000	1000	1000
AllUsers - Category 1	590	589	590	590	590
AllUsers - Category 2	800	800	800	800	800
AllUsers - Category 3	920	920	920	920	920
AllUsers - Category 4	900	900	892	900	900
All Filtered	350	346	350	350	350
Crowdsourced Test set	80	163	337	159	57

sample 1000 users from collected user lists of each of the five communities to form a balanced training dataset (denoted *AllUsers*). However, to determine what effect the four categories of potentially undesirable users have on classification performance, we produce alternative datasets (denoted *AllUsers - Category X*) that do not contain users from one of the identified categories, and adjust the number in each community to form balanced datasets. Finally, we create another dataset (denoted *AllFiltered*), by removing users of all the identified categories. Details about each dataset are provided in Table 3. We train classification models based on all 6 datasets using three types of learner, namely: Support Vector Machine (SVM), Multinomial Naive Bayes (NB) and Multilayer Perceptron (MLP).

Gold Standard. Having defined the training datasets, we next need a gold standard that we can evaluate against. To create this, we randomly sample another 1000 Twitter users (who do not appear in any of the training datasets) and use crowdsourced workers to manually label each user's community affiliation by examining that user's profile and his/her 8 most recent tweets. Three workers labelled each user and a majority vote is used to produce the final label for a user. If a majority could not be reached for a user, then more workers labelled that user until a majority was obtained (7.3% of the users required such additional labels to reach a majority).[1] Details about the test set are provided in Table 3.

Classifier Configuration. For the purposes of building the user classification models, we use the 20k most frequently occurring terms across all user's tweets and profile descriptions as features, after applying stopword removal and stemming. Each term is represented by its TF-IDF score. The configuration settings for the three learned models are: Multinomial NB $\alpha = 0.01$; for SVM we use a

---

[1] ~20% of accounts have been removed from Twitter, and are excluded from our test dataset.

Table 4. The F1 scores with different training data.

Classifier	Training Dataset	ACA	BSN	CTZ	MDA	PLT	Micro
RDN	AllUsers	0.10	0.15	0.24	0.15	0.08	0.18
NB	AllUsers	0.45	0.47	0.59	0.35	0.46	0.49
	AllUsers - Category 1	**0.47▲**	0.47	**0.61**	**0.41▲**	0.37	**0.51**
	AllUsers - Category 2	0.43	0.44	0.59	**0.40**	0.38	0.49
	AllUsers - Category 3	**0.46▲**	0.45	0.59	**0.42**	0.36	**0.50**
	AllUsers - Category 4	0.44	0.47	0.59	**0.37**	**0.47**	0.49
	AllFiltered	**0.49**	0.43	**0.62**	**0.39**	0.34	**0.50**
SVM	AllUsers	0.45	0.42	0.61	0.34	0.45	0.49
	AllUsers - Category 1	**0.48**	**0.53**	**0.64**	**0.42▲**	0.34	**0.54▲**
	AllUsers - Category 2	0.42	**0.52**	**0.63**	**0.35**	**0.46▲**	**0.52**
	AllUsers - Category 3	0.45	**0.49▲**	**0.63**	**0.41**	**0.46▲**	**0.53**
	AllUsers - Category 4	0.45	0.41	0.60	**0.36**	**0.47**	0.49
	AllFiltered	0.40	0.39	**0.62**	**0.38**	0.37	0.49
MLP	AllUsers	0.44	0.43	0.60	0.33	0.43	0.48
	AllUsers - Category 1	**0.46▲**	**0.49▲**	**0.63**	**0.41▲**	0.28	**0.51**
	AllUsers - Category 2	0.44	**0.47**	0.59	**0.34**	**0.48▲**	**0.50**
	AllUsers - Category 3	**0.45**	**0.45▲**	**0.62▲**	**0.36**	**0.45**	**0.50**
	AllUsers - Category 4	0.44	0.43	**0.61**	**0.34**	**0.44**	**0.49**
	AllFiltered	0.44	0.41	**0.62**	0.31	**0.34▲**	0.48

Linear kernel, L2 penalty, $C = 1.0$, $\gamma = 0.001$, and $multi\_class$ = one-vs-all[2]; for MLP we use one hidden layer with 500 neurons. All of the above parameters are obtained using a 10-fold cross-validation on the training dataset.

Baseline. To provide a basis for comparison, we also report the performance of a Random Classifier using uniform distribution (denoted as RDN) as a baseline.

Metric. We report F1 for classes, and Micro F1 across all classifiers and datasets.

3.2   Results

In this section, we report the outcome of our comparison between models trained on the AllUsers dataset and all other datasets. Table 4 reports classification performances for three learned models across 6 datasets. Scores highlighted in bold indicate increased performance over AllUsers. "▲" denotes statistically significant increases in performance (McNemar's test, $p < 0.05$) over AllUsers.

First, in Table 4, we observe that all the classifiers across all tested datasets achieve Micro F1 scores higher than 0.48, which is markedly higher than the RDN classifier (0.18). This indicates that the models produced are able to distinguish

---

[2] One-vs-all is the recommended setup for multi-class classification using SVM [12].

between the user classes. Next, comparing the classification models produced on each dataset, we observe that, for all three classifiers tested using all 5 filtered datasets, Micro F1 performance is greater than or equal to (by up to 10%) than that of AllUsers. Hence, the user categories identified in Sect. 2 do have negative impacts on community classification when used as training examples.

Among the four categories proposed in Sect. 2, we see that for all three classifiers, using dataset AllUsers - Category 1 as training set provides the highest Micro F1 scores, and obtains a significantly benefited SVM classifier compared to using AllUsers (McNemar's test, $p < 0.05$). As described above, AllUsers - Category 1 is the list that excludes ambiguous users, who mostly tweet about other communities. Indeed, by excluding users that appear to overlap with other communities, the classifiers perform better, as the difference between classes is clearer. Surprisingly, using the most sanitised dataset, namely AllFiltered, does not improve the result significantly. One reason can be that, as the size of dataset AllFiltered is only a third of the AllUsers dataset, the variety of text for the classifier to learn from is reduced, resulting in lower performance.

For the most difficult community observed, namely MDA, excluding ambiguous users (i.e. Category 1) results in an up-to 24.2% increase in the F1 score across almost all datasets and models. However, excluding the other 3 categories does not demonstrate a consistent benefit to F1 across the classifiers. Indeed, it is clear that ambiguous users are the most harmful for classifying MDA users.

4   Conclusions

In this paper, we examined how to construct a robust community classification dataset for Twitter and investigated challenges associated with selecting users as training examples. In particular, we first collected Twitter lists representing target communities and collected associated posts and profiles from each user. We analysed these lists, with the aim of performing a failure analysis, thereby identifying four categories of user that might be problematic and make poor training examples for classification. Therefore, we produced various datasets, by excluding users from each of the identified categories. We then trained several supervised community classification models based on the original and filtered datasets, and compared their performance when tested on a separate human-labelled gold-standard. We showed that public Twitter lists can be used as training data when analysing Twitter users, as all classifiers using AllUsers dataset achieved at least 0.48 Micro F1. On the other hand, the 4 categories of users we identified can be problematic, as Micro F1 scores increased by up to 10% when excluding each category in turn from the training dataset. Removing Category 1 (ambiguous users) in particular cause the largest increase in performance. Future studies are needed to develop automatic methods to identify and exclude such users for more effective community classification training datasets.

References

1. Purcell, K., Rainie, L., Mitchell, A., Rosenstiel, T.: Understanding the participatory news consumer. Pew Internet Am. Life Proj. **1**, 19–21 (2010)
2. Erikson, R., MacKuen, M., Stimson, J.: The Macro Polity. Cambridge University Press, Cambridge (2002)
3. Culotta, A., Kumar, N., Cutler, J.: Predicting the demographics of Twitter users from website traffic data. In: Proceedings of AAAI (2015)
4. Pennacchiotti, M., Popescu, A.: A machine learning approach to Twitter user classification. In: Proceedings of ICWSM (2011)
5. De Choudhury, M., Diakopoulos, N., Naaman, M.: Unfolding the event landscape on Twitter: classification and exploration of user categories. In: Proceedings of CSCW (2012)
6. Sachan, M., Dubey, A., Srivastava, S., Xing, E.P., Hovy, E.: Spatial compactness meets topical consistency: jointly modeling links and content for community detection. In: Proceedings of the ICWSDM (2014)
7. Chen, X., Wang, Y., Agichtein, E., Wang, F.: A comparative study of demographic attribute inference in Twitter. In: Proceedings of ICWSM (2015)
8. Fang, A., Ounis, I., Habel, P., Macdonald, C., Limsopatham, N.: Topic-centric classification of Twitter user's political orientation. In: Proceedings of SIGIR (2015)
9. Feng, V., Hirst, G.: Detecting deceptive opinions with profile compatibility. In: Proceedings of IJCNLP (2013)
10. Bergsma, S., Dredze, M., Van Durme, B., Wilson, T., Yarowsky, D.: Broadly improving user classification via communication-based name and location clustering on Twitter. In: Proceedings of HLT-NAACL (2013)
11. Bagdouri, M., Oard, D.: Profession-based person search in microblogs: using seed sets to find journalists. In: Proceedings of CIKM (2015)
12. Rifkin, R., Klautau, A.: In defense of one-vs-all classification. J. Mach. Learn. Res. **5**, 101–141 (2004)

Entity Retrieval via Type Taxonomy Aware Smoothing

Xinshi Lin[✉] and Wai Lam

Department of Systems Engineering and Engineering Management,
The Chinese University of Hong Kong, Sha Tin, Hong Kong
{xslin, wlam}@se.cuhk.edu.hk

Abstract. We investigate the task of ad-hoc entity retrieval from a knowledge base. We propose a type taxonomy aware smoothing method that exploits the hierarchical type information of a knowledge base and integrates into an existing language modelling framework. Unlike most existing type-aware retrieval models, our approach does not require an explicit inference of query type. Instead, it directly encodes the type information into a term-based retrieval model by considering the occurrence of query terms in multi-fielded pseudo documents of entities whose types have connections in the type taxonomy. We conduct experiments on a recent public benchmark dataset with the Wikipedia category information. Preliminary experiment results show that our framework improves the performance of existing models.

Keywords: Entity retrieval · Structure-aware smoothing · Type taxonomy

1 Introduction

An intelligent search engine should answer either precise or vague queries from users as well as handle diverse types of returned results: documents or entities. An entity is usually described by subject-predicate-object (SPO) triples in structured knowledge base or knowledge graph, as an object that has names, attributes, types and meaningful relations to other entities. Over the past decade, the availability of large-scale knowledge base such as DBpedia or Freebase significantly motivates research in ad-hoc entity retrieval.

Most previous approaches employ standard document retrieval methods which still work under the new setting of entity retrieval via predicate folding [1] that groups fields into several predefined categories based on the type of predicates and generate a pseudo document by aggregating SPO triples in each category. These term-based methods include BM25 [2], language modeling (LM) based ones [1, 3] and its multi-fielded extensions [4]. In recent years, Markov random field based retrieval models that take term dependence into account with well-designed or adaptive fielded representation have shown their effectiveness. These works include sequential dependence model (SDM) [5], fielded sequential dependence model (FSDM) [5, 6] and its parameterized

The work described in this paper is substantially supported by a grant from the Research Grant Council of the Hong Kong Special Administrative Region, China (Project Code: 14203414).

© Springer International Publishing AG, part of Springer Nature 2018
G. Pasi et al. (Eds.): ECIR 2018, LNCS 10772, pp. 773–779, 2018.
https://doi.org/10.1007/978-3-319-76941-7_75

version (PFSDM) [7]. A term-based model has stable empirical performance due to its solid theoretical foundation. However, it is also difficult to improve a term-based model due to its intrinsic limitation in capturing query intents.

It has been shown in the prior work that exploiting type information improves entity retrieval [8]. These studies based on TREC Entity Track [9] and INEX Entity Ranking Track [10] benchmarking platforms assume that the type information is provided by users as part of the definition of the information need. Recent work in this direction such as [11] focused on inference of the query type in a reasonable situation that type information is usually not given explicitly.

An important feature of these type-aware entity retrieval models is that they simply contain two separate component models: type matching model and term matching model. They assume conditional independence between the term-based and type-based components. A type matching model captures the relevance between a query type and an entity type while a term matching model measures the semantic relevance between the query and the pseudo documents of an entity. This paradigm of loose coupling setting has both advantages and disadvantages. On one hand, it simplifies the technical difficulty in a way that the type matching model and term matching model can be developed in different ways. On the other hand, the overall model would generate a biased estimation if any one of them makes an incorrect guess especially when the type matching model receives unreasonable types or an empty type as its input. This strategy ignores the connection between the type information and the text information associated with an entity.

To overcome the disadvantages of term-based and type-aware retrieval models, our model considers the text information and the type information in a unified way. It directly encodes the type information into a term-based retrieval model by considering the occurrence of query terms in multi-fielded pseudo documents of entities whose types have connections in the type taxonomy.

Besides the type-aware entity retrieval, our work is related to forum post retrieval task. The idea of exploiting structure information to improve retrieval accuracy has been proposed in [12], which inspires our model.

2 Model Description

In this section, we first introduce the Mixture of Language Models (MLM) for entity retrieval task which is based on the language modelling framework. Then we discuss our proposed extension for exploiting the type information. After that, we discuss the representation and processing of Wikipedia category information, which is the type taxonomy we consider in the experiment.

2.1 Mixture of Language Models

The Mixture of Language Models (MLM) is one of the state-of-the-art language modelling based models for the entity retrieval task [4]. Given an entity E and a query Q, the goal of MLM is to score the pseudo documents D of the entity E by the ranking function $f(Q, D)$:

$$f(Q, D) = \sum_{q_i \in Q} f_T(q_i, D) \tag{1}$$

where $f_T(\cdot)$ is a feature functions for individual terms formulated as follows:

$$f_T(q_i, D) = \log \sum_f w_f^T \frac{tf_{q_i, D_f} + \mu_f \frac{cf_{q_i f}}{|C_f|}}{|D_f| + \mu_f} \tag{2}$$

where q_i is the i-th query term and D_f is the pseudo document in the field f. tf_{q_i, D_f} denotes the term frequency of q_i in the D_f. The parameters w_f^T are weights for each field such that $\sum_f w_f^T = 1$. The smoothing parameter μ_f is set to the average length of the collection in field f. When each entity has only one field, the Mixture of language models degenerates into a standard language model (LM).

2.2 Smoothing by Exploiting Type Information

Following but also simplifying the entity retrieval assumption made in [13], we assume that a user generates an entity-targeted query via selecting several words to describe the types of entities plus some descriptive words. To encode the type information into the language modelling framework, we first consider a scoring function concerning types and find a maximum estimation:

$$f(Q, D) = \max_{c \in types(D)} \tilde{f}(Q, D, c) \tag{3}$$

where D is the pseudo documents of the entity E. $types(D)$ is the collection of types that belongs to the entity E.

Due to the nature of language, a query may not always be accurate enough to describe the most specific type of targeted entities. It is natural to consider the ancestors of the type i.e. a path in the type taxonomy. Therefore, we consider path as a variable in the feature functions, which are set as follows:

$$\tilde{f}(Q, D, c) = \max_{p \in Paths(c)} \sum_{q_i \in Q} f_T(q_i, D, p) \tag{4}$$

where $Paths(c)$ is the collection of directed paths starting from type c in the type taxonomy.

To incorporate the relatedness into the term-based retrieval model between a query term and a type path, we add a path-aware smoothing component to the feature function formulated as follows:

$$f_T(q_i, D, p) = \log \sum_f w_f^T \frac{tf_{q_i,D_f} + \mu_f \frac{cf_{q_i f}}{|C_f|} + n(\alpha, p) \sum_{i \in idx(V(p))} \alpha^i \mu_{v_i f} \frac{cf_{q_i f, v_i}}{|C_{f, v_i}|}}{|D_f| + \mu_f + n(\alpha, p) \sum_{i \in idx(V(p))} \alpha^i \mu_{v_i f}} \quad (5)$$

where $V(p)$ is the vertex set of the path p. The Function $idx(\cdot)$ returns the indexing set of the vertex set i.e. the set of indices $1 \ldots |V(p)|$. α is the reduction coefficient. $n(a, p)$ is used to normalize the weights $\{\alpha^i\}$, which equals to $(1 - \alpha)/(1 - \alpha^{|V(p)|})$. v_i represents the i-th vertex in the path p. $cf_{.f, v_i}$ is the collection frequency of a query term in the collection of documents in the field f of entities that share the same type v_i and the smoothing parameter $\mu_{v_i f}$ is set to the average length of the collection.

2.3 Wikipedia Category Representation and Processing

A Wikipedia category is intended to group together articles on a similar subject. The Wikipedia category system is periodically maintained by editors following the Wikipedia key policies and guidelines to achieve high quality.

By the end of 2015, there are about 1.4 million Wikipedia categories. It is commonly believed that the Wikipedia category system can be represented by a directed acyclic graph (DAG) where each vertex represents a category that connects to its parental categories. However, it is in fact a directed graph that contains many small strongly connected components (SCCs) due to the existence of disambiguation categories and editing errors. To process the category system, two different methods have been proposed in previous literature. The first method simply forces each category connects to single parent category using a heuristic strategy, which results in a tree representation [11]. The second method represents each category as a pair of head word and qualifiers, which transforms the system into a forest by defining a partial order on those pairs [14]. Both of them simplify the structure and show their effectiveness to some extent. In our model, we propose a graph-based approach that keeps the DAG structure of the Wikipedia category system by removing SCCs. We divide the graph into SCCs. For each SCC that is not a single vertex, we find its elementary cycles. A SCC is then reduced by removing common edges shared by intersected elementary cycles or one edge of a sole circle. To reduce computational complexity, a big SCC that contains more than one thousand vertices is dropped directly i.e. all edges in this SCC are removed.

3 Experimental Evaluation

3.1 Experiment Setup

We use DBpedia 2015-10 as our knowledge base along with the DBpedia-entity v2 test collection [15] which contains four query sets (INEX_LD, SemSearch ES, ListSearch and QALD2) with 467 queries in total. We use the same entity representation scheme with five fields: *names, attributes, categories, similar entity names* and *related entity names* as described in [5]. We use Python 3.6 and PyLucene 6.5 to implement our model and construct the index. The index contains an extra *catchall* field that simply aggregates contents of all fields. We use NLTK toolkit to remove stop words of all contents in the index. All index terms are stemmed using the Snowball stemmer. When constructing the graph representation of the Wikipedia category system, we consider the top 5 parental categories for each category in lexicographical order of their names. For each entity, we consider the first 30 categories in Wikipedia editors' order and the first 500 paths in length 3 in order of depth first search. Moreover, we consider the first 300 articles in lexicographical order of their titles in the collection that shares the same category.

For all experiments, we employ a two-stage retrieval method: First we use Lucene's default search engine to retrieve top 1000 results as candidates. After that we rank the candidates with the specific retrieval model. The reduction coefficient α is set to 0.75. We report results for standard language model (LM), MLM-tc concerning the field *names* and the field *attributes* with weights 0.2 and 0.8 as well as their variants that consider the type taxonomy aware smoothing method (denoted by identifier 'TAS'). We also report results from a variant of fielded sequential dependence model (FSDM*) [6] for reference. Following the guideline of the test collection in [15], we report normalized discounted cumulative gain at rank 10 (NDCG@10) as the main evaluation metric using 5-fold cross validation. To measure statistical significance, we employ a two-tailed paired t-test and denote differences at the 0.05 level.

3.2 Experimental Results

Table 1 summarizes retrieval results on the test collection. Our approach helps improve the LM model on all datasets except SemSearch ES. MLM-tc + TAS achieves improvements on all datasets except SemSearch ES. The negative effects brought by our method on SemSearch ES is probably because that most of the queries in this dataset are short and vague that tend to describe a large class of entities related to an entity instead of some distinguishing facts. Consequently, exploring the type paths may not help our model capture their actual intents. Compared to FSDM* which is one of the state-of-the-art models, results of our framework are reasonable. LM + TAS outperforms FSDM* on INEX-LD and their results on ListSearch and QALD2 are almost tied.

Table 1. Entity retrieval results on different query sets measured by NDCG@10. The numbers in parentheses show the relative improvements. Significance is tested against the corresponding baseline model. Significant improvements are bold.

	INEX-LD	SemSearch ES	ListSearch	QALD2
LM	0.3701	0.5767	0.3908	0.3482
LM + TAS	0.4071 (**+10.0%**)	0.5405 (−6.3%)	0.3969 (+1.6%)	0.3571 (**+2.6%**)
MLM-tc	0.3439	0.6114	0.3252	0.3118
MLM-tc + TAS	0.3946 (**+14.7%**)	0.5367 (−12.2%)	0.3684 (**+13.3%**)	0.3504 (**+16.1%**)
FSDM*	0.3882	0.6671	0.4070	0.3564

4   Conclusions

We propose a type taxonomy aware smoothing method that exploits the hierarchical type information of a knowledge base for entity retrieval. Experiment results show our framework improves the performance of existing models. Future work may include investigating the extension of our method to the MRF based framework and reducing the computational complexity.

References

1. Neumayer, R., Balog, K., Nørvåg, K.: On the modeling of entities for ad-hoc entity search in the web of data. In: Baeza-Yates, R., de Vries, A.P., Zaragoza, H., Cambazoglu, B.B., Murdock, V., Lempel, R., Silvestri, F. (eds.) ECIR 2012. LNCS, vol. 7224, pp. 133–145. Springer, Heidelberg (2012). https://doi.org/10.1007/978-3-642-28997-2_12
2. Tonon, A., Demartini, G., Cudré-Mauroux, P.: Combining inverted indices and structured search for ad-hoc object retrieval. In: Proceedings of the 35th International ACM SIGIR Conference on Research and Development in Information Retrieval, pp. 125–134 (2012)
3. Elbassuoni, S., Blanco, R.: Keyword search over RDF graphs. In: Proceedings of the 20th ACM International Conference on Information and Knowledge Management, pp. 237–242 (2011)
4. Neumayer, R., Balog, K., Nørvåg, K.: When simple is (more than) good enough: effective semantic search with (almost) no semantics. In: Baeza-Yates, R., de Vries, A.P., Zaragoza, H., Cambazoglu, B.B., Murdock, V., Lempel, R., Silvestri, F. (eds.) ECIR 2012. LNCS, vol. 7224, pp. 540–543. Springer, Heidelberg (2012). https://doi.org/10.1007/978-3-642-28997-2_59
5. Zhiltsov, N., Kotov, A., Nikolaev, F.: Fielded sequential dependence model for ad-hoc entity retrieval in the web of data. In: Proceedings of the 38th International ACM SIGIR Conference on Research and Development in Information Retrieval, pp. 253–262 (2015)
6. Hasibi, F., Balog, K., Bratsberg, S. E.: Exploiting entity linking in queries for entity retrieval. In: Proceedings of the 2016 ACM on International Conference on the Theory of Information Retrieval, pp. 209–218 (2016)
7. Nikolaev, F., Kotov, A., Zhiltsov, N.: Parameterized fielded term dependence models for ad-hoc entity retrieval from knowledge graph. In: Proceedings of the 39th International ACM SIGIR Conference on Research and Development in Information Retrieval, pp. 435–444 (2016)
8. Balog, K., Neumayer, R.: Hierarchical target type identification for entity-oriented queries. In: Proceedings of the 21st ACM International Conference on Information and Knowledge Management, pp. 2391–2394 (2012)
9. Balog, K., de Vries, A., Serdyukov, P., Westerveld, T., Thomas, P.: Overview of the TREC 2009 entity track (2009)
10. de Vries, A.P., Vercoustre, A.-M., Thom, J.A., Craswell, N., Lalmas, M.: Overview of the INEX 2007 entity ranking track. In: Fuhr, N., Kamps, J., Lalmas, M., Trotman, A. (eds.) INEX 2007. LNCS, vol. 4862, pp. 245–251. Springer, Heidelberg (2008). https://doi.org/10.1007/978-3-540-85902-4_22
11. Garigliotti, D., Balog, K.: On type-aware entity retrieval. arXiv preprint arXiv:1708.08291. (2017)

12. Duan, H., Zhai, C.: Exploiting thread structures to improve smoothing of language models for forum post retrieval. In: Clough, P., Foley, C., Gurrin, C., Jones, Gareth J.F., Kraaij, W., Lee, H., Mudoch, V. (eds.) ECIR 2011. LNCS, vol. 6611, pp. 350–361. Springer, Heidelberg (2011). https://doi.org/10.1007/978-3-642-20161-5_35

13. Lu, C., Lam, W., Liao, Y.: Entity retrieval via entity factoid hierarchy. In: ACL, vol. 1, pp. 514–523 (2015)

14. Chen, Y., Gao, L., Shi, S., Du, X., Wen, J.R.: Improving context and category matching for entity search. In: AAAI, pp. 16–22 (2014)

15. Hasibi, F., Nikolaev, F., Xiong, C., Balog, K., Bratsberg, S.E., Kotov, A., Callan, J.: DBpedia-entity v2: a test collection for entity search. In: Proceedings of SIGIR, vol. 17 (2017)

A Data Collection for Evaluating
the Retrieval of Related Tweets
to News Articles

Axel Suarez[1,2], Dyaa Albakour[2(✉)], David Corney[2], Miguel Martinez[2],
and José Esquivel[2]

[1] School of Computer Science and Electronic Engineering,
University of Essex, Colchester, UK
[2] Signal Media Ltd., 32-38 Leman Street, London E1 8EW, UK
dyaa.albakour@signal.uk.com, research@signalmedia.co

Abstract. Nowadays, social media users react in real-time to local and global events. Therefore, social media can be used to measure the impact of particular topics or events and to analyze public opinion. To this end, identifying and ranking social media posts, such as tweets, associated with a news article is an important information retrieval task. In this paper, we devise a new data collection to evaluate approaches for the task of related-tweet retrieval for news articles. Using two sets of (a) mainstream news articles and (b) tweets from curated newsworthy sources from the same period, we use a TREC-like pooling approach to associate news articles with relevant tweets. We also provide a benchmark for the related-tweet retrieval task by evaluating a number of retrieval approaches on this new data collection.

1 Introduction

In recent years, the way people produce and consume news has radically changed [7]. Where people used to read printed newspapers, many now read news websites and blogs. Along with these sites, social media platforms, such as Twitter, include content from many mainstream news sources, such as the BBC and The New York Times. However, mainstream news editors are not the only source of news, as individual social media users can report local or global events in real-time or comment on them afterwards. Recent studies have shown that social media posts can help understanding public opinion and sensing the state of the world. Indeed, Twitter has been used to replace opinion polls [10], to predict stock market movements [2], and to discover local events in a city [1]. Therefore, finding tweets related to news articles can be useful to analyze an event's context. For example, Fig. 1 shows two tweets that are related to a news article on the topic of Obamacare. In the first tweet, the user is sharing a link to the article to inform their followers about the event that Trump criticized Obamacare. In the second tweet, the user is commenting on the event and informing followers on their views about it. Finding, ranking and aggregating tweets about a particular

© Springer International Publishing AG, part of Springer Nature 2018
G. Pasi et al. (Eds.): ECIR 2018, LNCS 10772, pp. 780–786, 2018.
https://doi.org/10.1007/978-3-319-76941-7_76

news article may be helpful to understand the popularity and virality of the article, and also to analyze what the public thinks about the topic.

In this paper, we consider the task of *related-tweet retrieval*, where the aim is to identify and rank tweets that are related to a published news article. Examples of previous related work include the tasks of classifying tweets associated with news articles according to their subjectivity [6] and associating news articles with relevant twitter hashtags [11]. However, the data collections used for these tasks are not suitable for evaluating the retrieval of related tweets to news articles. Therefore, we create and present a new TREC-like data collection of relevance judgments for evaluating the task of related-tweet retrieval. In particular, we use the Signal "One-Million News Articles Dataset" [5], which contains articles from multiple sources, and a collection of tweets, created by Brigadir *et al.* [3], from a curated list of newsworthy sources. We follow a pooling approach, where we select a sample of 100 news articles, and propose a number of retrieval methods to create a pool of tweets for each news article in the sample. The pool is then annotated with relevance judgments. Furthermore, we use our created data collection to evaluate the retrieval methods used in the pooling process to provide a benchmark for the related-tweet retrieval task. We make the resulting data collection publicly available for research purposes[1] and together with the results of this paper, we aim at encouraging further research on this task.

Fig. 1. Examples of tweets related to a news article published in the New York Times and titled: *"Let Obamacare Fail", Trump says as G.O.P Health Bill Collapses.*

The rest of the paper is structured as follows. Section 2 describes our methodology to collect the relevance judgments using a pooling approach. Section 3 describes the data collection and gives insights from the annotation process and the results of retrieval evaluation. Finally, Sect. 4 summarizes our conclusions.

2 Methodology

2.1 The Pooling Approach

There are 3 components in a test data collection for a document retrieval task [4]: (i) *the corpus*; (ii) the set of information need statements, i.e. *information needs*;

---

[1] The data collection can be downloaded through this link:
http://research.signalmedia.co/datasets/signal1m-tweetir.html.

and (iii) *the relevance judgments* that indicate which documents should be presented for a particular information need. Generally, the most expensive component to produce, in terms of time and effort, are the judgments. Therefore, we follow a pooling approach to reduce the number of annotations, as is common in the TREC evaluation framework [12]. In our case, the corpus is a set of tweets, while the information needs are a random subset of articles selected from a larger set of news articles. To collect relevance judgments, we propose a variety of retrieval methods to retrieve related tweets. We then merge the top k tweets ranked by each retrieval method to create a pool of diverse tweets (see Fig. 2).

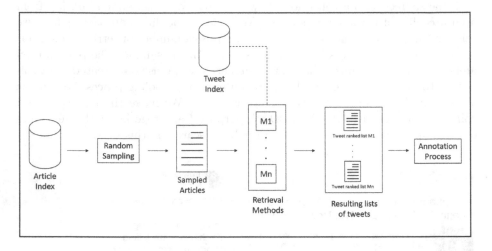

Fig. 2. Illustration of the pooling process

Each proposed related-tweet retrieval method generates a query Q. Using this query, the tweets are ranked with Lucene's Practical Scoring Function[2], which is the sum of the boosted and normalized tf-idf score for all terms in the query Q. We propose and compare eight retrieval methods:

M1. **Title search**: The title of the article is used as a query. The intuition is that the title represents a condensed version of the important subjects of the article.

M2. **Summary search**: We generate a summary for each article and use it as a query. We use the first two sentences of the article as a summary. This is an intuitive, yet effective, summarization approach that reflects the way journalists typically structure news articles.

M3. **Content search**: We use the full content of the article as a query.

---

[2] https://lucene.apache.org/core/4_6_0/core/org/apache/lucene/search/similarities/TFIDFSimilarity.

M4. **Summary + date search**: Same as M2, but the results are filtered so that only tweets posted on the day that the article is published are retrieved.

M5. **Bi-gram phrase search**: We train a bi-gram phrase recognition model, developed in [8], on a large collection of news articles. Using the trained model, we extract all the phrases in the summary of an article (as in M2) and use them as a query.

M6. **Named entity search**: With this approach, the intuition is to retrieve tweets about people, organizations and places mentioned in the article. We extract named entities using the Stanford NER Tagger configured with the default English 3-class model trained without part-of-speech tagging. We generate a query that consists of all terms representing the people, organizations and places mentioned in the summary.

M7. **Semantic summary search**: This method uses the 10 most 'semantically significant' terms in the news article as a query. To find these, we train a word2vec model [9] on a large collection of news articles to get the vector representation of each term in the article's summary, and compute the summary's centroid as follows:

$$\mu(c) = \frac{1}{|W_{summary}|} \sum_{w \in W_{summary}} v(w) \tag{1}$$

where $W_{summary}$ is the set of all the terms in the summary, and $v(w)$ is the vector representation of the term w using the word2vec model. After computing the centroid, we rank the terms by their cosine similarity to the centroid and select the top 10 terms as a query. The intuition here is that the closer a term is to the centroid, the more likely that it is relevant to the topic of the article.

M8. **Query expansion search**: This method uses the same 10 terms generated by M7 and expands the query with 10 different terms. In particular, for each of the 10 original terms t_i, we expand the query with the term in the word2vec space that has the highest cosine similarity to t_i. The intuition here is to fill the vocabulary gap where the related tweets do not mention the exact terms in the article.

2.2 Annotation for Relevance Judgments

In order to determine the performance of each of the aforementioned eight methods, we ask human annotators to provide relevance judgments. In particular, each tweet in the generated pool is annotated according to its relatedness to the corresponding article by labelling it with one of three different labels (grades of relevance):

1. **Non-relevant**: The tweet is about a completely different topic, or covers a similar topic but focusses on aspects not covered by the news article.
2. **Somewhat-relevant**: When the tweet refers to an event closely related to the main event of the article, or when it talks about a secondary theme mentioned at least once in the article.

3. **Completely-relevant**: When the tweet talks about the main topic of the article or directly mentions the article itself.

3   Data Collection and Experiments

3.1   Datasets

The Signal One Million News Articles (Signal-1M) dataset[3] contains meta-data about each article, such as title, content, and publication date. The Twitter dataset, by Brigadir *et al.* [3], consists of over 3.2 million tweets from a curated list of major news sources and journalists, which cover all their posts from the same range of dates as Signal-1M[4]. For pooling, we randomly selected 100 articles from Signal-1M, which were then passed into the eight retrieval methods (Sect. 2.1). Each retrieval method generated a ranked list of tweets for each article. We used a cut-off point $k = 10$ on each ranked list, and merge the tweets, whilst removing duplicates, resulting in 62.3 distinct tweets per article on average. We asked 10 undergraduate students to annotate the pool of tweets, as per Sect. 2.2. Finally, we train the phrase model used in M5, and the word2vec model used in M7, with a collection of 17 million news articles collected from the same sources of Signal-1M.

3.2   Annotation Agreement

We performed an annotator agreement experiment, where we gave the same tweets associated with 10 different news articles to 3 different annotators. Table 1 reports the agreement results. In the first row, we report the pair-wise agreement between annotators when considering all three labels of relevance. We see that the task is not trivial, as annotators only agree on the exact label 70.48% of the time on average. We also consider binary labels of relevance (rows 2 and 3), by merging the 'somewhat relevant' label with 'completely relevant' and 'non-relevant' respectively. Even with binary labels, the agreement is not perfect, but it is line with various retrieval tasks reported in TREC [12]. Next, we use the binary labels obtained for all 100 articles to evaluate the proposed retrieval methods.

3.3   Retrieval Results

Table 2 summarizes the retrieval results using binary relevance judgments obtained when merging the 'completely relevant' label with 'somewhat relevant'. Retrieval methods (M1–M4) perform well despite their simplicity. In particular, using the summary as a query (M2) outperforms all other methods in terms of MAP, P@5 and P@10. Perhaps surprisingly, M4 shows the worst performance among these four (MAP = 0.41), although it is only a slight variation of the best

---

[3] http://research.signalmedia.co/newsir16/signal-dataset.html.

[4] https://github.com/igorbrigadir/newsir16-data/tree/master/twitter/curated.

Table 1. Agreement rates between the three annotators, A, B and C.

Agreement between annotators	A and B	A and C	B and C	Average
3 Labels (Exact match)	68.05%	75.87%	67.53%	70.48%
2 Labels (Relevant = Somewhat relevant)	73.09%	81.94%	74.82%	76.62%
2 Labels (Somewhat relevant = Non-relevant)	**87.15%**	**88.02%**	**84.54%**	**86.57%**

performing retrieval method (M2). In M4, date filtering is used to select tweets posted just before or after the article is published. However, our results suggest that sometimes social media posts may discuss events and topics before they make it to mainstream media, or longer after they are published in mainstream media. The retrieval performance of the more complex methods (M5–M8) is worse than simpler methods (M1–M4), as their MAP scores are markedly lower. It is noteworthy however, that the Query Expansion method (M8) produces a slightly higher MAP score than the non-expanded version (Semantic Summary, M7).

Table 2. Retrieval scores for the proposed retrieval methods: Mean Average Precision (MAP); Precision at 5 (P@5); Precision at 10 (P@10)

Method		MAP	P@5	P@10
M1	Title search	0.62	0.52	0.48
M2	Summary search	**0.67**	**0.59**	**0.55**
M3	Content search	0.63	0.53	0.51
M4	Summary + sate search	0.48	0.40	0.36
M5	Bi-gram phrase search	0.41	0.32	0.31
M6	Named entity search	0.44	0.35	0.31
M7	Semantic summary search	0.37	0.28	0.29
M8	Query expansion search	0.40	0.28	0.27

4   Conclusion

We have created a new data collection that combines two existing datasets (news articles and tweets) and adds value to both. Annotating tweets with relevance judgments, on their relatedness to a news article, yields interesting insights, as we have observed that the inter-annotator agreement is not perfect. Furthermore, we have used our collection to evaluate a number of retrieval methods for the task of related-tweet retrieval. Our results show that simple approaches, e.g. using the terms in the title or the summary of the article as a query, can be very effective for this task. More complex approaches, such as phrase and entity search, failed to perform well on this task. This opens opportunities to consider more elaborate approaches to effectively bridge the gap between the vocabulary used in mainstream media and social media. To this end, the created data collection and the results presented in this paper will foster developing such approaches. For example, as future work, we aim to use the relevance judgments in our data collection to develop a learning-to-rank model for related-tweet retrieval.

References

1. Albakour, M., Macdonald, C., Ounis, I., et al.: Identifying local events by using microblogs as social sensors. In: Proceedings of OAIR, pp. 173–180 (2013)
2. Bollen, J., Mao, H., Zeng, X.: Twitter mood predicts the stock market. J. Comput. Sci. **2**(1), 1–8 (2011)
3. Brigadir, I., Greene, D., Cunningham, P.: Detecting attention dominating moments across media types. In: Proceedings of NewsIR 2016 Workshop at ECIR (2016)
4. Buckley, C., Dimmick, D., Soboroff, I., Voorhees, E.: Bias and the limits of pooling for large collections. Inf. Retrieval **10**(6), 491–508 (2007)
5. Corney, D., Albakour, D., Martinez-Alvarez, M., Moussa, S.: What do a million news articles look like? In: Proceedings of NewsIR 2016 Workshop at ECIR, pp. 42–47 (2016)
6. Kothari, A., Magdy, W., Darwish, K., Mourad, A., Taei, A.: Detecting comments on news articles in microblogs. In: Proceedings of ICWSM (2013)
7. Martinez-Alvarez, M., Kruschwitz, U., Kazai, G., Hopfgartner, F., Corney, D., Campos, R., Albakour, D.: First international workshop on recent trends in news information retrieval (NewsIR 2016). In: Proceedings of ECIR, pp. 878–882 (2016)
8. Mikolov, T., Sutskever, I., Chen, K., Corrado, G.S., Dean, J.: Distributed representations of words and phrases and their compositionality. In: Proceedings of NIPS, pp. 3111–3119 (2013)
9. Mikolov, T., Yih, W.T., Zweig, G.: Linguistic regularities in continuous space word representations. In: Proceedings of NAACL, pp. 746–751 (2013)
10. O'Connor, B., Balasubramanyan, R., Routledge, B.R., Smith, N.A.: From tweets to polls: linking text sentiment to public opinion time series. In: Proceedings of ICWSM (2010)
11. Shi, B., Ifrim, G., Hurley, N.: Be in the know: connecting news articles to relevant twitter conversations. arXiv:1405.3117 (2014)
12. Voorhees, E.M., Harman, D.K., et al.: TREC: Experiment and Evaluation in Information Retrieval, vol. 1. MIT Press, Cambridge (2005)

Content Based Weighted Consensus Summarization

Parth Mehta[✉] and Prasenjit Majumder

Dhirubhai Ambani Institute of Information and Communication Technology,
Gandhinagar, India
{parth_me,p_majumder}@daiict.ac.in

Abstract. Multi-document summarization has received a great deal of attention in the past couple of decades. Several approaches have been proposed, many of which perform equally well and it is becoming increasingly difficult to choose one particular system over another. An ensemble of such systems that is able to leverage the strengths of each individual systems can build a better and more robust summary. Despite this, few attempts have been made in this direction. In this paper, we describe a category of ensemble systems which use consensus between the candidate systems to build a better meta-summary. We highlight two major shortcomings of such systems: the inability to take into account relative performance of individual systems and overlooking content of candidate summaries in favour of the sentence rankings. We propose an alternate method, content-based weighted consensus summarization, which address these concerns. We use pseudo-relevant summaries to estimate the performance of individual candidate systems, and then use this information to generate a better aggregate ranking. Experiments on DUC 2003 and DUC 2004 datasets show that the proposed system outperforms existing consensus-based techniques by a large margin.

1 Introduction

A plethora of summarization techniques have been proposed in last two decades, but few attempts have been made to combine various summarization techniques to build a meta-summarizer. A study in [1] shows that several state-of-art systems with apparently similar performance in terms of ROUGE score, in fact, have very little overlap in terms of content. Essentially these systems seem to be picking out equally good, but different, information. It is possible to leverage this fact, to build a meta-system that combines all the *good* information across summaries and results in a better coverage.

A meta-summary can be created either before creating individual summaries or post-summarization. In the first case generally, the ranking algorithm is modified to encompass features from several different summarizers and directly generate the aggregate ranking [2]. In contrast, the latter systems use sentence rankings or summaries generated from individual systems and combine them

© Springer International Publishing AG, part of Springer Nature 2018
G. Pasi et al. (Eds.): ECIR 2018, LNCS 10772, pp. 787–793, 2018.
https://doi.org/10.1007/978-3-319-76941-7_77

to form a meta-summary [3–5]. The first type of ensembles depends on carefully combining various aspects of the individual systems, which is not only non-trivial but is also not possible in several cases. In contrast, the second approach can use the existing systems as it is without any modifications, which makes it possible to include as many candidate systems as required, without any overhead. The systems proposed in [6,7] looks into combining several sentence similarity scores to generate a more robust summary. These approaches show that using various combinations of ranking algorithms and sentence similarity metrics generally outperforms individual systems. In this work, we focus on generating an aggregate ranking of sentences from individual rankings rather than individual summaries. When using just the summaries to generate an ensemble, there is an upper bound on the overall performance [1], since the choice of sentences is limited to the existing summaries rather than entire documents. Both [4] and [5] focus on combining the sentence rankings from candidate systems using weighted linear combinations. While the former relies on a supervised approach that uses SVM-rank to learn relative rankings for all sentence pairs, the latter uses an unsupervised approach based on consensus between the candidate rankings. Existing summarization datasets are too small to train a generic supervised model. In this work, we focus on consensus-based methods to generate aggregates. While our approach is similar in principle to Weighted consensus summarization (WCS) [5], the way in which we define consensus differs. Unlike WCS, we do not consider sentence rankings to compare two systems. Rather we analyze the overlap in content selected by these systems to measure the consensus between them. We also take into account the relative performance of these systems for individual documents, thus ensuring that best performing system gets more weight compared to the ones with weaker performance.

2   Consensus Based Summarization

Consensus-based summarization is a type of ensemble system that *democratically* selects common content from several candidate systems by taking into account the individual rankings of candidate systems. As opposed to this, the *first past the post* types of ensembles select the highest ranked content from each individual system, even if they are ranked lower in other systems. In case of consensus-based systems, the sentences that are broadly accepted by several systems tend to be ranked higher rather than those championed by only some. Examples are Borda Count[1], and Weighted Consensus summarization [5]. Borda count assigns, to each sentence in the original rank lists, a score equal to their rank, i.e. sentence ranked 1st is given a score 1, the one ranked 2nd is given a score 2 and so on. The aggregate score is computed by averaging the score of a sentence in all the rank lists. One major problem with such techniques is their failure to take into account variance in performance of candidate systems across documents. A single system that performs very poorly, can limit the overall performance of the ensemble.

---

[1]   https://en.wikipedia.org/wiki/Borda_count.

The weighted consensus summarization [5] creates an aggregate ranking that is as close as possible to the individual rankings. As it is impossible to know beforehand, which candidate system will work best for a given document, the weighted consensus model gives equal importance to all the candidate systems. It then iteratively finds out the aggregate ranking that is as close as possible to each individual ranking. Like other consensus-based methods, WCS fails to take into account the variance in system performance. Another major issue is the manner in which difference between ranked lists is computed. WCS uses L2 Norm to compute the concordance between the aggregate and individual rankings. WCS minimizes Eq. 1 where, r* is the aggregate rank list, r_i are the individual rankings and w_i are the relative weights assigned to each system. The constraint on $||w||^2$ ensures that the weights are as uniform as possible.

$$(1 - \lambda) \sum_{i=1}^{K} w_i ||r^* - r_i||^2 + \lambda ||w||^2 \tag{1}$$

The constraint of minimizing the distance between entire rank lists, instead of the top-k sentences which form the summary, is unnecessary. As long as candidate systems agree in the top-k sentences, which are to be considered for the summary, any additional constraint on lower ranked sentences can adversely affect the performance. Besides that, considering the nature of documents, there will always be more than one sentence which will convey the same information. As a matter of fact, DUC 2003 corpus has on average 34 sentences per document cluster, that are repeated at least once, while DUC 2004 has 26 such sentences on an average per cluster. There are many more sentences that have near similar information. Simply comparing rank lists of the sentence does injustice, in cases where different sentences with very similar information were selected by different systems. To overcome these two problems, we propose a content-based consensus summarization method, which improves upon the existing WCS method. We use inter-system ROUGE scores to measure the similarity between rankings of two systems. The consensus is then achieved on content, rather than sentence rankings. Under certain constraints, this also takes into account the relative performance of individual systems, when computing the aggregate ranking. The method is described in detail in next section.

3   Proposed Approach

As in any consensus-based approach, the idea is to find a weighted combination of individual sentence rankings from the candidate systems to form an aggregate ranking. The problem boils down to finding the best combination of weights that maximizes the ROUGE score. In the proposed approach we define a new method for assigning weights to different candidate systems. We call this approach *Content based Weighted Consensus Summarization (C-WCS)*. Ideally, a better performing system should contribute more to the aggregate summary compared to a system with lower ROUGE scores. Of course in a practical setup, where the benchmark summaries are not available apriori, it is impossible to know which system will perform better. In theory, it is possible to train a system that can predict this information, by looking at the input document. But in practice, the utility of such a system would be limited by the amount of training data available. Instead of this approach, we propose using *pseudo relevant summaries*. For a given candidate summary S_i, each of the remaining $N - 1$ candidate systems, $S_j : j\epsilon\{1...N\}, j \neq i$, are considered to be *pseudo-relevant* summaries. We then estimate relative performance of the individual system from the amount of content it shares with these *pseudo relevant summaries*. Weights of a candidate system i is computed as shown in Eq. 2. $Sim(S_i, S_j)$ is defined as ROUGE-1 recall computed considering S_j as the benchmark summary used to evaluate S_i.

$$w_i = \frac{1}{N - 1} \sum_{j \neq i} Sim(S_i, S_j) \tag{2}$$

The underlying assumption in this proposed approach is that the systems performing poorly for a given document are much less in number than the ones performing well. This is not a weak constraint, but we show that this is generally true. In general, a given candidate system tends to perform well on more number of documents compared to the ones on which it performs poorly. Out of the six candidate systems that we experimented with, only one performed below average in more than 30% cases. The number of documents for which more than 50% systems performed below average, was 20%. Given this information, we assert that the number of systems performing well for a given document is generally larger than the ones that perform poorly. We present a hypothesis that for a summarization task in general, the *relevant* content in a document cluster is much lower compared to *non-informative* content. Under this assumption, two good or *informative* summaries would have a higher overlap in content, compared to two poor summaries. Simply because the good summaries will have lesser content to choose from, so they are bound to end up with higher overlap. Based on this we argue that the probability of a candidate summary, that has higher overlap with peers, performing better is high.

The limitation of this approach is the assumption that *good* summaries will have higher overlap amongst themselves, compared to the *bad* summaries. This condition will not be satisfied, if two systems that perform poorly, also generate very similar rankings. But this is not true in general and we show that there is a very good co-relation between rankings generated using Original ROUGE scores (based on handwritten summaries) and the pseudo ROUGE-scores (based on comparison with peers). While the scores themselves differ very much, the system rankings based on these two scores have a Kendal's Tau of 0.7. This indicates that in absence of handwritten summaries, a collection of several peer summaries can serve as a good reference.

4   Experimental Setup

The DUC 2003 and DUC 2004 datasets were used for evaluating the experiments. We report ROUGE-1, ROUGE-2 and ROUGE-4 recall. We experiment with six popular and well accepted extractive techniques as the candidate systems for our experiments: Lexrank [8], Textrank, Centroid [9], FreqSum [10], TopicSum [11] and Greedy-KL [12]. We use three baseline aggregation techniques against which the proposed method is compared. Besides Borda Count and WCS, we also compare the results with the *choose-best* Oracle technique. In case of the Oracle method, we assume that the performance of each candidate system, in terms of ROUGE score, is known to us. For each document, we directly select summary generated by the system that scored highest for that particular document and call it the meta-summary. This is a very strong baseline, and average ROUGE-1 score for this meta-system, on the DUC 2003 dataset, was 0.394 compared to a maximum ROUGE-1 of 0.357 for the best performing LexRank system. We further compare the results with two state of the art extractive summarization systems Determinantal Point Processes [13] and Submodular [14]. The results are shown in Table 1 below.

Table 1. System performance comparison

System	DUC 2003			DUC 2004		
	R-1	R-2	R-4	R-1	R-2	R-4
LexRank	0.357	0.081	0.009	.354	0.075	0.009
TexRank	0.353	0.072	0.010	0.356	0.078	0.010
Centroid	0.330	0.067	0.008	0.332	0.059	0.005
FreqSum	0.349	0.080	0.010	0.347	0.082	0.010
TsSum	0.344	0.750	0.008	0.352	0.074	0.009
Greedy-KL	0.339	0.074	0.005	0.342	0.072	0.010
Borda	0.351	0.080	0.0140	0.360	0.0079	0.015
WCS	0.375	0.088	0.0150	0.382	0.093	0.0180
C-WCS	0.390	**0.109**[†]	0.0198	**0.409**[†]	**0.110**	**0.0212**
Oracle	**0.394**	0.104	**0.0205**[†]	0.397	0.107	0.0211
Submodular	0.392	0.102	0.0186	0.400	**0.110**	0.0198
DPP	0.388	0.104	0.0154	0.394	0.105	0.0202

Figures in bold indicate the best performing system
[†] indicates significant difference with $\alpha = 0.05$

In all cases, the proposed C-WCS system outperforms other consensus-based techniques, Borda and WCS by a significant margin. It performs at par with the current state of art Submodular and DPP systems. In several cases, C-WCS even outperformed the Oracle system, which relies on apriori knowledge about which system will perform the best. We conducted a two-sided sign test to compare the C-WCS system with other systems. [†] indicates that the best performing system is significantly better than the next best performing system.

5   Conclusion

In this work, we propose a novel method for consensus-based summarization, that takes into account content of the existing summaries, rather than the sentence rankings. For a given candidate summary we treat other peer summaries as pseudo relevant model summaries and use them to estimate the performance of that candidate. Each candidate is weighted based on their expected performance when generating the meta-ranking. The proposed C-WCS system outperforms other consensus-based aggregation methods by a large margin and performs at par with the state-of-art techniques.

References

1. Hong, K., Conroy, J.M., Favre, B., Kulesza, A., Lin, H., Nenkova, A.: A repository of state of the art and competitive baseline summaries for generic news summarization. In: Proceedings of Language Resources and Evaluation Conference, pp. 1608–1616 (2014)
2. Mogren, O., Kågebäck, M., Dubhashi, D.: Extractive summarization by aggregating multiple similarities. In: Proceedings of Recent Advances in Natural Language Processing, pp. 451–457 (2015)
3. Hong, K., Marcus, M., Nenkova, A.: System combination for multi-document summarization. In: Proceedings of the 2015 Conference on Empirical Methods in Natural Language Processing, pp. 107–117. Association for Computational Linguistics, Lisbon, September 2015
4. Pei, Y., Yin, W., Fan, Q., Huang, L.: A supervised aggregation framework for multi-document summarization. In: Proceedings of 24th International Conference on Computational Linguistics: Technical Papers, pp. 2225–2242 (2012)
5. Wang, D., Li, T.: Weighted consensus multi-document summarization. Inf. Process. Manage. 48(3), 513–523 (2012)
6. Mehta, P.: From extractive to abstractive summarization: a journey. In: Proceedings of the ACL 2016 Student Research Workshop, Germany, pp. 100–106. ACL (2016)
7. Mehta, P., Majumder, P.: Effective aggregation of various summarization techniques. Inf. Process. Manage. 54(2), 145–158 (2018)
8. Erkan, G., Radev, D.R.: Lexrank: graph-based lexical centrality as salience in text summarization. J. Artif. Intell. Res. 22, 457–479 (2004)
9. Radev, D.R., Jing, H., Styś, M., Tam, D.: Centroid-based summarization of multiple documents. Inf. Process. Manage. 40(6), 919–938 (2004)
10. Nenkova, A., Vanderwende, L., McKeown, K.: A compositional context sensitive multi-document summarizer: exploring the factors that influence summarization. In: Proceedings of the 29th Annual International ACM SIGIR Conference on Research and Development in Information Retrieval, pp. 573–580. ACM (2006)
11. Lin, C.Y., Hovy, E.: The automated acquisition of topic signatures for text summarization. In: Proceedings of the 18th Conference on Computational Linguistics, vol. 1, pp. 495–501. Association for Computational Linguistics (2000)
12. Haghighi, A., Vanderwende, L.: Exploring content models for multi-document summarization. In: Proceedings of Human Language Technologies: The 2009 Annual Conference of the North American Chapter of the Association for Computational Linguistics, pp. 362–370. Association for Computational Linguistics (2009)

13. Kulesza, A., Taskar, B., et al.: Determinantal point processes for machine learning. Found. Trends Mach. Learn. **5**(2–3), 123–286 (2012)
14. Lin, H., Bilmes, J.: Learning mixtures of submodular shells with application to document summarization. In: Proceedings of the Twenty-Eighth Conference on Uncertainty in Artificial Intelligence, pp. 479–490. AUAI Press (2012)

Demonstrations

NOA: A Search Engine for Reusable Scientific Images Beyond the Life Sciences

Jean Charbonnier[1]📷, Lucia Sohmen[2]📷, John Rothman[1], Birte Rohden[1], and Christian Wartena[1(✉)]📷

[1] Hochschule Hannover, Expo Plaza 12, 30539 Hannover, Germany
{Jean.Charbonnier,christian.wartena}@hs-hannover.de
[2] Technische Informationsbibliothek, Welfengarten 1B, 30167 Hannover, Germany
lucia.sohmen@tib.eu

Abstract. NOA is a search engine for scientific images from open access publications based on full text indexing of all text referring to the images and filtering for disciplines and image type. Images will be annotated with Wikipedia categories for better discoverability and for uploading to WikiCommons. Currently we have indexed approximately 2,7 Million images from over 710 000 scientific papers from all fields of science.

Keywords: Open access · Image retrieval

1 Reusing Scientific Images

Images play an important role in scientific publications. In some cases images are specific for the paper, but in many cases images are general illustrations that could be reused in several papers or for illustrating a presentation. In order to effectively find reusable images one would need a search engine that allows for filtering scientific disciplines and image types, and that searches in scientific images only, or allows to filter images from scientific journals. NOA (*Nachnutzung von Open Access Bildern*, Reuse of Open Access Images) is such a search engine. A first version of this search engine is available at http://noa.wp.hs-hannover.de/.

2 Sustainability: Uploading Images to Wikimedia

The goal of the NOA project is to build a freely accessible corpus of images from open access articles and upload them to Wikimedia Commons which is a collection of freely reusable images that has existed for many years. Thus, access to the images will be secured even after the project is over and an already established user base will be able to make use of the images. In addition we will make them available through a dedicated search engine.

© Springer International Publishing AG, part of Springer Nature 2018
G. Pasi et al. (Eds.): ECIR 2018, LNCS 10772, pp. 797–800, 2018.
https://doi.org/10.1007/978-3-319-76941-7_78

3   Data

Since Wikimedia Commons only accepts images with a Creative Commons-Attribution or a Creative Commons-Attribution-Share Alike license, our article aggregation has focused on publishers using these licenses, including Hindawi, Frontiers, Copernicus, Springer Open and parts of PubMed Central.

Currently we have collected 2,7 Million images from over 710 000 papers in over 5 000 journals. 2 Million more images will be added in the upcoming months.

4   System Description

We retrieved all articles from the mentioned publishers through their public APIs or as a complete dump. We store the articles and their metadata in a MongoDB instance. This information is enriched and fed to an Apache Solr instance which delivers all data to a web frontend. Currently the system is hosted on a virtual machine with 32 GB RAM and 5 TB SSD Space. 32 GB are used to store the data in MongoDB, 237 GB are used to store all images on the file system.

4.1   Text Based Image Retrieval

Images in scientific journals often show specific and abstract objects, graphs, and drawings. Image recognition will not be effective in indexing these images. Instead, we use the text from the article and the metadata for retrieval. We add the image caption, the title of the paper, the journal title and the author names to the index. Words from the caption get the highest weight using Solr's eDisMax relevance score. Thus, we will get results based on matches from different fields, but usually ranked below images that have the query terms in the caption.

Yu [10] argues that information about an image in a scientific paper is found all over the paper. However, they try to generate a complete explanation of the picture. For indexing we need words that directly refer to concepts shown in the image. Moreover it is important that users understand why an image was found. This is easy in case the query terms are found in the meta data that are displayed in the result list, but not if these terms are found somewhere in the paper. Nevertheless, we will use the information from text regions with references to the image indirectly, as explained below.

Often image captions use specific abbreviations. In our collection we find on average almost one acronym per caption. We try to expand acronyms and add the definitions to the search index as well. Thus a query for *Fast Fourier Transformation* will also return images annotated solely with *FFT*.

Definitions for abbreviations are searched in the corresponding paper. If no definition was found, we take the definition found in another paper from the same journal, but only when the abbreviation is unambiguous within all papers from that journal. We found 2 838 713 occurrences of words written in all capitals that thus are likely to be abbreviations or acronyms. For 25 336 abbreviations with a total of 643 231 occurrences we found a definition in the paper. For 379 509 more we could find an unambiguous definition in another paper from the same journal. Thus a total of about 36% of all potential abbreviations could be expanded.

4.2 User Interface

The user interface has a typical design with a search field on top and search results below. Results are either displayed in a grid or as a list. The user can easily switch between the views and store their preference. In the first view (Fig. 1) only the images are displayed and the metadata is shown when an image is clicked. In the second view (Fig. 2) the metadata is shown next to the images.

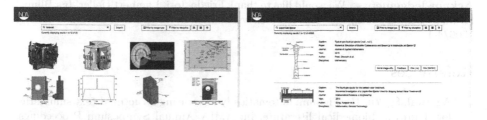

Fig. 1. Grid view   **Fig. 2.** List view

In the meta data pane the title of the paper contains a hyperlink to the official publication. Similarly, the original image can be downloaded. Finally, we offer the possibility to download the metadata of the paper in RIS and BibTeX format to enable easy import to common reference management programs.

The search results can be filtered by discipline and by image type.

5 Related Work

Other search engines for scientific images have been developed in the past, although they usually only index biomedical images extracted from articles in the PMC corpus. An early example of this is Figsearch [6], a prototypical search engine from 2004 that lets users search figure legends. BioText [2], which was developed in 2007, is a very basic search interface for figures and articles. The developers of the Yale image finder [9] from 2008 did research on text extraction from images [8,9] and made this text searchable. FigureSearch [1] from 2009 has a focus on image classification [3,4,7] and automatically generated text summaries for the images [1,10,11]. The newest search engine is Viziometrics [5] which implements automatic classification and crowd-source tagging of images from the PMC corpus. Only FigureSearch (http://figuresearch.askhermes.org/) and Viziometrics (http://viziometrics.org/) still have working instances.

6 Future Work

For most components basic algorithms have been used. Much of future work will deal with improving image classification, keyword extraction, abbreviation expansion, etc. We also plan to use OCR to index text in the images. In order to integrate the figures into Wikimedia Commons and enhance retrieval, we are currently working on annotating images with categories from the English Wikipedia using image captions and relevant sections from the papers.

Currently the database is static. In the upcoming months we will implement continuous updating with recently published papers. Another major topic will be evaluation of the service including what potential functions are useful for users.

Acknowledgements. We would like to thank Frieda Josi, Lambert Heller, Ina Blümel for many helpful comments. This research was funded by the DFG under grant no. WA 1506/4-1.

References

1. Agarwal, S., Yu, H.: FigSum: automatically generating structured text summaries for figures in biomedical literature. In: AMIA Annual Symposium Proceedings. AMIA Symposium 2009, pp. 6–10, November 2009
2. Hearst, M.A., Divoli, A., Guturu, H., Ksikes, A., Nakov, P., Wooldridge, M.A., Ye, J.: BioText search engine: beyond abstract search. Bioinformatics **23**(16), 2196–2197 (2007). https://academic.oup.com/bioinformatics/article-lookup/doi/10.1093/bioinformatics/btm301
3. Kim, D., Ramesh, B.P., Yu, H.: Automatic figure classification in bioscience literature. J. Biomed. Inf. **44**(5), 848–858 (2011). http://www.sciencedirect.com/science/article/pii/S1532046411000943
4. Kim, D., Yu, H.: Hierarchical image classification in the bioscience literature. In: AMIA Annual Symposium Proceedings 2009, pp. 327–331 (2009). http://www.ncbi.nlm.nih.gov/pmc/articles/PMC2815366/
5. Lee, P., West, J.D., Howe, B.: Viziometrics: analyzing visual information in the scientific literature. IEEE Trans. Big Data (2017)
6. Liu, F., Jenssen, T.K., Nygaard, V., Sack, J., Hovig, E.: FigSearch: a figure legend indexing and classification system. Bioinformatics **20**(16), 2880–2882 (2004). https://academic.oup.com/bioinformatics/article/20/16/2880/236814/FigSearch-a-figure-legend-indexing-and
7. Rafkind, B., Lee, M., Chang, S.F., Yu, H.: Exploring text and image features to classify images in bioscience literature. In: Proceedings of the Workshop on Linking Natural Language Processing and Biology: Towards Deeper Biological Literature Analysis, BioNLP 2006, pp. 73–80. Association for Computational Linguistics, Stroudsburg (2006). http://dl.acm.org/citation.cfm?id=1567619.1567632
8. Xu, S., Krauthammer, M.: A new pivoting and iterative text detection algorithm for biomedical images. J. Biomed. Inf. **43**(6), 924–931 (2010). http://www.ncbi.nlm.nih.gov/pmc/articles/PMC3265968/
9. Xu, S., McCusker, J., Krauthammer, M.: Yale Image Finder (YIF): a new search engine for retrieving biomedical images. Bioinformatics **24**(17), 1968–1970 (2008). http://www.ncbi.nlm.nih.gov/pmc/articles/PMC2732221/
10. Yu, H.: Towards answering biological questions with experimental evidence: automatically identifying text that summarize image content in full-text articles. In: AMIA Annual Symposium Proceedings, AMIA Symposium, pp. 834–838 (2006)
11. Yu, H., Lee, M.: BioEx - a novel user-interface that accesses images from abstract sentences (2006)

A Micromodule Approach for Building Real-Time Systems with Python-Based Models: Application to Early Risk Detection of Depression on Social Media

Rodrigo Martínez-Castaño[1]([✉])(iD), Juan C. Pichel[1](iD), David E. Losada[1](iD),
and Fabio Crestani[2]

[1] Centro de Investigación en Tecnoloxías da Información (CiTIUS),
Universidade de Santiago de Compostela, Santiago de Compostela, Spain
{rodrigo.martinez,juancarlos.pichel,david.losada}@usc.es
[2] Faculty of Informatics, Università della Svizzera Italiana (USI),
Lugano, Switzerland
fabio.crestani@usi.ch

Abstract. In this work we introduce CATENAE, a new library whose main goal is to provide an easy-to-use solution for scalable real-time deployments with Python micromodules. To demonstrate its potential, we have developed an application that processes social media data and alerts about early signs of depression. The architecture has the following modules: (1) a crawler for extracting users and content, (2) a classifier pipeline that processes new user contents, (3) an HTTP API for alert management and access to users' submissions, and (4) a web interface.

1 Introduction

Early risk detection [2] is a challenging and increasingly important research area. People are exposed to a wide range of threats and risks, and many of them exteriorize on social media. Some risks might come from criminals and offenders (e.g., stalkers, or offenders with sexual motivations). Other risks are not originated by external actors, but by the individuals themselves. For example, depression may lead to an eating disorder such as bulimia or anorexia or even to suicide. In this demo, we present the architecture of a system able to massively track online data and support risk assessment. The system is adaptable to multiple scenarios of early risk prediction, but we focus here on the case of depression. Such an application may be useful, for instance, to health agencies seeking a tool for analyzing the impact of depression on society.

CATENAE is a Python library for building topologies in the shape of directed acyclic graphs. Graph nodes represent points of data transformation and edges symbolize data flowing between nodes. Nodes can be connected to multiple nodes both to send and receive data. The communication between nodes is managed in

© Springer International Publishing AG, part of Springer Nature 2018
G. Pasi et al. (Eds.): ECIR 2018, LNCS 10772, pp. 801–805, 2018.
https://doi.org/10.1007/978-3-319-76941-7_79

the form of message queues by Apache Kafka,[1] a distributed message broker for high-throughput, low-latency handling of real-time data feeds. Each node can be instantiated multiple times in such a way that if one type of node becomes a bottleneck, it is only necessary to replicate it. In addition, unlike popular batch processing frameworks where resources are assigned to the whole topology, CATENAE assigns individually the hardware resources to each node.

We have developed a system for real-time prediction of signs of depression with CATENAE. The system uses the social network Reddit as data source. Following the lessons learned in [1], we implemented a dynamic strategy that works with a *depression classifier* (built from the training split detailed in [1]) and incrementally analyzes the stream of texts written by each user. To meet this aim, we defined micromodules as tiny, loosely coupled software modules (nodes of the topology) that can scale horizontally and be deployed independently. The micromodule approach has some advantages over batch processing architectures when dealing with real-time independent events. This is the case in retrieving user texts (posts or comments) from Reddit in real time, where they can be processed in parallel as they are being collected. Our system is oriented to early detection and, thus, it is more reasonable to make the alerts as soon as there is evidence of a potential risk (rather than accumulating cases and making batch processing). The system is explained in detail below.

2   Early Risk Detection of Depression in Real Time

2.1   The Reddit Crawler

In order to maximize the number of tracked users, we have built a web crawler for Reddit following the rules expressed in their `robots.txt` file. The crawler also uses the CATENAE library, as it is composed of multiple horizontal-scalable micromodules in a pipeline (see Fig. 1):

- **Submission and comment crawlers.** They retrieve all new submissions and comments and extract their author nicknames.
- **New user filter.** Nicknames will be checked to avoid repeated users. Aerospike[2] is used to deal with this task efficiently as it is a memory-based store. Those users who pass the filter will be sent to a queue of new users.
- **User content crawlers.** In this stage, all texts (submissions/comments) written on Reddit by the users that passed the filter are extracted as far as possible. Collecting all submissions of a user requires n calls, where n is the number of posts to retrieve (with a maximum of 100). On the other hand, retrieving the newest comments made by a user requires a single call. On every iteration, the system only retrieves the new texts available (it stores the identifiers of the last submission and comment for each user). The user content crawlers obtain user identifiers from different queues, ordered by priority.

---

[1] https://kafka.apache.org/.
[2] https://www.aerospike.com.

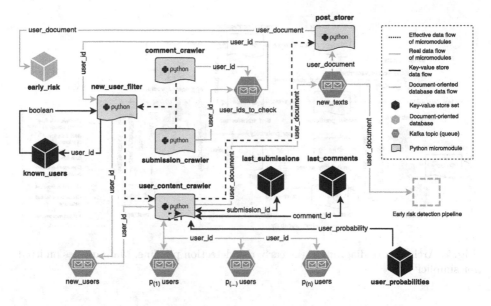

Fig. 1. Architecture diagram of the Reddit crawler.

Based on the estimated probability of risk and the activity since the previous iteration, users can be relocated in different priority queues. A single output queue receives the extracted texts.

- **Post storer.** It is in charge of storing texts in a document-oriented database. In addition, these texts will also feed the early prediction pipeline.

2.2 The Early Risk Detection Pipeline

The early prediction pipeline is a Logistic Regression classifier with L1 regularization, implemented in Python with *scikit-learn*.[3] The classifier is built with a training set of 486 users (83 positive, 403 negative) [1]. Users are represented with a single document, consisting of the concatenation of all their writings. The prediction process has four micromodules (see Fig. 2):

- **Text Vectorizer.** It transforms an input text into a vector of token counts.
- **Aggregator.** We accumulate a vector of token counts that represents all the texts of each user. The aggregator merges the current vector of counts with the vector obtained from any new submission or comment.
- **Tf–idf Transformer.** It transforms the aggregated vector of counts to a normalized tf-idf representation.
- **Model Predictor.** It produces the probability of risk of depression for users given their tf–idf representation. In addition, it produces an individual probability for each document which will be stored by the Post Updater micromodule.

---

[3] http://scikit-learn.org.

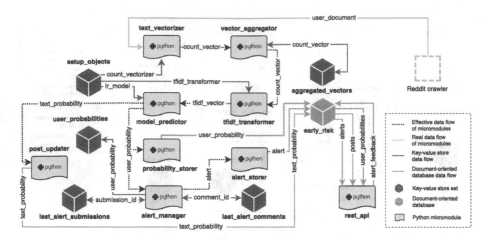

Fig. 2. Architecture diagram of the early risk detection pipeline. Main queues omitted for simplicity.

Both the Probability Storer and the Alert Manager micromodules are fed with user probabilities by the Model Predictor, storing alerts and generating them over a certain threshold, respectively.

The aggregated vectors and the Python objects (count vectorizer, tf–idf transformer, and the classification model) are stored in the Aerospike store. In this way, the vector updates and the micromodule initializations are fast.

2.3   The HTTP API and the User Interface

The user interface is a web application that retrieves information from the Alert Manager HTTP API and provides access to the generated alerts. Among its main functionalities are retrieving and processing those alerts, and retrieving the users' texts and historical records. Each alert has associated a confidence score (as produced by the classifier) and alerts are presented to the user by decreasing recency or in decreasing order of confidence. For each alert, the users' texts are presented ordered by decreasing probability of depression.

Acknowledgements. This work has been supported by MINECO (TIN2014-54565-JIN, TIN2015-64282-R), Xunta de Galicia (ED431G/08) and European Regional Development Fund.

References

1. Losada, D.E., Crestani, F.: A test collection for research on depression and language use. In: Fuhr, N., Quaresma, P., Gonçalves, T., Larsen, B., Balog, K., Macdonald, C., Cappellato, L., Ferro, N. (eds.) CLEF 2016. LNCS, vol. 9822, pp. 28–39. Springer, Cham (2016). https://doi.org/10.1007/978-3-319-44564-9_3
2. Losada, D.E., Crestani, F., Parapar, J.: eRISK 2017: CLEF Lab on early risk prediction on the Internet: experimental foundations. In: Jones, G.J.F., Lawless, S., Gonzalo, J., Kelly, L., Goeuriot, L., Mandl, T., Cappellato, L., Ferro, N. (eds.) CLEF 2017. LNCS, vol. 10456, pp. 346–360. Springer, Cham (2017). https://doi.org/10.1007/978-3-319-65813-1_30

YAKE! Collection-Independent Automatic Keyword Extractor

Ricardo Campos[1,2(✉)] , Vítor Mangaravite[2] , Arian Pasquali[2] ,
Alípio Mário Jorge[2,3] , Célia Nunes[4] , and Adam Jatowt[5]

[1] Polytechnic Institute of Tomar, Tomar, Portugal
ricardo.campos@ipt.pt
[2] LIAAD – INESC TEC, Porto, Portugal
{vima,arrp}@inesctec.pt
[3] DCC – FCUP, University of Porto, Porto, Portugal
amjorge@fc.up.pt
[4] University of Beira Interior, Covilhã, Portugal
celian@ubi.pt
[5] Kyoto University, Kyoto, Japan
adam@dl.kuis.kyoto-u.ac.jp

Abstract. In this paper, we present YAKE!, a novel feature-based system for multi-lingual keyword extraction from single documents, which supports texts of different sizes, domains or languages. Unlike most systems, YAKE! does not rely on dictionaries or thesauri, neither it is trained against any corpora. Instead, we follow an unsupervised approach which builds upon features extracted from the text, making it thus applicable to documents written in many different languages without the need for external knowledge. This can be beneficial for a large number of tasks and a plethora of situations where the access to training corpora is either limited or restricted. In this demo, we offer an easy to use, interactive session, where users from both academia and industry can try our system, either by using a sample document or by introducing their own text. As an add-on, we compare our extracted keywords against the output produced by the IBM Natural Language Understanding (IBM NLU) and Rake system. YAKE! demo is available at http://bit.ly/YakeDemoECIR2018. A python implementation of YAKE! is also available at PyPi repository (https://pypi.python.org/pypi/yake/).

Keywords: Keyword extraction · Information extraction · Text mining

1 Introduction

While considerable progress has been made over the last few years, the task of extracting meaningful keywords is yet to be solved, as the effectiveness of existing algorithms is still far from the ones in many other core areas of computer science. Most traditional approaches follow a supervised methodology, which largely depends on having access to training annotated text corpora. One of the first approaches has been proposed by Turney [5] who developed a custom-designed algorithm named GenEx. The great majority of the approaches developed so far, relied however, on supervised

© Springer International Publishing AG, part of Springer Nature 2018
G. Pasi et al. (Eds.): ECIR 2018, LNCS 10772, pp. 806–810, 2018.
https://doi.org/10.1007/978-3-319-76941-7_80

methods such as Naïve Bayes as a way to select relevant keywords. Arguably the most widespread implementation of such an approach is KEA [7] which uses the Naïve Bayes machine learning algorithm for keyword extraction. Despite their often-superior effectiveness, the main limitation of supervised methods is their relatively long training time process. This contrasts with general unsupervised algorithms [3, 4, 6], which may be quickly applied to documents across different languages or domains in a short time span and which demand reduced effort due to their plug and play nature. In this paper, we describe YAKE!, an online keyword extraction demo which builds upon text statistical features extracted from a single document to identify and rank the most important keywords. While keywords extraction systems have been extensively studied over the last few years, multilingual online single document tools are still very rare. YAKE! is an attempt to fill this gap. Overall, it provides a solution which does not need to be trained on a particular set of documents, and thus can be easily applied to single texts, regardless of the existence of a corpus, dictionary or any external collection. In an era of massive but likely unlabeled collections, this can be a great advantage over other approaches, particularly supervised ones. Another important feature is that YAKE! does not use NER nor PoS taggers, which makes the system to be language-independent, except for the use of different but static lists of stopwords for each language. This enables an easy adaptation of YAKE! to other languages other than English, especially, to minor languages for which open source language processing tools are scarce. It is an advantage over supervised methods, which demand training a custom model beforehand. Finally, the fact that YAKE! relies only on statistical features extracted from the text itself allows for easily scaling to vast collections.

2 Keyword Extraction Pipeline

The proposed system has six main components: (1) Text pre-processing; (2) Feature extraction; (3) Individual terms score; (4) Candidate keywords list generation; (5) Data Deduplication; and (6) Ranking. In the following we will provide a concise description of each of the six steps as a detailed discussion of them is beyond the scope of this paper and can be found on [1]. First, we apply a pre-processing step which splits the text into individual terms whenever an empty space or a special character (e.g., line breaks, brackets, comma, period, etc.) delimiter is found. Second, we devise a set of five features to capture the characteristics of each individual term. These are: (1) *Casing*; (2) *Word Positional*; (3) *Word Frequency*; (4) *Word Relatedness to Context*; and (5) *Word DifSentence*. The first one, *Casing*, reflects the casing aspect of a word. *Word Positional* values more those words occurring at the beginning of a document based on the assumption that relevant keywords often tend to concentrate more at the beginning of a document. *Word Frequency* indicates the frequency of the word, scoring more those words that occur more often. The fourth feature, *Word Relatedness to Context*, computes the number of different terms that occur to the left (resp. right) side of the candidate word. The more the number of different terms that co-occur with the candidate word (on both sides), the more meaningless the candidate word is likely to be. Finally, *Word DifSentence* quantifies how often a candidate word appears within different sentences. Similar to *Word Frequency*, *Word DifSentence* values more those words that often occur in different sentences. Both features however, are combined with *Word Relatedness to Context*, meaning that the more they occur in different sentences the better, as long as they do not occur frequently with different words on the right or left side (which would resemble a behavior close to the one of stopwords). In the third step, we heuristically combine all these features into a single measure such that each term is assigned a score S(w). This weight will feed the process of generating keywords which is to be taken in the fourth step. Here, we consider a sliding window of 3-grams, thus generating a contiguous sequence of 1, 2 and 3-gram candidate keywords. Each candidate keyword will then be assigned a final S(kw), such that the smaller the score the more meaningful the keyword will be. Equation 1 formalizes this:

$$S(kw) = \frac{\prod_{w \in kw} S(w)}{TF(kw) * \left(1 + \sum_{w \in kw} S(w)\right)} \tag{1}$$

where S(kw) is the score of a candidate keyword, determined by multiplying (in the numerator) the score S(w) of the first term of the candidate keyword by the subsequent scores of the remaining terms. This is divided by the sum of the S(w) scores to average out with respect to the length of the keyword, such that longer n-grams do not get benefited just because they have a higher n. The result is further divided by TF(kw)-term frequency of the keyword - to penalize less frequent candidates. In the fifth step, we eliminate similar candidates coming from the previous steps. For this, we use the *Levenshtein distance* [2]. Finally, the system will output a list of relevant keywords, formed by 1, 2, 3-grams, such that the lower the S(kw) score the more important the keyword will be.

3   Demonstration Overview

In this demonstration, we highlight some of the major features of YAKE!, in particular, its independence with regards to a training corpus, dictionary, size of the text, languages and domains. The online demo of YAKE! can be accessed at http://bit.ly/YakeDemoECIR2018. A python implementation of YAKE! is also available at https://pypi.python.org/pypi/yake/ meaning that our method can already be used and imported as a library. During the demonstration, we will showcase the behavior of our system on different kinds of datasets with various types of settings and we will show the audience how to interact with YAKE! All the results are immediately put side by side with the IBM NLU commercial solution and Rake [4] system for a comparison. In a nutshell, users can interact and test YAKE! under 3 different scenarios. For the first one, they can try the system by selecting a pre-chosen text from six datasets (500 N-KPCrowd-v1.1, INSPEC, Nguyen 2007, SemEval 2010, PubMed, 110-PT-BN-KP), the first five in English, the latter in Portuguese. Texts are from the scientific domain (including short abstracts, medium and large-size texts), TV broadcast and news articles. All the results can be compared against a ground-truth. Therefore, users can analyze the effectiveness of our system with regards to different domains, sizes and languages of the input text. For the second scenario, we arbitrarily choose as input to our system, a set of sample texts from different domains (politics, culture, history, tourism, technology, religion, economy, sports, education and biography) and languages (English, Portuguese, German, Italian, Netherland, Spanish, French, Turkish, Polish, Finnish and Arabic), thus enriching our demo with a range of text examples with characteristics different than those of formal datasets. Finally, we offer the user the chance to test YAKE! in an online environment and in real time with his/her own text, to see how it responds to different scenarios/texts and languages/domains. Users can input their text either by hand (copy/paste), or by referring to its URL. PDF files are also accepted. As a rule of thumb, the maximum size of n-grams is set to 3. However, this parameter can be adjusted by the user on the opening page. As an add-on, we also offer researchers access to a web service (under the API tab) so that our system can be used for research purposes. A screenshot of the results obtained for a document extracted from the INSPEC dataset is shown in Fig. 1.

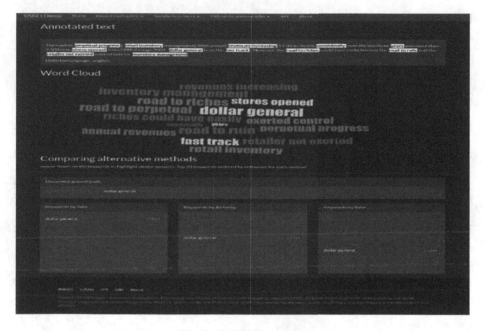

Fig. 1. YAKE! interface.

The results of our demo can be explored through three different functionalities: (1) annotated text; (2) word cloud; and (3) comparing YAKE! with alternative methods. The first one shows the text annotated with the top 10 keywords retrieved by YAKE!. The second, uses the relevance score of each keyword retrieved by YAKE!, to generate a word cloud, where more important keywords are given a higher size. Finally, we compare the results of YAKE! against IBM NLU and Rake [4]. Each result is assigned a relevance ranking value reflecting the importance of the keyword within the text. Note that in the case of YAKE!, the lower the value, the more important the keyword is. As a further additional feature, we offer the user to explore the results by means of a mouse-hover feature which highlights both the keyword selected as well as similar keywords identified within the three systems. In the example in Fig. 1, we highlight a keyword from the ground-truth ("dollar general") and observe its disposition within the three systems. Though anecdotal, this example shows that Yake! is able to list this keyword in the 1st position. A formal evaluation however is needed to take valid conclusions.

Acknowledgements. This work is partially funded by the ERDF through the COMPETE 2020 Programme within project POCI-01-0145-FEDER-006961, and by National Funds through the FCT as part of project UID/EEA/50014/2013 and of project UID/MAT/00212/2013. It was also financed by MIC SCOPE (171507010) and by Project "TEC4Growth - Pervasive Intelligence, Enhancers and Proofs of Concept with Industrial Impact/NORTE-01-0145-FEDER-000020" which is financed by the NORTE 2020, under the Portugal 2020, and through the ERDF.

References

1. Campos, R., Mangaravite, V., Pasquali, A., Jorge, A., Nunes, C., Jatowt, A.: YAKE! collection-independent automatic keyword extractor. In: Pasi, G., Piwowarski, B., Azzopardi, L., Hanbury, A. (eds.) ECIR 2018, LNCS, vol. 10772, pp. 806–810. Springer, Cham (2018)
2. Levenshtein, V.: Binary codes capable of correcting deletions, insertions, and reversals. Sov. Phys. Dokl **10**(8), 707–710 (1966)
3. Mihalcea, R., Tarau, P.: TextRank: bringing order into texts. In: EMNLP 2004, pp. 404–411, Barcelona, Spain, 25–26 July 2004
4. Rose, S., Engel, D., Cramer, N., Cowley, W.: Automatic keyword extraction from individual documents. In: Text Mining: Theory and Applications (2010)
5. Turney, P.: Learning algorithms for keyphrase extraction. Inf. Retr. J. **2**(4), 303–336 (2000)
6. Wan, X., Xiao, J.: Single document keyphrase extraction using neighborhood knowledge. In: AAAI 2008, pp. 855–860, 13–17 July 2008
7. Witten, I., Paynter, G., Frank, E., Gutwin, C., Nevill-Manning, C.: KEA: practical automatic keyphrase extraction. In: JCDL 2004, pp. 254–255, 7–11 June 1999

SIREN - Security Information Retrieval and Extraction eNgine

Lalit Mohan Sanagavarapu(✉) ⓘ, Neeraj Mathur, Shriyansh Agrawal,
and Y. Raghu Reddy

International Institute of Information Technology, Gachibowli, Hyderabad, India
{lalit.mohan,neeraj.mathur,shriyansh.agrawal}@research.iiit.ac.in,
raghu.reddy@iiit.ac.in

Abstract. Domain specific search engines (DSSE) are gaining popularity because of better search relevance and domain specificity. The growth of IT and internet led to the increase of cyber attacks, however, lack of DSSE for Security is making users refer multiple sites for security information. We demonstrate SIREN, a search engine for 'Information and Cyber Security' with coverage and classification at subdomain level and ranking search results based on site credibility. As part of our demonstration, we automated identification of seed URLs (34,007) and related child URLs (400,726) of the security domain using Artificial Bee Colony algorithm. Also, we evaluated functional and non-functional parameters of available open source software stack that can be used for building other DSSEs.

Keywords: Security · Domain specific search · Seed URL · Credibility

1 Introduction

There are 3.76+ Billion internet users accessing 1.27+ Billion websites and this is expected to increase with technology innovations and affordable computing devices. This large number of websites are indexed and accessed using search engines. However, the relevance of generic search engine results is an area of concern due to ambiguity, bias, content filtering and other issues. For improved search relevance, Domain Specific Search Engines (DSSE) are gaining wider acceptance. It is a well established fact that DSSEs have better Precision due to limited scope and focused corpus resulting in less load on network, storage and processor. Some of the challenges impeding extensive adoption of DSSEs are (i) Selection of seed URL is manual and requires thorough investigation for determining seed URL relevance (ii) All subdomains of a domain may not be represented in the extracted content, (iii) Search results to be ranked based on their domain credibility (iv) Additional processing is required for removal of unrelated content that is not specific to the domain and (v) Uniqueness/purpose

© Springer International Publishing AG, part of Springer Nature 2018
G. Pasi et al. (Eds.): ECIR 2018, LNCS 10772, pp. 811–814, 2018.
https://doi.org/10.1007/978-3-319-76941-7_81

of a domain needs to be well established. To address some of these identified challenges, we implement an approach to build security domain specific search engine.

With increased adoption of IT and internet, the concerns of security (confidentiality, integrity and availability) have also increased. Every day, thousands of websites are being compromised and millions of dollars are lost (hack, cyber extortion and other internet frauds) despite growing investment in security awareness and products. To address the concerns on security, there are increasing number of security related websites providing details on product comparisons, vulnerabilities, incidents, threats, controls, etc. However, with this growing internet information on security, relying on generic search engines that have issues of relevance and ambiguity can be detrimental to security knowledge. Thus, arises the need for a security search engine that could provide details on (i) vulnerabilities, threats, incidents, controls and advisories (ii) disambiguated and relevant search results ranked based on the credibility. Our interactions with Chief Information Security officers of Banking sector and other security experts reiterated the need for a security information search site. We also draw the inspiration of building SIREN (**S**ecurity **I**nformation **R**etrieval and **E**xtraction e**N**gine) from the established PubMed[1] search engine for Health domain (security is critical for non-bodily health of an organization/individual). Having a security search engine will benefit (i) Citizens - Security awareness; (ii) Government and related Bodies - To share advisories and controls leading to policy enablement and better collaboration with security agencies; (iii) Organizations using IT - reduces dependencies on paid security threat feeds; and (iv) Researchers - Access to knowledge for furthering research in security related topics.

2   Demonstration

In the industry, existing work on security search has been in the form of threat feeds, public IP scan and not as a security webpage content. With the established need of building a SIREN, we reviewed existing literature and evaluated software components for building a search engine. Our study of existing literature on search engine components and the related gaps for building a domain specific search engine are shown in Fig. 1. The functional and non-functional parameters comparison of the opensource software components for building search engine is available at [2]. Our effort in building SIREN includes

– **Seed URL and Crawling** - Twitter and Wikipedia are becoming URL repositories with increased crowdsourcing and Social Media usage. However, the current methods for identification of seed URL are manual. Inspired by Nature's Optimization algorithms, we extended Artificial Bee Colony (ABC) algorithm for automated identification of seed URLs [6]. Using the algorithm, we extracted 34,007 security domain related seed URLs. These initial seed URLs are provided to Apache Storm-crawler for crawling the content from

---

[1] https://www.ncbi.nlm.nih.gov/pubmed/.

Components	Existing Work	Gap
Seed URL Identification	• Repositories to get URLs • Metrics to measure URL relevance	1) Domain Coverage 2) Metrics does not factor redundancy
Crawling	• Focussed Crawlers – Ontology, Click through, Keyword, Anchor Text, etc	3) Crawler Efficiency
Noise reduction	• Stop word removal and Lemmatization	4) Domain Specific Acronyms
Classification	• Supervised and Unsupervised Learning • Reinforced Learning and Deep Learning	5) Ontology extraction from Web Text 6) Classification of Vulnerabilities, Threats, etc.
Indexing and Ranking	• TF-IDF, Okapi BM25, LSA/pLSA, Citation Indices and Multi-variate measurement	7) Domain based Credibility
Search Results		8) Domain Specific trends 9) Search specific Visualization

Fig. 1. Search engine components

400,726 child URLs [5]. To obtain domain coverage, we used Groups and Controls of ISO 27001:2013 standard, widely accepted in research and industry for identifying security gaps and implementing controls. We extended Claude Shannon Diversity Index [7] to validate domain coverage, the Diversity Index measure of 2.1 states that all security groups/subdomains are represented.

- **Noise Reduction** - Apache jsoup is used for removal of HTML tags of the crawled content. We used Python NLTK package for stopword removal and stemming to reduce the noise. A Security Acronyms list [3] is used to avoid loss of Security words while removing stopwords.

- **Classification** - For classifying security related content in a webpage, we compared the results of some of the widely used text processing algorithms such as SVM, Naive Bayes and Phrase2Vec and shortlisted Phrase2Vec [1]. The flexibility to parse ISO 27001:2013 manuals for security related phrases and the accuracy score (cut-off >0.75 similarity score on a scale 0–1, with value 1 being exact match of phrase) are the reasons for deciding on Phrase2Vec algorithm. The recall score for >0.75 similarity score is 0.393.

- **Indexing and Ranking** - Apache SOLR is used for indexing the classified content. Though there are prevalent pageranking algorithms, we combined the indexing with subdomain credibility of the page while displaying search results. Our credibility assessment to reduce search bias is based on Structural (Broken links, Page Loadtime, Spell errors and others) and Functional (Information source, Text cohesion, Content similarity, Webpage Genre and others) layers. A demo of credibility assessment is available at WebCred[2]. We normalized the values of credibility dimensions and assign a credibility score for each retrieved webpage.

- **Unique Features** - DSSEs are expected to have unique features beyond search results. SIREN uses ISO 27001:2013 manuals and related Security Ontology [4] to identify Vulnerabilities, Threats, Incidents and Controls in the crawled web content.

The demonstration of SIREN[3] to Chief Information Security Officers of Banking sector and the Department of Science and Technology, Government of India is well received. SIREN (component diagram is shown in Fig. 2) is deployed on a cloud platform to perform distributed crawling, faster classification and load balancing to cater user search queries.

---

[2] https://serc.iiit.ac.in/WEBCred/.

[3] https://serc.iiit.ac.in/Bhompoo/infosec.html.

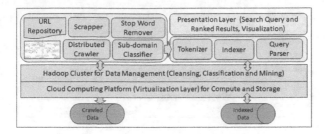

Fig. 2. SIREN - component diagram

3   Conclusions

SIREN, a security search engine using available open source software has crawled 400,726 URLs and can become one stop site for security related information. The usage of ABC algorithm and the diversity metric for subdomain coverage improves the relevance of the search engine. The approach for credibility assessment in a domain reduces the search bias leading to increased usage of search engine. The evaluation of software stack for building search engine can be used as a package for building other domain specific search engines. We plan to enhance SIREN to classify crawled content based on evolving security ontology, explore Deep web content, provide automated threat feeds and display visualization based on search criteria. The classified web content may be used for various security related analytics and recommendations.

References

1. Phrase2Vec Algorithm. https://github.com/zseymour/phrase2vec
2. Search Engine Software Comparison. http://tinyurl.com/SearchToolComp
3. Security Acronyms. http://tinyurl.com/SecurityAcronym
4. Ekelhart, A., Fenz, S., Klemen, M.D., Weippl, E.R.: Security ontology: simulating threats to corporate assets. In: Bagchi, A., Atluri, V. (eds.) ICISS 2006. LNCS, vol. 4332, pp. 249–259. Springer, Heidelberg (2006). https://doi.org/10.1007/11961635_17
5. Sanagavarapu, L.M., Sarangi, S., Reddy, Y.R., Varma, V.: Fine grained approach for domain specific seed URL extraction. In: HICSS. IEEE (2018, to be published)
6. Sanagavarapu, L., Sarangi, S., Reddy, Y.R.: ABC algorithm for URL extraction. In: ICWE, Practi-O-Web Workshop. Springer (2017, to be published)
7. Spellerberg, I.F., Fedor, P.J.: A tribute to claude shannon (1916–2001) and a plea for more rigorous use of species richness, species diversity and the shannon-wienerindex. Glob. Ecol. Biogeogr. **12**(3), 177–179 (2003)

CITEWERTs: A System Combining Cite-Worthiness with Citation Recommendation

Michael Färber[1]([⊠])(iD), Alexander Thiemann[1], and Adam Jatowt[2]

[1] University of Freiburg, Freiburg im Breisgau, Germany
michael.faerber@cs.uni-freiburg.de, mail@athiemann.net
[2] Kyoto University, Kyoto, Japan
adam@dl.kuis.kyoto-u.ac.jp

Abstract. Due to the vast amount of publications appearing in the various scientific disciplines, there is a need for automatically recommending citations for text segments of scientific documents. Surprisingly, only few demonstrations of citation-based recommender systems have been proposed so far. Moreover, existing solutions either do not consider the raw textual context or they recommend citations for predefined citation contexts or just for whole documents. In contrast to them, we propose a novel two-step architecture: First, given some input text, our system determines for each potential citation context, which is typically a sentence long, if it is actually "cite-worthy." When this is the case, secondly, our system recommends citations for that context. Given this architecture, in our demonstration we show how we can guide the user to only those sentences that deserve citations and how to present recommended citations for single sentences. In this way, we reduce the user's need to review too many sentences and recommendations.

Keywords: Citation recommendation · Citation context
Digital libraries · Recommender systems

1 Motivation

The number of published papers within the different scientific disciplines has increased dramatically in the last years: More than 100,000 new computer science papers are published every year [1]. A similar trend can be observed in other disciplines. This overload of information leads to the fact that scientists (and other people who need to cite facts, such as authors of news articles and editors of encyclopedias) cannot be aware of all publications at any given time and hence have difficulties during citing. To assist users in the process of citing, citation recommendation approaches have been proposed that recommend citations based on a given text fragment, which is called the *citation context* and which can be a sentence in a paper. However, most approaches (such as [2–4], to name only the most recent) do not explicitly incorporate the question

© Springer International Publishing AG, part of Springer Nature 2018
G. Pasi et al. (Eds.): ECIR 2018, LNCS 10772, pp. 815–819, 2018.
https://doi.org/10.1007/978-3-319-76941-7_82

of whether a given citation context for which citations are to be recommended, actually requires citations. As a consequence, citations are recommended for each potential citation context (e.g., sentence), even though the context may not "need" any citation in the first place.

This also holds in the case of presented demonstrations of citation recommendation systems. The RefSeer system [5], which is the most related demonstration to ours (though apparently no longer available online), recommends one citation for each sentence in the input text. Besides [5], to the best of our knowledge, only paper recommendation systems exist, i.e., systems that do not use any citation context, but, for instance, just use a citation graph [6]. *TheAdvisor* [7] and *FairScholar* [8] are further examples of paper recommender system demonstrations.

In this paper, we demonstrate an end-to-end system that is, to the best of our knowledge, the first one to combine two steps that have been considered separately so far, namely (1) determining the cite-worthiness of every potential citation context (e.g., sentence) and (2) recommending citations for "approved" citation contexts. As a consequence, our system does not require knowing a priori the locations in which a citation should be inserted; instead, they will be determined automatically (up to the sentence level). This two-step approach is not only more user-friendly, as it hides unnecessary recommendations, but it also reduces the number of costly recommendation computations.

The user of our system, which can be any scientist who is writing or reading text, can use the front-end to be guided to sentences that apparently need citations; he can then review the recommended citations for the single citation contexts. By training our approach on the publications hosted at arXiv.org (until August 2017), we are able to recommend papers published until recently. The source code of our system is available online.[1]

2   System Overview

Before describing the user interface and a typical workflow of the system, we outline the back-end. This consists of two main components:

Sentence Classification. This component makes a binary classification whether the given potential citation context needs a citation or not. We implement this classification step by means of a convolutional recurrent neural network (CRNN). The full architecture of the CRNN consists of four convolutional layers with 128 hidden states and with filter sizes of 1, 2, 3, and 5. Next is a concatenation step followed by max pooling. After the convolutional part, the recurrent part consists of three gated recurrent unit (GRU) layers with a (recurrent) dropout of 0.2. Finally, a densely connected layer with a softmax activation function and two outputs provides the final classification. An evaluation of this classification approach is provided in [9] (Fig. 1).

---

[1] See https://github.com/agrafix/grabcite and https://github.com/agrafix/grabcite-net.

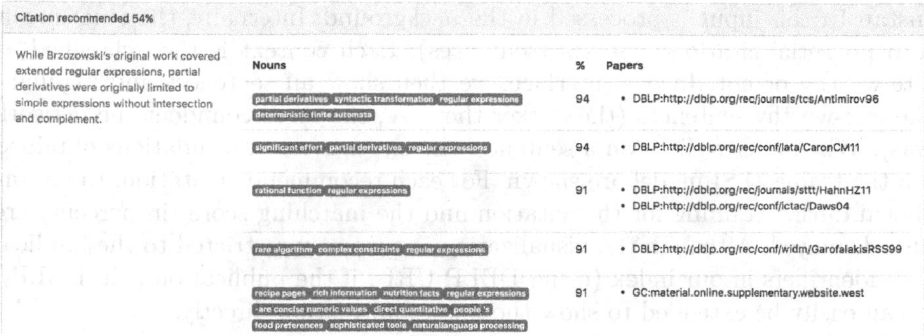

Fig. 1. Our system's user interface after selecting a cite-worthy sentence.

Citation Recommendation. By manually evaluating 1,500 randomly chosen sentences w.r.t. the characteristics of the citation contexts and w.r.t. their cite-worthiness,[2] we discover that citations mainly relate to a specific noun phrase mentioned in the citation context. This noun phrase is often either the name(s) of the cited paper's author(s) (e.g., "Carignan et al. [X] have"), or it is a specifically named technique or method (e.g., "discrete exterior calculus (DEC) [X]"), or an abstract concept (e.g., "recurrent neural networks"). Based on these insights, our idea is to extract all nouns from a sentence and to train a latent semantic index (LSI) [11] via TF-IDF on our data set.[3] Since LSI is only based on the citation contexts of the citing papers, no content of the *cited* papers is needed, making our approach applicable in many scenarios. To build the LSI, we only consider sentences that contain at least one citation.

We use the Python library TextBlob to extract the noun phrases. After splitting the data set into a training set (90%) and a test set (10%), the LSI model is computed with $n = 200$ factors via a fast truncated SVD [12]. In the testing phase, we convert the sequence of extracted nouns into LSI vector space and query the LSI index by computing similarities. Finally, we use a metadata mapping to recover contained references to the most similar entries of the LSI index, which are then returned as recommendations to the user.

3   Interface Usage

The interface of our system is available online at http://citewerts.citation-recommendation.org. A prototypical user inputs some text (e.g., a text paragraph without citations) into the input text box and presses "Analyze."

---

[2] We use the documents of the DRI corpus [10], as they have already been used for other citation analysis studies. The assessments, conducted by the authors, are available online at http://citation-recommendation.org/publications.

[3] Note that our intention was not to build a novel approach for citation recommendation that outperforms existing methods but to show the usefulness of combining citation recommendation with citation context classification.

Instantly, this input is processed in the background: Internally, the text is split into potential citation contexts (sentences). Each context is then classified as cite-worthy or not. In our interface, we then show all sentences and highlight the cite-worthy sentences (the darker they are, the more confident the network was). When a user clicks on a sentence, the citation recommendations obtained via the trained LSI model are shown. For each recommended citation, the terms shown during training for this citation and the matching score (in percent) are also displayed. Although the visualization is currently restricted to the publication identifiers in our index (using DBLP URIs, if the publication is in DBLP), it can easily be extended to show the meta-information directly.

4   Conclusions

In this paper, we demonstrated a system that not only recommends citations but firstly identifies cite-worthy contexts in the input text. Our system is user-friendly, since it hides unnecessary recommendations, and it reduces the number of costly recommendation computations. It can be used not only for scientific texts in different disciplines but also, for instance, for texts from encyclopedias or news articles.

Acknowledgements. Michael Färber is an International Research Fellow of the Japan Society for the Promotion of Science (JSPS). The work was partially supported by MIC SCOPE (171507010).

References

1. Küçüktunç, O., Saule, E., Kaya, K., Çatalyürek, Ü.V.: Diversifying citation recommendations. ACM Trans. Intell. Syst. Technol. 5(4), 55:1–55:21 (2014)
2. Ebesu, T., Fang, Y.: Neural citation network for context-aware citation recommendation. In: SIGIR 2017, pp. 1093–1096 (2017)
3. Jiang, Z., Liu, X., Gao, L.: Chronological citation recommendation with information-need shifting. In: CIKM 2015, pp. 1291–1300 (2015)
4. Huang, W., Wu, Z., Chen, L., Mitra, P., Giles, C.L.: A neural probabilistic model for context based citation recommendation. In: AAAI 2015, pp. 2404–2410 (2015)
5. Huang, W., Wu, Z., Mitra, P., Giles, C.L.: RefSeer: a citation recommendation system. In: JCDL 2014, pp. 371–374 (2014)
6. Huynh, T., Hoang, K., Do, L., Tran, H., Luong, H.P., Gauch, S.: Scientific publication recommendations based on collaborative citation networks. In: CTS 2012, pp. 316–321 (2012)
7. Küçüktunç, O., Saule, E., Kaya, K., Çatalyürek, Ü.V.: TheAdvisor: a webservice for academic recommendation. In: JCDL 2013, pp. 433–434 (2013)
8. Anand, A., Chakraborty, T., Das, A.: FairScholar: balancing relevance and diversity for scientific paper recommendation. In: Jose, J.M., Hauff, C., Altıngovde, I.S., Song, D., Albakour, D., Watt, S., Tait, J. (eds.) ECIR 2017. LNCS, vol. 10193, pp. 753–757. Springer, Cham (2017). https://doi.org/10.1007/978-3-319-56608-5_76
9. Färber, M., Thiemann, A., Jatowt, A.: To cite, or not to cite? Detecting citation contexts in text. In: ECIR 2018 (2018)

10. Fisas, B., Ronzano, F., Saggion, H.: A multi-layered annotated corpus of scientific papers. In: LREC 2016 (2016)
11. Deerwester, S.C., Dumais, S.T., Landauer, T.K., Furnas, G.W., Harshman, R.A.: Indexing by latent semantic analysis. JASIS **41**(6), 391–407 (1990)
12. Boutsidis, C., Magdon-Ismail, M.: Faster SVD-Truncated Least-Squares Regression. CoRR abs/1401.0417 (2014)

Elastic ChatNoir: Search Engine
for the ClueWeb and the Common Crawl

Janek Bevendorff[1,2], Benno Stein[1,2], Matthias Hagen[1,2],
and Martin Potthast[1,2(✉)]

[1] Bauhaus-Universität Weimar, Weimar, Germany
{janek.bevendorff,benno.stein,matthias.hagen}@uni-weimar.de
[2] Leipzig University, Leipzig, Germany
martin.potthast@uni-leipzig.de

Abstract. Elastic ChatNoir (Search:www.chatnoir.eu Code:www.gith
ub.com/chatnoir-eu) is an Elasticsearch-based search engine offering a
freely accessible search interface for the two ClueWeb corpora and the
Common Crawl, together about 3 billion web pages. Running across
130 nodes, Elastic ChatNoir features subsecond response times comparable to commercial search engines. Unlike most commercial search
engines, it also offers a powerful API that is available free of charge to
IR researchers. Elastic ChatNoir's main purpose is to serve as a baseline
for reproducible IR experiments and user studies for the coming years,
empowering research at a scale not attainable to many labs beforehand,
and to provide a platform for experimenting with new approaches to web
search.

1 Introduction

At heart, information retrieval is the art of building the perfect search engine. As
a special interest of computer science since 40 years, information retrieval (IR)
laid the foundation for technology that society takes for granted today—most
notably thanks to the emergence of the world wide web as IR's single most important application domain in terms of deployment scale and commercial success.
A large body of work examines nearly every aspect of web search and retrieval
under countless practical and theoretical scenarios, contributing as many insights
into how (better) web search engines can be built (in the future). For all its right-
ful claims to fame, there is one thing the scientific community has not yet built
for itself: an actual web-scale research search engine.

Wait, what? There are the Googles [1], Lucenes [5], Terriers [7], Indris [10],
Galagos [2], and even the non-elastic ChatNoirs [8] of this world, our esteemed
readers will interject. True, but none of them have been deployed at web
scale, *and* optimized for fast retrieval, *and* made publicly available for free,
and kept that way. Google was on the right track, but turned commercial and
hence opaque; Indri and the non-elastic ChatNoir offer search interfaces to the
ClueWeb, but they are not quite as fast as one would have liked them to be, nor
capable to withstand a high load of traffic. These shortcomings render the end

© Springer International Publishing AG, part of Springer Nature 2018
G. Pasi et al. (Eds.): ECIR 2018, LNCS 10772, pp. 820–824, 2018.
https://doi.org/10.1007/978-3-319-76941-7_83

user search experience unrealistic, since commercial search engines set the bar of users' expectations. Nevertheless, the validity of user studies hinges on realistic user interactions, so that many researchers either index much smaller corpora (e.g., the ClueWeb12-B) at the cost of generalizability, or resort to commercial search engines at the cost of reproducibility and influence over or knowledge about the underlying retrieval model. This applies likewise to experiments where a search engine is automatically queried to reach a higher-level retrieval goal, such as source retrieval for text reuse detection [6]; in fact, commercial search engines hardly offer (affordable) APIs.

Elastic ChatNoir aims at filling this gap by (1) hosting a freely accessible search engine, (2) indexing the two IR reference corpora ClueWeb09 and ClueWeb12 for compatibility with TREC, (3) indexing at least one instance of the Common Crawl for recency, (4) maintaining at least one baseline retrieval model long-term for reproducibility, (5) offering a free-of-charge API to IR researchers, and (4) publishing its full source code under a permissive open source license.

2 Background and Related Work

The rapid progress of retrieval technology in the 1990s—not least due to the corresponding TREC tracks—were soon adopted by several open-source and research-oriented search projects. Alongside the commercial success of Google [1], modern retrieval models were made available to the wider research community as working software within the popular Lucene library [5], followed by Terrier [7] and Lemur, the latter combining the contributions of the Indri [10] and the Galago [2] teams. In many experiments published at IR conferences and in IR journals, the aforementioned libraries or search systems serve as the underlying retrieval architecture to this day. However, with the rise of ever larger web corpora at TREC, most notably the ClueWeb09 with its more than 500 million English pages, many IR labs lacked the facilities necessary to process and index them. Today, the Common Crawl compiles more than 3 billion pages, with new and potentially even larger versions being published on a monthly basis.

Foreseeing this problem, the Indri team, who crawled the ClueWeb09 and ClueWeb12, also offered on-demand API access to a search engine indexing those corpora. Alas, the search interface has a rather slow retrieval time, rendering it non-realistic, and our demand for API requests swiftly outgrew the quotas. We hence took matters into our own hands and started developing the first version of ChatNoir in 2011 [8], which uses a custom implementation of the BM25F retrieval model [9] as a baseline and indexes the English portion of the ClueWeb09. Sharded across 40 search nodes, we were able to bring down retrieval times to a few seconds for most queries. For five years, the first ChatNoir has supported user studies at scale, shared tasks with dozens of participants, and has served as a teaching subject, answering millions of queries until today. Now the time is ripe for an overhaul and an upgrade.

Twenty years later, our goals do not differ much from Brin's and Page's [1]: "With ~~Google~~ ChatNoir, we have a strong goal to push more development and understanding into the academic realm. Another important design goal was to build systems that reasonable numbers of people can actually use. [..] Our final design goal was to build an architecture that can support novel research activities on large-scale web data, [..] to set up an environment where other researchers can come in [..] and produce interesting results that would have been very difficult to produce otherwise. [..] Another goal we have is to set up a Spacelab-like environment where researchers or even students can propose and do interesting experiments [..]". It goes without saying that today, thanks to technologies that simply did not exist back then, our task is rendered a lot easier.

3   A Scalable Search Engine Based on Elasticsearch

With Elastic ChatNoir, we depart from our custom implementation underlying the "old" ChatNoir, adopting Elasticsearch as a well-known, battle-tested, open source search backend, employed by many companies. At the time of writing, Elastic ChatNoir indexes the ClueWeb09, the ClueWeb12, and a 2015 instance of the Common Crawl. Regarding the latter, we plan on updating to the newest version at regular intervals.

For each corpus, we parse the plain WARC files, heuristically deduce content type and encoding of each entry (since corresponding meta data may be incorrect or missing), assign a deterministic name-based UUID to each entry, and create HDFS map files, mapping UUIDs to a JSON document containing headers and content, and URLs to UUIDs. The map files are input to a Hadoop MapReduce indexing job. In its map phase, from each raw HTML document, the main content is extracted, its language is detected, and meta data such as URLs, keywords, host names, headings, etc. are extracted. Additionally, external meta data (e.g., spam ranks) are mapped to their document IDs. Documents for which no useful content could be extracted, are discarded. During the reduce phase, all information belonging to an individual web page is collected in a multi-field JSON document, which is fed into the Elasticsearch index. As retrieval model for plain text fields, we employ BM25 with various filters and tokenizers for preprocessing. At 124 data nodes and 40 primary shards, an indexing throughput of at least 20,000 documents per second is achieved. For production, each shard is replicated two times.

The web frontend uses the Java Transport API to communicate with Elasticsearch. Every query is first run as a basic filtered Boolean query with AND semantics without expensive operations like proximity, phrasal search, or fuzzy matching. Each shard tries to find up to 70,000 results and then rescores the top-400 results per shard (parameters determined in pilot experiments) with a more complex query, taking into account many more factors (e.g., proximity boosting, additional boosts for Wikipedia articles or home pages, potential penalties for other factors, etc.). Results from the same website are visually grouped and potentially reordered.

Table 1 shows key figures of Elastic ChatNoir and its underlying hardware. Shards are distributed across 8 hard disks per data node. Search efficiency is

Table 1. Key figures of Elastic ChatNoir. Corpora include the ClueWeb09 (cw09), the ClueWeb12 (cw12), and the Common Crawl 11/2015 (cc1511).

Criterion	Corpus			Σ
	cw09	cw12	cc1511	
Indexed documents	734.8m	638.8m	1.6b	3.0b
Primary shards	40	40	40	120
Shard size	90.0 GB	77.5 GB	242.5 GB	–
Document map files (full)	6.1 TB	6.2 TB	28.8 TB	41.1TB
Index (full)	3.6 TB	3.1 TB	9.7 TB	16.4 TB
Replication	3	3	3	–
Document map files (replicated)	18.3 TB	18.6 TB	86.4 TB	123.3 TB
Index (replicated)	10.7 TB	9.4 TB	29.2 TB	49.3 TB
Total size on disk	29.0 TB	28.0 TB	115.6 TB	172.6 TB

Cluster	Betaweb
Nodes	130
Nodes (data)	124
Nodes (master / coord.)	5
Nodes (monitoring)	1
RAM (idle)	5–10 GB
RAM (max)	24 GB
RAM (cache)	150 GB
Avg. query time (warm)	~600 ms
Avg. query time (cold)	~2100 ms

optimized by continuous "warming," using the AOL log. This way, Elastic-search's node query and request caches are populated, and important parts of the index are served primarily from the host systems' page cache, guaranteeing fast response times in the order of a few hundred milliseconds for most real-world queries. "Cold" index regions (e.g., non-English queries) are of course slower. The nodes are deployed as Docker containers with different configurations for data and master roles and monitored with Check_MK and Kibana+X-Pack.

Finally, our system comes with a powerful API, accessible with API keys with adjustable quota and access restrictions. Usage of the API and the web interface are logged, allowing for post hoc analyses of user studies on a per-subject basis. API keys are issued for free to interested research parties. All of the above, including code, configuration, and documentation is available open source in our public GitHub repository.

4 Evaluation

We evaluated ChatNoir's search effectiveness on the TREC Web tracks of 2013 and 2014 which used the ClueWeb12. One run was conducted using ChatNoir "out of the box," with default parameters, and another run with field weights

Table 2. Evaluation results for the TREC Web track as ERR@20/nDCG@20 scores [3,4].

Participant (selection)	rank	TREC Web track (default) 2013	rank	2014
udel_fang	2	0.176 / 0.310	1	**0.233** / 0.325
ICTNET	4	0.158 / 0.236	2	**0.208** / 0.261
uogTr	3	0.160 / 0.259	5	**0.195** / 0.324
ChatNoir	–	0.130 / 0.193	–	**0.195** / 0.249
Terrier Base	–	–	6	**0.189** / 0.260
udel	5	0.157 / 0.246	7	**0.179** / 0.261
webis / BUW	12	0.101 / 0.181	9	**0.174** / 0.258
wistud	8	0.134 / 0.225	10	**0.174** / 0.291
ut	6	0.152 / 0.228	11	**0.172** / 0.226
Indri Base	14	0.096 / 0.168	12	**0.153** / 0.243

Participant (selection)	rank	TREC Web track (optimized) 2013	rank	2014
udel_fang	2	0.176 / 0.310	1	**0.233** / 0.325
ChatNoir	–	0.134 / 0.197	–	**0.213** / 0.254
ICTNET	4	0.158 / 0.236	2	**0.208** / 0.261
uogTr	3	0.160 / 0.259	5	**0.195** / 0.324
Terrier Base	–	–	6	**0.189** / 0.260
udel	5	0.157 / 0.246	7	**0.179** / 0.261
webis / BUW	12	0.101 / 0.181	9	**0.174** / 0.258
wistud	8	0.134 / 0.225	10	**0.174** / 0.291
ut	6	0.152 / 0.228	11	**0.172** / 0.226
Indri Base	14	0.096 / 0.168	12	**0.153** / 0.243

optimized against previous TREC Web tracks. Table 2 compares our results to a selection of other participants: the default parameters yield a medium rank (though still above the baselines Terrier and Indri), whereas with parameter optimization the performance can be substantially boosted (main optimization criterion was to considerably increase the importance of title matches compared to matches in the text body or other fields).

5 Conclusion

With Elastic ChatNoir, we provide a modern Elasticsearch-based retrieval system for important reference corpora like the ClueWebs and the Common Crawl. ChatNoir is freely available and features a powerful API. In its current version, ChatNoir uses the BM25 retrieval model and achieves subsecond answer times that are similar to commercial search engines, while offering a reasonable retrieval effectiveness as demonstrated by TREC experiments.

In the future, we plan to incorporate further versions of the Common Crawl, so that experiments and user studies with up-to-date web crawls are possible for everyone in a reproducible manner, instead of resorting to commercial search engines as black boxes. Furthermore, we plan to also provide other retrieval models, API functionality of user-defined weighting schemes, and possibly plugin support.

References

1. Brin, S., Page, L.: The anatomy of a large-scale hypertextual web search engine. In: Proceedings of WWW (1998)
2. Cartright, M.A., Huston, S., Field, H.: Galago: A modular distributed processing and retrieval system. In: Proceedings of SIGIR 2012 Workshop on Open Source Information Retrieval, pp. 25–31 (2012)
3. Collins-Thompson, K., Bennett, P.N., Diaz, F., Clarke, C., Voorhees, E.M.: TREC 2013 web track overview. In: Proceedings of TREC (2013)
4. Collins-Thompson, K., Macdonald, C., Bennett, P.N., Diaz, F., Voorhees, E.M.: TREC 2014 web track overview. In: Proceedings of TREC (2014)
5. Goetz, B.: The Lucene search engine: powerful, flexible, and free. In: JavaWorld (2000)
6. Hagen, M., Potthast, M., Adineh, P., Fatehifar, E., Stein, B.: Source retrieval for web-scale text reuse detection. In: Proceedings of CIKM 2017, pp. 2091–2094 (2017)
7. Ounis, I., Amati, G., Plachouras, V., He, B., Macdonald, C., Johnson, D.: Terrier information retrieval platform. In: Losada, D.E., Fernández-Luna, J.M. (eds.) ECIR 2005. LNCS, vol. 3408, pp. 517–519. Springer, Heidelberg (2005). https://doi.org/10.1007/978-3-540-31865-1_37
8. Potthast, M., Hagen, M., Stein, B., Graßegger, J., Michel, M., Tippmann, M., Welsch, C.: ChatNoir: a search engine for the ClueWeb09 corpus. In: Proceedings of SIGIR 2012, p. 1004 (2012)
9. Robertson, S.E., Zaragoza, H., Taylor, M.J.: Simple BM25 extension to multiple weighted fields. In: Proceedings CIKM 2004, pp. 42–49 (2004)
10. Strohman, T., Metzler, D., Turtle, H., Croft, W.B.: Indri: a language model-based search engine for complex queries. In: Proceedings of ICIA 2005, pp. 2–6 (2005)

Workshops

Bibliometric-Enhanced Information Retrieval: 7th International BIR Workshop

Philipp Mayr[1(✉)], Ingo Frommholz[2], and Guillaume Cabanac[3]

[1] GESIS – Leibniz-Institute for the Social Sciences, Cologne, Germany
philipp.mayr@gesis.org
[2] Institute for Research in Applicable Computing,
University of Bedfordshire, Luton, UK
ifrommholz@acm.org
[3] Computer Science Department, University of Toulouse, IRIT UMR 5505,
Toulouse, France
guillaume.cabanac@univ-tlse3.fr

The Bibliometric-enhanced Information Retrieval (BIR) workshop series has started at ECIR in 2014 [1] and serves as the annual gathering of IR researchers who address various information-related tasks on scientific corpora and bibliometrics [2]. The workshop features original approaches to search, browse, and discover value-added knowledge from scientific documents and related information networks (e.g., terms, authors, institutions, references). We welcome contributions elaborating on dedicated IR systems, as well as studies revealing original characteristics on how scientific knowledge is created, communicated, and used. The first BIR workshops set the research agenda by introducing the workshop topics, illustrating state-of-the-art methods, reporting on current research problems, and brainstorming about common interests. For the fourth workshop, co-located with the ACM/IEEE-CS JCDL 2016, we broadened the workshop scope and interlinked the BIR workshop with the natural language processing (NLP) and computational linguistics field [3]. This 7th full-day BIR workshop at ECIR 2018[1] aims to foster a common ground for the incorporation of bibliometric-enhanced services (including text mining functionality) into scholarly search engine interfaces. In particular we address specific communities, as well as studies on large, cross-domain collections. This workshop strives to feature contributions from core bibliometricians and core IR specialists who already operate at the interface between scientometrics and IR. Workshop proceedings will be published in open access with the CEUR workshop proceedings publication service.

References

1. Mayr, P., Scharnhorst, A., Larsen, B., Schaer, P., Mutschke, P.: Bibliometric-enhanced information retrieval. In: de Rijke, M., Kenter, T., de Vries, A.P., Zhai, C.X., de Jong, F., Radinsky, K., Hofmann, K. (eds.) ECIR 2014. LNCS, vol. 8416, pp. 798–801. Springer, Cham (2014). https://doi.org/10.1007/978-3-319-06028-6_99

---

[1] http://bit.ly/bir2018.

© Springer International Publishing AG, part of Springer Nature 2018
G. Pasi et al. (Eds.): ECIR 2018, LNCS 10772, pp. 827–828, 2018.
https://doi.org/10.1007/978-3-319-76941-7

2. Mayr, P., Scharnhorst, A.: Scientometrics and information retrieval: weak-links revitalized. Scientometrics **102**(3), 2193–2199 (2015)
3. Cabanac, G., Chandrasekaran, M.K., Frommholz, I., Jaidka, K., Kan, M.Y., Mayr, P., Wolfram, D.: Report on the joint workshop on bibliometric-enhanced information retrieval and natural language processing for digital libraries (BIRNDL 2016). SIGIR Forum **50**(2), 36–43 (2016)

BroDyn'18: Workshop on Analysis of Broad Dynamic Topics over Social Media

Tamer Elsayed[1(✉)], Walid Magdy[2], Mucahid Kutlu[1], Maram Hasanain[1], and Reem Suwaileh[1]

[1] Qatar University, Doha, Qatar
{telsayed,mucahidkutlu,maram.hasanain,reem.suwaileh}@qu.edu.qa
[2] University of Edinburgh, Edinburgh, UK
wmagdy@inf.ed.ac.uk

In recent years, users developed a widespread perception of social media as news sources that they follow (almost all day long) to get updated on topics of interest. Many of those topics attract *long-standing* user interest (i.e., stay active for long periods of time), are very *broad* (i.e., cover many subtopics), and are *dynamic* (i.e., develop and change focus over time with subtopics becoming obsolete and new subtopics emerging). Such kind of broad and dynamic topics can run for few weeks, such as crisis events (e.g., "Hurricane Irma"), or up to years, such as "the Syrian conflict". Other examples of such topics include "Refugees in Europe", "Qatar Crisis", "Brexit" to name a few.

Effective exploitation of content posted on social media that is relevant to broad dynamic topics requires *adaptive* techniques to effectively capture the changing aspects of topics. The techniques should also be *scalable* and *real-time* to cope with the large, rapidly flowing stream, and *reliable* to perform effectively over the long duration of the topic. Moreover, new evaluation frameworks for such domain are also needed with novel evaluation measures that capture the nature of topics (and thus systems) and new large reusable datasets that enable running meaningful and representative experiments.

BroDyn workshop[1] aims at building a community interested in developing and exchanging ideas and methods for analyzing social media for broad dynamic topics. It also aims at understanding the limitations of existing techniques in answering emerging information needs for such topics, and proposing new techniques, evaluation methods, and test collections to address these limitations. It is designed to bring together audience at all levels, including researchers from academia and industry as well as potential users (e.g., journalists and social scientists), to create a forum for discussing recent advances in this area.

The half-day workshop (held in conjunction with ECIR'18) covers several topics of interest including (but not limited to) adaptive filtering/topic tracking, adaptive summarization, multilingual topic/event detection, online/dynamic topic modeling, retrospective generation of timelines, cross-media filtering, real-time/scalable techniques of processing high-volume streams, and evaluation

---

[1] https://sites.google.com/view/brodyn2018.

© Springer International Publishing AG, part of Springer Nature 2018
G. Pasi et al. (Eds.): ECIR 2018, LNCS 10772, pp. 829–830, 2018.
https://doi.org/10.1007/978-3-319-76941-7

techniques and novel test collections. To encourage research on such tasks, we released two datasets: (1) GE2017, a dataset of around 18M tweets on the British General Elections 2017, and (2) USPresElect2016, a dataset of 3,450 labelled tweets on the US Presidential Elections 2016.

Second International Workshop on Recent Trends in News Information Retrieval (NewsIR'18)

Miguel Martinez[1], Dyaa Albakour[1(✉)], David Corney[5], Julio Gonzalo[2], Barbara Poblete[3], and Andreas Vlachos[4]

[1] Signal Media, London, UK
{miguel.martinez,dyaa.albakour}@signal.uk.com
[2] UNED, Madrid, Spain
julio@lsi.uned.es
[3] University of Chile, Santiago, Chile
barbara@poblete.cl
[4] University of Sheffield, Sheffield, England
a.vlachos@sheffield.ac.uk
[5] Factmata, London, UK
david.corney@factmata.com
https://www.signal.uk.com

Abstract. The news industry has undergone a revolution in the past decade, with substantial changes continuing to this day. News consumption habits are changing due to the increase in both the volume of news and the variety of sources. Readers need new mechanisms to cope with this vast volume of information in order to not only find a signal in the noise, but also to understand what is happening in the world given the multiple points of view describing every event. These challenges in journalism relate to IR and NLP fields such as: verification of a source's reliability; the integration of news with other sources of information; real-time processing of both news content and social streams; de-duplication of stories; and entity detection and disambiguation. Although IR and NLP have been applied to news for decades, the changing nature of the space requires fresh approaches and a closer collaboration with our colleagues from the journalism environment. The goal of this workshop is to stimulate such discussion between the communities and to share interesting approaches to solve real user problems.

© Springer International Publishing AG, part of Springer Nature 2018
G. Pasi et al. (Eds.): ECIR 2018, LNCS 10772, p. 831, 2018.
https://doi.org/10.1007/978-3-319-76941-7

Workshop on Social Aspects in Personalization and Search (SoAPS 2018)

Ludovico Boratto[1(✉)] and Giovanni Stilo[2]

[1] Digital Humanities, Eurecat Camí Antic de València 54, 08005 Barcelona, Spain
ludovico.boratto@acm.org
[2] Dipartimento di Informatica, Sapienza Università di Roma,
Via Salaria 113, 00198 Rome, Italy
stilo@di.uniroma1.it

In order to improve the web experience of the users, classic personalization technologies (e.g., recommender systems) and search engines usually rely on static schemes. Indeed, users are allowed to express ratings in a fixed range of values for a given catalogue of products, or to express a query that usually returns the same set of webpages/products for all the users.

With the advent of communication systems (social media platforms, instant messaging systems, speech recognition and transcription tools, etc.), users have been allowed to create new content and to express opinions and preferences in new forms (e.g., likes, textual comments, and audio feedbacks). Moreover, the social interactions can provide information on who influences whom. Being able to mine usage and collaboration patterns that arise thanks to social aspects and to analyze the collective cooperations, opens new frontiers in the generation of personalization services and in the improvement of search engines. Moreover, recent technological advances, such as deep learning, are able to provide a context to the analyzed data (e.g., word embeddings provide a vector representation of the words in a corpus, considering the context in which a word has been used).

Our workshop solicited contributions in all topics related to employing social aspects for personalization and search purposes, focused (but not limited) to the following list:

- Recommender systems
- Search and tagging
- Query expansion
- User modeling and profiling
- Advertising and ad targeting
- Content classification, categorization, and clustering
- Using social network features/community detection algorithms for personalization and search purposes
- Employing speech transcription in personalization and search
- Building benchmarking datasets
- Novel evaluation methodologies in the social context

© Springer International Publishing AG, part of Springer Nature 2018
G. Pasi et al. (Eds.): ECIR 2018, LNCS 10772, p. 832, 2018.
https://doi.org/10.1007/978-3-319-76941-7

First International Workshop on Narrative Extraction from Texts: Text2Story 2018

Alípio Mário Jorge[1,2]([✉]) [iD], Ricardo Campos[1,3] [iD], Adam Jatowt[4] [iD],
and Sérgio Nunes[1,2] [iD]

[1] INESC TEC, Porto, Portugal
amjorge@fc.up.pt
[2] University of Porto, Porto, Portugal
[3] Polytechnic Institute of Tomar, Tomar, Portugal
[4] Kyoto University, Kyoto, Japan

Keywords: Information extraction · Narrative extraction

1 Synopsis

The increasing availability of text information in the form of news articles, comments or posts poses new challenges for those who aim to understand the storyline of an event. Although understanding natural language text has improved over the last couple of years with several research works emerging on the grounds of information extraction and text mining, the problem of constructing consistent narrative structures is yet to be solved. We have a challenging path ahead of us for the development and improvement of algorithms that automatically identify, interpret and relate the different elements of a narrative which will be likely spread among different sources. In this first workshop on this topic, held at the 40th European Conference on Information Retrieval (ECIR 2018), we aim to foster the discussion of recent advances in the link between Information Retrieval (IR) and formal narrative representations from texts. More specifically, we aim to capture a wide range of multidisciplinary issues related to the text-to-narrative-structure and to its various related tasks. These include event identification; narrative representation language; sentiment and opinion detection; argumentation mining; narrative summarization; storyline visualization; temporal aspects of storylines; story evolution and shift detection; causal relation extraction and arrangement; evaluation methodologies for narrative extraction; big data applied to narrative extraction; resources and dataset showcase; personalization and recommendation; user profiling and user behavior modeling; credibility; fact checking and bots influence. This workshop features a diversity of tasks and techniques with promising results for an exciting event.

Acknowledgments. This work is financed by the ERDF – European Regional Development Fund through the Operational Programme for Competitiveness and Internationalisation - COMPETE 2020 Programme within project «POCI-01-0145-FEDER-006961», and by National Funds through the FCT – Fundação para a Ciência e a Tecnologia (Portuguese Foundation for Science and Technology) as part of project UID/EEA/50014/2013.

Tutorials

Neural Networks for Information Retrieval

Tom Kenter[1]([✉]), Alexey Borisov[2], Christophe Van Gysel[3], Mostafa Dehghani[3],
Maarten de Rijke[3], and Bhaskar Mitra[4]

[1] Booking.com, Amsterdam, Netherlands
tom.kenter@gmail.com
[2] Yandex, Moscow, Russia
alborisov@yandex-team.ru
[3] University of Amsterdam, Amsterdam, Netherlands
{cvangysel,dehghani,derijke}@uva.nl
[4] Microsoft, University College London, London, UK
bmitra@microsoft.com

Abstract. Machine learning plays a role in many aspects of modern IR systems, and deep learning is applied in all of them. The fast pace of modern-day research has given rise to many approaches to many IR problems. The amount of information available can be overwhelming both for junior students and for experienced researchers looking for new research topics and directions. The aim of this full-day tutorial is to give a clear overview of current tried-and-trusted neural methods in IR and how they benefit IR.

Prompted by the advances of deep learning in computer vision, neural networks (NNs) have resurfaced as a popular machine learning paradigm in many other directions of research, including IR. Recent years have seen NNs being applied to all key parts of the typical modern IR pipeline, such as click models, core ranking algorithms, dialogue systems, entity retrieval, knowledge graphs, language modeling, question answering, and text similarity. A key advantage that sets NNs apart from many learning strategies employed earlier, is their ability to work from raw input data. Where designing features used to be a crucial aspect and contribution of newly proposed IR approaches, the focus has shifted to designing network architectures instead. As a consequence, many different architectures and paradigms have been proposed, such as auto-encoders, recursive networks, recurrent networks, convolutional networks, various embedding methods, and deep reinforcement learning. The aim of this tutorial is to provide an overview of the main network architectures currently applied in IR and to show how they relate to previous work. The tutorial covers methods applied in industry and academia, with in-depth insights into the underlying theory, core IR tasks, applicability, key assets and handicaps, efficiency and scalability concerns, and tips & tricks. We expect the tutorial to be useful both for academic and industrial researchers and practitioners who either want to develop new neural models, use them in their own research in other areas or apply the models described here to improve actual IR systems.

© Springer International Publishing AG, part of Springer Nature 2018
G. Pasi et al. (Eds.): ECIR 2018, LNCS 10772, p. 837, 2018.
https://doi.org/10.1007/978-3-319-76941-7

Tutorial on Semantic Search on Text

Lynda Tamine[1]([✉]) and Lorraine Goeuriot[2]

[1] University of Toulouse UPS, IRIT, 118 route de Narbonne,
31062 Toulouse Cedex 9, France
tamine@irit.fr
[2] University of Grenoble Alpes, CNRS, Grenoble INP, LIG,
621 avenue Centrale, 38400 Saint-Martin-d'Hères, 38000 Grenoble, France
lorraine.goeuriot@imag.fr

1 Tutorial Abstract

This tutorial will first explore the peculiarities of medical and health-related queries with respect to various facets (eg. vocabulary, users expertise, task) with the attempt of better understanding the underlying human intent. Second, as envisioned in semantic search, we will focus on the techniques and theoretical models that go beyond lexical matching to drive the search. We will cover both the symbolic semantics through the use of external resources (eg. UMLS, MeSH, Gene Ontology) and the distributional semantics relying on words collocations in the corpus including recent representation learning approaches of concepts and documents. Third, we will develop a roadmap on the main evaluation frameworks used in medical IR and then particularly examine and compare the effectiveness of semantic-based IR approaches. Finally, we summarize the research findings in the area and outline the key open research questions. To sum up, the goals of the tutorial are the following:

- Summarize the lessons that can be drawn from studies investigating the peculiarities of medical-related information needs;
- Present state-of-the art semantic search models supporting medical IR processes;
- Describe the major medical search evaluation benchmarks used in the IR community and report the key result trends achieved by the application of semantic IR models.

2 Organizers

Lynda Tamine is a Professor of Computer Science at the Paul Sabatier university in Toulouse and member of the Institut de Recherche en Informatique de Toulouse (IRIT). Her research interests include modelling and evaluation of medical, contextual, collaborative and social information retrieval. Lorraine Goeuriot is an associate professor in Universit Grenoble Alpes. Her research interests include medical information retrieval and the evaluation of information retrieval.

© Springer International Publishing AG, part of Springer Nature 2018
G. Pasi et al. (Eds.): ECIR 2018, LNCS 10772, p. 838, 2018.
https://doi.org/10.1007/978-3-319-76941-7

Extreme Multi-label Classification for Information Retrieval

Krzysztof Dembczyński[1] and Rohit Babbar[2(✉)]

[1] Poznan University of Technology, Poznań, Poland
[2] Aalto University, Helsinki, Finland
rohit.babbar@aalto.fi

The goal in extreme multi-label classification is to learn a classifier which can assign a small subset of relevant labels to an instance from an extremely large set of target labels. It has been shown that this framework can be applied to effectively address the challenges in ranking, recommendation and automatic tagging systems. Extreme classification simultaneously exhibits two seemingly contrary challenges – data abundance as a whole for all the labels on one hand, and data scarcity for individual labels on the other hand. The former poses a computational challenge while the latter posses a statistical challenge.

In the tutorial we will motivate extreme classification as an active and rapidly growing research area with many potential applications in information retrieval. We will present three main strands for addressing the challenges which consist of (i) label embedding methods [2, 8], (ii) tree-based methods [3, 6], and (iii) smart one-vs-rest approaches [1, 7]. Futhermore, we will discuss extreme classification solutions in the domain of deep networks [4, 5]. Finally, we will highlight open benchmark datasets derived from sources such as Wikipedia, Amazon and Delicious,[1] and also live demonstrate open source code by running it on these benchmark datasets.

References

1. Babbar, R., Schölkopf, B.: DiSMEC: distributed sparse machines for extreme multi-label classification. In: WSDM 2017, pp. 721–729. ACM (2017)
2. Bhatia, K., Jain, H., Kar, P., Varma, M., Jain, P.: Sparse local embeddings for extreme multi-label classification. In: NIPS 2015, pp. 730–738. Curran Associates Inc. (2015)
3. Jasinska, K., Dembczyński, K., Busa-Fekete, R., Pfannschmidt, K., Klerx, T., Hüllermeier, E.: Extreme F-measure maximization using sparse probability estimates. In: ICML 2016, vol. 48, pp. 1435–1444. PMLR (2016)
4. Joulin, A., Grave, E., Bojanowski, P., Mikolov, T.: Bag of tricks for efficient text classification. CoRR abs/1607.01759 (2016)
5. Liu, J., Chang, W.C., Wu, Y., Yang, Y.: Deep learning for extreme multi-label text classification. In: SIGIR 2017, pp. 115–124. ACM (2017)
6. Prabhu, Y., Varma, M.: FastXML: a fast, accurate and stable tree-classifier for extreme multi-label learning. In: KDD 2014, pp. 263–272. ACM (2014)

---

[1] http://manikvarma.org/downloads/XC/XMLRepository.html.

© Springer International Publishing AG, part of Springer Nature 2018
G. Pasi et al. (Eds.): ECIR 2018, LNCS 10772, pp. 839–840, 2018.
https://doi.org/10.1007/978-3-319-76941-7

7. Yen, I.E., Huang, X., Dai, W., Ravikumar, P., Dhillon, I., Xing, E.: PPDSparse: a parallel primal-dual sparse method for extreme classification. In: KDD 2017, pp. 545–553. ACM (2017)
8. Yu, H.F., Jain, P., Kar, P., Dhillon, I.: Large-scale multi-label learning with missing labels. In: ICML 2014, vol. 32, pp. 593–601. PMLR (2014)

Author Index

Printed in the United States
By Bookmasters